Springer Geology

The book series Springer Geology comprises a broad portfolio of scientific books, aiming at researchers, students, and everyone interested in geology. The series includes peer-reviewed monographs, edited volumes, textbooks, and conference proceedings. It covers the entire research area of geology including, but not limited to, economic geology, mineral resources, historical geology, quantitative geology, structural geology, geomorphology, paleontology, and sedimentology.

More information about this series at http://www.springer.com/series/10172

Rajesh K. Srivastava · Richard E. Ernst
Peng Peng
Editors

Dyke Swarms of the World: A Modern Perspective

Springer

Editors
Rajesh K. Srivastava
Department of Geology
Institute of Science, Banaras Hindu University
Varanasi, India

Richard E. Ernst
Department of Earth Sciences
Carleton University
Ottawa, ON, Canada

and

Department of Geology and Geography
Tomsk State University
Tomsk, Russia

Peng Peng
Institute of Geology and Geophysics, Chinese Academy of Sciences
Beijing, China

ISSN 2197-9545 ISSN 2197-9553 (electronic)
Springer Geology
ISBN 978-981-13-1665-4 ISBN 978-981-13-1666-1 (eBook)
https://doi.org/10.1007/978-981-13-1666-1

Library of Congress Control Number: 2018957044

This Springer imprint is published by the registered company Springer Nature Singapore Pte Ltd.
The registered company address is: 152 Beach Road, #21-01/04 Gateway East, Singapore 189721, Singapore

Preface

Detailed multidisciplinary studies on mafic dyke swarms play a crucial role in solving geodynamic problems of Earth's history. Professor Henry C. Halls recognized the importance of dyke swarms long ago and organized the first International Dyke Conference focusing on geological, geochemical, and geophysical aspects of dykes and related units in Toronto, Canada, in 1985. Given the new insights arising during that conference, it was decided that International Dyke Conferences (IDCs) should be held every 5 years. Consequently, IDC-2 was held in Australia in 1990, IDC-3 in Israel in 1995, IDC-4 in South Africa in 2001, IDC-5 in Finland in 2005, and IDC-6 in India in 2010. Each IDC also produced a proceedings volume (apart from IDC-4 whose contributions were published as part of the IDC-5 volume) (Halls and Fahrig 1987; Parker et al. 1990; Baer and Heimann 1995; Hanski et al. 2006; Srivastava 2011).

Continuing the tradition, the Seventh International Dyke Conference (IDC-7) was hosted by the State Key Laboratory of Lithospheric Evolution, Institute of Geology and Geophysics, Chinese Academy of Sciences, and held on August 18–20, 2016, at the Friendship Hotel in Beijing. Approximately, 140 dyke researchers from 19 countries attended this conference which had the theme "Dyke Swarms: Keys to Paleogeographic Reconstruction," which was further divided into 10 sub-themes to cover almost every aspect of modern research on dykes and related units:

1. Regional maps/reviews of dyke swarms and related units
2. The role of giant dyke swarms in the reconstruction of supercontinents/paleocontinents: progress, problems, and potential
3. Mapping of dykes using remote-sensing techniques: aeromagnetic data, Landsat, radar, etc.
4. Geochronology of dyke swarms
5. Petrology, geochemistry, and petrogenesis of dykes
6. Emplacement mechanism of dykes
7. Dyke swarms on planetary bodies

8. Links to mineralization and resources
9. Miscellaneous: synplutonic mafic dykes and alkaline dykes, etc.
10. Oceanic dyke complexes: seafloor spreading, oceanic plateaus, or juvenile arcs?

An abstract volume (containing 133 abstracts) was provided at the conference, and these abstracts were subsequently published in Acta Geologica Sinica (Peng et al. 2016).

Group photograph of the IDC-7 participants

As a tradition of IDCs, it is customary to publish a volume based mainly on presentations during the conference but also including additional key research on dykes. The large volume of abstracts at IDC-7 and the varied themes led to the planning of two volumes. The first one, titled "Dyke Swarms: Keys for Precambrian Paleogeographic Reconstruction," will be published as a special issue of Precambrian Research (Editors: Peng Peng, Richard E. Ernst, Ulf Söderlund, and Michael Hamilton), and the second one is this volume. Originally, about 30 participants expressed interest to submit their work to this volume. Ultimately, 15 manuscripts were received and reviewed. All the submitted manuscripts were reviewed by at least two reviewers, in addition to reviews from the guest editors. Thirteen manuscripts were accepted for this volume. A brief description of each contribution is presented below.

It is well established that mafic dyke swarms in shield areas provide the most complete record of short-lived, mantle-generated large igneous province (LIP) events through time and space, which provides valuable information on geodynamics, locating mantle plume centers, evolution of the mantle, paleogeographic reconstructions, links with metallogeny, and links to climate change including mass extinction events (e.g., Halls 1982; Halls and Fahrig 1987; Parker et al. 1990; Baer and Heimann 1995; Hanski et al. 2006; Srivastava 2011; Ernst 2014; Ernst and Jowitt 2013, 2017; Ernst and Youbi 2017). Therefore, systematic study and information on mafic dyke swarms can be very useful for solving many geological problems as illustrated in this volume.

The volume starts with two review chapters: Chapter "Giant Circumferential Dyke Swarms: Catalogue and Characteristics" on the giant circumferential dyke swarms (**Buchan and Ernst**) and Chapter "Magma Transport Pathways in Large Igneous Provinces: Lessons from Combining Field Observations and Seismic Reflection Data" on using seismic reflection data to understand magma plumbing systems (**Magee et al.**). A dominant feature of LIP plumbing systems is their regional radiating mafic dyke swarms usually interpreted to focus on the center of a mantle plume responsible for the LIP (e.g., Halls 1982; Ernst and Buchan 1997, 2001; Ernst 2014). However, a new class of giant swarms is now recognized, termed giant circumferential swarms which with a circular or elliptical geometry (Ernst and Buchan 1998; Buchan and Ernst 2018). **Buchan and Ernst** (Chapter "Giant Circumferential Dyke Swarms: Catalogue and Characteristics") provide a comprehensive catalogue and characteristics of giant circumferential dyke swarms from around the world (most newly recognized and presented here for the first time). Most of these appear to be linked to giant radiating dyke swarms, LIPs, and mantle plumes. A comparison is made between these identified terrestrial examples and possible analogues on Venus (coronae) and Mars.

Magee et al. (Chapter "Magma Transport Pathways in Large Igneous Provinces: Lessons from Combining Field Observations and Seismic Reflection Data") present seismic reflection data on sill complexes and dyke swarms in order to better constrain their structure and emplacement. This review provides insights into the connectivity of and magma flow pathways within extensive sill complexes and how sill complexes are spatially accommodated. This work also reveals changes in dyke width with height and how dyke-induced normal faults and pit chain craters can be used to locate sub-vertical dykes offshore. These observations provide insights into properties of LIP magma plumbing systems both on Earth and other planetary bodies.

Hollanda et al. (Chapter "The Mesozoic Equatorial Atlantic Magmatic Province (EQUAMP)") provide $^{40}Ar/^{39}Ar$ ages, geochemical and airborne magnetic data on mafic dyke swarms and sills belonging to the Equatorial Atlantic Magmatic Province (EQUAMP), a newly identified early Cretaceous $\sim$135–120 Ma LIP in South America. They suggest that the Rio Ceará Mirim dykes (which can now be traced for over 1000 km along an arcuate trajectory) and Sardinha sills are derived from melting of a subcontinental lithospheric mantle source.

Teixeira et al. (Chapter "Intraplate Proterozoic Magmatism in the Amazonian Craton Reviewed: Geochronology, Crustal Tectonics and Global Barcode Matches") provide a detailed evaluation of geochronology (including U–Pb baddeleyite ages) and geochemistry of the Proterozoic mafic dyke swarms and sills of the Amazonian Craton. Such data are provided for a number of LIPs/SLIPs (silicic LIPs) that include the major Orocaima (1.98–1.96 Ga), Uatumã (1.89–1.87 Ga), Avanavero (1.79 Ga), and the Rincón del Tigre–Huanchaca (1.11 Ga) events. Other magmatic events are reported from the Central Brazil and Guiana shields, for example the Mata-Matá (1.57 Ga), Salto do Céu (1.44 Ga), and Nova Floresta (1.22 Ga) mafic sills and the

Cachoeira Seca troctolite dykes and laccoliths (1.19 Ga). Petrogenesis and paleogeographic reconstruction of all these LIPs/SLIPs are also discussed, and the authors suggest that the Proterozoic intracratonic LIP/SLIP events were part of the breakup of the Columbia supercontinent.

The next three chapters focus on dyke swarms and their associated LIPs from the African shield. The first of these, **de Kock et al.** (Chapter "The Precambrian Mafic Magmatic Record, Including Large Igneous Provinces of the Kalahari Craton and Its Constituents: A Paleogeographic Review"), is a review of the Precambrian mafic magmatic record of the Kalahari Craton. They summarized the available precise U–Pb crystallization ages and paleopoles of mafic rocks with a particular focus on reclassification of the mafic dyke swarms, identification of LIPs, and implications for paleogeographic reconstructions. They also identified gaps in our knowledge of the Precambrian mafic record of the Kalahari Craton to be filled by multidisciplinary studies combining the latest advances in U–Pb geochronology along with both paleomagnetism and geochemistry.

Wabo et al. (Chapter "Constraining the Chronology of the Mashishing Dykes from the Eastern Kaapvaal Craton in South Africa") provide new geochronological, geochemical, and paleomagnetic data for NNE-trending dykes near Mashishing in the eastern Kaapvaal Craton. These data indicate that these Mashishing dykes belong to four different events: (i) a ~2.25–2.20 Ga pre-Bushveld dyke swarm, (ii) a dyke related to the ~2.06 Ga Dullstroom Lavas (part of the Bushveld LIP), (iii) the ~1.87–1.83 Ga Back Hills dyke swarm, and (iv) the ~1.11 Ga Umkondo dyke swarm.

Baratoux et al. (Chapter "New U–Pb Baddeleyite Ages of Mafic Dyke Swarms of the West African and Amazonian Cratons: Implication for Their Configuration in Supercontinents Through Time") present 14 new high precision new U–Pb TIMS ages ranging between 1790 and 200 Ma from doleritic dykes in the southern part of the West African Craton. The following swarms are distinguished: 1791 Ma N010° trending Libiri, 1764 Ma N035° trending Kédougou, 1575 Ma N100° trending Korsimoro, ~1525–1529 Ma N130° trending Essakane, 915 Ma N070° trending Oda, ~870 Ma N355° trending Manso, and 202 Ma N040° trending Hounde swarm, and also a 200 Ma for the mafic sills in the Taoudeni basin. All these Proterozoic swarms were previously unknown, this chapter provides the first robust LIP bar code for the southern West African Craton, and the results are used in a new paleogeographic Proterozoic reconstruction. The newly discovered 200 Ma Hounde swarm has trend oblique to the previously recognized giant radiating swarm of the CAMP LIP.

Central Asia is a major part of the Asian–European continent and occupies most of the western Central Asian Orogenic Belt. **Feng et al.** (Chapter "Spatial and Temporal Distribution Patterns of Mafic Dyke Swarms in Central Asia: Results from Remote-Sensing Interpretation and Regional Geology") present spatial–temporal distribution patterns of the mafic dyke swarms on the basis of detailed mapping with the help of remote-sensing interpretation and regional geology. The majority of these mafic dykes were emplaced in the Eastern Tianshan and Beishan,

Western Mongolian Altai, Eastern Junggar, North and West Bank of Balkhash, Western Junggar, and Chingis-Taerbahatai regions.

Samal et al. (Chapter "Neoarchean-Mesoproterozoic Mafic Dyke Swarms of the Indian Shield Mapped Using Google Earth™ Images and ArcGIS™, and Links with Large Igneous Provinces") present a comprehensive study of the Neoarchean–Mesoproterozoic mafic dyke swarms of the Indian Shield including mapping them using Google Earth™ images and ArcGIS™. They identify 24 Neoarchean–Mesoproterozoic mafic dyke swarms, which belong to 14 distinct mafic magmatic events in the Indian Shield (Dharwar, Bastar, Singhbhum, Bundelkhand, and Aravalli Cratons) and suggested that these swarms represent the exposed plumbing system for large igneous provinces (LIPs). This work also suggests a connection of the Indian Shield with Kenorland/Superia (~2.75–2.07 Ga), Columbia/Nuna (1.90–1.38 Ga), and Rodinia (1.20–0.72 Ga) supercontinents. They note that additional U–Pb geochronology and associated paleomagnetism are required to fully constrain the timing and pattern of the assembly of the various Indian Cratons and any post-assembly rotations between these cratons.

Sesha Sai et al. (Chapter "Petrology and Mineral Chemistry of a Porphyritic Mafic Dyke, Jonnagiri Schist Belt, Eastern Dharwar Craton, India: Implications for Its Magmatic Origin") share petrological and mineral chemistry of a Paleoproterozoic porphyritic mafic dyke from the eastern Dharwar craton, India. They estimated temperature and oxygen fugacity for the coexisting magnetite–ilmenite solid solution pairs which yielded an equilibration temperature of ~756 °C and $10^{-15.6}$ atm f_{O_2}.

Proterozoic mafic dykes from the Bomdila area, NE Lesser Himalaya, India, have been studied for geochemical characteristics and petrogenesis and tectonic significance (**Rashid et al.**) (Chapter "Geochemistry, Petrogenesis and Tectonic Significance of the Proterozoic Mafic Dykes from the Bomdila Area, NE Lesser Himalaya, India"). Geochemistry of these intrusive mafic rocks suggests their derivation from an enriched lithospheric mantle source (rather than being affected by crustal contamination) and emplacement in a continental rift tectonic environment.

Torkian (Chapter "Petrology and Tectonic Setting of Dyke Swarms Emplaced in the Upper Jurassic Qorveh Granitoid Complex (Majidaba and Kangareh), Kurdistan Province, Iran") discusses petrological characteristics of mafic dykes from the Kurdistan Province, Iran. These mafic rocks show a tholeiitic to calc-alkaline nature and probable emplacement in an active continental margin tectonic setting.

The final Chapter "From Ophiolites to Oceanic Crust: Sheeted Dike Complexes and Seafloor Spreading" is focused on sheeted dyke complexes (**Karson**). He has provided a thorough review of sheeted dyke complexes in ophiolites and also includes sheeted dike complexes on the seafloor. He demonstrates how sheeted dyke complexes are key to understanding the fundamentals of the tectonics, magma plumbing networks, and hydrothermal/biological systems at mid-ocean ridges.

This IDC-7 volume demonstrates significant progress since the last dyke conference (IDC-6 in India in 2010) and shows that the field of dyke swarm research remains rich in research ideas. This volume is also an acknowledgment of the vision of Henry Halls who launched the field of modern dyke swarm studies with Halls (1982) and IDC-1 in 1985 (Halls and Fahrig 1987). Current research foci include mafic dyke swarms associated with LIPs, with ophiolites, and on planetary bodies, and also the implications for the continental breakup, climate change, and resource exploration. These research areas remain important, but going forward there is also opportunity in addressing: mafic dyke swarms in arc systems, and also addressing silicic and kimberlite dyke swarms.

Varanasi, India — Rajesh K. Srivastava
Ottawa, Canada — Richard E. Ernst
Beijing, China — Peng Peng

References

Baer G, Heimann A (1995) Physics and chemistry of dykes. A. A. Balkema, Rotterdam

Buchan KL, Ernst RE (2018) A giant circumferential dyke swarm associated with the High Arctic Large Igneous Province (HALIP). Gondwana Res 58:39–57

Ernst RE (2014) Large igneous provinces. Cambridge University Press, UK

Ernst RE, Buchan KL (1997) Giant radiating dyke swarms: their use in identifying pre-mesozoic large igneous provinces and mantle plumes. In: Mahoney J, Coffin M (eds) Large igneous provinces: continental, oceanic, and planetary volcanism. American Geophysical Union, Geophysical monograph series, vol 100, pp 297–333

Ernst RE, Buchan KL (1998) Arcuate dyke swarms associated with mantle plumes on earth: implications for venusian coronae. In: Lunar and planetary science conference #29, Houston, Texas, CD-ROM, Abstract #1021

Ernst RE, Buchan KL (2001) Large mafic magmatic events through time and links to mantle-plume heads. In: Ernst RE, Buchan KL (eds) Mantle plumes: their identification through time. Geological Society of America, special paper 352, 483–575

Ernst RE, Jowitt SM (2013) Large igneous provinces (LIPs) and metallogeny. In: Colpron M, Bissig T, Rusk BG, Thomspon JFH (eds) Tectonics, metallogeny, and discovery: the North American Cordillera and similar accretionary settings. Society of economic geologists, special Publication, vol 17, pp 17–51

Ernst, RE, Jowitt, SM (2017) Multi-commodity, multi-scale exploration targeting using the large igneous province record. GSWA record 2017/6 target 2017, Perth, Australia, Abstracts volume, pp 41–44

Ernst RE, Youbi N (2017) How large igneous provinces affect global climate, sometimes causemass extinctions, and represent natural markers in the geological record. Palaeogeogr Palaeoclimatol Palaeoecol 478:30–52

Halls HC (1982) The importance and potential of mafic dyke swarms in studies of geodynamic process. Geosci Canada 9:145–154

Halls HC, Fahrig WF (1987) Mafic dyke swarms. Geol Assoc Canada Spl Paper 34:503

Hanski E, Mertanen S, Ramo T, Vuollo J (2006) Dyke swarms: time markers of crustal evolution. Taylor & Francis, London

Parker AJ, Rickwood PC, Tucker DH (1990) Mafic dykes and emplacement mechanism. A. A. Balkema, Rotterdam

Peng P, Ernst RE, Hou G, Soderlund U, Zhang S, Hamilton M, Xu Y, Denyszyn S, Mege D, Pisarevsky S, Srivastava RK, Kusky TM (2016) Dyke swarms: keys to paleogeographic reconstructions. Sci Bull 61:1669–1671

Srivastava RK (2011) Dyke swarms: keys for geodynamic interpretation. Springer, Heidelberg

Contents

Giant Circumferential Dyke Swarms: Catalogue and Characteristics

Kenneth L. Buchan and Richard E. Ernst

Abstract Giant circumferential dyke swarms have a primary geometry that is quasi-circular or quasi-elliptical. Examples and possible examples described previously or identified in this study have outer diameters that range from ~450 to ~2500 km. There has been little study of these features. Here, we present a global catalogue of giant circumferential dyke swarms and discuss their characteristics. All of the identified giant circumferential swarms are of mafic composition. Many, but not all, are associated with a roughly coeval giant radiating dyke swarm whose focus is at or near the centre of the circumferential system. As giant radiating swarms are usually interpreted to focus above mantle plume centres and form a key component of the plumbing system of large igneous provinces (LIPs), it is likely that giant circumferential swarms linked to radiating systems are also plume and LIP related. The largest giant circumferential swarms have diameters comparable to the diameters postulated for the flattened heads of plumes that have risen from the core-mantle boundary, suggesting that they may be associated with the outer edge of a flattening or flattened mantle plume head. Smaller giant circumferential swarms could be linked with small plumes from the mid-mantle or with the edge of a magmatic underplate above a plume head. Giant circumferential dyke swarms on Earth may be analogues of coronae on Venus and similar features on Mars. Coronae are large tectono-magmatic features that typically consist of a quasi-circular or quasi-elliptical graben-fissure system and associated topography (central uplift or depression, and circular rim or

K. L. Buchan (✉)
Geological Survey of Canada, Natural Resources Canada,
601 Booth Street, Ottawa K1A 0E8, Canada
e-mail: kenneth.buchan@canada.ca

R. E. Ernst
Department of Earth Sciences, Carleton University,
Ottawa ON, Canada
e-mail: richard.ernst@ernstgeosciences.com

R. E. Ernst
Department of Geology and Geography, Tomsk State University,
36 Lenin Ave, Tomsk 634050, Russia

R. K. Srivastava et al. (eds.), *Dyke Swarms of the World: A Modern Perspective*,
Springer Geology, https://doi.org/10.1007/978-981-13-1666-1_1

moat). In some instances, they are linked to a giant radiating graben-fissure system and LIP-scale volcanism. Both radiating and circumferential graben on Venus and Mars have been interpreted to be underlain by dykes.

Keywords Giant circumferential dyke swarm · Giant radiating dyke swarm Large igneous province · Mantle plume · Corona

1 Introduction

Giant radiating dyke swarms, which are mainly of mafic composition, are an important component of the plumbing systems of large igneous provinces (LIPs). They are usually thought to be emplaced from a focal region above a mantle plume (e.g., Halls 1982; Ernst and Buchan 1997, 2001), mainly because of their large scale, radiating geometry, evidence for lateral flow and the presence in many cases of an associated domal uplift. Several authors, however, support alternative models, such as those in which radiating dyke swarms form as a result of plate tectonic processes (e.g., McHone et al. 2005). For example, dykes could conceivably be emplaced into fanning fracture systems that develop as the result of indentor-style plate convergence, or could be emplaced vertically along the arms of radially oriented rift systems without an associated plume.

Ernst and Buchan (1998, 2001) identified three large mafic dyke swarms with a circular or elliptical geometry associated with the 250 Ma Siberian Traps, 1110–1185 Ma Keweenawan (North America) and ca. 980–930 Ma Rogaland (Scandinavia) magmatic events, and proposed that they may circumscribe mantle plume centres. They further suggested that such circumferential swarms may be eroded versions of tectono-magmatic structures, called coronae, on Venus. Coronae usually exhibit a quasi-circular or quasi-elliptical annulus of graben and fissures, volcanics (sometimes of LIP-scale) and topographic features such as a central uplift, central depression, uplifted rim or moat (e.g., Stofan et al. 1992). However, in some cases the geometry of coronae is more complicated. Occasionally, the annulus is characterized by compressional structures termed wrinkle ridges.

Giant circumferential dyke swarms are large swarms with a primary geometry that is quasi-circular or quasi-elliptical. Examples and possible examples that are catalogued in this study have diameters that range from ~450 to ~2500 km.

It should be noted that some giant dyke swarms have an arcuate pattern as the result of secondary deformation, or as the result of deflection from a linear or radiating geometry in the presence of a regional stress field (Ernst et al. 1995). Such swarms are not considered to be giant circumferential swarms.

A number of the catalogued giant circumferential swarm are centred near the focus of a roughly coeval giant radiating swarm. Assuming that the radiating swarm is linked to a mantle plume, then it is likely that the circumferential swarm is also linked to the same plume. However, other giant circumferential swarms, without an associated radiating swarm, may or may not be plume related.

In addition to giant circumferential swarms that are the subject of this study, there are much smaller circumferential swarms with diameters that range from a few kilometres to several tens of kilometres or more. The intrusions of these swarms are usually referred to as ring dykes if vertical or outward dipping, and cone sheets if inward dipping (e.g., Neuendorf et al. 2005). They are found in a variety of setting. Examples include circumferential swarms related to active volcanoes of the Galapagos Islands (Chadwick and Howard 1991; Chadwick et al. 2011) and the active Niuafo'ou volcano in a back-arc basin setting behind the Tonga Arc (Newhall and Dzurisin 1988), circumferential swarms within and around the margins of the central igneous complexes (deeply eroded roots of large volcanoes) of the 65 Ma British Tertiary Igneous Province of the North Atlantic LIP (Emeleus 1982; Burchardt et al. 2013), the 65 km diameter Meugueur-Meugueur troctolite ring dyke/cone sheet of northwestern Africa (Moreau et al. 1995), and the circumferential dyke swarm on the periphery of the deformed, late Archean Blake River Group megacaldera complex of North America with an original diameter of 80–90 km (Pearson and Daigneault 2009).

Since the original identification of giant circumferential dyke swarms by Ernst and Buchan (1998), further examples have been described in the literature. Mäkitie et al. (2014), and Ruotoistenmäki (2014) described a ca. 1380 Ma giant circumferential swarm at Lake Victoria in Africa which is linked to the Kunene-Kibaran LIP (Mäkitie et al. 2014; Ernst 2014; Buchan and Ernst 2016). Buchan and Ernst (2018) reconstructed a giant circumferential dyke swarm related to the ca. 135–75 Ma High Arctic LIP (HALIP).

In addition, there are some studies where it has been suggested that individual dykes or small dyke swarms, which are too short to display an arcuate pattern but are oriented perpendicular to a roughly coeval giant radiating dyke swarm, may be part of a giant circumferential system. Pehrsson et al. (1993) described an example in which a 1267 ± 3 Ma (U-Pb) dyke, located within and parallel to the Great Slave Lake shear zone along the southern margin of the Slave craton, is oriented perpendicular to the coeval Mackenzie giant radiating dyke swarm of North America, and is located ~1000 km from the swarm focus. Denyszyn et al. (2009) proposed that the ~150 km long 713–716 Ma (U-Pb) Clarence Head dyke swarm of North America, which is slightly younger than and crosscuts the ca. 720 Ma Franklin giant radiating dyke swarm at right angles ~1250 km from the swarm focus, may represent a segment of a giant circumferential system. In the discussion that follows we only include the cases where a swarm of dykes (e.g., Clarence Head swarm), rather than individual dykes, has been identified perpendicular to a radiating swarm.

In this paper, we catalogue and briefly describe the giant circumferential dyke swarms and possible segments of circumferential swarms that have been reported to date from around the globe, and identify a number of additional examples. The additional examples include swarms associated with the 65 Ma Deccan (India) and 135–120 Ma Paraná-Etendeka (South America and Africa) LIPs, and possible swarms linked to the 62–54 Ma North Atlantic, ca. 92–88 Ma Madagascar (Madagascar and India), ca. 183 Ma Karoo (Africa), ca. 370 Ma Yakutsk-Vilyui (Siberia), ca. 1110 Ma Umkondo (Africa), ca. 1210 Ma Marnda Moorn (Australia) and ca. 1780 Ma

Xiong'er-Taihang (northern China) magmatic events. We briefly summarize the characteristics of giant circumferential swarms and compare these characteristics with those of coronae on Venus and similar structures on Mars.

It is hoped that our catalogue of giant circumferential swarms and the preliminary analysis of their characteristics herein will stimulate further studies to identify addition examples and to better understand these little-known features.

2 Global Catalogue of Giant Circumferential Dyke Swarms

Giant circumferential dyke swarms and possible swarms or swarm segments are listed by age in Table 1, along with basic characteristics including outer diameter, arc length, width and the presence or absence of a coeval radiating swarm. It should be noted that all of the identified swarms are entirely or almost entirely of mafic composition. One example of a possible circumferential system of wrinkle ridges is also included. The global distribution of giant circumferential systems is illustrated in Fig. 1. Each swarm is briefly described below.

2.1 *Giant Circumferential Dyke Swarms*

Dykes of Deccan LIP, India (ca. 67–65 Ma): Dykes of the ca. 67–65 Ma (Renne et al. 2015; Schoene et al. 2015) Deccan LIP are summarized in Fig. 2. The best defined and described dykes form dense subswarms parallel to the ENE-trending Narmada-Tapi and S-trending West Coast rift systems (e.g., Bondre et al. 2006; Vanderkluysen et al. 2011). Dyke orientations in the region between these subswarms are more complicated and have variously been described as random (Hooper 1990) or, in some areas, as having a dominant NE trend (e.g., Bondre et al. 2006). Vanderkluysen et al. (2011) analysed dyke trends and geochemistry, and concluded that lower portions of the Deccan lava pile were likely fed by the ENE- and S-trending dykes, whereas upper portions were fed by the dykes without a preferred orientation. Dykes in more northern regions of the Deccan LIP are less well studied.

We interpret the Narmada-Tapi and West Coast subswarms as part of a giant radiating swarm (cf. Ernst and Buchan 1997), along with dykes that trend NW across the Saurashtra (Kathiawar) region (Auden 1949). A sparse set of dykes has been mapped at approximately right angles to each of these radiating subswarms (e.g., Clark 1880; Auden 1949; Sant and Karanth 1990; Dessai and Viegas 1995; Ray et al. 2007). In the Saurashtra region the dykes that intersect the radiating subswarm display an arcuate pattern as described by Auden (1949). We interpret the dykes that are perpendicular to the subswarms of the radiating system, including the arcuate dykes of Saurashtra, and the NE dykes located between the Narmada-Tapi and West Coast subswarms, as part of a giant circumferential dyke swarm with an outer diameter of >600 km and an arc of ~220° (Fig. 2). The centre of the proposed circumferential

Table 1 Catalogue of giant circumferential dyke swarms and possible segments of giant circumferential swarms

Name of event/swarm	Age (Ma)	Outer diameter (km)	Arc (°)	Width[a] (km)	Associated radiating swarm	References
Giant circumferential swarms						
Deccan LIP (India)	ca. 67–65	620	220	200	Yes	Herein
High Arctic LIP (Franz Josef Land, Svalbard, Greenland, N. America)	135–75	1600	220	300	Yes	Buchan and Ernst (2015, 2017, 2018)
Paraná-Etendeka (S. America, Africa)	135–120	1000	130	250	Yes	Herein
Kochikha swarm of Siberian Traps LIP (Siberia)	250	600–300 (ellipse)	~150	130	Yes	Ernst and Buchan (1998, 2001)
Blekinge-Dalarna swarm (Scandinavia)	978–946	1100	70	≥100	?	Ernst and Buchan (1998), Buchan and Ernst (2016)
Keweenawan LIP (N. America)	1110–1085	700–530 (ellipse)	140	≥50	?	Ernst and Buchan (1998)
Lake Victoria swarm of Kunene-Kibaran LIP (Africa)	ca. 1370	650	160	140	No	Mätikie et al. (2014), Ruotoistenmäki (2014), Ernst et al. (2014)
Possible giant circumferential swarms and swarm segments[b]						
British-Irish portion of North Atlantic LIP (Britain, Ireland)	62–54	≥1700	–	≥150?	Yes	Herein
Madagascar LIP (Madagascar, India)	92–88	~2400	70?	450	Yes	Herein

(continued)

Table 1 (continued)

Name of event/swarm	Age (Ma)	Outer diameter (km)	Arc (°)	Width[a] (km)	Associated radiating swarm	References
Karoo LIP (Africa)	183	460	300?	120	?	Chevallier and Woodford (1999), Chevallier et al. (2001), herein
Yakutsk-Vilyui LIP (Siberia)	380–360	1600	150?	300?	Yes	Herein
Clarence Head swarm of Franklin LIP (N. America)	716–713	~2500	–	≥70	Yes	Denyszyn et al. (2009)
Umkondo LIP (Africa)	ca. 1110	2300	–	70	Yes	Herein
Marnda Moorn LIP (Australia)	ca. 1210	1040	90 (N) 100 (S)	250	?	Wang et al. (2014), herein
Lüliang swarm of Xiong'er-Taihang LIP (China)	ca. 1780	1200	–	350	Yes	Peng (2015), Buchan and Ernst (2016)
Circumferential wrinkle ridges						
Yakima folds of Columbia River LIP (N. America)	17	1200–1400	–	≥150	Yes	Mège and Ernst (2001)

[a]Width of a circumferential dyke swarm refers to the distance between the inner and outer diameters of the swarm

[b]Includes cases where dyke distributions are somewhat complicated or dating is not yet sufficient to confirm that all proposed circumferential dykes are of similar age, as well as cases where short swarm segments are both coeval with and perpendicular to a radiating swarm

swarm is near the focus of the radiating swarm in the Gulf of Cambay and falls along the Réunion mantle plume track (Misra et al. 2014).

In some areas the radiating dykes are observed to crosscut the circumferential dykes (e.g., Krishnamurthy 1972; Dessai and Viegas 1995), whereas in other areas the opposite relationship is reported (e.g., Krishnamacharlu 1972). This suggests multiple pulses of circumferential and/or radiating dyke emplacement. Precise radio-

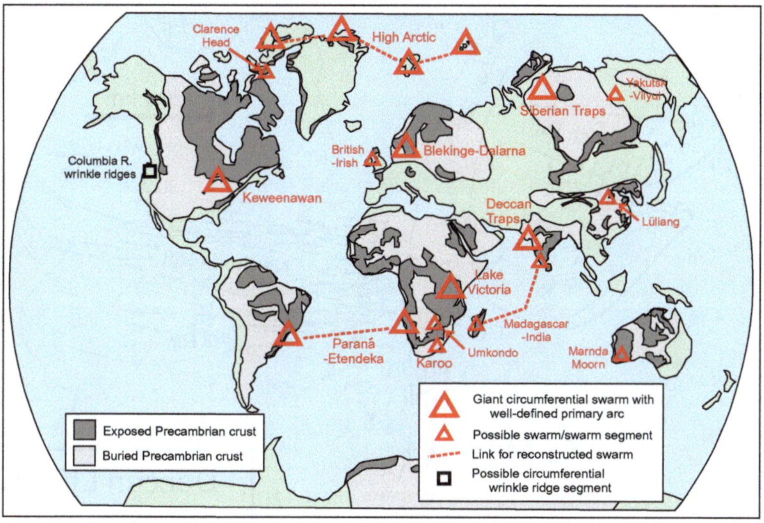

Fig. 1 Global distribution of giant circumferential dyke swarms and possible segments of giant circumferential dyke swarms. Characteristics of each arc listed in Table 1 and discussed in the text. Possible compressional wrinkle ridges associated with the Columbia River LIP of North America are also shown. The geological background is modified after Goodwin (1996)

metric dating of the circumferential and radiating dykes from the various regions is needed to clarify the relative ages of the two systems.

Dykes of High Arctic LIP (HALIP), Canadian Arctic islands, northern Greenland, Svalbard and Franz Josef Land (135–75 Ma): Maher (2001) and Buchan and Ernst (2006) described a giant radiating dyke swarm associated with the High Arctic LIP in a reconstruction of the high Arctic region. Buchan and Ernst (2015, 2017) identified a HALIP giant circumferential dyke swarm using the reconstruction of the radiating swarm. Buchan and Ernst (2018) have utilized a more up-to-date reconstruction of the Arctic region to refine the geometry of the circumferential swarm (Fig. 3). The reconstruction involves closing the Eurasia Basin and Baffin Bay (following the reconstruction of Gion et al. 2017), and undoing deformation of the Eurekan orogeny in the eastern Queen Elizabeth Islands of northern Canada based on the radiating dyke pattern and paleomagnetic constraints. Both radiating (black) and circumferential (red) dykes of Fig. 3 are located in each of the four separate regions of the high Arctic that are involved in the reconstruction. The radiating swarm consists of the Queen Elizabeth Islands dykes, the Erlandsen Land dykes of northern Greenland and unnamed dykes of Svalbard and Franz Josef Land, as well as associated aeromagnetic anomalies in the four regions. The circumferential swarm comprises the Surprise Fiord dykes of the Queen Elizabeth Islands, the J. P. Koch dykes of northern Greenland, and unnamed dykes of Svalbard and Franz Josef Land. In the reconstruction of Buchan and Ernst (2018), the circumferential swarm has a roughly circular geometry, an outer diameter of 1600 km, an arc of 220° and a centre that is

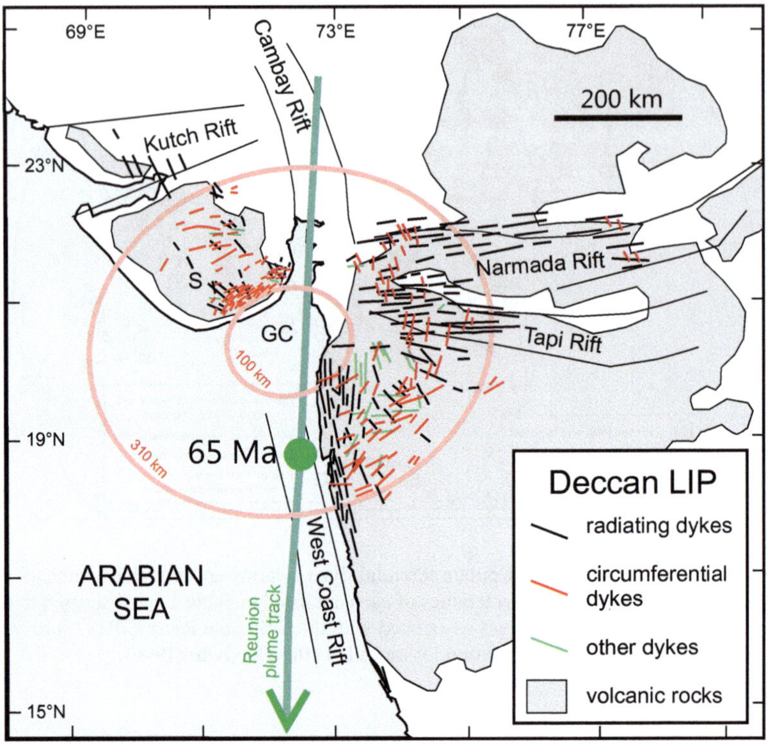

Fig. 2 Circumferential (red) and radiating (black) dykes of the ca. 65 Ma Deccan LIP, India, as described in the text. Dykes of other trends are green. The radiating dykes in the vicinity of the Narmada-Tapi and West Coast rift systems are dense, and have been thinned on the figure for the sake of clarity. Pink circles indicate the approximate outer and inner dimensions of the circumferential swarm. The centre of the circumferential swarm and the focus of the radiating swarm are similar and fall roughly on the (green) Reunion plume track (from Misra et al. 2014). S = Saurashtra (Kathiawar) region; GC = Gulf of Cambay

near, and perhaps slightly offset from, the focus of the radiating swarm. The ages of the circumferential and radiating swarms are poorly constrained. Only a single U-Pb age of 95 ± 1 Ma is available from the radiating swarm, from a probable dyke in the Queen Elizabeth Islands (Kingsbury et al. 2018). Ar–Ar ages of uncertain reliability range between 128 and 86 Ma, and may indicate multiple pulses of radiating dyke emplacement (see summary in Buchan and Ernst 2018). Only a single U-Pb age of 82 ± 2 Ma (Thórarinsson et al. 2015) and a single Ar–Ar age of 82 ± 1 Ma (Kontak et al. 2001), both from northern Greenland, are available from the circumferential swarm, so that it is not possible to speculate on the age span of the overall swarm. However, in Franz Josef Land, proposed circumferential dykes both crosscut and are cut by radiating dykes (Dibner 1998), suggesting that there may have been multiple pulses of circumferential and/or radiating dyke emplacement. HALIP-related topography (summarized in Maher 2001) in the vicinity of the circumferential swarm

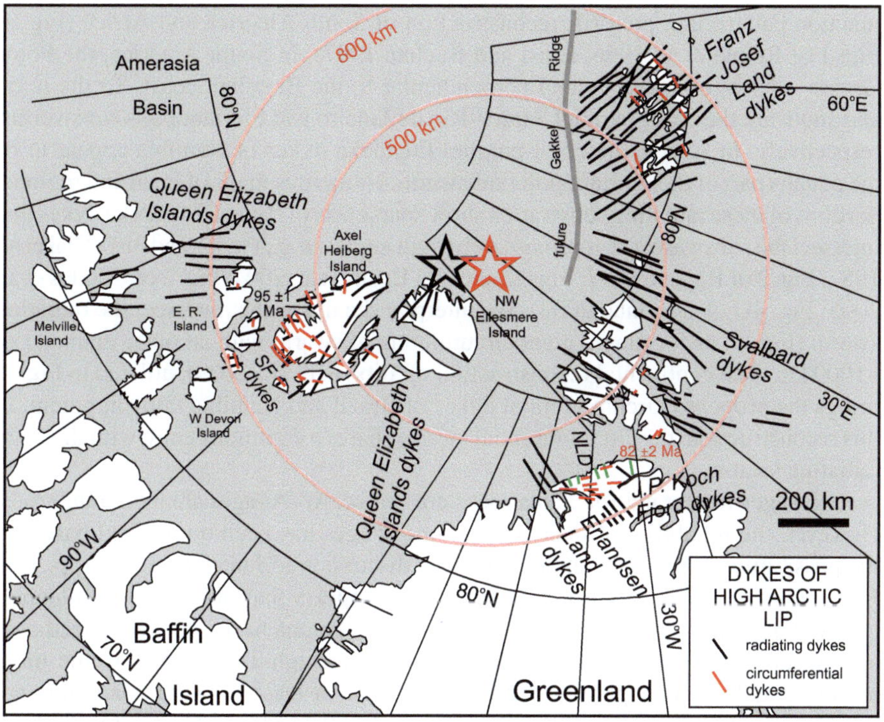

Fig. 3 Circumferential (red) and radiating (black) dykes of the reconstructed 135–75 Ma High Arctic LIP (HALIP) of the Canadian Arctic islands, northern Greenland, Svalbard and Franz Josef Land (modified after Buchan and Ernst 2018). The radiating swarm is dense in some areas, and has been thinned on the figure for clarity. The reconstruction is based on closing the Eurasia Basin, which opened in the Paleogene along the Gakkel Ridge, and Baffin Bay (following Gion et al. 2017), and undoing deformation in the Queen Elizabeth Islands (as described by Buchan and Ernst 2018). Pink circles indicate the approximate outer and inner dimensions of the circumferential swarm. The red star is the centre of the circumferential swarm. The black star is the focus of the radiating swarm. U-Pb ages for a circumferential dyke (Thórarinsson et al. 2015) and a probable radiating dyke (Kingsbury et al. 2018) are indicated with red and black lettering respectively. SF dykes = Surprise Fiord dykes; E.R. Island = Ellef Ringnes Island; NLD = Nansen Land dykes. Nansen Land dykes form a dense N-trending swarm along the northern Greenland coast. They are coloured green, as they may not be part of the HALIP radiating swarm, but rather may be related to rifting that led to the opening of the Eurasia Basin (e.g., Thórarinsson et al. 2015). In addition, although a U-Pb age of 81 Ma has been reported for these dykes (Thórarinsson et al. 2015), paleomagnetic data suggests that some Nansen Land dykes may be Early Carboniferous rather than Cretaceous in age (Abrahamsen et al. 1997)

on Franz Joseph Land and Svalbard could be related to the edge of a domal uplift associated with plume arrival (Maher 2001), or perhaps with a circular uplifted rim (Buchan and Ernst 2018) that is typical of many Venusian coronae.

Dykes of Paraná-Etendeka LIP, Brazil and Namibia (135–120 Ma): Radiating dykes of the Paraná-Etendeka LIP form prominent subswarms that display a triple junction pattern in a pre-drift reconstruction of South America and Africa (Fig. 4; Fig. 1 of Renne et al. 1996; Ernst and Buchan 1997). In South America, the Ponta Grossa subswarm strikes inland perpendicular to the Brazilian coast. To the north and south are the coast-parallel Santos-Rio de Janeiro and Florianópolis subswarms, respectively. In Africa, the coast-parallel Etendeka dykes of Namibia appear to be the counterpart of the Florianópolis subswarm. Numerous maps of local and regional portions of these radiating subswarms show sparse sets of Paraná-Etendeka dykes that intersect the subswarms at approximately right angles (e.g., Figs. 1 and 3b of Sial et al. 1987; Fig. 2 of Raposo 1997; Figs. 2 and 3 of Ewart et al. 2004; Fig. 2 of Guedes et al. 2005; Fig. 4U of Coutinho 2008). We interpret the dykes that intersect the radiating swarm (Fig. 4) as forming a giant circumferential swarm with an outer diameter of ~1000 km. The continental reconstruction of Torsvik et al. (2009), utilized in Fig. 4, brings the proposed circumferential dykes of Brazil and Namibia into alignment. In this reconstruction, the circumferential swarm shares a common centre with the giant radiating swarm.

U-Pb ages for Paraná-Etendeka rocks are sparse. Ar–Ar ages are more numerous. However, the reliability of many of the Ar–Ar ages has been questioned based on the possibility of Ar excess and/or loss (see discussion in Janasi et al. 2011, p. 149; Florisbal et al. 2014, p. 148). Nevertheless, it appears that older low-Ti volcanics outcropping mainly in the southeastern part of the Paraná basin were emplaced over a very short interval at 135–134 Ma, and younger high-Ti volcanics in the more northerly and westerly parts of the basin between at least 134–131 Ma (summary in Florisbal et al. 2014). U-Pb ages of 134.7–133.9 Ma for radiating dykes of the Florianópolis subswarm that are chemically similar to the younger volcanics suggest emplacement at the beginning of the second volcanic phase (Florisbal et al. 2014). Ar–Ar ages for high-Ti radiating dykes of the Ponta Grossa subswarm (Renne et al. 1996) fall in the narrow range between 133 and 131 Ma, consistent with emplacement during the second volcanic phase. There are no U-Pb ages for the proposed circumferential dykes. However, in some cases Ar–Ar ages appear to be younger than for the radiating set. For example, Renne et al. (1996) obtained 122 Ma ages for a pair of high-Ti dykes intersecting the Ponta Grossa radiating subswarm at right angles, suggesting that they could have been intruded long after the main phases of volcanism (Florisbal et al. 2014). Much more precise dating is needed to clarify the ages of the proposed circumferential and radiating swarms.

Kochikha dykes of Siberian Traps LIP (250 Ma): A coupled giant circumferential and radiating system (Fig. 5) was proposed by Ernst and Buchan (1998, 2001) on the basis of available mapping of dykes associated with the Siberian Traps LIP. The Kochikha circumferential swarm has a pronounced elliptical shape (maximum and minimum outer diameters of ~600 and 300 km) and shares a common centre with the radiating swarm. The circumferential swarm spans an arc of ~150°. The radiating swarm (comprising the Ebekhaya and Maimecha subswarms and linear feeder zones to volcanic flows of the Noril'sk area) extends far beyond the circumferential system to a distance of 900 km from its focus. The circumferential dykes crosscut Siberian Traps volcanics, whereas the radiating dykes do not, indicating that the circumferential dykes are younger that the radiating set.

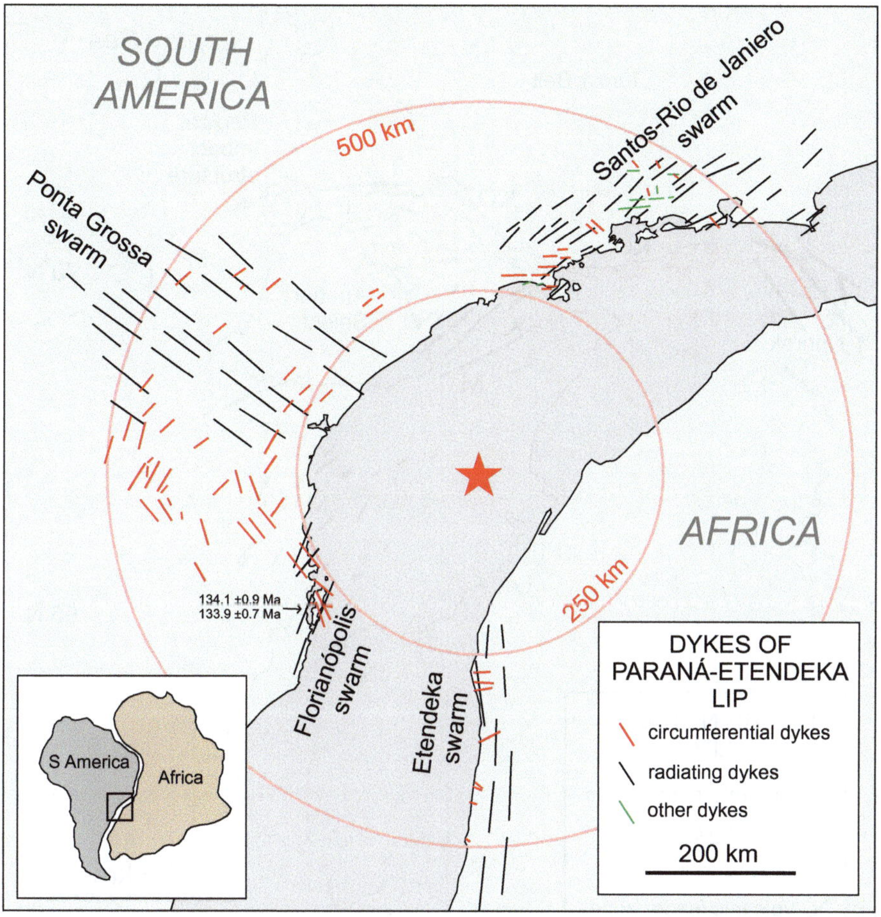

Fig. 4 Circumferential (red) and radiating (black) dykes of the reconstructed ca. 135–120 Ma Paraná-Etendeka LIP of Brazil and Namibia. The radiating swarm is dense in some areas, and has been thinned on the figure for the sake of clarity. Circumferential dykes are from various sources listed in the text. The reconstruction is from Torsvik et al. (2009). Pink circles indicate the approximate outer and inner dimensions of the circumferential swarm. The red star is the centre of the circumferential swarm and the focus of the radiating swarm. U-Pb ages for radiating Florianópolis dykes are shown in black lettering

Blekinge-Dalarna dykes of Rogaland event, Sweden (ca. 978–946 Ma): The Blekinge-Dalarna swarm (Fig. 6; Gorbatschev et al. 1987; Bylund 1992; Söderlund et al. 2005) follows the eastern margin of the high-grade Protogine Zone, which separates the Svecofennian and Sveconorwegian domains of the Baltic Shield in Sweden. It has a length of 700 km, an arc of ~60°, and is truncated to the north by the Caledonian orogeny and overlain to the south by Phanerozoic cover rocks. Ernst and Buchan (1998) proposed that the swarm may represent a giant circumferential

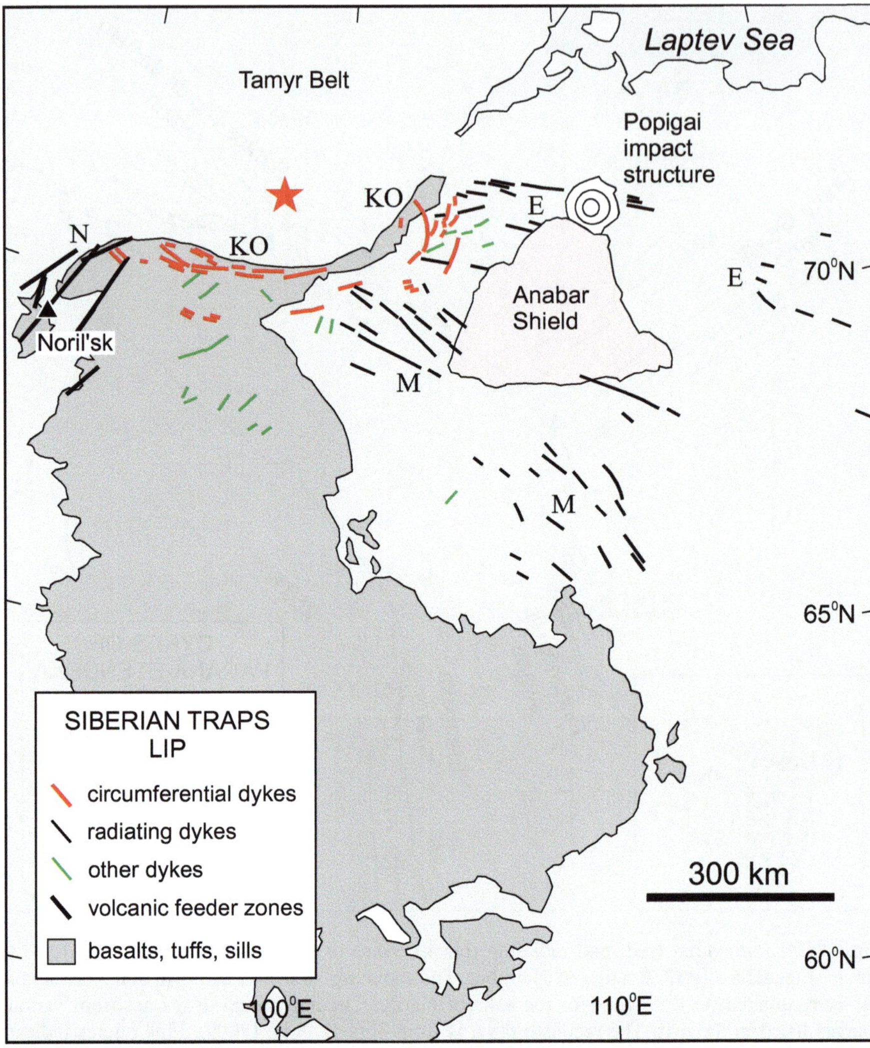

Fig. 5 Circumferential (red) dykes and radiating (black) dykes and volcanic feeder zones of the ca. 250 Ma Siberian Trap LIP (modified after Ernst and Buchan (2001), with additional circumferential dykes from Ryabov et al. (2014, Fig. 2.30)). The red star locates the approximate focus of the radiating swarm and the centre of the circumferential swarm. E = Ebekhaya dykes; KO = Kochikha dykes; M = Maimecha dykes; N = Noril'sk feeder zones to volcanic flows. The circumferential dykes cut the volcanics, whereas the radiating dykes do not

system (outside diameter of 1100 km) about a centre near the prominent 932–929 Ma (Schärer et al. 1996) Rogaland complex. More recently, precise U-Pb dating has confirmed that the dykes along the length of the swarm are of approximately the same age (978–946 Ma; Söderlund et al. 2005). Consistent paleomagnetic directions along

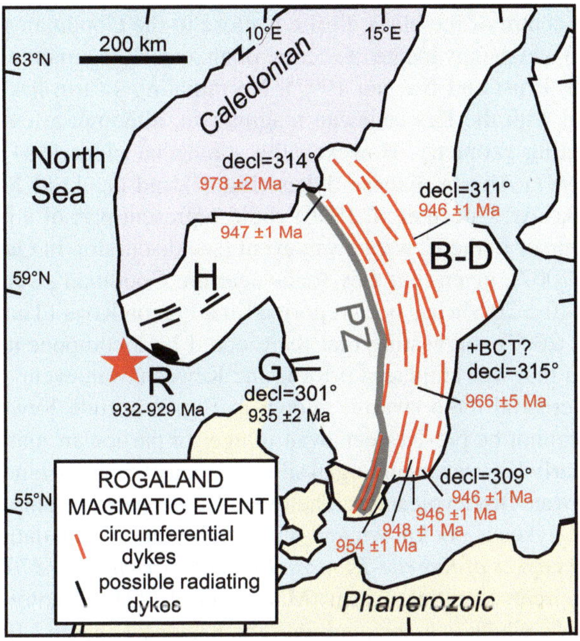

Fig. 6 Circumferential (red) dykes of the ca. 978–946 Ma Blekinge-Dalarna (B-D) swarm of the Rogaland magmatic event, Sweden. The red star locates the centre of the swarm. The slightly younger (ca. 935 Ma) Göteborg (G) dykes (black) could represent part of a radiating swarm. Dykes of the Hunnedalan (H) swarm (black) could also be part of a radiating swarm, but are poorly dated. Paleomagnetic declinations obtained at various locations along the Blekinge-Dalarna swarm (Pisarevsky and Bylund 2006) are labelled 'decl.' The label '+BCT?' locates a probable positive baked contact test indicating the remanences are primary. U-Pb ages for the Blekinge-Dalarna dykes (Söderlund et al. 2005) are in red lettering. U-Pb ages for the Göteborg dykes (Hellström et al. 2004) and the Rogaland complex (Schärer et al. 1996) are in black lettering

the swarm (summarized in Fig. 6, from the compilation in Pisarevsky and Bylund 2006) indicate that its arcuate geometry is primary. Dykes of the small Göteborg (Tuve) swarm (Fig. 6) have been dated at 935 ± 2 Ma (U-Pb; Hellström et al. 2004), somewhat younger than the Blekinge-Dalarna swarm, and could represent part of a radiating system in the interior of the circumferential swarm.

Keewenawan dykes of Mid-Continent Rift, central North America (ca. 1110–1085 Ma): Ernst and Buchan (1998) described dykes of the ca. 1110–1085 Ma Keweenawan LIP that ring the northern shore of Lake Superior (Fig. 7) as a giant circumferential dyke swarm, and suggested that opening of the Mid-Continent Rift may have occurred along the extensional belt defined by the circumferential swarm. The northern portion of the black ellipse shown in Fig. 7 roughly matches the geometry of the rift system in the Lake Superior region, as well as the curvature of the dyke swarms along the northern Lake Superior shore (cf., red ellipse segment), provided that the dyke swarms are shifted ~80 km southward to take account of the opening of the rift. The centre of the black ellipse is close to the Goodman swell (Peterman and Sims 1988) which may locate the centre of an underlying mantle plume (Hutchison et al. 1990; Ernst and Buchan 1997). No radiating swarm has been identified which is coeval with the Keweenawan magmatism, although a few isolated dykes may fit a radiating geometry. However, the somewhat older 1141 ± 2 Ma (U-Pb, Krogh et al. 1987) Abitibi diabase dykes (Fig. 7) and ca. 1144 Ma lamprophyre dykes (U-Pb, Ar–Ar; Queen et al. 1996) could represent part of a radiating swarm that was a precursor to the Keweenawan event (see discussion in Queen et al. 1996; Heaman et al. 2007), given that they focus near the Goodman swell. An imprecise U-Pb age of 1140 ± 20 Ma has been reported for a Pigeon River (Thunder Bay) dyke (Heaman et al. 2007), suggesting that there could be a component of the circumferential swarm that was emplaced prior to the Keweenawan event. However, some of the circumferential dyke swarms shown in Fig. 7 intrude Keweenawan rocks, and therefore cannot be pre-Keweenawan in age. In particular, many Pigeon River dykes intrude early Keweenawan Logan sills and many Copper Island dykes crosscut early Keweenawan Olser volcanics. The overall time span for emplacement of the circumferential dykes is not known and will require more U-Pb dating.

Lake Victoria dykes of Kunene-Kibaran LIP, east Africa (ca. 1370 Ma): The Lake Victoria giant circumferential swarm (Mäkitie et al. 2014; Ruotoistenmäki 2014; Ernst et al. 2014), which is associated with the Kunene-Kibaran LIP (Mäkitie et al. 2014; Ernst 2014), has a circular geometry over an arc of 160° and an outer diameter of ~650 km (Fig. 8). It is a prominent feature on aeromagnetic maps of the region. The swarm is only imprecisely dated at 1374 ± 42 Ma and 1368 ± 41 Ma (Sm–Nd; Mäkitie et al. 2014). There is no associated giant radiating dyke swarm. However, numerous roughly coeval sills, layered intrusions and granitoids are located within the circumferential swarm. It is uncertain whether the "dykes" are vertical. Geophysical modelling suggests inward-dipping cone sheets (Ruotoistenmäki 2014), whereas field descriptions indicate the presence of some subvertical dykes (Mäkitie et al. 2014).

2.2 *Possible Giant Circumferential Dyke Swarms and Swarm Segments*

There are a number of cases in which a set of dykes appears to form a circumferential pattern, but the overall distribution of dykes in the region is rather complicated, or the possibility of secondary deformation has not been ruled out. In other cases, dating of the dykes of an apparent circumferential system is currently inadequate and further study is needed to confirm that the dykes are all related to the same magmatic event. In addition, there are examples in which a small dyke swarm without a clear arcuate geometry is perpendicular to coeval dykes of a giant radiating swarm, suggesting that the small swarm could represent a segment of a giant circumferential swarm with its

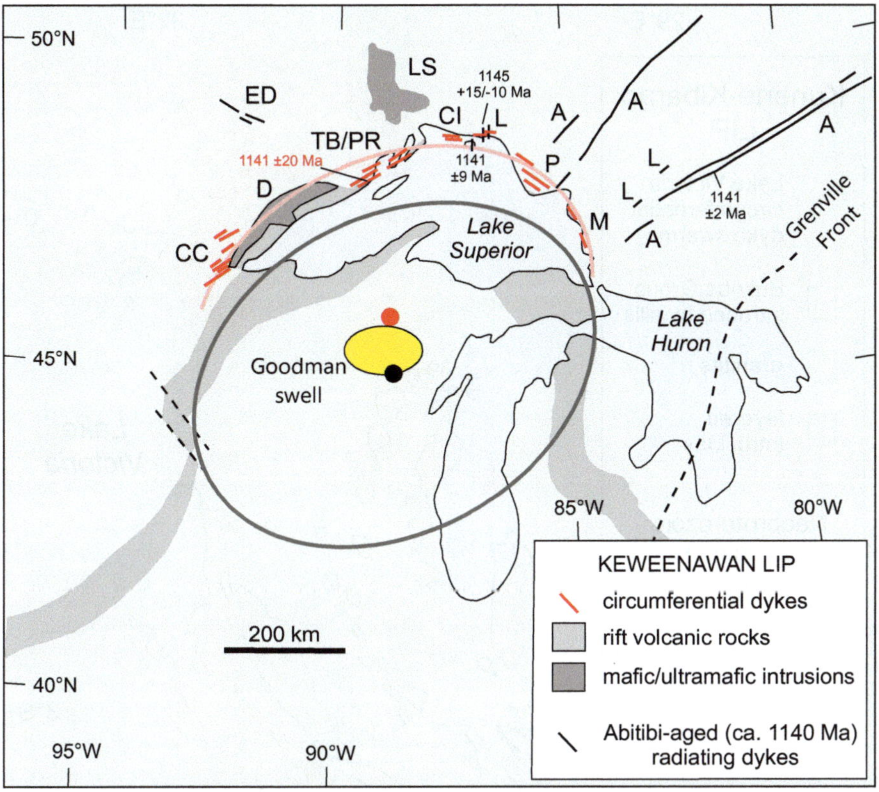

Fig. 7 Circumferential (red) and radiating dykes (black) of the ca. 1110–1085 Ma Keweenawan LIP and ca. 1140 Ma Abitibi magmatic event associated with the Mid-Continent Rift of North America. Rift-parallel dyke swarms: CC = Carleton County, TB/PR = Thunder Bay (Pigeon River), CI = Copper Island, P = Pukaskwa, M = Mamainse Point. Other mafic/ultramafic intrusions: D = Duluth anorthosite complex, LS = Logan (Nipigon) sills. Radiating dykes: A = Abitibi dykes, ED = Eye Dashwa dykes. L = lamprophyre dykes. Dykes are modified from Buchan and Ernst (2004). The Goodman swell (Peterman and Sims 1988) may locate the centre of a Keweenawan mantle plume as described in the text. The northern portion of the black ellipse (centred at the black dot) roughly matches the geometry of the rift system in the Lake Superior region. If shifted north ~80 km (as the pink arc centred at the red dot) to reflect opening of the rift, it roughly matches the curvature of the dyke swarms along the northern Lake Superior shore. Linear Keweenawan Baraga and Mellen-Gogebic dyke swarms (e.g., Buchan and Ernst 2004) that occur south of Lake Superior are not shown. Their trends are not consistent with either the radiating or circumferential systems. U-Pb ages for radiating Abitibi (Krogh et al. 1987) and lamprophyre (Queen et al. 1996) dykes are shown in black lettering. A U-Pb age for a Pigeon River circumferential dyke (Heaman et al. 2007) is shown in red. This age suggests emplacement well before the Keweenawan event. However, many dykes of the circumferential system (including some Pigeon River dykes) are observed to cut Keweenawan sills and volcanics and hence cannot be pre-Keweenawan in age

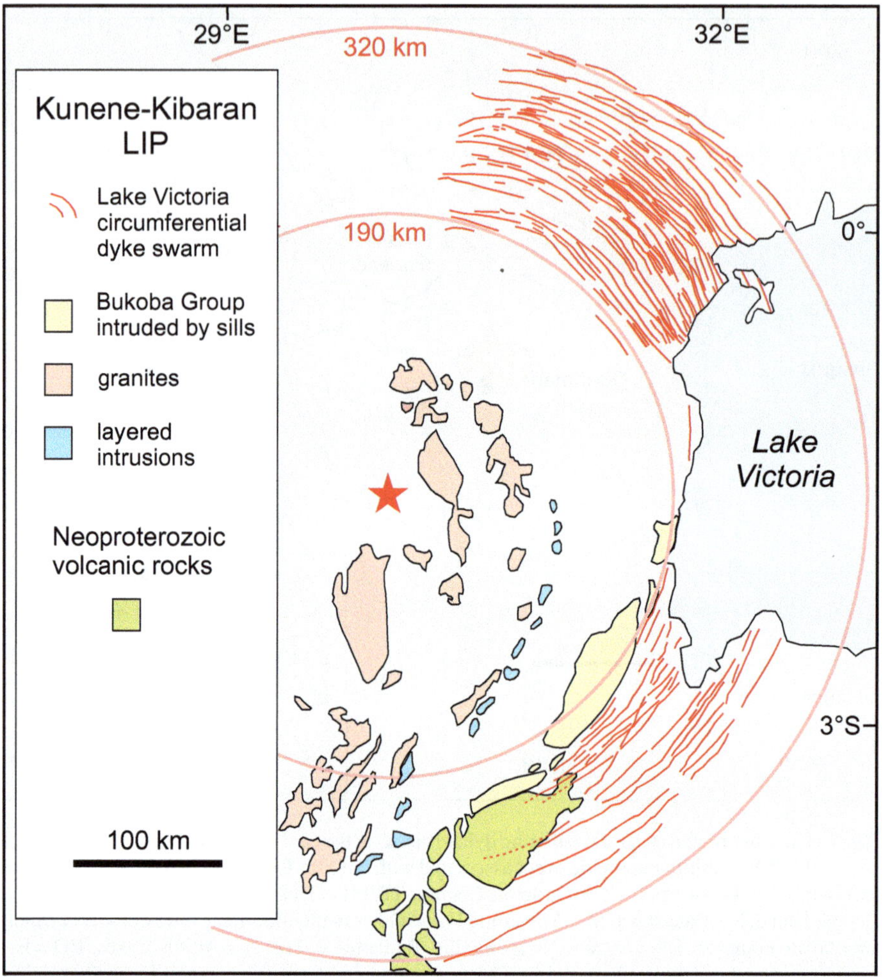

Fig. 8 Ca. 1370 Ma giant circumferential Lake Victoria dyke swarm of the Kunene-Kibaran LIP of eastern Africa, traced mainly from aeromagnetic data. Dashed pink circles indicate the approximate outer and inner dimensions of the swarm. The red star locates the centre of the swarm. The figure is modified after Mäkitie et al. (2014), with the Bukoba Group after Tack et al. (2010)

centre at the focus of the radiating swarm. Several examples of possible swarms are described briefly below.

Dykes of British-Irish portion of North Atlantic LIP (62–54 Ma): SSE- to SE-trending dykes associated with the North Atlantic LIP (or North Atlantic Igneous Province, NAIP) form a dense swarm across Britain and Ireland (Fig. 9; e.g., Speight et al. 1982; Cooper et al. 2012) and may represent part of a giant radiating swarm focused above the Iceland mantle plume. See a reconstruction of the North Atlantic LIP in the inset to Fig. 9. Maclennan and Jones (2006) proposed a model of an ellipse-shaped transient uplift (Fig. 9 inset) associated with the arrival of the plume based on observations in sedimentary basins bordering the North Atlantic Ocean. In addition to the dominant SSE-trending dyke swarm of Britain and Ireland, dykes are locally associated with a number of central igneous complexes. These dykes typically occur within a few tens of km or less of the complexes and can be parallel to the SSE swarm, perpendicular to this swarm or form a fan about the complexes (Speight et al. 1982). In addition, small ring dyke or cone sheet swarms occur within or around the edge of the complexes (Emeleus 1982). Finally, dyke swarms at a high angle to the dominant SSE swarm, but not close to central complexes, occur in Ireland, both within the Republic of Ireland and Northern Ireland. They include the ENE Mayo (or West Connacht) swarm (Mohr 1982, 1988; Preston 2001), similarly trending dykes in the Malin Head region (Geological Survey of Northern Ireland 1977) and Belfast region (Walker 1959; Geological Survey of Northern Ireland 1971), as well as the E-trending Inischrone (or North Connacht) swarm (Mohr 1987, 1988; Preston 2001). It has been suggested that the Mayo dyke swarm may emanate from the offshore Brendan complex (e.g., Mohr 1982) shown on Fig. 9. However, this swarm extends nearly 200 km from the centre of the Brendan complex, many times farther that the dykes associated with other central complexes. We suggest that the Mayo swarm and the similarly trending dykes of the Malin Head and Belfast regions may belong to a giant circumferential dyke swarm that parallels the outer edge of the elliptical uplift described by Maclennan and Jones (2006), and that is related to the outer edge of the flattened plume head. The Inischrone swarm, which comprises a dense system of very narrow dykes (usually <2 m) (Mohr 1987), may have a more local source. It is not clear if the proposed circumferential swarm extends into Scotland, but we suggest that two long ENE-trending dyke segments (labelled SUF in Fig. 9) adjacent to the Southern Upland Fault, that are at right angles to the dominant SSE dyke swarm, and have been interpreted as right-angle bends in SSE-trending dykes (MacGregor 1949; MacDonald et al. 2014), may actually be part of the circumferential swarm. The Mayo and Inischrone swarms of western Ireland are interpreted to belong to the North Atlantic LIP event based on K-Ar and Ar–Ar dating (e.g., Thompson 1985; Mitchell and Mohr 1986). However, the relative ages of the proposed circumferential and radiating swarms is not clear and will require more detailed studies.

Dykes of Madagascar LIP, Madagascar and southern India (ca. 92–88 Ma): The 92–88 Ma Madagascar LIP in Madagascar is characterized by widespread flood basalts and dykes associated with the breakup of Madagascar and India (e.g., Storey et al. 1995; Melluso et al. 2009; Cucciniello et al. 2013, 2015). Ernst and Buchan (1997) proposed that the dominant dyke sets form a giant radiating swarm focused on the Marion plume centre off the southeast corner of the island (Fig. 10). Subsidiary dyke sets intersect the dominant system roughly at right angles (Fig. 10). We suggest that they may represent components of a giant circumferential swarm with outer diameter of ~2400 km. In addition, coeval dykes have been identified in southern India (Fig. 10), paralleling the west coast (Radhakrishna et al. 1990; Radhakrishna and Joseph 2012), and inland perpendicular to the coast (Kumar et al. 2001). How the southern India dykes relate to the pattern of dykes in Madagascar is uncertain because the exact reconstruction of Madagascar and India is controversial. Nevertheless, using

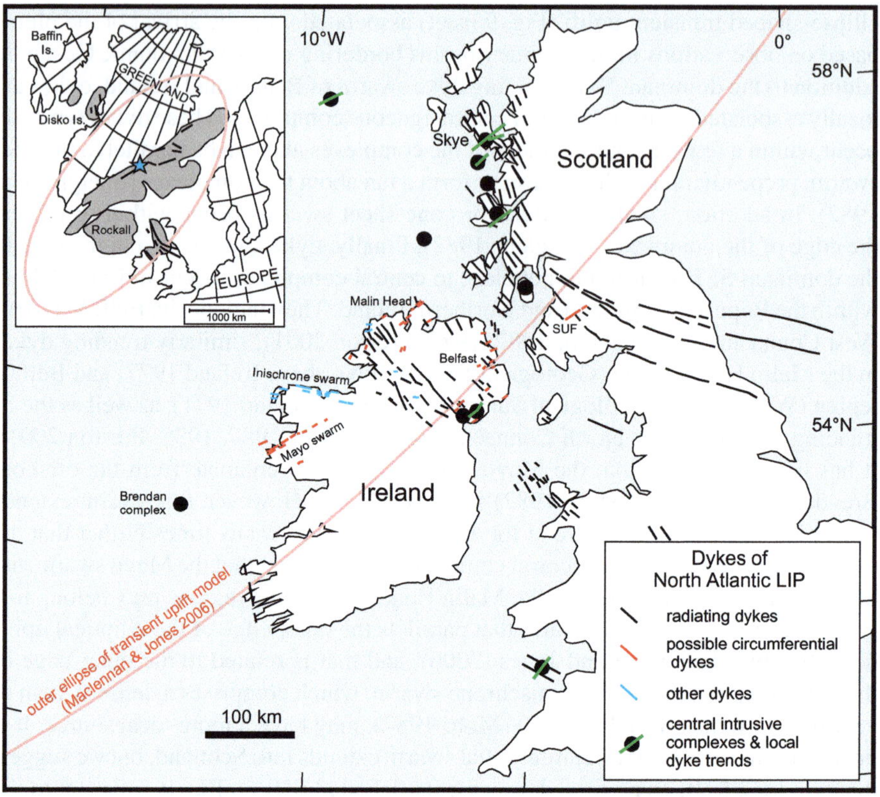

Fig. 9 Possible circumferential (red) and radiating (black) dykes of the British-Irish portion of the 62–54 Ma North Atlantic LIP. The radiating swarm is dense in some areas, and has been thinned on the figure for the sake of clarity. Central complexes (black circles) are often associated locally with dykes of the overall radiating system, dyke sets of other trends (in green), and internal ring dykes or cone sheets. Possible circumferential dykes are from various sources described in the text. The inset shows a reconstruction of the North Atlantic LIP (modified after Ernst and Buchan 1997, Fig. 4 and references therein), with volcanics in grey, and the blue star locating the approximate plume centre. The pink ellipse is the outer contour of the uplift associated with plume arrival described by Maclennan and Jones (2006)

the reconstruction of Torsvik et al. (2000, Fig. 3), the dykes parallel to the coast of India fit the radiating pattern and the inland dykes fit the circumferential pattern as illustrated in Fig. 10. Precise dating is needed to establish the relative ages of the proposed circumferential and radiating swarms.

Dykes of the eastern Karoo Basin, southern Africa (ca. 183 Ma): Dykes that crosscut sediments, as well as sills and volcanics of the ca. 183 Ma Karoo LIP, within the Karoo Basin form a dense array with multiple trends (Fig. 1 of Chevallier and Woodford 1999). Although usually assumed to be of Karoo age, reliable radiometric dating of the dykes is very sparse, so that dykes of younger ages may also be present. Chevallier and Woodford (1999) have proposed that certain Karoo "megaswarms"

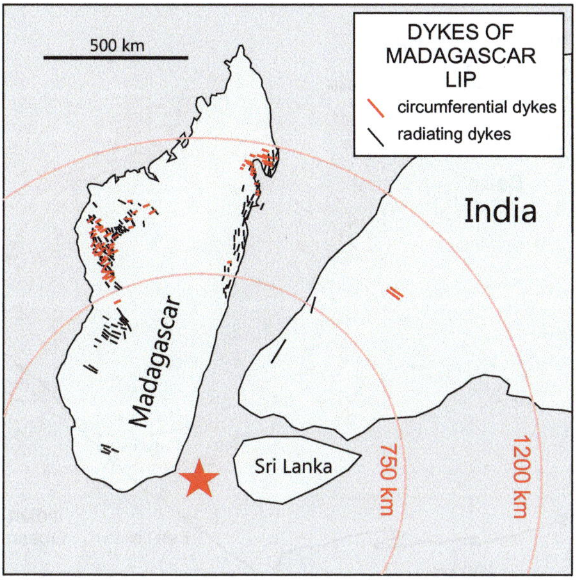

Fig. 10 Possible circumferential (red) and radiating (black) dykes of the reconstructed 92–88 Ma Madagascar LIP of Madagascar and southern India. Dykes of Madagascar are from Storey et al. (1995) and Melluso et al. (2009). Dykes of India are from Radhakrishna et al. (1990), Kumar et al. (2001) and Radhakrishna and Joseph (2012). The reconstruction is from Torsvik et al. (2000). Pink circles indicate the approximate outer and inner dimensions of the circumferential swarm. The red star is the focus of the radiating swarm, centre of the circumferential swarm, and approximate location of the Marion plume

in the eastern Karoo Basin form a giant dyke swarm radiating from a triple junction and mantle plume centre off the east coast (Fig. 11b; Fig. 1a, b of Chevallier and Woodford 1999). In particular, Chevallier et al. (2001) described one prominent set of dykes as "a major curvi-linear swarm of extensive and thick dykes". The southern portion of this swarm trends E-W near the coast (see the Gap dykes of Fig. 11a) and curves gradually to the north farther inland. The well-studied Gap dykes (e.g., Mountain 1943; Moore 1965) have precise U-Pb zircon ages of 182.5 ± 0.3 and 182.6 ± 0.3 Ma (Svensen et al. 2012) and typical widths of 100–200 m (occasionally up to 290 m). Farther north, in the northeastern part of the Basin, Chevallier et al. (2001) identified a second arcuate set of dykes, trending NW close to the coast "in the Transkei region" and curving to an E-W trend farther inland "in the Free State" (broadly from north of Port St. Johns to Bloemfontein in Fig. 11). Together these two arcuate swarms were thought to radiate from the offshore triple junction and plume centre (Fig. 11b).

However, dykes of a giant radiating dyke swarm linked to a mantle plume typically display a radiating geometry in the region of plume-generated uplift, and curve into a more linear pattern beyond the uplift in response to regional stresses (e.g., Ernst and Buchan 2001). This is unlike the pattern proposed by Chevallier and Woodford (1999)

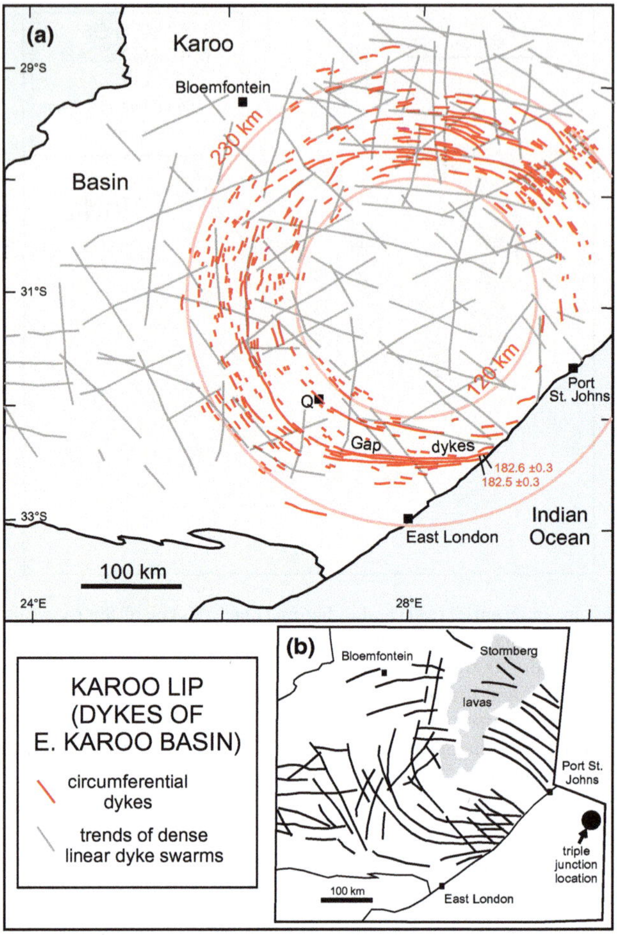

Fig. 11 **a** Possible circumferential dykes (red) of the ca. 183 Ma Karoo LIP in the eastern Karoo Basin of southern Africa. **b** Earlier interpretation of arcuate Karoo dykes in the same area (modified from Fig. 1a of Chevallier and Woodford 1999, with triple junction location from Fig. 1b of Chevallier and Woodford 1999). In this earlier interpretation, the southern to western arc (including the Gap dykes) and a modified version of the northwestern to northern arc of the proposed circumferential system were recognized by Chevallier and Woodford (1999, Fig. 1a) and Chevallier et al. (2001), but linked to an offshore triple junction, rather than a circumferential swarm (see discussion in text). Dense linear dyke swarms of several trends also occur throughout the eastern Karoo Basin. Grey lines in (**a**) indicate only their generalized distribution and approximate trends. They often overlap in trend with dykes interpreted to belong to the circumferential swarm. The pink circles indicate the approximate outer and inner dimensions of the proposed circumferential swarm. The dyke distribution is simplified from Chevallier and Woodford (1999, Fig. 1). With the exception of the 183 Ma (U-Pb; Svensen et al. 2012) Gap dykes, most dykes in the Basin are poorly dated. Q = Queenstown

for the dykes radiating from a triple junction and mantle plume off the east coast (Fig. 11b), where the dykes of the northern and southern sets appear to arc towards one another far from the hypothesized plume centre. Therefore, we suggest an alternative explanation for the more southerly arcuate dyke set—namely that it could represent the southern to western portion of a giant circumferential dyke swarm with an outer diameter of ~460 km and an arc of >100° (Fig. 11). More speculatively, the northern arcuate dyke set (or at least the northern portion of it) could form the northeastern to northern portion of this giant circumferential swarm. However, because of the high density of NW- to W-trending dykes in this northern region it is difficult to assess which specific dykes might belong to the circumferential system and which might belong to various other crosscutting and potentially younger swarms. Figure 11a illustrates an interpretation of a circumferential system, assuming that all the dykes selected are actually of Karoo age.

Dykes of Yakutsk-Vilyui LIP, Siberia (ca. 380–360 Ma): The Yakutsk-Vilyui LIP of Siberia is thought to consist of two magmatic pulses that peaked at ca. 377 and 364 Ma based on Ar–Ar dating (e.g., Ricci et al. 2013, Fig. 4). A giant radiating dyke swarm belonging to this event is composed of subswarms that parallel prominent triple-junction rifts and associated volcanics (Fig. 12; Shpount and Oleinikov 1987; Kiselev et al. 2012; Ricci et al. 2013). Their focus has been interpreted to locate a mantle plume centre (Ernst and Buchan 1997). Minor dykes, thought to belong to the event, are oriented at approximately right angles to the dominant subswarms in several areas (Fig. 12; Kiselev et al. 2012). We suggest that they may be components of a giant circumferential swarm that circumscribes the plume centre, with an outer diameter of ~1600 km. Ages constraints on the possible circumferential dykes and the radiating system are very poor, so that it is not yet possible to establish the relative age of the swarms or to attempt to link them to the main magmatic pulses.

Clarence Head dykes of Franklin LIP, northern Canada (ca. 716–713 Ma): The ca. 720 Ma Franklin LIP includes a giant radiating dyke swarm, small subswarms with discordant trends, as well as volcanics and sills (Fig. 7 of Buchan et al. 2010). Based on U-Pb dating, Denyszyn et al. (2009) demonstrated that the 716–713 Ma Clarence Head subswarm, which crosscuts the slightly older (ca. 720 Ma) Franklin radiating swarm at right angles (Fig. 2 of Denyszyn et al. 2009) ~1250 km east of the focus of the radiating swarm, is part of the Franklin LIP. Although the subswarm is too short (~150 km) to show an arcuate pattern, they suggested that it could be a component of a giant circumferential dyke swarm. Its diameter would be ~2500 km, the largest yet proposed on Earth, and comparable in size to the largest Venusian corona (~2600 km diameter).

Denyszyn et al. (2009) also proposed that the Lasard River subswarm, located south of the radiating swarm focus and dated at ca. 715 Ma (U-Pb; M. Hamilton in Buchan et al. 2010), could be part of the circumferential swarm. Similarly, Denyszyn et al. (2016) have suggested that the Tatonduk dykes located southwest of the radiating swarm focus, which they dated at 714 ± 1 Ma (U-Pb), may be part of the circumferential system. However, the Lasard River subswarm is oriented within ~30° of nearby dykes of the radiating swarm (Fig. 7 of Buchan et al. 2010), and the Tatonduk dykes have been folded so that their primary orientation is unknown

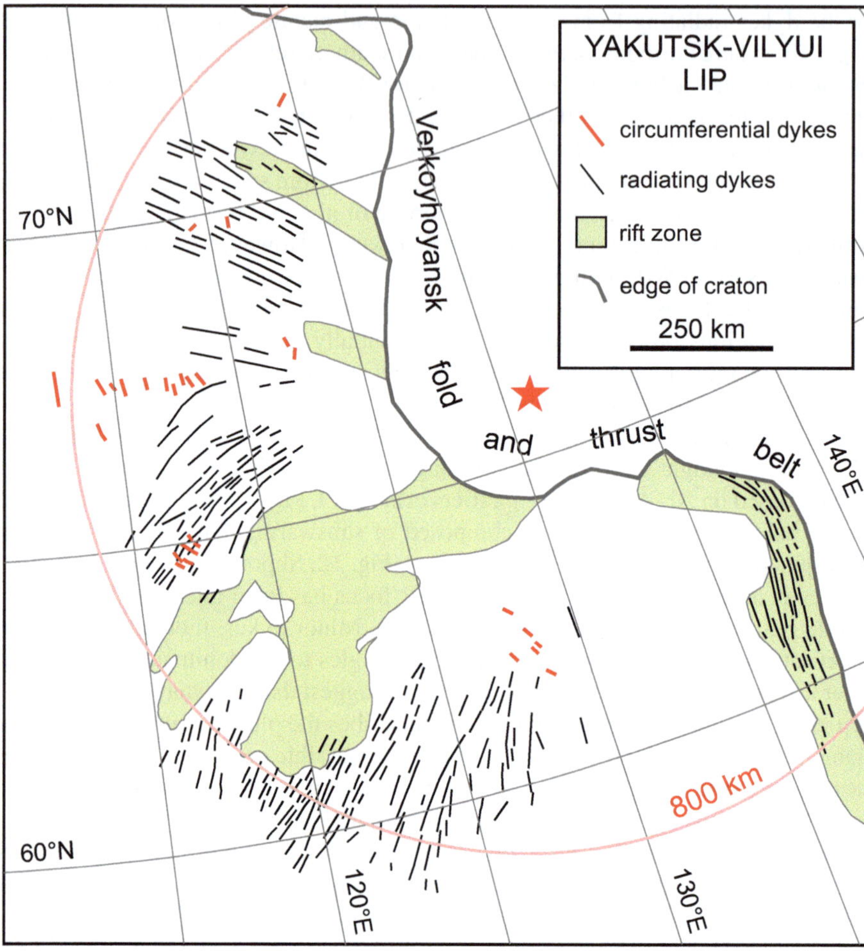

Fig. 12 Possible circumferential (red) and radiating (black) dykes of the ca. 370 Ma Yakutsk-Vilyui LIP, Siberia. The dyke distribution is simplified from Kiselev et al. (2012). The pink circle indicates the approximate outer limit of the circumferential dykes. The black star locates the focus of the radiating swarm and proposed plume centre (Ernst and Buchan 1997)

(Cox et al. 2017). In addition, Cox et al. (2017) have proposed that the Tatonduk dykes and related volcanics may be linked to rifting along the northwestern margin of Laurentia, rather than to the Franklin LIP. Given these uncertainties, we conclude that the Clarence Head dykes may be part of a circumferential swarm, but that there is no clear evidence linking the Lasard River or Tatonduk dykes to a circumferential swarm.

Dykes of Umkondo LIP, southern Africa (ca. 1110 Ma): To explain the complicated pattern in the trends of Umkondo-aged dykes, de Kock et al. (2014, Fig. 1) postulated up to three coeval centres of magmatism linked to a single mantle plume, each with a radiating dyke swarms. They considered the magmatic centre along the present-day NW margin of the Kalahari craton as the most robust, given its association with the significant "NW Botswana Rift", its bimodal volcanism and the presence nearby of the Xade and Tsetseng layered magmatic complexes. De Kock et al. (2014) suggested that the multiple magmatic centres could be generated from a single mantle plume if flow is channeled laterally to thinspots in the overlying lithosphere, a process occasionally suggested for other LIP events (e.g., Sleep 2006; Bright et al. 2014).

We propose a simpler interpretation of the Umkondo dyke orientations (Fig. 13) in terms of a single giant radiating swarm (black dykes) and a single giant circumferential swarm (red dykes), both centred along the present-day NW margin of the Kalahari craton. This interpretation eliminates the need for multiple magmatic centres and multiple radiating swarms associated with the Umkondo plume. The relative age of the proposed circumferential and radiating dykes is unclear. Only a single, rather imprecise U-Pb age of 1110 ± 19 Ma (Divuli Ranch dyke of Fig. 13; de Kock et al. 2014) is available from the circumferential swarm, and would allow these dykes to be older or younger than the somewhat better dated radiating dykes (Fig. 13; de Kock et al. 2014). The proposed circumferential Mutare M2 and M3 dykes of Fig. 13 are only linked to the Umkondo event based on their geochemistry (see summary in de Kock et al. 2014).

Dykes of Marnda Moorn LIP, Australia (ca. 1210 Ma): The ca. 1210 Ma Marnda Moorn LIP of Australia is comprised of several coeval dyke swarms, most of which "exhibit regional arcuate trends around the margin of the Yilgarn craton" (Pirajno and Hoatson 2012), and fit within the so-called "outer zone" between a pair of concentric circles drawn by Wang et al. (2014, Fig. 1) and reproduced in Fig. 14. Along the craton's northwest margin, the Muggamurra dykes arc from a NE trend to an ENE trend (Fig. 1 of Wang et al. 2014). Along the northern craton margin this arcuate system continues from ENE to E based on aeromagnetic anomalies that have been interpreted to represent Muggamurra dykes (Fig. 19 of Wyche et al. 2013). The Gnowangerup (or Ravensthorpe) dykes arc from WNW to ENE along the southern margin of the craton (Myer 1990; Wang et al. 2014). They have been mapped farther east by Spaggiari et al. (2009, Fig. 3) to link with the Fraser dykes which trend NE parallel to the craton's SE margin (Wingate et al. 2000; Wang et al. 2014). Precise U-Pb ages have been obtained at several locations in these swarm (Fig. 14; Wang et al. 2014; age references summarized in Wingate and Pidgeon 2005).

We suggest that the Marnda Moorn dykes that parallel the craton margin may constitute a giant circumferential dyke swarm with an outer diameter of ~1050 km. However, there are some complications related to such an interpretation. The dykes are mapped mainly from aeromagnetic data so that multiple ages of dyke emplacement may be present in some or in all of these swarms (e.g., Wang et al. 2014). Indeed, recent U-Pb dating has identified a much older 1.89 Ga Boonadgin swarm of WNW trend (Stark et al. 2018) in the vicinity of the similar-trending western portion of the Gnowangerup swarm (Fig. 14). In addition, two dyke swarms, the Boyagin and Wheatbelt swarms with Marnda Moorn U-Pb ages (Fig. 14; Wang et al. 2014, Fig. 1), do not readily fit the circumferential pattern, with the possible exception of some Boyagin dykes north of Perth. Wheatbelt dykes are mainly of W trend (with

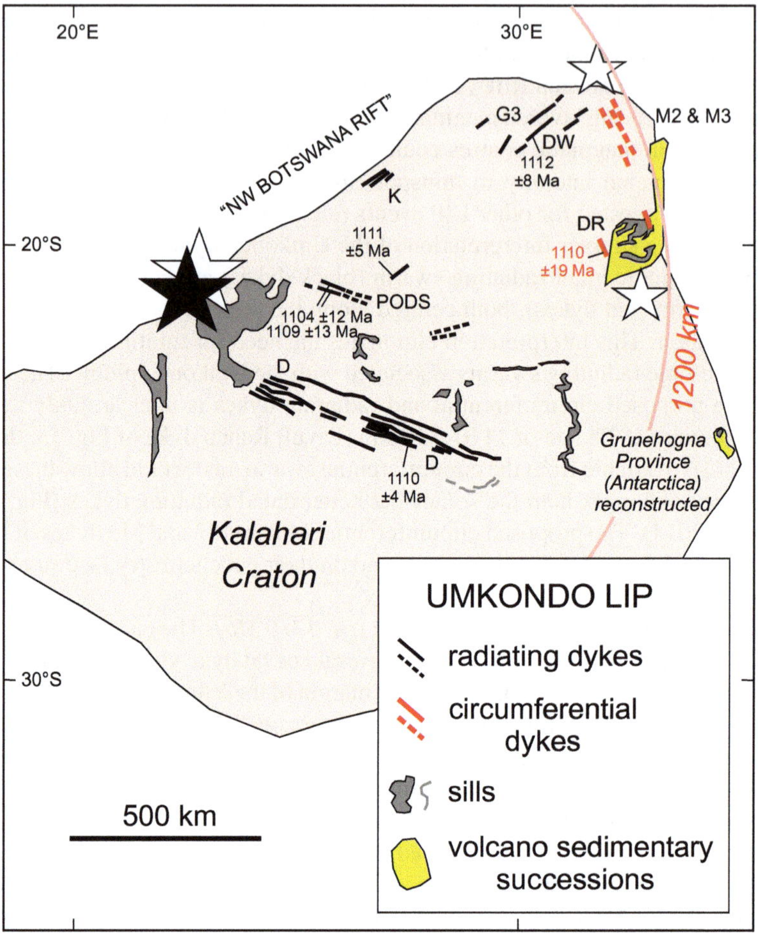

Fig. 13 Possible circumferential (red) and radiating (black) dykes of the ca. 1110 Ma Umkondo LIP, southern Africa. Individual dykes and dyke swarms are from de Kock et al. (2014), who interpreted them in terms of three radiating swarms. White stars locate the three proposed radiating centres of de Kock et al. (2014), with the larger star considered by these authors as the most robust. The black star is the approximate focus of the radiating dykes, as reinterpreted herein. The pink arc indicates the approximate outer edge of the circumferential swarm (red), assuming its centre is close to the focus of the radiating swarm (i.e., black star). D = Dibete swarm, DR = Divuli Ranch dyke, DW = Deweras dyke, G3 = Umkondo portion of Guruve swarm, K = Kamatavi swarm, M2 and M3 = Umkondo portion of Mutare swarm, PODS = Umkondo (Proterozoic) portion of Okavango swarm. U-Pb ages and corresponding sampling locations are from de Kock et al. (2014, Fig. 2 and Table 1), with red and black lettering for circumferential and radiating dykes respectively. Note that it appears that the Dewares dyke sample label [ZM10/21] and the Duval Ranch dyke sample label [ZM10/4] were accidently switched in Fig. 1 of de Kock et al. (2014). The Grunehogna Province of Antarctica is shown in its reconstructed location with respect to the rest of the Kalahari craton

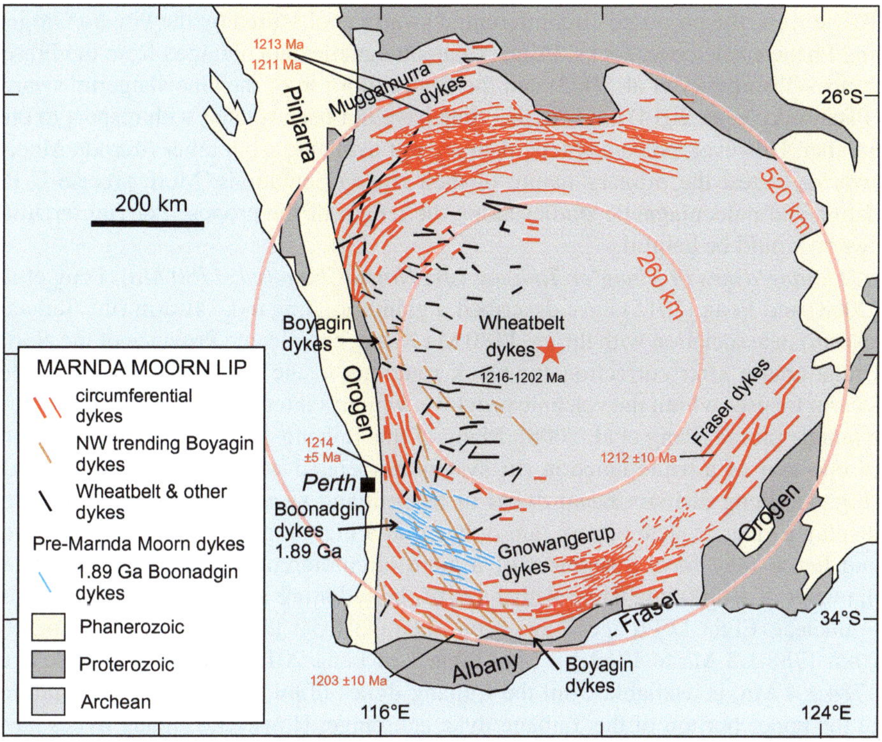

Fig. 14 Possible circumferential dyke swarm of the ca. 1210 Ma Marnda Moorn LIP of south-western Australia. Thick red dykes interpreted herein to be part of the circumferential swarm are from Wang et al. (2014). Thin red dykes are from Wyche et al. (2013) and Spaggiari et al. (2009). Dykes of the Boyagin swarm (brown) and Wheatbelt swarm and other dykes (black) are from Wang et al. (2014). Some Boyagin dykes northeast of Perth have been reinterpreted herein as possibly part of the circumferential swarm. Pink circles are from Wang et al. (2014) and encompass the dyke sets that are parallel to the Yilgarn craton margin. They also correspond to the approximate outer and inner dimensions of the proposed circumferential swarm. The red star locates the centre of the proposed circumferential swarm. U-Pb ages (in red) for Marnda Moorn dykes are after Wang et al. (2014), with primary references given in Table 1 of Wingate and Pidgeon (2005). 1.89 Ga Boonadgin dykes (blue) are from Stark et al. (2018). Assignment of dykes to specific swarms in the western region close to the Pinjara Orogen is very difficult due to limited dating, overlapping trends of dykes of different ages and limited exposure. Background geology modified after Wang et al. (2014)

various other trends also observed), at a high angle to the craton margin and the circumferential system. Wheatbelt dykes that are perpendicular to the craton margin and to the circumferential swarm could represent part a radiating system, but an overall radiating pattern is not observed. The Boyagin dykes are interpreted to form a linear NW trending swarm that intersects the swarms along the craton margin. Finally, orogenic belts parallel much of the proposed circumferential swarm (Fig. 14). In the south, the Albany-Fraser Orogeny parallels the Fraser and Gnowangerup swarms.

To the west, the proposed circumferential swarm is obscured by the Pinjara Orogen and Phanerozoic cover rocks. Primary paleomagnetic data obtained from the Fraser swarm (Pisarevsky et al. 2003) and the central portion of the Gnowangerup swarm (Pisarevsky et al. 2014) confirm that they have not been rotated with respect to one another. However, paleomagnetic data are not available from other Marnda Moorn swarms to test the primary nature of their dyke orientations. More precise U-Pb dating and paleomagnetic studies along the length of the proposed circumferential swarm would be helpful.

Lüliang dykes of Xiong'er-Taihang LIP, North China (ca. 1780 Ma): Peng et al. (2008) and Peng (2015) have described a giant radiating dyke swarm (the Taihang swarm) in association with the ca. 1780 Ma Xiong'er Volcanic Province of the North China craton after correction for block rotations in the Mesozoic (Fig. 15). The swarm focuses within the volcanic province, which is interpreted to be located above a mantle plume (Peng et al. 2008). Many of the Taihang dykes are oriented parallel to one arm of a triple-junction rift system associated with the volcanic province (Fig. 15; Peng 2015). Coeval dykes of the Lüliang swarm, which is too short to display a clear arcuate pattern, are oriented at a high angle to the radiating swarm and hence may form a segment of a giant circumferential swarm with an outer diameter of ~1200 km. The relative age of the radiating and circumferential dykes is unclear. Eight U-Pb ages have been obtained from the Taihang dykes ranging from 1788 ± 5 Ma to 1769 ± 3 Ma (Table 1 of Peng 2015). Only one precise age, 1784 ± 4 Ma, is available from the Lüliang dykes (Table 1 of Peng 2015), falling in the upper portion of the Taihang dyke age range. However, Lüliang dykes have been observed cutting Taihang dykes (Peng et al. 2008) indicating that at least some Lüliang dykes are younger than some Taihang dykes.

2.3 Possible Circumferential Wrinkle Ridges

Circumferential wrinkle ridges occur at a number of coronae on Venus and similar features on Mars (Squyres et al. 1992; Stofan et al. 1992; Mège and Ernst 2001). A possible analogue is associated with the Columbia River LIP of western North America.

Yakima folds of the Columbia River LIP, North America (ca. 17 Ma): Mège and Ernst (2001) noted that the Yakima folds or ridges (e.g., Fig. 3 of Watters 1989) that are associated with the ca. 17 Ma onset of volcanic activity in the Columbia River LIP of western North America are roughly perpendicular to dykes of a giant radiating dyke swarm (Ernst and Buchan 1997). They proposed that the folds form a segment of a circumferential arc at a distance of ~600–700 km from the focus of the radiating swarm and probable mantle plume centre.

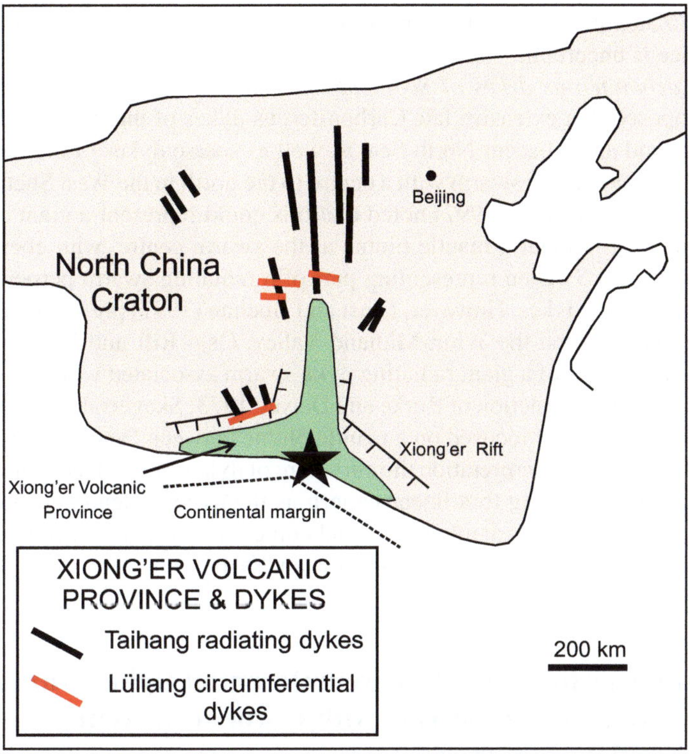

Fig. 15 Possible circumferential (red) and radiating (black) dykes of the ca. 1780 Ma Xiong'er-Taihang LIP, North China craton. The dyke distribution is modified from Peng et al. (2008). The black star locates the focus of the radiating dykes in the region of the Xiong'er Volcanic Province and associated triple-junction rifting (Peng et al. 2008)

2.4 Unsubstantiated Circumferential Swarms

In addition to the examples discussed above and catalogued in Table 1, there are cases in the literature where a circumferential geometry has been proposed for a large dyke set, but in subsequent studies the interpretation has not been confirmed or an alternative interpretation is preferred.

Dykes of Ethiopian Plateau, Afar (≤30 Ma): In an early study, Mohr (1971) interpreted many of the dykes in the Ethiopia Plateau region to form the "eastern half of a circle" (diameter ~800 km) and speculated that either a "magmatic source lay within the semicircle, under the central Ethiopian plateau" or that the dyke orientations might "reflect a warped protomargin of the Nubian crustal plate .... against Arabia and the Indian Ocean". However, subsequent studies by the same author (e.g.,

Mohr and Zanettin 1988) do not appear to mention this arcuate swarm and, hence, its existence is uncertain.

Late Carboniferous dykes of Scotland and Norway (~300 Ma): Smythe et al. (1995) proposed that extensive late Carboniferous dykes of the Whin-Midland Valley of Scotland and adjacent North Sea, as well as sparse dykes of the Oslo Rift of Norway, form an arcuate swarm with a centre to the north in the West Shetland Shelf region. Ernst and Buchan (1997) noted that this could represent a giant circumferential dyke swarm about a mantle plume at the swarm centre, with coeval Scania dykes in southern Sweden representing part of a radiating swarm perpendicular to the circumferential dykes. However, Ernst and Buchan (1997) preferred an alternate interpretation in which the Whin-Midland Valley, Oslo Rift and Scania dyke sets constitute three arms of a giant radiating dyke swarm associated with triple junction rifting (Jutland triple junction of Burke and Dewey 1973; Skagerrak-Centered LIP of Torsvik et al. 2008) and focused on a mantle plume between Denmark and southern Norway. In this latter interpretation the curvature of dykes in Scotland simply reflects a change from a radiating to a linear pattern as the dykes extend beyond the zone of domal uplift above the mantle plume and come under the influence of a regional stress field (Ernst et al. 1995; Ernst and Buchan 2001).

3 Characteristics of Giant Circumferential Dyke Swarms on Earth and Comparison with Coronae on Venus and Similar Features on Mars

In this section we summarize the basic characteristics of the terrestrial giant circumferential dyke swarms described in this paper, many of which are common to coronae on Venus and similar features on Mars.

Most giant circumferential swarms are approximately circular in shape (e.g. Lake Victoria of Fig. 8). Occasionally, they are elliptical (e.g., Siberian Traps of Fig. 5). Similarly, most Venusian coronae and similar features on Mars are quasi-circular (e.g., Stofan et al. 1992). However, some coronae are elliptical, and others have more irregular geometries.

Circumferential dyke swarms exhibit a very wide range in size, with those catalogued herein having outer diameters from ~450 to ~2500 km (Table 1). In comparison, coronae on Venus range mainly from 60 to 1100 km, with one larger example (Artemis corona) having a diameter of ~2600 km (Stofan et al. 1992). On Mars, features that are similar to coronae range in diameter from 500 to 1000 km (Ernst et al. 2001).

Giant circumferential swarms exhibit a broad range in widths (Table 1). With the exception of the largest swarms, which appear to have highly variable widths, there may be a general increase in swarm width with swarm diameter. However, more data are needed to draw any definitive conclusion. The graben-fissure systems of coronae also show a broad range in widths, with a tendency for widths to "increase somewhat with corona size" (Stofan et al. 1992).

Although the presence of dykes is a defining characteristic of giant circumferential swarms, dykes are not observed directly in the case of coronae on Venus or similar features on Mars. Nevertheless, evidence suggests that dykes do underlie circumferential graben-fissure systems on Venus (McKenzie et al. 1992; Ernst et al. 2001; Bethell et al. 2016) and Mars (Zimbelman and Edgett 1992; Montési 2001; Ernst et al. 2001). For example, channelized volcanic flows can be seen emanating from circumferential graben on both planets (e.g., Fig. 6 of Zimbelman and Edgett 1992; Fig. 2A of Bethell et al. 2016). There is also substantial evidence for dykes beneath radiating graben on the two planets (McKenzie et al. 1992; Grosfils and Head 1994; Mège and Masson 1996; Ernst et al. 2001, 2003; Wilson and Head 2002; Studd et al. 2011). For example, magma is observed to be fed from radiating graben. In addition, radiating graben extend far beyond their associated domal uplift. This cannot be explained by uplift-driven extension, but requires a magmatic driver (e.g., Grosfils and Head 1994; Wilson and Head 2002).

Graben, such as those observed in the annuli of Venusian coronae, have not yet been described above dykes in any of the terrestrial giant circumferential swarms discussed herein. This is likely because they have been lost to erosion, as graben have occasionally been observed above dykes of giant radiating swarms (Balkwill and Fox 1982; Brown et al. 1994; Aspler and Ernst 2003) and in association with modern-day volcanic rift zones both on land and along mid-ocean ridges (e.g., Mastin and Pollard 1988; Rubin 1992; Chadwick and Embley 1998).

Dykes density (i.e., the spacing of dykes) around the circumference of a circumferential swarm is not well constrained in most cases. However, there are examples, where the density appears to vary considerably. For example, in the Lake Victoria swarm (Fig. 8) the dyke density decreases to the south, suggesting that the dykes may have been emplaced laterally from a source region along the arc to the north (Mäkitie et al. 2014). Graben density, inferred to reflect dyke density, also varies substantially around the circumference of some coronae.

Widths of individual dykes in circumferential swarms are quite variable. In some circumferential swarms (e.g., Lake Victoria, Keweenawan, Karoo, High Arctic, Clarence Head) the maximum widths are 100 m or more, comparable to the maximum widths observed in giant radiating dyke swarms (Ernst et al. 1995). Maximum dyke widths on Venus and Mars are thought to be roughly similar to or perhaps somewhat greater (up to several hundred metres) than maximum dyke widths on Earth, based on analysis of the cross sectional geometry (depth and width) of circumferential (and radiating) graben on these planets (e.g., Ernst et al. 1995; Mège and Masson 1996; Wilson and Head 2002).

Although data are quite limited in some cases, it appears that most circumferential swarms consist of vertical or sub-vertical dykes, as such dykes have been described, at least locally, from the High Arctic, Blekinge-Dalarna, Paraná-Etendeka, Deccan, Keweenawan, Marnda Moorn and Clarence Head circumferential or possible circumferential swarms. On the other hand, Ruotoistenmäki (2014) interpret the Lake Victoria circumferential "dykes" as dipping inward at ~30–40° (as cone sheets) based on analysis of magnetic profiles, although field studies have also reported sub-vertical dykes (Mäkitie et al. 2014). There is no information on the dip of dykes that may underlie the circumferential graben of coronae.

Most giant circumferential dyke swarm have an associated giant radiating dyke swarm. However, in several instances a radiating swarm is absent or problematic (Table 1). On Venus, coronae occur both with and without radiating graben-fissure systems.

Most coupled giant radiating and circumferential dyke swarms appear to share a common centre. However, there may be cases where the centres are offset from one another. For example, in the case of the High Arctic LIP (Fig. 3) the centres are close to one another, but may be slightly offset (Buchan and Ernst 2018). Offset centres have been observed with several coupled radiating-circumferential graben-fissure systems on Venus (e.g., Odudova corona of Fig. 15 in Krassilnikov and Head 2003).

The relative age of giant circumferential and radiating dykes in coupled systems is, in most cases, poorly constrained due to a paucity of precise dating. U-Pb ages are often of high precision. Ar–Ar ages can also be of high precision in some instances (e.g., see dating of Deccan volcanics in Renne et al. 2015). Unfortunately, in other cases, Ar–Ar ages are likely not sufficiently reliable to resolve the age difference between circumferential and radiating dykes (see discussion of Ar–Ar age reliability related to the Paraná-Etendeka and High Arctic events in Janasi et al. 2011; Senger et al. 2014; Polteau et al. 2016). Despite these difficulties, there are a few cases where relative age information is available based on U-Pb ages or crosscutting relationships. Circumferential dykes appear to be younger than radiating dykes in the Siberian Traps and Franklin (Clarence Head) events as discussed in the individual event descriptions above. On the other hand, possible radiating dykes of the Rogaland event are significantly younger than the Blekinge-Dalarna circumferential dykes. In a number of cases (e.g., the High Arctic, Deccan and Lüliang-Taihang events) there is evidence of multiple pulses of circumferential and/or radiating dykes based on conflicting crosscutting relationships or overlapping U-Pb ages for the two systems. On Venus the relative age of coronae and radiating graben-fissure systems is the subject of discussion. In many cases, it appears that the radiating systems are older than the circular systems (e.g., Janes et al. 1992; Squyres et al. 1992; Koch and Manga 1996; Krassilnikov and Head 2003). In other cases, there is evidence from crosscutting relationships that there were alternating pulses of the two systems (McKenzie et al. 1992; Aittola and Kostama 2002; Kostama and Aittola 2004).

The ages of giant circumferential dyke swarms span much of the geological record on Earth, from the Paleoproterozoic to the Cenozoic (Table 1). However, the age range of Venusian coronae is poorly defined, because of a lack of radiometric dating and the uncertainty in the timing and style of planetary resurfacing on Venus (e.g., Davaille et al. 2017 and reference therein).

Most coronae have a topographic expression which can include a central domal uplift or central depression, as well as an outer rim and/or moat (Smrekar and Stofan 1997). The rim and moat often coincide roughly with the annulus of circumferential graben. On Earth, any topography associated with giant circumferential dyke swarms will have been removed by erosion, but can be inferred, in some cases, from the sedimentary record. Thus, domal uplift has been reported in relation to the arrival of many mantle plumes and associated giant radiating dyke swarms (summary in Ernst 2014), including several cases (e.g., High Arctic LIP, Maher 2001; North Atlantic LIP, Maclennan and Jones 2006; Yakutsk-Vilyui LIP, Kiselev et al. 2012; Franklin LIP, Rainbird and Ernst 2001) where there are also giant circumferential swarms. It is not yet known if circumferential rims or moats were present with circumferential swarms on Earth, although Buchan and Ernst (2018) have suggested that the topography identified in associated with the High Arctic LIP could represent either the edge of a domal uplift (Maher 2001) or a raised circumferential rim.

Finally, many giant circumferential dyke swarms discussed herein occur in association with LIP-scale volcanism, although it is not known whether volcanic flows are fed from the circumferential dykes or only from other components of the LIP plumbing system, such as giant radiating dykes. On Venus, LIP-scale volcanism is observed in association with a significant number of coronae (Magee and Head 2001; Aittola and Kostama 2002).

In summary, it appears that giant circumferential dyke swarms share a number of characteristics with Venusian coronae and similar features on Mars, and so could be analogues that were generated by a similar process or processes as has been suggested by Ernst and Buchan (1998), Buchan and Ernst (2016), and Bethell et al. (2016).

4 Possible Models for Generation of Giant Circumferential Dyke Swarms and Coronae

Given the very limited recognition of giant circumferential dyke swarms on Earth prior to the present study, there has been relatively little discussion of models for their generation. Indeed, with their wide range in size (Table 1), it is possible that they may not all be generated by a single process.

The larger giant circumferential swarms and possible swarms (diameters of 1500–2500 km) appear to be coupled with giant radiating swarms, likely indicating the presence of mantle plumes. The diameters of these circumferential swarms are comparable to those estimated from theoretical modelling and geological evidence for thermal plumes that are generated from the core-mantle boundary and have flattened at the base of the lithosphere (Griffiths and Campbell 1991; Ernst and Buchan 2002; Campbell 2007; Ernst 2014). Circumferential swarm diameters are also comparable to the size of the domal uplifts that are predicted above such plumes (e.g., Griffiths and Campbell 1991). Therefore, it seems plausible that these very large circumferential swarms are linked to the edges of flattened (or flattening) plume heads and are located along the margins of the associated domal uplifts (Fig. 16a). For example, the circumferential dykes of the High Arctic LIP (Fig. 3; Buchan and Ernst 2018) and the North Atlantic LIP (Fig. 9) are located near the outer edge of proposed domal uplifts above plumes, lending credence to this interpretation. Models for generating the circumferential graben-fissure systems of Venusian coronae at the outer edge of plumes or diapirs have been considered in the literature and are discussed briefly later in the section.

Giant circumferential swarms with diameters less than 1500 km could also be generated at the outer edge of flattened plume heads if the plumes are of small size. Theoretical models suggest that small plumes could be spawned at the 670 km transition zone or could be generated as secondary plumes from the top of larger plumes that have stalled there temporarily (e.g., Thompson and Tackley 1998; Arndt 2000; Courtillot et al. 2003; Farnetani and Samuel 2005; Campbell 2007; Wang et al. 2016). Dobretsov et al. (2008) have proposed a type of thermochemical plume that rises from the core-mantle boundary but does not develop a significant plume head. Recent seismic tomography studies (e.g., French and Romanowicz 2015; Bozdağ et al. 2016) have imaged deep mantle plumes of probable thermochemical origin that rise to a depth of ~1000 km, where there is a prominent increase in viscosity (Rudolph et al. 2015), and where they may be deflected and/or decrease in diameter (Rickers et al. 2013; French and Romanowicz 2015). Alternatively, circumferential swarms with diameters <1500 km could be generated from the edge of large plume heads that are in the process of flattening. A third option is that such swarms are produced along the edge of a magmatic underplate located at the base of the crust above a mantle plume head (Fig. 16b), as has been suggested by Mäkitie et al. (2014, Fig. 13e) for the Lake Victoria circumferential swarm (Fig. 8). Wang et al. (2014) have proposed that the Marna Moorn dykes (Fig. 14) were emplaced along the margins of the Yilgarn craton because a plume rising beneath thick (>200 km) lithosphere at the centre of the craton was deflected to the craton margins where the lithosphere is thinner (100–125 km), resulting in enhanced melting and dyke emplacement there (cf. Thompson and Gibson 1991; Sleep 2006). It should be noted that some circumferential swarms with diameters <1500 km do not have a clear association with a radiating swarm (Table 1; e.g. Blekinge-Dalarna, Lake Victoria, Marnda Moorn) and, hence, might have a non-plume origin.

In the extensive literature on Venusian coronae, a number of authors have discussed models for the generation of these features including their circumferential, and where present, radiating fracture systems (with surface expression as graben) (e.g., Squyres et al. 1992; Koch and Manga 1996; Smrekar and Stofan 1997; Krassilnikov and Head 2003; Dombard et al. 2007; Davaille et al. 2017). Most of these models link coronae with mantle plumes or diapirs. For example, Squyres et al. (1992) proposed a conceptual model in which a rising plume generates domal uplift with an associated radiating fracture system, and then flattens at the base of the lithosphere. As the flattened plume head cools, gravitational relaxation of the topography produces a central depression as well as a rim, moat and circumferential fracture system near the edge of the plume. Using a boundary integral method, Koch and Manga (1996, Figs. 2 and 3) also described a model in which a plume generates a central dome and radiating fractures as it approaches the level of neutral buoyancy. As the plume head flattens, a central depression surrounded by an elevated rim evolves and expands in diameter, with circumferential fractures and compressional features (wrinkle ridges) adjacent to the rim. On the other hand, Smrekar and Stofan (1997) used numerical experiments to propose that the variety of corona topographic fea-

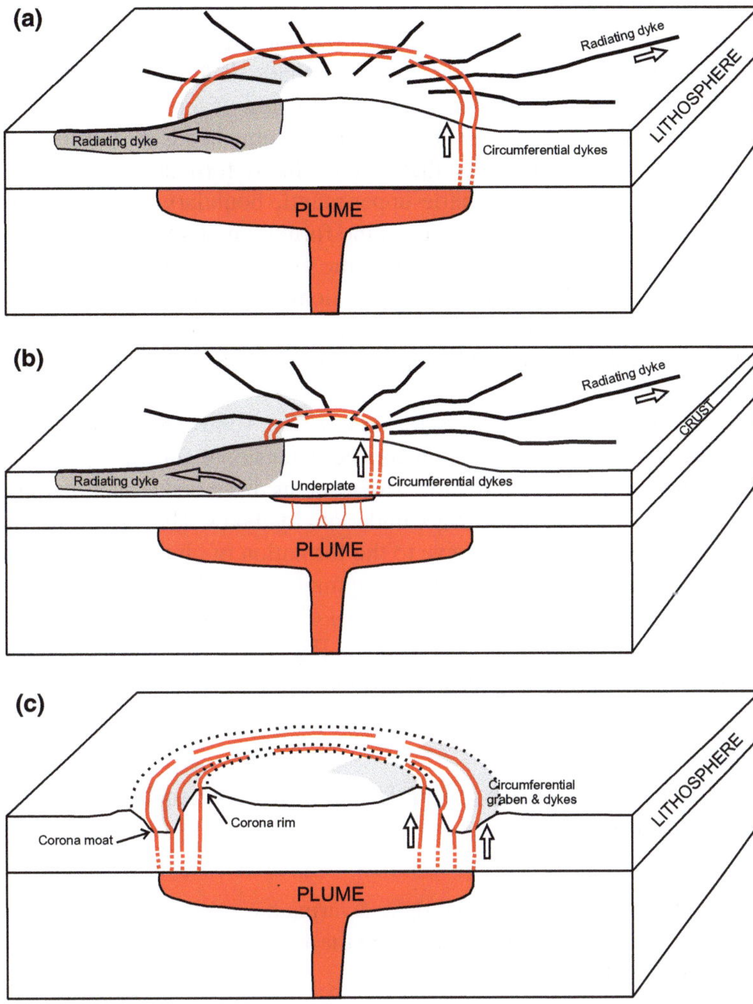

Fig. 16 **a** Model of giant radiating dykes emplaced during topographic uplift above a plume and giant circumferential dykes emplaced above the edge of the plume head which has flattened at the base of the crust (not to scale). Modified after Ernst and Buchan (1998). **b** Model of giant radiating dykes emplaced during topographic uplift above a plume and giant circumferential dykes emplaced above the edge of a magmatic underplate (located at the base of the crust above the middle of the plume head). **c** Model of a Venusian corona generated above the edge of a plume that has flattened at the base of the lithosphere showing a raised rim, a moat and an annulus of graben underlain by dykes (not to scale). Modified after Ernst and Buchan (1998). As discussed in the text, some coronae also have associated radiating graben thought to be underlain by dykes. In **a–c** the dashed lower portion of circumferential dykes indicates the uncertainty in how magma propagates from the edge of the plume or magmatic underplate to reach the upper crust

tures are best explained by lithospheric delamination around the edge of the plume. Dombard et al. (2007, Fig. 3) proposed that the wide range in coronae size can be best accounted for if a corona is related to a magmatic underplate located at the base of the crust above a zone of partial melting at the top of a plume head. Given that the average size of coronae is smaller than large plume heads on Earth, Stofan and Smrekar (2005, Fig. 12) concluded that coronae likely form above small secondary plumes generated at the lower mantle-upper mantle boundary from the top of large deep mantle plumes. This is analogous to the formation of small secondary plumes described earlier in the section for some plume models on Earth (e.g., Arndt 2000). Models for circumferential fracture/graben systems on Mars are broadly divided into those associated with loading (either of a volcanic edifice or from emplacement of mafic-ultramafic intrusions) or with uplift (e.g., Öhman and McGovern 2014 and references therein).

In general, the published models for coronae on Venus and similar features on Mars provide a purely tectonic origin for the extensional fractures/graben, but do not address the issue of dyke intrusion. However, as discussed earlier there is ample evidence to suggest that the graben are underlain by dykes. The Venus-Mars models provide stress distributions conducive to the formation of circumferential fractures, and should also be broadly applicable if magma is available. The magma would fill the fractures as dykes, and could reach a level of neutral buoyancy below the paleosurface, with a surface expression as graben. Figure 16c illustrates circumferential dykes of a corona being generated from the edge of a mantle plume.

5 Identifying and Characterizing Giant Circumferential Dyke Swarms

Until recently, giant circumferential dyke swarms have gone largely unrecognized on Earth. To better identify and characterize these features we suggest the following important considerations.

(1) Giant circumferential swarms with a primary circular/elliptical geometry are most easily identified if they can be observed over a long arc length either in outcrop or on aeromagnetic surveys. However, many circumferential swarm are much less visible on aeromagnetic maps than has typically been the case for radiating swarms. This could be explained if the dykes of circumferential swarms are much less dense or narrower than in radiating swarms, so that the aeromagnetic signature of the radiating dykes is dominant. The Lake Victoria circumferential swarm (Fig. 8), which does not have a companion radiating swarm, is unusual because it can be readily traced from aeromagnetic maps.

(2) Circumferential swarms with a relatively short arc length can be difficult to distinguish from arcuate swarms that result from secondary deformation or from deflection from a linear or radiating geometry in the presence of a regional stress field.

(3) Where secondary deformation may have occurred, a paleomagnetic study along the arc of a potential giant circumferential swarm can determine whether or not its geometry is primary (cf. Blekinge-Dalarna swarm of Fig. 6).
(4) Evidence that a dyke swarm is located along the edge of a contemporaneous domal uplift or circular uplifted rim or moat can indicate that the swarm is circumferential and likely linked to an underlying plume responsible for the uplift or rim.
(5) If a short dyke swarm intersects a roughly coeval giant radiating swarm at right angles, it may be a segment of a circumferential swarm and likely linked to an underlying plume responsible for the radiating swarm. In some cases, a plate tectonic reconstruction may be required to link short circumferential dyke segments and reveal the overall extent and geometry of the swarm.
(6) It should be noted some dykes associated with a given magmatic event may have orientations that are not consistent with the pattern of the radiating or circumferential swarms. For example, short dyke segments that feed sills or volcanic flows within a supracrustal package may be oriented under the influence of a local stress field or the structure of local host rocks (e.g. Muirhead et al. 2012) (see further discussion in Sect. 3 of Buchan and Ernst 2018).
(7) Precise U-Pb dating of dykes of different trends is critical to determining the relative age of circumferential and radiating dyke swarms, and whether they have been emplaced in pulses.
(8) Geochemistry of dykes of known trend is helpful in correlating dykes of different swarms and subswarms of circumferential and radiating systems.
(9) If the circumferential and/or radiating swarms are part of a LIP, precise dating and geochemistry may help identify specific sill and/or flood basalt packages that they have fed.
(10) Recording dyke dips is helpful in determining if circumferential dykes are emplaced vertically as ring dykes, or as cone sheets.

6 Conclusions

We have catalogued the giant circumferential dyke swarms and possible components of giant circumferential swarms from around the globe and over a broad time range from the Paleoproterozoic to the Cenozoic. Many are identified for the first time herein. They display a very wide range in diameters (~450–2500 km). Several are only evident in plate tectonic reconstructions. Many appear to be linked to giant radiating dyke swarms, LIPs and mantle plumes. Furthermore, the diameters of the circumferential swarms may reflect the outer edges of flattened or flattening plume heads or perhaps the edges of magmatic underplates above plumes. Giant circumferential dyke swarms may be analogues of coronae on Venus and similar circular structures on Mars.

Acknowledgements We thank Erin Bethell, Benoit Saumur, Sally Pehrsson, Eric Grosfils, Daniel Mège, Marie-Claude Williamson and Laurent Montési for discussions on circumferential dykes on Earth, coronae on Venus and/or potential analogues on Mars. Sally Pehrsson, Michiel de Kock and Franco Pirajno provided helpful reviews of the manuscript. R. Ernst has been partially supported from Mega-Grant 14.Y26.31.0012 of the Russian Federation. This is Natural Resources Canada contribution #20180022.

References

Abrahamsen N, Bengaard H-J, Friderichsen JD, Van der Voo R (1997) Palaeomagnetism of three dyke swarms in Nansen Land, northern Greenland (83°N). Geol Mijnbouw 76:83–95

Aittola M, Kostama V-P (2002) Chronology of the formation process of Venusian novae and the associated coronae. J Geophys Res 107(E1):5112

Arndt N (2000) Hot heads and cold tails. Nature 407:458–461

Aspler LB, Ernst RE (2003) Dyke-induced graben on Venus and Mars: analogues for Earth's rock record? In: Lunar and Planetary Science Conference #34, abstract #1711

Auden JB (1949) Dykes in western India: a discussion of their relationship with the Deccan traps. Trans Nat Inst Sci India 3:123–157

Balkwill HR, Fox FG (1982) Incipient rift zone, western Sverdrup Basin, Arctic Canada. In: Embry AS, Balkwill HR (eds) Arctic geology and geophysics. Canadian Society of Petroleum Geologists, Memoir 8, pp 171–187

Bethell E, Ernst RE, Samson C, Buchan KL (2016) Circumferential graben-fissure systems of Venusian coronae as possible analogues of giant circumferential dyke swarms on Earth. In: Lunar and Planetary Science Conference #57, abstract #1471

Bondre NR, Hart WK, Sheth HC (2006) Geology and geochemistry of the Sangamner mafic dyke swarm, Western Deccan Volcanic Province, India: implication for regional stratigraphy. J Geol 114:155–170

Bozdağ E, Peter D, Lefevre M, Komatitsch D, Tromp J, Hill J, Podhorszki N, Pugmire D (2016) Global adjoint tomography: first-generation model. Geophys J Int 207:1739–1766

Bright RM, Amato JM, Denyszyn SW, Ernst RE (2014) U-Pb geochronology of 1.1 Ga diabase in the southwestern United States: testing models for the origin of a post-Grenville Large Igneous Province. Lithosphere 6:135–156

Brown G, Platt NH, McGrandle A (1994) The geophysical expression of Tertiary dykes in the southern North Sea. First Break 12(3):137–146

Buchan KL, Ernst RE (2004) Diabase dyke swarms and related units in Canada and adjacent regions. Geological Survey of Canada, Map 2022A, scale: 1:500 000

Buchan KL, Ernst RE (2006) Giant dyke swarms and the reconstruction of the Canadian Arctic islands, Greenland, Svalbard and Franz Josef Land. In: Hanski E, Mertanen S, Rämö T, Vuollo J (eds) Dyke swarms—time markers of crustal evolution. Taylor and Francis Group, London, pp 27–48

Buchan KL, Ernst RE (2015) A giant circumferential dyke swarm associated with the High Arctic Large Igneous Province (HALIP)—a possible analogue for coronae on Venus. American Geophysical Union-Geological Association of Canada-Mineralogical Association of Canada-Canadian Geophysical Union Joint Assembly, Abstract P42A-05

Buchan KL, Ernst RE (2016) Giant circumferential dyke swarms on Earth as possible analogues of coronae on Venus. In: Lunar and Planetary Science Conference #57, abstract #1183

Buchan KL, Ernst RE (2017) Giant circumferential and radiating dyke swarms of the High Arctic Large Igneous Province. In: Williamson M-C (ed) GEM 2 High Arctic Large Igneous Province (HALIP) activity: workshop report. Geological Survey of Canada, Open File 8151, pp 9–10

Buchan KL, Ernst RE (2018) A giant circumferential dyke swarm associated with the High Arctic Large Igneous Province (HALIP). Gondwana Res 58:39–57

Buchan KL, Ernst RE, Bleeker W, Davis WJ, Villeneuve M, van Breeman O, Hamilton M, Söderlund U (2010) Proterozoic magmatic events of the Slave craton, Wopmay orogeny and environs. Geological Survey of Canada, Open File 5985, poster and 25 page report

Burchardt S, Troll VR, Mathieu L, Emeleus HC, Donaldson CH (2013) Ardnamurchan 3D cone-sheet architecture explained by a single elongate magma chamber. Sci Rep 3:2891. https://doi.org/10.1038/srep02891

Burke K, Dewey JF (1973) Plume-generated triple junctions: key indicators in applying plate tectonics to old rocks. J Geol 81:406–433

Bylund G (1992) Palaeomagnetism, mafic dykes and the Protogine Zone, southern Sweden. Tectonophysics 201:49–63

Campbell IH (2007) Testing the plume theory. Chem Geol 241:153–176

Chadwick WW, Embley RW (1998) Graben formation associated with recent dike intrusions and volcanic eruptions on the mid-ocean ridge. J Geophys Res 103(B5):9807–9825

Chadwick WW, Howard KA (1991) The pattern of circumferential and radial eruptive fissures on the volcanoes of Fernandina and Isabela islands, Galapagos. Bull Volc 53:259–275

Chadwick WW, Jónsson S, Geist DJ, Poland M, Johnson DJ, Batt S, Harpp KS, Ruiz A (2011) The May 2005 eruption of Fernandina volcano, Galápagos: the first circumferential dike intrusion observed by GPS and InSAR. Bull Volc 73:679–697

Chevallier L, Woodford A (1999) Morpho-tectonics and mechanisms of emplacement of the dolerite rings and sills of the western Karoo, South Africa. S Afr J Geol 102:43–54

Chevallier L, Goedhart M, Woodford AC (2001) The influences of dolerite sill and ring complexes on the occurrence of groundwater in Karoo fractured aquifers: a morpho-tectonic approach. Water Research Commission, Report 937/1/01

Clark ET (1880) On volcanic foci of eruption in the Konkan. Rec Geol Surv India 13:69–73 and accompanying map

Cooper MR, Anderson H, Walsh JJ, Van Dam CL, Young ME, Earls G, Walker A (2012) Palaeogene Alpine tectonics and Icelandic plume-related magmatism and deformation in Northern Ireland. J Geol Soc Lond 169:29–36

Courtillot V, Davaille A, Besse J, Stock J (2003) Three distinct types of hotspots in the Earth's mantle. Earth Planet Sci Lett 205:295–308

Coutinho JMV (2008) Dyke swarms of the Paraná triple junction, southern Brazil. Geologia USP, Série Cientifica 8(2):29–52

Cox GM, Halverson GP, Denyszyn S, Foden J, Macdonald FA (2017) Cryogenian magmatism along the north-western margin of Laurentia: plume or rift? Precambr Res. https://doi.org/10.1016/j.precamres.2017.09.025

Cucciniello C, Melluso L, Jourdan F, Mahoney JJ, Meisel T, Morra V (2013) ^{40}Ar–^{39}Ar ages and isotope geochemistry of Cretaceous basalts in northern Madagascar: refining eruption ages, extent of crustal contamination and parental magmas in a flood basalt province. Geol Mag 150(1):1–17

Cucciniello C, Melluso L, Morra V (2015) The Madagascar large igneous province. Large Igneous Provinces Commission website, LIP of the Month for August 2015. http://lips.iavceivolcano.org/15aug. Accessed 19 Apr 2018

Davaille A, Smrekar SE, Tomlinson S (2017) Experimental and observational evidence for plume-induced subduction on Venus. Nat Geosci 10:349–355

de Kock MO, Ernst RE, Söderlund U, Jourdan F, Hofmann A, Le Gall B, Bertrand H, Chisonga BC, Beukes N, Rajesh HM, Moseki LM, Fuch R (2014) Dykes of the 1.11 Ga Umkondo LIP, southern Africa: clues to a complex plumbing system. Precambr Res 249:129–143

Denyszyn SW, Davis DW, Halls HC (2009) Paleomagnetism and U-Pb geochronology of the Clarence Head dykes, Arctic Canada: orthogonal emplacement of mafic dykes in a large igneous province. Can J Earth Sci 46:155–167

Denyszyn S, Cox G, Halverson G (2016) Geochemistry and geochronology of circumferential dykes of the Franklin LIP: a rotated perspective on plate reconstruction. Acta Geol Sinica (English Edition) 90(supp. 1):30

Dessai AG, Viegas AAA (1995) Multi-generation mafic dyke-swarm related to Deccan magmatism, south of Bombay: implications on the evolution of the western Indian continental margin. Geol Soc India, Memoir 33:435–451

Dibner VD (ed) (1998) Geology of Franz Josef Land. Norsk Polarinstitutt, Meddelelse 146, 190 p

Dobretsov NL, Kirdyashkin AA, Kirdyashkin AG, Vernikovsky VA, Gladkov IN (2008) Modelling of geochemical plumes and implications for the origin of the Siberian traps. Lithos 100:66–92

Dombard AJ, Johnson CL, Richards MA, Solomon SC (2007) A magnetic loading model for coronae on Venus. J Geophys Res 112:E04006. https://doi.org/10.1029/2006JE002731

Emeleus CH (1982) The central complexes. In: Sutherland DS (ed) Igneous rocks of the British Isles. Wiley, Chichester, pp 369–425

Ernst RE (2014) Large igneous provinces. Cambridge University Press, Cambridge

Ernst RE, Buchan KL (1997) Giant radiating dyke swarms: their use in identifying pre-Mesozoic large igneous provinces and mantle plumes. In: Mahoney JJ, Coffin MF (eds) Large igneous provinces, continental, oceanic, and planetary flood volcanism. American Geophysical Union Geophysical Monograph 100, pp 297–333

Ernst RE, Buchan KL (1998) Arcuate dyke swarms associated with mantle plumes on Earth: implications for Venusian coronae. In: Lunar and Planetary Science Conference #29, abstract #1021

Ernst RE, Buchan KL (2001) The use of mafic dyke swarms in identifying and locating mantle plumes. In: Ernst RE, Buchan KL (eds) Mantle plumes: their identification through time. Geological Society of America, Special Paper 352, pp 247–265

Ernst RE, Buchan KL (2002) Maximum size and distribution in time and space of mantle plumes: evidence from large igneous provinces. J Geodyn 34:309–342

Ernst RE, Buchan KL, Palmer HC (1995) Giant dyke swarms: characteristics, distribution and geotectonic applications. In: Baer G, Heimann A (eds) Physics and chemistry of dykes. Balkema, Rotterdam, pp 3–21

Ernst RE, Grosfils EB, Mège D (2001) Giant dike swarms: Earth, Venus and Mars. Ann Rev Earth Sci 29:489–534

Ernst RE, Desnoyers DW, Head JW, Grosfils EB (2003) Graben–fissure systems in Guinevere Planitia and Beta Regio (264°–312°E, 24°–60°N), Venus, and implications for regional stratigraphy and mantle plumes. Icarus 164:282–316

Ernst RE, Buchan KL, Mäkitie H, Klausen MB, Samson C, Gosfils EB (2014) Giant circumferential mafic dyke swarms on Earth, Mars and Venus. In: Geological Society of America annual meeting, abstract #260-13

Ewart A, Marsh JS, Milner SC, Duncan AR, Kamber BS, Armstrong RA (2004) Petrology and geochemistry of Early Cretaceous bimodal continental flood volcanism of the NW Etendeka, Namibia. Part 1: introduction, mafic lavas, and re-evaluation of mantle source components. J Petrol 45:59–105

Farnetani CG, Samuel H (2005) Beyond the thermal plume paradigm. Geophys Res Lett 32:L07311

Florisbal LM, Heaman LM, Janasi VA, Bitencourt MF (2014) Tectonic significance of the Florianópolis dyke swarm, Paraná-Etendeka magmatic province: a reappraisal based on precise U-Pb dating. J Volcanol Geoth Res 289:140–150

French SW, Romanowicz B (2015) Broad plumes rooted at the base of the Earth's mantle beneath major hotspots. Nature 525:95–99 and extended data file

Geological Survey of Northern Ireland (1971) Geology of Belfast and district (special engineering and geology sheet, solid and drift). Scale 1:21 120

Geological Survey of Northern Ireland (1977) Geological map of Northern Ireland (solid edition). Department of Commerce, Northern Ireland, scale 1:250 000

Gion AM, Williams SE, Müller RD (2017) A reconstruction of the Eurekan Orogeny incorporating deformation constraints. Tectonics 36:304–320. https://doi.org/10.1002/2015TC004094

Goodwin AM (1996) Principles of Precambrian geology. Academic Press, London

Gorbatschev R, Lindh A, Solyom Z, Laitakari I, Aro K, Lobach-Zhuchenko SB, Markov MS, Ivliev AI, and Bryhni I (1987) Mafic dyke swarms of the Baltic Shield. In: Halls HC, Fahrig WF (eds) Mafic dyke swarms. Geological Association of Canada, Special Paper 34, pp 361–372

Griffiths RW, Campbell IH (1991) Interaction of mantle plume heads with the Earth's surface and onset of small-scale convection. J Geophys Res 96(B11):18, 295-18, 310

Grosfils EB, Head JW (1994) The global distribution of giant radiating dike swarms on Venus: implications for the global stress state. Geophys Res Lett 21:701–704

Guedes E, Heilbron M, Vasconcelos PM, Valeriano CM, Almeida JCH, Teixeira W, Thomaz Filho A (2005) K-Ar and $^{40}Ar/^{39}Ar$ ages of dikes emplaced in the onshore basement of the Santos Basin, Resnede area, SE Brazil: implications for the south Atlantic opening and Tertiary reactivation. J S Am Earth Sci 18:371–382

Halls HC (1982) The importance and potential of mafic dyke swarms in studies of geodynamic processes. Geosci Can 9:145–154

Heaman LM, Easton RM, Hart TR, Hollings P, MacDonald CA, Smyk M (2007) Further refinement to the timing of Mesoproterozoic magmatism, Lake Nipigon region, Ontario. Can J Earth Sci 44:1055–1086

Hellström FA, Johansson Å, Larsen SÅ (2004) Age and emplacement of late Sveconorwegian monzogabbroic dykes, SW Sweden. Precambr Res 128:39–55

Hooper PR (1990) The timing of crustal extension and the eruption of continental flood basalts. Nature 345:246–249

Hutchison DR, White RS, Cannon WF, Schulz KJ (1990) Keweenawan hot spot: geophysical evidence for a 1.1 Ga mantle plume beneath the Midcontinent Rift system. J Geophys Res 95:10, 869-10, 884

Janasi VA, Freitas VA, Heaman LH (2011) The onset of flood basalt volcanism, northern Paraná basin, Brazil: a precise U-Pb baddeleyite/zircon age for a Chapecó-type dacite. Earth Planet Sci Lett 302:147–153

Janes DM, Squyres SW, Bindschadler DL, Baer G, Schubert G, Sharpton VL, Stofan ER (1992) Geophysical models for the formation and evolution of coronae on Venus. J Geophys Res 97(E10):16, 055-16, 067

Kingsbury CG, Kamo SL, Ernst RE, Söderlund U, Cousens BL (2018) U-Pb geochronology of the plumbing system associated with the Late Cretaceous Strand Fiord Formation, Axel Heiberg Island, Canada: part of the 130-90 Ma High Arctic large igneous province. J Geodyn. https://doi.org/10.1016/j.jog.2017.11.001

Kiselev AI, Ernst RE, Yarmolyuk VV, Egorov KN (2012) Radiating rifts and dyke swarms of the middle Paleozoic Yakutsk plume of eastern Siberia. J Asian Earth Sci 45:1–16

Koch DM, Manga M (1996) Neutrally buoyant diapirs: a model for Venus. Geophys Res Lett 23:25–228

Kontak DJ, Jensen SM, Dostal J, Archibald DA, Kyser TK (2001) Cretaceous mafic dyke swarm, Peary Land, northernmost Greenland: geochronology and petrology. Can Mineral 39:997–1020

Kostama V-P, Aittola M (2004) Arcuate graben of Venusian volcano-tectonic structures: the last phase of tectonic activity? Astron Astrophys 428:235–240. https://doi.org/10.1051/0004-6361:200400061

Krassilnikov AS, Head JW (2003) Novae on Venus: geology, classification, and evolution. J Geophys Res 108(E9):5108. https://doi.org/10.1029/2002JE001983

Krishnamacharlu T (1972) Dykes around Dadiapada, Broach District, Gujarat. Bull Volc 35:947–956

Krishnamurthy P (1972) Petrology of the dyke rocks of the western portions of Rajpipla Hills, Broach District, Gujarat, India. Bull Volc 35:930–946

Krogh TE, Corfu F, Davis DW, Dunning GR, Heaman LM, Kamo SL, Machado N (1987) Precise U-Pb isotopic ages of diabase dykes and mafic to ultramafic rocks using trace amounts of baddeleyite and zircon. In: Halls HC, Fahrig WF (eds) Mafic dyke swarms. Geological Association of Canada, Special Publication 34, pp 331–348

Kumar A, Pande K, Bhashkar Venkatesan TR, Rao YJ (2001) The Karnataka Late Cretaceous dykes as products of the Marion hot spot at the Madagascar-India breakup event: evidence from ^{40}Ar-^{39}Ar geochronology and geochemistry. Geophys Res Lett 28:2715–2718
MacDonald R, Bagiński B, MacInnnes DA, MacGillivray JC, Fettes DJ (2014) The Palaeogene Bracken Bay-Straiton dyke: composition and controls on intrusion. Scott J Geol 50:57–69
MacGregor AG (1949) Dykes of the post-Carboniferous age. In: Eyles VA, Simpson JB, MacGregor AG (eds) Geology of Central Ayrshire (explanation of one-inch sheet 14). Memoirs of the Geological Survey of Scotland, pp 118–123
Maclennan J, Jones SM (2006) Regional uplift, gas hydrate dissociation and the origins of the Paleocene-Eocene thermal maximum. Earth Planet Sci Lett 245:65–80
Magee KP, Head JW (2001) Large flow fields on Venus: implications for plumes, rift associations, and resurfacing. In: Ernst RE, Buchan KL (eds) Mantle plumes: their identification through time. Geological Society of America, Paper 352, pp 81–101
Maher HD Jr (2001) Manifestation of the Cretaceous High Arctic Large Igneous Province in Svalbard. J Geol 109:91–104
Mäkitie H, Data G, Isabirye E, Mäntärri I, Huhma H, Klausen MB, Pakkenen L, Virransaio P (2014) Petrology, geochronology and emplacement model of the giant 1.37 Ga arcuate Lake Victoria Dyke Swarm on the margin of a large igneous province in eastern Africa. J Afr Earth Sci 97:273–296
Mastin LG, Pollard DD (1988) Surface deformation and shallow dike intrusion processes at Inyo craters, Long Valley, California. J Geophys Res 93(B11):13, 221-13, 235
McHone JG, Anderson DL, Beutel EK, Fialko YA (2005) Giant dikes, rifts, flood basalts, and plate tectonics: a contention of mantle models. In: Foulger GR, Natland JH, Presnall DC, Anderson DL (eds) Plates, plumes and paradigms. Geological Society of America, Special Paper 388, pp 401–420
McKenzie D, McKenzie JM, Saunders RS (1992) Dike emplacement on Venus and Earth. J Geophys Res 97(E10):15, 977-15, 990
Mège D, Ernst RE (2001) Contractional effects of mantle plumes on Earth, Mars, and Venus. In: Ernst RE, Buchan KL (eds) Mantle plumes: their identification through time. Geological Society of America, Special Paper 352, pp 103–140
Mège D, Masson P (1996) A plume tectonics model for the Tharsis province, Mars. Planet Space Sci 44:1499–1546
Melluso L, Sheth HC, Mahoney JJ, Morra V, Petrone CM, Storey M (2009) Correlations between silicic volcanic rocks of the St Mary's Islands (southwestern India) and eastern Madagascar: implications for Late Cretaceous India-Madagascar reconstructions. J Geol Soc Lond 166:283–294
Misra AA, Bhattacharya G, Mukherjee S, Bose N (2014) Near N-S paleoextension in the western Deccan region, India: does it link strike-slip tectonics with India-Seychelles rifting? Indian J Earth Sci 103:1645–1680
Mitchell JG, Mohr P (1986) K-Ar systematics in Tertiary dolerites from West Connacht, Ireland. Scott J Geol 22:225–240
Mohr PA (1971) Ethiopian Tertiary dike swarms. Smithsonian Astrophysical Observatory, Special Report 339
Mohr P (1982) Tertiary dolerite intrusions of west-central Ireland. Proc Roy Ir Acad 82(B):53–82
Mohr P (1987) The Cill Ala dike swarm, Cos Sligo and Mayo: physical parameters. Ir Nat J 22:328–334
Mohr P (1988) The analcime-olivine dolerites of West Connaught, Ireland: classification and genetic problems. Ir J Earth Sci 9:133–140
Mohr P, Zanettin B (1988) The Ethiopian flood basalt province. In: Macdougall JD (ed) Continental flood basalts. Kluwer Academic Publishers, Dordrecht, pp 63–110
Montési LGJ (2001) Concentric dikes on the flanks of Pavonis Mons: implications for the evolution of martian shield volcanoes and mantle plumes. In: Ernst RE, Buchan KL (eds) Mantle plumes: their identification through time. Geological Society of America, Special Paper 352, pp 165–181
Moore A (1965) The North Gap dyke of the Transkei. Trans Geol Soc S Afr 68:89–120

Moreau C, Ohnenstetter D, Diot H, Demaiffe D, Brown WL (1995) Emplacement of the Meugueur-Meugueur cone-sheet (Niger, West Africa), one of the world's largest igneous ring-structures. In: Baer G, Heimann A (eds) Physics and chemistry of dykes. Balkema, Rotterdam, pp 41–49

Mountain ED (1943) The dikes of the Transkei Gaps. Trans Geol Soc S Afr 46:55–73

Muirhead JD, Airoldi G, Roweland JV, White JDL (2012) Interconnected sills and inclined-sheet intrusions control shallow magma transport in the Ferrar Large Igneous Province, Antarctica. Geol Soc Am Bull 124:162–180

Myers JS (1990) Precambrian tectonic evolution of part of Gondwana, southwestern Australia. Geology 18:537–540

Neuendorf KKE, Mehl JP Jr, Jackson JA (2005) Glossary of geology, 5th edition. American Geological Institute, Alexandria, Virginia, USA

Newhall CG, Dzurisin D (1988) Historical unrest at large calderas of the world. U.S. Geol Surv Bull 1855

Öhman T, McGovern PJ (2014) Circumferential graben and the structural evolution of Alba Mons, Mars. Icarus 233:114–125

Pearson V, Daigneault R (2009) An Archean megacaldera complex: the Blake River Group, Abitibi greenstone belt. Precambr Res 168:66–82

Pehrsson SJ, van Breemen O, Hanmer S (1993) Ages of diabase dyke intrusions, Great Slave Lake shear zone, Northwest Territories. In: Radiogenic age and isotope studies: Report 5, Geological Survey of Canada, Paper 93-2, pp 23–28

Peng P (2015) Precambrian mafic dyke swarms in the North China Craton and their geological implications. Sci China Earth Sci 58:649–675

Peng P, Zhai M, Ernst RE, Guo J, Liu F, Hu B (2008) A 1.78 Ga large igneous province in the North China craton: the Xiong'er Volcanic Province and the North China dyke swarm. Lithos 101:260–280

Peterman ZE, Sims PK (1988) The Goodman swell: a lithospheric flexure caused by crustal loading along the Midcontinent Rift system. Tectonics 7:1077–1090

Piragno F, Hoatson DM (2012) A review of Australia's large igneous provinces and associated mineral systems: implications for mantle dynamics through geological time. Ore Geol Rev 48:2–54

Pisarevsky S, Bylund G (2006) Palaeomagnetism of 935 Ma mafic dykes in southern Sweden and implications for the Sveconorwegian Loop. Geophys J Int 166:1095–1104

Pisarevsky SA, Wingate MTD, Harris LB (2003) Late Mesproterozoic (ca. 1.2 Ga) palaeomagnetism of the Albany-Fraser orogeny: no pre-Rodinia Australia-Laurentia connection. Geophys J Int 155:F6–F11

Pisarevsky SA, Wingate MTD, Li Z-X, Wang X-C, Tohver E, Kirkland CL (2014) Age and paleomagnetism of the 1210 Ma Gnowangerup-Fraser dyke swarm, Western Australia, and implications for Mesoproterozoic paleogeography. Precambr Res 246:1–15

Polteau S, Hendriks BWH, Planke S, Ganerød M, Corfu F, Faleide JI, Midtkandal I, Svensen HS, Myklebust R (2016) The Early Cretaceous Barents Sea sill complex: distribution, 40Ar/39Ar geochronology, and implications for carbon gas formation. Palaeogeogr Palaeoclimatol Palaeoecol 441:83–95

Preston J (2001) Tertiary igneous activity. In: Holland CH (ed) The geology of Ireland. Dunedin Academic Press, Edinburgh, pp 353–373

Queen M, Heaman LM, Hanes JA, Archibald DA, Farrer E (1996) $^{40}Ar/^{39}Ar$ phlogopite and U-Pb perovskite dating of lamprophyre dykes from the eastern Lake Superior region: evidence for a 1.14 Ga magmatic precursor in Midcontinent Rift volcanism. Can J Earth Sci 33:958–965

Radhakrishna T, Joseph M (2012) Geochemistry and paleomagnetism of Late Cretaceous mafic dikes in Kerala, southwest coast of India in relation to large igneous provinces and mantle plumes in the Indian Ocean region. Geol Soc Am Bull 124:240–255

Radhakrishna T, Joseph M, Thampi PK, Mitchell JG (1990) Phanerozoic mafic dyke intrusions from the high grade terrain of southwestern India: K-Ar isotope and geochemical implications. In: Parker AJ, Rickwood PC, Tucker DH (eds) Mafic dykes and emplacement mechanisms. Balkema, Rotterdam, pp 363–372

Rainbird RH, Ernst RE (2001) The sedimentary record of mantle-plume uplift. In: Ernst RE, Buchan KL (eds) Mantle plumes: their identification through time. Geological Society of America, Special Paper 352, pp 227–245
Raposo MIB (1997) Magnetic fabric and its significance in the Florianópolis dyke swarm, southern Brazil. Geophys J Int 131:159–170
Ray R, Sheth HC, Mallik J (2007) Structure and emplacement of the Nandurbar-Dhule mafic dyke swarm, Deccan Traps, and the tectonomagnatic evolution of flood basalts. Bull Volc 69:537–551
Renne PR, Deckart K, Ernesto M, Féraud G, Piccirillo EM (1996) Age of the Ponta Grossa dike swarm (Brazil), and implications to Paraná flood volcanism. Earth Planet Sci Lett 144:199–211
Renne PR, Sprain CJ, Richards MA, Self S, Vanderkluysen L, Pande K (2015) State shift in Deccan volcanism at the Cretaceous-Paleogene boundary, possibly induced by impact. Science 350:76–78
Ricci J, Quidelleur X, Pavlov V, Orlov S, Shatsillo A, Courtillot V (2013) New $^{40}Ar/^{39}Ar$ and K-Ar ages of the Viluy traps (Eastern Siberia): further evidence for a relationship with the Frasnian-Famennian mass extinction. Palaeogeogr Palaeoclimatol Palaeoecol 386:531–540
Rickers F, Fichtner A, Trampert J (2013) The Iceland-Jan Mayen plume system and its impact on mantle dynamics in the North Atlantic region: evidence from full-waveform inversion. Earth Planet Sci Lett 367:9–51
Rubin AM (1992) Dike-induced faulting and graben subsistence in volcanic rift zones. J Geophys Res 97(B2):1839–1858
Rudolph ML, Lekié V, Lithgow-Bertelloni C (2015) Viscosity jump in Earth's mid-mantle. Science 350(6266):1349–1352
Ruotoistenmäki T (2014) Geophysical characteristics of Asza shear, Nagasongola discontinuity and ring dyke complex in Uganda. J Afr Earth Sci 93:23–41
Ryabov VV, Shevko AY, Gora MP (2014) Trap magmatism and ore formation in the Siberian Noril'sk region, Trap petrology, vol 1. Springer, Dordrecht
Sant DA, Karanth RV (1990) Emplacement of dyke swarms in the Lower Narmada Valley, western India. In: Parker AJ, Rickwood PC, Tucker DH (eds) Mafic dykes and emplacement mechanisms. Balkema, Rotterdam, pp 383–389
Schärer U, Wilmart E, Duchesne J-C (1996) The short duration and anorogenic character of anorthosite magmatism: U-Pb dating of the Rogaland complex, Norway. Earth Planet Sci Lett 139:335–350
Schoene B, Samperton KM, Eddy MP, Keller G, Adatte T, Bowring SA, Khadri SFR, Gertsch B (2015) U-Pb geochronology of the Deccan Traps and relation to the end-Cretaceous mass extinction. Science 347:182–184
Senger K, Tveranger J, Ogata K, Braathen A, Planke S (2014) Late Mesozoic magmatism in Svalbard: a review. Earth Sci Rev 139:123–144
Shpount BR, Oleinikov BV (1987) A comparison of mafic dyke swarms from the Siberian and Russian platforms. In: Halls HC, Fahrig WF (eds) Mafic dyke swarms. Geological Association of Canada, Special Paper 34, pp 379–383
Sial AN, Oliveira EP, Choudhuri A (1987) Mafic dyke swarms of Brazil. In: Halls HC, Fahrig WF (eds) Mafic dyke swarms. Geological Association of Canada, Special Paper 34, pp 467–481
Sleep NH (2006) Mantle plumes from top to bottom. Earth Sci Rev 77:231–271
Smrekar SE, Stofan ER (1997) Corona formation and heat loss on Venus by coupled upwelling and delamination. Science 277:1289–1294
Smythe DK, Russell MJ, Skuce AG (1995) Intra-continental rifting from the major late Carboniferous quartz-dolerite dyke swarm of NW Europe. Scott J Geol 31:151–162
Söderlund U, Isachsen CE, Bylund G, Heaman LM, Patchett PJ, Vervoort JD, Andersson UB (2005) U-Pb baddeleyite ages and Hf, Nd isotope chemistry constraining repeated mafic magmatism in the Fennoscandian Shield from 1.6 to 0.9 Ga. Contrib Miner Petrol 150:174–194
Spaggiari CV, Bodorkos S, Barquero-Molina M, Tyler IM, Wingate MTD (2009) Interpreted bedrock geology of the south Yilgarn and central Albany–Fraser Orogen, Western Australia. Geological Survey of Western Australia, Record 2009/10

Spreight JM, Skelhorn RR, Sloan T, Knaap RJ (1982) The dykes swarms of Scotland. In: Sutherland DS (ed) Igneous rocks of the British Isles. Wiley, Chichester, pp 449–459

Squyres SW, Janes DM, Baer G, Bindschadler DL, Schubert G, Sharpton VL, Stofan ER (1992) The morphology and evolution of coronae on Venus. J Geophys Res 97(E8):13, 611-13, 634

Stark JC, Wang X-C, Denyszyn SW, Li Z-X, Rasmussen B, Zi J-W, Sheppard S, Liu Y (2018) Newly identified 1.89 Ga mafic dyke swarm in the Archean Yilgarn Craton, Western Australia suggests a connection with India. Precambr Res. https://doi.org/10.1016/j.precamres.2017.12.036

Stofan ER, Smrekar SE (2005) Large topographic rises, coronae, large flow fields, and large volcanoes on Venus: evidence for mantle plumes? In: Foulger GR, Natland JH, Presnall DC, Anderson DL (eds) Plates, plumes and paradigms. Geological Society of America Special Paper 288, pp 841–861

Stofan ER, Sharpton VL, Schubert G, Baer G, Bindschadler DL, Janes DM, Squyres SW (1992) Global distribution and characteristics of coronae and related features on Venus: implications for origin and relation to mantle processes. J Geophys Res 97(E8):13, 347-13, 378

Storey M, Mahoney JJ, Saunders AD, Duncan RA, Kelley SP, Coffin MF (1995) Timing of hot spot-related volcanism and the breakup of Madagascar and India. Science 267:852–855

Studd D, Ernst RE, Samson C (2011) Radiating graben-fissure systems in the Ulfrun Regio area, Venus. Icarus 215:279–291

Svensen H, Corfu F, Polteau S, Hammer Ø, Planke S (2012) Rapid magma emplacement in the Karoo large igneous province. Earth Planet Sci Lett 325–326:1–9

Tack L, Wingate MTD, De Waele B, Meert J, Belousova E, Griffin B, Tahon A, Fernandez-Alonso M (2010) The 1375 Ma "Kibaran event" in Central Africa: prominent emplacement of bimodal magmatism under extensional regime. Precambr Res 180:63–84

Thompson P (1985) Dating the British Tertiary Igneous Province in Ireland by the $^{40}Ar/^{39}Ar$ stepwise degassing method. Unpublished Ph.D. thesis. University of Liverpool, United Kingdom

Thompson RN, Gibson SA (1991) Subcontinental mantle plumes, hotspots and pre-existing thinspots. J Geol Soc Lond 148:973–977

Thompson PF, Tackley PJ (1998) Generation of mega-plumes from the core-mantle boundary in a compressible mantle with temperature-dependent viscosity. Geophys Res Lett 25:1999–2002

Thórarinsson SB, Söderlund U, Døssing A, Holm PM, Ernst RE, Tegner C (2015) Rift magmatism on the Eurasia basin margin: U-Pb baddeleyite ages of alkaline dyke swarms in North Greenland. J Geol Soc 172:721–726

Torsvik TH, Tucker RD, Ashwal LD, Carter LM, Jamtveit B, Vidyadharan KT, Venkataramana P (2000) Late Cretaceous India-Madagascar fit and timing of breakup related magmatism. Terra Nova 12:220–224

Torsvik TH, Smethurst MA, Burke K, Steinberger B (2008) Long term stability in deep mantle structure: evidence from the ~300 Ma Skagerrak-Centred Large Igneous Province (the SCLIP). Earth Planet Sci Lett 267:444–452

Torsvik TH, Rousse S, Labails C, Smethurst MA (2009) A new scheme for the opening of the South Atlantic Ocean and the dissection of an Aptian salt basin. Geophys J Int 177:1315–1333

Vanderkluysen L, Mahoney JJ, Hooper PR, Sheth HC, Ray R (2011) The feeder system of the Deccan Traps (India): insights from dike geochemistry. J Petrol 52:315–343

Walker GPL (1959) Some observations on the Antrim basalts and associated dolerite intrusions. Proc Geol Assoc 70:179–205

Wang W-C, Li Z-X, Li J, Pisarevsky SA, Wingate MTD (2014) Genesis of the 1.21 Ga Marnda Moorn large igneous province by plume-lithosphere interaction. Precambr Res 241:85–103

Wang X-C, Wilde SA, Xu B, Pang C-J (2016) Origin of arc-like continental basalts: implications for deep-Earth fluid cycling and tectonic discrimination. Lithos 261:5–45

Watters TR (1989) Periodically spaced anticlines of the Columbia Plateau. In: Reidel SP, Hooper PR (eds) Volcanism and tectonism in the Columbia River flood-basalt province. Geologcial Society of America, Special Paper 239, pp 283–292

Wilson L, Head JW (2002) Tharsis-radial graben systems as the surface manifestation of plume-related dike intrusion complexes: models and implications. J Geophys Res 107(E8):5057. https://doi.org/10.1029/2001JE001593

Wingate MTD, Pidgeon RT (2005) The Marnda Moorn LIP, a late Mesproterozoic large igneous province in the Yilgarn craton, Western Australia. Large Igneous Provinces Commission website, LIP of the Month for July 2005. http://lips.iavceivolcano.org/05jul. Accessed 19 Apr 2018

Wingate MTD, Campbell IH, Harris LB (2000) SHRIMP baddeleyite age for the Fraser dyke swarm, southeastern Yilgarn craton, Western Austrtalia. Aust J Earth Sci 47:309–313

Wyche S, Pawley MJ, Chen SF, Ivanic TJ, Zibra I, VanKranendonk MJ, Spaggiari CV, Wingate MTD (2013) Geology of the northern Yilgarn craton. In: Wyche S, Ivanic TJ, Zibra I (compilers) Youanmi and Southern Carnarvon seismic and magnetotelluric (MT) workshop 2013. Geological Survey of Western Australia, Record 2013/6, pp 33–63

Zimbelman JR, Edgett KS (1992) The Tharsis Montes, Mars: comparison of volcanic and modified landforms. Proc Lunar Planet Sci 22:31–44

Magma Transport Pathways in Large Igneous Provinces: Lessons from Combining Field Observations and Seismic Reflection Data

Craig Magee, Richard E. Ernst, James Muirhead, Thomas Phillips and Christopher A.-L. Jackson

Abstract Large Igneous Province (LIP) formation involves the generation, intrusion, and extrusion of significant volumes (typically > 1 Mkm^3) of mainly mafic magma and is commonly associated with episodes of mantle plume activity and major plate reconfiguration. Within LIPs, magma transport through Earth's crust over significant vertical (up to tens of kilometres) and lateral (up to thousands of kilometres) distances is facilitated by dyke swarms and sill-complexes. Unravelling how these dyke swarms and sill-complexes develop is critical to: (i) evaluating the spatial and temporal distribution of contemporaneous volcanism and hydrothermal

---

C. Magee (✉) · T. Phillips · C. A.-L. Jackson
Basins Research Group, Department of Earth Science and Engineering,
Imperial College London, London SW7 2BP, UK
e-mail: c.magee@imperial.ac.uk

T. Phillips
e-mail: t.phillips13@imperial.ac.uk

C. A.-L. Jackson
e-mail: c.jackson@imperial.ac.uk

C. Magee
School of Earth and Environment, University of Leeds, Leeds LS2 9JT, UK

R. E. Ernst
Department of Earth Sciences, Carleton University, Ottawa
ON, Canada
e-mail: Richard.Ernst@ernstgeosciences.com

R. E. Ernst
Department of Geology and Geography, Tomsk State University,
36 Lenin Ave, Tomsk 634050, Russia

J. Muirhead
Department of Earth Sciences, Syracuse University, Syracuse, NY 13210, USA
e-mail: james.muirhead@fulbrightmail.org

T. Phillips
Department of Earth Sciences, Durham University, Durham DH1 3LE, UK

R. K. Srivastava et al. (eds.), *Dyke Swarms of the World: A Modern Perspective*,
Springer Geology, https://doi.org/10.1007/978-981-13-1666-1_2

venting, which can drive climate change; (ii) determining melt source regions and volume estimates, which shed light on the mantle processes driving LIP formation; and (iii) assessing the location and form of associated economic ore deposits. Here, we review how seismic reflection data can be used to study the structure and emplacement of sill-complexes and dyke swarms. We particularly show that seismic reflection data can reveal: (i) the connectivity of and magma flow pathways within extensive sill-complexes; (ii) how sill-complexes are spatially accommodated; (iii) changes in the vertical structure of dyke swarms; and (iv) how dyke-induced normal faults and pit chain craters can be used to locate sub-vertical dykes offshore.

1 Introduction

Large Igneous Provinces (LIPs) comprise large volume (>0.1 Mkm^3 but frequently >1 Mkm^3), predominantly mafic (-ultramafic) intrusion networks, extrusive sequences (e.g. flood basalts), and crustal magmatic underplate emplaced either over a relatively rapid timeframe (<5 Myr) or through multiple, short-lived pulses over a few 10s of millions of years (Ernst 2014; Ernst and Youbi 2017; Coffin and Eldholm 1994, 2005; Bryan and Ferrari 2013; Bryan and Ernst 2008). LIPs occur in intraplate continental or oceanic settings, and many appear to result from the impingement of a mantle plume at the base of the lithosphere (e.g. Ernst 2014; Ernst et al. 1995; Coffin and Eldholm 1992). Transport of magma through Earth's crust during LIP development is facilitated by sheet-like conduits including dykes, sills, and inclined sheets. Dykes are traditionally considered to be the most important component of LIP magma plumbing systems and typically form impressive swarms that can extend laterally for >2000 km (Fig. 1) (e.g. Ernst 2014; Ernst and Baragar 1992; Ernst et al. 1995, 2001; Halls 1982). Giant circumferential (sub-circular to elliptical) swarms up to 2000 km in diameter have also been recently recognized (e.g. Buchan and Ernst 2018a,b). However, field observations reveal that sill-complexes, i.e. extensive interconnected networks of sills and inclined sheets (e.g. the Siberian Trap sill-complex is exposed over $>1.5 \times 10^6$ km^2), form the primary volumetric elements of many LIP plumbing systems (Fig. 1) (e.g. Svensen et al. 2012; Leat 2008; Elliot 1992; Elliot and Fleming 2004; Burgess et al. 2017). Analysing and integrating our understanding of these two key parts of LIP plumbing systems (dyke swarms and sill-complexes) is crucial to: (i) estimating LIP melt volumes, which provide critical insight into mantle processes; (ii) reconstructing palaeogeographies, particularly during the Precambrian (e.g. Ernst et al. 2013; Bleeker and Ernst 2006); (iii) locating economic exploration targets associated with LIP formation (e.g. Ernst and Jowitt 2013; Jowitt et al. 2014); (iv) quantifying LIP-related greenhouse gas production, and its role in driving mass extinction (e.g. Ernst and Youbi 2017; Svensen et al. 2004); and (v) understanding the emplacement mechanics and surface expression of sheet intrusions, which can allow similar features to be identified and interpreted on other planetary bodies (e.g. Head and Coffin 1997; Ernst 2014).

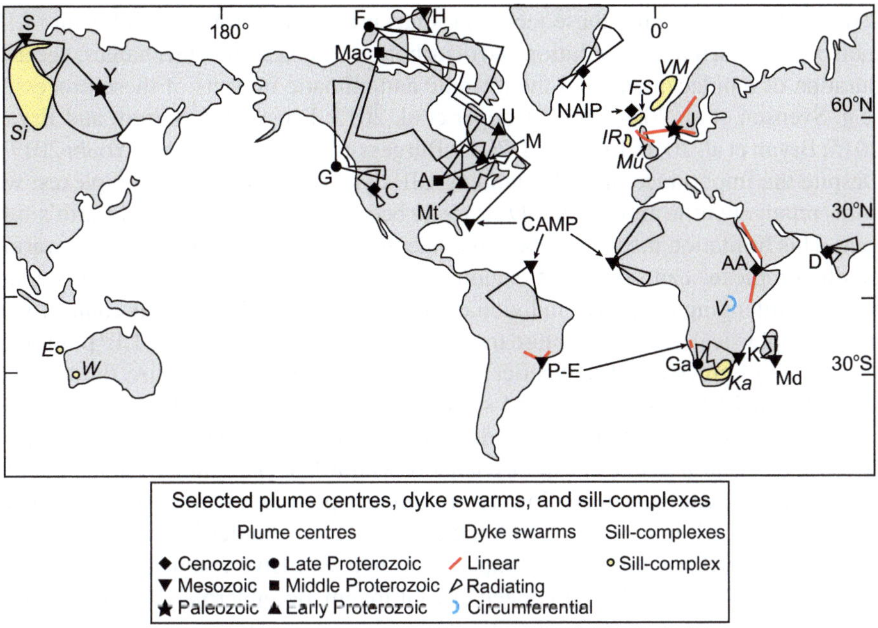

Fig. 1 Map showing selected plume centres and associated dyke swarms and sill-complexes, with each representing part of the plumbing system of a Large Igneous Province (LIP) (modified from Ernst 2014). Dyke swarms: A = 1140 Ma Abitibi swarm, precursor to the 1115–1085 Ma Keweenawan LIP; AA = 30–0 Ma Afar-Arabian swarms; C = 17–0 Ma Columbia River swarms; CAMP = 201 Ma Central Atlantic Magmatic Province swarm; D = 66 Ma Deccan swarm; F = 725–716 Ma Franklin swarm; G = 779 Ma Gunbarrel swarms; Ga = 799 Ma Gannakouriep swarm; H = 130–90 Ma High Arctic LIP (HALIP) swarm; J = 301 Ma Skagerrak (Jutland) swarms; K = 183 Ma Karoo swarms; M = 2510 Ma Mistassini swarm; Mac = 1267 Ma Mackenzie swarm; Md = 89 Ma Madagascar swarm; Mt = 2480–2450 Ma Matachewan swarm; NAIP = 62–55 Ma North Atlantic Igneous Province swarms (e.g. *Mu* is the Mull dyke swarm); P-E = ~135–128 Ma Paraná-Etendeka dyke swarms; S = 251 Ma Siberian Traps swarm; U = 2217–2210 Ma Ungava swarm; Y = 370 Ma Yakutsk-Vilyui swarm. Also show is the 1385 Ma Victoria dyke swarm (*V*) (Mäkitie et al. 2014). Selected sill-complexes: E = ~150 Ma Exmouth Sub-basin sills; F = ~55 Ma Faroe-Shetland Basin sill-complex; IR = ~65–50 Ma Irish Rockall Basin sill-complex; Ka = 183 Ma Karoo sill-complex; Si = 251 Ma Siberian Traps sill-complex; VM = Vøring-Mørebasins sill-complex; W = 1070 Ma Warakurna sill-complex. The ~183 Ma Ferrar sill-complex in Antarctica is shown in Fig. 2a

To date, our understanding of magma transport through LIP plumbing systems has primarily been driven by the mapping of dyke swarm and sill-complex extents and geometries using traditional field approaches and geophysical techniques (e.g. aeromagnetic data), coupled with a range of petrological, geochemical, and geochronological analyses (see Ernst 2014 and references therein). Furthermore, application of different modelling approaches has helped evaluate emplacement mechanics of dyke swarms and sill-complexes (e.g. Townsend et al. 2017; Rivalta et al. 2015; Bunger et al. 2013; Malthe-Sørenssen et al. 2004; Galland 2012; Kavanagh et al.

2015, 2018). Combining these techniques has provided profound insights into LIP melt conditions, magma evolution and distribution, emplacement mechanics, age and duration of emplacement, and the tectonic and climatic impacts of these processes (e.g. Svensen et al. 2004, 2012; Cooper et al. 2012; Ernst 2014; Jerram and Bryan 2015; Bryan et al. 2010; Ernst et al. 2005; Burgess et al. 2017; Ernst and Youbi 2017). Despite the importance of understanding LIP dyke swarms and sill-complexes, we know relatively little about their 3D structure because of the 2D nature of Earth's surface. This limitation means that both the lateral and vertical extents of dyke swarms or sill-complexes can seldom be examined simultaneously. Furthermore, it is commonly challenging to resolve individual intrusions and determine their connectivity using geophysical data that capture the subsurface expression of these LIP plumbing systems (e.g. gravity and magnetics). Without being able to visualise the true 3D geometry of LIP dyke swarms and sill-complexes, it is difficult to answer a wide range of questions, including: How does magma travel such large lateral distances (>100 km)? Can magma transport channels within LIP plumbing systems remain open and facilitate sustained flow for protracted periods of time? How do dyke swarms and/or sill-complexes feed flood basalts? How volumetrically important are intrusive networks within LIPs compared to their extrusive counterparts? What is their relationship to underlying, deep crustal magmatic underplate? How do sheet intrusions relate to stress field conditions?

Reflection seismology is the only technique that can image intrusion networks in 3D, at a relatively high-resolution (i.e. metres to tens of metres) over vast areas, allowing individual magma bodies and associated structures to be mapped (e.g. Smallwood and Maresh 2002; Thomson and Hutton 2004; Planke et al. 2005; Cartwright and Hansen 2006; Schofield et al. 2017; Magee et al. 2016; Planke et al. 2015). For example, mapping of magma flow indicators across regionally extensive sill-complexes imaged in seismic reflection data reveal that these intrusion networks can transport melt over significant vertical and lateral distances (e.g. Magee et al. 2016; Schofield et al. 2017; Reynolds et al. 2017). These magma flow observations are further supported by field, magnetic, and geochemical analyses of LIP sill-complexes (e.g. Airoldi et al. 2016; Leat 2008; Aspler et al. 2002; Marsh 2004). One challenge with reflection seismology, which relies on measuring arrival times of acoustic energy reflected back from geological features, is that the technique favours imaging of sub-horizontal to moderately inclined structures (e.g. sills). Dykes and dyke swarms are thus rarely imaged in seismic reflection data (e.g. Thomson 2007), except when they have been rotated post-emplacement to shallower dips (e.g. Phillips et al. 2017; Abdelmalak et al. 2015). Several studies have also shown, however, that dykes can be detected in seismic reflection data through the recognition of dyke-related host rock structures and/or geophysical artefacts (e.g. Wall et al. 2010; Malehmir and Bellefleur 2010; Zaleski et al. 1997; Bosworth et al. 2015; Kirton and Donato 1985; Ardakani et al. 2017).

In addition to imaging the 3D structure of dyke swarms and sill-complexes, seismic reflection data also provide unprecedented insight into the host rock deformation mechanisms that spatially accommodate magma emplacement and associated hydrothermal venting (e.g. Hansen and Cartwright 2006b; Magee et al. 2013, 2014;

Reeves et al. 2018; Trude et al. 2003; Jackson et al. 2013; Svensen et al. 2004; Hansen 2006). Being able to examine how intrusion induces deformation and the style of structures formed is critical because it provides insight into the processes driving magmatism (e.g. Pollard et al. 1983; Ernst et al. 2001; Magee et al. 2017; Galland 2012; van Wyk de Vries et al. 2014; Pollard and Johnson 1973). Furthermore, understanding how subsurface magma movement and accumulation translates into ground deformation, through 3D analysis of ancient intrusions and associated deformation, inform show surficial intrusion-induced structures on Earth and other planetary bodies relate to the underlying plumbing system (e.g. Pollard et al. 1983; Ernst et al. 2001; Magee et al. 2017; Galland 2012; Wyrick and Smart 2009; Wyrick et al. 2004; Manga and Michaut 2017).

We consider seismic reflection to be a powerful tool in the study of magma plumbing systems, and one that is yet to be fully utilised by the LIP community. To highlight the potential use of seismic reflection data, we briefly examine seismically imaged sill-complexes and dyke swarms and discuss how insights into their 3D geometry inform our understanding of LIP magma plumbing systems. We particularly aim to promote future integration and collaboration between of sectors of the LIP community using reflection seismology and those who apply more conventional analytical methods.

2 Magma Transport Pathways in LIPs

Here, we summarise our current understanding of the roles that sill-complexes and dyke swarms play in the development of LIPs using selected examples from field and seismic reflection studies. In addition to examining the 3D structure of sill-complexes and dyke swarms, we establish how this knowledge can be combined with analyses of host rock and free surface deformation to unravel the emplacement mechanics of these intrusive networks. LIP plumbing systems also comprise local, geometrically complicated intrusive complexes (e.g. laccoliths and layered intrusions) and, whilst these features are poorly imaged in seismic reflection data (e.g. Archer et al. 2005), their development can be linked to dyke swarm and sill-complex emplacement (e.g. Menand 2008); we therefore briefly describe insights into these intrusive complexes derived from field observations. Overall, this short review of previous work illustrates how seismic reflection data has advance our understanding of LIP magma plumbing systems, and provides a key tool for future studies addressing unresolved questions related to LIP emplacement.

2.1 *Field Observations of Sill-Complexes*

2.1.1 Sill-Complex Geometry and Connectivity

Prior to the application of seismic reflection data to study sill-complexes associated with LIPs, our understanding of these plumbing system components was largely driven by analyses of the Ferrar and Karoo sill-complexes (Figs. 1 and 2a) (e.g. du Toit 1920; Gunn and Warren 1962; Grapes et al. 1974; Chevallier and Woodford 1999; Leat 2008; Elliot and Fleming 2017; Elliot et al. 1999; Svensen et al. 2015; Polteau et al. 2008a, b). The Karoo sill-complex is relatively well-exposed across ~0.5×10^6 km^2 in South Africa (Chevallier and Woodford 1999). The Ferrar sill-complex can be traced across an apparently linear belt that, based on palaogeographic reconstructions of the Early Jurassic, likely extended for >4000 km from the Theron Mountains in Antarctica to Australia and New Zealand (Fig. 2a) (Elliot and Fleming 2017). Both of the Ferrar and Karoo sill-complexes consist of numerous sills, which have tabular or saucer-shaped morphologies, and inclined sheets intruded into relatively unstructured sedimentary basins at depths of <5 km (e.g. Fig. 2b, c) (e.g. Elliot and Fleming 2017; Muirhead et al. 2014; Airoldi et al. 2011; Svensen et al. 2012; Galerne et al. 2008, 2011; Chevallier and Woodford 1999). High-precision U-Pb zircon and baddeleyite dating of the Ferrar and Karoo sill-complexes yield ages of 182.779 ± 0.033–182.59 ± 0.079 Ma and 183 ± 0.5–182.3 ± 0.6 Ma, respectively, indicating emplacement likely occurred simultaneously over <0.5 Myr (Svensen et al. 2012; Burgess et al. 2015). Geochronological and geochemical data suggest a common source for both the Ferrar and Karoosill-complexes, perhaps a plume centre located in the Weddell Sea (Encarnación et al. 1996; Elliot and Fleming 2000; Burgess et al. 2015). Linear dyke swarms and picrite distribution associated with the Karoo sill-complex suggest the province may have alternatively been sourced from onshore southern Africa in the Mwenezi (Nuanetsi) area (Fig. 2a) (Hastie et al. 2014; Burke and Dewey 1973; Ernst and Buchan 1997a). Unlike many LIPs (e.g. the Mackenzie LIP; Ernst and Baragar 1992; Baragar et al. 1996), no major sub-parallel or giant radiating dyke swarms are associated with the Ferrar LIP (Figs. 1 and 2a) (e.g. Elliot and Fleming 2004, 2017; Muirhead et al. 2014). Here, we briefly discuss how observations from the Ferrar and Karoo sill-complexes shed light on magma transport within these extensive LIP plumbing systems.

The Ferrar sill-complex is spectacularly exposed in the McMurdo Dry Valleys of Antarctica, where networks of sills and inclined sheets can be observed and measured on >1 km-high cliff faces (Fig. 2b). These outcrops reveal a conduit system of inter-connected, stacked sills that facilitated magma ascent and possibly fed voluminous lava outpourings (Elliot and Fleming 2004; Muirhead et al. 2012, 2014). Although these cliff face outcrops limit access to the 3D geometry of individual sills, extensive exposures along much of the Ferrar sill-complex length mean that magma transport within the system can be regionally assessed (Fig. 2a) (Elliot and Fleming 2017). Key observations from the Ferrar sill-complex include: (i) recognized sill-sill feeding relationships (e.g. Fig. 2b) (Muirhead et al. 2012, 2014); (ii) decreasing Mg# and MgO

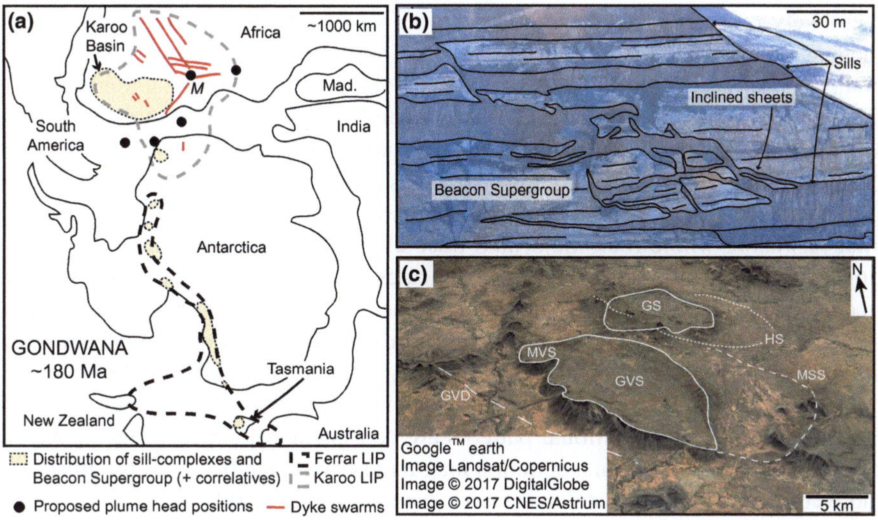

Fig. 2 **a** Paleogeographic reconstruction showing the distribution of intrusive elements within the Karoo and Ferrar LIPs (modified from Leat 2008). **b** Interpreted photograph of interconnected sills and inclined sheets, part of the Ferrar sill-complex, intruded into the Beacon Supergroup sedimentary rocks (image courtesy of James White). **c** Interpreted Google Earth image of the Golden Valley Sill and surrounding intrusions, located in the Karoo Basin, South Africa. GVS = Golden Valley Sill; MVS = MV Sill; GS = Glen Sill; HS = Harmony Sill; MSS = Morning Sun Sill; GVD = Golden Valley Dyke

contents consistent with fractional crystallization away from the Weddell Sea region (Leat 2008; Elliot and Fleming 2000; Elliot et al. 1999); and (iii) the consistent orientation of broken bridge structure long axes and magnetic lineations within sills and interconnected sheets across Antarctica (e.g. in the Theron Mountains, Antarctica) (Hutton 2009; Dragoni et al. 1997; Airoldi et al. 2012). These observations indicate that interconnected sills facilitated long distance magma transport (e.g. Leat 2008; Elliot and Fleming 2017). However, it remains contentious as to whether magma: (i) entered the sill-complex only at the plume head region (i.e. magma flowed within the sill-complex for >4000 km; Leat 2008); or (ii) if dykes injected laterally from the plume head region within the crust and periodically ascended to feed the sill-complex (Elliot and Fleming 2017).

In contrast to the Ferrar sill-complex, exposure styles within the Karoo sill-complex commonly allow the broad 3D geometry of individual intrusions to be assessed (e.g. Fig. 2c). At deeper levels within the sill-complex, the sills are up to 200 m thick and tabular, with those intruded closer to the palaeosurface, considered to be marked by the base of the Drakensburg Group lavas, displaying saucer-shaped morphologies with thicknesses <100 m (e.g. Fig. 2c) (Svensen et al. 2012); intrusion of these sills is predicted to have released at least 27,400 Gt CO_2 from hydrothermal vents, contributing to Toarcian global warming (e.g. Svensen et al. 2006, 2007; Jamtveit et al. 2004). The change in sill geometry with proximity to the palaeosurface likely reflects the increasing interaction between the local stress field around the magma body and the free surface (Malthe-Sørenssen et al. 2004). It is, however, commonly difficult from field observations alone to determine how individual sills are fed, i.e. are sills interconnected or are they fed by dykes? For example, within the vicinity of the Golden Valley Sill, it may be assumed that the close spatial association of five major sills and one major dyke indicates that the intrusions are interconnected (Fig. 2c) (Galerne et al. 2011). However, Forward Stepwise-Discriminant Function Analysis using whole rock major and trace element geochemical data, which distinguishes compositional populations, indicates that different magma pulses can be tracked across the intrusions (Fig. 2c) (Galerne et al. 2011). Three sills appear connected and fed by a single magma batch (i.e. the Golden Valley, MV, and Glen sills), whereas the Morning Sun Sill, Harmony Sill and Golden Valley Dyke are not connected and were fed by discrete magma batches (Fig. 2c) (Galerne et al. 2011). Furthermore, the compositional patterns are consistent with each of these magma batches being derived from a common magma reservoir that becomes progressively more contaminated by crustal host rocks (Neumann et al. 2011).

2.1.2 Surface Expression of Sills and Sill-Complexes

The growth of broadly sub-horizontal, tabular sills and laccoliths emplaced at shallow-levels, particularly in sedimentary basins, is typically accommodated by overburden uplift (e.g. Fig. 3) (e.g. Jackson and Pollard 1988; Koch et al. 1981; Johnson and Pollard 1973; Pollard and Johnson 1973; Morgan et al. 2008; Wilson et al. 2016; Galland 2012; Agirrezabala 2015; Le Gall et al. 2010). This doming of the overburden, termed forced folding, commonly instigates deformation of the free surface (e.g. Fig. 3) (e.g. Magee et al. 2017; Pollard and Johnson 1973). Although erosion and exposure of ancient sill-complexes involves removal of the overburden, making it difficult to determine how entire sill-complexes are accommodated, geodetic and remote sensing data from the Danakil Depression, Ethiopia document the growth of forced folds above several sill intrusions (e.g. Fig. 3) (Magee et al. 2017). In particular, analysis of ground deformation at and lava flow disposition around the Alu dome, in the Danakil Depression, indicate that sill growth can occur over protracted time spans through discrete injection and eruption cycles driving uplift and subsidence (Fig. 3) (Magee et al. 2017). Identifying forced folds at the surface thus allows the geometry and locations of subsurface sills and sill-complexes to be inferred (Magee et al. 2017; van Wyk de Vries et al. 2014). It should, however, be noted that inelastic deformation of the host rock can also accommodate sill emplacement (e.g. fluidisation), which can partly or fully inhibit expression of intrusions at the contemporaneous surface (e.g. Schofield et al. 2012b, 2014).

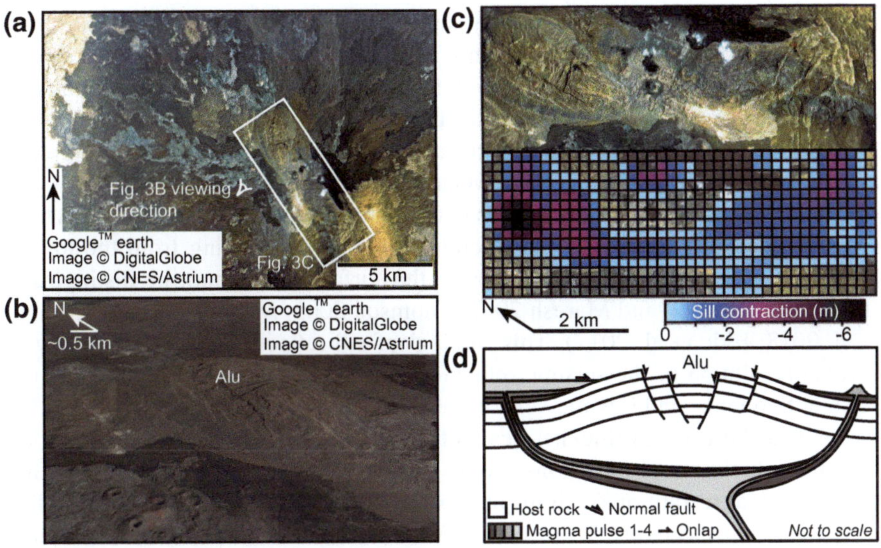

Fig. 3 **a**, **b** Plan-view and 3D view (vertical exaggeration = 2) of the Alu dome and surrounding lava flows. **c** Sill contraction area and magnitude inverted from InSAR data collected during a basalt eruption in 2008 (Pagli et al. 2012). Note that the area of sill contraction underlies the Alu dome. **d** Schematic of interpreted plumbing system beneath Alu. Images modified from Magee et al. (2017)

2.2 *Imaging and Identifying Sill-Complexes in Seismic Reflection Data*

Seismic reflection data have revolutionized our understanding of sill-complexes because these intrusion networks can be imaged in 3D at resolutions on the order of tens of meters (e.g. Magee et al. 2014; Smallwood and Maresh 2002; Cartwright and Hansen 2006; Thomson and Hutton 2004; Planke et al. 2015). Mafic sills are typically well imaged in seismic reflection data, particularly within sedimentary basins, where they are expressed as positive polarity, high-amplitude reflections (Fig. 4) (e.g. Smallwood and Maresh 2002; Thomson 2005; Planke et al. 2005, 2015). This high-amplitude, positive polarity reflection configuration occurs because the high density and seismic velocity of mafic igneous rocks, relative to the host sequence, produces a high acoustic impedance contrast at intrusion contacts that reflects more acoustic energy (e.g. Brown 2004; Magee et al. 2015; Eide et al. 2017); more felsic intrusions have lower densities and seismic velocities and thus, depending on the host rock lithology, may not be easy to identify in seismic reflection data (Mark et al. 2017). Mapping sills in seismic reflection data reveals that they can display a range of geometries, from strata-concordant to saucer-shaped (Fig. 4a–c) (e.g. Planke et al. 2005). These sill-related reflections commonly display minor vertical offsets that radiate outwards from the deepest portion of the sill (Fig. 4d), or where the sill appears connected to an inclined sheet (e.g. Magee et al. 2014; Schofield et al.

2012a; Reynolds et al. 2017). Comparisons between sills observed in outcrop and reflection seismic reveal that these minor vertical offsets are likely intrusive steps or bridge structures (e.g. Fig. 4a, d), which form during and shed light on initial sheet propagation direction (Schofield et al. 2012b). Mapping these seismically resolvable magma flow indictors thus provides critical insights into how entire sill-complexes are emplaced (Schofield et al. 2017; Magee et al. 2014, 2016).

One limitation of seismic reflection data is that sills are typically imaged as tuned reflection packages, whereby the discrete reflections emanating from the top and base of a sill converge on their return to the surface and cannot be distinguished (Fig. 4) (Smallwood and Maresh 2002; Thomson 2005; Magee et al. 2015; Rabbel et al. 2018; Eide et al. 2017). This 'tuning' of sill reflections is controlled by the vertical resolution of the seismic reflection data, itself a function of the dominant seismic frequency and seismic velocity of the host rock interval (Brown 2004), and means that the true sill thickness cannot be extracted. Instead, the maximum and minimum thickness of intrusions producing tuned reflection packages are defined by the vertical resolution (or limit of separability) and the detection limit (or limit of visibility), respectively (Brown 2004; Magee et al. 2015; Smallwood and Maresh 2002; Eide et al. 2017; Rabbel et al. 2018). Intrusions with thicknesses above the vertical resolution of the seismic reflection data will produce discrete top and base reflections.

2.2.1 North Atlantic Igneous Province (~61–50 Ma): Irish Rockall Basin

The North Atlantic Igneous Province formed ~61–50 Ma in response to the impingement of a mantle plume (i.e. now centred on Iceland) and break-up of Pangaea (e.g. Storey et al. 2007). Here, we specifically focus on a sill-complex imaged in seismic reflection data from the NE Irish Rockall Basin and particularly consider how its emplacement deformed the host rock and contemporaneous surface (Figs. 1 and 5). Within a 748 km^2 region of the sill-complex, which is covered by 3D seismic reflection data, the imaged plumbing system comprises 82 seismically resolved, stacked saucer-shaped sills and inclined sheets (Figs. 5a, b) (Magee et al. 2014). Overall, the sill-complex was intruded into a ~2 km thick sequence of Cretaceous marine shale and Paleogene volcaniclastic sandstone (Magee et al. 2014). Individual intrusions have diameters of ~0.3–7.1 km and may transgress up to 1.79 km of strata (Magee et al. 2014). Mapping of magma flow indicators (e.g. intrusive steps and bridge structures; Fig. 4d) and comparison to vertical seismic sections reveal that individual intrusions propagated away from apparent connection sites to underlying sills (Magee et al. 2014). These inferred flow patterns imply that individual sills are connected and facilitated lateral magma flow, from a melt source to the NW likely corresponding to the Hebridean Terrace Igneous Complex (i.e. an intrusive complex), across a broad area without dykes (Magee et al. 2014).

The Irish Rockall Basin sill-complex is overlain by a series dome-shaped forced folds (e.g. Fig. 5a) (Magee et al. 2014). Two types of forced fold can be recognized

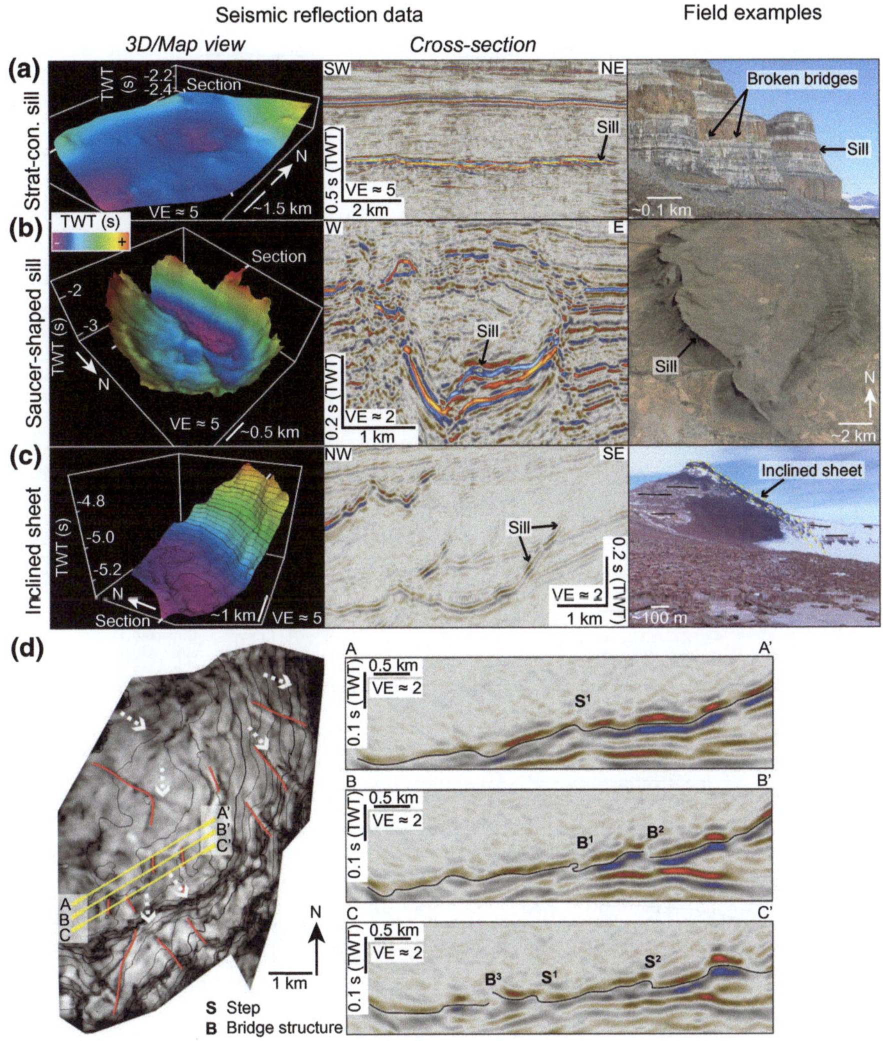

Fig. 4 **a–c** Examples of strata-concordant (strat-con.) and saucer-shaped sills, as well as inclined sheets, observed in seismic reflection data (i.e. in cross-section and 3D views) and field outcrop. Field photographs courtesy of Donny Hutton (**a**), Google Earth (**b**), and James White (**c**). Figure modified from Magee et al. (2016). **d** Intrusive step and broken bridge traces (red lines), which allow magma flow patterns to be inferred (dashed white arrows) observed on a plan-view variance map of a sill located in the Irish Rockall Basin (see Magee et al. 2014). Cross-sections through the sill, depicting the different magma flow indicators, are also shown

(Magee et al. 2014): (i) individual forced folds, the outline of which corresponds to the lateral tips of underlying sills; and (ii) compound forced folds, which consist of and result from the amalgamation of individual forced folds (Fig. 5c). These

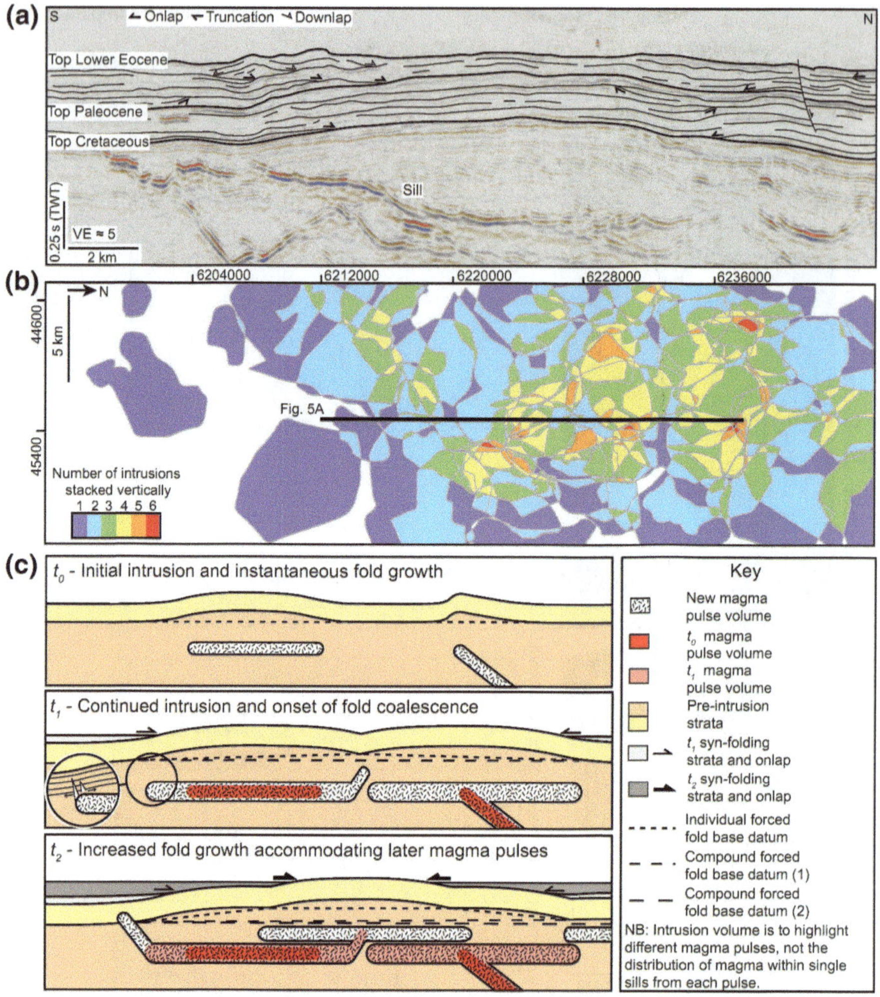

Fig. 5 **a** Seismic section from the Irish Rockall Basin showing an interconnected network of high-amplitude reflections, interpreted to be sills and inclined sheets, overlain by domed strata (Magee et al. 2014). Seismic-stratigraphic onlap, downlap, and truncation patterns occur throughout the folded strata. **b** Map of the 82 seismically resolved sills interpreted by Magee et al. (2014), with colours depicting how the sills are vertically stacked. **c** Schematic of the evolution of sill emplacement and forced fold growth, over a protracted period of time, in the Irish Rockall Basin (Magee et al. 2014)

compound forced folds indicate that the entire sill-complex was, at least partly, accommodated by overburden uplift (Magee et al. 2014). Importantly, throughout the folded Cretaceous-to-Lower Eocene sequence and particularly at the Top Palaeocene, seismic-stratigraphic onlap of strata onto the forced folds is observed (Fig. 5a) (Magee et al. 2014). These onlap relationships form when sediment is deposited around a topobathymetric high, demonstrating that the forced folds were expressed at the contemporaneous surface (Magee et al. 2014). Furthermore, biostratigraphic dating of the onlapping strata indicates that forced fold growth, and thus sill emplacement, occurred incrementally over 15 Myr between ~65 and 54 Ma; prominent onlap onto the Top Palaeocene horizon suggests a major pulse of sill-complex emplacement occurred at ~56 Ma (Fig. 5a) (Magee et al. 2014). Recognition of these forced folds and onlap onto them indicates that the surface expression of sill-complexes may be complicated, with individual injection events shaping the deformation history of a broad area over a protracted period of time (e.g. Fig. 5c) (Magee et al. 2014). The geometry and history of intrusion-induced forced folds currently expressed at the surface of Earth or other planetary bodies may thus not relate to evolution of a single intrusion, as is commonly assumed (Magee et al. 2014).

2.2.2 North Atlantic Igneous Province (~61–50 Ma): Faroe-Shetland Basin

Within the Faroe-Shetland Basin, to the north of the Rockall Basin, seismic reflection data image a ~61–52 Myr old, extensive sill-complex (>22,500 km^2) emplaced into Upper Cretaceous shale (Figs. 1 and 6) (Passey and Hitchen 2011; Schofield et al. 2017). Over 300 seismically resolved saucer-shaped sills, with diameters up to tens of kilometers, have been mapped (Fig. 6) (Schofield et al. 2017). Reconstruction of flow patterns within the Faroe-Shetland Basin sill-complex implies that magma was fed into the basin via dykes intersecting basement-involved faults, from which sills propagated laterally for up to ~25 km into the hanging wall (Fig. 6) (Schofield et al. 2017). Individual intrusions transgress up to 2 km of the sedimentary succession and appear to feed overlying sills and lava flows (Fig. 6b) (Schofield et al. 2017).

2.2.3 North Atlantic Igneous Province (~61–50 Ma): Vøring and Møre Basins

Similar to the Irish Rockall and Faroe-Shetland basins, the Vøring and Møre basins offshore Norway host an extensive sill-complex (i.e. >80,000 km^2) emplaced during the break-up of the North Atlantic (Fig. 7a) (e.g. Jamtveit et al. 2004; Skogseid et al. 1992; Planke et al. 2005; Schmiedel et al. 2017; Cartwright and Hansen 2006; Hansen and Cartwright 2006a, b; Hansen et al. 2004). Associated with the sill-complex are >735 hydrothermal vent complexes; i.e. pipe-like structures that emanate from the lateral tips of sills and extend upwards to the syn-emplacement seabed, where they fed crater-, dome-, or eye-shaped vents (e.g. Fig. 7b) (e.g. Hansen 2006; Svensen et al. 2004). Biostratigraphic dating suggests the majority of the hydrothermal vent complex was active 55.8–55 Ma, coincident with the onset of the 200 Kyr-long Eocene Thermal Maximum global warming event (Svensen et al. 2004). It is estimated that sill-complex emplacement heated host rock organic shales, rapidly releasing 3–0.3 × 10^{18} g of methane through the hydrothermal vent complex and thereby instigating

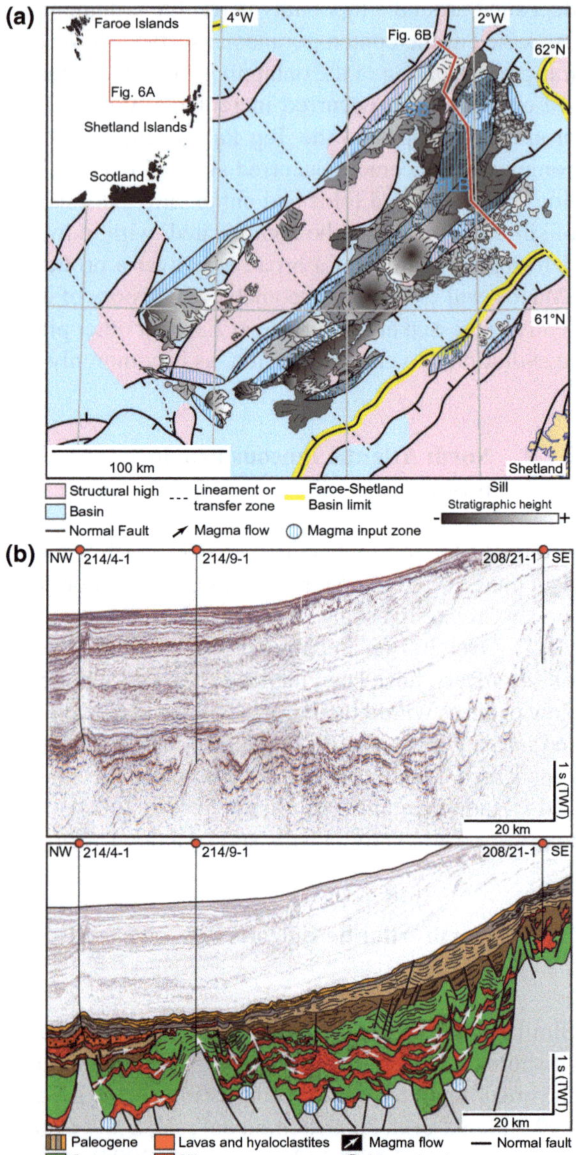

Fig. 6 **a** Map of seismically resolved sills interpreted across the Faroe-Shetland Basin, showing magma flow patterns and interpreted zones of magma input into the basin (Schofield et al. 2017). **b** Uninterpreted and interpreted seismic section through the Faroe-Shetland Basin, highlighting the distributions, structure, and connectivity of sills within the Cretaceous strata (Schofield et al. 2017). See Fig. 6a for location

a global carbon isotope excursion of $-1‰$, which may have promoted the Eocene Thermal Maximum (Svensen et al. 2004).

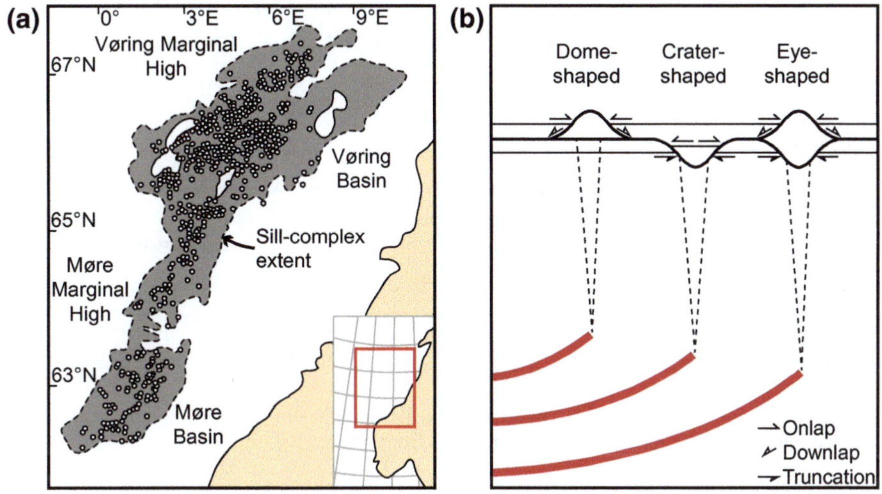

Fig. 7 **a** Map showing the areal coverage of the sill-complex and associated hydrothermal vent complexes (filled circles) spanning the Vøring and Møre basins, offshore Norway (modified from Svensen et al. 2004). **b** Schematic showing the different types of hydrothermal vent complex, which are classified based on the morphology of their upper section; i.e. dome-, crater-, or eye-shaped

2.2.4 Warakurna LIP (~1070 Ma): Northern Yilgarn Craton, Western Australia

Within the ~3.05–2.62 Ga Youanmi Terrane in the Yilgarn Craton of Western Australia (Fig. 1), three 2D seismic reflection profiles image an interconnected network of highly reflective sills that extend for several hundreds of kilometers (e.g. Fig. 8) (Ivanic et al. 2013a). In contrast to the vast majority of sill-complexes described from seismic reflection data, which occur in sedimentary basins (see Magee et al. 2016 and references therein), these sills within the Youanmi Terrane were emplaced into crystalline rocks comprising greenstone belts, granites, and granitic gneiss (Fig. 8) (Wyche et al. 2013). Radiometric dating of outcropping intrusions within this sill-complex, including the Mount Homes Gabbro Sill that has a baddeleyite SHRIMP U-Pb age of 1070 ± 18 Ma, suggests emplacement occurred during formation of the ~1070 Ma Warakurna LIP (Wyche et al. 2013; Ivanic et al. 2013a; b; Wingate et al. 2004; Pirajno and Hoatson 2012). Basement sills are also potentially associated with other LIP events, such as the ~1080 Ma Southwestern Laurentia LIP, the ~1270 Ma Mackenzie LIP, and the ~1740 Ma Cleaver LIP (e.g. Mandler and Clowes 1997; Ernst 2014). Although the relatively low resolution of the seismic reflection data imaging this Warakurna sill-complex mean that its overall geometry and contemporaneous surface expression cannot fully be assessed, these data indicate that interconnected sill-complexes can facilitate long-distance lateral magma transport within crystalline continental crust (Fig. 8) (Ivanic et al. 2013a; b; Magee et al. 2016).

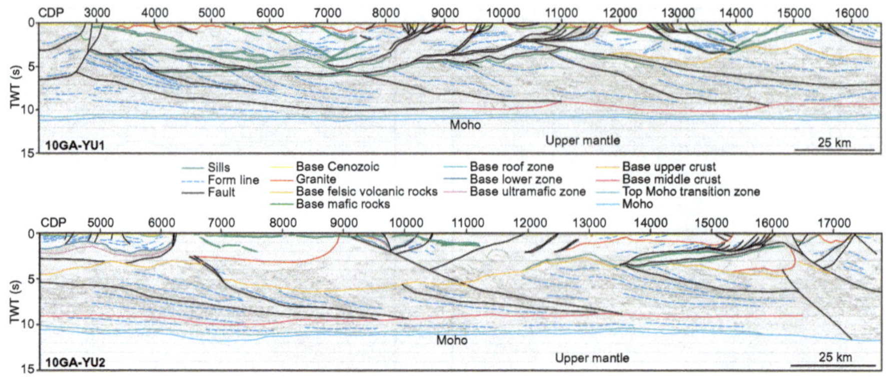

Fig. 8 Two seismic lines from the Yilgarn Craton, Western Australia, showing a network of sills intruded into crystalline basement rocks (modified from Ivanic et al. 2013a, b)

2.3 Field Observations of Dyke Swarms

2.3.1 Dyke Swarm Geometry and Connectivity

Key to the mapping of dyke swarms has been the combination of field observations and regional and local, high-resolution aeromagnetic maps (e.g. Ernst 2014; Cooper et al. 2012; Reeves 2000; Glazner et al. 1999; Buchan and Ernst 2004, 2013). Aeromagnetic mapping of dyke swarms has proved particularly important because these data allow dykes to be traced beneath areas of thin sedimentary cover (Fig. 9) (e.g. Cooper et al. 2012; Mäkitie et al. 2014; Buchan and Ernst 2004). An excellent example of how aeromagnetic data can illuminate otherwise unseen dyke swarm extents and properties is provided by the Tellus regional airborne geophysical survey of Northern Ireland, which imaged several Palaeocene dyke swarms related to the formation of the North Atlantic Igneous Province (Fig. 9) (Cooper et al. 2012). Subtle changes in dyke orientation, cross-cutting relationships, and variable offset across strike-slip faults revealed by the Tellus aeromagnetic data highlight dyke swarm emplacement occurred periodically during north-south Alpine-related compression (Fig. 9) (Cooper et al. 2012). Overall, field and aeromagnetic mapping reveal that individual dykes within dyke swarms can display a range of lengths, from metres to thousands of kilometres, and thicknesses of typically 10–50 m, but up to >100 m and >1 km in the case of dyke-like layered intrusions. It should, however, be noted that it can be challenging to determine dyke lengths if dykes are segmented and linked across en echelon offsets; i.e. the actual length of the dyke could be much greater than the length of measured segments (e.g. Ernst 2014; Ernst et al. 1995; Townsend et al. 2017). In particular, whilst the en echelon arrangement of dyke segments is clear in some swarms (e.g. 700 km long 1140 Ma Great Abitibi dyke of the Superior craton), connections between segments can be difficult to recognise within dense swarms.

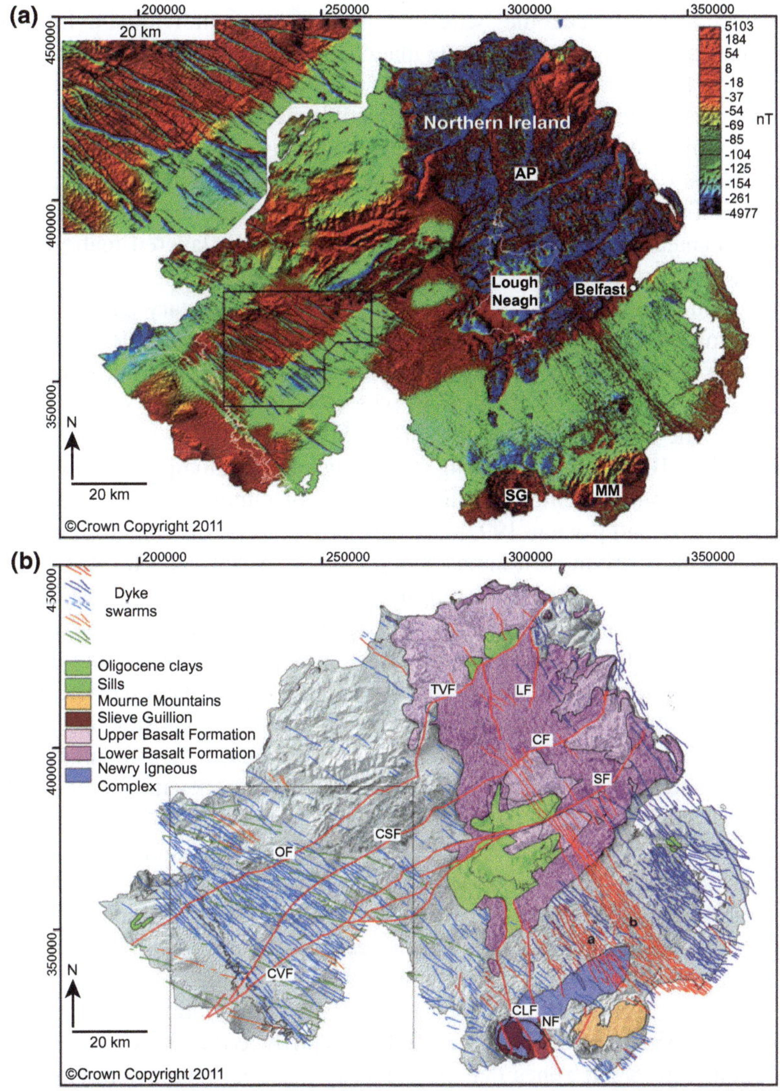

Fig. 9 **a**, **b** Uninterpreted and interpreted total magnetic intensity anomaly map acquired during the Tellus airborne geophysical survey of Northern Ireland (modified from Cooper et al. 2012). Linear magnetic anomalies are attributed to different dyke swarms, with larger anomalies corresponding to the Antrim Plateau (AP) basalt lavas and the Slieve Guillian (SG) and Mourne Mountains (MM) intrusive complexes. Offsets in the dyke-related linear anomalies mark the position of strike-slip faults: CF = Carnlough Fault; CLF = Camlough Fault; CVF = Clogher Valley Fault; LF = Loughguile Fault; NF = Newry Fault; OF = Omagh Fault; SF = Sixmilewater Fault; TSF = Tempo-Sixmilecross Fault; TVF = Tow Valley Fault

Geochemical and palaeomagnetic studies suggest that individual dykes within a swarm may represent a single, unique magma pulse (Ernst 2014). For example, the 700 km long 1140 Ma Great Abitibi dyke of the Superior craton displays distinct and internally consistent ratios of mantle incompatible elements (i.e. expressed as parallel curves on multi-element diagrams), which are slightly but clearly different from the shapes of multi-element diagrams for other nearby dykes within the swarm (Ernst 2014). It is likely that individual dykes within a swarm are injected from a magma reservoir located in the plume centre region, probably a layered mafic-ultramafic intrusion, in separate injection events (e.g. Baragar et al. 1996).

Whilst sill-complexes are typically restricted to sedimentary basins, dykes and dyke swarms can vertically span lower- to supra-crustal structural levels and extend along-strike across various rock types. Dyke swarm geometries are thus primarily controlled by regional stress conditions, whereby dykes typically orient orthogonal to σ_3 and in the σ_1-σ_2 plane. A range of overall dyke swarm geometries have been identified (Fig. 10): (i) giant radiating swarms that have radii up to 2000 km, diverge away from the plume centre, and attain widths of hundreds of kilometres (Ernst et al. 2001; Ernst and Buchan 1997a); (ii) linear dykes swarms, which are commonly associated with overlying rift zones, extend along-strike for up to >2000 km, and have widths of tens of kilometres (e.g. Ernst and Buchan 1997a; Ernst 2014); and (iii) giant circumferential swarms that have diameters of up to 2000 km (Buchan and Ernst 2018a, b). All three dyke swarm geometries can develop around and appear to emanate from a mantle plume centre (Ernst et al. 2001). Although the plan-view geometry of a dyke swarm can be assessed at the level of exposure and under thin cover using field observations and remote sensing data, it is difficult to estimate dyke height and determine whether dyke swarm width varies with height. Modelling by Wilson and Head (2002) suggests that the bottom of laterally emplaced dykes could extend into the upper lithospheric mantle. However, this interpretation differs from studies that suggest a lower density of dykes emplaced in high grade terranes, compared lower grade terranes, probably reflects the presence of a lower boundary to laterally propagating dykes within the lower crust (Fahrig 1987). Here, we provide a brief summary of several major dyke swarms.

Linear dyke swarms for many LIPs develop when uplift induced by mantle plume impingement localise mechanical (i.e. faulting) and dyke-driven extension of Earth's crust within narrow rift zones that extend away from the plume centre, potentially forming a rift-rift-rift triple junction (Fig. 10a). For example, a major linear dyke swarm parallels the Red Sea rift for ~2500 km (Fig. 1) (Bosworth et al. 2015). The dyke swarm and rifting are associated with the Afro-Arabian LIP, which is in turn related to the Afar plume (Fig. 1). The ~30 Ma Ethiopian flood basalts and related magmatism was focused above the inferred location of the Afar plume and triple junction rifting extends along the Red Sea, the Gulf of Yemen, and southward along the East Africa rift system (Ebinger and Sleep 1998). Initial flood basalt magmatism occurred at 30 Ma (Hofmann et al. 1997), but the major pulse of rift-parallel diking in the Red Sea occurred at ~23 Ma (e.g. Bosworth et al. 2015).

If extension at the level of dyke emplacement is instead accommodated solely by intrusion, giant radiating dyke swarms may form, whereby dyke-spacing increases

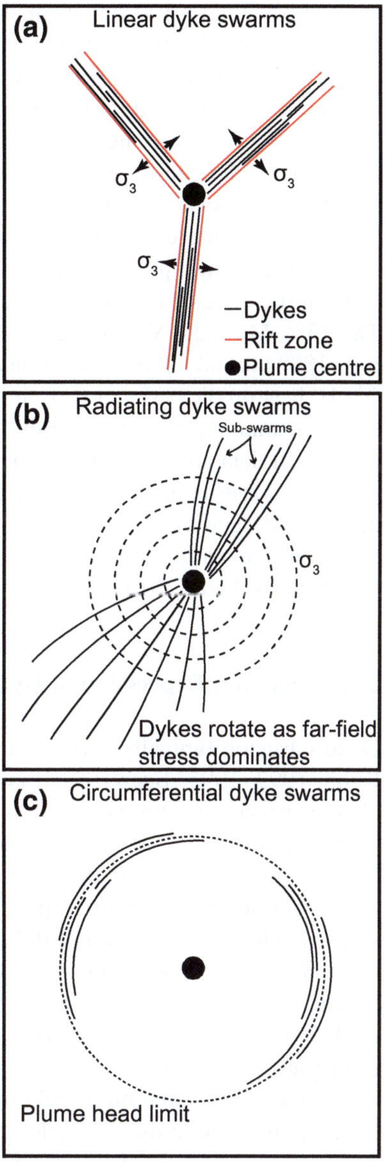

Fig. 10 Schematics showing differences in the geometry of linear, radial, and circumferential dyke swarms. See text for discussion

away from the plume centre (e.g. Fig. 10b). The 1267 Ma Mackenzie dykes exhibit a radiating pattern that spans about 100° arc and extends at least 2400 km from the inferred plume centre at the focus of the swarm (Fig. 1) (e.g. Baragar et al. 1996; Buchan and Ernst 2013; Ernst and Bleeker 2010). Average dyke widths are ~30 m and magnetic fabric measurements demonstrate vertical magma flow dominated within ~500 km of the plume centre (Ernst and Baragar 1992). All dykes analysed beyond ~500 km from the plume centre (i.e. sample sites at 600, 800, 1000, 1400 and 2100 km) display magnetic fabrics consistent with lateral injection (Ernst and Baragar 1992). Extrusive lavas of the Coppermine River flood basalts are only present in isolated remnants and their original extent remains uncertain (Ernst 2014). Associated sills are present in scattered sedimentary basins and it is predicted that some of these sill-complexes and also layered mafic intrusions were fed by the radiating dyke swarm (Ernst and Buchan 1997b).

Giant circumferential dyke swarms likely originate from and define the outer edge of the mantle plume head, which flattens against the base of the lithosphere, or from the edge of magmatic underplate (e.g. Fig. 10c) (Buchan and Ernst 2019). The 1385 Ma Lake Victoria giant dyke swarm has an arcuate geometry extending for ~180° around an arc, which has a diameter of ~600 km (Fig. 1) (Mäkitie et al. 2014). It is predicated that this circumferential pattern continues for the full 360°, but this cannot be tested at present because no aeromagnetic data and only limited geological data are available for the western region (Mäkitie et al. 2014). The Lake Victoria swarm is interpreted to mark a plume centre for the Kunene-Kibaran LIP that extends across the Congo craton (Ernst et al. 2013). The magmatic-tectonic conditions responsible for the circumferential pattern are still being investigated (Buchan and Ernst 2018a, b).

2.3.2 Surface Expression of Dykes and Dyke Swarms

Dyke intrusion typically involves tensile failure and opening of the host rock (e.g. Rubin 1992, 1995; Baer 1991; Lister and Kerr 1991), although emplacement may also occur via viscous indentation in some instances (e.g. Spacapan et al. 2017; Abdelmalak et al. 2012; Donnadieu and Merle 1998; Guldstrand et al. 2017). Because the majority of dykes do not breach the surface and instead arrest at depth, the act of generating space for dyke intrusion by host rock extension or viscous indentation can induce deformation of the overlying rock (e.g. Fig. 11) (e.g. Gudmundsson and Brenner 2004; Gudmundsson 2003; Mastin and Pollard 1988; Pollard et al. 1983; Donnadieu and Merle 1998; Abdelmalak et al. 2012; Guldstrand et al. 2017; Kavanagh et al. 2018). For example, the development of an extension discrepancy between the level of dyke intrusion and the overlying rock may be resolved by the formation of dyke-induced normal faults that dip towards the upper dyke tip and bound dyke-parallel graben (Fig. 11) (e.g. Mastin and Pollard 1988; Rubin 1992; Pollard et al. 1983; Hjartardóttir et al. 2016; Hofmann 2013; Mège et al. 2003; Wilson and Head 2002). Dyke-induced normal faults have been observed on Earth (e.g. Iceland and Hawaii) and other planetary bodies (e.g. Mars and Venus) and are considered to accommodate the same extension as the underlying dyke, i.e. their cumulative heave is expected to equal the dyke thickness (Fig. 11) (e.g. Mastin and Pollard 1988; Pollard et al. 1983; Rubin 1992; Mège et al. 2003; Ernst et al. 2001; Hjartardóttir et al. 2016; Muirhead et al. 2016). Because the geometry and growth of dyke-induced normal faults is kinematically linked to dyke propagation, it is expected that the surface expression of these faults can be used as a proxy for dyke parameters (e.g. Rowland et al. 2007; Dumont et al. 2016, 2017; Hofmann 2013; Mastin and Pollard 1988). To this end, four hypotheses have been developed to explain dyke-induced normal fault growth and explain their surface expression (Fig. 11): (i) dyke-induced normal faults nucleate at the surface and propagate downwards (Acocella et al. 2003); (ii) they propagate upwards from the dyke tip (Grant and Kattenhorn 2004); (iii) a combination of both (Rowland et al. 2007; Tentler 2005); or (iv) that they nucleate between and grow towards the dyke tip and surface (Mastin and Pollard 1988). It is, however, difficult to rigorously test these hypotheses due to a lack of sufficient 3D field exposures or geophysical images that detail dyke-induced fault structure. An important distinction though between the four hypotheses is that, whilst they all produce similar final fault geometries, the different fault growth patterns predict unique fault displacement trends (Fig. 11). Quantitatively analysing displacement across dyke-induced normal faults could thus allow us to evaluate the kinematic relationship between dykes and dyke-induced normal faults and determine how dyke-induced normal faults can be distinguished from rift-related tectonic normal faults (Fig. 11). Instead of dyke-induced normal faulting, some experimental and numerical models have also demonstrated that space for dyke intrusion may be generated by antiformal folding (i.e. uplift) of the overlying rock volume and/or reverse faulting, particularly if the dyke behaves as a viscous indenter (Abdelmalak et al. 2012; Wyrick and Smart 2009; Guldstrand et al. 2017).

Satellite imagery suggests that inferred dykes on Mars and Venus appear to occasionally be overlain by pit chain craters (e.g. Wyrick et al. 2004; Ferrill et al. 2004; Patterson et al. 2016). Pit chain craters have been observed on Earth, although they are typically smaller than those on other planetary surfaces (e.g. Ferrill et al. 2011). Overall, these elliptical to circular craters commonly occur in linearly aligned arrays, typically have long axes up to 4.5 km, and are several hundred metres deep (e.g. Wyrick et al. 2004; Ferrill et al. 2004). No eruptive material appears to be associated with pit chain craters. Whilst pit chain craters are generally considered to develop through cavity collapse instigated by dyke intrusion, we cannot validate the various mechanisms put forth to explain their formation, including (Fig. 12): (i) tensile fracturing; (ii) dilational faulting; (iii) void collapse following volatile or magma removal (Wyrick et al. 2004); or (iv) pulsed movement of the dyke front during lateral propagation (Patterson et al. 2016). It is, however, plausible that pit chain craters may also develop through processes not associated with dyke intrusion such as tectonic faulting, karst dissolution or collapse of lava tubes (Fig. 12) (Wyrick et al. 2004).

Regardless of dyke emplacement mechanics, associated surface deformation structures (e.g. faults and pit chain craters; Figs. 11 and 12) represent an important record of ongoing and historical dyke intrusion events. Characterising the surface expression of dyke-induced deformation thus provides crucial insight into the distribution and geometry of subsurface dykes and dyke swarms (e.g. Ernst et al. 2001; Mastin and Pollard 1988). However, to accurately invert ground deformation patterns and recover the properties and dynamics of underlying dykes, it is critical to understand how dyke intrusion translates to elastic and inelastic surface deformation in four-dimensions (e.g. Kavanagh et al. 2018). Whilst dyke-induced ground defor-

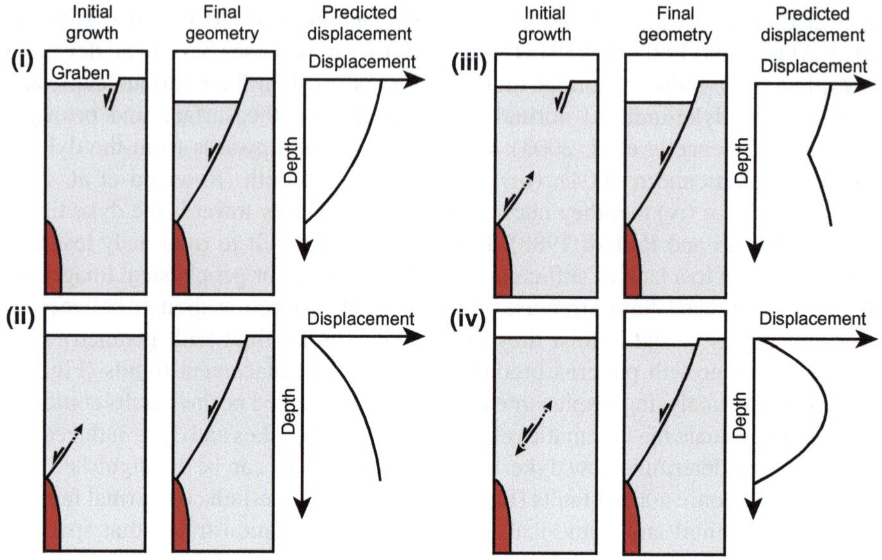

Fig. 11 Sketches depicting how different inferred models of dyke-induced fault growth produce graben observed above dykes on Earth and other planetary bodies: (i) faults nucleate at the surface and propagate downwards (Acocella et al. 2003); (ii) faults nucleate at the dyke tip and propagate upwards (Grant and Kattenhorn 2004); (iii) faults nucleate at the surface and dyke tip, propagating towards each other and becoming dip-linked (Tentler 2005; Rowland et al. 2007); and (iv) dykes nucleate between the dyke tip and surface, propagating towards each (Mastin and Pollard 1988). Also shown are the predicted displacement-depth plots for the different fault models

mation can be monitored in near real-time at active volcanoes (e.g. Hjartardóttir et al. 2016), we cannot assess the 3D structure of natural systems from surface data alone.

2.4 *Imaging and Identifying Dyke Swarms in Seismic Reflection Data*

One may infer from a global map of known dyke swarms that many should continue offshore onto the continental passive margins (Fig. 1). Furthermore, dyke swarms may be expected to occur along magma-rich margins where continental break-up was potentially driven by intense dyke intrusion, as is currently observed along magmatic segments of the East African Rift (e.g. Kendall et al. 2005; Karson and Brooks 1999; Abdelmalak et al. 2015; Muirhead et al. 2015). However, with the exception of scant evidence from aeromagnetic and limited seismic reflection surveys for the extension of dyke swarms offshore, few have been observed along continental passive margins (cf. Kirton and Donato 1985; Wall et al. 2010; Abdelmalak et al. 2015). This is primarily because the sub-vertical orientation of dykes makes them difficult to image

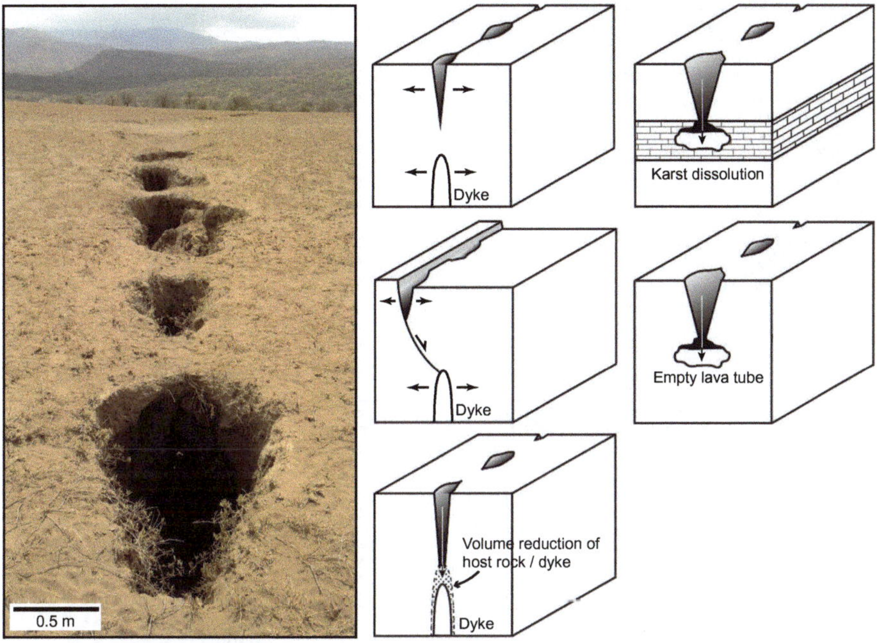

Fig. 12 Photograph of a linear array of pit chain craters in the Natron Basin, Tanzania that occur above a dilated normal fault and formed in relation to a dyke intrusion event in 2007 (see Muirhead et al. 2015 for further detail). Also shown are schematics of different models for pit chain crater formation (modified from Wyrick et al. 2004)

in seismic reflection data, as the vast majority of acoustic energy encountering a dyke contact is reflected downwards and does not return to the surface (Thomson 2007). Where dykes have been interpreted in seismic reflection data, they have been attributed to either: (i) vertical discontinuities in the data, such that the dykes are not imaged per se but rather correspond to zones from which little acoustic energy is returned (e.g. Wall et al. 2010; Bosworth et al. 2015; Ardakani et al. 2017); or (ii) local inclined reflections representing dykes that have been rotated post-emplacement (e.g. Phillips et al. 2017; Abdelmalak et al. 2015). Here, we describe how seismic reflection data can be used to constrain the offshore limits of dyke swarms.

2.4.1 Skagerrak (Jutland) LIP (~300–280 Ma): Farsund Dyke Swarm, Offshore Southern Norway

Phillips et al. (2017) recently recognised and described a dyke swarm (i.e. the Farsund Dyke Swarm) imaged in 2D and 3D seismic reflection data from offshore southern Norway (e.g. Fig. 13a). The dyke swarm is expressed in seismic data as a series of high-amplitude, tuned reflection packages, which dip northwards at ~35–50°N

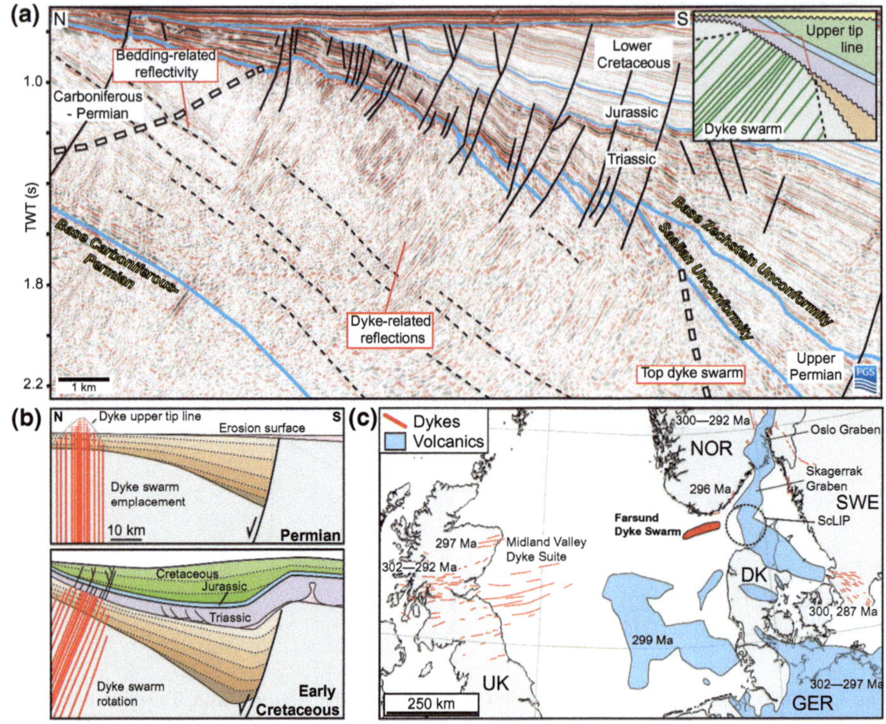

Fig. 13 **a** Seismic section from the Farsund Basin, offshore southern Norway that images part of a dyke-swarm (modified from Phillips et al. 2017). Note that the dyke swarm cross-cuts Carboniferous-Permian strata, but is truncated by the Saalian-Altmark and Base Zechstein unconformities (Phillips et al. 2017). **b** Schematic detailing how basin flexure rotated the originally sub-vertical dykes, allowing them to be imaged in seismic reflection data (modified from Phillips et al. 2017). **c** Map showing the modern day distribution of dyke swarms and volcanic rocks related to the Skagerrak (Jutland) LIP (Phillips et al. 2017)

and cross-cut but do not offset gently (~10–20°) S-dipping, Carboniferous-Permian (<320 Ma) strata (Fig. 13a) (Phillips et al. 2017). Post-emplacement flexure of the passive margin during Mesozoic extension rotated the dykes to their current inclination, presumably from an initial sub-vertical orientation, facilitating their imaging in seismic reflection data (Fig. 13b) (Phillips et al. 2017). The tuned reflection packages within the inferred dyke swarm (Fig. 13a) display complex wave-trains, suggesting they represent an amalgamation of reflections emanating from two or more intrusion-host rock contacts that cannot be distinguished seismically (Phillips et al. 2017). These tuned reflection packages, thus, do not represent individual dykes but rather likely result from interference between the margins of dykes with thicknesses and/or spacings <75 m (Phillips et al. 2017). No borehole data is available to test this hypothesis.

The dyke swarm is mapped across multiple seismic datasets within a ~100 km long, ~25 km wide zone that trends WSW-ENE (Fig. 13c) (Phillips et al. 2017). In the centre of the dyke swarm, reflections extend from a depth of ~2.5 s TWT (i.e. ~4 km), below which seismic imaging deteriorates, up to ~0.8 s TWT (i.e. ~1 km) where they are truncated by the Saalian-Altmark (~290–270 Ma) and Base Zechstein (~260 Ma) unconformities (Fig. 13a, b) (Phillips et al. 2017). Towards the margins of the dyke swarm, the spacing of the tuned reflection packages increases and their height progressively decreases as the dykes arrest at lower stratigraphic levels (Fig. 13a, b) (Phillips et al. 2017). Because the dyke swarm cross-cuts Carboniferous-Permian (<320 Ma) strata and is truncated by the 290–270 Ma Saalian-Altmark unconformity, it is likely that emplacement occurred ~320–270 Ma (Phillips et al. 2017). This inferred emplacement age is broadly coincident with the 300–280 Ma Skagerrak LIP, also termed the Skagerrak (Jutland) LIP, which is marked by convergent, linear dyke swarms that place the plume centre at the southern end of the Oslo graben (Figs. 1 and 13c) (e.g. Phillips et al. 2017; Ernst and Buchan 1997a; Torsvik et al. 2008). The Farsund Dyke Swarm occurs proximal to the proposed source of the Skagerrak LIP and its WSW-trend suggests that it links to the Midland Valley Dyke Suite in the UK (Fig. 13c) (Phillips et al. 2017; Torsvik et al. 2008).

Variation in dyke height across the swarm, with dyke-related reflections terminating at lower stratigraphic levels towards its margins (Fig. 13a, b), implies that the measured width of dyke swarms exposed at Earth's surface are dependent on the erosion level (Phillips et al. 2017). For example, the width of the Farsund Dyke Swarm is ~12 km at the Saalian-Altmark unconformity, but at deeper levels it increases to ~20 km (Fig. 13a) (Phillips et al. 2017). This change in swarm width with height has implications for calculations of intrusion volume and associated extension. For example, assuming the imaged portion of the dyke swarm has a 1:1 dyke to host rock ratio, with dyke heights of 3 km, the variable width of the swarm with depth means that estimates of its volume and the extension it accommodates range from 900–1500 km^3 and 6–10 km, respectively (Phillips et al. 2017). These observations and approximate calculations highlight that the exposed width of a dyke swarm cannot necessarily be used to accurately determine melt volumes and extension across ancient dyke swarms.

2.4.2 North Atlantic Igneous Province (~61–50 Ma): Mull Dyke Swarm, UK

The NW-striking Mull Dyke Swarm extends south-westwards from the Mull Central Complex in NW Scotland, which represents the deeply eroded roots of an ancient volcano, for >300 km onshore (Figs. 1 and 14a) (e.g. Speight et al. 1982; Macdonald et al. 2010; Ishizuka et al. 2017). Geochemical and rock magnetic data indicate that dykes intruded laterally away from the central complex, with K-Ar dating highlighting emplacement occurred ~58 Ma; i.e. as part of the North Atlantic Igneous Province (e.g. Macdonald et al. 2009, 2015). Continuation of the Mull Dyke Swarm offshore, into the southern North Sea, has been postulated based on interpretation of NW-striking, linear features identified in both 3D seismic reflection and aeromagnetic data as dykes (Fig. 14a) (Kirton and Donato 1985; Wall et al. 2010). In the seismic reflection data, these inferred dykes correspond to vertical zones of discontinuous host rock reflections, which are 0.5–2 km wide and up to ~40 km long (Fig. 14b) (Wall et al. 2010). Where the seismic reflection data overlaps with high-resolution aeromagnetic data, the inferred dykes correspond to linear, negative magnetic anomalies (Kirton and Donato 1985; Wall et al. 2010). Seismic reflection data further reveal that overlying and extending along the length of the dykes are a series of circular, ellipsoidal, and linear craters (e.g. Fig. 14b) (Wall et al. 2010). At their top, which occurs at the Top Cretaceous horizon, these craters are up to 2 km wide and infilled by overlying Paleogene strata, implying crater formation occurred to the Early Paleogene (Fig. 14b) (Wall et al. 2010). The craters typically have inward-dipping walls, which extend down to the top of the inferred dykes (Fig. 14b) (Wall et al. 2010). These craters are interpreted to be pit chain craters formed in response to explosive volatile release as the propagating dyke approached the contemporaneous seabed (Wall et al. 2010). The age of the pit chain craters, which formed during dyke emplacement, and the trend of the imaged dykes suggests that the Mull Dyke Swarm extends for >600 km (Kirton and Donato 1985; Wall et al. 2010). Where the dykes and pit chain craters are observed in a single 2D seismic section, their appearance is reminiscent of some hydrothermal vents developed above the lateral tips of sills, which can correspond to a roughly cylindrical, vertical zone of dim or chaotic reflections overlain by a crater (Fig. 7b) (e.g. Jamtveit et al. 2004; Hansen 2006); to differentiate between hydrothermal vents and pit chain craters above a dyke, it is necessary to evaluate the along-strike extent of the structure using 3D seismic reflection data.

2.4.3 Dyke and Dyke-Induced Faults Imaged in Egypt

Subtle vertical disruptions to host rock reflections within 3D seismic data from the Abu Gharadig Basin, onshore Egypt, have been inferred to correspond to dykes (Fig. 15) (Bosworth et al. 2015). Similar to the pit chain craters observed above the dykes in the North Sea, these Egyptian dykes are also overlain by crater-like features up to several hundred meters in diameter (Fig. 15) (Bosworth et al. 2015). However, the dykes in the Abu Gharadig Basin are also associated with overlying, dyke-parallel normal faults that dip towards and extend upwards from the dyke tip for ~0.3 s TWT (Fig. 15) (Bosworth et al. 2015). These normal faults bound graben and extend for >100 km and have offsets of only a few tens of meters maximum (Bosworth et al. 2015). The association with underlying dykes suggests that these graben-bounding faults nucleated and grew in response to dyke emplacement, i.e. they are dyke-induced normal faults (Bosworth et al. 2015). Such dyke-induced normal faults have not been recognised in other seismic datasets, but these data from Egypt indicate that (Bosworth et al. 2015): (i) dyke-induced normal faults can be more easily recognised in seismic reflection data than dykes (Fig. 15); and (ii) seismic reflection data can allow displacement across dyke-induced normal faults to be quantitatively assessed in 3D. This latter observation is important because mapping

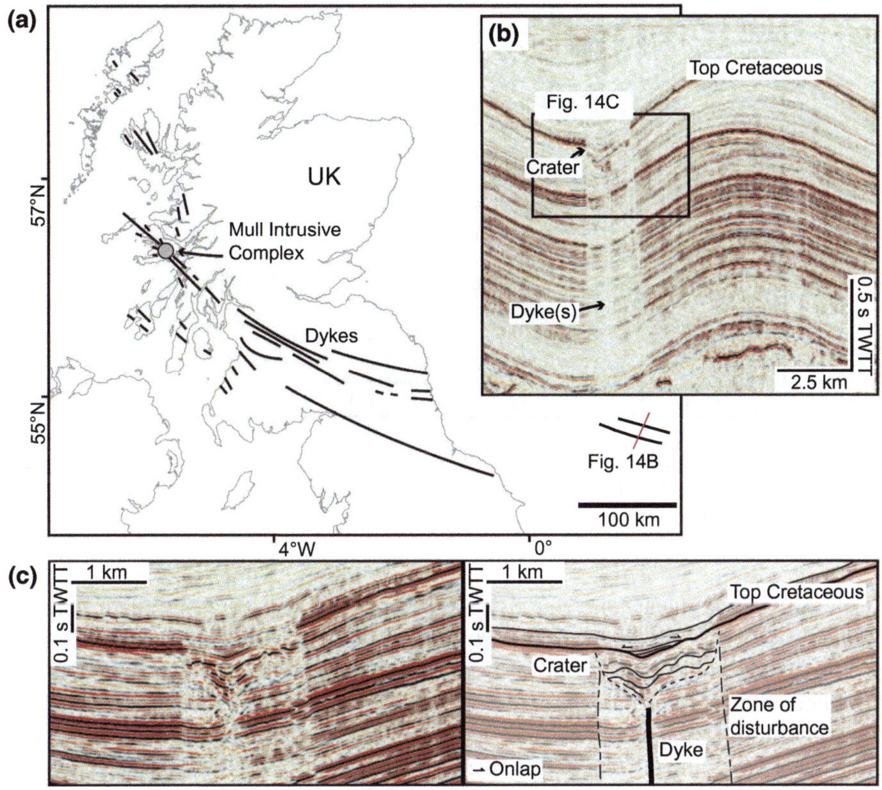

Fig. 14 **a** Map of the Mull dyke swarm, which emanates from the Mull Central Complex (based on Wall et al. 2010; Macdonald et al. 2015; Ishizuka et al. 2017). **b**, **c** Seismic sections from Wall et al. (2010) showing the dykes, which correspond to vertical zones of low-amplitude, disturbed reflections, and the pit chain craters. See Fig. 14a for location

displacement patterns across dyke-induced normal faults could allow their kinematic history to be resolved, providing a means of testing existing hypotheses concerning dyke-induced normal fault growth; i.e. where dyke-induced normal faults nucleate, they are likely to be active for longer and therefore accrue more displacement, producing diagnostic displacement-depth trends.

2.5 Field Observations of Intrusive Complexes

Intrusive complexes, also referred to as central complexes, represent the deeply eroded roots of major volcanic centres within LIPs (e.g. Emeleus and Bell 2005; Jerram and Bryan 2015; Ewart et al. 2002). These complexes are likely vertically extensive and comprise a dense network of closely spaced magma conduits (e.g.

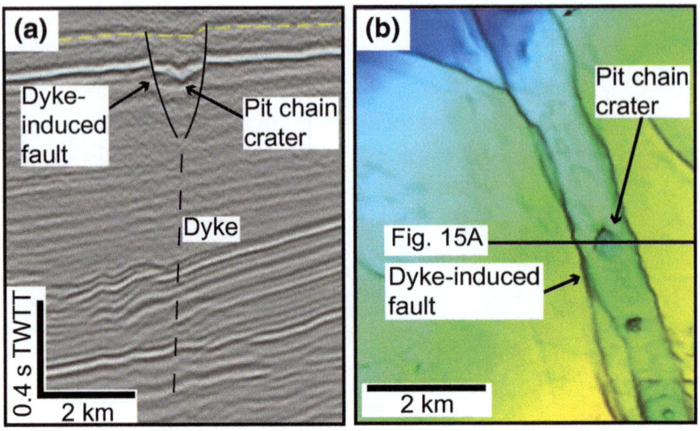

Fig. 15 Seismic section and map showing a dyke, dyke-induced faults (and graben), and pit chain craters imaged in data from Egypt (Bosworth et al. 2015). Figure 15b shows the location of the seismic section in (**a**)

dykes, sills, cone sheets, and ring dykes) and major plutons (e.g. Emeleus and Bell 2005; Jerram and Bryan 2015; Ewart et al. 2002). Such intrusive complexes are perhaps most well-known from onshore the British and Irish Palaeogene Igneous Province, which forms part of the North Atlantic Igneous Province, where they are typically circular in plan-view and up to ~15 km in diameter (Emeleus and Bell 2005). The structure of these intrusive complexes and the geometry of individual components is partly controlled by their composition, with many comprising a combination of ultramafic-mafic layered intrusions and granitoid plutons (e.g. laccoliths) (e.g. Emeleus and Bell 2005; Jerram and Bryan 2015; Bryan et al. 2010; Ewart et al. 2002). Whilst intrusive complexes provide a site for assimilation-fractional crystallisation processes, they can also feed dyke swarms and sill-complexes (e.g. Emeleus and Bell 2005; Macdonald et al. 2015; Preston et al. 1998; Jolly and Sanderson 1995; Bryan et al. 2010). For example, the Mull Intrusive Complex on the Isle of Mull, Scotland fed a laterally propagating dyke swarm (i.e. the Mull dyke swarm), which extends >600 km to the SW (Fig. 14), and the Loch Scridain sill-complex (e.g. Emeleus and Bell 2005; Ishizuka et al. 2017; Macdonald et al. 2015; Preston et al. 1998; Jolly and Sanderson 1995; Speight et al. 1982).

3 Discussion

Dyke swarms and sill-complexes facilitate the transport of magma through Earth's crust to the surface during the formation of LIPs, controlling the distribution of eruption sites and associated economic ore deposits. Whilst field observations coupled with petrological, geochemical, geochronological, structural, and geophysical analyses have advanced our understanding of how magma is emplaced and evolves within these systems, the 3D structure of dyke swarms and sill-complexes has largely remained enigmatic. We have demonstrated that, over the last two decades, seismic reflection data has provided unprecedented insights into the 3D structure of LIP plumbing systems. For example, reflection seismology has revealed: (i) sill-complexes commonly transport magma over significant lateral (>10s km) and vertical (up to 12 km) distances (e.g. Cartwright and Hansen 2006; Schofield et al. 2017; Magee et al. 2016); (ii) sill-complexes can feed extensive networks of hydrothermal vents, which may contribute to and/or drive climate change (e.g. Svensen et al. 2004, 2007); (iii) dykes and dyke swarms can be identified and mapped across continental margins (e.g. Phillips et al. 2017; Abdelmalak et al. 2015; Wall et al. 2010); and (iv) the width of dyke swarms may decrease with height, implying erosion level may control measured widths, and thus volume estimates, of dyke swarms observed at Earth's surface (Phillips et al. 2017). Despite the advances reflection seismology has offered, here we select and discuss four outstanding questions and/or problems.

3.1 Plumbing System Volumes

Many sheet intrusions imaged in seismic reflection data correspond to tuned reflection packages; i.e. reflections from the top and base contact of the intrusion interfere on their way back to the surface and cannot be distinguished (e.g. Smallwood and Maresh 2002; Thomson 2005; Peron-Pinvidic et al. 2010; Magee et al. 2015; Eide et al. 2017; Rabbel et al. 2018; Phillips et al. 2017). This tuning effect is dependent on the thickness of the body relative to the limit of separability (i.e. vertical resolution) of the seismic reflection data, which itself is a function of data frequency and the seismic velocity of the intrusion (Brown 2004). Intrusions with thicknesses below the limit of separability but above the limit of visibility are imaged as tuned reflection packages. The thickness of intrusions expressed as tuned reflection packages is thus below the limit of separability but above the limit of visibility and cannot be constrained further (Brown 2004). Without being able to accurately assess intrusion thickness, magma volume estimates of LIP plumbing systems imaged in seismic reflection data are compromised. Whilst higher resolution seismic reflection data will help to further constrain intrusive volumes within LIPs, the following would also be useful: (i) a better understanding of how intrusions are expressed in seismic reflection data, which can be achieved through synthetic seismic forward modelling (e.g. Magee et al. 2015; Peron-Pinvidic et al. 2010; Eide et al. 2017; Rabbel et al. 2018); (ii) constraints on the seismic velocity of sills, perhaps through analysis of borehole data or geophysical techniques such as full-waveform inversion (e.g. Morgan et al. 2013); and (iii) a comprehensive analyses of intrusion scaling relationships within exposed and seismically imaged sill-complexes (cf. Cruden et al. 2017 and references therein).

3.2 Lateral Magma Flow in Sill-Complexes

Seismic reflection data, coupled with field observations, clearly show connections between individual intrusions within sill-complexes can define tortuous magma flow pathways (Magee et al. 2016). Although it is difficult to accurately constrain the duration of sill-complex formation, seismic-based studies suggest that emplacement may occur incrementally over millions of years (e.g. Reeves et al. 2018; Magee et al. 2014). However, most intrusions within sill-complexes are emplaced at shallow-levels (>3 km depths) within relatively cold sedimentary strata and are expressed as tuned reflection packages, implying their thickness is below the limit of separability for any given seismic volume (i.e. typically <100 m) (see Magee et al. 2016 and references therein). It may therefore be expected that individual intrusions will solidify rapidly, raising the question as to how magma transits through long-lived sill-complexes. Several studies have demonstrated that sheet accretion or magma channelization within discrete intrusions can facilitate sill-complex emplacement (e.g. Annen et al. 2015; Holness and Humphreys 2003; Magee et al. 2016), but whether these processes can maintain elevated host rock temperatures and thereby inhibit solidification remains unknown. To understand how magma moves through these intrusion networks it is critical integrate rock magnetic, petrological, geochemical, and high-resolution geochronological data to track flow pathways, residence times, and magma evolution across exposed sill-complexes.

3.3 Sill-Complex and Dyke Swarm Connectivity

Understanding the 4D connectivity and evolution of intrusive components in LIP plumbing systems is integral to evaluating controls on the accumulation and location of volumetrically large magma bodies, the eruption of which (e.g. to form flood basalts) can trigger catastrophic climatic and environmental changes. Magma transport within LIPs is dominated by sill-complexes and/or dyke swarms, yet it remains unclear how these two intrusion networks relate to each other. For example, it is traditionally assumed that dykes bring magma from mantle source areas vertically into the lower crust, and upward from the magmatic underplate into the upper brittle crust (see Ernst 2014 and references therein). Given that sill-complexes are predominantly observed within sedimentary basins and appear to have been emplaced at relatively shallow-levels (>5 km depth), it may be expected that they are fed by dyke swarms. This potential relationship LIP plumbing system components is supported by the apparent feeding of the 2.21 Ga Nipissing sill-complex by the contemporaneous Senneterre dyke swarm (e.g. Ernst and Buchan 1997a; Buchan et al. 1993). However, the mechanics and implications of the transition from dyke swarm to sill-complex have not been assessed. Furthermore, field- and seismic-based observations demonstrate sill-complexes can transport magma up to 12 km vertically, from lower crustal areas, and develop within crystalline rocks (Cartwright and Hansen 2006;

Ivanic et al. 2013a; McBride et al. 2018). Whilst these observations of basement-hosted sill-complexes questions the volumetric importance of dykes at depth, it also raises the possibility that sill-complexes could feed dyke swarms. In particular, the repeated and rapid input of magma into a sill-complex could promote the development of a large intrusive centre (e.g. Cawthorn 2012; Ernst and Buchan 1997a; Annen et al. 2015; Annen 2011), potentially capable of feeding dyke swarms (e.g. the Mull Intrusive Complex; Emeleus and Bell 2005; Ishizuka et al. 2017; Macdonald et al. 2015; Preston et al. 1998; Jolly and Sanderson 1995; Speight et al. 1982).

Our poor of understanding of sill-complex and dyke swarm connectivity is partly due to: (i) a lack of exposed dyke and sill feeding relationships within LIPs; and (ii) difficulties in imaging dykes in seismic reflection. Resolving possible feeder dyke relationships within sill-complexes using seismic reflection data is particularly challenging because their imaging results in a significant reduction in data quality directly beneath the complex as acoustic energy is reflected back or scattered by the sills (e.g. Planke et al. 2005; Eide et al. 2017). From a reflection seismology perspective, one way to potentially circumvent this problem is to map magma flow indicators across sill-complexes to determine the location and geometry of magma input zones (e.g. Schofield et al. 2017; Magee et al. 2016; Schofield et al. 2012a; Thomson and Hutton 2004), which may provide insight into whether sills are fed by dykes (e.g. Goulty and Schofield 2008). Mapping flow patterns and identifying feeder locations within seismically imaged flood basalts spatially associated with sill-complexes may also provide a method for distinguishing the role of dykes and dyke swarms within LIPs.

4 Conclusions

Large Igneous Provinces (LIPs) have become an important focus for research in recent years due to their use in paleocontinental reconstructions, in exploration targeting, as planetary analogues, and as a result of their links to dramatic climate change. A key frontier within the study of LIPs concerns deciphering the 3D geometry of their plumbing systems and, in particular, identifying how dyke swarms and sill-complexes grow and interact. Seismic reflection data uniquely reveal the 3D geometry of LIP magma plumbing systems, highlighting that dyke swarms and sill-complexes dictate melt distribution. Here, we review several sill-complexes and dyke swarms either exposed at Earth's surface or analysed using seismic reflection data. In particular, we highlight that: (i) mapping sill-complexes in 3D and recognising magma flow indicators can be used to unravel their source regions, sill connectivity, and overall growth history; (ii) sill-complexes can facilitate significant vertical and lateral magma transport both within sedimentary basins and crystalline crust, potentially instigating widespread hydrothermal venting; and (iii) dyke swarm width varies with dyke height, implying that the observed width of dyke swarms exposed at Earth's surface is a function of erosion level and may not represent the true swarm size. Furthermore, we also show that seismic reflection data allows the visualisation and quantification of host rock deformation, including deformation of the contemporaneous surface, induced by dyke swarm and sill-complex emplacement. By understanding how dyke swarms and sill-complexes are expressed at the surface we can estimate properties of LIP magma plumbing systems that are unexposed on Earth (e.g. East African Rift System) or other planetary bodies (e.g. Mars). Overall, building our understanding of LIP magma plumbing system geometries, emplacement mechanics, and relative timings through analysis of seismic reflection data is critical to evaluating associated volcanic hazards, either in active systems or where magmatism (both extrusive and intrusive) may have driven mass extinctions, and accumulation of economic resources.

Acknowledgements CM is funded by a Junior Research Fellowship at Imperial College London. REE was partially supported from Mega-Grant 14.Y26.31.0012 of the government of the Russian Federation. JDM is supported by National Science Foundation grant EAR-1654518. We would like to thank Rajesh Srivastava for editorial handling and an anonymous reviewer and Dougal Jerram for their comments.

References

Abdelmalak MM, Mourgues R, Galland O, Bureau D (2012) Fracture mode analysis and related surface deformation during dyke intrusion: results from 2D experimental modelling. Earth Planet Sci Lett 359:93–105

Abdelmalak MM, Andersen TB, Planke S, Faleide JI, Corfu F, Tegner C, Shephard GE, Zastrozhnov D, Myklebust R (2015) The ocean-continent transition in the mid-Norwegian margin: insight from seismic data and an onshore Caledonian field analogue. Geology 43(11):1011–1014

Acocella V, Korme T, Salvini F (2003) Formation of normal faults along the axial zone of the Ethiopian Rift. J Struct Geol 25(4):503–513

Agirrezabala LM (2015) Syndepositional forced folding and related fluid plumbing above a magmatic laccolith: insights from outcrop (Lower Cretaceous, Basque-Cantabrian Basin, western Pyrenees). Geol Soc Am Bull B31192. 31191

Airoldi G, Muirhead JD, White JD, Rowland J (2011) Emplacement of magma at shallow depth: insights from field relationships at Allan Hills, south Victoria Land, East Antarctica. Antarct Sci Inst Subscription 23(3):281

Airoldi G, Muirhead JD, Zanella E, White JD (2012) Emplacement process of Ferrar Dolerite sheets at Allan Hills (South Victoria Land, Antarctica) inferred from magnetic fabric. Geophys J Int 188(3):1046–1060

Airoldi GM, Muirhead JD, Long SM, Zanella E, White JD (2016) Flow dynamics in mid-Jurassic dikes and sills of the Ferrar large igneous province and implications for long-distance magma transport. Tectonophysics 683:182–199

Annen C (2011) Implications of incremental emplacement of magma bodies for magma differentiation, thermal aureole dimensions and plutonism–volcanism relationships. Tectonophysics 500(1):3–10

Annen C, Blundy JD, Leuthold J, Sparks RSJ (2015) Construction and evolution of igneous bodies: towards an integrated perspective of crustal magmatism. Lithos 230:206–221

Archer SG, Bergman SC, Iliffe J, Murphy CM, Thornton M (2005) Palaeogene igneous rocks reveal new insights into the geodynamic evolution and petroleum potential of the Rockall Trough, NE Atlantic Margin. Basin Res 17(1):171–201

Ardakani EP, Schmitt DR, Currie CA (2017) Geophysical evidence for an igneous dike swarm, Buffalo Creek, Northeast Alberta. Geol Soc Am Bull

Aspler LB, Cousens BL, Chiarenzelli JR (2002) Griffin gabbro sills (2.11 Ga), Hurwitz Basin, Nunavut, Canada: long-distance lateral transport of magmas in western Churchill Province crust. Precambr Res 117(3–4):269–294

Baer G (1991) Mechanisms of dike propagation in layered rocks and in massive, porous sedimentary rocks. J Geophys Res Solid Earth 96(B7):11911–11929

Baragar W, Ernst R, Hulbert L, Peterson T (1996) Longitudinal petrochemical variation in the Mackenzie dyke swarm, northwestern Canadian Shield. J Petrol 37(2):317–359

Bleeker W, Ernst R (2006) Short-lived mantle generated magmatic events and their dyke swarms: the key unlocking earth's paleogeographic record back to 2.6 Ga. Dyke swarms—time markers of crustal evolution, pp 3–26

Bosworth W, Stockli DF, Helgeson DE (2015) Integrated outcrop, 3D seismic, and geochronologic interpretation of Red Sea dike-related deformation in the Western Desert, Egypt-The role of the 23 Ma Cairo "mini-plume". J Afr Earth Sci 109:107–119

Brown AR (2004) Interpretation of three-dimensional seismic data, vol 42. AAPG Memoir 42, SEG investigations in geophysics no. 9, 6th edn. AAPG and SEG, Oklahoma, USA

Bryan SE, Ernst RE (2008) Revised definition of large igneous provinces (LIPs). Earth Sci Rev 86(1–4):175–202

Bryan SE, Ferrari L (2013) Large igneous provinces and silicic large igneous provinces: progress in our understanding over the last 25 years. GSA Bull 125(7–8):1053–1078

Bryan SE, Peate IU, Peate DW, Self S, Jerram DA, Mawby MR, Marsh JG, Miller JA (2010) The largest volcanic eruptions on Earth. Earth Sci Rev 102(3–4):207–229

Buchan KL, Ernst R (2004) Diabase dyke swarms and related unites in Canada and adjacent regions. Geol Surv Can

Buchan K, Ernst R (2013) Diabase dyke swarms of Nunavut, Northwest Territories and Yukon, Canada. Geol Surv Can, Open File 7464(10.4095):293149

Buchan KL, Ernst RE (2018a) A giant circumferential dyke swarm associated with the High Arctic Large Igneous Province (HALIP). Gondwana Res 58:39–57

Buchan KL, Ernst RE (2018b) Giant circumferential dyke swarms: catalogue and characteristics. In Srivastava RK, Ernst RE, Peng P (eds) Dyke swarms of the world – a modern perspective. Springer (in press)

Buchan K, Mortensen J, Card K (1993) Northeast-trending early proterozoic dykes of southern superior province: multiple episodes of emplacement recognized from integrated paleomagnetism and U-Pb geochronology. Can J Earth Sci 30(6):1286–1296

Bunger AP, Menand T, Cruden A, Zhang X, Halls H (2013) Analytical predictions for a natural spacing within dyke swarms. Earth Planet Sci Lett 375:270–279

Burgess S, Bowring S, Fleming T, Elliot D (2015) High-precision geochronology links the Ferrar large igneous province with early-Jurassic ocean anoxia and biotic crisis. Earth Planet Sci Lett 415:90–99

Burgess S, Muirhead J, Bowring S (2017) Initial pulse of Siberian Traps sills as the trigger of the end-Permian mass extinction. Nature Commun 8(1):164

Burke K, Dewey J (1973) Plume-generated triple junctions: key indicators in applying plate tectonics to old rocks. J Geol 81(4):406–433

Cartwright J, Hansen DM (2006) Magma transport through the crust via interconnected sill complexes. Geology 34(11):929–932

Cawthorn R (2012) Multiple sills or a layered intrusion? Time to decide. S Afr J Geol 115(3):283–290

Chevallier L, Woodford A (1999) Morpho-tectonics and mechanism of emplacement of the dolerite rings and sills of the western Karoo, South Africa. S Afr J Geol 102(1):43–54

Coffin MF, Eldholm O (1992) Volcanism and continental break-up: a global compilation of large igneous provinces. In: Storey BC, Alabaster T, Pankhurst RJ (eds) Magmatism and the causes of continental break-up, vol 68. vol 1. Geological Society of Special Publications, pp 17–30

Coffin MF, Eldholm O (1994) Large igneous provinces: crustal structure, dimensions, and external consequences. Rev Geophys 32(1):1–36

Coffin MF, Eldholm O (2005) Large igneous provinces. Encycl Geol 315–323

Cooper M, Anderson H, Walsh J, Van Dam C, Young M, Earls G, Walker A (2012) Palaeogene alpine tectonics and Icelandic plume-related magmatism and deformation in Northern Ireland. J Geol Soc 169(1):29–36

Cruden A, McCaffrey K, Bunger A (2017) Geometric scaling of tabular igneous intrusions: implications for emplacement and growth

Donnadieu F, Merle O (1998) Experiments on the indentation process during cryptodome intrusions: new insights into Mount St. Helens deformation. Geology 26(1):79–82

Dragoni M, Lanza R, Tallarico A (1997) Magnetic anisotropy produced by magma flow: theoretical model and experimental data from Ferrar dolerite sills (Antarctica). Geophys J Int 128(1):230–240

du Toit AL (1920) The Karroo dolerites of South Africa: a study in hypabyssal injection. S Afr J Geol 23(01):1–42

Dumont S, Socquet A, Grandin R, Doubre C, Klinger Y (2016) Surface displacements on faults triggered by slow magma transfers between dyke injections in the 2005–2010 rifting episode at Dabbahu–Manda–Hararo rift (Afar, Ethiopia). Geophys J Int 204(1):399–417

Dumont S, Klinger Y, Socquet A, Doubre C, Jacques E (2017) Magma influence on propagation of normal faults: Evidence from cumulative slip profiles along Dabbahu-Manda-Hararo rift segment (Afar, Ethiopia). J Struct Geol 95:48–59

Ebinger CJ, Sleep N (1998) Cenozoic magmatism throughout east Africa resulting from impact of a single plume. Nature 395(6704):788

Eide CH, Schofield N, Lecomte I, Buckley SJ, Howell JA (2017) Seismic interpretation of sill-complexes in sedimentary basins: The 'sub-sill imaging problem'. J Geol Soc 175(2):193–209

Elliot DH (1992) Jurassic magmatism and tectonism associated with Gondwanaland break-up: an Antarctic perspective. Geol Soc Lond Special Publications 68(1):165–184

Elliot DH, Fleming TH (2000) Weddell triple junction: the principal focus of Ferrar and Karoo magmatism during initial breakup of Gondwana. Geology 28(6):539–542

Elliot DH, Fleming TH (2004) Occurrence and dispersal of magmas in the Jurassic Ferrar large igneous province, Antarctica. Gondwana Res 7(1):223–237

Elliot DH, Fleming TH (2017) The Ferrar Large Igneous Province: field and geochemical constraints on supra-crustal (high-level) emplacement of the magmatic system. Geol Soc Lon Special Publications 463:SP463

Elliot DH, Fleming TH, Kyle PR, Foland KA (1999) Long-distance transport of magmas in the Jurassic Ferrar large igneous province, Antarctica. Earth Planet Sci Lett 167(1):89–104

Emeleus CH, Bell B (2005) The Palaeogene volcanic districts of Scotland, vol 3. British Geological Survey

Encarnación J, Fleming TH, Elliot DH, Eales HV (1996) Synchronous emplacement of Ferrar and Karoo dolerites and the early breakup of Gondwana. Geology 24(6):535–538

Ernst RE (2014) Large igneous provinces. Cambridge University Press

Ernst RE, Baragar WRA (1992) Evidence from magnetic fabric for the flow pattern of magma in the Mackenzie giant radiating dyke swarm. Nature 356(6369):511–513

Ernst RE, Bleeker W (2010) Large igneous provinces (LIPs), giant dyke swarms, and mantle plumes: significance for breakup events within Canada and adjacent regions from 2.5 Ga to the Present This article is one of a selection of papers published in this Special Issue on the theme Lithoprobe—parameters, processes, and the evolution of a continent. Lithoprobe Contribution 1482. Geological Survey of Canada Contribution 20100072. Can J Earth Sci 47(5):695–739

Ernst RE, Buchan KL (1997a) Giant radiating dyke swarms: their use in identifying pre-Mesozoic large igneous provinces and mantle plumes. Large igneous provinces: continental, oceanic, and planetary flood volcanism, pp 297–333

Ernst RE, Buchan K (1997b) Layered mafic intrusions: a model for their feeder systems and relationship with giant dyke swarms and mantle plume centres. S Afr J Geol 100(4):319–334

Ernst RE, Jowitt SM (2013) Large igneous provinces (LIPs) and metallogeny. In: Colpron M, Bissig T, Rusk BG, Thompson JFH (eds) Tectonics, metallogeny, and discovery: the North American Cordillera and similar accretionary settings, vol 17. Society of Economic Geologists Special Publication, pp 17–51

Ernst RE, Youbi N (2017) How large igneous provinces affect global climate, sometimes cause mass extinctions, and represent natural markers in the geological record. Palaeogeogr Palaeoclimatol Palaeoecol 478:30–52

Ernst RE, Head J, Parfitt E, Grosfils E, Wilson L (1995) Giant radiating dyke swarms on Earth and Venus. Earth Sci Rev 39(1):1–58

Ernst RE, Grosfils E, Mege D (2001) Giant dike swarms: earth, venus, and mars. Annu Rev Earth Planet Sci 29(1):489–534

Ernst RE, Buchan KL, Campbell IH (2005) Frontiers in large igneous province research. Lithos 79(3):271–297

Ernst RE, Bleeker W, Söderlund U, Kerr AC (2013) Large igneous provinces and supercontinents: toward completing the plate tectonic revolution. Lithos 174:1–14

Ewart A, Milner SC, Duncan AR, Bailey M (2002) The Cretaceous Messum igneous complex, S.W. Etendeka, Namibia: reinterpretation in terms of a downsag-cauldron subsidence model. J Volcanol Geoth Res 114(3–4):251–273

Fahrig WF (1987) The tectonic setting of continental mafic dyke swarms: failed arm and early passive margin. In: Halls HC, Fahrig WF (eds) Mafic dyke swarms. Geological Association of Canada Special Paper 34, pp 331–348

Ferrill DA, Wyrick DY, Morris AP, Sims DW, Franklin NM (2004) Dilational fault slip and pit chain formation on Mars. GSA Today 14(10):4

Ferrill DA, Wyrick DY, Smart KJ (2011) Coseismic, dilational-fault and extension-fracture related pit chain formation in Iceland: analog for pit chains on Mars. Lithosphere 3(2):133–142

Galerne CY, Neumann E-R, Planke S (2008) Emplacement mechanisms of sill complexes: information from the geochemical architecture of the Golden Valley Sill Complex, South Africa. J Volcanol Geoth Res 177(2):425–440

Galerne CY, Galland O, Neumann E-R, Planke S (2011) 3D relationships between sills and their feeders: evidence from the Golden Valley Sill Complex (Karoo Basin) and experimental modelling. J Volcanol Geoth Res 202(3–4):189–199

Galland O (2012) Experimental modelling of ground deformation associated with shallow magma intrusions. Earth Planet Sci Lett 317:145–156

Glazner AF, Bartley JM, Carl BS (1999) Oblique opening and noncoaxial emplacement of the Jurassic Independence dike swarm, California. J Struct Geol 21(10):1275–1283

Goulty NR, Schofield N (2008) Implications of simple flexure theory for the formation of saucer-shaped sills. J Struct Geol 30(7):812–817

Grant JV, Kattenhorn SA (2004) Evolution of vertical faults at an extensional plate boundary, southwest Iceland. J Struct Geol 26(3):537–557

Grapes R, Reid D, McPherson J (1974) Shallow dolerite intrusion and phreatic eruption in the Allan Hills region, Antarctica. NZ J Geol Geophys 17(3):563–577

Gudmundsson A (2003) Surface stresses associated with arrested dykes in rift zones. Bull Volcanol 65(8):606–619

Gudmundsson A, Brenner L (2004) Local stresses, dyke arrest and surface deformation in volcanic edifices and rift zones. Ann Geophys 47(4)

Guldstrand F, Burchardt S, Hallot E, Galland O (2017) Dynamics of surface deformation induced by dikes and cone sheets in a cohesive Coulomb brittle crust. J Geophys Res Solid Earth 122(10):8511–8524

Gunn B, Warren G (1962) Geology of Victoria Land between the Mawson and Mulock Glaciers: New Zealand geological survey. Bulletin 71:133–135

Halls H (1982) The importance and potential of mafic dyke swarms in studies of geodynamic processes. Geosci Can 9(3):145–154

Hansen DM (2006) The morphology of intrusion-related vent structures and their implications for constraining the timing of intrusive events along the NE Atlantic margin. J Geol Soc 163:789–800
Hansen DM, Cartwright J (2006a) Saucer-shaped sill with lobate morphology revealed by 3D seismic data: implications for resolving a shallow-level sill emplacement mechanism. J Geol Soc 163(3):509–523
Hansen DM, Cartwright J (2006b) The three-dimensional geometry and growth of forced folds above saucer-shaped igneous sills. J Struct Geol 28(8):1520–1535
Hansen DM, Cartwright JA, Thomas D (2004) 3D seismic analysis of the geometry of igneous sills and sill junction relationships. In: Davies RJ, Cartwright J, Stewart SA, Lappin M, Underhill JR (eds) 3D seismic technology: application to the exploration of sedimentary basins, vol 29, vol 1. Geological Society, London, Special Publications, pp 199–208
Hastie WW, Watkeys MK, Aubourg C (2014) Magma flow in dyke swarms of the Karoo LIP: implications for the mantle plume hypothesis. Gondwana Res 25(2):736–755
Head JW, Coffin MF (1997) Large igneous provinces: a planetary perspective. Large igneous provinces: continental, oceanic, and planetary flood Volcanism, pp 411–438
Hjartardóttir ÁR, Einarsson P, Gudmundsson MT, Högnadóttir T (2016) Fracture movements and graben subsidence during the 2014 Bárðarbunga dike intrusion in Iceland. J Volcanol Geoth Res 310:242–252
Hofmann B (2013) How do faults grow in magmatic rifts? LiDAR and InSAR observations of the Dabbahu rift segment, Afar. University of Leeds, Ethiopia
Hofmann C, Courtillot V, Feraud G, Rochette P, Yirgu G, Ketefo E, Pik R (1997) Timing of the Ethiopian flood basalt event and implications for plume birth and global change. Nature 389(6653):838
Holness M, Humphreys M (2003) The Traigh Bhàn na Sgùrra sill, Isle of Mull: flow localization in a major magma conduit. J Petrol 44(11):1961–1976
Hutton DHW (2009) Insights into magmatism in volcanic margins: bridge structures and a new mechanism of basic sill emplacement—Theron Mountains, Antarctica. Petrol Geosci 15(3):269–278
Ishizuka O, Taylor RN, Geshi N, Mochizuki N (2017) Large-volume lateral magma transport from the Mull volcano: an insight to magma chamber processes. Geochem Geophys Geosyst 18(4):1618–1640
Ivanic T, Korsch R, Wyche S, Jones L, Zibra I, Blewett R, Jones T, Milligan P, Costelloe R, Van Kranendonk M (2013a) Preliminary interpretation of the 2010 Youanmi deep seismic reflection lines and magnetotelluric data for the Windimurra Igneous Complex. In: Youanmi and Southern Carnarvon seismic and magnetotelluric (MT) workshop, Geological Survey of Western Australia, Record
Ivanic T, Zibra I, Doublier M, Wyche S (2013b) Geological interpretation of the Youanmi and Southern Carnarvon seismic lines 10GA-YU1, 10GA-YU2, 10GA-YU3, and 11GA-SC1. Youanmi and Southern Carnarvon seismic and magnetotelluric (MT) workshop 2013. Geological Survey of Western Australia
Jackson MD, Pollard DD (1988) The laccolith-stock controversy: new results from the southern Henry Mountains, Utah. Geol Soc Am Bull 100(1):117–139
Jackson CA-L, Schofield N, Golenkov B (2013) Geometry and controls on the development of igneous sill–related forced folds: a 2-D seismic reflection case study from offshore southern Australia. Geol Soc Am Bull 125(11–12):1874–1890
Jamtveit B, Svensen H, Podladchikov YY, Planke S (2004) Hydrothermal vent complexes associated with sill intrusions in sedimentary basins. Phys geol High-Level Mag Syst 234:233–241
Jerram DA, Bryan SE (2015) Plumbing systems of shallow level intrusive complexes. In: Physical geology of shallow magmatic systems. Springer, Cham,, pp. 39–60
Johnson AM, Pollard DD (1973) Mechanics of growth of some laccolithic intrusions in the Henry mountains, Utah, I: field observations, Gilbert's model, physical properties and flow of the magma. Tectonophysics 18(3):261–309

Jolly R, Sanderson DJ (1995) Variation in the form and distribution of dykes in the Mull swarm, Scotland. J Struct Geol 17(11):1543–1557

Jowitt SM, Williamson M-C, Ernst RE (2014) Geochemistry of the 130 to 80 Ma Canadian High Arctic large igneous province (HALIP) event and implications for Ni-Cu-PGE prospectivity. Econ Geol 109(2):281–307

Karson JA, Brooks CK (1999) Structural and magmatic segmentation of the Tertiary East Greenland volcanic rifted margin. Geol Soc London Special Publications 164(1):313–338

Kavanagh JL, Boutelier D, Cruden A (2015) The mechanics of sill inception, propagation and growth: experimental evidence for rapid reduction in magmatic overpressure. Earth Planet Sci Lett 421:117–128

Kavanagh JL, Burns AJ, Hazim SH, Wood E, Martin SA, Hignett S, Dennis DJ (2018) Challenging dyke ascent models using novel laboratory experiments: implications for reinterpreting evidence of magma ascent and volcanism. J Volcanol Geoth Res

Kendall J-M, Stuart G, Ebinger C, Bastow I, Keir D (2005) Magma-assisted rifting in Ethiopia. Nature 433(7022):146–148

Kirton S, Donato J (1985) Some buried Tertiary dykes of Britain and surrounding waters deduced by magnetic modelling and seismic reflection methods. J Geol Soc 142(6):1047–1057

Koch F, Johnson A, Pollard D (1981) Monoclinal bending of strata over laccolithic intrusions. Tectonophysics 74(3):T21–T31

Le Gall B, Daoud MA, Maury RC, Rolet J, Guillou H, Sue C (2010) Magma-driven antiform structures in the Afar rift: The Ali Sabieh range, Djibouti. J Struct Geol 32(6):843–854

Leat PT (2008) On the long-distance transport of Ferrar magmas. Geol Soc London Special Publications 302(1):45–61

Lister JR, Kerr RC (1991) Fluid-mechanical models of crack propagation and their application to magma transport in dykes. J Geophys Res Solid Earth 96(B6):10049–10077

Macdonald R, Bagiński B, Upton BGJ, Dzierżanowski P, Marshall-Roberts W (2009) The Palaeogene Eskdalemuir dyke, Scotland: long-distance lateral transport of rhyolitic magma in a mixed-magma intrusion. Mineral Mag 73(3):285–300

Macdonald R, Bagiński B, Upton BGJ, Pinkerton H, MacInnes DA, MacGillivray JC (2010) The Mull Palaeogene dyke swarm: insights into the evolution of the Mull igneous centre and dyke-emplacement mechanisms. Mineral Mag 74(4):601–622

Macdonald R, Fettes D, Bagiński B (2015) The Mull Paleocene dykes: some insights into the nature of major dyke swarms. Scott J Geol 51(2):116–124

Magee C, Briggs F, Jackson CA-L (2013) Lithological controls on igneous intrusion-induced ground deformation. J Geol Soc 170(6):853–856

Magee C, Jackson CL, Schofield N (2014) Diachronous sub-volcanic intrusion along deep-water margins: insights from the Irish Rockall Basin. Basin Res 26(1):85–105

Magee C, Maharaj SM, Wrona T, Jackson CA-L (2015) Controls on the expression of igneous intrusions in seismic reflection data. Geosphere GES01150. 01151

Magee C, Muirhead JD, Karvelas A, Holford SP, Jackson CA, Bastow ID, Schofield N, Stevenson CT, McLean C, McCarthy W (2016) Lateral magma flow in mafic sill complexes. Geosphere GES01256. 01251

Magee C, Bastow ID, de Vries BvW, Jackson CA-L, Hetherington R, Hagos M, Hoggett M (2017) Structure and dynamics of surface uplift induced by incremental sill emplacement. Geology 45(5):431–434

Mäkitie H, Data G, Isabirye E, Mänttäri I, Huhma H, Klausen MB, Pakkanen L, Virransalo P (2014) Petrology, geochronology and emplacement model of the giant 1.37 Ga arcuate Lake Victoria Dyke Swarm on the margin of a large igneous province in eastern Africa. J Afr Earth Sc 97:273–296

Malehmir A, Bellefleur G (2010) Reflection seismic imaging and physical properties of base-metal and associated iron deposits in the Bathurst Mining Camp, New Brunswick, Canada. Ore Geol Rev 38(4):319–333

Malthe-Sørenssen A, Planke S, Svensen H, Jamtveit B (2004) Formation of saucer-shaped sills. In: Breitkreuz C, Petford N (eds) Physical geology of high-level magmatic systems. Geological Society, London, Special Publications, vol 234. Geological Society, London, Special Publications, pp 215–227

Mandler HAF, Clowes RM (1997) Evidence for extensive tabular intrusions in the Precambrian shield of western Canada: a 160-km-long sequence of bright reflections. Geology 25(3):271

Manga M, Michaut C (2017) Formation of lenticulae on Europa by saucer-shaped sills. Icarus 286:261–269

Mark N, Schofield N, Pugliese S, Watson D, Holford S, Muirhead D, Brown R, Healy D (2017) Igneous intrusions in the Faroe Shetland basin and their implications for hydrocarbon exploration; new insights from well and seismic data. Marine and Petroleum Geology

Marsh B (2004) A magmatic mush column rosetta stone: the McMurdo Dry Valleys of Antarctica. Eos Trans Am Geophys Union 85(47):497–502

Mastin LG, Pollard DD (1988) Surface deformation and shallow dike intrusion processes at Inyo Craters, Long Valley, California. J Geophys Res Solid Earth 93(B11):13221–13235

McBride JH, William Keach R, Leetaru HE, Smith KM (2018) Visualizing Precambrian basement tectonics beneath a carbon capture and storage site, Illinois Basin. Interpretation 6(2):T257–T270

Mège D, Cook AC, Garel E, Lagabrielle Y, Cormier MH (2003) Volcanic rifting at Martian grabens. J Geophys Res Planets 108(E5)

Menand T (2008) The mechanics and dynamics of sills in layered elastic rocks and their implications for the growth of laccoliths and other igneous complexes. Earth Planet Sci Lett 267(1–2):93–99. https://doi.org/10.1016/j.epsl.2007.11.043

Morgan S, Stanik A, Horsman E, Tikoff B, de Saint Blanquat M, Habert G (2008) Emplacement of multiple magma sheets and wall rock deformation: Trachyte Mesa intrusion, Henry Mountains, Utah. J Struct Geol 30(4):491–512

Morgan J, Warner M, Bell R, Ashley J, Barnes D, Little R, Roele K, Jones C (2013) Next-generation seismic experiments: wide-angle, multi-azimuth, three-dimensional, full-waveform inversion. Geophys J Int 195(3):1657–1678

Muirhead JD, Airoldi G, Rowland JV, White JD (2012) Interconnected sills and inclined sheet intrusions control shallow magma transport in the Ferrar large igneous province, Antarctica. Geol Soc Am Bull 124(1–2):162–180

Muirhead JD, Airoldi G, White JD, Rowland JV (2014) Cracking the lid: sill-fed dikes are the likely feeders of flood basalt eruptions. Earth Planet Sci Lett 406:187–197

Muirhead JD, Kattenhorn SA, Le Corvec N (2015) Varying styles of magmatic strain accommodation across the East African Rift. Geochem Geophys Geosyst 16(8):2775–2795

Muirhead J, Kattenhorn S, Lee H, Mana S, Turrin B, Fischer T, Kianji G, Dindi E, Stamps D (2016) Evolution of upper crustal faulting assisted by magmatic volatile release during early-stage continental rift development in the East African Rift. Geosphere 12(6):1670–1700

Neumann ER, Svensen H, Galerne CY, Planke S (2011) Multistage evolution of dolerites in the Karoo large igneous province, Central South Africa. J Petrol 52(5):959–984

Pagli C, Wright TJ, Ebinger CJ, Yun S-H, Cann JR, Barnie T, Ayele A (2012) Shallow axial magma chamber at the slow-spreading Erta Ale Ridge. Nature Geosci 5(4):284–288

Passey S, Hitchen K (2011) Cenozoic (igneous). Geology of the Faroe-Shetland Basin and adjacent areas. British Geological Report Survey Research Report, RR/11/01; Jarofeingi Research Report RR/11/01

Patterson CW, Ernst RE, Samson C (2016) Pit Chains belonging to radiating graben-fissure systems on venus: model for formation during lateral dyke injection. Acta Geol Sin (English Edition) 90(s1):143–144

Peron-Pinvidic G, Shillington DJ, Tucholke BE (2010) Characterization of sills associated with the U reflection on the Newfoundland margin: evidence for widespread early post-rift magmatism on a magma-poor rifted margin. Geophys J Int 182(1):113–136

Phillips TB, Magee C, Jackson CA-L, Bell RE (2017) Determining the three-dimensional geometry of a dike swarm and its impact on later rift geometry using seismic reflection data. Geology 46(2):119–122

Pirajno F, Hoatson DM (2012) A review of Australia's Large Igneous Provinces and associated mineral systems: implications for mantle dynamics through geological time. Ore Geol Rev 48:2–54

Planke S, Rasmussen T, Rey SS, Myklebust R (2005) Seismic characteristics and distribution of volcanic intrusions and hydrothermal vent complexes in the Vøring and Møre basins. In: Doré AG (ed) Petroleum geology: north-west Europe and global perspectives—proceedings of the 6th petroleum geology conference. Geological Society, London, pp 833–844

Planke S, Svensen H, Myklebust R, Bannister S, Manton B, Lorenz L (2015) Geophysics and remote sensing. Springer, Berlin, Heidelberg, pp 1–16

Pollard DD, Johnson AM (1973) Mechanics of growth of some laccolithic intrusions in the Henry Mountains, Utah, II: bending and failure of overburden layers and sill formation. Tectonophysics 18(3):311–354

Pollard DD, Delaney PT, Duffield WA, Endo ET, Okamura AT (1983) Surface deformation in volcanic rift zones. Tectonophysics 94(1–4):541–584

Polteau S, Ferré EC, Planke S, Neumann ER, Chevallier L (2008a) How are saucer-shaped sills emplaced? Constraints from the Golden Valley Sill, South Africa. J Geophys Res 113(B12)

Polteau S, Mazzini A, Galland O, Planke S, Malthe-Sørenssen A (2008b) Saucer-shaped intrusions: occurrences, emplacement and implications. Earth Planet Sci Lett 266(1):195–204

Preston RJ, Bell BR, Rogers G (1998) The Loch Scridain xenolithic sill complex, Isle of Mull, Scotland: fractional crystallization, assimilation, magma-mixing and crustal anatexis in subvolcanic conduits. J Petrol 39(3):519–550

Rabbel O, Galland O, Mair K, Lecomte I, Senger K, Spacapan JB, Manceda R (2018) From field analogues to realistic seismic modelling: a case study of an oil-producing andesitic sill complex in the Neuquén Basin, Argentina. J Geol Soc jgs2017–2116

Reeves C (2000) The geophysical mapping of Mesozoic dyke swarms in southern Africa and their origin in the disruption of Gondwana. J Afr Earth Sc 30(3):499–513

Reeves J, Magee C, Jackson CAL (2018) Unravelling intrusion-induced forced fold kinematics and ground deformation using 3D seismic reflection data. Volcanica 1(1):17

Reynolds P, Holford S, Schofield N, Ross A (2017) The shallow depth emplacement of mafic intrusions on a magma-poor rifted margin: an example from the Bight Basin, Southern Australia. Mar Pet Geol 88:605–616

Rivalta E, Taisne B, Bunger A, Katz R (2015) A review of mechanical models of dike propagation: schools of thought, results and future directions. Tectonophysics 638:1–42

Rowland J, Baker E, Ebinger C, Keir D, Kidane T, Biggs J, Hayward N, Wright T (2007) Fault growth at a nascent slow-spreading ridge: 2005 Dabbahu rifting episode, Afar. Geophys J Int 171(3):1226–1246

Rubin AM (1992) Dike-induced faulting and graben subsidence in volcanic rift zones. J Geophys Res Solid Earth 97(B2):1839–1858

Rubin AM (1995) Propogation of magma-filled cracks. Annu Rev Earth Planet Sci 23:49

Schmiedel T, Kjoberg S, Planke S, Magee C, Galland O, Schofield N, Jackson CA-L, Jerram DA (2017) Mechanisms of overburden deformation associated with the emplacement of the Tulipan sill, mid-Norwegian margin. Interpretation 5(3):SK23–SK38

Schofield N, Heaton L, Holford SP, Archer SG, Jackson CA-L, Jolley DW (2012a) Seismic imaging of 'broken bridges': linking seismic to outcrop-scale investigations of intrusive magma lobes. J Geol Soc 169(4):421–426

Schofield NJ, Brown DJ, Magee C, Stevenson CT (2012b) Sill morphology and comparison of brittle and non-brittle emplacement mechanisms. J Geol Soc 169(2):127–141

Schofield N, Alsop I, Warren J, Underhill JR, Lehné R, Beer W, Lukas V (2014) Mobilizing salt: magma-salt interactions. Geology G35406. 35401

Schofield N, Holford S, Millett J, Brown D, Jolley D, Passey SR, Muirhead D, Grove C, Magee C, Murray J, Hole M, Jackson CAL, Stevenson C (2017) Regional magma plumbing and emplace-

ment mechanisms of the Faroe-Shetland sill complex: implications for magma transport and petroleum systems within sedimentary basins. Basin Res 29(1):41–63
Skogseid J, Pedersen T, Eldholm O, Larsen BT (1992) Tectonism and magmatism during NE Atlantic continental break-up: the Voring Margin. Geol Soc London, Special Publications 68(1):305–320
Smallwood JR, Maresh J (2002) The properties, morphology and distribution of igneous sills: modelling, borehole data and 3D seismic from the Faroe-Shetland area. In: Jolley DW, Bell BR (eds) The north atlantic igneous province: stratigraphy, tectonic, volcanic and magmatic processes, vol 197, vol 1. Geological Society, London, Special Publications, pp 271–306
Spacapan JB, Galland O, Leanza HA, Planke S (2017) Igneous sill and finger emplacement mechanism in shale-dominated formations: a field study at Cuesta del Chihuido, Neuquén Basin, Argentina. J Geol Soc 174(3):422–433
Speight J, Skelhorn R, Sloan T, Knaap R (1982) The dyke swarms of Scotland. Igneous rocks of the British Isles 449–459
Storey M, Duncan RA, Tegner C (2007) Timing and duration of volcanism in the North Atlantic Igneous Province: implications for geodynamics and links to the Iceland hotspot. Chem Geol 241(3–4):264–281
Svensen H, Planke S, Malthe-Sorenssen A, Jamtveit B, Myklebust R, Rasmussen Eidem T, Rey SS (2004) Release of methane from a volcanic basin as a mechanism for initial Eocene global warming. Nature 429(6991):542–545
Svensen H, Jamtveit B, Planke S, Chevallier L (2006) Structure and evolution of hydrothermal vent complexes in the Karoo Basin, South Africa. J Geol Soc 163:11
Svensen H, Planke S, Chevallier L, Malthe-Sørenssen A, Corfu F, Jamtveit B (2007) Hydrothermal venting of greenhouse gases triggering Early Jurassic global warming. Earth Planet Sci Lett 256(3–4):554–566
Svensen H, Corfu F, Polteau S, Hammer Ø, Planke S (2012) Rapid magma emplacement in the Karoo Large Igneous Province. Earth Planet Sci Lett 325–326:1–9
Svensen HH, Polteau S, Cawthorn G, Planke S (2015) Sub-volcanic intrusions in the Karoo basin, South Africa
Tentler T (2005) Propagation of brittle failure triggered by magma in Iceland. Tectonophysics 406(1):17–38
Thomson K (2005) Volcanic features of the North Rockall Trough: application of visualisation techniques on 3D seismic reflection data. Bull Volcanol 67(2):116–128
Thomson K (2007) Determining magma flow in sills, dykes and laccoliths and their implications for sill emplacement mechanisms. Bull Volcanol 70(2):183–201
Thomson K, Hutton D (2004) Geometry and growth of sill complexes: insights using 3D seismic from the North Rockall Trough. Bull Volcanol 66(4):364–375
Torsvik TH, Smethurst MA, Burke K, Steinberger B (2008) Long term stability in deep mantle structure: Evidence from the ~300 Ma Skagerrak-Centered Large Igneous Province (the SCLIP). Earth Planet Sci Lett 267(3):444–452
Townsend MR, Pollard DD, Smith RP (2017) Mechanical models for dikes: a third school of thought. Tectonophysics 703:98–118
Trude J, Cartwright J, Davies RJ, Smallwood JR (2003) New technique for dating igneous sills. Geology 31:4
van Wyk de Vries B, Márquez A, Herrera R, Bruña JG, Llanes P, Delcamp A (2014) Craters of elevation revisited: forced-folds, bulging and uplift of volcanoes. Bull Volcanol 76(11):1–20
Wall M, Cartwright J, Davies R, McGrandle A (2010) 3D seismic imaging of a Tertiary Dyke Swarm in the Southern North Sea, UK. Basin Res 22(2):181–194
Wilson L, Head JW (2002) Tharsis-radial graben systems as the surface manifestation of plume-related dike intrusion complexes: models and implications. J Geophys Res Planets 107(E8)
Wilson PI, McCaffrey KJ, Wilson RW, Jarvis I, Holdsworth RE (2016) Deformation structures associated with the Trachyte Mesa intrusion, Henry Mountains, Utah: Implications for sill and laccolith emplacement mechanisms. J Struct Geol 87:30–46

Wingate MT, Pirajno F, Morris PA (2004) Warakurna large igneous province: a new Mesoproterozoic large igneous province in west-central Australia. Geology 32(2):105–108
Wyche S, Pawley M, Chen S, Ivanic T, Zibra I, Van Kranendonk M, Spaggiari C, Wingate M (2013) Geology of the northern Yilgarn Craton. In: Youanmi and Southern Carnarvon Seismic and Magnetotelluric (MT) Workshop 2013: extended abstracts, Compiled by S Wyche, TJ Ivanic and I Zibra. Geological Survey of Western Australia. Record
Wyrick DY, Smart KJ (2009) Dike-induced deformation and Martian graben systems. J Volcanol Geoth Res 185(1–2):1–11
Wyrick D, Ferrill DA, Morris AP, Colton SL, Sims DW (2004) Distribution, morphology, and origins of Martian pit crater chains. J Geophys Res 109(E6):E06005
Zaleski E, Eaton DW, Milkereit B, Roberts B, Salisbury M, Petrie L (1997) Seismic reflections from subvertical diabase dikes in an Archean terrane. Geology 25(8):707–710

The Mesozoic Equatorial Atlantic Magmatic Province (EQUAMP)

A New Large Igneous Province in South America

M. H. B. M. Hollanda, C. J. Archanjo, A. A. Macedo Filho, H. Fossen, R. E. Ernst, D. L. de Castro, A. C. Melo and A. L. Oliveira

Abstract Mafic dike swarms and sills intruding, respectively, the Precambrian Borborema and Paleozoic Parnaíba provinces (NE Brazil) constitute two major magmatic events related to the continental break-up that formed the Equatorial Atlantic. Available whole rock K–Ar determinations and a few plagioclase $^{40}Ar/^{39}Ar$ ages suggest that these events occurred approximately between 135 and 120 Ma. Airborne magnetic data indicates that the main dike swarm, the Rio Ceará-Mirim swarm, can be traced for about 1,000 km along an arcuate trajectory between the Cretaceous Potiguar rift near the Atlantic coastline, and the northern margin of the São Francisco craton. The dikes of the such a giant swarm form isolated (often en-echelon) segments that can reach up to 40 km in length and have a mean width of about 70 m. The sills, known as Sardinha magmatism, are intercalated between the Paleozoic sedimentary strata of the Parnaíba basin having major expression in subsurface rather than on surface. Geochemical data indicate that the parental tholeiitic magmas of dikes and sills would largely derive from melting of subcontinental lithospheric sources. However, a subordinate OIB-type component identified in some dikes would indicate contribution of a mantle plume as a melt source. In this paper we propose that the Sardinha and Rio Ceará-Mirim magmatic products, which are found over an area of about 700,000 km^2, all together represent a newly recognized Cretaceous LIP in South America here named Equatorial Atlantic Magmatic Province (EQUAMP).

M. H. B. M. Hollanda (✉) · C. J. Archanjo · A. A. Macedo Filho · A. L. Oliveira
Instituto de Geociências, Universidade de São Paulo, São Paulo 05508–080, Brazil
e-mail: hollanda@usp.br

H. Fossen
Museum of Natural History, University of Bergen, N-50, Bergen, Norway

R. E. Ernst
Department of Earth Science, Carleton University, Ottawa, Canada

R. E. Ernst
Department of Geology and Geography, Tomsk State University,
36 Lenin Ave, Tomsk 634050, Russia

D. L. de Castro · A. C. Melo
Programa de Pós-Graduação em Geodinâmica e Geofísica,
Universidade Federal do Rio Grande do Norte, Natal 58078–970, Brazil

R. K. Srivastava et al. (eds.), *Dyke Swarms of the World: A Modern Perspective*,
Springer Geology, https://doi.org/10.1007/978-981-13-1666-1_3

1 Introduction

During the time interval spanning the Late-Triassic to Mid-Cretaceous, one of the major continental land masses existing on Earth, the Gondwana supercontinent, was gradually dismembered into South America, Africa (Madagascar), India, Australia and Antarctica after the efficient combination of global-scale plate boundary stresses and deep mantle dynamics including mantle plumes. As a result, large igneous provinces (LIPs) (Ernst 2014) characterized by extensive mafic and/or felsic lava flows and related plumbing systems formed. Lava flows, sills and dikes are preserved in the northern Brazil, Bolivia and northwestern Africa representing part of the Triassic-Jurassic Central Atlantic Magmatic Province (CAMP; Marzoli et al. 1999a, b; Knight et al. 2004; Verati et al. 2005, 2007; Merle et al. 2011; Bertrand et al. 2014). The Cretaceous LIPs include Karoo-DML (Dronning Maud Land)-Ferrar related to the break-up of the East Gondwana involving Africa, Antarctica and Australia (183.0 $\pm$ 0.5–182.3 $\pm$ 0.6 Ma for Karoo, Svensen et al. 2012, 182.78 $\pm$ 0.03–182.43 $\pm$ 0.04 Ma for Ferrar, Burgess et al. 2015), and the Paraná-Etendeka event related to West Gondwana fragmentation to form Africa and South America in the Early Cretaceous (ca. 135 Ma; Thiede and Vasconcelos 2010; Florisbal et al. 2014; Janasi et al. 2011). At the Cretaceous-Paleogene boundary, the separation of India from the Seychelles micro-continent was intimately linked to the Deccan traps (e.g., Hooper et al. 2010; Renne et al. 2015), while the Afro-Arabian province represents the youngest LIP of Cenozoic age (30–5 Ma) along the East-African rift system (e.g., Ukstins Peate et al. 2005; Riisager et al. 2005). One common feature found in all these LIPs is the contemporaneity of the eruptive and intrusive igneous rocks emplaced either as a short-term single pulse or as multiple pulses (e.g., Courtillot et al. 2003; Ernst 2014).

Extensive flood basalt volcanism related to continental break-up is often connected to plume (or hot spot) activity, while dike swarms are at first associated with variations in the orientation and magnitude of local or regional stress fields in intraplate settings (e.g., Pollard et al. 1975). Having strike lengths in excess of 300 km, giant dike swarms are usually considered to be an important component of the plumbing system of LIPs (Ernst and Buchan 1997) and can be interpreted as sub-volcanic feeder systems of the lava flows. The spatial geometry of such giant swarms, particularly the recognition of giant radiating dike swarms (Halls 1982; Fahrig 1987) led Ernst and Buchan (1997) to propose a close link with LIPs originated by plumes, but also with either the thermal and mechanical rheology of the crustal section they are intruding, such as preexisting crustal discontinuities along with the regional stress field and its modification by domal uplift associated with mantle plume arrival at the base of the lithosphere. More recently, increasing field observations and geophysical modeling have shown that plumbing systems can also develop at very shallow levels, particularly in sedimentary basins, as extensive sill complexes (Thomson and Hutton 2004; Polteau et al. 2008; Svensen et al. 2012; Magee et al. 2016). These intrusive complexes are mapped often over several tens of kilometers and are dominated by an interconnected network of mafic dikes and sills, which frequently exhibit saucer-shaped morphologies associated with a small proportion of dikes (e.g., Muirhead et al. 2012; Magee et al. 2016). Such particular sill morphology along with lateral magma flow patterns found in regional dike swarms enable magma transport over many hundreds of kilometers and in some cases more than 2,000 km and, therefore, have broad implications for the interpretation of geochemical data obtained from large-scale magmatic areas (e.g., Meade et al. 2009). Hence, detailed structural, geophysical, geochemical and geochronological examination of giant plumbing systems (dikes plus sills) has become an essential tool to support models integrating plate tectonics and mantle dynamics in continental settings.

This paper intends to formally propose a new Cretaceous LIP in South America formed during the Equatorial Atlantic Ocean opening and, therefore, related to the break-up of the West Gondwana supercontinent. This igneous province, herein coined as Equatorial Atlantic Magmatic Province (EQUAMP), differs from other Gondwanan LIPs by encompassing dominantly (or exclusively) intrusive instead of eruptive rocks, thus satisfying the criteria for plumbing system-type continental LIPs of Bryan and Ernst (2008; also Ernst 2014) and the definition of LPPs (Large Plutonic Province) of Sheth (2007). The EQUAMP LIP consist of dikes and sills that, although having been considered as separated intrusive components in previous works, can now be investigated within the context of a single magmatic province on the basis of chemical and geochronological similarities. Some aspects of the magmatic products making up this hitherto previously unknown LIP are summarized here.

2 Geological Framework

The main EQUAMP magmatic components are the Rio Ceará-Mirim dike swarm (RCM) and the Sardinha sill province, which are intrusive into two contrasting geological settings in northeast Brazil, the Precambrian Borborema Province and the Paleozoic-Mesozoic Parnaíba Basin. The Borborema Province is a major crustal block formed by convergence between the West African and Congo-São Francisco cratons to assemble West Gondwana during the Neoproterozoic Brasiliano/Pan-African orogeny (e.g., Frimmel and Frank 1998; Caby 2003; Gray et al. 2008; Brito Neves et al. 2014). As a result of the convergence, the Borborema Province shares a number of geological features with terranes of the northwestern Africa, amongst which a dominant Paleoproterozoic gneissic-migmatitic basement, Neoproterozoic (passive margin and intracontinental) mobile belts, and voluminous (acid to mafic) plutonism dated at ca. 630–530 Ma (see Santos et al. 2010, 2014; Van Schmus et al. 2008 for a general overview of the Precambrian geology of the Borborema province). A major tectonic feature of the Brasiliano/Pan-African orogeny concerns a continental-scale network of strike-slip shear zones developed coevally with high-temperature deformation and metamorphism, and crustal anatexis (Vauchez et al. 1995; Archanjo et al. 2002a, 2013). These shear zones consist of mylonite belts that split up the Borborema Province into terranes with contrasting tectono-lithological histories, some of them accommodating late brittle fault reactivations (normal, strike-slip or locally reverse faults) related to Cambrian-Ordovician extensional events that led to the development of small intra-continental rifted basins (e.g., Sénant and Popoff 1991; Françolin et al. 1994). In the Cretaceous, the Borborema Province was affected by widespread rifting associated with the Equatorial Atlantic opening, during which the RCM and related sub-swarms and sills were emplaced.

The Neoproterozoic shear zones of the Borborema Province are covered to the west by one of the major Paleozoic sag basins of the South America, the Parnaíba Basin. The basin features a ~5-km-deep predominantly siliciclastic succession that covers approximately 600,000 km^2 overlying the Precambrian rocks of the Amazonian craton and Araguaia belt (to the west), Borborema Province (to the east) and São Francisco craton to the south (e.g., Góes and Feijó 1984). Sandstones, subordinate shales and conglomerates, as well as local limestones and evaporites, are grouped into three supersequences that record successive transgressive-regressive cycles separated by regional erosive unconformities: Silurian (Serra Grande Group), Middle Devonian-Early Carboniferous (Canindé Group) and Late Carboniferous-Early Triassic (Balsas Group) (Góes and Feijó 1984; Vaz et al. 2007). Long-term denudation of neighboring Precambrian source rocks and intrabasinal recycling were likely concurrent processes controlling the ca. 250 m.y. of sedimentation in the basin (Hollanda et al. 2018). This wide sedimentary succession, in turn, hosts two magmatic events. The older magmatic activity is represented by equivalents of the CAMP-related flood basalts (locally named Mosquito Formation; Bellieni et al. 1990; Fodor et al. 1990; Marzoli et al. 1999a; Merle et al. 2011), and the younger is represented by mafic sills and subordinate dikes grouped as the Sardinha Formation (Bellieni et al. 1990; Fodor et al. 1990). Moreover, and maybe corresponding to the lateral equivalent of the Sardinha sills, the RCM intrudes, over hundreds of kilometers in length, the Precambrian basement of the Borborema Province and the basal Silurian succession of the Parnaiba Basin.

In the following sections we outline some characteristics of the magmatic components of the EQUAMP LIP, specially their elemental and isotope geochemistry, geochronology and structural aspects. We also bring up some research issues that will be the focus of future investigations of this newly recognized magmatic province.

3 The EQUAMP Components

3.1 Dikes

In contrast to the large volumes of magma erupted during the opening of the South Atlantic, the Equatorial margin is featured by relatively modest volume of magmatism (e.g., Benkhelil 1989). The more prominent magmatic activity is the RCM dike swarm, which has long been recognized as an EW-trending swarm approximately 400 km in length and formed prior to and concurrent with the Equatorial Atlantic rifting (e.g., Matos et al. 1992, 2000). The RCM dikes are characterized by relatively thick, tholeiitic mafic bodies on the order of 20–190 m in width and 1–15 km in length (e.g., Archanjo et al. 2000, 2002b), the longer dikes being generally emplaced as en-echelon arrays. In recent years, the Brazilian Geological Survey has made available high-resolution airborne magnetic data that reveal a southwestward continuation of the RCM dikes that increases the overall swarm length by at least 600 km. The swarm changes in trend, turning from EW towards SW at approximately 38°W. Both the total length of the swarm (about 1,000 km) and its change in trend make the RCM dikes an arcuate giant swarm in the sense of Ernst and Buchan (1997). Contrasting with the well-studied dikes of the EW-trending portion, the NE-trending dikes have not yet been studied in terms of geochemical, chronological and structural aspects.

In addition to the arcuate swarm, airborne magnetic anomalies also discriminate two other sub-sets of dikes (Fig. 1). One of them is a somewhat diffuse, curved NW-trending dikes exposed nearly parallel to the Equatorial Atlantic coastline—the Canindé dike swarm, while the other is located to the southeast of the RCM swarm consisting of a NE-trending en-echelon dike system—the Riacho do Cordeiro dike swarm. This latter extends over ca. 300 km in length from the São Francisco craton to the south until intersecting with the Pernambuco lineament (Fig. 1). A local, unpublished K–Ar age of 119 ± 2 Ma in the Riacho do Cordeiro dikes suggests their temporal correlation with the RCM dikes (see the Geochronology section). As for the NE-trending portion of the RCM (NE-RCM), no petrological and age data are available for Canindé and Riacho do Cordeiro dikes.

3.2 *Sills*

The second EQUAMP component is the Sardinha magmatism. It consists of sheet-like bodies (sills) and local dikes that intruded the Parnaíba Basin to the east of 45°W, in the opposite side of the main area of occurrence of the CAMP lava flows (Fig. 1) (e.g., Fodor et al. 1990; Bellieni et al. 1990). Although not exclusive, the emplacement of sills as seen in the field is dominantly controlled by rheological differences between the host-rock lithologies of the Canindé Group, i.e. along the contact between the regressive-transgressive sedimentary successions. A recent study combining seismic, magnetic and surface geologic data showed that the occurrence of both CAMP and Sardinha magmatic events are much more extensive in subsurface than those mapped on surface (Mocitaiba et al. 2017). An integrated modeling using geochemical (major element) and geophysical (magnetic susceptibility) data suggest that the CAMP magmatism is not restricted to the west (and to the south) areas, but might occur to the east interstratified with Sardinha sills (de Castro et al. 2018).

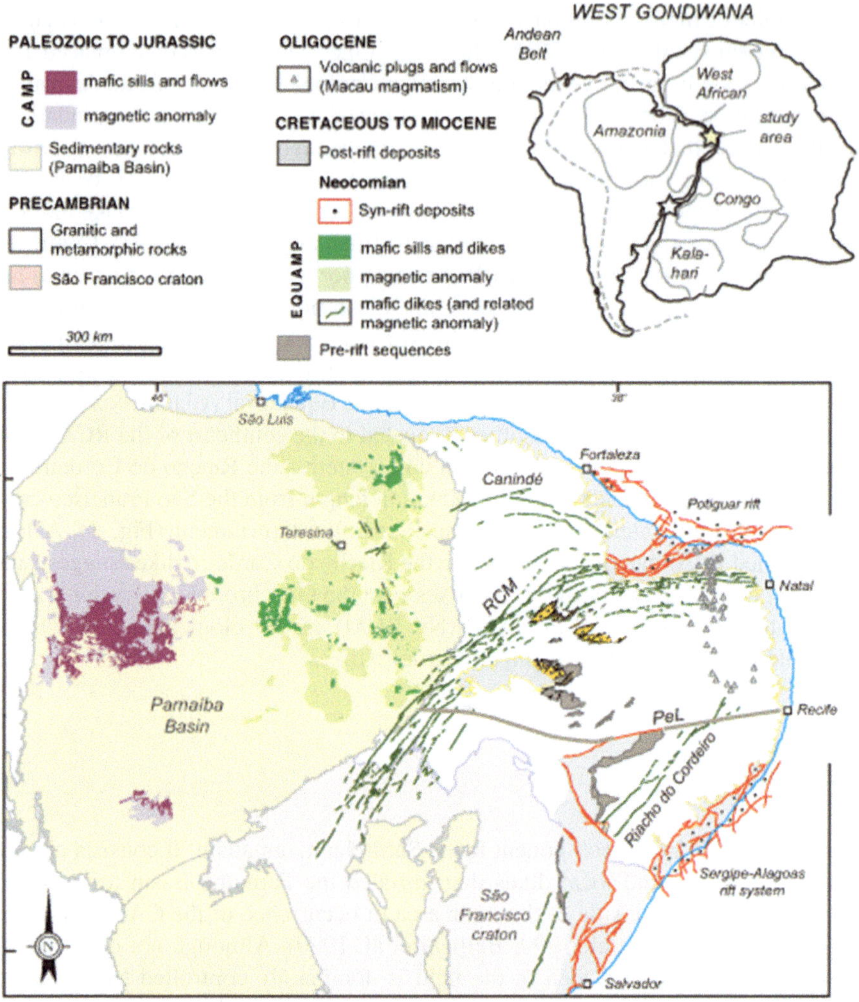

Fig. 1 Simplified map integrating the Borborema Province and Parnaíba Basin. Dikes that are associated to the Rio Ceará-Mirim magmatic event are shown as green lines. The Sardinha magmatic event is mostly represented by sills within the basin; the dark green areas are the exposures of sills on surface, while the light green areas illustrate the subsurface distribution based on geophysical data (after Mocitaiba et al. 2017). The remnant CAMP flood basalts are highlighted in dark purple (surface exposures) and light purple (subsurface interpretation). The inset on the top is a schematic picture of West Gondwana and related cratonic blocks, with location of the study area. RCM refers to the Rio Ceará-Mirimdikeswarm and PeL refers to the Pernambuco Lineament, whereas PEMP (see inset) is the abbreviation to Paraná-Etendeka Magmatic Province

3.3 Size of the Dike Swarm

Physical parameters (length and width) of the RCM dike swarm were determined using Google Earth™ imagery combined with regional magnetic anomalies map. The semi-arid environment of NE Brazil offers good contrast between mafic dikes and their host granite-gneissic rocks, with dikes forming well-defined dark, ribbon-like patches cutting across the fabric of the Precambrian basement. As the regional relief is nearly flat along dike exposures and most these occur as vertical bodies, dimensions measured on satellite image approach to their real length and thickness. Image sources provided by Google Earth™ in the NE Brazil combines Spot Image scenes acquired between 2011 and 2012 and Digital Globe Imagery (LANDSAT 7 and 8) acquired from 2002 to 2010. Image resolution is typically about 15 m per pixel, which means that mafic dikes below about 30 m in width and 100 m in length become increasingly difficult to detect on Google Earth™. Hence, the study on dike dimensions is focussed mostly to the larger RCM dikes. Some dikes with widths lower than 30 m included on Table 1 were identified in field work.

Dikes of the EW- and NE-trending portions of the RCM dike swarm are mostly linear consisting of simple isolated bodies or arrays of two or more dike segments connected by steps. Both simple and connected dike segments show left- or right-lateral en-echelon arrangements on satellite image. In a few places where vertical sections are available, the dike walls are vertical or show steep dips. We did not observe steps connecting dikes in vertical sections, although they are necessary to interconnect the dike system from their more continuous underlying sources up to the actual level of exposure. Columnar joints resulting from shrinkage on cooling were not detected, although such features, if present, may have been masked by the common onion-skin weathering that affect the mafic dikes. Furthermore, and possibly due to pervasive physical alteration that usually mask the contact between dikes and their wall-rocks, chilled margins were recorded in just a few dikes.

Dike length (L) was estimated by measuring the distance between the most remote points on the dikes exposure. When connected by steps, the length is the total distance between the most remote points of the dike array. Widths were measured perpendicular to dike wall and its maximum width (Wm) refers to the widest segment recorded on the satellite image. Lengths of 71 dike segments of the EW-RCM range from 0.1 to 14.8 km (mean L = 3.3 km). On the other hand, maximum widths range from 18 to 193 m (mean Wm = 84.7 m). Similar dimensions are recorded in dikes of the NE-trending portion (NE-RCM). Their lengths vary from 0.4 to 37.7 km (mean L = 4.9 km) with widths that vary from 16.5 to 193 m (mean Wm = 68.3 m). In a logarithmic L – Wm diagram (Fig. 2) the RCM dikes plot above the Oligocene mafic dike swarm that intrudes the extended Precambrian crust of East Africa (Schultz et al. 2008) and radial composite dikes distributed around an eroded volcanic neck intrusive in shallow crust of North America (Shiprock dikes; Delaney and Pollard 1981). RCM and Ethiopian dike swarms show equivalent dike lengths, and both are longer than Shiprock dikes. However, the RCM dikes are thicker than Ethiopian and Shiprock dikes, which we attribute to the deeper crustal exposure of the RCM dikes compared to those swarms.

Table 1 Length (L) and maximum width (Wm) of single dikes and dikes connected by steps of the Rio Ceará Mirim dike swarm (geographic coordinates in decimal degrees)

E-trending sub-swarm single dike		L (m)	Wm (m)	E-trending sub-swarm connected dike		L (m)	Wm (m)	NE-trending sub-swarm single dike		L (m)	Wm (m)	NE-trending sub-swarm connected dike		L (m)	Wm (m)
−5.580729	−36.063835	1568.6	100.9	−5.582922	−36.106884	2371.8	109.8	−5.515651	−38.898315	3627.3	84.1	−5.478237	−38.658944	9110	130.6
−5.586941	−36.143577	686.1	59.8	−5.586154	−36.126771	1413.9	96.5	−5.479401	−38.638361	2736.2	83.8	−5.562484	−39.025787	3690.2	108.7
−5.586059	−36.155305	1453.7	96.7	−5.584593	−36.310338	1954.2	80.4	−5.653879	−39.149624	1897	78.8	−5.611699	−39.089268	6651.6	124.5
−5.582604	−36.325942	378.3	62.8	−5.586278	−36.490779	2246.2	70.3	−5.669289	−39.169342	1957.5	70.5	−6.522031	−40.159948	2534.4	48.9
−5.581192	−36.333218	780.9	75.4	−5.615733	−36.135447	4337.1	77.6	−5.720268	−39.215706	708.5	67.5	−5.704575	−39.207428	4929.4	84.5
−5.583323	−36.340984	706.7	65.7	−5.605185	−36.175568	3234.9	38	−5.746920	−39.234538	674.5	43.4	−5.792078	−39.276657	3705.1	101.7
−5.583684	−36.346782	326.9	46.2	−5.631748	−36.214467	3542.1	159.9	−5.750573	−39.237466	397.4	35.3	−6.535719	−40.130091	2733.1	75.1
−5.582820	−36.349681	173.5	44.7	−5.628287	−36.260285	2968.8	112	−5.757382	−39.245056	2008.3	59.8	−6.508134	−40.211957	15625.7	62.8
−5.581147	−36.359331	1194.8	97	−5.628834	−36.298031	5821.5	146.9	−5.922960	−39.395905	833.9	31	−6.449727	−40.137205	4478.7	51.8
−5.588746	−36.379231	999.9	80.5	−5.628308	−36.348808	3862.5	90.3	−5.950754	−39.438298	1252.2	56.6	−6.616423	−40.243631	13593.1	193.1
−5.586448	−36.390985	479.2	73.8	−5.623948	−36.430519	9736.7	165.4	−5.963765	−39.467555	2643	83.7	−6.178690	−38.831951	2464.7	57.5
−5.581983	−36.401008	622.7	62.6	−5.629994	−36.503285	2255.5	131.4	−6.419035	−40.010023	4259.5	82.3	−5.768205	−38.906234	7318.9	78.2
−5.578341	−36.413192	1243.6	40.3	−5.636220	−36.528466	2895.7	115.5	−6.444844	−40.128876	2952.8	55.1	−5.818030	−38.999919	1621.6	20.3
−5.586691	−36.483759	545.7	62.6	−5.652184	−36.597987	8558.8	154.8	−6.582735	−40.275499	2903.3	33.1	−7.615401	−41.279334	7595.4	52.9
−5.586969	−36.508836	678.5	64.5	−5.654612	−36.759833	6382.3	97.5	−6.655130	−40.396479	6019.5	91.4	−7.744713	−41.354655	10573.8	82.5
−5.590312	−36.715075	1267.5	67.5	−5.712730	−36.200267	14785.2	82.7	−6.656107	−40.390006	4142.1	47.9				
−5.591448	−36.725798	1008.4	52.5	−5.694811	−36.390401	6492.3	54.5	−6.565351	−40.160537	6444.2	53.9				
−5.626569	−36.102989	2083.5	74.7	−5.707974	−37.005634	8541.4	193.5	−6.807680	−40.013111	5959.8	116.4				
−5.621491	−36.150741	4040.4	88.1	−5.702631	−37.061279	4796.6	170.9	−6.008898	−38.704992	1539.5	77.8				
−5.622041	−36.167189	2101.3	61.3	−5.706217	−37.081675	3289.9	107.3	−6.025138	−38.725893	4093.8	109.7				

(continued)

Table 1 (continued)

E-trending sub-swarm single dike		L (m)	Wm (m)	E-trending sub-swarm connected dike		L (m)	Wm (m)	NE-trending sub-swarm single dike		L (m)	Wm (m)	NE-trending sub-swarm connected dike		L (m)	Wm (m)
−5.626001	−36.184523	2932	101.8	−5.702291	−37.127776	11606.3	46.6	−6.092764	−38.776757	6651.7	105.1				
−5.623021	−36.390696	2032	107.1	−5.705699	−37.206947	7973.7	40.6	−5.686260	−38.697981	37744.4	84.3				
−5.639188	−36.546479	2088.7	108.5	−5.721650	−37.122460	4961.2	65.8	−6.145310	−38.811325	893.3	38.8				
−5.651099	−36.673277	2014.4	128.6	−5.726393	−37.138106	7684.2	69.5	−6.212479	−38.863307	4176.1	59.1				
−5.651418	−36.689752	1728	99.5	−5.760599	−37.273365	10731.9	110.9	−5.721989	−38.764520	2125.6	16.5				
−5.654194	−36.712985	1368.2	70.4	−5.759746	−37.293428	3782.7	120	−5.730039	−38.790512	1835.9	33.1				
−5.692279	−36.330835	439	44.2	−5.757640	−37.333447	4822.6	146.9	−5.741068	−38.831904	5771	68.3				
−5.693060	−36.333754	100.9	41.4	−5.767072	−37.383816	8048.9	129.2	−5.812867	−38.972914	1224.8	29.4				
−5.695595	−36.354311	2124.7	42.4	−5.764409	−36.972563	2891.9	69.8	−5.802497	−38.972086	1039.8	21.6				
−5.695944	−36.357476	1156.5	27.1	−5.784362	−37.186460	11338.8	69.7	−5.823231	−39.012125	1038.7	17.9				
−5.699989	−36.527563	1714.1	104.4					−6.792033	−40.562761	2761.5	73.1				
−5.702113	−36.574401	2307.2	40.3					−6.825990	−40.595860	14827.4	67.4				
−5.702919	−36.583349	2575.9	62.6					−7.013061	−40.867360	2239.9	59.6				
−5.707797	−36.615090	2873.2	52					−7.630816	−41.305865	1531.1	28.1				
−5.670239	−36.794425	7620.9	104					−7.754221	−41.416877	1111.8	31.3				
−5.674502	−36.836615	2365.3	134.4					−7.803599	−41.489832	3918.6	106.2				
−5.731334	−36.826275	1709	67.7					−7.823625	−41.519608	3820.6	86.5				
−5.779134	−37.036340	1608.7	30					−7.850077	−41.548152	3148.8	45.5				
−5.776622	−37.151489	3553.4	64.4					−7.872843	−41.572725	5014.8	78.2				
−5.584905	−36.356969	170	25					−7.945703	−41.641504	14978.5	70.2				
−5.795622	−37.769662	313	60												

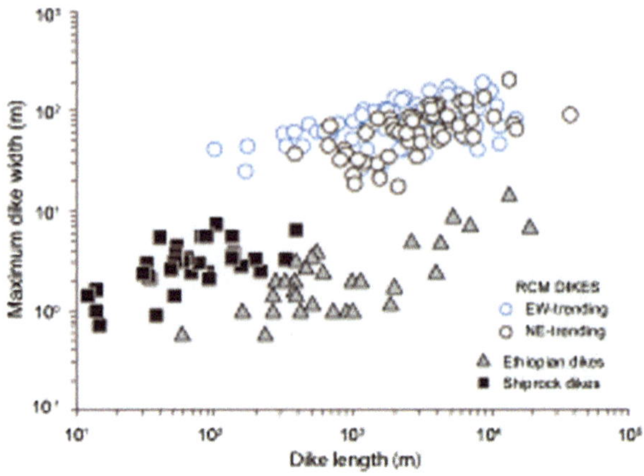

Fig. 2 Relationship between the length (L) and maximum width (Wm) for mafic dikes of the RCM dike swarm (see details in the text). Data from Shiprock dikes and Ethiopian mafic swarm are available in Delaney and Pollard (1981) and Schultz et al. (2008)

3.4 AMS Data from RCM Dikes

According to Archanjo et al. (2002b) and Archanjo and Launeau (2004), normal-type anisotropy of magnetic susceptibility (AMS) fabrics of the EW-RCM dikes are consistent with a lateral magma flow along the central and western sectors of the swarm. AMS magnitude is weak, mostly between 2 and 5%, and provided by multidomain titanomagnetite grains. In contrast, the low-field magnetic susceptibility of the mafic dikes is strong, about 10^{-2} SI, which is consistent with the intensity of the aeromagnetic anomalies that define the trace of the dikes in the regional maps. To investigate the relationship between the opening of fissures to emplace the basaltic magma and the orientation of magnetic fabric we sampled two mafic dikes for which the wall-rock contacts are exposed in field. These dikes are situated in the middle and western part of the EW-RCM dike swarm and are respectively, 18 m (CM30) and 150 m (CM43) in width. Although the susceptibility and anisotropy parameters of these samples have already been published in Archanjo et al. (2002b), we introduce here a new interpretation of AMS based on the obliquity between the dike fabric and the wall-rock orientation (Fig. 3).

Shape ellipsoids vary from oblate to prolate in one dike (CM30) and are dominantly oblate in the other (CM43). Strongly prolate ellipsoids ($T < -0.5$) indicate a linear fabric while oblate ellipsoids ($T > 0.5$) indicate a planar fabric. The majority of samples in one dike (CM30) shows dominant prolate ellipsoids, with lineations oblique (samples F, G, H and I) to a high angle (A, C) to the dike wall. In samples with neutral ($T \sim 0$) to oblate ($T > 0$) ellipsoids, the foliations tend to be perpendicular to the dike plane; one sample (D) shows a well-defined foliation ($T = 0.66$) nearly parallel to the dike.

In the larger dike (CM45), the magnetic foliations typically are oblique to the dike trend. The strike of the magnetic foliation is usually in a NE-SW directions and dips are variable, but usually steep to the NW or SE. Lineations, in contrast, tend to disperse to different directions in agreement with the oblate shape of the AMS ellipsoids. The magnetic fabric that transects the dike, and particularly the foliation

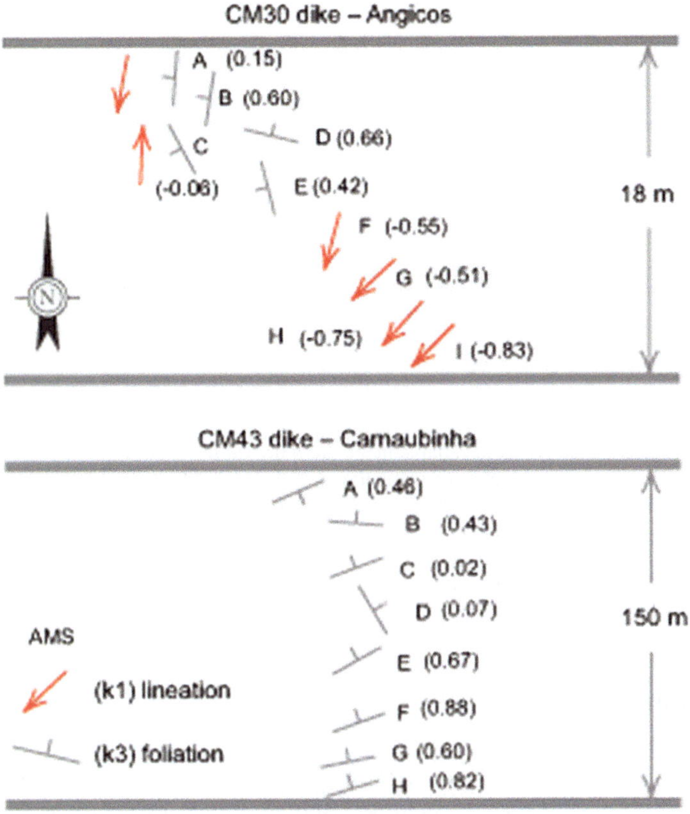

Fig. 3 Anisotropy of magnetic susceptibility (AMS) of two dikes (CM30 and CM45) whose contacts are exposed with the host regional rocks. Magnetic fabric of dike CM30 shows prolate (T < 0) and oblate (T > 0) AMS shape ellipsoids while oblate fabrics dominate the AMS fabric in the dike CM45. Oblique to highly-oblique lineations in the dike with neutral to prolate ellipsoids (CM30) and well-defined asymmetrical magnetic foliations in the larger dike (CM45) suggest that the fracture opening that accommodated the mafic magma included a component of lateral strike slip displacement parallel to the dike wall. Note: capital letters refer to the measurement site and respective AMS shape parameter (value given in parentheses) (see details in the text)

along the larger dike, would be formed by the combined influence of magma flux and syn-emplacement shearing before magma freezing (Correa-Gomes et al. 2001; Féménias et al. 2004; Clemente et al. 2007). The AMS of the dike with neutral to prolate ellipsoids (CM30) would be carried by needle-like titanomagnetite grains that would grow preferentially along the principal direction of local extension (Knight and Walker 1988). Late magmatic shearing parallel to the dike wall, in turn, would provide a transected magnetic foliation in the thicker dike. AMS results suggest, therefore, that the opening of fractures that accommodated the EW-RCM dikes combined a component of normal dilatation and a component of lateral strike slip offset of the wall-rocks.

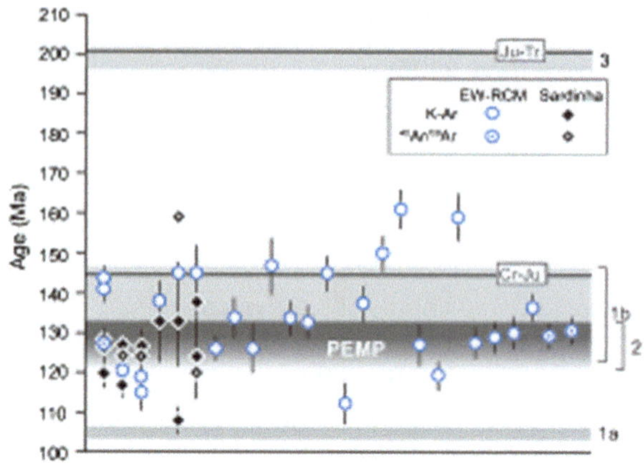

Fig. 4 K–Ar and $^{40}Ar/^{39}Ar$ ages available in the literature for the EQUAMP diabases (see sources in the text). Gray areas are shown for comparison: (1a, 1b) Northern Benue Trough ($^{40}Ar/^{39}Ar$ ages—Maluski et al. 1995); (2) Paraná-Etendeka ($^{40}Ar/^{39}Ar$ ages—Renne et al. 1992, 1996a, b; Raposo et al. 1998; Ernesto et al. 1999; Marzoli et al. 1999a, b; Mincato 2000; Thiede and Vasconcelos 2010; U–Pb ages—Janasi et al. 2011; Florisbal et al. 2014; Almeida et al. 2017); and (3) CAMP-Parnaíba ($^{40}Ar/^{39}Ar$ ages—Baksi and Archibald 1997; Marzoli et al. 1999a; Merle et al. 2011)

4 Geochronology

Diabases of the EW-RCM dikes and Sardinha sill magmatic events have been dated since the 90s using the K–Ar method (Bellieni et al. 1990, 1992; Fodor et al. 1990; Misusaki et al. 2002), while only a few $^{40}Ar/^{39}Ar$ ages were published for the sills (Baksi and Archibald 1997) and dikes (Smith et al. 2001; Ngonge et al. 2016a) (Fig. 4). The ages are mainly scattered between 160 and 110 Ma, although younger and older K–Ar ages are also reported. Paleomagnetic data obtained from diabases of the EW-RCM showed that the dikes were emplaced at different times and under reverse and normal geomagnetic polarities, but with magnetization directions acquired mainly in the Early Cretaceous (the sub-swarms I, II, IV and V of Bellieni et al. 1992). Conversely, another set of dikes gave paleopoles corresponding to the Jurassic (the sub-swarm II). The authors suggest that the different magnetic signatures of the EW-RCM dikes, would reflect rifting processes related to the opening of the Equatorial (Cretaceous) and Central (Jurassic) domains of the Atlantic Ocean. If the duration of the RCM magmatic event as revealed by the K–Ar ages is trustworthy, this would imply that the dike swarm was emplaced over a time span of at least 55 m.y. and, therefore, would represent one longer-lasting magmatic activity when compared to other Gondwanan LIPs (ca. 135 Ma Paraná-Etendeka and 200 Ma CAMP) in South America. This lifetime is approximately coeval with the volcanic and plutonic activity in the Benue Trough (Maluski et al. 1995; Coulon et al. 1996), which is thought to be the African failed rift of a triple junction linked to the Atlantic opening (e.g., Matos 1992, 2000).

Defining the timing and duration of LIPs requires a precise and accurate geochronological survey that certainly does not involve the use of the K–Ar system. This is mainly because of limitations in recognizing either Ar loss or excess radiogenic Ar, which can lead to age misinterpretations. Taking into account only the $^{40}Ar/^{39}Ar$ results available for both sills and dikes, the EQUAMP lifespan would be nearly coincident with the main peak of the Paraná-Etendeka (flood lavas and intrusives) magmatic activity, between 135 and 120 Ma. Nevertheless, the poor precision associated to the current dataset of ages makes still difficult to determine the time span of the EQUAMP. Hence, much remains to be done to constrain the temporal history of the EQUAMP LIP.

5 Petrological Aspects

The data available for the EW-RCM and the Sardinha magmatic events reveal that both are dominantly represented by fine- to medium-grained, tholeiitic basalts and basaltic andesites, while the RCM dikes also includes trachy-basalts and basaltic trachy-andesites (Bellieni et al. 1990, 1992; Fodor et al. 1990; Hollanda et al. 2006; Ngonge et al. 2016a) (Fig. 5a, b). Plagioclase, two pyroxenes (augite and pigeonite) and Fe–Ti oxides make up the main mineral assemblage, while olivine occurs as a main constituent in some dikes of the EW-RCM dikes or as very subordinate minerals in Sardinha sills.

The EQUAMP dikes are dominantly (~80% vol.) high-Ti tholeiites with $MgO < 5$ wt% and $TiO_2 \geq 1.5$ wt% ($Ti/Y \sim 360$), whereas low-Ti tholeiites ($TiO_2 \leq 1.5$ wt%; $Ti/Y \leq 360$) and a set of high-Ti olivine tholeiitic dikes are minor components (Fig. 5c, d). Tholeiites of intermediate and acid compositions are absent in the EW-RCM dikes. The Sardinha intrusives, in turn, have been described essentially as high-Ti tholeiites ($MgO < 5$ wt% and $TiO_2 \geq 2$ wt%; Bellieni et al. 1990; Fodor et al. 1990), while low-Ti tholeiitic diabases would represent a minor component. As aforementioned, it should be acknowledged the current uncertainty in distinguishing sills belonging to Sardinha and/or CAMP events at the east side of the Parnaíba Basin. In general, both RCM and Sardinha tholeiites are strongly affected by fractional crystallization.

The distinction in terms of enrichment in incompatible trace elements between the three tholeiite groups of the EW-RCM follows the Ti-based definition, i.e. the evolved high-Ti tholeiites have higher incompatible element (Rb, Ba, Th, U, LREE, Sr, Zr, Hf) contents relative to the low-Ti tholeiites, whereas the olivine tholeiites are lesser enriched (Ngonge et al. 2016a). A negative Nb–Ta anomaly is more prominent in the low-Ti tholeiites than in the evolved high-Ti tholeiites, contrasting with the OIB-type abundances found for the olivine tholeiites. Such contrasting trace element signatures are evidence for a non-parentage relationship between the olivine tholeiitic and the other magma types. In fact, the initial ($t \sim 130$ Ma) isotopic compositions of the olivine tholeiites reveal likely contribution of a FOZO-type asthenospheric component ($^{87}Sr/^{86}Sr = 0.70339$–$0.70373$, $^{143}Nd/^{144}Nd = 0.512518$–$0.512699$ and $^{206}Pb/^{204}Pb > 19.1$; Ngonge et al. 2016a), while the high- and low-Ti tholeiites have

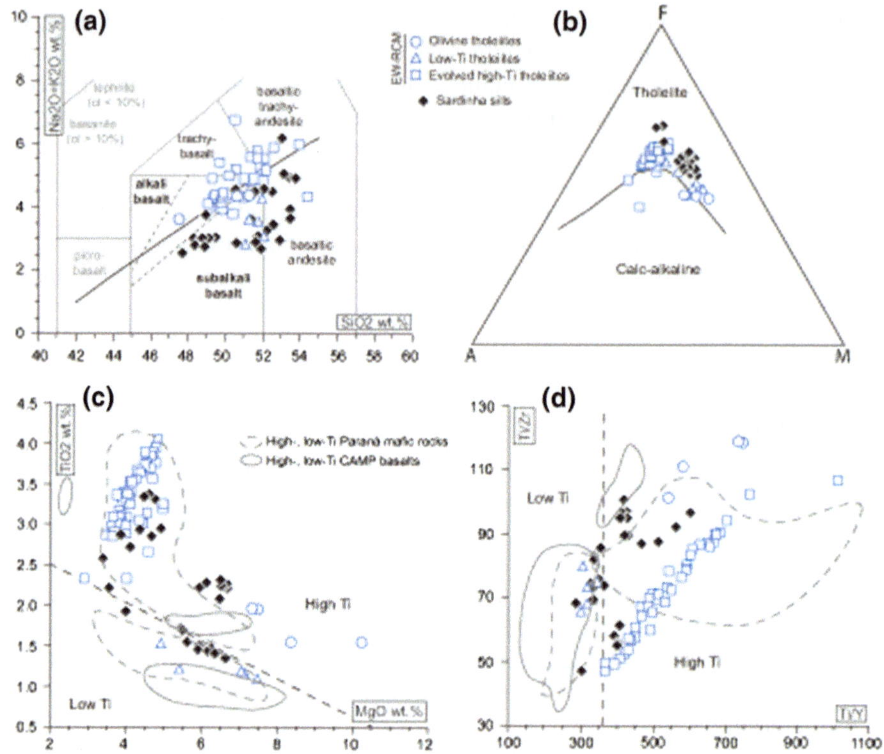

Fig. 5 Total oxide and element contents of the EQUAMP tholeiites from database of the literature (Bellieni et al. 1990, 1992; Fodor et al. 1990; Hollanda et al. 2006; Ngonge et al. 2016a). **a** Total alkali-silica (TAS) diagram after (Le Maitre 2002); **b** AFM (Alkalis–FeO–MgO) plot with the dividing line of Irvine and Baragar (1971); **c** TiO_2–MgO and **d** Ti/Y–Ti/Zr variation diagrams. Fields of Paraná-Etendeka (Peate 1997; Rocha-Júnior et al. 2013) and CAMP-Parnaíba (Bellieni et al. 1990; Fodor et al. 1990; Merle et al. 2011) are shown for comparison

variable Sr–Nd and relatively non-radiogenic $^{206}Pb/^{204}Pb$ initial ratios indicating that the primary melts of the RCM magmas were segregated from a heterogeneous EM1-type mantle (Hollanda et al. 2006; Ngonge et al. 2016a). The low-Ti magmas, however, display some degree of crustal contamination.

Although not studied to the same level of detail as the EW-RCM dikes, the geochemical signature of the Mg-evolved, Sardinha tholeiites is also featured by enrichment in incompatible elements and a negative Nb anomaly. From the current geochemical information and the (very limited) isotope dataset (i.e., radiogenic Sr, and non-radiogenic Nd and Pb compositions), the origin of the Sardinha parental magmas has been attributed to variable degrees of melting of an enriched and strongly heterogeneous, spinel-garnet peridotite source (Bellieni et al. 1990; Fodor et al. 1990; Silva et al. 2017).

6 Discussion

6.1 Dike Physical Parameters and AMS Data

The RCM swarm, before restricted to EW-trending dikes situated to the south of the Potiguar rift, has increased dramatically in length based on mapping from high-resolution aeromagnetic data collected for northeast Brazil. To the west, the swarm changes its direction to reach, to the south of Permanbuco Lineament, the border of the São Francisco craton (Fig. 1). According to aeromagnetic and remote sensing criteria, the arcuate RCM dikes can be traced for about 1,000 km in length, hence defining a giant dike swarm. The mafic dikes appear as isolated, en-echelon segments sometimes connected by steps in map view. Mean lengths of these segments reach 3.3 and 4.7 km in the EW- and NE-trending portions, whereas maximum widths range from 60 to 80 m, respectively. The thickness of the RCM mafic dikes are, therefore, higher than those in the Ethiopian swarm, which is emplaced in a rifting setting. We can speculate that the RCM dikes would be closer to their reservoirs compared to the Ethiopian dikes, although the relation between dike thickness and geological setting needs further investigation.

According to Archanjo et al. (2000, 2002b), AMS studies in the EW-RCM dikes indicated a possible feeder zone located in the east sector of the swarm. Such a feeder zone was characterized by vertical magnetic lineations found at the intersection of the dike swarm and the Cenozoic volcanic centers of the Macau magmatism (Fig. 1; Ngonge et al. 2016b). In contrast, AMS of the central and west sectors of the EW-trending swarm indicates lateral flow of magma that, in some cases, may include magnetic fabrics of tectonic origin, i.e. formed in the final stages of crystallization of the melt in a distensive stress field. Transected magnetic foliations recorded in some dikes are suggestive of a lateral (strike slip) displacement of the host-rock walls which, in turn, would indicate that the mode I-type fracture opening that assisted the emplacement of the dikes would include a component of shearing. These structural findings, observed just in the EW-trending portion of the swarm, have to be compared to AMS of the NE-RCM dikes in order to propose a consistent regional tectonic model for the crustal fracturing and intrusion of dikes. Moreover, a detailed AMS investigation is required in the connection of the RCM and Canindé swarms since they regionally form a radial arrangement whose focus appears to be centered next to the Cretaceous Potiguar rift (see Fig. 1). A modeled geoid anomaly in NE Brazil indicates a low-density (thermal) zone situated at between 17 and 78 km in depth (Ussami et al. 1993), which agrees with our estimated melting depths (between 60 and 75 km; Ngonge et al. 2016a) for the RCM tholeiites, i.e., in the garnet-spinel facies of the transition zone.

6.2 Preliminary Comparison with the ca. 135 Ma Paraná-Etendeka

Taking into account that a complete geochemical survey is still unavailable for the entire RCM dikes and Sardinha sills, a comparison between EQUAMP and the approximately coeval Paraná-Etendeka LIP can only be made at some extent. The Paraná-Etendeka LIP is located approximately 1,500 km to the south of the exposed area of the EQUAMP (Fig. 1, inset), being asymmetrically split into South America (southern Brazil) and Africa (Namibia and Angola). The volcanic rocks are mainly tholeiitic basalts and basaltic andesites, with subordinate occurrence of silicic rocks that span the dacite-trachyte-rhyolite compositional fields (e.g., Piccirillo and Melfi 1988; Peate et al. 1992; Ewart et al. 1998a, b, 2004a, b; Nardy et al. 2008). From the published ages, the Paraná-Etendeka volcanism began at ca. 135 Ma and lasted 133–132 Ma (see references in Fig. 4), although ages attributed to the related plumbing systems can progress up to 131 Ma (Raposo et al. 1998; Mincato 2000). Flood basalt lavas and associated intrusive rocks are geochemically classified into low-Ti and high-Ti types based on an arbitrary threshold of 2 wt% of TiO_2 content with corresponding Ti/Y values of ~330–350 (e.g., Peate et al. 1992). On both the Brazilian side, these two major chemical groups are spread into two areas referenced as southern (dominated by low-Ti lavas and some dike swarms) and northern (dominated by high-Ti lavas, sills and dikes) domains, whereas one central domain brings together both geochemical groups. Conversely, high- and low-Ti rocks of the EQUAMP are not geographically controlled and, at least from the current knowledge, the low-Ti types are very subordinate relative to those of high-Ti compositions, representing only 10% of volume of the EW-RCM dikes, while even more subordinate in the Sardinha sills.

The two igneous provinces are formed under nearly similar tectonic frameworks. Paraná-Etendeka lavas and Sardinha sills of EQUAMP were formed, respectively, by flooding and intruding Paleozoic sag-type basins, while the RCM and the subsidiary Canindé and Riacho do Cordeiro swarms of EQUAMP are intrusive into the Precambrian basement, as well as a number of dike swarms related to the Paraná-Etendeka LIP (e.g., Ponta Grossa, Florianópolis, São Paulo-Rio de Janeiro and their equivalents in the African side; e.g., Renne et al. 1996a; Deckart et al. 1998; Raposo et al. 1998; Marzoli et al. 1999b; Florisbal et al. 2014; Almeida et al. 2017; Raposo 2017). The spatial distribution of dike swarms in both these provinces suggests a triple junction configuration developed during the stepwise opening of the South Atlantic in the Cretaceous. The intimate association of triple junction geometry and with voluminous and short-lived magmatism as seen for the Paraná-Etendeka LIP is strongly suggestive of plume-related provinces, whether the plume had participated as an active source for melt generation or providing heat to trigger melting of the lithospheric mantle. Differences in Ti contents in the Paraná magmas are followed by distinctive major and trace element concentrations and isotope compositions, and they have been explained as a function of either heterogeneities in their lithospheric mantle sources (e.g., Milner et al. 1995; Peate 1997; Marzoli et al.

1999b; Rocha-Júnior et al. 2012, 2013) or plume (Tristan)-lithosphere interactions (e.g., Gibson et al. 1995; Ewart et al. 1998a, b; Thompson et al. 2001). No definitive OIB compositions has been reported for the flood lavas of the Paraná-Etendeka LIP, hence favoring the idea of Tristan plume was mainly a heat instead a melt source.

Chemical mantle heterogeneity is also an explanation for the coexistence of high- and low-Ti compositions in the RCM and Sardinha tholeiites of the EQUAMP (Fodor et al. 1990; Ngonge et al. 2016a). Based on the correlation between negative Nb-anomaly, incompatible element enrichment and isotope modeling, a subduction-modified subcontinental lithospheric mantle has been proposed as the more likely source for the Cretaceous magmatism in NE Brazil. As the ultimate pre-Andean subduction history in the South America dates back the Brasiliano/Pan-African orogeny, which was accounted for the convergence of major cratonic blocks and ocean consumption to assembly the West Gondwana at the Ediacaran-Cambrian. Subduction-related metasomatism might be certainly an ubiquitous signature preserved in the lithospheric mantle beneath the Precambrian terranes in South America. Melt production from potentially enriched lithospheric mantle reservoir will mainly depend on the rheology and fertility of the lithosphere, partial melting degree, and the thermal gradient of the heat source. Otherwise, evidence for OIB-type (FOZO-like) signature shown by the olivine tholeiitic dikes of the EW-RCM suggests that a deep mantle source has also contributed with melt provenance during the Equatorial Atlantic geodynamic evolution.

Given the current level of understanding about petrogenesis of the Gondwanan LIPs in South America, the 135 Ma Paraná-Etendeka, 200 Ma CAMP and even the 'newborn' ca. 130 Ma EQUAMP, much remains to be investigated concerning to the extent of partial melting, crust assimilation and fractional crystallization processes in governing the compositional variations observed in continental basalts from these provinces. As a direct consequence of obtaining such information, the contribution of lithosphere and convective (asthenosphere or plume) mantle reservoirs can be determined. Future investigations on the EQUAMP tholeiites will be focused on quantitative and numerical modeling based on a priori trace element and isotope data in order to define the nature and the thermomechanical conditions of the sub-continental mantle beneath NE Brazil during the breakup of the Gondwana along the Equatorial Atlantic margin.

7 Conclusions

We have reviewed the evidence from two magmatic suites in NE Brazil—mafic sills which are intrusive in the Parnaíba Basin and dike swarms located mostly in the Precambrian basement of the Borborema Province. We propose the dikes and sills comprise a new Cretaceous LIP in South America, which we are labeling as the EQUAMP (EQUatorial Atlantic Magmatic Province). The approximate ages of ca. 135–120 Ma suggest a close association with the opening of the Equatorial Atlantic, making the study of this LIP a thought-provoking subject in the context of supercontinent cycle. The relationship with the approximately coeval ca. 135 Ma Parana-Etendeka LIP centred about 1,500 km further south, also requires investigation.

Acknowledgements The authors Maria Helena B. M. Hollanda, Carlos J. Archanjo and David L. de Castro thanks to the Conselho Nacional de Pesquisa e Desenvolvimento (CNPq) for grants 304979/2016-3, 305824/2014-7 and 303015/2013-6, respectively. Antomat A. Macedo, Alanny C. C. Melo and Alisson L. Oliveira thanks to FAPESP and CAPES, respectively, for their PhD scholarships. REE was partially supported from Mega-Grant 14.Y26.31.0012 of the government of the Russian Federation. The authors are grateful to the referees (Dr. David Peate and Dr. Elson Oliveira) for the comments and corrections that helped to improve the final version of the manuscript. This paper is the first scientific contribution funded by FAPESP 2017/08423-9.

References

Almeida VV, Janasi VA, Heaman LM, Shaulis BJ, Hollanda MHBM, Renne PR (2017) Contemporaneous alkaline and tholeiitic magmatism in the Ponta Grossa Arch Paraná-Etendeka Magmatic Province: constraints from U-Pb zircon/baddeleyite and ^{40}Ar/^{39}Ar phlogopite dating of the José Fernandes Gabbro and mafic dikes. J Volcanol Geoth Res (accepted). https://doi.org/10.1016/j.volgeores.2017.01.018

Archanjo CJ, Launeau P (2004) Magma flow inferred from preferred orientations of plagioclase of the Rio Ceará-Mirim dike swarm (NE Brazil) and its AMS significance. In: Martin-Hernandez F, Lunemburg CM, Aubourg C, Jackson M (eds) Magnetic fabric: methods and applications. Geological Society, London, Special Publications, vol 238, pp 285–298. https://doi.org/10.1144/gsl.sp.2004.238.01.17

Archanjo CJ, Trindade RIF, Macedo JWP, Araújo MGS (2000) Magnetic fabric of a basaltic dike swarm associated with Mesozoic rifting in northeastern Brazil. J S Am Earth Sci 13:179–189. https://doi.org/10.1016/S0895-9811(00)00023-7

Archanjo CJ, Trindade RI, Bouchez JL, Ernesto M (2002a). Granite fabrics and regional-scale strain partitioning in the Seridó belt (Borborema Province, NE Brazil). Tectonics. https://doi.org/10.1029/2000tc001269

Archanjo CJ, Araújo MGS, Launeau P (2002b) Fabric of the Rio Ceará-Mirim mafic dike swarm (Northeastern Brazil) determined by anisotropy of magnetic susceptibility and image analysis. J Geophys Res 107(B3):1–13, 2046. https://doi.org/10.1029/2001jb000268

Archanjo CJ, Viegas LG, Hollanda MHBM, Souza LC, Liu D (2013) Timing of the HT/LP transpression in the Neoproterozoic Seridó belt (Borborema Province, Brazil): constraints from U-Pb (SHRIMP) geochronology and implications for the connections between NE Brazil and West Africa. Gondwana Res 23:701–714. https://doi.org/10.1016/j.gr.2012.05.005

Baksi AK, Archibald DA (1997) Mesozoic igneous activity in the Maranhão province, northern Brazil: ^{40}Ar/^{39}Ar evidence for separate episodes of basaltic magmatism. Earth Planet Sci Lett 151(3–4):139–153

Bellieni G, Piccirillo EM, Cavazzini G, Petrini R, Comin-Chiaramonti P, Nardy AJR, Civetta L, Melfi AJ, Zantedeschi P (1990) Low- and high-TiO_2 Mesozoic tholeiitic magmatism of the Maranhão basin (NE-Brasil): K-Ar age, geochemistry, petrology, isotope characteristics and relationships with Mesozoic low- and high-TiO_2 flood basalts of the Paraná basin (SE-Brazil). Neues Jahrbuch Miner Abh 162(1):1–33

Bellieni G, Macedo MHF, Petrini R, Picirillo EM, Cavazzini G, Comin-Chiaramonti P, Ernesto M, Macedo JWP, Martins G, Melfi AJ, Pacca IG, De Min J (1992) Evidence of magmatic activity related to Middle Jurassic to early cretaceous rifting from northeastern Brazil (Ceará-Mirim): K/Ar age, palaeomagnetism, petrology and Sr-Nd isotope characteristics. Chem Geol 97:9–32

Benkhelil J (1989) The origin and evolution of the cretaceous Benue trough (Nigeria). J Afr Earth Sci 8(2/3/4):251–282

Bertrand H, Fornari M, Marzoli A, García-Duarte R, Sempere T (2014) The Central Atlantic Magmatic Province extends into Bolivia. Lithos 188:33–43. https://doi.org/10.1016/j.lithos.2013.10.019

Brito Neves BB, Fuck RA, Pimentel MM (2014) The Brasiliano collage in South America: a review. Braz J Geol 44(3):493–518. https://doi.org/10.5327/Z2317-48892014000300010

Bryan S, Ernst RE (2008) Revised definition of Large Igneous Provinces (LIPs). Earth Sci Rev 86:175–202. https://doi.org/10.1016/j.earscirev.2007.08.008

Burgess SD, Bowring SA, Fleming TH, Elliot DH (2015) High-precision geochronology links the Ferrar large igneous province with early-Jurassic ocean anoxia and biotic crisis. Earth Planet Sci Lett 415:90–99. https://doi.org/10.1016/j.epsl.2015.01.037

Caby R (2003) Terrane assembly and geodynamic evolution of central-western Hoggar: a synthesis. J Afr Earth Sci 37:133–159. https://doi.org/10.1016/j.afrearsci.2003.05.003

Clemente CS, Amorós EB, Crespo MG (2007) Dike intrusion under stress: effects on magnetic and vesicle fabrics in dikes from rift zones of Tenerife (Canary Islands). J Struct Geol 29:1931–1942. https://doi.org/10.1016/j.jsg.2007.08.005

Correa-Gomes LC, Souza Filho CR, Martins CFJN, Oliveira EP (2001) Development of symmetrical and asymmetrical fabric in sheet-like igneous bodies: the role of magma flow and wall-rock displacements in theoretical and natural cases. J Struct Geol 23:1415–1428. https://doi.org/10.1016/S0191-8141(01)00007-4

Coulon C, Vidal P, Dupuy C, Baudin P, Popoff M, Maluski H, Hermitte D (1996) The Mesozoic to early Cenozoic magmatism of the Benue trough (Nigeria); geochemical evidence for the involvement of the St Helena plume. J Petrol 37:341–358. https://doi.org/10.1093/petrology/37.6.1341

Courtillot V, Davaille A, Besse J, Stock J (2003) Three distinct types of hotspots in the Earth's mantle. Earth Planet Sci Lett 205:295–308

Deckart K, Féraud G, Marques LS, Bertrand H (1998) New time constraints on dike swarms related to the Paraná-Etendeka magmatic province, and subsequent South Atlantic opening, southeastern Brazil. J Volcanol Geoth Res 80:67–83. https://doi.org/10.1016/S0377-0273(97)00038-3

de Castro DL, Oliveira DC, Hollanda MHBM (2018) Geostatistical interplay between geophysical and geochemical data: mapping litho-structural assemblages of Mesozoic igneous activities in the Parnaíba Basin (NE Brazil). Surv Geophys 39:683–713. https://doi.org/10.1007/s10712-018-9463-5

Delaney PT, Pollard DD (1981) Deformation of host rocks and flow of magma during growth of minette dikes and breccia-bearing intrusions near Ship Rock, New Mexico. US Geol Surv Prof Pap 1202:61p

Ernesto M, Raposo MIB, Marques LS, Renne PR, Diogo LA, de Min A (1999) Paleomagnetism, geochemistry and $^{40}Ar/^{39}Ar$ dating of the northeastern Paraná Magmatic Province: tectonic implications. J Geodyn 28:321–340. https://doi.org/10.1016/S0264-3707(99)00013-7

Ernst RE (2014) Large Igneous Province. Cambridge Un Press, United Kingdom, 653 pp

Ernst RE, Buchan KL (1997) Giant radiating dike swarms: their use in identifying pre-Mesozoic large igneous provinces and mantle plumes. AGU Geophys Monogr 100:297–333

Ewart A, Milner SC, Armstrong RA, Duncan AR (1998a) Etendeka volcanism of the Goboboseb Mountains and Messum Igneous complex, Namibia. Part I: Geochemical evidence of Early Cretaceous Tristan plume melts and the role of crustal contamination in the Parana-Etendeka CFB. J Petrol 39:191–225. https://doi.org/10.1093/petroj/39.2.191

Ewart A, Milner SC, Armstrong RA, Duncan AR (1998b) Etendeka volcanism of the Goboboseb Mountains and Messum Igneous Complex, Namibia. Part II: voluminous quartz latite volcanism of the Awahab magma system. J Petrol 39:227–253. https://doi.org/10.1093/petroj/39.2.227

Ewart A, Marsh JS, Milner SC, Duncan AR, Kamber BS, Armstrong RA (2004a) Petrology and geochemistry of Early Cretaceous bimodal continental flood volcanism of the NW Etendeka, Namibia. Part 1: introduction, mafic Lavas and re-evaluation of mantle source components. J Petrol 45:59–105. https://doi.org/10.1093/petrology/egg083

Ewart A, Marsh JS, Milner SC, Duncan AR, Kamber BS, Armstrong RA (2004b) Petrology and geochemistry of Early Cretaceous bimodal continental flood volcanism of the NW Etendeka, Namibia. Part 2: Characteristics and petrogenesis of the high-Ti latite and high-Ti and low-Ti voluminous quartz latite eruptives. J Petrol 44:107–138. https://doi.org/10.1093/petrology/egg082

Féménias O, Diot H, Berza T, Gauffriau A, Demaiffe D (2004) Asymmetrical to symmetrical magnetic fabric of dikes: paleo-flow orientations and paleo-stresses recorded on feeder-bodies from the Motru Dike Swarm (Romania). J Struct Geol 26:1401–1418. https://doi.org/10.1016/j.jsg.2003.12.003

Florisbal LM, Heaman LM, Janasi VA, Bitencourt MF (2014) Tectonic significance of the Florianópolis dike swarm, Paraná-Etendeka Magmatic Province: a reappraisal based on precise U-Pb dating. J Volcanol Geoth Res 289(2014):140–150. https://doi.org/10.1016/j.jvolgeores.2014.11.007

Françolin JBL, Cobbold PR, Szatmari P (1994) Faulting in the early cretaceous Rio do Peixe basin (NE Brazil) and its significance for the opening of the Atlantic. J Struct Geol 16(5):647–661. https://doi.org/10.1016/0191-8141(94)90116-3

Frimmel HE, Frank W (1998) Neoproterozoic tectono-thermal evolution of the Gariep Belt and its basement, Namibia and South Africa. Precambr Res 90:1–28. https://doi.org/10.1016/S0301-9268(98)00029-1

Fahrig WF (1987) The tectonic setting of continental mafic dike swarms: failed arm and early passive margin. In Halls HC, Fahrig WF (eds) Mafic dike swarms. Geological Association of Canada, Special Publication, St John's, NL, vol 34, pp 331–348

Fodor RV, Sial AN, Mukasa SB, McKee EH (1990) Petrology, isotope characteristics, and K-Ar ages of the Maranhão, northern Brazil, Mesozoic basalt province. Contrib Miner Petrol 104:555–567

Gibson SA, Thompson RN, Dickin AP, Leonardos OH (1995) High-Ti and low-Ti mafic potassic magmas: key to plume-lithospheric interactions and continental flood-basalt genesis. Earth Planet Sci Lett 136:149–165. https://doi.org/10.1016/0012-821X(95)00179-G

Góes AMO, Feijó FJ (1984) Bacia do Parnaíba. Boletim de Geociências da Petrobras 8(3):57–67

Gray DR, Foster DA, Meert JG, Goscombe BD, Armstrong R, Trouw RAJ, Passchier CW (2008) A Damara orogen perspective on the assembly of southwestern Gondwana. In: Pankhurst RJ, Trouw RAJ, Brito Neves BB, De Wit MJ (eds), West Gondwana: pre-cenozoic correlations across the South Atlantic Region. Geological Society of London, Special Publications, vol 294, pp 399–412. https://doi.org/10.1144/sp294.5

Knight MD, Walker GPL (1988) Magma flow directions in dikes of the Koolau Complex, Oahu, determined from magnetic fabric studies. J Geophys Res 93:4301–4319. https://doi.org/10.1029/JB093iB05p04301

Halls HC (1982) The importance and potential of mafic dike swarms in studies of geodynamic process. Geosci Can 9:145–154

Hollanda MHBM, Pimentel MM, Oliveira DC, Jardim de Sá EF (2006) Lithosphere-asthenosphere interaction and the origin of cretaceous tholeiitic magmatism in Northeastern Brazil: Sr–Nd–Pb isotopic evidence. Lithos 86:34–49. https://doi.org/10.1016/jlithos.2005.04.004

Hollanda MHBM, Góes AN, Negri FA (2018) Sedimentary provenance of sandstones of the Parnaíba Basin through detrital zircon geochronology. In: Daly MC, Fuck RA, Julià J, Watts AB, Andrade V (eds) Cratonic basin formation: a case of the Parnaíba Basin of Brazil. Geological Society, London, Special Publications, 472. https://doi.org/10.1144/SP472.16
Hooper P, Widdowson M, Kelley S (2010) Tectonic setting and timing of the final Deccan flood basalt eruptions. Geology 38(9):839–842. https://doi.org/10.1130/G31072.1
Irvine TN, Baragar WRA (1971) A guide to the chemical classification of the common volcanic rocks. Can J Earth Sci 8:523–548
Janasi VA, Freitas VA, Heaman LH (2011) The onset of flood basalt volcanism, Northern Paraná Basin, Brazil: a precise U-Pb baddeleyite/zircon age for a Chapecó-type dacite. Earth Planet Sci Lett 302(1–2):147–153. https://doi.org/10.1016/j.epsl.2010.12.005
Knight KB, Nomade S, Renne PR, Marzoli A, Bertrand H, Youbi N (2004) The Central Atlantic Magmatic Province at the Triassic-Jurassic boundary: paleomagnetic and $^{40}Ar/^{39}Ar$ evidence from Morocco for brief, episodic volcanism. Earth Planet Sci Lett 228:143–160. https://doi.org/10.1016/j.epsl.2004.09.022
Le Maitre RW (2002) Igneous rocks—a classification and glossary of terms. In: Recommendations of the IUGS subcommission on the systematics of igneous rocks, 2nd edn. Cambridge University Press, Cambridge
Magee C, Muirhead JD, Karvelas A, Holford SP, Jackson CAL, Bastow ID, Schoeld N, Stevenson CTE, McLean C, McCarthy W, Shtukert O (2016) Lateral magma flow in mafic sill complexes. Geosphere 12(3):809–841. https://doi.org/10.1130/GES01256.1
Maluski H, Coulon C, Popoff M, Baudin P (1995) $^{40}Ar/^{39}Ar$ chronology, petrology and geodynamic setting of Mesozoic to early Cenozoic magmatism from the Benue trough, Nigeria. J Geol Soc Lond 152:311–326. https://doi.org/10.1144/gsjgs.152.2.0311
Marzoli A, Renne PR, Piccirillo EM, Ernesto M, Bellieni G, De Min A (1999a) Extensive 200-million-year-old continental flood basalts of the Central Atlantic Magmatic Province. Science 284:616–618. https://doi.org/10.1126/science.284.5414.616
Marzoli A, Melluso L, Morra V, Renne PR, Sgrosso I, D'Antonio M, Duarte Morais L, Morais EAA, Ricci G (1999b) Geochronology and petrology of Cretaceous basaltic magmatism in the Kwanza basin (western Angola), and relationships with the Paraná-Etendeka continental flood basalt province. J Geodyn 28:341–356. https://doi.org/10.1016/s0264-3707(99)00014-9
Matos RMD (1992) The northeast Brazilian rift system. Tectonics 11(4):766–791. https://doi.org/10.1029/91TC03092
Matos RMD (2000) Tectonic evolution of the Equatorial South Atlantic. In: Webster M, Talwani M (eds) Atlantic rifts and continental margins, American Geophysical Union, vol 115, pp 331–354
Meade FC, Chew DM, Troll VR, Ellam RM, Page L (2009) Magma ascent along a major terrane boundary: crustal contamination and magma mixing at the Drumadoon intrusive complex, Isle of Arran, Scotland. J Petrol 50:2345–2374. https://doi.org/10.1093/petrology/egp081
Merle R, Marzoli A, Bertrand H, Reisberg L, Verati C, Zimmermann C, Chiaradia M, Belienni G, Ernesto M (2011) $^{40}Ar/^{39}Ar$ ages and Sr-Nd-Pb-Os geochemistry of CAMP tholeiites from Western Maranhão basin (NE Brazil). Lithos 122:137–151. https://doi.org/10.1016/j.lithos.2010.12.010
Milner SC, Duncan AR, Whittingham AM, Ewart A (1995) Trans-atlantic correlation of eruptive sequences and individual silicic volcanic units within the Paraná-Etendeka igneous province. J Volcanol Geoth Res 69:137–157. https://doi.org/10.1016/0377-0273(95)00040-2
Mincato RL (2000) Metalogenia dos elementos do grupo da Platina com base na estratigrafia e geoquímica da Província Ígnea Continental do Paraná. Ph.D. thesis, Universidade de Campinas, São Paulo, 172 pp

Misusaki AMP, Thomaz-Filho A, Milani EJ, Césero P (2002) Mesozoic and Cenozoic igneous activity and its tectonic control in Northeastern Brazil. J S Am Earth Sci 15:183–198. https://doi.org/10.1016/S0895-9811(02)00014-7

Mocitaiba LSR, de Castro DL, Oliveira DC (2017) Cartografia geofísica regional do magmatismo mesozoico na Bacia do Parnaíba. Geologia USP Série Científica 17(2):169–192. https://doi.org/10.11606/issn.2316-9095.v17-455

Muirhead JD, Airoldi G, Rowland JV, White JDL (2012) Interconnected sills and inclined sheet intrusions control shallow magma transport in the Ferrar large igneous province, Antarctica. Geol Soc Am Bull 124(1/2):162–180. https://doi.org/10.1130/B30455.1

Nardy AJR, MAchado FB, Oliveira MAF (2008) As rochas vulcânicas mesozóicas ácidas da Bacia do Paraná: litoestratigrafia e considerações geoquímico-estratigráficas. Revista Brasileira de Geociências 38(1):178–195

Ngonge ED, Hollanda MHBM, Archanjo CJ, Oliveira DC, Vasconcelos PMP, Muñoz PRM (2016a) Petrology of continental tholeiitic magmas forming a 350-km-long Mesozoic dike swarm in NE Brazil: constraints of geochemical and isotopic data. Lithos 258–259:228–252. https://doi.org/10.1016/j.lithos.2016.04.008

Ngonge ED, Hollanda MHBM, Pimentel MM, Oliveira DC (2016b) Petrology of the alkaline rocks of the Macau Volcanic Field, NE Brazil. Lithos 206–207:453–470. https://doi.org/10.1016/j.lithos.2016.10.008

Peate DW (1997) The Paraná-Etendeka province. In: Large igneous provinces: continental, oceanic and planetary flood volcanism. AGU Geophysical Monographs, vol 100, pp 217–245

Peate DW, Hawkesworth CJ, Mantovani MSM (1992) Chemical stratigraphy of the Paraná lavas (South America): classification of magma types and their spatial distribution. Bull Volc 55:119–139

Piccirillo EM, Melfi AJ (1988) The Mesozoic flood volcanism of the Paraná Basin: petrogenetic and geophysical aspects. Universidade de São Paulo, São Paulo, 600 pp

Pollard DD, Müller OH, Dockstader DR (1975) The form and growth of fingered sheet intrusions. Geol Soc Am Bull 86(3):351–363

Polteau S, Mazzini A, Galland O, Planke S, Malthe-Sørenssen A (2008) Saucer-shaped intrusions: occurrences, emplacement and implications. Earth Planet Sci Lett 266:195–204. https://doi.org/10.1016/j.epsl.2007.11.015

Raposo MIB (2017) Magnetic fabrics of the Cretaceous dike swarms from São Paulo coastline (SE Brazil): its relationship with South Atlantic Ocean opening. Tectonophysics. https://doi.org/10.1016/j.tecto.2017.10.023

Raposo MIB, Ernesto M, Renne PR (1998) Paleomagnetism and dating of the early Cretaceous Florianópolis dike swarm (Santa Catarina Island), Southern Brazil. Phys Earth Planet Inter 108:275–290. https://doi.org/10.1016/S0031-9201(98)00102-2

Renne PR, Ernesto M, Pacca IG, Coe RS, Glen JM, Prevot M, Perrin M (1992) The age of Paraná flood volcanism, rifting of Gondwanaland, and the Jurassic-Cretaceous boundary. Science 258:975–979. https://doi.org/10.1126/science.258.5084.975

Renne PR, Deckart K, Ernesto M, Féraud G, Piccirillo EM (1996a) Age of the Ponta Grossa dike swarm (Brazil), and implications to Paraná flood volcanism. Earth Planet Sci Lett 144:199–211. https://doi.org/10.1016/0012-821X(96)00155-0

Renne PR, Glen JM, Milner SC, Duncan AR (1996b) Age of Etendeka flood volcanism and associated intrusions in southwestern Africa. Geology 24:659–662. https://doi.org/10.1130/0091-7613(1996)024.00659

Renne PR, Sprain CJ, Richards MA, Self S, Vanderkluysen L, Pande K (2015) State shift in Deccan volcanism at the Cretaceous-Paleogene boundary, possibly induced by impact. Science 350(62556):76–78. https://doi.org/10.1126/science.aac7549

Riisager P, Knight KB, Baker JA, Ukstins Peate I, Al-Kadasi M, Al-Subbary A, Renne PR (2005) Paleomagnetism and $^{40}Ar/^{39}Ar$ geochronology of Yemeni Oligocene volcanics: implications for timing and duration of Afro-Arabian traps and geometry of the Oligocene paleomagnetic field. Earth Planetary Sci Lett 237:647–672. https://doi.org/10.1016/j.epsl.2005.06.016

Rocha-Júnior ERV, Putchel IS, Marques LS, Walker RJ, Machado FB, Nardy AJR, Babinski M, Figueiredo AMG (2012) Re-Os isotope and highly siderophile element systematics of the Paraná continental flood basalts (Brazil). Earth Planet Sci Lett 337–338:164–173. https://doi.org/10.1016/j.epsl.2012.04.050

Rocha-Júnior ERV, Marques LS, Babinski M, Nardy AJR, Figueiredo AMG, Machado FB (2013) Sr–Nd–Pb isotopic constraints on the nature of the mantle sources involved in the genesis of the high-Ti tholeiites from northern Paraná Continental Flood Basalts (Brazil). J S Am Earth Sci 46:9–25. https://doi.org/10.1016/j.jsames.2013.04.004

Santos EJ, Van Schmus WR, Kozuch M, Brito Neves BB (2010) The Cariris Velhos tectonic event in Northeast Brazil. J S Am Earth Sci 29:61–76. https://doi.org/10.1016/j.jsames.2009.07.003

Santos EJ, Souza Neto JA, Silva MRR, Beurlen H, Cavalcanti JAD, da Silva MG, Dias VM, Costa AF, Santos LCML, Santos RB (2014) Metalogênese das porções norte e central da Província Borborema. In: Silva MG, Rocha Neto MB, Jost H, Kuyumijan RM (eds) Metalogênese das Províncias Tectônicas Brasileiras, pp 343–388

Sénant J, Popoff M (1991) Early cretaceous extension in Northeast Brazil related to the South Atlantic opening. Tectonophysics 198:35–46. https://doi.org/10.1016/0040-1951(91)90129-G

Schultz RA, Mège D, Diot H (2008) Emplacement conditions of igneous dikes in Ethiopian traps. J Volcanol Geoth Res 178:683–692. https://doi.org/10.1016/j.jvolgeores.2008.08.012

Sheth HC (2007) 'Large Igneous Provinces (LIPs)': definition, recommended terminology, and a hierarchical classification. Earth Sci Rev 85:117–124. https://doi.org/10.1016/j.earscirev.2007.07.005

Silva AG, Almeida CN, Valente SC, Almeida LFB (2017) The petrogenesis of tholeiitic diabases in eastern Parnaíba Basin: evidence for geochemical heterogeneities in the subcontinental lithospheric mantle in NE Brazil. Braz J Geol 47(1):109–126. https://doi.org/10.1590/2317-4889201720160041

Smith PE, Evensen NM, York D, Szatmari P, Oliveira DC (2001) Single crystal ^{40}Ar-^{39}Ar dating of pyrite: no fool's clock. Geology 29:403–406. https://doi.org/10.1130/0091-7613(2001)029

Svensen H, Corfu F, Polteau S, Hammer Ø, Planke S (2012) Rapid magma emplacement in the Karoo Large Igneous Province. Earth Planet Sci Lett 325–326:1–9. https://doi.org/10.1016/j.epsl.2012.01.015

Thiede D, Vasconcelos PM (2010) Paraná flood basalts: rapid extrusion hypothesis confirmed by new $^{40}Ar/^{39}Ar$ results. Geology 38(8):747–750. https://doi.org/10.1130/G30919.1

Thompson RN, Gibson SA, Dickin AP, Smith PM (2001) Early cretaceous basalt and picrite dikes of the Southern Etendeka region, NW Namibia: windows into the role of the Tristan mantle plume in Paraná-Etendeka magmatism. J Petrol 42:2049–2081. https://doi.org/10.1093/petrology/42.11.2049

Thomson K, Hutton D (2004) Geometry and growth of sill complexes: insights using 3D seismic from the North Rockall Trough. Bull Volc 66:364–375. https://doi.org/10.1007/s00445-003-0320-z

Ukstins Peate I, Baker JA, Al-Kadasi M, Al-Subbary A, Knight KB, Riisager P, Thirlwall MF, Peate DW, Renne PR, Menzies MA (2005) Volcanic stratigraphy of large-volume silicic pyroclastic eruptions during Oligocene Afro-Arabian flood volcanism in Yemen. Bull Volc 68:135–156. https://doi.org/10.1007/s00445-005-0428-4

Ussami N, Sá NC, Molina EC (1993) Gravity Map of Brazil II: regional and residual gravity anomalies and their a correlation with major tectonic provinces. J Geophys Res 98:2199–2208

Van Schmus WR, Oliveira EP, Silva Filho AF, Toteu SF, Penaye J, Guimarães IP (2008) Proterozoic links between the Borborema Province, NE Brazil, and the Central African Fold Belt. In: Pankhurst RJ, Trouw RAJ, Brito Neves BB, De Wit MJ (eds) West Gondwana: pre-cenozoic correlations across the South Atlantic Region. Geological Society of London, Special Publications, vol 294, pp 69–99. https://doi.org/10.1144/sp294.5

Vauchez A, Neves S, Caby R, Corsini M, Egydio-Silva ME, Arthaud M, Amaro V (1995) The Borborema Province shear zone system. J S Am Earth Sci 8(3/4):247–266. https://doi.org/10.1016/0895-9811(95)00012-5

Vaz PT, Rezende NGAM, Wanderley Filho JR, Travassos WAS (2007) Bacia do Parnaíba. Boletim de Geociências DA Petrobras 15(2):253–263

Verati C, Bertrand H, Féraud G (2005) The farthest record of the Central Atlantic Magmatic Province into West Africa craton: Precise $^{40}Ar/^{39}Ar$ dating and geochemistry of Taoudenni basin intrusives (northern Mali). Earth Planet Sci Lett 235:391–407. https://doi.org/10.1016/j.epsl.2005.04.012

Verati C, Rapaille C, Féraud G, Marzoli A, Bertrand H, Youbi N (2007) $^{40}Ar/^{39}Ar$ ages and duration of the Central Atlantic Magmatic Province volcanism in Morocco and Portugal and its relation to the Triassic-Jurassic boundary. Palaeogeogr Palaeoclimatol Palaeoecol 244(2007):308–325. https://doi.org/10.1016/j.palaeo.2006.06.033

Intraplate Proterozoic Magmatism in the Amazonian Craton Reviewed: Geochronology, Crustal Tectonics and Global Barcode Matches

W. Teixeira, N. J. Reis, J. S. Bettencourt, E. L. Klein and D. C. Oliveira

Abstract We review geochronological data including U-Pb baddelyite ages of Proterozoic mafic dyke swarms and sills of the Amazonian Craton, as well as their geochemical character and geological settings, in order to arrive at an integrated tectonic interpretation. The information together with the characteristics of coeval volcanic-plutonic suites indicates a cyclicity of the mafic-felsic activity through time and space. At least four LIP/SLIP events are apparent, and each one appears to accompany the stepwise accretionary crustal growth of Amazonia. The oldest two, the Orocaima (1.98–1.96 Ga) and Uatumã (c. 1.89–1.87 Ga) SLIPs, comprise calc-alkaline I-type and subordinate A-type plutonic and volcanic rocks. Synchronous mafic intraplate activity occurs across the Guiana and Central-Brazil Shields. These two events may be caused by interaction between subduction-related processes and mantle plumes with synchronous lithosphere extension during the two time periods. The Avanavero (1.79 Ga) LIP event mostly consists of mafic dykes and sills which are intrusive into the Roraima platform cover, in the Guiana Shield. They show tholeiitic chemistry and similarities with E-MORB and subcontinental lithospheric mantle-derived basalts, whereas the REE pattern suggests affinity with intraplate settings. The age of the Avanavero rocks is identical to the Crepori Diabase, located ca. 1800 km away to the south (Central-Brazil Shield). The youngest LIP event (1.11 Ga), the Rincón del Tigre-Huanchaca, has the Rio Perdido Suite as a component in the Rio Apa Terrane, which is ca. 300 km away from the Rincón del Tigre Complex, located in the SW portion of the Amazonian Craton. Furthermore, the Central-Brazil and Guiana Shields boasts widespread intraplate mafic activity, highlighted by the Mata-Matá (1.57 Ga),

W. Teixeira (✉) · J. S. Bettencourt
Instituto de Geociências, Universidade de São Paulo, São Paulo, SP, Brazil
e-mail: wteixeir@usp.br

N. J. Reis
CPRM – Serviço Geológico do Brasil, Manaus, AM, Brazil

E. L. Klein
CPRM – Serviço Geológico do Brasil, Brasília, DF, Brazil

D. C. Oliveira
Instituto de Geociências, Universidade Federal do Pará, Belém, PA, Brazil

R. K. Srivastava et al. (eds.), *Dyke Swarms of the World: A Modern Perspective*, Springer Geology, https://doi.org/10.1007/978-981-13-1666-1_4

Salto do Céu (1.44 Ga) and Nova Floresta (1.22 Ga) mafic sills and the Cachoeira Seca Troctolite (1.19 Ga). Contemporaneous A-type, rapakivi granites with roughly similar ages also occur elsewhere. These particular episodes are extension specific steps of the Mesoproterozoic Amazonia, and the quite large distribution is consistent with LIP events. In a broader perspective, the intermittent Proterozoic intracratonic activity has a barcode that matches LIP/SLIP events in Columbia and Rodinia.

Keywords Amazonian craton · Proterozoic dykes and sills
Large Igneous Provinces · Intraplate magmatism · U-Pb geochronology

1 Introduction

As widely known, mafic dyke swarms are one of the key tectonic elements for understanding the geodynamic evolution of the Earth in the Precambrian time, including timing of assembly and breakup of supercontinents. Dyke swarms are particularly important in paleogeographic reconstructions and coupled paleomagnetic poles. Also, potential barcode matches may be found with LIP/SLIP events and/or mantle plume events on other continents through time (e.g., Halls 1982; Fahrig 1987; Halls and Fahrig 1987; Bleeker and Ernst 2006; Bryan and Ernst 2008a, b; Ernst et al. 2013; Ernst 2014; Peng 2015; Shumlyanskyy et al. 2016a, b; Antonio et al. 2017). In this regard, Amazonia/West Africa discloses coherent connections with Baltica and eastern Laurentia through the Paleo- and Mesoproterozoic time frame, supported by the paleomagnetic evidences, geologic correlation of the accretionary belts along the active margins, as well periods of intraplate magmatism (e.g., Rämö and Haapala 1995; Bettencourt et al. 1999; Åhäll et al. 2000; Dall'Agnol et al. 2012; Gower and Krogh 2002; Högdahl et al. 2004; Tohver et al. 2006; Cordani et al. 2010a; Cordani and Teixeira 2007; Johansson 2009, 2014; Bogdanova et al. 2013; Bispo-Santos et al. 2014a, b; D'Agrella-Filho et al. 2012, 2016; Teixeira et al. 2016).

Precambrian mafic dyke swarms and sills (and coeval plutons) like in most cratons are common in the Amazonian Craton, where they highlight a recurrent activity in association with alkaline complexes and A-type plutonic and volcanic rocks (e.g., Santos et al. 2002a; Klein et al. 2012; Teixeira et al. 2015). Furthermore, the Amazonian Craton hosts scattered Mesozoic mafic dykes, such as the c. 200 Ma Cassiporé dyke swarm which aligns with the CAMP Large Igneous Province (LIP) event (e.g., Priem et al. 1968; Amaral 1974; Sial et al. 1987; Gibbs 1987; Teixeira et al. 2018), and this not focused here.

Since the studies in the 1960s, achievements have been made on the geochronology, geochemistry and petrology of dykes and sills over the Amazonian Craton, such as synthetized by Teixeira (1978, 1990), Gibbs (1987), Sial et al. (1987), Santos et al. (2002a), among others. These studies have aimed to characterize the main geologic features of such igneous episodes and their tectonic significance. Further paleomagnetic studies have provided insights on the paleogeographic context of the Amazonian Craton (see for review D'Agrella-Filho et al. 2016). For most of the mafic dyke swarms, only K-Ar, $^{40}Ar/^{39}Ar$ and Rb-Sr ages have been used as estimates of crystallization ages (e.g., Snelling and McConnell 1969; Hebeda et al. 1973; Amaral 1974; Teixeira 1990). More recently an increasing number of U-Pb baddeleyite ages of dykes and sills and associated rocks have defined prominent magmatic episodes in time and space (e.g., Vasquez et al. 2002; Santos et al. 2002a; Klein et al. 2012; Reis et al. 2013a; Silva et al. 2016; Teixeira et al. 2015, 2016; Antonio et al. 2017).

This review deals with the Paleo- to Mesoproterozoic dyke swarms and associated rocks in the Amazonian Craton, whose ages, geochemical characteristics and geology are reassessed for an integrated tectonic interpretation. The roughly coeval plutonic-volcanic events (A-type and calc-alkaline affinity) associated with extensional settings through time are also considered. In addition, we emphasize the importance of U–Pb baddeleyite geochronology in mafic dykes for constraining the timing of recurrent intracratonic mafic activity and to understand its overall tectonic significance. To complete this review we compiled more than 300 published U-Pb ages (in a GIS database) of Precambrian dykes, sills, layered mafic-ultramafic intrusions and associated silicic rocks (e.g., A-type rapakivi complexes) to identify the geologic correlations (Teixeira and Albrez 2017).

2 The Amazonian Craton

The Amazonian Craton (Fig. 1) is a result of a long-lived crustal growth by addition of continental and/or oceanic arcs in Paleo- to Mesoproterozoic times to one or two Archean nuclei in the northeast. The accretionary regime possibly involved intervening older continental blocks such as the Imataca Block which accreted to the northern portion of the Craton (Guiana Shield) at the time of the Transamazonian orogeny (2.25–1.95 Ga), whereas the late Paleoproterozoic Paraguá and Rio Apa Terranes subsequently accreted to the SW portion of Amazonia around 1.35 and 1.00 Ga, respectively (Teixeira et al. 1989, 2010; Santos et al. 2000; Tassinari and Macambira 2004; Boger et al. 2005; Cordani and Teixeira 2007; Cordani et al. 2010b). According to Cordani and Teixeira (2007) the tectonic framework comprises the Central Amazonian (>2.6 Ga) province and the Maroni-Itacaiunas (2.45–1.95 Ga) province in the northeast, as well as four adjoining provinces to the SW, namely Ventuari-Tapajós (2.01–1.80 Ga), Rio Negro-Juruena (1.82–1.60 Ga), Rondonian-San Ignacio (1.59–1.30 Ga) and Sunsas-Aguapeí (1.20–0.95 Ga) (Fig. 1). Whereas this early model is roughly followed in the review, we take into account recent geophysical information, systematic U-Pb dating and geochemical and isotopic constraints that have shed more light on the internal organization of some provinces and the nature of particular tectonic-magmatic events (e.g., Bettencourt et al. 2010; Fernandes et al. 2011; Santos et al. 2004; Rizzotto et al. 2013, 2014; Kroonenberg et al. 2016; Scandolara et al. 2017).

The Amazonian Craton is the host of four LIP-scale magmatic events (see Fig. 1), discriminated by the Orocaima (1.98–1.96 Ga), Uatumã (1.89–1.87 Ga), Avanavero (1.79 Ga) and Rincón del Tigre (1.11 Ga) events, among other intraplate activity through time and space—to be dealt with here. In particular, the two oldest events have

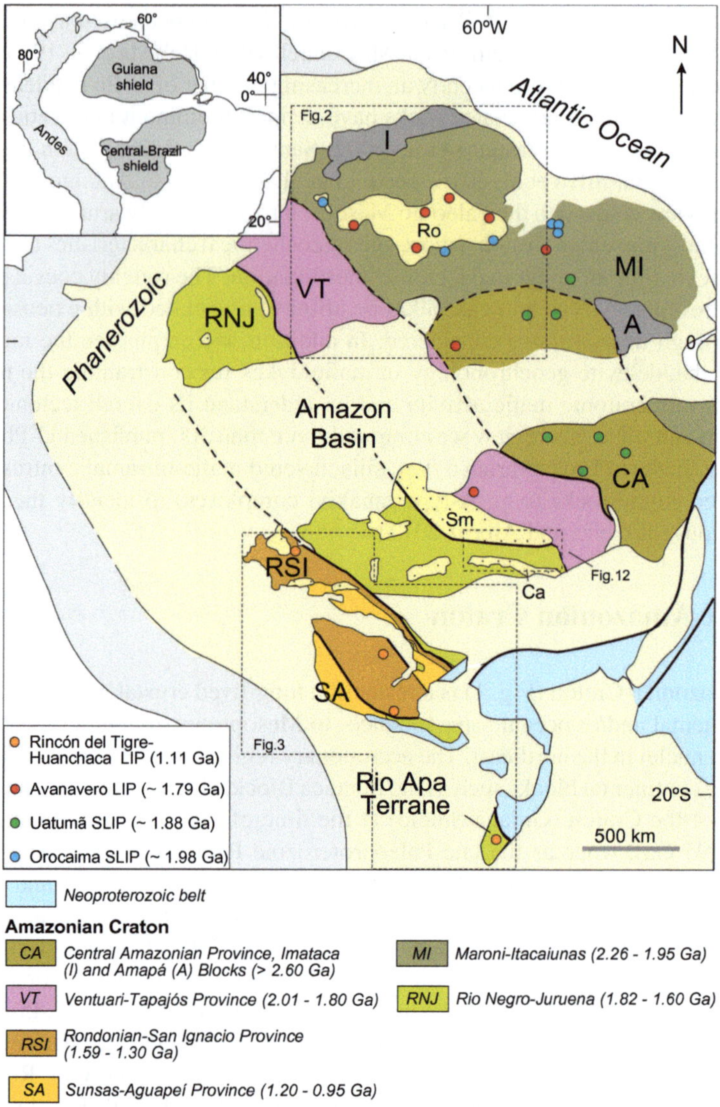

Fig. 1 Tectonic framework of the Amazonian Craton (adapted from Cordani and Teixeira 2007; Kroonenberg et al. 2016; Fraga et al. 2017). Proterozoic sedimentary covers (yellow doted domains): Ro (Roraima Supergroup), Sm (Sumaúma Supergroup; Cachimbo Graben), Cb (Caiabis Group; Caibis Graben). The distinct Proterozoic LIP/SLIP components (colourful dots) are also shown. See text for discussion

been termed with different names based on regional correlations with similar-looking volcanic-plutonic rocks. We summarize below the pertinent tectonic-geologic aspects of the Amazonian Craton that have important bearing for the scope of our review.

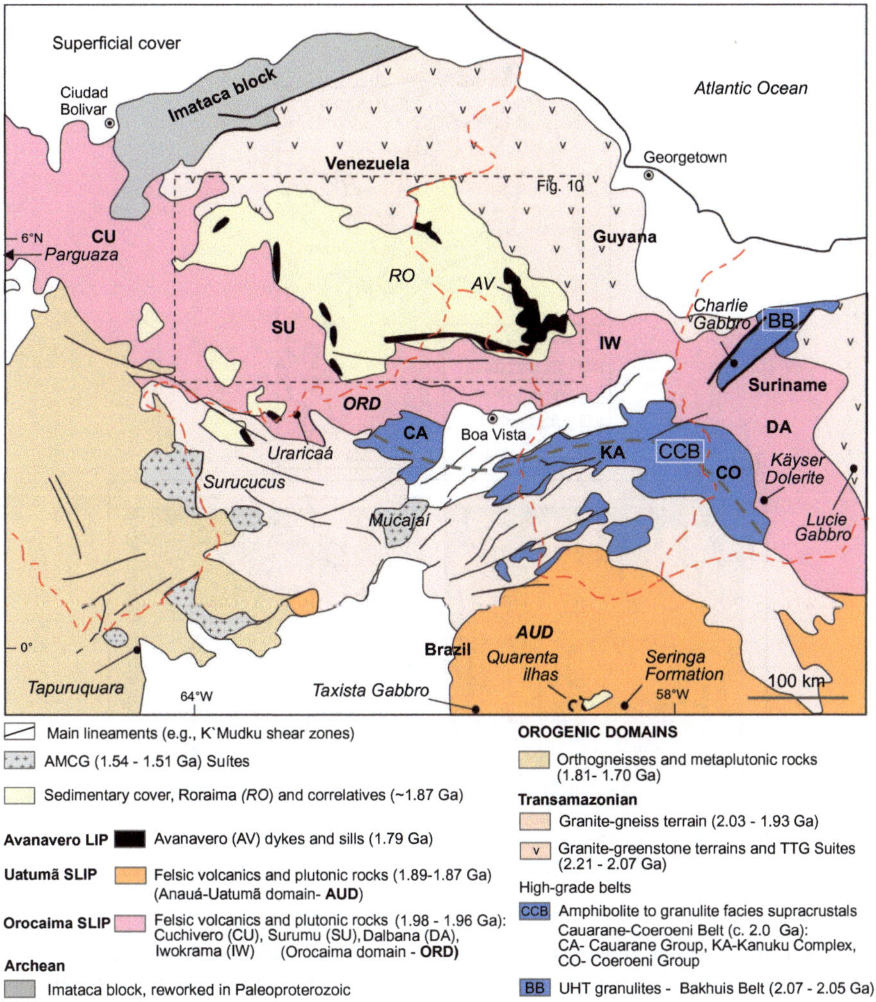

Fig. 2 Geologic outline of the central portion of the Guyana shield, including portions of the Imataca Block and the Central Amazonian, Maroni-Itacaiunas and Ventuari-Tapajos provinces. The distribution of the Orocaima and Uatumã SLIP and Avanavero LIP magmatisms are also shown. Adapted from Reis et al. (2017a, b) and Fraga et al. (2009, 2017). See text for details

The Maroni-Itacaiunas province (Fig. 2) where components of the Orocaima, Uatumã and Avanavero events occurred (see Sect. 4) is a product of successive accretionary arcs with later collision during the Orosirian. The most prominent tectonic components are the granite-greenstone and granite-gneissic terranes, as well as two roughly coeval high-grade metamorphic belts (e.g., Gibbs and Barron 1993; Delor et al. 2003; Cordani and Teixeira 2007; Vasquez et al. 2008; Fraga et al. 2009; Reis et al. 2004, 2013a). The granulitic belts in the Guiana Shield comprise: (i) the

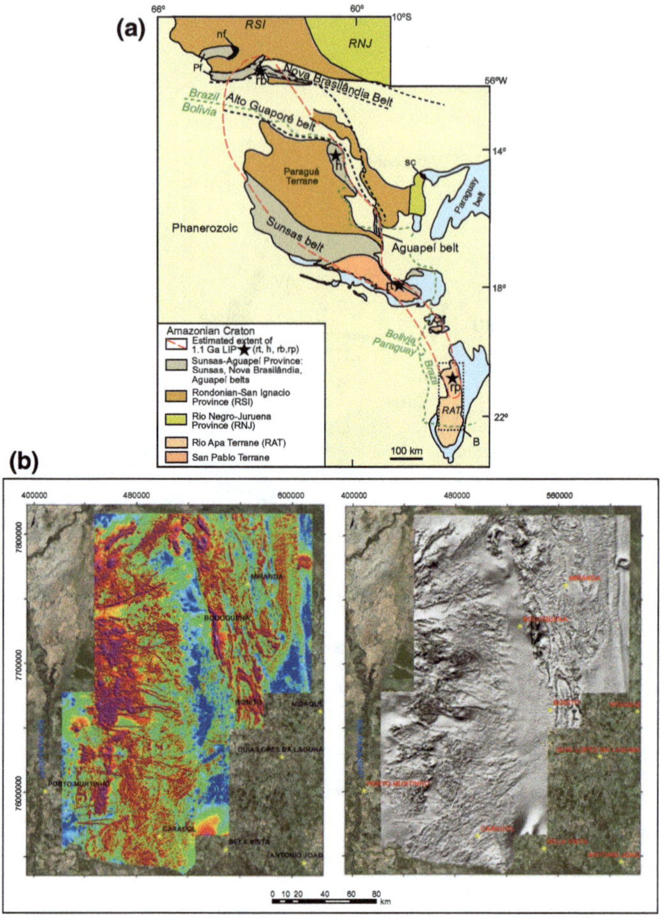

Fig. 3 **a** Tectonic outline of the SW portion of the Amazonian Craton, including parts of the Rio Negro-Juruena, Rondonian-San Ignacio and Sunsas-Aguapeí provinces. The possible correlative Rio Apa Terrane is also shown (adapted from Bettencourt et al. 2010; Rizzotto et al. 2013; Faleiros et al. 2016; Scandolara et al. 2017). Interpreted nodes of the 1.11 Ga Large Igneous Province are indicated by stars: rt (Rincón del Tigre), h (Huanchaca sill), rb (Rio Branco Suite in Mato Grosso), rp (Rio Perdido Suite), and its approximate outline indicated by the red dashed line. Other intraplate (≥1.20 Ga) mafic intrusions: Salto do Céu (sc) and Nova Floresta (nf) sills which is intrusive into the Palmeiral Formation (Pf) rift basin. Dashed black lines are inferred tectonic limits for the Nova Brasilândia and Alto Guaporé belts, respectively. **b** Aerogeophysical (total magnetic field) and radar image of a portion of the Rio Apa Terrane. The observed ~NW- and E-W trends highlight the orientations of the Rio Perdido Suite, supported by geologic mapping

2.07–2.05 Ga Bakhuis belt (fault bounded) whose granulites show peak conditions at 950–1050 °C and 8½–9 Kb (Kroonenberg et al. 2016 and references therein) and; (ii) the arcuate 2.01–1.98 Ga Cauarane-Coeroeni belt which is composed of amphibolite to granulite facies supracrustal rocks of Cauarane Group in Brazil, the

Kanuku Complex in Guyana and Coeroeni Group in Suriname (e.g., Fraga et al. 2009). The Paleoproterozoic crust is intruded in places by S-type granitic plutons (1.76–1.75 Ga), as well as by anorthosite-mangerite-charnockite-granite (AMCG) plutons (1.55–1.51 Ga), such as the Parguaza rapakivi batholith (1.51 Ga; Tassinari et al. 1996) and the 1.54 Ga Mucajaí anorthosite-mangerite-rapakivi granite (AMG) association (see Fig. 2) (Fraga et al. 2009). Restricted parts of the Maroni-Itacaiunas province are likely to be Archean (Amapá block; Fig. 1) and/or reclycled crust intruded by rich-K granites during the Orosirian times, such as in parts of the Imataca block (e.g., Tassinari et al. 2004; Tassinari and Macambira 2004; Cordani and Teixeira 2007; Kroonenberg et al. 2016; Fraga et al. 2017).

Large portions of the continental crust in the Central-Brazil Shield consolidated in Orosirian times, similar to the Maroni-Itacaiunas province. In particular, the southern portion (i.e., Tapajós region) of the Ventuari-Tapajós province accreted through the Tapajonic arcs, active between 2.01 and 1.90 Ga (Fernandes et al. 2011 and references therein). According to these authors these arcs involved northward flat subduction beneath the Archean/Paleoproterozoic continental margin, which eventually triggered nearby post-collisional activity over the already stable area, and linked with the Uatumã SLIP event to be discussed in Sect. 4.2.

After the consolidation of the Ventuari-Tapajós province, multiple orogenic belts subsequently accreted to the SW portion of the proto-Amazonian Craton. For instance, recent structural-tectonic studies, supported by geophysical interpretation (Scandolara et al. 2017) in the southern segment of the Rio Negro-Juruena province (Fig. 3a), characterized the Juruena accretionary orogeny (1.82–1.74 Ga). Contemporaneous calc-alkaline, granitic and (K-high) felsic volcanic rocks such as the Teles Pires Suite (1.78–1.76 Ga) occur within the adjoining portion of the Ventuari-Tapajós province, which in turn is age-equivalent to the Avanavero LIP event to be dealt with here. The Juruena orogenic belts underwent metamorphism at c. 1.69–1.63 and 1.42–1.37 Ga (Bettencourt et al. 2010; Scandolara et al. 2017). Furthermore the Rio Negro-Juruena province hosts several generations of A-type granitoids such as the Serra da Providência Suite (1.60–1.51 Ga; Table 2). Other A-type plutons with documented U-Pb ages from 1.40 to 1.31 Ga, as well as Mesoproterozoic tectonic basins are also present (Payolla et al. 2002; Bettencourt et al. 2010; Scandolara et al. 2013).

The Rondonian-San Ignacio province (Fig. 3a) is a result of a composite orogeny initiated at c. 1.59 Ga, which evolved through multiple arc accretion (e.g., 1.59–1.52, 1.51–1.48, 1.48–1.42 Ga), oceanic closure and final microcontinent–continent collision against the continental margin of the Rio Negro-Juruena province at c. 1.35 Ga (Cordani and Teixeira 2007; Bettencourt et al. 2010). As a matter of fact widespread reworking of the western portion of the Rio Negro-Juruena crust occurred during the time. The stepwise evolution was accompanied by intraplate mafic-felsic activity within the already stable parts of the proto-Craton (see Sects. 4 and 5).

The Alto Guaporé orogen (e.g., Rizzotto et al. 2014) which is the most prominent geologic-tectonic element of the Rondonian-San Ignacio province evolved in two steps, from 1.47–1.43 (accretionary phase) to 1.35–1.33 Ga (collisional phase). Strong magnetic anomalies and associated thrusts and shear zones highlight a WNW/ESE fossil suture zone preserved in the Alto Guaporé belt (see Fig. 3a), which is exemplified by syntectonic tonalitic-plagiogranitic plutons (c. 1.43 Ga) intruded into the ophiolites, marking the accretionary orogenic phase (Rizzotto et al. 2013 and references therein). The collisional phase marks the suturing of the Paraguá Terrane (1.82–1.66 Ga) against the proto-continental margin (i.e., Rio Negro-Juruena province) and eventually the consolidation of the Rondonian-San Ignacio province (e.g., Boger et al. 2005; Cordani and Teixeira 2007; Santos et al. 2008; Bettencourt et al. 2010; Rizzotto et al. 2014; Scandolara et al. 2017). The latter province is also the host of penecontempaneous late- to post-tectonic intraplate magmatism related to the collisional stage, exemplified by the 1.34–1.36 Ga Alto Candeias and the 1.31–1.30 Ga São Lourenço-Caripunas suites—as depicted in Table 3 (Bettencourt et al. 2010).

Finally, both the Rondonian-San Ignacio and the Rio Negro-Juruena provinces experienced later collision-related deformation which is considered to be late offshoots of the c. 1.10–1.00 Ga Sunsas orogeny, such as shear zones, tectonothermal overprints, metasedimentary belts (e.g., Nova Brasilândia), as well as magmatism. Post-tectonic to an orogenic stages took place after c. 1.00 Ga, evidenced by the occurrences of granites, pegmatites (0.99–0.97 Ga) (e.g., Rizzotto et al. 2002; Tohver et al. 2004, 2006; Teixeira et al. 2010). This orogeny which is the youngest and southwesternmost of the orogenic events recorded along the SW portion of the Amazonian Craton built the Sunsas-Aguapeí province (e.g., Teixeira et al. 2010; Bettencourt et al. 2010). This province hosts the 1.11 Ga Rincón del Tigre-Huanchaca LIP (Darbyshire 1979; Teixeira et al. 2015 and references therein) to be discussed in Sect. 4.4.

In particular, the Nova Brasilândia metasedimentary belt (see Fig. 3a) within the Rondonian-San Ignacio province (reworked portion of the Rio Negro-Juruena province; Bettencourt et al. 2010) originated in a precursor intracontinental rift setting (c. 1.35 Ga) which evolved to passive margin with narrow oceanic opening and subsequent crustal shortening and thrusting during which coeval bimodal magmatism (Rio Branco Suite) emplaced into the strata (Rizzotto et al. 2001, 2002). The Aguapeí metasedimentary belt (aborted rift) is roughly coeval with the Nova Brasilândia belt, as indicated by the comparable detrital zircon ages (<1230 and <1215 Ma), respectively (e.g., Teixeira et al. 2010 and references therein). During the onset of the Sunsas orogeny intraplate activity, reactivation of structures and tectonic basins occurred in places of the Central-Brazil and Guiana Shields. Specifically the Palmeiral rift basin which occurs in the Rondonian-San Ignacio province (Fig. 3) is intruded by 1.2 Ga mafic sills to be dealt with later.

3 Characteristics of the Paleo- to Mesoproterozoic Intraplate Activity

The most important Paleo- to Mesoproterozoic dyke swarms and sills, associated intraplate felsic volcanic-plutonic units and other mafic-ultramafic intrusions crop out in relatively low relief areas in the Guiana and Central-Brazil Shields, overlain by the

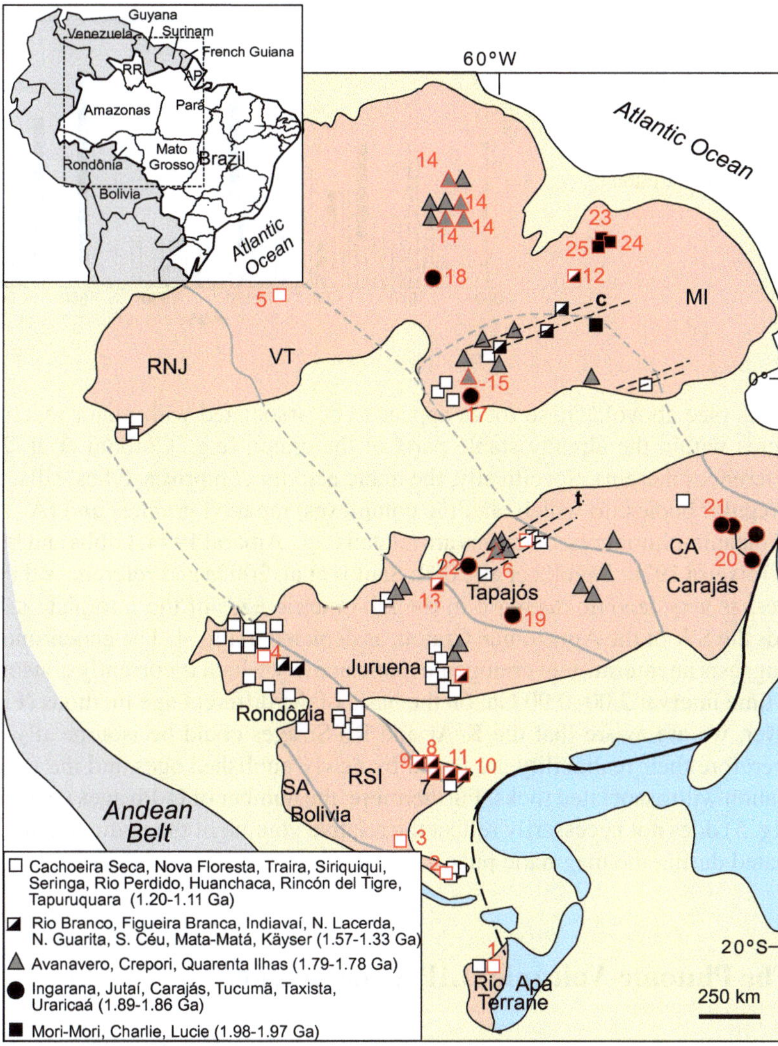

Fig. 4 Map of the Amazonian Craton, with symbols showing mafic dykes and sills and mafic-ultramafic complexes with published age determinations, grouped in different time groups (adapted from Teixeira 1990; Santos et al. 2002a). Some symbols represent multiple dated samples. Red frames (symbols) with numbers highlight key U-Pb ages used in the text. Boundaries between the tectonic provinces (abbreviations as in Fig. 1) are shown by gray dashes. 1. Major shear zones/lineaments are also outlined (black dashes): c = Cachorro; t = Tocantins. Inset: Brazilian Federal States (Amazon region): RR = Roraima, AP = Amapá, MT = Mato Grosso, RO = Rondônia, AM = Amazonas

Phanerozoic strata of the Amazon Basin. Within the Amazonian Craton the profuse mafic magmatic episodes of intracontinental setting are roughly contemporaneous with the Paleo- to Mesoproterozoic stepwise accretionary evolution of the tectonic

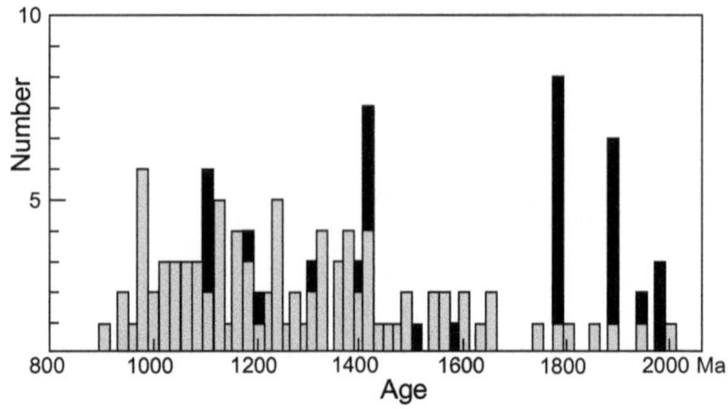

Fig. 5 Age-histogram (K-Ar, Ar-Ar and Rb-Sr ages: gray bars; U-Pb ages: black bars) of Proterozoic dykes and sills, and mafic-ultramafic complexes in the Amazonian Craton

provinces (see above). These rocks appear to be associated with major reactivated structures within the already stable parts of the craton (e.g., Cordani et al. 2010a and references therein). Specifically, the mafic activity comprises dykes, sills, flows and irregular bodies, to which alkaline complexes, rapakivi granites and (A- and I-type) volcanic-plutonic associations are related (e.g., Amaral 1974; Gibbs and Barron 1993; Teixeira 1978; Payolla et al. 2002; Santos et al. 2002a and references therein).

There is a systematic decrease in the radiometric ages of the intraplate activity towards the SW of the Amazonian Craton, as depicted in Fig. 4. The geochronologic data suggests an intermittent cratogenic mafic activity which is currently constrained in the time interval 2.00–0.90 Ga, on the basis of the different age methods (Fig. 5). However, we are aware that the K-Ar and Rb-Sr ages could be isotopically reset, and therefore their realiability are tested by newly published ages and the geologic correlation with associated rocks. Furthermore, the number of U-Pb ages (see Table 1 and Fig. 5) does not necessarily reflect the relative volume of the mafic magmas that originated during the magmatic pulses.

4 The Plutonic-Volcanic (LIP Scale) Events

At least four magmatic events (SLIP or LIP scale) occurred in the Amazonian Craton, well-defined by U-Pb ages, namely Orocaima (1.98–1.96 Ga), Uatumã (1.89–1.87 Ga), Avanavero (1.79 Ga) and Rincón del Tigre-Huanchaca (1.11 Ga), respectively (dots in Fig. 1). Their geologic-tectonic characteristics are overviewed, as suggested by the geochronologic and geochemical information. Morphology, dimension and geologic setting of each SLIP/LIP are discriminated when available from the literature, otherwise inferred by the integrated information.

The terminology of the lithostratigraphic units along the text follows the published information (e.g., suite, association, complex, swarm, mafic intrusion, Group, Formation) in a flexible way because of the review scope. However, we are aware that in some cases the standard classification may not have been strictly applied. Table 1 presents the available U-Pb ages for the mafic dykes, sills and mafic-ultramafic

Table 1 Compilation of U-Pb ages of intraplate activity (basaltic rocks and mafic-ultramafic complexes) in the Amazonian Craton

Stratigraphic unit (N)	Description	Location	U-Pb age (Ma)	References
Rio Perdido Suite (1)	Dykes (gabbros and diabases)	Rio Apa Block (SE Amazonia; Mato Grosso do Sul state)	[1]1111 ± 1 (b)	Teixeira et al. (2018)
Rincón del Tigre (2)	Layered complex (*granophyre*)	Bolivian Precambrian Shield	[1]*1110* ± *2* (b)	Teixeira et al. (2015)
Huanchaca Suite (3)	Gabbro sills and dykes	Bolivian Precambrian Shield (Paragua Terrane)	[1]1112 ± 2 (b)	Teixeira et al. (2015)
[a]Rio Branco Suite (4)	(Meta) leucogabbro	Rondonia-San Ignacio Province (Rondonia state)	[1]1110 ± 10 (z)	Rizzotto et al. (2002)
Tapuruquara Suite (5)	Elongated to semi-rounded stocks (*gabbro*)	Ventuari-Tapaós Province (NW Amazonas state)	[2]*1172* ± *8* (*b*)	Santos et al. (2006)
Cachoeira Seca Troctolite (6)	Lacoliths and dykes of alkalic gabbro-troctolite	Ventuari-Tapaós Province (SW Para state)	[2]1186 ± 12 (b)	Santos et al. (2002a)
Nova Lacerda (7)	Mafic dykes of basaltic composition and tholeiitic affinity	Rio Negro-Juruena Province (Mato Grosso state)	[1]1387 ± 17 (b)	Teixeira et al. (2015)
Indiavaí Suite (8)	Mafic intrusion	Rio Negro-Juruena Province (Mato Grosso state)	[1]1416 ± 7 (b)	Teixeira et al. (2011)
Figueira Branca Suite (9)	Mafic-ultramafic layered complex and gabbroic stocks	Rio Negro-Juruena Province (Mato Grosso state)	[1]1426 ± 8 (b)	Teixeira et al. (2011)
Rio Branco Suite (10)	Mafic member (gabbro, tholeiitic dykes and porphyritic basalt). Felsic member (rapakivi granite)	Rio Negro-Juruena Province (Mato Grosso state)	[1]1423 ± 2 (z)	Geraldes et al. (2004)

(continued)

Table 1 (continued)

Stratigraphic unit (N)	Description	Location	U-Pb age (Ma)	References
Salto do Ceu Gabbro (11)	Tholeiitic, gabbro sills	Rio Negro-Juruena Province (Mato Grosso state)	[1]1439 ± 4 (b)	Teixeira et al. (2016)
Kayser Diabase (12)	Alkaline diabase dykes	Maroni-Itacaiunas Province (Suriname)	[3]1528 ± 2 (b)	De Roever et al. (2014)
Mata-Matá Gabbro (13)	Tholeiitic, diabase sill	Ventuari-Tapajós Province (SE Amazonas state)	[2]1576 ± 4 (b)	Betiollo et al. (2009)
[b]Avanavero Diabase (14)	Tholeiitic dykes, sills, ring structures: Pedra Preta, Cipo, Cotingo, Manga Brava (Brazil); Omai, Tumatumari (Guyana)	Central Amazonian Province (Roraima state; Venezuela, Guyana)	[1]1793 ± 1 (b); [1]1794 ± 2 (b); [1]1794 ± 2 (b, zr) [2]1787 ± 14 (b); [2]1782 ± 3 (b); [2]1786 ± 3 (b)	Reis et al. (2013a), Santos et al. (2004), Norcross et al. (2000)
Quarenta Ilhas Diabase (15)	Tholeiitic sill (elliptical cone sheet structures); Ring dyke	Central Amazonian Province (NE Amazonas state)	[2]1780 ± 3 (b)	Santos et al. (2003)
[b]Crepori Diabase (16)	Sills and dykes of tholeiitic gabbro, diabase, andesite and granophyre	Ventuari-Tapajós Province (SW Para state)	[2]1780 ± 7 (b)	Santos et al. (2002a)
[c]Taxista Gabbro (17)	Dyke intrusive into the Agua Clara granitoid	Central Amazonian Province (NE Amazonas state)	[2]1859 ± 15 (b)	Simões et al. (2015)
[a]Uraricaá Suite (18)	Gabbro, diabase, and minor hornblendite	Central Amazonian Province (NW Roraima state)	[2]1882 ± 4 (b)	Fraga et al. (2017)

(continued)

Table 1 (continued)

Stratigraphic unit (N)	Description	Location	U-Pb age (Ma)	References
Ingarana Suite (19)	Gabbro, anorthosite, diabase dykes, syenite: Ingarana, Jutaí	Ventuari-Tapajós Province (SW Para state)	^{2}<1893 ± 10; 1879 ± 3; 1878 ± 9 (b-z)	Santos et al. (2000, 2001, 2002a)
[c]Tucumã (20)	Felsic and mafic dykes (bimodal), including those in Carajas	Central Amazonian Province (Para state)	^{2}1881 ± 7; ^{2}1882 ± 9	Silva et al. (2016), Teixeira et al. (2018)
[c]Santa Rosa (21)	Rhyolite dyke	Central Amazonian Province (Para state)	^{2}1895 ± 11 (z)	Antonio et al. (2017)
[c]Teodorico Diabase (22)	Tholeiitic gabbro dykes (Ingarana Suite?)	Ventuari-Tapajós Province (SW Para state)	^{2}1893 ± 10 (z)	Santos et al. (2002a)
[d]Charlie Gabbro (23)	Gabbro	Maroni-Itacaiunas Province (Suriname)	^{1}1971 ± 15 (b)	Klaver et al. (2016)
[d]Moi-Moi Gabbro (24)	(Meta)gabbro	Maroni-Itacaiunas Province (Suriname)	^{2}1984 ± 4 (z)	Klaver et al. (2016)
[d]Lucie Gabbro (25)	Gabbro	Maroni-Itacaiunas Province (Suriname)	^{3}1985 ± 4 (z)	Kroonenberg et al. (2016)

See text for explanation
U-Pb ages: [1]Isotopic dilution-thermal ionization mass spectrometer (TIM S); [2]sensitive high-resolution ion microprobe; [3]Kober method (single zircon Pb evaporation) 207/206 age. Components of LIP/SLIP events: [a]Rincón del Tigre-Huanchaca; [b]Avanavero; [c]Uatuma; [d]Orocaima. Age dating corresponding with particular rocks are shown in italics in lines 2 and 5 (see columns 2 and 4). Minerals: zircon (zr), baddeleyite (b)

complexes, used as key-time markers for the tectonic events. The accompanying Table 2 shows the age-equivalent plutonic-volcanic associations and plutonic rocks (selected) as well as their tectonic settings. The potential barcode matches between the Amazonian SLIP/LIP events and intraplate activity on other continental blocks through time are shown in Table 3.

Table 2 Compilation of U-Pb ages of selected plutonic rocks in the Amazonian Craton and their interpreted tectonic characteristics

Stratigraphic unit	Assemblage	U-Pb age (Ma)	Interpreted tectonic setting	References
Sunsas Granites	Taperas granite	1076 ± 18	Arogenic	1
Rio Branco Suite	Mafic to felsic members: gabbro, metagabbro, amphibolite, metadiabase, granite.	1110–1098	A-type, within-plate	2, 3
Santa Clara Suite	Monzogranite, syenogranite, granite	1082–1074	Anorogenic	4
Seis Lagos Complex	Carbonatite, carbonite breccia, siderite carbonatite	1154–1328	Anorogenic	5
Sao Lourengo-Caripunas Suite	Alkali-feldspar granite, syenogranite, quartz- syenite, and rhyolite porphyry	1314–1309	Post-collisional	4, 12, 17
Alto Candeias Suite	Monzogranite, syenogranite, charnockite and syenite	1346–1338	Late-tectonic	4, 12, 17
Rio Branco Suite	Felsic to mafic members: syenite, rapakivi granite, gabbro, diabase, basalt	1423 ± 2	Anorogenic	6, 7
Serra Grande Suite	Co-magmatic charnockites and rapakivi granites	1434–1425	Anorogenic	8
Mucajaí	Granite, mangerite, anorthosite (AMG) complex	1538 ± 5	Within-plate	9
Parguaza and Surucucus Suites	Rapakivi granites	1510;1551 ± 5	Anorogenic	4, 10, 11, 12
Serra da Providência Suite	Rapakivi granites, charnockitic/mangeritic granitoids, gabbronorite, gabbro	1600–1570, 1570–1560, 1550–1540; 1510	Anorogenic; within-plate	4, 12, 13
Teles Pires Suite	Granitic plutons and felsic volcanic rocks	1780–1760	A-type; post-collisional	14, 15, 16, 17
Madeira Suite	Albite-rich granite, per-alkaline granite	1839–1822	Post-collisional	18
Maloquinha Suite	Aluminous alkaline granites, syenogranites	1877–1865	A-type, post-collisional	19, 20
Parauari Suite	Calc-alkaline, monzogranite, granodiorite (mainly)	1883–1879	Post-collisional	19, 20

(continued)

Table 2 (continued)

Stratigraphic unit	Assemblage	U-Pb age (Ma)	Interpreted tectonic setting	References
Rio Pardo	Rapakivi granites, charnockites, enderbites and a gabbroic complex	1882–1860	Within-plate	21
Mapuera Suite	Granite, granodiorite, syenite, charnockite	1880–1871	Within-plate	22, 23
Kabalebo	Charnockites	1992–1984	Post-collisional	24
Pedra-Pintada-Surumu	Calc-alkaline (mainly), I-type plutonic and volcanic rocks	1985–1982	Post-collisional	25
Aricamã-Cachoeira da Ilha Suite	Syenogranite, K-feldspar granite, ignimbrite	1983 ± 4; 1973 ± 8	A-type, post-collisional	9, 25

Bold ages (Serra da Providência suite) are considered peak magmatism. See text for explanation
[1]Boger et al. (2005); [2]Rizzotto et al. (2001); [3]Tohver et al. (2004); [4]Bettencourt et al. (1999, 2010); [5]Rossoni et al. (2017); [6]Geraldes et al. (2004); [7]Teixeira et al. (2016); [8]Santos et al. (2011); [9]Fraga et al. (2009a); [10]Tassinari et al. (2006); [11]Santos et al. (2003); [12]Payolla et al.(2002); [13]Scandolara et al. (2013); [14]Santos et al. (2001); [15]Santos et al. (2008); [16]Neder et al. (2002); [17]Scandolara et al. (2014; 2017); [18] Bastos Neto et al. (2014); [19]Santos et al. (2000); [20]Vasquez et al. (2002); [21]Simoes et al. (2015); [22]Ferron et al. 2010; [23]Reis et al. (2017b); [24]Klaver et al. (2015); [25]CPRM (2010)

4.1 The Orocaima (1.98–1.96 Ga) Magmatism

The Orocaima volcanic-plutonic magmatism (Reis et al. 2003) is the oldest late Paleoproterozoic intraplate event of LIP scale across the Guiana Shield (Reis et al. 2003; Fraga et al. 2009 and references therein). The minimum areal extent of the Orocaima volcanic-plutonic magmatism is c. 200,000 km^2, only considering the exposure in the Guiana Shield.

The Orocaima rocks (e.g., Reis et al. 2004) mainly comprise high-K, calc-alkaline I-type rocks belonging to the Pedra Pintada-Surumi plutonicvolcanic association (1982 ± 3; 1984 ± 7; 1985 ± 1 Ma). This association is coeval with subordinate A-type volcanic rocks (Cachoeira da Ilha Formation) and rapakivi granites such as the Aricamã Suite (1986 ± 4 Ma; Fraga et al. 2009—see Table 2) in the Roraima block, northwestern portion of the Central-Amazonian province (Cordani and Teixeira 2007). This block is the host of the Roraima clastic sedimentary rocks, which is well-constrained in age to 1873 ± 3 Ma, on the basis of zircon U-Pb age determinations in interbedded tuff layers (Santos et al. 2003; Reis et al. 2017a). This sedimentary basin (Fig. 2) is intruded by mafic sills of the Avanavero Suite (Gibbs and Barron 1993; Reis et al. 2017a)—to be dealt with here.

The Surumu Group dominantly consists of ignimbrites with rhyolitic to andesitic composition, and subordinate subvolcanic intrusions and sedimentary layers (CPRM

Table 3 Potential barcode matches with the studied LIP/SLIP events. See pertinent sections for explanation

LIP/SLIP events	Potential Barcode matches	Paleotectonic connections
Rincón del Tigre–Huanchaca LIP (1.11 Ga)	Initiation of Keweenawan magmatism of Central Laurentia (Midcontinent Rift); Southwestern USA diabase province and its extent to northern Mexico (Tohver et al. 2006; Ernst et al. 2013; Stein et al. 2014; Bright et al. 2014); intraplate magmatism throughout the Kalahari, Congo and India cratons (Ernst 2014 and references therein)	Grenville Orogen
Avanavero LIP (1.79 Ga)	Ropruchey sills (Fedotova et al. 1999), Shosksha Formation (Pisarevsky and Sokolov 2001), Hoting gabbro (Elming et al. 2009) and Smaland-Varmland Intrusions in the Fennoscandia shield (Högdahl et al. 2004; Pisarevsky and Bylund 2010); Ultramafic-mafic dykes, coeval tholeiitic and jutonitic dykes and associated AMCG complexes in the Ukranian shield (Shumlyanskyy et al. 2016a, b, 2017); Xiong'er volcanics and Taihang dyke swarm in the North China Craton (Peng 2015)	Columbia
Uatumã SLIP (1.88–1.87 Ga)	Circum-Superior LIP of the Superior Craton (Jowitt and Ernst 2013); Widespread intraplate magmatism throughout the Slave Craton (Buchan et al. 2010; Ernst et al. 2013; Ernst 2014 and references therein). Dykes and sills in the Bastar and Dharwar, Siberia, Kaapvaal and Zimbabwe cratons (Ernst 2014)	Columbia
Orocaima SLIP (1.98–1.96 Ga)	Pechenga-Onega LIP of the Karelian-Kola Craton (Lubnina et al. 2016); Xiwangshan dyke swarm of the North China Craton (Peng 2015); Khajuraho-Jhansi dyke swarm of the Bhundelkhand Craton (Pradhan et al. 2012)	Columbia

2010 and references therein), mainly exposed in the Guiana Shield, Brazil (see Fig. 2). The correlative lithostratigraphic units occur in Venezuela (Cuchivero Group), Guyana (Iwokrana Group) and Suriname (Dalbana Group), and are predominantly composed of ignimbrites with varied composition similar to the Surumu rocks (e.g., Gibbs and Barron 1993; Delor et al. 2003 and references therein) (Fig. 6). However, we are aware that more robust geochronology coupled with petrologic studies is necessary to confirm the correlation between these lithostratigraphic associations. The fact that the Orocaima rocks are predominantly acid to intermediate in composition and their large geographic occurrence is consistent with a Silicic Large Igneous Province (SLIP) (cf. Bryan 2007; Bryan and Ernst 2008a, b).

Subordinate mafic activity (see Table 1) synchronous with the Orocaima SLIP event is apparent in Suriname, such as the Charlie gabbros and Moi-Moi metagabbros (1971 and 1984 Ma; see Table 1) which occur as km-sized bodies completely

Fig. 6 Field aspects of the Orocaima SLIP: **a** Lapilli-tuff (Surumu Formation, Brazil); **b** Ignimbrite (Iwokrama Formation, Guyana); **c** Dacitic Ignimbrite (Dalabana Formation, Suriname); **d** Aricamã rapakivi granite (Brazilian Roraima state) (photos by N. J. Reis)

enclosed by the Kabalebo charnockites (1984–1992 Ma), in the Bakhuis UHT granulitic belt (see Fig. 2). One anorthosite in the center of this belt was dated by zircon Pb evaporation at 1980 Ma, whereas the Lucie Gabbro in the eastern portion of Suriname yielded a similar zircon Pb evaporation age (1985 $\pm$ 4 Ma). These (ACG) rocks post-date the UHT event (Transamazonian orogeny) (Klaver et al. 2015, 2016 and references therein; Kroonenberg et al. 2016 and references therein).

The fact that the Orocaima volcanic-plutonic event roughly accompanies the arcuate exposure of the Cauarane-Coeroneni collisional belt (2.01–1.98 Ga) similarly suggests a connection between the orogen and extensional dynamics that gave rise to this particular event. Furthermore, the calc-alkaline characteristics of the volcanic-plutonic association are consistent with subduction-related arc magmatism, whereas the contemporaneous A-type volcanics, the Aricamã Suite (Roraima) and the ACG suites of Suriname (see above) are post-collisional in character (Reis et al. 2003, 2004; Delor et al. 2003; CPRM 2010; Klaver et al. 2015, 2016). Additional petrogenetic studies are needed for contraining the mantle and/or lithospheric sources from with the Orocaima rocks derived. Finally, from a global perspective the 1.98–1.96 Ga SLIP has potential age-equivalent units in many continental blocks, as summarized in Table 3.

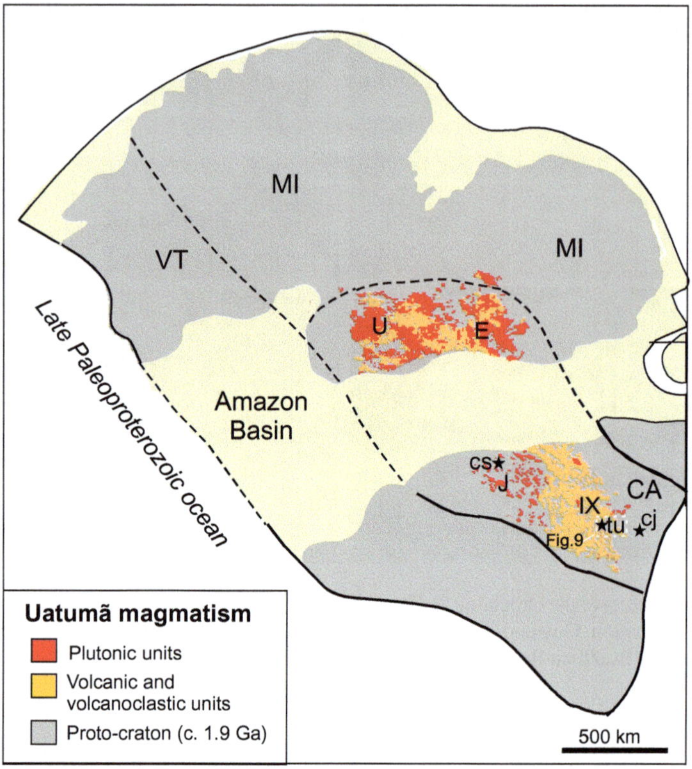

Fig. 7 Geologic cartoon showing the Uatumã SLIP in the proto-Amazonian Craton at c. 1.8 Ga. Tectonic domains within the Central Amazonian province: Anauá-Uatumã; E = Erepecuru-Trombetas; J = Jamanxim; I = Iriri-Xingu. Keys: Cachoeira Seca troctolite (cs); Tucumã bimodal swarm (tu), Carajás dykes (cj). Tectonic provinces (CA, MI and VT) are delineated as in Fig. 1

4.2 The Uatumã (1.89–1.87 Ga) Magmatism

The Uatumã SLIP (Klein et al. 2012 and references therein) has an estimated exposure of ~400,000 km^2 across the already stable continental crust of the Central Amazonian province, although part of the rocks within this area (Fig. 7) is possibly related to the older Orocaima magmatism. Similar to this, the Uatumã event is predominantly acid to intermediate, with peak ages between 1.89 and 1.87 Ga. This SLIP event has age match with the Uraricaá Suite (1882 ± 4 Ma; see Table 1 and Fig. 2) which is composed of low-Titanium tholeiitic gabbro with chemistry akin to basalts, basaltic-andesites and picrobasalts. Hornblendites are subordinate (Fraga et al. 2017). The A-type Rio Pardo Association (1882–1860 Ma; Table 2), which is composed of rapakivi granites, charnockites, enderbites and a gabbroic complex located in the Guiana Shield, is age equivalent with the Uatumã SLIP event (Simões et al. 2015). According to these authors, this particular A-type activity may include the spatially related Taxista Gabbro (1859 ± 15 Ma; see Table 1) which is hosted by the Água Branca granite (c. 1890 Ma) in the Anauá-Uatumã domain (see Fig. 2).

The Uatumã event, according to Klein et al. (2012), comprises an association of calc-alkaline I-type and A-type volcanic and plutonic rocks, as discriminated below:

(i) The effusive rocks where erupted rhyolites predominate. These rocks have been included into several stratigraphic units (Iriri and Iricoumé Groups; Bom Jardim, Sobreiro and Santa Rosa Formations) with different units exposed in different parts of the Central-Brazil Shield (Klein et al. 2012; Roverato et al. 2017; Antonio et al. 2017). The Iricoumé (1896–1882 ± 11 Ma; Santos et al. 2002a; Ferron et al. 2010) and Iriri (1870 Ma; Santos et al. 2001) groups have been correlated with similar-looking, dominantly silicic volcanic sequences in southern Guyana (Kuyuwini Group)—see Fig. 2 (Santos et al. 2004; Klein et al. 2012; Barreto et al. 2013; Silva et al. 2016; Fraga et al. 2017). Field aspects of the volcanic lithotypes of the Uatumã LIP are presented in Fig. 8.
(ii) The A-type (post-collisional and anorogenic) granitic plutons in the Central-Brazil Shield such as the Maloquinha Suite (1877–1865 Ma; Lamarão et al. 2005), as well as the I-type Água Branca Suite (1890–1980 Ma) and the Parauari Suite (1890–1870 Ma). Penecontemporaneous granitic intrusions occur in the Guiana Shield, such as the Mapuera (1880–1871 Ma) and Madeira Suites (1839–1822 Ma) in the Uatumã-Anauá Domain (Fig. 2), encompassed by the Iricoumé volcanics. The A-type granites are metaluminous to peralkaline, whereas the Parauari rocks show high-K, calc-alkaline affinity (e.g., Vasquez et al. 2002; Santos et al. 2002a; Lamarão et al. 2005; Dall'Agnol and Oliveira 2007; Ferron et al. 2010; Bastos Neto et al. 2014; Reis et al. 2017b).

The Uatumã SLIP event also includes the NW-oriented Tucumã bimodal dyke swarm (1881 ± 7 and 1882 ± 9 Ma; Silva et al. 2016 and references therein) (Fig. 9), as well as NW- and NW trending, mafic to intermediate dykes with identical ages (Teixeira et al. 2018); see Table 1) in the Carajás-Rio Maria region to the east. Locally these dykes form composite mafic-felsic bodies. All these dykes are correlated to the Uatumã SLIP on the basis of tight age match and similar intraplate setting (see Table 1).

According to Silva et al. (2016), the Tucumã swarm is comprised of three lithotypes, intrusive into the Archean granite-gneissic rocks and the associated greenstone belt (see Fig. 9): felsic (porphyritic rhyolites), mafic (basaltic andesites and subordinate basalts) and intermediate (andesite, dacite) rock. In particular, the rhyolites show porphyric, rapakivi-textures that suggest a close relationship with the nearby granitic plutons of the Maloquinha Suite (see Fig. 8e, f), intrusive into the Archean crust of the Central Amazonian province (e.g., Silva et al. 2016 and references therein). The felsic dykes are peraluminous to slightly metaluminous and akin to A2, ferroan and reduced granites (see Fig. 9 in Silva et al. 2016), whereas intermediate and mafic dykes are metaluminous and belong to the tholeiitic series. The emplacement of the Maloquinha plutons roughly follows the NW-SE general trend of the Tucumã dykes, whereas their sheeted-like geometry can be explained as a direct response to the extensional tectonic regime (Oliveira et al. 2008). The Tucumã dykes also have age

Fig. 8 Field aspects of the Uatumã SLIP: **a** Ignimbrite and rhyolitic tuff (Iricoumé Group) showing steeply dipping bedding; **b** Fine-grained porphyritic granite associated with rhyolites and tuffs of the Uatumã-Anauá domain (Photos by E. Klein); **c** Outcrop view of the Tucumã swarm (Photo by P. J. Antonio); **d** General view of one Tucumã dyke intrusive into local Archean greenstone belt sequence; **e** Macroscopic appearance of porphyritic rhyolitic dyke (Tucumã area) showing locally plagioclase mantled K-feldspar phenocrysts; **f** Dacitic dyke with K-feldspar phenocrysts (Tucumã area). See Fig. 7 for geographic location of the Tucumã swarm (Photos D. Oliveira)

match with some Au-bearing mafic intrusions that further west in the Central-Brazil Shield (Ventuari-Tapajós province), such as the Ingarana Suite (1881 ± 4 Ma; Santos et al. 2002a) which includes the Teodorico gabbro (1893 ± 10 Ma) and the Jutaí gabbro-anorthosite (1879 ± 7 Ma; Santos et al. 2001)—see Table 1. Nevertheless,

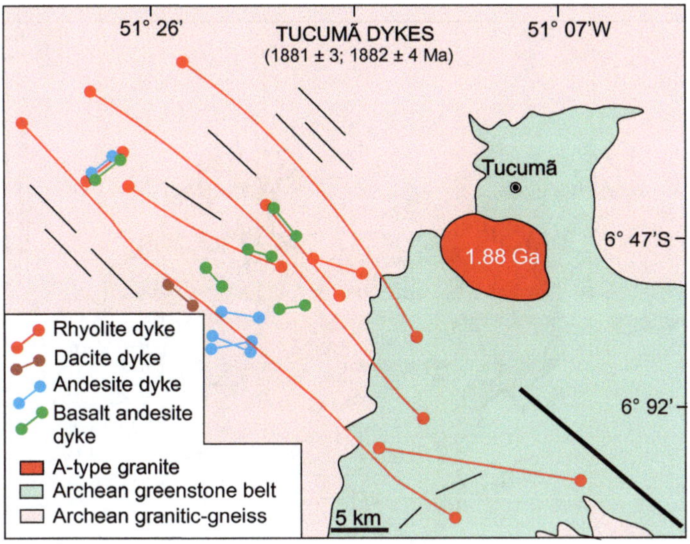

Fig. 9 Geologic sketch map of the Tucumã bimodal dyke swarm, intrusive into Archean lithostratigraphic units in the Central Amazonian province (after Silva et al. 2016). Uatumã SLIP component: 1.88 Ga granite. See Fig. 7 for geographic location of the Tucumã swarm and coeval dykes in the Carajás region

these particular intrusions may be more directly related to subduction going on in that area, and then not part of the Uatumã SLIP.

From a geodynamic perspective, the Uatumã SLIP event has been suggested to be related to thermal perturbations in the upper mantle and mafic underplating, of plume-like character (e.g., Silva et al. 2016). The felsic volcanic-plutonic activity (1.89–1.87 Ga) could have been derived by large crustal fusion, as suggested by available Nd-Sr isotopic constraints. The associated mafic magmatism should presumably be directly mantle-derived, as usually assumed for intraplate bimodal magmatic suítes (Silva et al. 2016; Fraga et al. 2017; Antonio et al. 2017). The Uatumã magmatism is also age-equivalent with the final stages (in a post-collisional setting) of the Tapajonic arcs (2.01–1.90 Ga; Fernandes et al. 2011) after which large portions of continental crust in the Central-Brazil Shield (e.g., Central-Amazonian and Ventuari-Tapajós provinces) consolidated as a stable landmass (see Sect. 2). According to Fernandes et al. (2011), the orogenic process included a northward flat subduction process beneath the Archean/Paleoproterozoic continental margin, which eventually triggered felsic plutonic intrusions (e.g., Parauari Suite; 1883 ± 4 to 1879 ± 3 Ma), as well as the c. 1.88 Ga Uatumã volcanic rocks in a back arc setting (e.g., Fernandes et al. 2011; Teixeira et al. 2017). This hypothesis is coherent with the geochemical signature of the 1.88 Ga basaltic and andesitic dykes of Carajás/Rio Maria, located to the east of the Tucumã dykes, which shows N-MORB and E-MORB characteristics with variable metasomatic effects (e.g., LREE and LILE enrichment with respect to HFSE; Rivalenti et al. 1998).

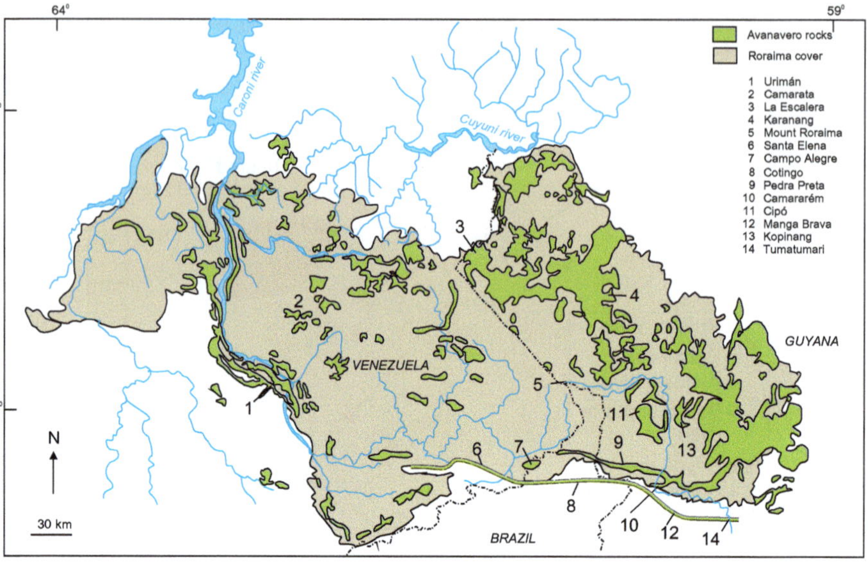

Fig. 10 Distribution of the Avanavero sills and dykes, emplaced into the Roraima cover within the Guiana shield (border area of Venezuela, Brazil and Guyana), northern Amazonian Craton. The dotted line indicates the country boundaries (after Reis et al. 2013a)

Noteworthy, the Uatumã SLIP event has economic importance, as shown by significant gold and tin deposits, in addition to tantalite-columbite and gemstones occurrences, and also has potential for Rare Earth Elements. Finally, Table 3 presents potential barcode matches (1.88–1.87 Ga) for this SLIP in the context of the Columbia supercontinent.

4.3 *The 1.79–1.78 Ga Avanavero LIP Event*

The Avanavero magmatism (Reis et al. 2013a) was active c. 100 my later than the Uatumã SLIP. It is the most widespread and voluminous intraplate mafic activity (Figs. 2 and 10) in the Guiana Shield, occurring over an extensive area (ca. 73,000 km^2) of Brazil, Venezuela and Guyana. The crystalline basement which hosts the Avanavero magmatism comprises distinct geological units developed during the Transamazonian orogeny, and is tectonically ascribed to the Maroni-Itacaiunas province (e.g., Gibbs and Barron 1993; Cordani and Teixeira 2007). However, much of the Avanavero magmatism is emplaced into the Roraima sedimentary cover (see Fig. 2) in at least four stratigraphic levels (Reis et al. 2013a; Beyer et al. 2015; Reis et al. 2017a and references therein).

Reis et al. (2013a) recently addressed the tectonic significance of the Avanavero LIP based on new and compiled U-Pb ages (see Table 1), geochemical data and geo-

Fig. 11 Field aspects of the Avanavero mafic magmatism (photos by N. J. Reis). **a** The Pedra Preta sill along the Uailã river (Roraima Supergroup), northern portion of the Brazilian Roraima state; **b** The Pedra Preta sill along the Quinô river, northern portion of the Roraima state; **c** Landscape view of the Pedra Preta sill along the Quinô river, northern portion of the Roraima state, Roraima Basin; **d** The Cotingo dyke crosscuting the Orocaima volcanics (Orosirian in age), northern portion of the Roraima state

logic information. This LIP is dominated by widespread mafic sills such as the locally termed Cotingo, Kopinang, Tumatumari, Santa Elena, Cipó, and Monte Roraima sills that are emplaced into the Roraima cover (Figs. 2, 10 and 11). Sills can be as much as 400 m thick whereas the associated mafic dykes can be several of kilometers in length, sometimes showing branching or bifurcating forms.

Precise U-Pb baddeleyite ages for the Avanavero rocks in the Guiana Shield are: 1794 ± 2, 1782 ± 3 and 1786 ± 3 Ma (see Table 1), respectively (Reis et al. 2013a; Santos et al. 2003). These ages indicate a short-lived character for the Avanavero magmatism, and also imply that all previous published K-Ar and Rb-Sr ages are meaningless (Table 2 in Reis et al. 2013a). In the northeast portion of the Amazonas state, coeval mafic rocks form saucer shaped sills, the Quarenta Ilhas Diabase (see Fig. 2), intrusive into the Urupi Formation (Reis et al. 2013a). This diabase yielded U-Pb crystallization ages of 1793 ± 1 and 1780 ± 3 Ma (Reis et al. 2013a; Santos et al. 2003). These are 300–400 m.y. older that the previous K-Ar ages which reflect partial or total Ar loss due to much younger overprints.

Geochemically, the Avanavero rocks are tholeiitic in composition and are similar to E-MORB and subcontinental lithospheric mantle basalts, whereas the REE pattern indicates an intraplate character. In contrast, the low Nb/LREE ratios suggest lithospheric mantle sources modified by crustal contamination and/or by variable enrich-

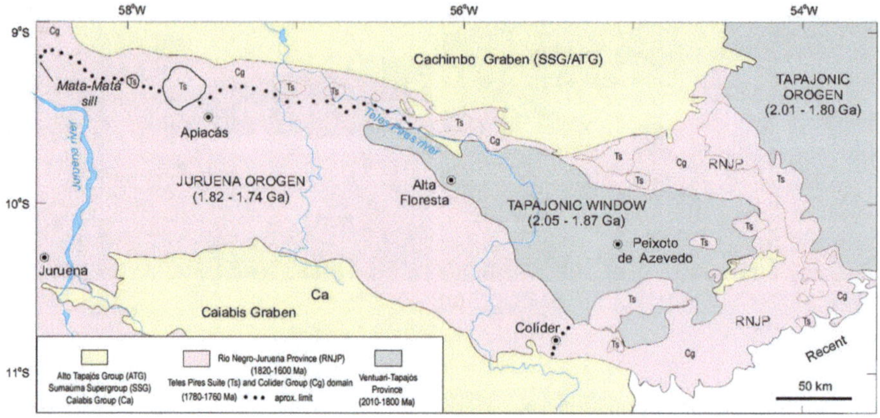

Fig. 12 Geologic outline of Teles Pires Suite and Colider Group (Alta Floresta Gold Province) in the southern portion of the Ventuari-Tapajós province (adapted from Cordani and Teixeira 2007; Scandolara et al. 2014; Bettencourt et al. 2016). See text and Fig. 1 for details

ment with slab fluids or melt from oceanic lithosphere (Reis et al. 2013a). According to these authors, such inherited chemical characteristics of arc-setting are genetically associated with the 1.98–1.95 Ga high-K calc-alkaline volcano-plutonism (Orocaima SLIP) whilst the mafic magmatism post-dates the evolution of the Maroni–Itacaiúnas Province as an intraplate activity.

Coeval rocks with the Avanavero magmatism occur elsewhere in the Amazonian Craton, as suggested by previous published U–Pb geochronology (e.g., Santos et al. 2002a and references therein). For instance, the Crepori Diabase (1780 ± 7 Ma; see Table 1) is ca. 1800 km away from the Avanavero-type area (see Fig. 4), discontinuously exposed 1.5–4.0 km south of the lower Crepori River in the Central-Brazil Shield. The documented U-Pb baddeleyite age is c. 200 my older than the previous K-Ar ages (see Santos et al. 2002a for review). The Crepori Diabase is composed of sills and dykes of tholeiitic gabbro and diabase, as well as differentiated rocks (andesite and granophyre). The sills and dykes stricking N55–75W and N10–15E are intrusive into the sedimentary rocks of the Palmares Group (<1.89 Ga) (Santos et al. 2001, 2002). From a chemical point of view (Almeida et al. 2000) the Crepori rocks are basalts, trachybasalts, tephrites and trachyandesites of alkaline affinity (alkali-basalts or alkali-olivine basalts).

The Crepori Diabase also has good age correlation with the Teles Pires Suite (1780–1760 Ma) which is located south of the Cachimbo Graben (Fig. 12). This suite consists of calc-alkaline granitic and (high-K) felsic volcanic rocks and unmetamorphosed volcanic-sedimentary rocks. The associated Colider Group is dominated by acid volcanics, showing medium- to high-K or shoshonitic calc-alkaline and subalkaline compositions, while the subordinate rocks with trachytic-rhyolitic composition and slightly alkalic (e.g., Santos et al. 2001, 2002b, 2004, 2008; Neder et al. 2002; Pinho et al. 2003; Bispo-Santos et al. 2008; Scandolara et al. 2014; CPRM 2014). The bulk plutonic-volcanic association continuously underlies the southern edge of the Cachimbo Graben (Figs. 1 and 12). These rocks contain significant Au-deposits in the mafic rocks (1.78–1.77 Ga), known as the Alta Floresta Gold Province (Bettencourt et al. 2016 and references therein).

The Cachimbo Graben comprises the basal Sumaúma Group, which is composed of two distinct lithostratigraphic units: the lower volcanosedimentary Vila do Carmo Group (1.76–1.74 Ga; rift phase) and the upper (>1.03 Ga; post-rift phase) sedimentary Beneficente Group (Reis et al. 2005, 2013b). The Vila do Carmo Group is intruded by the 1576 Ma (see Table 1) Mata-Matá gabbro sill (Betiollo et al. 2009), pointing to significantly younger intraplate activity than the Crepori Diabase within the southern portion of the Ventuari-Tapajós province. The Sumaúma Supergroup in turn is unconformably overlain by flat-lying, sedimentary strata of the Alto Tapajós Group (Paleozoic in age).

From a geodynamic perspective, the Teles Pires-Colider association has been considered to be subduction-related (Juruena orogeny) akin to a post-collisional (back arc) setting. The volcanic-plutonic rocks occur over the already stable tectonic environment east of the 1.81–1.74 Ga Juruena belt (i.e., southern border of the Ventuari-Tapajós province after Cordani and Teixeira 2007), similar as the tectonic setting of the older Uatumã rocks which occur east of the Ventuari-Tapajós province and are similarly considered to be subduction-related (see above). The age matches between the Teles Pires and Colider magmatisms with the Juruena accretionary orogeny are consistent with post-collisional manifestations (Scandolara et al. 2017). Whereas the Teles Pires and Colider rocks have similar ages with the Avanavero LIP, they may be rather formed by unrelated events because of the marked compositional and tectonic differences, as well as far distal exposures between them.

In summary, the 1.79 Ga Avanavero LIP likely marks a breakup attempt of the continental lithosphere due to mantle convection reorganization below the already stable Archean/late Paleoproterozoic continental lithosphere (proto-Amazonian Craton). This process occurred shortly after the cratonization of the Maroni-Itacaiunas province, and was resumed c. 100 my after the Uatumã SLIP event. The available ages for the mafic rocks of the Avanavero LIP suggest its evolution in two steps (1795–1780 Ma and c. 1780–1760 Ma), and therefore with a maximum life span of ~25 my (Reis et al. 2013a). This life span has been considered the result of a migrated mantle plume (Santos et al. 2002a), though it may be fortuitous due to the lack of more age data.

The potential barcode matches of the Avanavero LIP (see Table 3) around the world are consistent with many paleotectonic reconstructions which consider Amazonia, Baltica, Laurentia and West Africa possibly constituted a single landmass during much of the Proterozoic time (e.g., Zhao et al. 2004; Johansson 2009, 2014; Bispo-Santos et al. 2008; D'Agrella-Filho et al. 2016). From this paleogeographic perspective, there is also a plausible correlation between the Teles Pires Suite and Colider Group volcanics which occur along the Ventuari-Tapajós/Rio Negro-Juruena border (Proterozoic Amazonian proto-craton) and the Småland-Värmland Granitic-Porphyry Belt, particularly those which are roughly coeval to the TIB-1 phase (1.81–1.76 Ga) of the Transscandinavian Igneous Belt, and including the calc-alkaline c. 1.77 Ga Tving granitoids of Blekinge (see for review Högdahl et al. 2004; Pisarevsky and Bylund 2010). The latter rocks occupy a comparable position along the border between the Svecofennian and Gothian/Sveconorwegian provinces of Proterozoic Fennoscandia proto-craton, and are subduction-related (e.g., Johansson 2009), similar to the Teles Pires-Colider association. However, an extensional setting with mantle upwelling in a back-arc setting could also be envisaged for the Småland-Värmland rocks.

4.4 The 1.11 Ga Rincón del Tigre-Huanchaca LIP Event

Figure 3a outlines the geologic-tectonic framework of the SW portion of the Amazonian Craton, where the Rincón del Tigre-Huanchaca LIP occurs (Teixeira et al. 2015). This LIP has also a component (Rio Perdido Suite) in the distal Rio Apa Terrane, according to precise U-Pb age dating in baddeleyite (Teixeira et al. 2018). These LIP components can be characterized as follow: (i) mafic dykes and sills (Huanchaca and Rio Perdido Suites); (ii) the Rincón del Tigre layered Complex and; (iii) the Rio Branco Suite.

4.4.1 Huanchaca and Rio Perdido Suites

Mafic rocks of the Huanchaca Suite (Lima et al. 2012) crosscut the Paraguá Terrane (Bolivia; Fig. 3a) and the Huanchaca-Aguapeí sedimentary cover (1.17–1.15 Ga) (e.g., Litherland et al. 1986; Litherland and Powel 1989; Teixeira et al. 2010; Lima et al. 2012). The magmatism consists of large sills 50–200 m thick (trending N30 W; dipping ~10°SW) and associated (E-W oriented) dykes of ca. 150 km^2 (Fig. 13a, b). The sills have a documented U-Pb baddeleyite age of 1112 ± 2 Ma (Teixeira et al. 2015). Therefore all previously published K-Ar and $^{40}Ar/^{39}Ar$ ages between 1040 and 845 Ma (Litherland et al. 1986; Santos et al. 2002a; Lima et al. 2012) are isotopically reset. The Huanchaca Suite is classified as intraplate andesitic basalts of tholeiitic affinity and subalkaline character (Lima et al. 2012).

The Rio Perdido Suite (Faleiros et al. 2016 and references therein) comprises NW-SE and NWN-ESE oriented dykes (see Fig. 3a). One of these dykes (N85 W oriented; ~5 m thick) has a documented U-Pb baddeleyite age of 1110.1 ± 1.4 Ma (Teixeira et al. 2018). Subordinate E-W oriented dykes and NE-SW trending basaltic sills are also included in the Rio Perdido Suite, but for those we do not have any age constraints. The dykes (gabbros and diabases; Fig. 13c) are up to 30 m thick and crosscut mainly the southern exposure of the Rio Apa Terrane (1.95–1.72 Ga). The whole dyke swarm appears to be ca. 140 km wide with the traceable length as long as 200 km, according to aerogeophysical (total magnetic field) and radar images (Lacerda Filho 2015; Faleiros et al. 2016) (see Fig. 3b). Geochemical data suggest that the Rio Perdido dykes are andesitic basalts of tholeiitic affinity but with controversial

Fig. 13 Field aspects of the Huanchaca and Rio Perdido magmatism: **a** Landscape view of the Huanchaca sill (wheathered rock boulders in the foreground), intrusive into sedimentary strata of the Huanchaca-Aguapeí platform cover (top); **b** Block of mafic dyke (Huanchaca); **c** Rio Perdido dyke (Photos by A. S. Ruiz). **d** Outcrop view of fractured, pink granophyre (upper member of the Rincón del Tigre complex (Photo R. Matos))

signatures suggesting either intraplate (Lima 2016) or subduction-related magmatism (Remédio and Faleiros 2014).

The Rio Perdido dyke swarm is located ca. 300–1000 km SE from the other components of Rincón del Tigre-Huanchaca LIP (Teixeira et al. 2015) which implies a much larger scale for this LIP than previously thought. Furthermore, one may consider that distances could not be original because of the allochtonous character of the Rio Apa Terrane.

4.4.2 The Rincón del Tigre Layered Complex

This complex (e.g., Darbyshire 1979; Litherland et al. 1986 and references threrein) is a layered mafic-ultramafic sill (~4.8 km in thickness) within the Sunsas belt in Bolivia (see Fig. 3a). It consists of basal ultrabasic types (harzburgites, olivine bronzitites), intermediate gabbros and upper granophyres (Fig. 13d). The Rincón del Tigre Complex intruded along the contact between the Sunsas Group (below) and Vibosi Group (above), whose sedimentary strata experienced low-grade metamorphism and folding during the Sunsas collisional orogeny (e.g., Teixeira et al. 2010 and references therein)—see Sect. 2.

The Rincón del Tigre Complex is precisely dated at 1110 ± 2 Ma (see Table 1) by U-Pb baddeleyite age from the granophyres (Teixeira et al. 2015). There is a Ni-Cu-PGE ore potential for the Rincón del Tigre Complex, suggested by the geochemical anomalies and associated mineral occurrences. Specifically associated with the base of the magnetite gabbro portion, there occurs a precious metals zone with low grade sulfide and precious metals mineralization, 80–185 m thick (Prendergast 2000).

4.4.3 The Rio Branco Suite

The ~1.11 Ga Rio Branco magmatism (Brazilian state of Rondônia) is coeval with the crustal shortening/extension in the Nova Brasilândia belt (see Fig. 3a), contemporaneous with the Sunsas (Grenville) orogen (see Sect. 2) (e.g., Tohver et al. 2004, 2005, 2006). The magmatism comprises mafic and felsic plutonic and meta-plutonic intrusions: gabbro (1110 ± 8 Ma), metagabbro, amphibolite, metadiabase, and within-plate granite (1098 ± 10 Ma).

It is worth mentioning that roughly coeval mangerite-charnockite-rapakivi granite (MCG) plutonism is apparent in the state of Rondônia, SW portion of the Amazonian Craton (Fig. 12 in Bettencourt et al. 2016). For instance, the plutons of the Santa Clara Suite yields SHRIMP zircon U-Pb ages between 1082 and 1074 Ma (Bettencourt et al. 2010) (see Table 2). They are emplaced in major transpressional to extensional structures which transect the portion of the Rio Negro-Juruena province which was reworked at the time of the Sunsas collision (e.g., Bettencourt et al. 1999, 2016 and references therein).

Finally, the components of the 1.11 Ga Rincón del Tigre LIP broadly follow the arcuate structural trend of the Alto Guaporé and Sunsas belts (see Fig. 3a, and Sect. 2). This suggests a protracted causal relationship, controlled by a large-scale crustal weakness along the SW margin of the proto-Craton. The LIP components (intraplate character) have good age correlation with the timing of the Sunsas orogen, as illustrated in Fig. 14. From a geodynamic point of view, the nature of the Rincón del Tigre-Huanchaca LIP could involve the action of a subjacent mantle plume producing heat under the thickened crust—the stable foreland in relation to the Sunsas orogen. In a broader perspective this event has a remarkable match with prominent intraplate activity in other places around the world, as summarized in Table 3.

5 Other Mesoproterozoic Magmatism and Rifting

Intermittent magmatic episodes pre-dating the Rincón del Tigre LIP are apparent, mostly associated with basin tectonics and reactived structures in the SW portion of the Amazonian Craton (see Figs. 4 and 5). The intraplate activity can be tentatively grouped into two intervals, 1.57–1.37 and 1.20–1.17 Ga, given the age compilation,

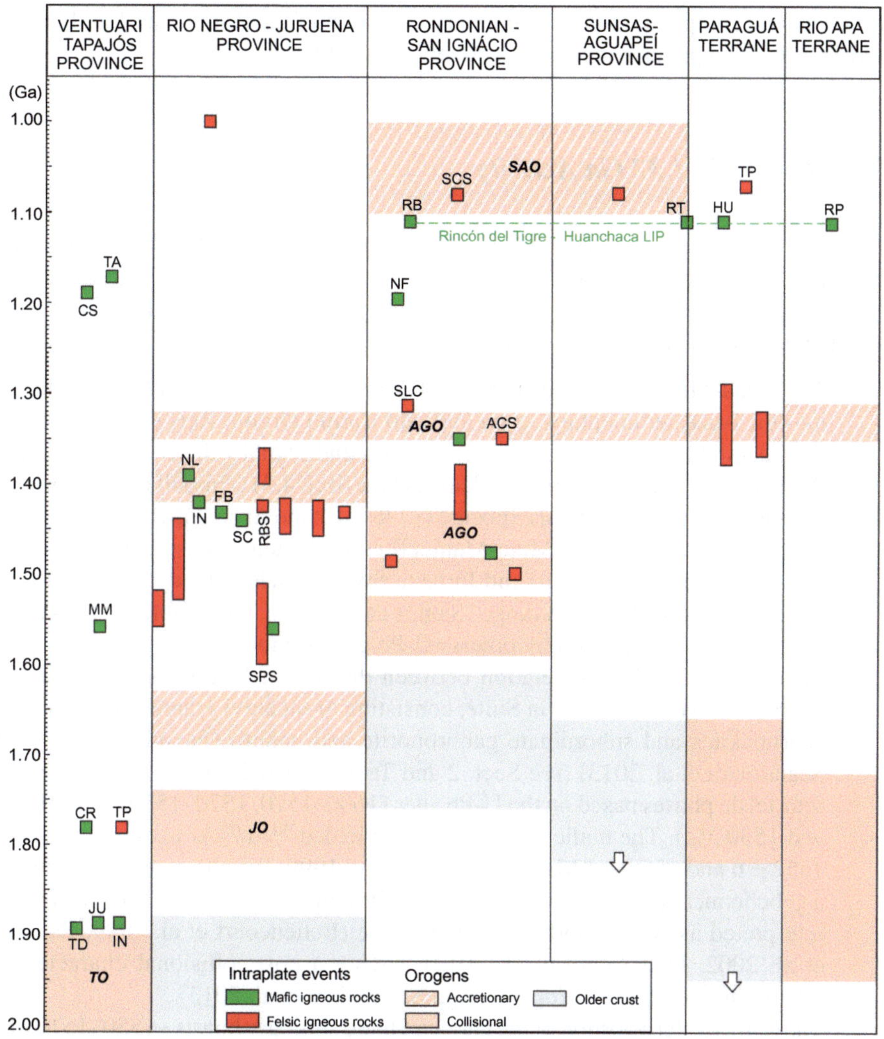

Fig. 14 Time-diagram comparing ages of anorogenic and post-collisional mafic and felsic magmatism in the Amazonian Craton with the timing of accretionary orogens along its active SW margin. Adapted from Bettencourt et al. (2010), Faleiros et al. (2016) and Scandolara et al. (2017). Green and red dots are individual U-Pb crystallization ages (data shown in Tables 1 and 2) of the intraplate (post-collisional and/or anorogenic) episodes. Mafic igneous rocks: RP = Rio Perdido Suite, HU = Huanchaca Suite, RT = Rincón del Tigre Complex, RB = Rio Branco Suite (Rondônia state), NF = Nova Floresta Gabbro, TA = Taparuquara Suite, CS = Cachoeira Seca Troctolite, NL = Nova Lacerda swarm, IN = Indiavaí Suite, FB = Figueira Branca Suite, SC = Salto do Céu Gabbro, MM = Mata Matá Gabbro, CR = Crepori Diabase, TD = Teodorico Diabase, Ingarana Suite (JU = Jutaí, IN = Ingarana). Felsic igneous rocks: TP = Teles Pires Suite, SPS = Serra da Providência Suite, SCS = Santa Clara Suite, SLC = São Lourenço-Caripunas Suite, TP = Taperas Suite. Other red bars: late- to post-orogenic and anorogenic granitic suites (data compiled from Fig. 5 in Bettencourt et al. (2010) and from Fig. 12 in Scandolara et al. (2017)). Shaded salmon bands: Proterozoic crustal growth events along the SW margin of Amazonia: SAO = Sunsas Orogen, AGO = Alto Guaporé Orogen, JO = Juruena Orogen, TO = Tapajonic Orogen. Older crust (gray bars): inferred from published Sm–Nd T_{DM} ages. See text for explanation

which can be compared to the accretionary evolution of the Rondonian-San Ignacio and Sunsas-Aguapeí provinces, respectively (see Table 1).

5.1 The 1.57–1.37 Ga Activity

This magmatism (e.g., dolerite dykes, bimodal magmatism, mafic-ultramafic complexes) is well-dated by U-Pb baddeleyite geochronology (Table 1). Representative examples are concentrated in the SE portion of the Rio Negro-Juruena province, as summarized below in three age groups:

(i) The Mata-Matá gabbro sill (1576 ± 4 Ma; Betiollo et al. 2009) is emplaced into the basal sedimentary strata (Vila do Carmo Group; Reis et al. 2013b) of the Cachimbo Graben (see Fig. 12) which is underlain by the c. 1790 Ma Teles Pires Suite and Colider Group volcanics (see Sect. 4.3). This sill is 250 m thick, cropping out as an elliptical exposure (~1 km^2) in the Aripuanã River (Brazilian state of Amazonas). The Mata-Matá sill could be tentatively related to some diabase dykes along the Iriri and Juruena rivers, which yielded minimum K-Ar ages in the 1470–1400 Ma range (Santos et al. 2002a; their Table 2). However, this remains to be proved by precise U-Pb geochronology.
There is a good age correlation between the Mata-Matá gabbro sill and the nearby Serra da Providência Suite, consisting of rapakivi granites, mangerites, charnockites and subordinate gabbronorite and gabbro (Payolla et al. 2002; Scandolara et al. 2013) (see Sect. 2 and Tables 1 and 2). This suite comprises four felsic phases based on the U-Pb ages (1600–1570, 1570–1560, 1550–1540 and 1530 Ma). The mafic representatives yielded $^{40}Ar/^{39}Ar$ ages of 1554 ± 6, 1552 ± 6 and 1558 ± 4 Ma (Bettencourt et al. 1999; Payolla et al. 2002). From a geochemical point of view the Serra da Providência (AMCG) Suite has been interpreted as A-type and within-plate type (Bettencourt et al. 1999; Payolla et al. 2002; Dall'Agnol et al. 2012), having a post-collisional character in relation to the Juruena orogeny (e.g., Scandolara et al. 2013).
The northwestern portion of the Guiana Shield similarly boasts ~1550–1510 Ma intraplate activity (see Table 2), such as the Parguaza and the Surucucus rapakivi granites in the Venezuela and Brazilian state of Roraima (e.g., Tassinari et al. 1996; Dall'Agnol et al. 2012; Santos et al. 2003. Cordani and Teixeira 2007 and references therein), as well as the Mucajaí mangerite/granite and the closely related Repartimento massive-type anorthosite in the Amazonas state (see Fig. 2). In a global scale, the Mata-Matá magmatism has a few barcode matches such as the Ålandand (buried) Riga rapakivi granites and Värmland dolerite intrusions in Fennoscandia and the Gawler Range event in Australia (Wahlgren et al. 1996; Fanning et al. 1988; Åhäll et al. 2000; Söderlund et al. 2005; Ernst 2014).

(ii) The Käyser NW-SE trending swarm of dykes occurs in southwestern Suriname (see Fig. 2), composed of olivine dolerites of alkaline character (Delor et al.

2003; De Roever et al. 2014 and references therein). The emplacement is precisely dated at 1528 ± 2 Ma (weighted mean $^{207}Pb/^{206}Pb$ age in baddeleyite), and is associated to major faults that crosscut the Paleoproterozoic crust of the Maroni-Itacaiunas province. The age of the Käyser Dolerite swarm provides an exact age match with the oldest dykes of the Essakane swarm of Burkina Faso, in the West African Craton (De Roever et al. 2014). According to these authors this barcode match supports that Amazonia and the West African Craton should be treated as a single crustal block at the time frame.

(iii) The Salto do Céu gabbro sill (1439 ± 4 Ma; Teixeira et al. 2016) is closely related with the anorogenic Rio Branco Suite (see Table 2), which comprises felsic members (syenites to rapakivi granites), and mafic members (gabbro, tholeiitic diabase dykes and porphyritic basalt). The gabbro sill (see Fig. 3a) is emplaced into a flat-lying platform sequence (<1.54 Ga) underlain by 1.82–1.79 Ga crystalline basement at the SE edge of the Rio Negro-Juruena province (e.g., Geraldes et al. 2004). In a regional perspective, this mafic event is roughly coeval with the syntectonic stage (1435 Ga) in the Alto Guaporé orogeny located to the west (see Sect. 2 and Fig. 3a). There is thus a coincidence in time between convergent-margin processes and the within-plate activity (i.e. Salto do Céu sill and coeval rocks) in the already stable parts of the SW Amazonian Craton. It is worth mentioning that similar-aged anorogenic activity is also present in the Guiana Shield, such as the 1.43–1.42 Ga co-magmatic charnockites and rapakivi granites of the Serra Grande Suite (Santos et al. 2011) which is ca. 2000 away from the Salto do Céu sill.

(iv) The Figueira Branca layered mafic–ultramafic Complex and the coeval Indiavaí gabbro and felsic-mafic plugs (1429 ± 3–1416 ± 7 Ma) occur the southwestern portion of the Rio Negro-Juruena province (Teixeira et al. 2011, 2016; D'Agrella-Filho et al. 2012). The age-equivalent NW-trending Nova Guarita dyke swarm (1418 ± 3 Ma; Bispo-Santos et al. 2012) composed of gabbro, microgabbro and diabase is located ca. 700 km northward in the Central-Brazil Shield. Contemporaneous anorogenic activity is also present within the Guiana Shield (such as the 1434–1425 Ma co-magmatic charnockites and rapakivi granites of the Serra Grande Suite (Santos et al. 2011). They are emplaced into the NW portion of the Central Amazonian province, ca. 2000 km away from the Figueira Branca Complex, suggesting therefore LIP scale magmatism.

(v) The NW-oriented Nova Lacerda dyke swarm (1387 ± 17 Ma) occurs not far from the Salto do Céu sill (Teixeira et al. 2011, 2016). The dykes likely derived from heterogenous mantle sources enriched by subduction-related metasomatism, according to the geochemical and isotopic data (Girardi et al. 2013). The Nova Lacerda swarm has age match with cratogenic granitic suites in the SW portion of the craton, such as the Alvorada, Teotônio, San Ramon and Santo Antônio plutons and associated dykes (Bettencourt et al. 2010 and references therein).

In summary, all these activities between 1.57 and 1.37 Ga can be considered as discrete extensional episodes within already stable parts, contemporaneous with the crustal growth steps along the active SW continental margin of Mesoproterozoic Amazonia. Figure 14 illustrates the potential age matches between the accretionary/collisional orogenies with the intraplate activity (felsic and mafic igneous events. The recurrent activity (Rb-Sr and K-Ar datings between 1.51 and 1.45 Ga, as well as few U-Pb ages; see Fig. 4), is particularly significant along reactived faults and/or shear zones (e.g., Cachorro and Tapajóslineaments). In addition, along the NW margin of the Amazonian Craton the recently dated Seis Lagos carbonatite Complex (1328–1154 Ma; Rossoni et al. 2017) is a further evidence of the withinplate magmatism. The carbonatite is emplaced into orthogneisses that are as old as 1.81–1.78 Ga (Almeida et al. 2007).

In a broader perspective, this intermittent Mesoproterozoic may have a linkage with global convergent dynamics, but could represent inboard extensional episodes within and overall convergent system. In this sense, the 1.57–1.37 Ga activity of Amazonia is roughly coeval with major intracontinental-related A-type felsic magmatism and mafic dykes (e.g., 1.47–1.42 Ga AMCG suites) such as in Eastern Laurentia and Fennoscandia, as well as in the West African Craton (e.g., Anderson and Bender 1989; McLelland 1989; Rämö and Haapala 1995; Åhäll et al. 2000; Gower and Krogh 2002; Högdahl et al. 2004; Brander and Söderlund 2009; Zariņš and Johansson 2009; Johansson et al. 2016). This particular Mesoproterozoic intraplate magmatism may be attempts of break-up in different parts of Columbia, penecontemporaneous with growing parts of the supercontinent (e.g., Rogers and Santosh 2002; Zhao et al. 2004; D'Agrella-Filho et al. 2016).

5.2 *The 1.20–1.17 Ga Activity*

Magmatic activity of this age range comprises mafic dykes and flows, laccoliths, anorthosite complexes with a regional distribution in the Amazonian Craton (see Figs. 4 and 5). This intraplate activity is associated with extension events (e.g., Teixeira 1990; Santos et al. 2002a) that also formed the Tonian rift basins such as in the SW portion of the Amazonian Craton (e.g., Palmeiral, Apiacás; see Figs. 3a and 12). The available ages of the mafic magmatism suggest a causal correlation between intraplate episodes and the Sunsas collision along the SW continental margin of the proto-Amazonian Craton (Tohver et al. 2006; Cordani et al. 2010a; Teixeira et al. 2010). Figure 14 illustrates the age correlation between the intraplate magmatism and orogenic events. The geodynamic scenario (1.10–1.00 Ga time frame) includes regional shearing and tectonic reactivation along crustal weakness zones such as the Ji-Paraná strike-slip shear network (1180–1150 Ma; Tohver et al. 2005) in Rondônia and the system of aborted rifts that evolved into the Huanchaca/Aguapeí basin in Bolivia (e.g., Teixeira et al. 2010 and references therein). Anorogenic granites, alkaline ring complexes, as well as pervasive isotopic resetting of the country rocks in some localized regions (e.g., K'Mudku tectonothermal episode in the Guiana Shield) are also reflective of the orogenic dynamics along the active SW margin of Amazonia (Cordani et al. 2010a for a review).

Mafic dykes with K-Ar ages in this range are concentrated in the SW edge of the Rio Negro-Juruena, Rondonian-San Ignacio and Sunsas-Aguapeí provinces, as shown in Fig. 4. This preferential geographic distribution is consistent with the general younging of the Amazonian Craton towards the southwest (current coordinates), with the dykes being related to inboard intraplate magmatism, according to the Cordani and Teixeira (2007) model. However, one may suspect that many K-Ar ages could have been reset (by Ar loss) due to the Mesoproterozoic polycyclic metamorphism and deformation, and thus they need to be confirmed by U-Pb dating. The representative examples of the 1.22–1.17 Ga intraplate mafic activity are summarized below:

The Nova Floresta sills (see Fig. 3a) display alkaline affinity and have documented $^{40}Ar/^{39}Ar$ ages of 1198 ± 3 to 1201 ± 2 Ma (Tohver et al. 2002). These ages are identical to a new U-Pb baddelyite dating age from a gabbro sill (M. Hamilton, written comm., 2017). The associated K-Ar ages are significantly younger, between 1118 and 1045 Ga (Santos et al. 2002a, see Table 2), reflecting Ar loss. Intraplate activity at c. 1.22 Ga is rare in the Amazonian Craton; one example could be the Canamã syenite (Rb-Sr age of c. 1217 Ma; Teixeira 1990), which occurs far north at the Cachorro lineament (Fig. 4) in the Guiana Shield. There are roughly coeval events like this globally, such as the Fraser Dyke Swarm in the Yilgarn Craton (Australia) and the dolerite dykes, associated syenites and granites along the Protogine Zone in Fennoscandia (e.g., Johansson 1991; Wingate et al. 2000; Pirajno and Hoatson 2012; Ernst 2014). The slightly older (c. 1.27 Ga) Mackenzie dyke swarm and associated sills and layered intrusions in Laurentia (e.g., Fahrig 1987), as well as the widespread 1.27–1.25 Ga Central Scandinavian Dolerite Group in north-central Sweden (Söderlund et al. 2005, 2006; Söderlund and Ask 2006) could be also associated.

The Cachoeira Seca lacoliths and dykes of troctolite and olivine gabbro crop out in Tapajós River region (Santos et al. 2002a), hosted by the Parauari Suite (1.89–1.87 Ga), which is tectonically ascribed to the Uatumã SLIP (see Sect. 4.2 and Fig. 7). The troctolite has a documented U-Pb baddeleyite crystallization age of 1186 ± 12 Ma (see Table 1). This age is c. 110–140 my older that the K-Ar ages for associated dykes (mean 1060 Ma). Potential contemporary mafic activity occurs elsewhere in the Central-Brazil Shield, as also suggested by a broad range of K-Ar ages from 1250 to 910 Ma (see Fig. 5). However, these dates need more precise age constraints. These mafic intrusions have dominantly tholeiitic affinity (e.g., Santos et al. 2002a and references therein).

The Cachoeira Seca magmatism also has age matches with the Tapuruquara Suite (see Fig. 2), comprised of several stocks of gabbros, as well as websterites and lherzolites, occurring far northward in the Guiana Shield (NW Amazonas state) (Reis et al. 2006). A representative gabbro from this suite yielded a SHRIMP baddeleyite age of 1172 ± 8 Ma, while two previous K-Ar datings gave 2250 and 2910 Ma ages, indicating excess Argon (Santos et al. 2006). The grabbroic rocks are high-alumina, showing strong subalkaline and silica-subsaturated character.

We also note that the Seis Lagos carbonatite Complex might be as young as 1187–1154 Ma (see previous section). If so there is a good age correlation with the nearby Tapuruquara mafic-ultramafic complex (1172 Ma; see Table 1). Both complexes could be related to the K'Mudku tectonothermal episode (e.g., Cordani et al. 2010a) that cut across the western portion of the Central Amazonian and Maroni-Itacaiunas provinces. In the Guiana Shield the alkaline sills and dykes of the Seringa Formation (see Fig. 2) have with minimum K-Ar ages from 1090 to 1080 Ma (Araujo Neto and Moreira 1976), but this needs to be confirmed by more robust geochronology.

Finally, the post-tectonic, A-type Taperas Granite (Sunsas Suite; see Table 2) which is intrusive into the Paraguá Terrane (see Fig. 3) yielded a zircon U-Pb age of 1076 ± 18 Ma (Boger et al. 2005). Whereas this granite is age-equivalent with the Cachoeira Seca troctolite, it could be associated with a particular extension episode within the Amazonian Craton that overlaps in time. Much younger occurrences of AMCG plutons (990–970 Ma) collectively known as the Rondônia Tin Province represent the anorogenic manifestations over in the Rio Negro-Juruena and Rondonian-San Ignacio provinces (Bettencourt et al. 1999; Payolla et al. 2002; Teixeira et al. 2010).

6 Summary and Conclusions

We have reviewed the geochronology, geochemistry and geologic setting of distinct mafic dyke swarms and sills and roughly coeval mafic and felsic plutonic and volcanic rocks in the Amazonian Craton. The conclusions and possible tectonic implications are summarized below:

- Four important and voluminous events of intracratonic magmatisms are apparent in the Amazonian Craton: the Orocaima (1.98–1.96 Ga) and Uatumã (1.89–1.87 Ga) SLIP events, and the Avanavero (1.79–1.78 Ga) and the Rincón del Tigre-Huanchaca LIP events (1.11 Ga).
- Both the Orocaima and Uatumã SLIP events consist of calc-alkaline I-type and subordinate A-type plutonic and volcanic rocks of widespread occurrence across the Central Amazonian and Maroni-Itacaiunas provinces. The Orocaima event has age matches with gabbros and associated charnockites and anorthosites which cross-cut the Bakhuis belt, in Suriname. The Uatumã event likely includes the Tucumã bimodal swarm and nearby mafic to intermediate dykes and intraplate granites in the Central-Brazil Shield, given the age matches. It is also age-correlative to the Uraricaá Suite, the Rio Pardo rapakivi Association and some A-type and I-type plutons in the Guiana Shield.
- From a geodynamic point of view both the Orocaima and Uatumã SLIPs appears to be triggered by subduction-related processes with consequent extension of the continental lithosphere during these time periods. Their formation could be related to thermal perturbations in the upper mantle with associated mafic underplating, plume-like.
- The Avanavero event comprises mostly mafic sills and dykes with widespread occurrence in the Guiana Shield. Coeval dykes (Crepori Diabase) occur in the

Central-Brazil Shield, as well as other age-equivalent units such as the I-type, plutonic-volcanic Teles Pires Suite and the associated Colider Group in the Ventuari-Tapajós province. These suites with predominantly calc-alkaline composition have been related to a back-arc setting akin to the Juruena orogeny, as illustrated in Fig. 14.

- The 1.11 Ga LIP event previously defined in the SW portion of the Amazonian Craton (Huanchaca sills and dykes, Rio Branco Suite, Rincón del Tigre Complex) has tight age match with the Rio Perdido dyke in the distal Rio Apa Terrane. There is roughly contemporary anorogenic magmatism with this LIP elsewhere in the Amazonian Craton. In a broader scale this event is associated with the Grenville orogeny that consolidated the southwestern margin of the craton.
- Other intracratonic mafic activities are apparent in the main Craton, exemplified by the precisely dated: Mata-Matá sill (1.57 Ga), Käyser Dolerite (1.53 Ga), Salto do Céu (1.44 Ga) and Nova Floresta (1.22 Ga) sills, Cachoeira Seca Troctolite (1.19 Ga) and Tapuruquara Suite (1.17 Ga). These events could be associated in time and space with reativated structures and emplacement of A-type, rapakivi granites which largely occur in the SW portion of the Amazonian Craton. Taken together these 1.57–1.17 Ga episodes are also coeval with the long-lived accretion regime and convergent episodes that progressively built in Amazonia Mesoproterozoic times, as depicted in Fig. 14. The large scale of these events and the long-time period of magmatism are suggestive of a model coupling several plumes with distinct subduction-related processes through time. This is a key issue for understanding the relationship between intraplate magmatism (including that of LIP scale) with a convergent setting.
- The Paleo- to Mesoproterozoic intraplate activity of the Amazonian Craton has barcode matches with LIP events on other blocks in the context of the Columbia and Rodinia supercontinents, as summarized in Table 3.

Acknowledgements This review is a contribution to "Large Igneous Provinces-Supercontinent Reconstruction-Resource Exploration Project" (www.supercontinent.org NSERC CRDPJ 419503-11; www.camiro.org/exploration/ongoing-projects CAMIRO Project 08E03). Useful comments from R. Ernst on an early version of the manuscript are appreciated. We also thank Åke Johansson and Leonid Shumlyanskyy for their thorough and constructive reviews, which resulted in significant improvements to this paper. WT and EK are thankful to the National Council of Technological and Scientific Development (CNPq) for the research productivity scholarship grants (303498/2014-5 and 306798/2016-6, respectively). F. Bispo-Santos, A. S. Ruiz, G. Lima, O. Lazcano, F. M. Faleiros, R. Matos and P. Y. J. Antonio provided the valuable photos and key outcrop descriptions, selected for this work.

References

Åhäll K-I, Connelly JN, Brewer TS (2000) Episodic rapakivi magmatism due to distal orogenesis? Correlation of 1.69–1.50 Ga orogenic and inboard, "anorogenic" events in the Baltic Shield. Geology 28:823–826

Almeida ME, Brito MFL, Ferreira A (2000) Projeto Especial Província Mineral do Tapajós: Geologia da folha SB.21-V-D- Mamãe Anã. Estado do Pará. Escala 1:250,000. Superintendência Regional de Manaus, 238 p

Almeida ME, Macambira MJB, Reis NJ, Luzardo R, da Pinheiro S (2007) Geologia, geoquímica multielementar e isotópica (Sm-Nd) das rochas do embasamento do extremo oeste da Província Rio Negro, NW do Amazonas, Brasil. In: SBG-Núcleo Norte, Anais do Simpósio de Geologia da Amazônia 10, Porto Velho, Brasil, 26–29

Amaral G (1974) Geologia Pré-cambriana da região Amazônica. Tese de Doutorado, Instituto de Geociências, Universidade de São Paulo, Brazil, 212 pp

Anderson JL, Bender EE (1989) Nature and origin of Proterozoic A-type granitic magmatism in the southwestern United States of America. Lithos 23:19–52

Antonio PYJ, D'Agrella-Filho MS, Trindade RIF, Nédélec A, Oliveira DC, Silva FF, Roverato M, Lana C (2017) Turmoil before the boring billion: Paleomagnetism of the 1880–1860 Ma Uatumã event in the Amazonian craton. Gondwana Res 49:106–129

Araujo Neto H, Moreira HL (1976) Projeto Estanho de Abonari. Relatório Final, DNPM/CPRM, Manaus, 2 v

Barreto CJS, Lafon JM, da Rosa-Costa LT, de Lima EF (2013) Vulcanismo félsicopaleoproterozoico do Grupo Iricoumé, Domínio Erepecuru-Trombetas, Província Amazônia Central: dados de campo, caracterização petrográfica e geocronologia Pb-Pb em zircão. Geologia USP, Série Científica 13(1):4–72

Bastos Neto AC, Ferron TMM, Chauvet A, Chemale F Jr, Lima EF, Barbanson L, Costa CFM (2014) U-Pb dating of the Madeira Suite and structural control of the albite-enriched granite at Pitinga (Amazonia, Brazil): evolution of the A-type magmatism and implications for the genesis of the Madeira Sn-Ta-Nb (REE, cryolite) world-class deposit. Precambr Res 243:181–196

Betiollo LM, Almeida ME, Reis NJ, Bahia RBC, Splendor F, Costa UA, Luzardo R (2009) Magmatismo máfico calimiano (sill Mata-Matá), Rio Aripuanã, Amazonas – implicações geológicas. XI Simpósio de Geologia da Amazônia. Resumos expandidos, CDOM

Bettencourt JS, Tosdal RM, Leite Junior WB, Payolla BL (1999) Mesoproterozoic rapakivi granites of Rondonia Tin Province, south-western border of the Amazonian Craton, Brazil—I. Reconnaissance U-Pb geochronology and regional implications. Precambr Res 95:41–67

Bettencourt JS, Leite W Jr, Ruiz B, Matos AS, Payolla RS, Tosdal RM (2010) The Rondonian-San Ignacio Province in the SW Amazonian Craton: an overview. J S Am Earth Sci 29:28–46

Bettencourt JS, Juliani C, Xavier RP, Monteiro LVS, Bastos Neto AC, Klein EL, Assis R, Leite WB Jr, Moreto CPN, Fernandes CMD, Pereira VP (2016) Metallogenetic systems associated with granitoid magmatism in the Amazonian Craton: an overview of the present level of understanding and exploration significance. J S Am Earth Sci 68:22–49

Beyer SR, Hiatt EE, Kyser K, Drever GL, Marlatt J (2015) Stratigraphy, diagenesis and geological evolution of the Paleoproterozoic Roraima Basin, Guyana: links to tectonic events on the Amazon Craton and assessment for uranium mineralization potential. Precambr Res 267:227–249

Bispo-Santos F, D'Agrella-Filho MS, Pacca IIG, Janikian L, Trindade RIF, Elming S-Å, Silva JA, Barros MAS, Pinho FEC (2008) Columbia revisited: Paleomagnetic results from the 1790 Ma Colider volcanics (SW Amazonian Craton, Brazil). Precambr Res 164:40–49

Bispo-Santos F, D'Agrella-Filho MS, Trindade RIF, Elming S-A, Janikian L, Vasconcelos PM, Perillo BM, Pacca IIG, Silva JA, Barros MAS (2012) Tectonic implications of the 1419 Ma Nova Guarita mafic intrusives paleomagnetic pole (Amazonian Craton) on the longevity of Nuna. Precambr Res 196–197:1–22

Bispo-Santos F, D'Agrella-Filho MS, Janikian L, Reis NJ, Trindade RIF, Reis MAA (2014a) Towards Columbia: Paleomagnetism of 1980-1960 Ma Surumu Volcanic Rocks, Northern Amazonian Craton. Precambr Res 244:123–138

Bispo-Santos F, D'Agrella-Filho MS, Trindade RIF, Janikian L, Reis NJ (2014b) Was there SAMBA in Columbia? Paleomagnetic evidence from 1790 Ma Avanavero mafic sills (northern Amazonian Craton). Precambr Res 244:139–155

Bleeker W, Ernst R (2006) Short-lived mantle generated magmatic events and their dyke swarms: the key unlocking earth's paleogeographic record back to 2.6 Ga. In Hanski E, Mertanen S, Rämö T, Vuollo J (eds) Dyke swarms—time markers of crustal evolution. Taylor and Francis/Balkema, London, pp 3–26

Bogdanova SV, Gintov OB, Kurlovich D, Lubnina NV, Nilsson M, Orlyuk MI, Pashkevich IK, Shumlyanskyy LV, Starostenko VI (2013) Late Palaeoproterozoic mafic dyking in the Ukrainian Shield (Volgo-Sarmatia) caused by rotations during the assembly of supercontinent Columbia. Lithos 174:196–216

Boger SD, Raetz M, Giles D, Etchart E, Fanning CM (2005) U-Pb age from the Sunsás region of eastern Bolivia, evidence for allochtonous origin of the Paragua Block. Precambr Res 139:121–146

Brander L, Söderlund U (2009) Mesoproterozoic (1.47–1.44 Ga) orogenic magmatism in Fennoscandia; baddeleyite U-Pb dating of a suite of massif type anorthosite in S, Sweden. Int J Earth Sci 98:499–516

Bright RM, Amato JM, Denyszyn SW, Ernst RE (2014) U-Pb geochronology of 1.1 Ga diabase in the southwestern United States: testing models for the origin of a post-Grenville large igneous province. Lithosphere 6:135–156 (Geological Society of America). https://doi.org/10.1130/l335.1

Bryan S (2007) Silicic large Igneous Provinces. Episodes 30(1):20–31

Bryan S, Ernst RE (2008a) Revised definition of Large Igneous Provinces (LIP). Earth Sci Rev 86:175–202

Bryan S, Ernst RE (2008b) Revised definition of Large Igneous Provinces (LIPs). Earth Sci Rev 86:961–964

Buchan KL, Ernst RE, Bleeker W, Davis WJ, Villeneuve M, van Breemen O, Hamilton MA, Söderlund U (2010) Proterozoic magmatic events of the Slave craton, Wopmay orogen and environs. Geological Survey of Canada, Open File 5989, CDROM, poster, and report, 25 p

Cordani UG, Teixeira W (2007) Proterozoic accretionary belts in the Amazonian Craton. In: Hatcher Jr RD, Carlson MP, McBride JH, Martinez Catalán JR (eds) The 4D framework of continental crust. GSA Memoir. Geological Society of America Book Editors 200, Boulder, CO, pp 297–320

Cordani UG, Fraga LM, Reis NJ, Tassinari CCG, Brito-Neves BB (2010a) On the origin and tectonic significance of the intra-plate events of Grenvillian-type age in South America: a discussion. J S Am Earth Sci 29:143–159

Cordani UG, Teixeira W, Tassinari CCG, Coutinho JMV, Ruiz AS (2010b) The Rio Apa craton in Mato Grosso do Sul (Brazil) and northern Paraguay: geochronological evolution, correlations and tectonic implications. Am J Sci 310:981–1023

CPRM (2010) Geology of Brazil Program. Geology and Mineral Resources of the Vila de Tepequém Sheet, NA.20-X-A-III, 1:100.000 scale. Roraima State. In: Fraga LM, Dreher AM (org) Manaus, Amazonas, CPRM, Geological Survey of Brazil, 182 pp, CD-ROM

CPRM 2014. Geology of Brazil Program. Geology and Mineral Resources of the Sumaúma Sheet, SB.20-Z-D, 1:250,000 scale. Amazonas State. In: Almeida ME, Costa UAP (org) Sumaúma-Roosevelt-Mutum Project. Manaus, Amazonas, CPRM, Geological Survey of Brazil, 297 pp

D'Agrella-Filho MS, Trindade RIF, Elming S-Å, Teixeira W, Yokoyama E, Tohver E, Geraldes MC, Pacca IIG, Barros MAS, Ruiz AS (2012) The 1420 Ma Indiavaí Mafic Intrusion (SW Amazonian Craton): paleomagnetic results and implications for the Columbia Supercontinent. Gondwana Res 22:956–973

D'Agrella-Filho MS, Bispo-Santos F, Trindade RIF, Antonio PJY (2016) Paleomagnetism of the Amazonian Craton and its role in paleocontinents. Braz J Geol 46(2):275–299

Dall'Agnol R, Frost CD, Rämö OT (2012) IGCP Project 510 A-type granites and related rocks through time: project vita, results, and contribution to granite research. Lithos 151:1–16

Dall'Agnol R, Oliveira DC (2007) Oxidized, magnetite-series, rapakivi-type granites of Carajás, Brazil: implications for classification and petrogenesis of A-type granites. Lithos 93:215–233

Darbyshire DPF (1979) Results of the age determination programme. Unpublished Report Eastern Bolivia Mineral Exploration Project, phase I, p 9. Unpublished Open file in Bolivia (Geobol) and the United Kingdom (BGS)

De Roever E, Söderlund U, Breecker W, Klaver M (2014) A precise U-Pb baddeleyite age for the Käiser dolerite swarm in Suriname: an exact age match with mafic dykes in West African Craton. Unpublished report A-178, 1–5. www.supercontinent.org

Delor C, De Roever EWF, Lafon J-M, Lahondère D, Rossi P, Cocherie A, Guerrot C, Potrel A (2003) The Bakhuis ultrahigh-temperature granulite belt (Suriname): II. Implications for late Transamazonian crustal stretching in a revised Guiana shield framework. Geology of France and surrounding areas—Special Guiana Shield, 2-3-4, pp 207–232

Elming S-Å, Moakhar MO, Layer P, Donadini F (2009) Uplift deduced from remanente magnetization of a proterozoic basic dyke and the baked country rock in the Hoting area, Central Sweden: a palaeomagnetic and $^{40}Ar/^{39}Ar$ study. Geophys J Int 179:59–78

Ernst RE (2014) Large Igneous Provinces, 1–653. Cambridge University Press, Cambridge

Ernst ER, Bleeker W, Soderlund U, Kerr CA (2013) Large Igneous Provinces and supercontinents: toward completing the plate tectonic revolution. Lithos 230:103–118

Fahrig WF (1987) The tectonic setting of continental mafic dyke swarms: failed arm and early passive margin. In: Halls HC, Fahrig WF (eds) Mafic Dyke Swarms: Geological Association of Canada Special Paper 34, pp 331–348

Faleiros FM, Pavan M, Remédio MJ, Rodrigues JB, Almeida VV, Caltabeloti FP, Pinto LGR, Oliveira AA, Pinto de Azevedo EJ, Costa VS (2016) Zircon U-Pb ages of rocks from the Rio Apa Cratonic Terrane (Mato Grosso do Sul, Brazil): new insights for its connection with the Amazonian Craton in pre-Gondwana times. Gondwana Res 34:187–204

Fanning CM, Flint RB, Parker AJ, Ludwig KR, Blisett AH (1988) Refined Proterozoic evolution of the Gawler Craton, South Australia, through U–Pb zircon geochronology. Precambr Res 40/41:363–386

Fedotova MA, Khramov NA, Pisakin BN, Priyatkin AA (1999) Early Proterozoic palaeomagnetism: new results from the intrusives and related rocks of the Karelian, Belomorian and Kola provinces, eastern Fennoscandian Shield. Geophys J Int 137:691–712

Fernandes CMD, Juliani C, Monteiro LVS, Lagler B, Echeverri-Misas CM (2011) High-K calc-alkaline to A-type fissure-controlled volcano-plutonism of the São Felix do Xingu region, Amazonian craton, Brazil: exclusively crustal sources or only mixed Nd model ages? J S Am Earth Sci 32:351–368

Ferron JMTM, Bastos Neto AC, Lima EF, Nardi LVS, Costi HT, Pierosan R, Prado M (2010) Petrology, geochemistry and geochronology of Paleoproterozoic volcanic and granitic rocks (1.89 to 1.87 Ga) of the Pitinga Province, Amazonian Craton, Brazil. J S Am Earth Sci 29(2):483–497

Fraga LM, Macambira MJB, Dall'Agnol R, Sena JB (2009) 1.94–1.93 Ga charnockitic magmatism from the central part of the Guyana Shield, Roraima, Brazil: single-zircon evaporation data and tectonic implications. J S Am Earth Sci 27:247–257

Fraga LM, Vasquez ML, Almeida ME, Dreher AM, Reis NJ (2017) A influência da orogenia Eo-Orosiriana na formação da SLIP Uatumã, parte central do Cráton Amazônico. Simpósio de Geologia da Amazônia 15:405–408

Geraldes MC, Heilbron MCP, Teixeira W (2004) Lithospheric versusasthenospheric source of the SW Amazonian craton A-type granites: the role of the Paleo- and Mesoproterozoic accretionary belts for their coevalcontinental suites. Episodes 27:1–5

Gibbs AK (1987) Contrasting styles of continental mafic intrusions in the Guiana Shield. In: Halls HC, Fahrig WF (eds) Mafic Dyke Swarms: Geological Association of Canada Special Paper, 34, pp 457–465

Gibbs AK, Barron CN (1993) The geology of the Guiana Shield. Oxford University Press, Claredon Press, New York, 245 pp

Girardi VAV, Teixeira W, Mazzucchelli M, Corrêa da Costa PC (2013) Sr-Nd constraints and trace-elements geochemistry of selected Paleo and Mesoproterozoic mafic dykes and related intrusions from the South American Platform: insights into their mantle sources and geodynamic implication. J S Am Earth Sci 41:65–82

Gower C, Krogh TE (2002) A U-Pb geochronological review of the Proterozoic history of the eastern Grenville Province. Can J Earth Sci 39:795–829

Halls HC (1982) The importance and potential of mafic dyke swarms in studies of geodynamic process. Geosci Can 9:145–154

Halls HC, Fahrig WF (eds) (1987) Mafic Dyke Swarms. St John's, NL. Geological Association of Canada, Special Publication 34

Hebeda EH, Boelrijk NAIM, Priem HNA, Verdurmen EATH, Verschure RA (1973) Excess radiogenic argon in the Precambrian Avanavero dolerite in western Surinam (South America). Earth Planet Sci Lett 20:189–200

Högdahl K, Andersson UB, Eklund O (eds) (2004) The Transscandinavian Igneous Belt (TIB) in Sweden: a review of its character and evalution. Geological Survey of Finland, Special Paper 37, 123 pp

Johansson Å (1991) Age of the Önnestad syenite and some gneissic granites along the southern part of the Protogine Zone, southern Sweden. In: Gower CF, Rivers T, Ryan B (eds) Mid-Proterozoic Laurentia-Baltica. Geological Association of Canada, Special Paper 38, pp 131–148

Johansson A (2009) Baltica, Amazonia and the SAMBA connection—1000 million years of neighbourhood during the Proterozoic? Precambr Res 175:221–234

Johansson Å (2014) From Rodinia to Gondwana with the "SAMBA" model: a distant view from Baltica towards Amazonia and beyond. Precambr Res 244:226–235

Johansson Å, Waight T, Andersen T, Simonsen SL (2016) Geochemistry and petrogenesis of Mesoproterozoic A-type granitoids from the Danish island of Bornholm, southern Fennoscandia. Lithos 244:94–108

Jowitt SM, Ernst RE (2013) Geochemical assessment of the metallogenic potential of Proterozoic LIPs of Canada. Lithos 174:291–307

Klaver M, De Roever EWF, Nanne JAM, Mason PRD, Davies GR (2015) Charnockites and UHT metamorphism in the Bakhuis Granulite Belt, western Suriname: evidence for two separate UHT events. Precambr Res 262:1–19

Klaver M, De Roever EWF, Thijssen ACD, Bleeker W, Söderlund U, Chamberlain K, Ernst R, Berndt J, Zeh A (2016) Mafic magmatism in the Bakhuis Granulite Belt (western Suriname): relationship with charnockite magmatism and UHT metamorphism. GFF 138(1):203–218

Klein EL, Almeida ME, Rosa-Costa LT (2012) The 1.89-1.87 Ga Uatumã Silicic Large Igneous Province, northern South America: Large Igneous Provinces Commission. http://www.largeigneousprovinces.org

Kroonenberg SB, De Roever EWF, Fraga LM, Reis NJ, Faraco MT, Lafon J-M, Cordani U, Wong TE (2016) Paleoproterozoic evolution of the Guiana Shield in Suriname: a revised model. Geol Mijnbouw 94(4):491–522

Lamarão CN, Dall'Agnol R, Pimentel M (2005) Nd isotopic composition of Paleoproterozoic volcanic and granitoid rocks of Vila Riozinho: implications for the crustal evolution of the Tapajos gold province, Amazon craton. J S Am Earth Sci 18(3–4):277–292

Lima GA (2016) Soleiras e enxames de diques máficos do sul-sudoeste do Cráton Amazônico. Tese de Doutorado, Instituto de Geociências, Universidade Federal do Pará, Brasil, 130 p

Lima GA, Sousa MZA, Ruiz AS, D'agrella Filho MS, Vasconcelos P (2012) Sills máficos da Suíte Intrusiva Huanchaca - SW do Cráton Amazônico: registro de magmatismo fissural relacionado à ruptura do Supercontinente Rodínia. Revista Brasileira de Geociências 42:111–129

Litherland M, Powel PEJ (1989) The geologic and geomorphologic evolution of Serrania Huanchaca, eastern Bolivia: the legendary "Lost World". J S Am Earth Sci 2(1):1–17

Litherland M, Annells RN, Appleton JD, Berrangé JP, Bloomfield K, Burton CCJ, Darbyshire DPF, Fletcher CJN, Hawkins MP, Klinck BA, Llanos A, Mitchell WI, O'Connor EA, Pitfield PEJ, Power G, Webb BC (1986) The geology and mineral resources of the Bolivian Precambrian shield. Br Geol Surv Overseas Memoir 9:153 p

Lubnina NV, Stepanova AV, Ernst RE, Nilsson M, Söderlund U (2016) New U-Pb baddeleyite age, and AMS and paleomagnetic data for dolerites in the Lake Onega region belonging to the 1.98–1.95 Ga regional Pechenga-Onega Large Igneous Province. GFF 138(1):54–78

McLelland JM (1989) Crustal growth associated with anorogenic, mid-Proterozoic anorthosite massifs in northeastern North America. Tectonophysics 161:331–341

Neder RD, Leite JAD, Figueiredo BR, McNaughton NJ (2002) 1.76 Ga volcano-plutonism in the southwestern Amazonian Craton, Aripuanã-MT, Brazil: Tectono-stratigraphic implications from SHRIMP U-Pb zircon data and rock geochemistry. Precambr Res 119:171–187

Norcross CE, Davies DW, Spooner ETC, Rust A (2000) U-Pb and Pb–Pb age constraints on Paleoproterozoic magmatism, deformation and gold mineralization in the Omai area, Guyana Shield. Precambr Res 102:69–86

Oliveira DC, Dall'Agnol R, Silva JBC, Almeida JAC (2008) Gravimetric, radiometric, and magnetic susceptibility study of the Paleoproterozoic Redenção and Bannach plutons: implications for architecture and zoning of A-type granites. J S Am Earth Sci 25:100–115

Payolla BL, Bettencourt JS, Kozuch M, Leite WB, Fetter AH, Van Schmus WR (2002) Geological evolution of the basement rocks in the east-central part of the Rondônia Tin Province, SW Amazonian Craton, Brazil: U-Pb and Sm–Nd isotopic constraints. Precambr Res 119:141–169

Peng P (2015) Precambrian mafic dyke swarms in the North China Craton and their geological implications. Sci China Earth Sci 58:649–675

Pinho MASB, Chemale Júnior F, Van Schmus WR, Pinho FEC (2003) U-Pb and Sm–Nd evidence for 1.76–1.77 Ga magmatism in the Moreru region, Mato Grosso, Brazil: implications for province boundaries in the SW Amazon Craton. Precambr Res 126:1–25

Pirajno F, Hoatson DM (2012) A review of Australia's large igneous provinces and associated mineral systems: implications for mantle dynamics through geological time. Ore Geol Rev 48:2–5

Pisarevsky SA, Bylund G (2010) Paleomagnetism of 1780–1770 Ma mafic and composite intrusions of Småland (Sweden): implications for the Mesoproterozoic supercontinent. Am J Sci 310:1168–1186

Pisarevsky SA, Sokolov SJ (2001) The magnetostratigraphy and a 1780 Ma palaeomagnetic pole from the red sandstones of Vazhinka River section, Karelia, Russia. Geophys J Int 146:531–538

Pradhan VR, Meert JG, Padit MK, Kamenov G, Mondal MdEAG (2012) Paleomagnetic and geochronological studies of the mafic dyke swarms of Bundelkhand craton, central India: implications for the tectonic evolution and paleogeographic reconstructions. Precambr Res 198–199:51–76

Prendergast MD (2000) Layering and precious metals mineralization in the Rincón del Tigre Complex, Eastern Bolivia. Econ Geol 95:113–130

Priem HNA, Boelrijk NAIM, Verschure RH, Hebeda EH (1968) Isotopic age determinations on Surinam rocks, 3. Proterozoic and Permo-Triassic basalt magmatism in the Guiana Shield. Geol Mijnbouw 47:17–20

Rämö OT, Haapala I (1995) One hundred years of Rapakivi Precambrian basement of the Gulf of Finland and granite. Mineral Petrol 52:129–185

Reis NJ, Fraga LM, de Faria MSG, Almeida ME (2003) Geologia do Estado de Roraima, Brasil. In: Rossi P, Lafon J-M, Vasquez ML (org) Geology of France and surrounding areas—special Guiana Shield. 2-3-4, BRGM, pp 121–134

Reis NJ, Faria MSG, Almeida ME, Oliveira MA (2004) Folhas NA.20-Boa Vista e NB.20-Roraima. In: Schobbenhaus C, Gonçalves JH, Santos JOS, Abram MB, Leão Neto R, Matos GMM, Vidotti RM, Ramos MAB, de Jesus JDA (eds) Carta Geológica do Brasil ao Milionésimo, Sistema de Informações Geográficas—SIG. Programa Geologia do Brasil. CPRM, Brasília. CD-ROM

Reis NJ, Riker SRL, da Pinheiro S, Nobre JN, da Cruz NM, Costi HT (2005) Geologia dos rios Tapajós, Juruena (Bararati e São Tomé) e Teles Pires, porção sul do Cráton Amazônico, em área limítrofe dos estados do Amazonas, Pará e Mato Grosso. In: Horbe AMC, da VS Souza (eds) Contribuições à Geologia da Amazônia, v 4, Manaus, SBG/Núcleo Norte, pp 55–68

Reis NJ, Almeida ME, Riker SRL, Ferreira AL (2006) Geologia e Recursos Minerais do Estado do Amazonas. In: Reis NJ, Ferreira AL (coord.). Escala 1:1.000.000. Manaus, CPRM, 2006 (Convênio CPRM/CIAMA), 125 p

Reis NJ, Teixeira W, Hamilton MA, Bispo-Santos F, Almeida ME, D'Agrella-Filho MS (2013a) Avanavero mafic magmatism, a late Paleoproterozoic LIP in the Guiana Shield, Amazonian Craton: U-Pb IDTIMS baddeleyite, geochemical and paleomagnetic evidence. Lithos 174:175–195

Reis NJ, Bahia RBC, Costa UAP, Betiollo LM, Oliveira AC, Oliveira AA, Splendor F (2013b) O Supergrupo Sumaúma no contexto geológico da Folha SB.20-Z-D (Sumaúma), sudeste do Amazonas: modo de ocorrência, discussão de idades em zircões detríticos e correlações no SW do Cráton do Amazonas. Contribuições à Geologia da Amazônia 8:199–222

Reis NJ, Nadeau S, Fraga LM, Betiollo LM, Faraco MTL, Reece J, Lachhman D, Ault R (2017a) Stratigraphy of the Roraima Supergroup along the Brazil-Guyana border in the Guiana shield, Northern Amazonian Craton—results of the Brazil-Guyana Geology and Geodiversity Mapping Project. Braz J Geol 47(1):43–57

Reis NJ, Cordani UG, Almeida ME, Fraga LM, Wahnfried I, Oliveira V, Maurer VC (2017b) Novas idades U-Pb SHRIMP da região Jauaperi-Mocidade-Demêni, sul do Escudo das Guianas, Cráton Amazônico. Anais do XV Simpósio de Geologia da Amazônia, Belém, PA, pp 464–467

Remédio MJ, Faleiros FM (2014) Programa Geologia do Brasil—PGB. Geologia e Recursos Minerais da Folha Fazenda Margarida—SF.21-X-C-IV, Estado de Mato Grosso do Sul, Escala 1:100,000. CPRM, São Paulo

Rivalenti G, Mazzucchelli M, Girardi VAV, Cavazzini G, Finatti C, Barbieri MA, Teixeira W (1998) Petrogenesis of the Paleoproterozoic basalt–andesite–rhyolite dyke association in the Carajás region, Amazonian craton. Lithos 43:235–265

Rizzotto GJ, Lima EF, Chemale Junior F (2001) Geologia do Grupo Nova Brasilândia, sudeste de Rondônia, acresção continental e implicações geotectônicas. In: Reis NJ, Monteiro MAS (eds) Contribuições à geologia da Amazônia: SBG, Manaus, Brazil, vol 2, pp 342–442

Rizzotto GJ, Bettencourt JS, Teixeira W, Pacca IG, D'Agrella-Filho MS, Vasconcelos PMP, Basei MAS, Onoe AT (2002) Geologia e geocronologia da Suíte Metamórfica Colorado e suas encaixantes SE de Rondônia: implicações para a evolução mesoproterozóica do SW do Cráton Amazônico. Geologia USP Série Científica 2:41–55

Rizzotto GJ, Santos JOS, Hartmann LA, Tohver E, Pimentel MM, McNaughton NJ (2013) The Mesoproterozoic Guaporé suture in the SW Amazonian craton: geotectonic implications based on field geology, zircon geochronology and Nd-Sr isotopic geochemistry. J S Am Earth Sci 48:271–295

Rizzotto GJ, Hartmann LA, Santos JOS, McNaughton NJ (2014) Tectonic evolution of the southern margin of the Amazonian craton in the late Mesoproterozoic based on field relationships and zircon U-Pb geochronology. An Acad Bras Ciênc 86:57–84

Rogers JJW, Santosh M (2002) Configuration of Columbia, a mesoproterozoic supercontinent. Gondwana Res 5:5–22

Rossoni MB, Bastos AC, Souza VS, Marques JC, Dantas E, Botelho NF, Giovannini AL, Pereira VP (2017) U-Pb zircon geochronological investigation of the Morro dos Seis Lagos Carbonatite Complex and associated Nb deposit (Amazonas, Brazil). J S Am Earth Sci 80:1–17

Roverato M, Juliani C, Fernandes CMD, Capra L (2017) Paleoproterozoic andesitic volcanism in the southern Amazonian craton, the Sobreiro Formation: New insights from lithofacies analysis of the volcaniclastic sequences. Precambr Res 289:18–30

Santos JOS, Hartmann LA, Gaudette HE, Groves DI, McNaughton NJ, Fletcher IR (2000) A new understanding of the provinces of the Amazon craton based on integration of field mapping and U-Pb and Sm–Nd geochronology. Gondwana Res 3(4):453–488

Santos JOS, Groves DI, Hartmann LA, McNaughton NJ, Moura MB (2001) Gold deposits of the Tapajós Province, Amazon Craton. Miner Deposita 36:278–299

Santos JOS, Hartmann LA, McNaughton NJ, Fletcher IR (2002a) Timing of mafic magmatism in the Tapajós Province (Brazil) and implications for the evolution of the Amazon Craton: evidence from baddeleyite and zircon U-Pb SHRIMP geochronology. J S Am Earth Sci 15:409–429

Santos JOS, Rizzotto G, Easton MR, Potter PE, Hartmann LA, McNaughton NJ (2002b) The Sunsas Orogen in Western Amazon Craton, South America and correlation with the Grenville Orogen of Laurentia, based on U-Pb isotopic study of detrital and igneous zircons. In: Geological Society of America, Precambrian Geology, pp 122–128

Santos JOS, Potter PE, Reis NJ, Hartmann LA, Fletcher IR, McNaughton NJ (2003) Age, source and regional stratigraphy of the Roraima Supergroup and Roraima-like sequences in northern South America, based on U-Pb geochronology. Geol Soc Am Bull 115:331–348

Santos JOS, Van Breemen OB, Groves DI, Hartmann LA, Almeida ME, McNaughton NJ, Fletcher IR (2004) Timing and evolution of multiple Paleoproterozoic magmatic arcs in the Tapajós Domain, Amazon Craton: constraints from SHRIMP and TIMS zircon, baddeleyite and titanite U-Pb geochronology. Precambr Res 131:73–109

Santos JOS, Faria MSG, Riker SRL, Souza MM, Hartmann LA, Almeida ME, McNaughton NJ, Fletcher IR (2006) A faixa colisional K'Mudku (Idade Grenvilleana) no norte do Cráton Amazonas: Reflexo intracontinental do Orógeno Sunsás na margem ocidental do cráton. 9th Simpósio de Geologia da Amazônia, Belém, Pará, Brazil. Sessão Temática III, 4 pp. CD-ROM

Santos JOS, Rizzotto GJ, Potterd PE, McNaughton NJ, Matos R, Hartmann LA, Chemale F Jr, Quadros MLES (2008) Age and autochthonous evolution of the Sunsás Orogen in West Amazon Craton based on mapping and U-Pb geochronology. Precambr Res 165:120–152

Santos JOS, Pinto V, McNaughton NJ, Silva LC (2011) Magmatismo Serra Grande em Roraima: formação cogenética de granito rapakivi e charnockito em ca. 1430 Ma [Serra Grande anorogenic suite (Roraima state, Brazil): coeval rapakavi granite and charnockite at ca. 1430 Ma]. In: Simpósio de Geologia da Amazônia, Brazil. CDROM

Scandolara JE, Fuck RA, Dall'Agnol R, Dantas EL (2013) Geochemistry and origin of the early Mesoproterozoic mangerite-charnockite-rapakivi granite association of the Serra da Providência suite and associated gabbros, central-eastern Rondônia, SW Amazonian Craton, Brazil. J S Am Earth Sci 4:166–193

Scandolara JE, Ribeiro PSE, Frasca AAS, Fuck RA, Rodrigues JB (2014) Geochemistry and geochronology of mafic rocks from the Vespor suite in the Juruena arc, Roosevelt-Juruena terrain, Brazil: Implications for Proterozoic crustal growth and geodynamic setting of the SW Amazonian craton. J S Am Earth Sci 53:20–49

Scandolara JE, Correa RT, Fuck RA, Souza VS, Rodrigues JB, Ribeiro PSE, Frasca AAS, Saboia AM, Lacerda Filho JV (2017) Paleo-Mesoproterozoic arc-accretion along the southwestern margin of the Amazonian craton: the Juruena accretionary orogen and possible implications for Columbia supercontinente. J S Am Earth Sci 73:223–247

Shumlyanskyy L, Ernst R, Söderlund U, Billström K, Mitrokhin O, Tsymbal S (2016a) New U-Pb ages for mafic dykes in the Northwestern region of the Ukrainian shield: coeval tholeiitic and jotunitic magmatism. GFF 138:79–85

Shumlyanskyy L, Mitrokhin O, Billström K, Ernst R, Vishnevska E, Tsymbal S, Cuney M, Soesoo A (2016b) The ca. 1.8 Ga mantle plume related magmatism of the central part of the Ukrainian shield. GFF 138:86–101

Shumlyanskyy L, Hawkesworth C, Billström K, Bogdanova S, Mytrokhyn O, Romer R, Dhuime B, Claesson S, Ernst R, Whitehouse M, Bilan O (2017) The origin of the Palaeoproterozoic AMCG complexes in the Ukrainian Shield: new U-Pb ages and Hf isotopes in zircon. Precambr Res 292:216–239

Sial AN, Oliveira EP, Choudhuri A (1987) Mafic dyke swarms of Brazil. In: Mafic Dyke Swarms (Halls HC, Fahrig WF editors). Geological Association of Canada Special Paper 34, pp 467–481

Silva FF, Oliveira DC, Antonio PYJ, D'Agrella Filho MS, Lamarão CN (2016) Bimodal magmatism of the Tucumã area, Carajás province: U-Pb geochronology, classification and processes. J S Am Earth Sci 72:95–114

Simões MS, de Lisboa TM, Almeida ME, de Souza AGH, Silva DPB, da Silva SRAS, Lombello JC, Bahia RBC (2015) Programa Geologia do Brasil. Programa Cartografia da Amazônia. Geologia e Recusos Minerais das folhas Igarapé Canoa (SA.20-X-D-VI), Santo Antônio do Abonari (SA.20-X-D-III) e Vila do Pitinga (SA.20-X-B-VI). Escala 1:100.000. Estado do Amazonas. In: Simões MS, de Lisboa TM, Almeida ME, da Silva SRA, Souza AGH (org) Projeto Uatumã-Abonari. Superintendência Regional de Manaus, Nota Explicativa, 50 p

Snelling NJ, McConnell RB (1969) The geochronology of Guyana: Geologie en Mijnbouw 48:201–213

Söderlund U, Ask R (2006) Mesoproterozoic bimodal magmatism along the Protogine Zone, S Sweden: three magmatic pulses at 1.56, 1.22 and 1.205 Ga, and regional implications. GFF 128:303–310

Söderlund U, Isachsen CE, Bylund G, Heaman LM, Patchett PJ, Vervoort JD, Andersson UB (2005) U-Pb baddeleyite ages and Hf, Nd isotope chemistry constraining repeated mafic magmatism in the Fennoscandian Shield from 1.6 to 0.9 Ga. Contrib Miner Petrol 150:174–194

Söderlund U, Elming S-Å, Ernst RE, Schissel D (2006) The Central Scandinavian Dolerite Group—protracted hotspot activity or back-arc magmatism? Constraints from U-Pb baddeleyite geochronology and Hf isotopic data. Precambr Res 150:136–152

Stein CA, Stein S, Merino M, Keller GR, Flesch LM, Jurdy DD (2014) Was the Midcontinent Rift part of a successful seafloor-spreading episode? Geophys Res Lett 41(5):1465–1470

Tassinari CCG, Macambira M (2004) A evolução tectônica do Craton Amazônico. In: Mantesso-Neto V, Bartorelli A, Carneiro CDR, Brito Neve BB (eds) Geologia do Continente Sul Americano: Evolução da obra de Fernando Flávio Marques Almeida. São Paulo, pp 471–486

Tassinari CCG, Cordani UG, Nutman AP, Van Schmus WR, Bettencourt JS, Taylor PN (1996) Geochronological systematics on basement rocks from the Rio Negro-Juruena province (Amazonian Craton) and tectonic implications. Int Geol Rev 38:161–175

Tassinari CCG, Munhá JMU, Teixeira W, Palacios T, Nutman A, Sosa SC, Santos AP, Calado BO (2004) The Imataca Complex, NW Amazonian Craton, Venezuela: crustal evolution and integration of geochronological and petrological cooling histories. Episodes 27:3–12

Teixeira W (1978) Significação tectônica do magmatismo anorogênico pré-Cambriano básico e alcalino na região Amazônica. Anais XXX Congresso Brasileiro de Geologia, Recife 1:491–505

Teixeira W (1990) The Proterozoic mafic dyke swarms and alcaline intrusions in the Amazonian Craton, South America, and their tectonic evolution based on Rb-Sr, K-Ar and ^{40}Ar-^{39}Ar geochronology. In: Parker AJ, Rickwood PC, Tucker DH (eds) Maifc dykes and emplacement mechanisms. Proceeding of the Second International Dyke Conference, Adelaide, Australia. Publication 23, International Geological Correlation Program, Project 257 pp 285–293

Teixeira W, Albrez E (2017) Dyke and LIPs Arc GIS compilation for the Precambrian shield of South America. Large Igneous Provinces—Supercontinent Reconstruction—Resource Exploration Project (www.supercontinent.org) (unpublished report)

Teixeira W, Tassinari CCG, Cordani UG, Kawashita K (1989) A review of the geochronology of the Amazonian Craton: tectonic implications. Precambr Res 42:213–227

Teixeira W, Geraldes MC, Matos R, Ruiz AS, Saes G, Vargas-Mattos G (2010) A review of the tectonic evolution of the Sunsas belt, SW Amazonian Craton. J S Am Earth Sci 29:47–60

Teixeira W, Geraldes MC, D'Agrella-Filho MS, Santos JOS, Sant'Ana Barros MA, Ruiz AS, Corrêa da Costa PC (2011) Mesoproterozoic juvenile mafic-ultramafic magmatism in the SW Amazonian Craton (Rio Negro-Juruena province): SHRIMP U–Pb geochronology and Nd–Sr constraints of the Figueira Branca Suite. J S Am Earth Sci 32:309–323

Teixeira W, Hamilton MA, Lima G, Ruiz AS, Matos R, Ernst RE (2015) Precise ID-TIMS U-Pb baddeleyite ages (1110–1112 Ma) for the Rincón del Tigre-Huanchaca large igneous province (LIP) of the Amazonian Craton: implications for the Rodinia supercontinent. Precambr Res 265:273–285

Teixeira W, Ernst RE, Hamilton MA, Lima G, Ruiz AS, Geraldes MC (2016) Widespread ca. 1.4 Ga intraplate magmatism and tectonics in a growing Amazonia. GFF 138:241–254

Teixeira W, Hamilton MA, Girardi VAV, Faleiros FM, Ernst RE (2018) U-Pb baddeleyite ages of key dyke swarms in the Amazonian craton (Carajás/Rio Maria and Rio Apa areas): tectonic implications for 1880, 1110 Ma, 535 Ma and 200 Ma events. Precambr Res (in press). http://doi.org/10.1016/j.precamres.2018.02.008
Tohver E, Van der Pluijm BA, Van der Voo R, Rizzotto GJ, Scandolara JE (2002) Paleogeography of the Amazon Craton at 1.2 Ga: early Grenvillian collision with the Llano segment of Laurentia. Earth Planet Sci Lett 199:185–200
Tohver E, Van der Pluijm BA, Mezger K, Essene E, Scandorala JE, Rizzotto GJ (2004) Significance of the Nova Brasilândia metasedimentary belt in western Brazil: redefining the Mesoproterozoic boundary of the Amazon Craton. Tectonics 23:TC6004. https://doi.org/10.1029/2003tc001563
Tohver E, Pluijm BVD, Mezger K, Scandolara JE, Essene E (2005) Two stage tectonic history of the SW Amazon craton in the late Mesoproterozoic: identifying a cryptic suture zone. Precambr Res 137:35–59
Tohver E, Teixeira W, van der Pluijm B, Geraldes MC, Bettencourt JS, Rizzotto G (2006) Restored transect across the exhumed Grenville orogen of Laurentia and Amazonia, with implications for crustal architecture. Geology 34(8):669–672
Vasquez ML, Klein EL, Ricci PSF (2002) Granitóides pós-colisionais da porção ao leste da Província Tapajós. In: Klein EL, Vasquez ML, Rosa-Costa LT (eds) Contribuições à Geologia da Amazônia v. 3. SBG, Belém, PA, Brazil, pp 67–84
Vasquez ML, Macambira MJB, Armstrong RA (2008) Zircon geochronology of granitoids from the western Bacajá domain, southeastern Amazonian craton, Brazil: Neoarchean to Orosirian evolution. Precambr Res 161:279–302
Wahlgren CH, Heaman LM, Kamo S, Ingvald E (1996) U-Pb baddeleyite dating of dolerite dykes in the eastern part of the Sveconorwegian orogen, south-central Sweden. Precambr Res 79:227–237
Wingate MTD, Campbell IH, Harris LB (2000) SHRIMP baddeleyite age for the Fraser dike swarm, southeast Yilgarn craton, Western Australia. Aust J Earth Sci 47:309–313
Zariņš K, Johansson Å (2009) U-Pb geochronology of gneisses and granitoids from the Danish island of Bornholm: new evidence for 1.47–1.45 Ga magmatism at the southern margin of the East European Craton. Int J Earth Sci 98:1561–1580
Zhao GC, Sun M, Wilde SA, Li SZ (2004) A Paleo-Mesoproterozoic supercontinent: assembly, growth and breakup. Earth Sci Rev 67:91–123

The Precambrian Mafic Magmatic Record, Including Large Igneous Provinces of the Kalahari Craton and Its Constituents: A Paleogeographic Review

M. O. de Kock, A. P. Gumsley, M. B. Klausen, U. Söderlund and C. Djeutchou

Abstract The study of Precambrian dyke swarms, sill provinces and large igneous provinces on the Kalahari craton in southern Africa has expanded greatly since the pioneering work initiated almost four decades ago. The main contributors to this progress have been a large number of precise U–Pb crystallization ages of mafic rocks, published in a number of recent papers. This information is compiled here into a series of maps that provide a nearly 3 billion year intraplate magmatic record of the Kalahari craton and its earlier constituents, the proto-Kalahari, Kaapvaal and Zimbabwe cratons. We also review their possible paleogeographic relations to other cratons or supercontinents. This review provides a more accessible overview of individual magmatic events, and mostly includes precise U–Pb ages of mafic dykes and sills, some of which can be linked to stratigraphically well-constrained volcanic rocks. The extrusion ages of these volcanic units are also starting to be refined by, among others, in situ dating of baddeleyite. Some mafic dyke swarms, previously characterized entirely on similarity in dyke trends within a swarm, are found to be temporally composite and sometimes consist of up to three different generations. Other mafic dyke swarms, with different trends, can now be linked to protracted volcanic events like the stratigraphically well preserved Mesoarchean Nsuze Group (Pongola Supergroup) and Neoarchean Ventersdorp Supergroup. Following upon these Archean events, shorter-lived Proterozoic large igneous provinces also intrude the Transvaal Supergroup, Olifantshoek Supergroup and Umkondo Group,

M. O. de Kock (✉) · C. Djeutchou
Department of Geology, University of Johannesburg, Auckland Park 2006, South Africa
e-mail: mdekock@uj.ac.za

A. P. Gumsley · U. Söderlund
Department of Geology, Lund University, 223 62 Lund, Sweden

A. P. Gumsley
Institute of Geophysics, Polish Academy of Sciences, 01-452, Warsaw, Poland

M. B. Klausen
Department of Earth Sciences, Stellenbosch University, Stellenbosch 7600, South Africa

U. Söderlund
Department of Geosciences, Swedish Museum of Natural History, Stockholm, Sweden

R. K. Srivastava et al. (eds.), *Dyke Swarms of the World: A Modern Perspective*, Springer Geology, https://doi.org/10.1007/978-981-13-1666-1_5

and include the world's largest layered intrusion, the Bushveld Complex. Longer-lived late Paleoproterozoic magmatic events are also preserved as mafic intrusions and lava units within the Waterberg and Soutpansberg groups as well as the granitic basement. Many gaps in our knowledge of the Precambrian mafic record of the Kalahari craton remain, but further multi-disciplinary studies combining the latest advances in U–Pb geochronology and both paleomagnetism and geochemistry will help solve the Precambrian paleogeographic puzzle.

1 Introduction

The recognition of intraplate igneous events, and especially large igneous provinces (LIPs), has increased dramatically in recent years and offer key temporal and spatial constraints for creating paleogeographic reconstructions back into Precambrian time. Such studies rely mainly on mafic volcanic rocks and their feeder systems, which are preserved in Precambrian terranes as regional mafic sill provinces and dyke swarms, as well as mafic-ultramafic layered intrusions.

Dyke swarms are defined as groups of dykes of similar age that may form linear, radiating or arcuate arrays (Ernst et al. 1995). It is important to note that dyke swarms of different ages—yet without other apparent petrographical or compositional differences—may have overlapping patterns that can only be separated on the basis of precise age constraints. The adjective "giant" can be added if swarms are longer than 300 km (Ernst et al. 1995). However, some smaller individual dyke swarm fragments may be part of "giant" swarms, when properly restored (or chronologically matched) together with other craton fragments (e.g., Bleeker and Ernst 2006). For this reason we avoid using the adjective "giant" in this contribution. Sill provinces are less rigorously defined into lesser and greater types and are herein simply regarded as any collection of coeval sills that share a similar stratigraphic host unit, and is confined within a certain geographical extent.

Numerous mafic dyke swarms and sill provinces cross-cut the Precambrian terranes of southern Africa, and many of these intrusions are only recently constrained as syn-magmatic with volcanic units preserved within remnants of supracrustal volcano-sedimentary successions. Many of these magmatic events form integral parts of recognized LIPs; whereas other events have not yet been linked to a specific LIP. However, apart from some pioneering regional studies (e.g., McElhinny and Opdyke 1964; Jones and McElhinny 1966; Hunter and Reid 1987; Wilson et al. 1987; Uken and Watkeys 1997), the Precambrian dyke swarms and sill provinces across southern Africa received little scientific attention before the early part of the twenty-first century. McElhinny and Opdyke (1964) were the first to show that there were two major Proterozoic sill provinces on the Zimbabwe craton of distinctly different ages, based on paleomagnetic signatures. The follow up work of Jones and McElhinny (1966) is one of the first examples of the use of paleomagnetism for correlating isolated occurrences of mafic units over large distances. Hunter and Reid (1987) and Wilson et al. (1987) began to characterize and group mafic dykes across southern Africa into different swarms. The identification of these swarms was primarily based on observable trends, degrees of deformation and metamorphism, and lithology. Ages were mostly based upon cross-cutting relationships, paleomagnetism and a limited number of K–Ar mineral and whole-rock as well as Rb–Sr whole-rock ages. Some swarms were tentatively linked to igneous units with better temporal constraints, including volcanic units, sills, and even larger layered intrusions. Uken and Watkeys (1997) advanced these early observations by interpreting the dyke swarm trends in terms of prominent structural lineaments, associating them more specifically with continental 'rifts' during the extrusion and deposition of various known volcano-clastic successions in the supracrustal record. With an increased development of analytical techniques employed for U–Pb geochronology, both on igneous zircon, and especially baddeleyite crystals (Krogh 1973; Heaman and LeCheminant 1993), and improved separation techniques for these minerals (Söderlund and Johansson 2002), the study of mafic dykes and sills in southern Africa has increased substantially; sometimes with far-reaching implications. In this review, we aim to highlight these recent advances in our knowledge of mafic magmatism across southern Africa and through Precambrian times, and thereby try to simplify what may otherwise be regarded as a complex spatial and temporal array of mainly subalkaline mafic lavas, sills, dykes and larger intrusions. In addition, some attention is given to relatively low-volume kimberlites, carbonatites and other alkaline igneous complexes, as these may also represent intraplate mafic magmatism that can be part of LIPs (e.g., Ernst 2014); albeit, often representing magmas derived through much lower degrees of partial mantle melting. Brief mention is also made of felsic igneous units that form part of some identified LIP or magmatic event. This review, however, focuses primarily on the precise U–Pb dates obtained directly from mafic units in recent years. Dates discussed in this contribution are U–Pb dates obtained during recent years, derived directly from igneous zircons and baddeleyites within mafic intrusions unless stated otherwise. It is acknowledged that this review could be complimented by an existing rich geochronological database based on several other isotopic systems, techniques and minerals from a range of other rock types, including igneous ages of granitic rocks and tuffaceous units, as well as detrital zircon studies of sedimentary cover successions that can be associated with a mafic magmatic event. A complete discussion of such a larger database is, however, beyond the scope of this contribution. Where no robust age data is available directly from a mafic unit, we do speculate about possible age assignments based on other isotopic systems (e.g., Rb–Sr or noble gas constraints).

The ultimate goal of this focused review is strictly to provide an updated overview of Precambrian mafic magmatic events across southern Africa, and thereby possibly group or subdivide these into lesser or greater events of shorter or longer durations. This naturally provide some new constraints on the architecture, modes of emplacement and magma petrogenesis within paleogeographic reconstructions that may ultimately shed more light on tectonic settings. We only make superficial reference to the geochemistry and petrology of these mafic magmatic events, however, unless composition may be used to further distinguish between different, yet spatially and structurally overlapping events. Instead, we aim to identify questions that remain to be resolved on (1) the paleogeographic and magmatic barcode record of southern Africa's Precambrian crustal blocks, back through time, (2) how these events were emplaced, and (3) which of these events classify as LIPs.

2 Crustal Architecture of Southern Africa

The Kalahari craton of southern Africa (originally defined by Clifford 1970) is a mosaic of crustal terranes of varying age. It is principally made up of a composite Archean core, which consists of the Kaapvaal and Zimbabwe cratons joined along the Limpopo Metamorphic Complex (Fig. 1). Along the western margin of this Archean Kaapvaal-Zimbabwe core there are progressively younger Paleoproterozoic accreted terranes and fold-and-thrust belts (i.e., the Kheis orogen, Magondi orogen, Okwa terrane, and Rehoboth terrane), which collectively make up the so-called proto-Kalahari craton. The proto-Kalahari craton is in turn surrounded by an accreted Mesoproterozoic rim, inclusing the Namaqua-Natal orogen, and the Konkiep and Choma Kaloma terranes. We in this context follow the definition of Hartnady et al. (1985), in which the Kalahari craton is constituted by the proto-Kalahari craton and its Mesoproterozoic rim. In addition, paleomagnetic data indicates that the Kaapvaal craton was likely contiguous with the Archean Grunehogna terrane, and associated Mesoproterozoic Maud orogen, of East Antarctica, at least during Mesoproterozoic to Phanerozoic times (e.g., Jones et al. 2003). Grunehogna, as well as the Mesoproterozoic basement of the Falkland Islands, and the Mesoproterozoic Haag Nunatak are therefore also considered a part of the Kalahari craton, but separated from their southern African counterparts during and since the break-up of Gondwana (Jacobs et al. 2008). Much of the focus of this review is on the Archean-Paleoproterozoic proto-Kalahari craton, however, and this is not intentional, but rather a combined artifact of preservation, outcrop, and accessibility of the involved terranes. Much of the western and southern margins of the Kalahari craton is, for example, covered by younger Phanerozoic (mainly Karoo Supergroup) rock successions and unconsolidated Quaternary (mainly Kalahari Group) sands, respectively; whereas, much older host rocks are remarkably well exposed within the Kaapvaal and Zimbabwe craton cores.

The Kaapvaal craton of South Africa, Botswana and Swaziland (Fig. 1) can be subdivided into separate structural domains that generally young towards the north and west (e.g., De Wit et al. 1992; Eglington and Armstrong 2004). The southeastern Swaziland block includes the oldest recorded basement rocks of the Kaapvaal craton, which probably merged with the Witwatersrand block along the Barberton greenstone belt. The 3.3–2.8 Ga Pietersburg block constitutes the northern part of the Kaapvaal craton, north of the prominent Thabazimbi-Murchison lineament (TML; Laurent et al. 2013). The east- to northeast trending and 3.6–2.8 Ga old basement structures of the Swaziland, Witwatersrand, and Pietersburg blocks are truncated by more north-trending structures of the 3.0 Ga Kimberley block in the west (De Wit et al. 1992; Tinker et al. 2002; Eglington and Armstrong 2004). Granitoid emplacement between 2.93 and 2.88 Ga places a minimum age constraint on this amalgamation

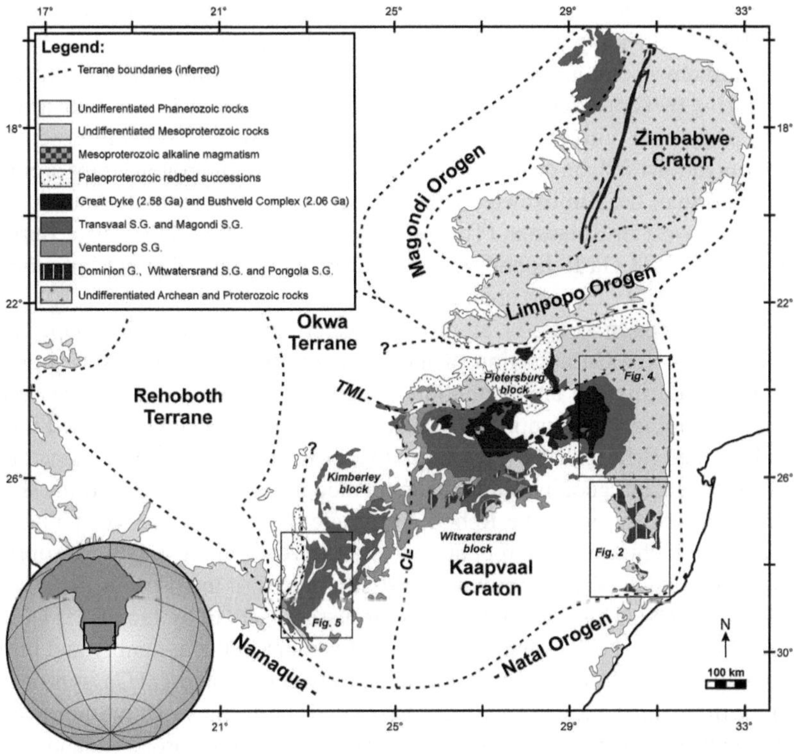

Fig. 1 Generalized Precambrian geology of the Kalahari craton of southern Africa after Hammerbeck and Allcock (1985) The Kalahari craton is made up of the Kaapvaal and Zimbabwe cratons with the intervening Limpopo Metamorphic Complex (LMC), as well as younger Proterozoic terranes and orogens to the west (e.g., the Rehoboth terrane). The Witwatersrand, Kimberley and Pietersburg blocks constitute the Kaapvaal craton, and these blocks are separated by the prominent Colesberg and Thabazimbi-Murchison lineaments (i.e., the CL and TML)

along the Colesberg lineament (CL; Mapeo et al. 2004a; Schmitz et al. 2004). Final stabilization of the Kaapvaal craton occurred with the intrusion of late-stage granites at about 2.7 Ga (Eglington and Armstrong 2004). The Zimbabwe craton borders the northern margin of the Kaapvaal craton (Fig. 1) along an intervening LMC, which is a polymetamorphic belt that experienced high-grade metamorphism between 2.7 and 2.6 Ga, and again at approximately 2.0 Ga (e.g., Kramers and Mouri 2011). The second episode of metamorphism and deformation is thought by some (e.g., Schaller et al. 1999; Söderlund et al. 2010) to represent the transpressive docking of the Zimbabwe and Kaapvaal cratons to form the Archean-Paleoproterozoic nucleus of the proto-Kalahari craton. Hanson et al. (2011b) have even argued for a final docking after 1.88 Ga. However, the idea that the LMC is mainly the product of continent-

continent collision between 2.7 and 2.6 Ga (e.g., Roering et al. 1992; Kramers et al. 2011) is the more widely accepted interpretation at present.

The Zimbabwe craton, which occurs mostly in Zimbabwe (Fig. 1), probably formed during two episodes of crustal growth between 3.8 and 3.2 Ga, with a voluminous and protracted third episode occurring between 3.0 and 2.6 Ga representing its final cratonization (Wilson et al. 1995). The 2.0 Ga Magondi orogen wraps around the western margin of the Zimbabwe craton, and separates it from the Rehoboth terrane further to the west in Botswana (Treloar 1988; Majaule et al. 2001; McCourt et al. 2001). The ca. 1.78 Ga Rehoboth terrane is a geophysically distinct but otherwise poorly exposed entity (Jacobs et al. 2008).

The Kheis orogen (Fig. 1) along the western margin of the Kaapvaal craton records post 1.9 Ga accretion of the Rehoboth terrane and crustal shortening (Moen 1999), during the amalgamation of parts of the proto-Kalahari craton into the Nuna/Columbia supercontinent. The Kheis orogen has been correlated with the Magondi orogen (Master 1991), but these two orogens are structurally separated by a poorly exposed, but geophysically distinct Paleoproterozoic Okwa terrane (Fig. 1). Farther to the southwest of the Rehoboth terrane the 2.0–1.7 Ga Richtersveld terrane was an allochtonous microcontinent that accreted along a late Mesoproterozoic Namaqua-Natal orogeny, around the southern half of the proto-Kalahari craton (Jacobs et al. 2008). This crustal growth is discussed in detail by Jacobs et al. (2008), and it records the proto-Kalahari craton's involvement in the assembly and ultimate amalgamation of the Rodinia Supercontinent. Between 1.2 and 1.0 Ga, syn-orogenic volcano-sedimentary successions were deposited along these active northwestern (e.g., Hanson et al. 2006), southern and eastern margins of the proto-Kalahari craton, culminating in continent-continent or continent-island arc collisions, which formed the Namaqua-Natal orogen that can be traced via the Falkland Islands, Haag nunatak and Maud orogen in paleogeographic reconstructions of Gondwanaland (Robb et al. 1999; Thomas et al. 2000; Jacobs et al. 2008).

3 Catalogue of (Mostly Mafic) Intraplate Magmatic Events

The Precambrian mafic dyke swarm and sill province record of the exposed eastern Kaapvaal craton and that of the exposed Zimbabwe craton are more apparent than the record of the more poorly exposed central and western Kaapvaal craton, as well as the remainder of the proto-Kalahari craton. The latter largely due to Archean to Paleoproterozoic volcano-sedimentary cover successions across the central and western Kaapvaal craton, as well as younger Phanerozoic Karoo Supergroup and post-Karoo cover (including the Quaternary sands of the Kalahari desert) across other parts. Geophysical constraints, scarce outcrops and drill core intersections provide most of the record for these more peripheral parts of the Kalahari craton outside of its Archean-Proterozoic crustal core. The Zimbabwe craton is essentially stripped of its supracrustal cover successions, making darker mafic dykes and sills more prominently exposed against the paler granitic basement.

Mafic dykes have only been sporadically mapped throughout southern Africa in the past, and traditionally been subdivided into more pristine "dolerite dykes" or metamorphosed "diabase dykes", tentatively assigning these to the Jurassic Karoo LIP or older (Precambrian) magmatic events, respectively. We use the more neutral, but less specific term "mafic" in this review's coverage of the Precambrian intrusions.

Since the latter half of the twentieth century, a large number of geochronological studies have improved upon our understanding of the different mafic dyke swarms and sill provinces across southern Africa. As an introduction to a systematic catalogue of mafic magmatic events presented here, it can be stated that the first major mafic igneous event following cratonic stabilization of the Zimbabwe craton was the intrusion of the Great Dyke and its satellites at 2.58–2.57 Ga (e.g., Wingate 2000). On the Kaapvaal craton, the oldest well-dated major mafic igneous event following initial cratonization formed the 2.99–2.98 Ga Usushwana Complex and associated 2.98–2.97 Ga Badplaas mafic dyke swarm (Olsson et al. 2010; Gumsley et al. 2015). Only from ca. 1.88 Ga onwards does the Kaapvaal and Zimbabwe cratons share igneous rocks derived from common magmatic events (Söderlund et al. 2010), like the Umkondo and Karoo LIPs (e.g., Erlank 1984; Hanson et al. 2004a) and coeval Waterberg-hosted and Mashonaland sill provinces (e.g., Hanson et al. 2011b). In the following, we review our current understanding of mafic magmatic events that formed within apparently more coherent episodes of the (1) Mesoarchean, (2) Neoarchean, (3) late Neoarchean to early Paleoproterozoic, (4) the middle Paleoproterozoic Bushveld Complex, (5) late Paleoproterozoic, (6) early Mesoproterozoic, (8) the late Mesoproterozoic Umkondo LIP, and, finally, (9) the Neoproterozoic.

3.1 Mesoarchean Mafic Magmatism Within the Witwatersrand-Pongola Basins

Following initial 3.1 Ga stabilization of the Witwatersrand block a bimodal volcanic pile of basalts and rhyolites of the Dominion Group (Marsh et al. 1989; Armstrong et al. 1991) erupted across its central part. The Dominion Group has been shown to be geochemically and stratigraphically similar to another bimodal volcanic pile within the south-eastern Swaziland block, known as the Nsuze Group of the Pongola Supergroup (e.g., Cole 1994; Gold 2006; Fig. 2). This is supported by the proposed stratigraphic correlation of the Pongola's Mozaan Group and the Witwatersrand Supergroup unconformably overlying the Dominion Group (Beukes and Cairncross 1991). However, the Nsuze Group has been dated to between 2985 and 2968 Ma (Hegner et al. 1994; Nhleko 2003; Mukasa et al. 2013), which is significantly younger than the available 3074 ± 6 Ma age for the Dominion Group (Armstrong et al. 1991). Even though the Witwatersrand Supergroup is predominantly sedimentary, its upper successions host the basaltic-andesitic Crown and Bird Member lavas (Fig. 2) that are stratigraphically comparable with the Mozaan Group's Tobolsk Formation and

Gabela Formation lavas, and where an age of 2914 ± 8 Ma has been determined for the Crown Member (Armstrong et al. 1991).

Uken and Watkeys (1997) first noted that the Nsuze Group is broadly coeval to mafic dykes within a 80 km wide, >100 km long southeast-trending swarm across the southeastern Kaapvaal craton, called the Badplaas dyke swarm by Olsson et al. (2010). Many of the dykes within this swarm were presumed to be of Archean age (Hunter and Reid 1987) based on cross-cutting relations to ca. 3.1 Ga granitoid batholiths in the region of the Barberton Greenstone Belt. None of these dykes appear to intrude the Neoarchean to Paleoproterozoic Transvaal Supergroup and many terminate against the ca. 2.7 Ga Mbabane granitoid batholith (Layer et al. 1989). Geochronological studies show that the Badplaas dyke swarm consists of at least two sub-parallel dyke generations. The two generations have been dated at 2980 ± 1 Ma and 2967–2966 Ma (Olsson et al. 2010; Gumsley et al. 2015).

The two main dyke-like limbs of the Mesoarchean Usushwana Complex—crossing the border between South Africa and Swaziland—conspicuously follow the same southeast trend as the Badplaas dyke swarm (Fig. 2). The Complex is a layered intrusion composed of both gabbro (Piet Retief Suite) and granophyre (Hlelo Suite), which both reportedly formed at ca. 2860 Ma (Hunter and Reid 1987). Using the same ID-TIMS dating methodology, however, Gumsley et al. (2015) showed that the mafic Piet Retief Suite of the Usushwana Complex is broadly coeval with the older (2990–2978 Ma) Nsuze Group and the ca. 2980 Ma generation of the Badplaas dyke swarm. Thus, the Piet Retief Suite of the Usushwana Complex and southeast-trending mafic dykes were likely part of a protracted event, which, in turn, are coeval with, and likely fed the 2985–2977 Ma Nhlebela Formation (also known as the Pypklipberg Formation; Fig. 2) lavas of the lower Nsuze Group (Hegner et al. 1994; Nhleko 2003; Mukasa et al. 2013; Gumsley et al. 2015). The other overlapping younger generation of sub-parallel dykes likely fed the 2968–2966 Ma Agatha Formation lavas of the upper Nsuze Group (Fig. 2). Paleomagnetic studies on the ca. 2980 Ma and 2967–2966 Ma Badplaas dyke swarm also link the dykes to the Nsuze Group lavas (Lubnina et al. 2010; Maré and Fourie 2012). Klausen et al. (2010), furthermore, found that coeval dykes and lavas roughly share similar basaltic andesite compositions (with clac-alkaline affinities) and incompatible element signatures.

Mafic sills dated at 2874–2866 Ma by U–Pb on baddeleyite using ID-TIMS (Gumsley et al. 2013, 2015) intrude the Pongola Supergroup, south of Swaziland in South Africa. This includes the layered mafic-ultramafic 2866 ± 2 Ma Hlagothi Complex near the base of the Nsuze Group (Gumsley et al. 2013), as well as potentially the layered mafic-ultramafic Thole Suite. Mafic sills of this age generation are pervasive throughout the Mozaan Group (Fig. 2), and may be coeval with the mentioned Tobolsk and Gabela Formation lavas within in the upper Pongola Supergroup, as well as Crown and Bird Member lavas within the Witwatersrand Supergroup (Beukes and Cairncross 1991; Gumsley et al. 2013); all of which are of basaltic to more evolved andesitic or even dacitic compositions. The Hlelo Suite granophyres of the Usushwana Complex may also relate to the same 2.86 Ga event, although further U–Pb geochronology is needed to confirm its 2860 ± 26 Ma whole-rock Pb–Pb date by Walraven and Pape (1994). The dacitic to rhyolitic volcanic rocks and granophyre of

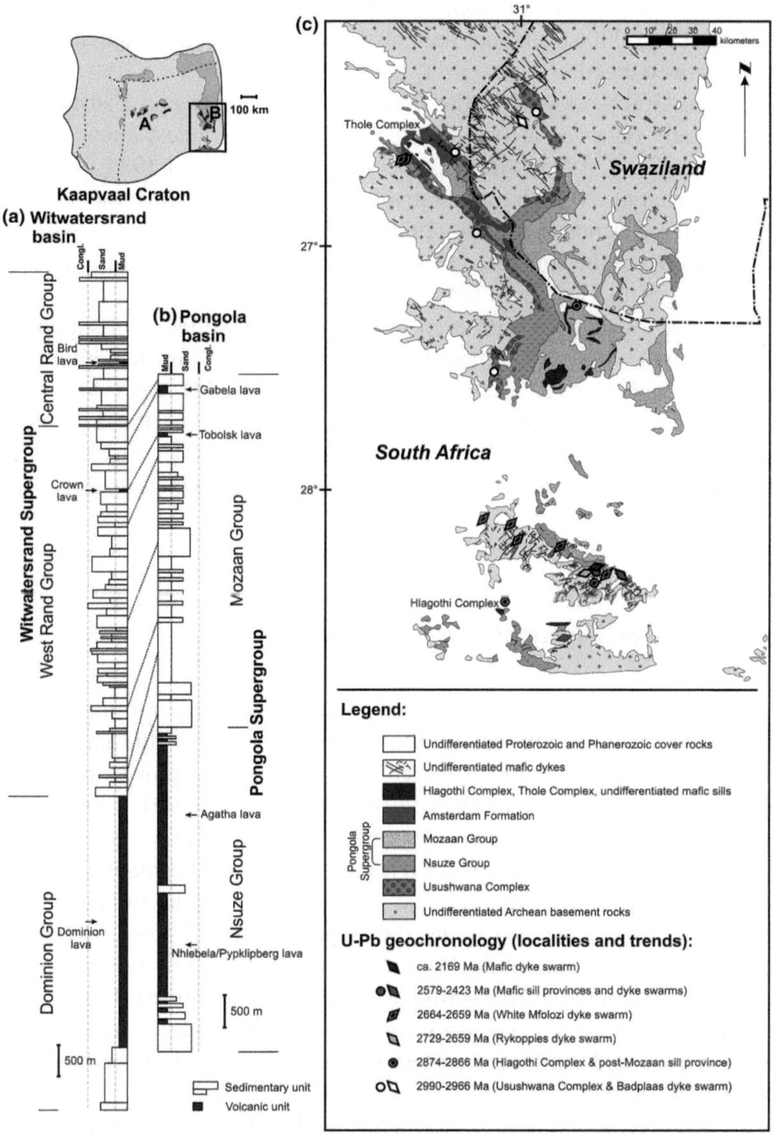

Fig. 2 Generalized lithological succession and correlation of the Witwatersrand basin (**a**) and the Pongola basin (**b**). **c** Geology and mafic dykes of the southeastern Kaapvaal craton. Unless stated otherwise, all maps in this and subsequent map figures are extracted from the Council of Geosciences' electronic 1:1,000,000 ArcGIS geology map of South Africa; whereas, dyke swarms, were digitized and added from a large number georeferenced 1:250,000 geological maps of the Republic of South Africa (Council of Geoscience). Geochronology symbols: diamonds = dated dyke and its trend; circles = dated sill, sheet, or plutonic body

the Amsterdam Formation, outcropping within the central parts of a major Pongola Supergroup (Gold 2006) syncline may also be related to this magmatic event, but have not yet been dated.

The Nsuze Group, as well as the Badplaas dyke swarm and Usushwana Complex is generally regarded as being emplaced within the oldest known continental rift (e.g., Burke et al. 1985). Many authors (Hunter and Reid 1987; Klausen et al. 2010; Gumsley et al. 2015) have tentatively suggested the involvement of a mantle plume for all three events (i.e., 2990–2978, 2968–2966 Ma, and the 2874–2866 Ma), typically centered at the southeastern end of the Badplaas dyke swarm and near the current margin of the Kaapvaal craton. However, there are few constraints on the tectonic setting for this rift, and even less conclusive evidence in support of any plume involvement. Coeval granites are associated with volcanism within the Nsuze Group, and include the 2973–2960 Ma Hlatikulu and 2981–2961 Ma Nhlangano plutons (Mukasa et al. 2013; Hofmann et al. 2015). However, these granitoids are difficult to relate to any orogenic event, and might rather reflect increased regional crustal anataxis during a period of anomalously high mantle melting (e.g., Hofmann et al. 2015).

3.2 Long-Lived Neoarchean Magmatism

After an apparent quiescence in mafic magmatism, crossing into the Neoarchean, volcanism resumed toward the central and northwestern parts of the Kaapvaal craton, as evidenced by the Ventersdorp Supergroup and associated groups and formations within related coeval sub-basins (Van der Westhuizen et al. 1991, 2006).

The oldest ages for compositionally bimodal volcanic rocks, associated with the Ventersdorp Supergroup, are determined for the Derdepoort Formation of northwestern South Africa as well as the bimodal volcanic rocks of the Lobatse Group and Kanye Formation, of Botswana all dated to approximately 2785–2781 Ma (Grobler and Walraven 1993; Moore et al. 1993; Walraven et al. 1996; Wingate 1998). An associated 2.78 Ga mafic dyke swarm has not yet been identified (Fig. 3), but these volcanic rocks are broadly coeval with both mafic and felsic plutons (e.g., the Modipe gabbro, Gabarone and Turfloop granites; Moore et al. 1993; Henderson et al. 2000; Denyszyn et al. 2013).

The Ventersdorp Supergroup conformably overlies the Witwatersrand Supergroup in the central part of the Kaapvaal craton (along the gold-rich and so-called Ventersdorp Contact Reef), but this contact becomes more unconformable towards the west. An up to 3 km-thick pile of flood basalts—overlying some basal conglomerate breccia and other clastic horizons interbedded by komatiitic lavas—dominate the Klipriviersberg Group of the lower Ventersdorp Supergroup. The Klipriviersberg Group is unconformably overlain by compositionally bimodal volcanic rocks and sedimentary successions of the Platberg Group, which, in turn is unconformably overlain by the clastic Bothaville Formation and overlying, predominantly basaltic andesite lavas of the Allanridge Formation (Van der Westhuizen et al. 1991). So-called proto-basinal

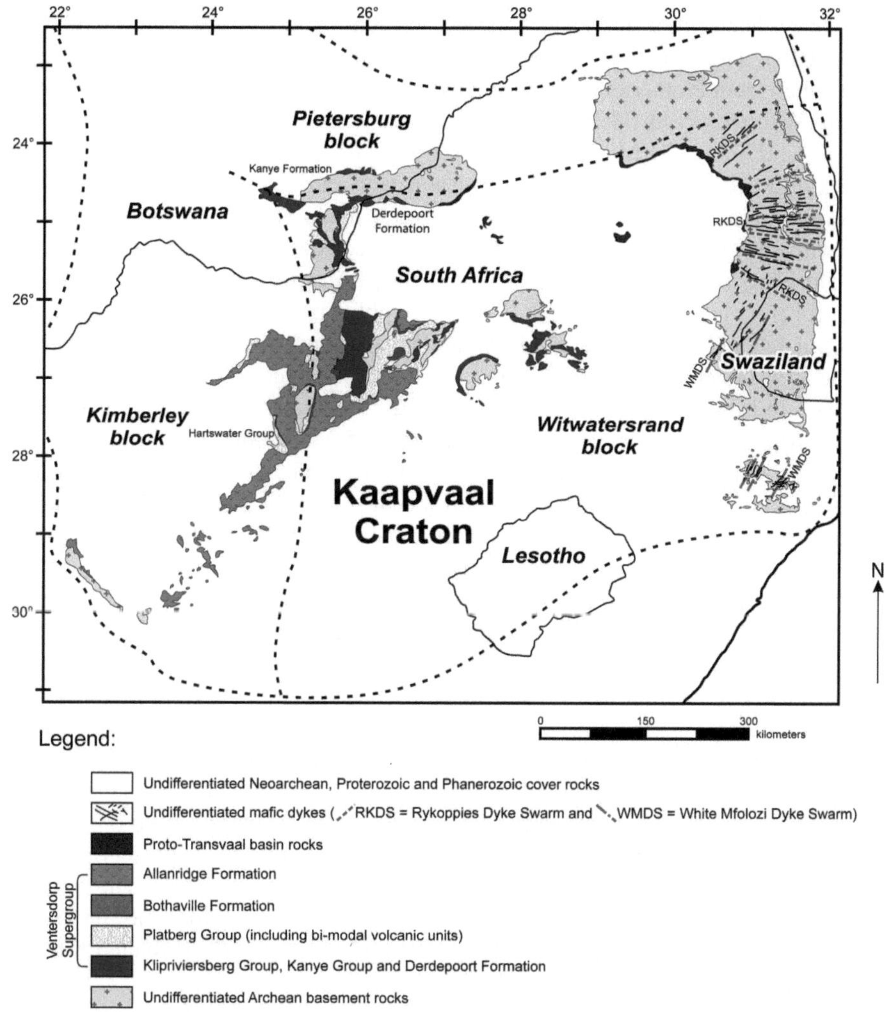

Fig. 3 Distribution of the Ventersdorp Supergroup and equivalent rock units on the Kaapvaal craton, as well as the prominent Neoarchean mafic dyke swarms of the eastern Kaapvaal craton (i.e., the radiating Rykoppies and the northeast-trending White Mfolozi swarms). Compiled as in Fig. 2

volcanic and sedimentary rocks (i.e., the Buffelsfontein Group, Godwan Formation and Wolkberg Group) were deposited unconformably on top of granite-greenstone basement, presumably during or after the Allanridge Formation lavas, forming the base of the Neoarchean-Paleoproterozoic Transvaal Supergroup, with the Buffelsfontein Group having been dated by Barton et al. (1995) to 2664 ± 1 Ma.

The Ventersdorp Supergroup and the proto-basinal rocks are generally thought to have been deposited within a relatively short time interval, between 2714 and 2664 Ma (Armstrong et al. 1991; Barton et al. 1995), but this has recently been questioned by older 2746 ± 9 Ma and 2720 ± 2 Ma ages for the Platberg Group (Cornell et al. 2018). These new ages, however, are in agreement with crystallization ages determined from Platberg Group correlatives such as the 2714 ± 3 Ma felsic Zoetlief Group (Walraven et al. 1991), the 2729 ± 3 Ma felsic Amalia Group (Poujol et al. 2005), the 2733–2724 Ma bimodal Hartswater Group (De Kock et al. 2012), and 2739 ± 39 Ma bimodal Sodium Group (Altermann and Lenhardt 2012)—see Fig. 3. These ages put a maximum age limit to the overlying Allanridge Formation (capping the Ventersdorp Supergroup), which has not yet been dated. Of greater implications, the Klipriviersberg Group must be older than the previously accepted 2714 Ma age by Armstrong et al. (1991). Wingate (1998) highlighted possible lead loss affecting the 2714 Ma date of Armstrong et al. (1991) and regarded it as a minimum age constraint for the Klipriviersberg Group. De Kock et al. (2012) suggested that the Klipriviersberg Group is perhaps better correlated with 2.78 Ga volcanic and magmatic units in the northwest of the craton, such as the 2782 ± 5 Ma Derdepoort Formation. It is interesting to note that Cornell et al. (2018) reports a 2781 ± 5 Ma date from a basaltic unit intersected in drill core near Kimberley that is usually assigned to the Platberg Group. This older sample more likely represents a correlative of the volcanics of the 2.78 Ga Derdepoort Formation and possibly the Klipriviersberg Group.

The development of the Platberg Group is broadly coeval with ca. 2.72 Ga felsic magmatism on the southeastern Kaapvaal craton (e.g., the Hlathikulu and Kwetta plutons; Mukasa et al. 2013). Although only observed in sub-outcrop mapping from mines and drill core, mafic dykes and sills have been documented within the Witwatersrand Supergroup, and have been geochemically linked to the overlying Klipriviersberg Group basalts in the central Kaapvaal craton (Meier et al. 2009). The age of the Allanridge Formation flood basalts remains unknown, but is bracketed by the extrusion of the 2720 ± 2 Ma quartz porphyries of the Platberg Group (Armstrong et al. 1991; Cornell et al. 2018) and the eruption of volcanic rocks preserved within the proto-basinal fill sequences at ca. 2664 Ma (Barton et al. 1995). The Ventersdorp Supergroup thus likely presents two pulses of continental flood basalt eruption at 2.78 and 2.70 Ga separated by a long intervening period of rift related bimodal volcanism and sedimentation.

Across the better exposed Archean granitoid-greenstone basement part of the Kaapvaal craton, a dominantly east-trending Rykoppies mafic dyke swarm radiates out eastward from beneath the Bushveld Complex and underlying Transvaal Supergroup (Olsson et al. 2010; Fig. 4). This dyke swarm was initially thought to be younger than the southeast-trending Badplaas dyke swarm, as several of the east-trending dykes cross-cut southeast-trending dykes (Hunter and Reid 1987; Uken and Watkeys 1997). Uken and Watkeys (1997) argued that these dykes are most likely of ca. 2.05 Ga Bushveld Complex age because they coincide conspicuously with the elongated outcrop of Bushveld's main eastern and western lobes and likely diverge into mafic sills within the Transvaal Supergroup. However, mafic dykes within this Rykoppies dyke swarm are now dated using ID-TIMS U–Pb on baddeleyite to between 2685 and 2683 Ma, and at 2662 ± 2 Ma (Olsson et al. 2010, 2011), consistent with the observation that these dykes do not intrude the Transvaal Supergroup. These often more andesitic dykes, with strong calc-alkaline affinities, commonly incorporate large quantities of partially digested country rock xenoliths (Klausen et al. 2010; Olsson et al. 2010) and have geochemical signatures that are best explained by the assimilation of large amounts of the tonalite-trondhjemite-granodiorite basement (Klausen et al. 2010; Gumsley et al. 2016). While Klausen et al. (2010) matched compositions of both Rykoppies dykes and Allanridge Formation lavas, geochemical signatures remain to be compared with more likely coeval proto-basinal volcanic rocks, for which there at present are no geochemical analysis including a significant number of trace elements.

Both to the north and south of the east-trending Rykoppies dyke swarm, broadly coeval dykes are found amongst a younger and the roughly parallel northeast-trending Black Hills dyke swarm (Sect. 3.6) and the southeast-trending Badplaas dyke swarm, respectively, which combine into a conspicuously radiating pattern (Olsson et al. 2010, 2011). However, this radiating swarm is found to be made up of two distinct generations, clustering between 2701–2692 Ma and 2662–2659 Ma, which both occur on either side of the Rykoppies dyke swarm (Olsson et al. 2010, 2011). Regardless, the radiating nature of these dyke swarms led Olsson et al. (2011) to propose a coeval mantle plume (after Hatton 1995) to coincide with the convergence center of this radiating dyke swarm, and thereby offering an alternate hypothesis for the much later emplacement of the Bushveld Complex, as discussed in Sect. 3.5.

The protracted, extensive and complex Ventersdorp magmatic event is further accentuated by a ca. 2729 Ma mafic dyke, trending east across the southeasternmost margin of the Kaapvaal craton (Larsson 2015) as indicated on Fig. 2. This dyke is likely related to Ventersdorp Supergroup volcanism (e.g., Hartswater Group), albeit located almost 500 km away (Fig. 2). A northeast-trending White Mfolozi dyke swarm also cuts across the southeastern part of the Kaapvaal craton, in the northern KwaZulu-Natal and southeastern Mpumalanga provinces, and has been dated using U–Pb on baddeleyite by ID-TIMS, producing a combined weighted mean age of 2662 ± 2 Ma for the entire swarm (Gumsley et al. 2016). This ca. 2662 Ma age is near-coeval with the youngest dykes of the 2701–2659 Ma radiating Rykoppies dyke swarm, as well as ca. 2664 Ma proto-basinal volcanic rocks of the Transvaal Supergroup (Barton et al. 1995). However, the White Mfolozi mafic dyke swarm not only cuts across the Rykoppies radiating dyke swarm, but is also made up of distinctly different plagioclase megacrystic dykes, with more depleted geochemical signatures (Klausen et al. 2010; Gumsley et al. 2016). It should in this context be mentioned that the radiating Rykoppies dyke swarm, overall, exhibits a significant correlation between geochemical signatures and the terrain they are hosted within (although, not exactly the same terrains as defined by Eglington and Armstrong 2004), rather than dyke ages or trends (Gumsley et al. 2016). This suggests yet undefined lithospheric control on magma compositions.

Finally, within the Central Zone of the LMC, the deformed mafic Causeway and Stockford dykes yielded emplacement ages of 2607 ± 4 Ma and 2604 ± 6 Ma, and younger metamorphic ages at ca. 2.0 Ga using zircon analyzed for U–Pb Sensitive High-Resolution Ion Microprobe or SHRIMP geochronology (Xie et al. 2017). Xie et al. (2017) suggested that the ca. 2605 Ma dykes intruded after an earlier continent-

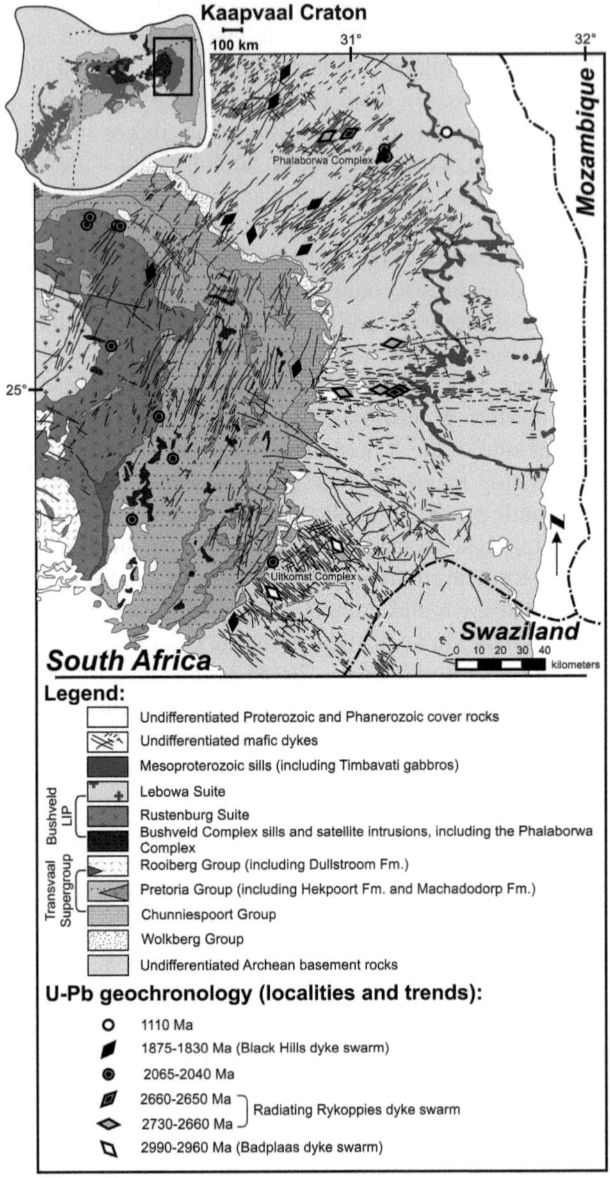

Fig. 4 Geological map emphasizing the distribution of mafic dykes across the eastern Kaapvaal craton. The affinity of undated northeast-trending dykes in the northwest of the mapped area and northwest-trending dykes to the north of the Badplaas dyke swarm remains to be confirmed. Compiled as in Fig. 2. Geochronology symbols: diamonds = dated dyke and its trend; circles = dated sill, sheet, or plutonic body

continent collision of the Zimbabwe and Kaapvaal cratons, during post-orogenic collapse and long before intracontinental transpressive deformation at 2.0 Ga.

3.3 Late Neoarchean to Early Paleoproterozoic Magmatic Events

Near the end of the Neoarchean, the Great Dyke of Zimbabwe is the first main mafic magmatic event following the stabilization of the Zimbabwe craton (Fig. 5). This 550 km-long, north to northeast-trending layered intrusion spans the entire craton (Wilson et al. 1987; Söderlund et al. 2010), and hosts world-class platinum-group element deposits. It was dated to 2574 ± 2 Ma using baddeleyite (Wingate 2000) and 2575 ± 1 Ma using zircon and rutile (Oberthür et al. 2002). Its sub-parallel East and Umvimeela satellite dykes have yielded near-coeval ages, with Söderlund et al. (2010) providing the first U–Pb ID-TIMS age on baddeleyite at 2575 ± 2 Ma for the Umvimeela dyke (Fig. 5).

U–Pb ID-TIMS baddeleyite ages of ca. 2580 Ma and ca. 2574 Ma are also reported for an east and a southeast-trending dyke, respectively, from the southeasternmost part of the Kaapvaal craton by Larsson (2015), which is broadly coeval with the Great Dyke and its satellite intrusions (Fig. 2). Granitoids of similar age have also been reported within the Vredefort Dome (Hart et al. 1999). However, these South African dykes and granitoids are located nearly 1500 km south of the Great Dyke of Zimbabwe, which, together with their very different compositions do not lend much support for any direct link between these events inside a coherent Kalahari craton at present.

Following upon the Great Dyke, the north to northeast-trending Sebanga mafic dykes constitute one of the most extensive swarms on the Zimbabwe craton (Fig. 5), being approximately 550 km wide and 350 km long (Wilson et al. 1987; Söderlund et al. 2010). These dykes were originally linked to the late Paleoproterozoic Mashonaland mafic sill province on the basis of similar paleomagnetic signatures (Jones et al. 1974). However, Söderlund et al. (2010) revealed much older ages within the swarm, spanning up to 100 Myr. The age of the most prominent Sebanga Poort dyke was determined by ID-TIMS on baddeleyite to 2408 ± 2 Ma (Söderlund et al. 2010), whereas the parallel-trending Crystal Springs dyke was dated to 2512 ± 2 Ma (Söderlund et al. 2010). In addition the Mtshingwe dyke, again of the same trend, was dated to 2470 ± 1 Ma (Söderlund et al. 2010). As no similar ages were at the time reported from the Kaapvaal craton, Söderlund et al. (2010) used this as further support for a later amalgamation of the Kaapvaal and Zimbabwe cratons along the ca. 2.0 Ga Central Zone of the LMC.

Unlike the Zimbabwe craton, and following upon the Ventersdorp Supergroup, proto-basinal volcanism and related intrusions, reviewed in Sect. 3.2, the Kaapvaal craton entered a period of relative magmatic quiescence until the eruption of the early Paleoproterozoic Ongeluk Formation basalts. The Kaapvaal craton preserves

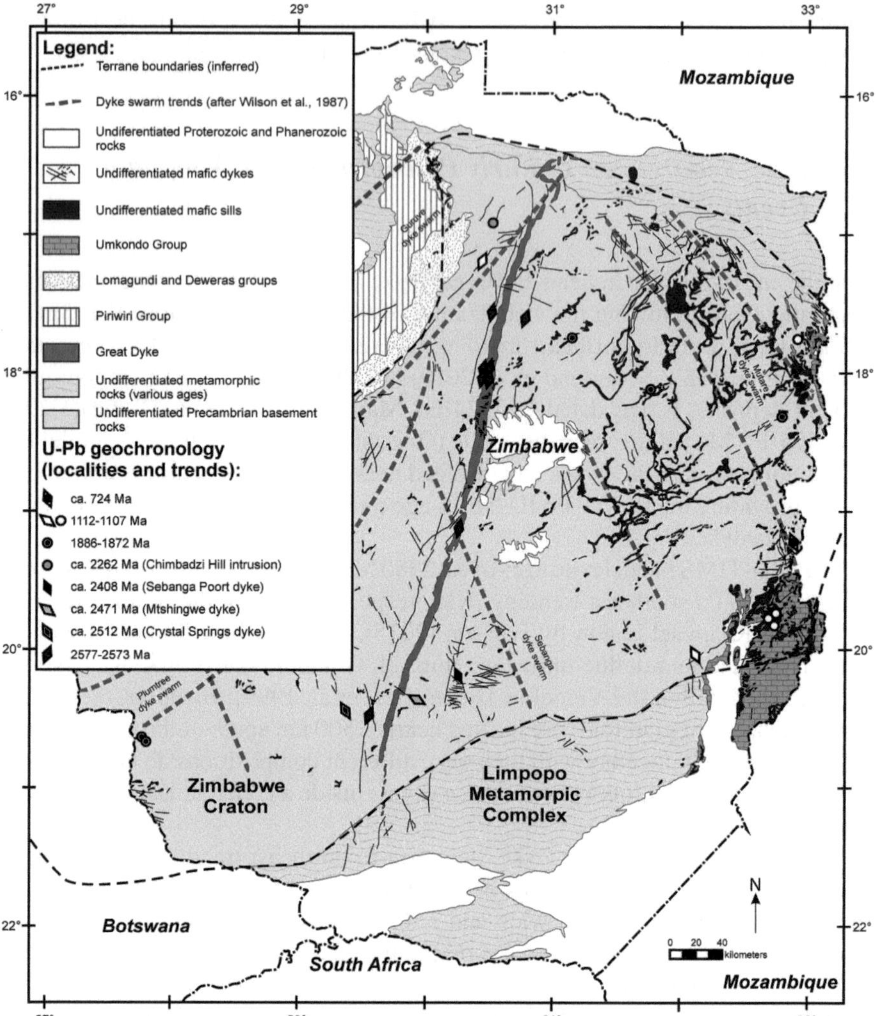

Fig. 5 The geology, including mafic dyke swarms and sills of the Zimbabwe, digitized from the 1:1,000,000 provisional geological map of Zimbabwe (1977) of the Zimbabwe Geological Survey. Note that although Botswana geology is not shown here, the Mashonaland sills extend across the border of Zimbabwe into Botswana. Geochronology symbols: diamonds = dated dyke and its trend; circles = dated sill, sheet, or plutonic body

a supracrustal record of being partly submerged during a major marine transgression, and records the deposition of shallower water carbonates and deeper water iron formations of the Transvaal Supergroup during the late Neoarchean (e.g., Eriksson et al. 2006). Subsequent uplift of the Kaapvaal craton, with accompanying glaciation, as testified by glacial diamictite of the Makganyene Formation of the Postmasburg Subgroup (Fig. 6), led to a deep erosional incision and a resulting unconformity across much of the craton as sea-levels fell. An exception is where the Makganyene Formation is preserved in the Griqualand-West sub-basin of the Transvaal Supergroup (e.g., Eriksson et al. 2006; Polteau et al. 2006). The 2426 ± 3 Ma Ongeluk Formation volcanic rocks erupted in close succession as sub-aqueous flood basalts on top of these diamictites (Cornell et al. 1996; Kampmann et al. 2015; Gumsley et al. 2017). These 2424 ± 24 Ma basalts (Gumsley et al. 2017) are coeval with the 2428–2426 Ma Westerberg sill province (Kampmann et al. 2015), and a 2423 ± 4 Ma north-trending mafic dyke along the western margin of the Kaapvaal craton (Gumsley et al. 2017; Fig. 6), where several other parallel dykes cluster (Fig. 6). The magmatic event also includes a ca. coeval 2423 Ma-old northeast dipping mafic sheet on the southeastern part of the Kaapvaal craton and almost 1000 km east of the Ongeluk Formation (Gumsley et al. 2017; Fig. 2). Finally, this recently discovered Ongeluk event on the Kaapvaal craton also falls within the protracted 2.51–2.41 Ga age-span of Zimbabwe's Sebanga dyke swarm, and thereby represents the first correlation of mafic magmatism on both cratons.

The Ongeluk Formation has previously been correlated stratigraphically with the compositionally similar Hekpoort Formation basaltic-andesites (Fig. 4), farther to the east within the main Transvaal sub-basin (e.g., Cornell et al. 1996; Eriksson et al. 2006). However, this correlation is now shown to be incorrect using the combined ID-TIMS and in situ secondary ion mass spectrometry (SIMS) age data on baddeleyite by Gumsley et al. (2017) and Kampmann et al. (2015) from the Ongeluk Formation and the Westerberg sill province. Detrital zircon population ages establish a likely minimum age of ≤2250–2240 Ma (Rasmussen et al. 2013; Schröder et al. 2016) for subaerial Hekpoort Formation volcanism, which also has a different paleomagnetic signature compared to the Ongeluk LIP (Evans et al. 1997; Gumsley et al. 2017; Humbert et al. 2017). Cornell et al. (1996) reported a Rb–Sr whole-rock age of ca. 2184 ± 76 Ma for the Hekpoort Formation, which is similar to a ca. 2168 Ma mafic dyke dated by Larsson (2015) using ID-TIMS on baddeleyite from the southeastern part of the Kaapvaal craton (Fig. 2). However, the large error bar on the Rb–Sr result and the possibility of open-system behavior in the Rb–Sr isotopic system does not presently allow a definite correlation between the two rock units.

Undated Bushy Bend Member andesitic lavas (Eriksson et al. 1994) underlie the Hekpoort Formation, but are of very limited extent, while higher up in the succession the undated volcanic rocks of the Machadodorp Member alkali basaltic and volcanoclastic rocks are more extensively exposed within the Transvaal Supergroup (cf., Fig. 4). Elsewhere in this volume, Wabo et al. (2019) report 2208–2276 Ma amphibole Ar–Ar ages for intrusions that east of the town of Mashising and close to the Transvaal-hosted Machadodorp Member. In Zimbabwe the Chimbadzi Hill intrusion yielded a U-Pb baddeleyite age of 2262 ± 2 Ma, but the extent of this Paleoporterozoic event is unkown as this is the only such reported age from that craton (Manyeruke et al. 2004).

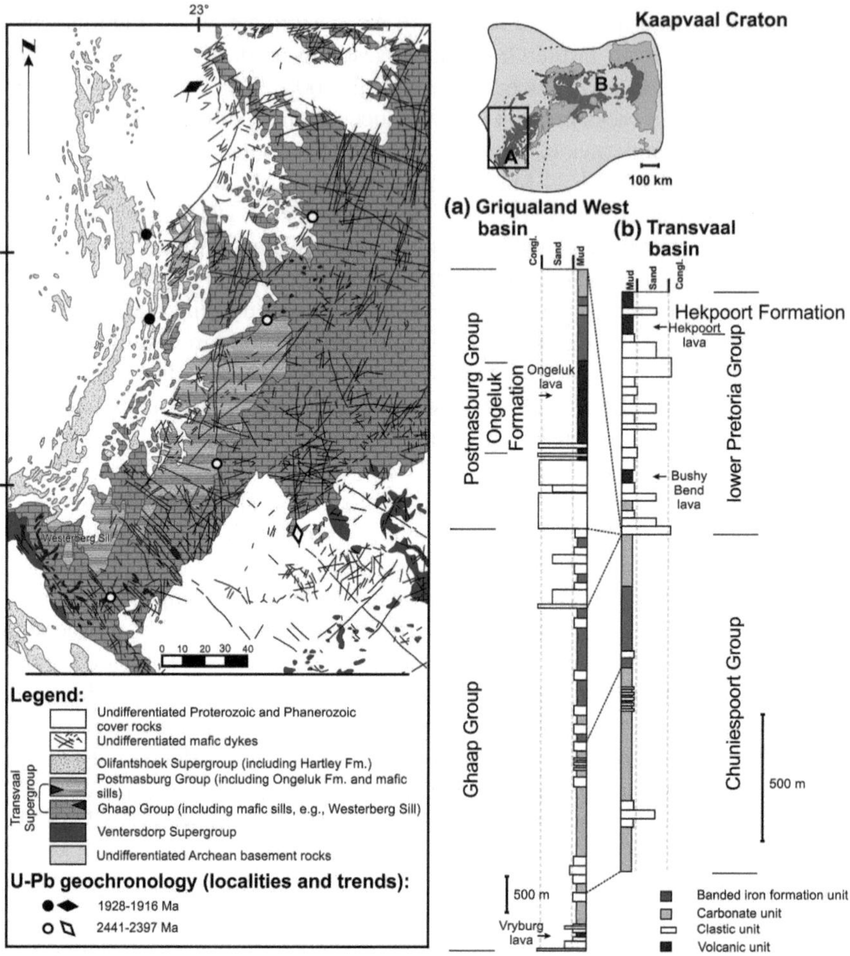

Fig. 6 Stratigraphic correlation between the **a** Griqualand West sub-basin and **b** the Transvaal sub-basin (after Gumsley et al. 2017). **c** Geology including mafic dyke swarms from the southwestern Kaapvaal craton. The affinity of southeast-trending dykes in the south of the mapped area, east-southeast-trending dykes in central and northern part of the mapped area, and north-northeast-trending dykes in the northern part of the mapped area remains to be determined. Compiled as in Fig. 2. Geochronology symbols: diamonds = dated dyke and its trend; circles = dated sill, sheet, or plutonic body

3.4 *The Bushveld Complex and Associated Sills, Complexes and Volcanic Rocks*

The Bushveld Complex was emplaced into the Kaapvaal craton closely following on from the intrusion of the alkaline and carbonatititic Phalaborwa Complex and Schiel Complex (Cawthorn et al. 2006; Fig. 4). The Phalaborwa Complex has U–Pb baddeleyite ages of ca. 2.06 Ga (e.g., Wu et al. 2011; and references therein). Walraven et al. (1992) report a Pb–Pb whole-rock errorchron date of 2059 ± 35 Ma for the Schiel Complex, but Barton et al. (1996) suggested that the complex is significantly youger. Barton et al. (1996) presented zircon dates for the Schiel Complex at 1853 ± 6 Ma and 2005 ± 7 Ma, the older population of dates being interpreted as xenocrystic. More recently Laurent and Zeh (2015) precisely dated zircon from rocks of Schiel Complex at 2054 ± 4 Ma and 2051 ± 6 Ma. The Phalaborwa Complex is broadly coeval with the ca. 2061 Ma bimodal volcanic Rooiberg Formation (Walraven 1997; Fig. 4), which partly underlies the most recognized part of the Bushveld Complex; namely, its layered ultramafic-mafic Rustenburg Layered Suite that hosts the world's largest deposits of platinum and chromium. The Bushveld Complex also incorporates the felsic Lebowa Suite (Cawthorn et al. 2006), many near-coeval satellite intrusions (e.g., De Waal et al. 2008), as well as numerous mafic sills mostly hosted in the Transvaal Supergroup (Cawthorn et al. 1981; Sharpe 1981, 1982). Cawthorn et al. (1981) identified these sills as having a pre-, syn-, or post-emplacement relationship with the Bushveld Complex based mostly on mineralogy and stratigraphic locations. The sills were geochemically characterized by Barnes et al. (2010); whereas, Wabo et al. (2015a) studied the sills geochemically and paleomagnetically, as well as presented two U–Pb ID-TIMS baddeleyite ages of ca. 2058 Ma (Fig. 4). These ages are broadly coeval with other satellite intrusions, including ca. 2058 Ma ages on both baddeleyites and zircons from the Uitkomst Complex (Wabo et al. 2015b; Maier et al. 2017); the ca. 2057 Ma Molopo Farms Complex (De Kock et al. 2016); and the ca. 2055 Ma Moshaneng Complex, Marble Hall diorite, Lindeques Drift intrusion and Roodekraal Complex (De Waal and Armstrong 2000; Mapeo et al. 2004b; De Waal et al. 2006; Fig. 7).

The Rustenburg Layered Suite is an up to 9 km thick layered sequence that arguably accumulated within an enormous lopolith (e.g., Cawthorn et al. 2006; Cole et al. 2014). Zeh et al. (2015) demonstrated with high-precision chemical abrasion ID-TIMS U–Pb zircon ages that the entire suite crystallized within less than 1 Myr from 2055.91 ± 0.26 Ma to 2054.89 ± 0.37 Ma. Additional high-precision U–Pb dating of zircons and baddeleyites support the rapid emplacement of the suite, where chromite layers may unexpectedly have been emplaced out-of-sequence within slightly older norites (Mungall et al. 2016). An older U–Pb ID-TIMS baddeleyite age of 2058 ± 2 Ma from the Marginal Zone of the Rustenburg Layered Suite (Olsson et al. 2010) does not possess equally high precision as those by Zeh et al. (2015) and Mungall et al. (2016). Although the Rustenburg Layered Suite formed very rapidly, Bushveld Complex magmatism likely continued with the emplacement of an overlying Lebowa Suite of granophyres and granites until ca. 2054 Ma (Walraven and Hattingh 1993; Van Tongeren et al. 2016).

The Bushveld Complex differs from other large mafic events on the Kaapvaal craton by apparently not having any associated regional dyke swarm, or any other type of exposed feeder, whereby one could explain the rapid mode of emplacement of such unusually large volumes of mafic magma (e.g., Olsson et al. 2011). An equally large, but currently hidden feeder, like the Great Dyke of Zimbabwe, could initially have fed marginal sills of the Bushveld Complex, which eventually coalesced into a

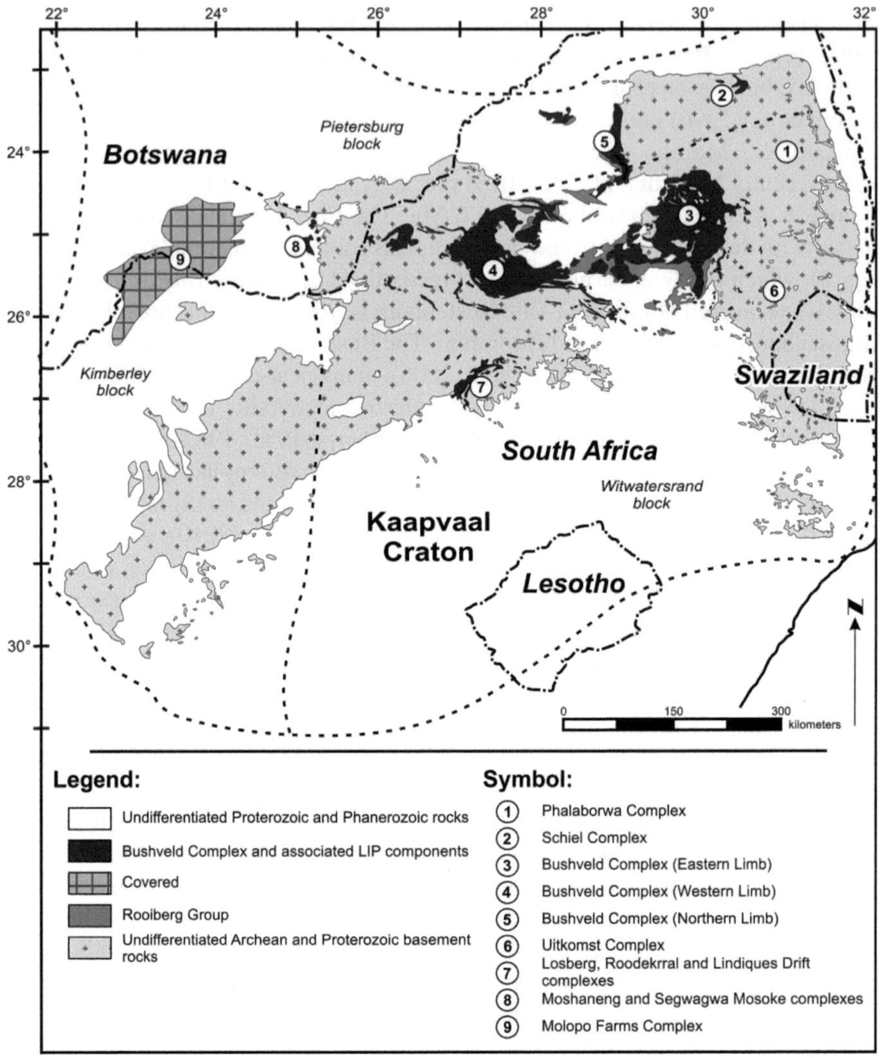

Fig. 7 Distribution of the Bushveld LIP across the Kaapvaal craton. Compiled as in Fig. 2

replenished large lopolith (e.g., Cawthorn et al. 2006), which ultimately crystallized into the Rustenburg Layered Suite. The most likely location of such a hypothetical feeder zone might be along the E–W trending TML (Clarke et al. 2009), which aligns sub-parallel to the LMC, and its Central Zone that is dominated by ages that conform to a major 2.05–2.03 Ga metamorphic event and subsequent dextral displacement between 2.03 and 1.9 Ga (Holzer et al. 1998).

3.5 Late Paleoproterozoic Events

Isolated remnants of volcanic rocks have been recognized within the extensive Paleoproterozoic red-bed successions on the Kaapvaal craton. SHRIMP zircon U–Pb dates indicate that the deposition of these red-bed successions spanned more than 200 Myr (Dorland et al. 2006; Geng et al. 2014). Minor quartz porphyry lavas near the bottom of the Waterberg Group immediately postdate the intrusion of the Bushveld Complex (Dorland et al. 2006). More prominent occurrences of younger volcanic rocks, including more mafic lavas, are preserved as the Hartley Formation, within the Olifantshoek Supergroup, and the Sibasa and Ngwanedzi formations, within the Soutpansberg Group.

The volcanic rocks within the Hartley Formation are compositionally bimodal basalts and rhyolites, which have been dated to 1916 ± 1 Ma using four different methods on a quartz porphyry flow unit (Cornell et al. 2016). A previous age of 1928 ± 4 Ma on the same quartz porphyry (Cornell et al. 1998) and a 1920 ± 4 Ma age on another quartz porphyry have also been reported (Alebouyeh Semami et al. 2016). An east-northeast-trending mafic dyke swarm, termed the Tsineng dyke swarm, was identified by Goldberg (2010) as stemming from a magmatic centre that was located outside the western margin of the Kaapvaal craton (Alebouyeh Semami et al. 2016; Figs. 6 and 8). As this up to 300 km long, and at least 100 km wide swarm appears to be truncated by the <1.9 Ga Kheis orogen, but cuts the Waterberg Group, Goldberg (2010) tentatively assigned it a ca. 1.9 Ga age. One mafic dyke of the Tsineng dyke swarm was dated to 1923 ± 6 Ma using ID-TIMS on baddeleyite (Alebouyeh Semami et al. 2016). Furthermore, broadly coeval mafic sills from Botswana (near Moshaneng) yield ages of ca. 1927 Ma (Hanson et al. 2004b). In addition, drill core samples from the buried Trompsburg Complex (Fig. 8)—a roughly circular and 50 km wide layered mafic intrusion beneath the Phanerozoic Karoo Supergroup sedimentary rocks in the southern part of the Kaapvaal craton—is dated to ca. 1915 Ma (Maier et al. 2003), and is thereby also a potential member of the same LIP.

Approximately 40 Myr later, the northeastern Kaapvaal craton was intruded by northeast-trending mafic dykes termed the Black Hills dyke swarm. Extensive dating by Olsson et al. (2016; 2011) has shown that dykes within this swarm are intermixed with the northeast-trending 2.66 Ga branch of the proposed radiating dyke swarm, described in Sect. 3.2 (Figs. 4 and 8). The ID-TIMS U–Pb baddeleyite ages of the Black Hills dyke swarm (sensu stricto) range from 1871 to 1839 Ma (Olsson et al. 2016), where age resolutions allow these authors to distinguish between an earlier and compositionally more primitive suite, and a later more evolved suite. Mafic sills further west within the Waterberg Group are geochemically indistinguishable from the Black Hill dyke swarm, and thereby extend the magmatic age range back to ca. 1886 Ma (Hanson et al. 2004b; Olsson et al. 2016; Fig. 8). Paleomagnetic data are available for many Black Hills dykes and post-Waterberg sills, and give consistent results that are supported by positive baked-contact tests from two dykes (Letts et al. 2005; De Kock et al. 2006; Letts et al. 2010; Lubnina et al. 2010; Maré and Fourie 2012). The post-Waterberg sills across the northeastern part the Kaapvaal

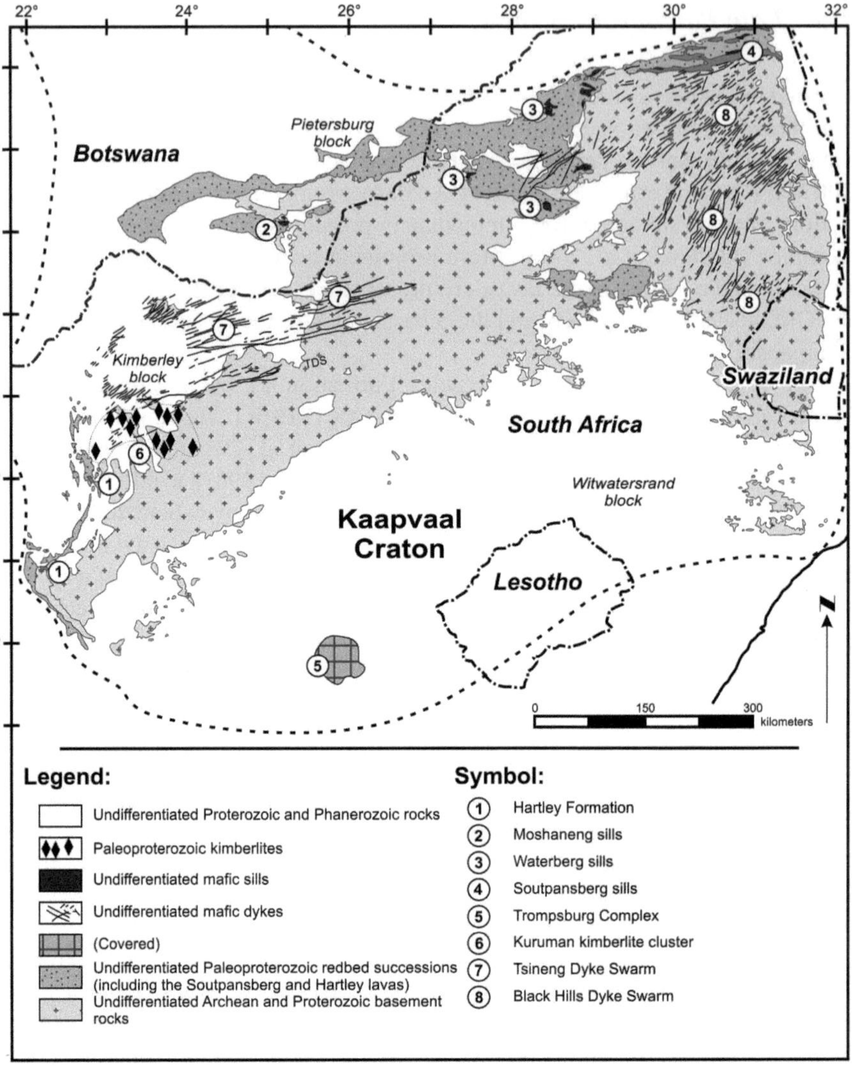

Fig. 8 Late Paleoproterozoic magmatic units on the Kaapvaal craton. These are the 1.93–1.92 Ga Hartley LIP (including the east-northeast-trending Tsineng dyke swarm, Trompsburg Complex, and Moshaneng sills), the 1.88–1.83 Ga Black Hills dyke swarm, post-Waterberg sills, and Soutpansberg Group volcanic units. Compiled as in Fig. 2

craton are furthermore broadly coeval with the Mashonaland sill province within the Zimbabwe craton (Wilson et al. 1987; Söderlund et al. 2010; Fig. 5), representing the first well-established magmatic barcode link between the Kaapvaal craton and the Zimbabwe craton. Based on significantly different paleomagnetic poles of the Mashonaland and Waterberg sill provinces, Hanson et al. (2011b) so suggested that these two sill provinces—and thereby their respective host cratons—must have been as much as 3000 km apart, before being brought closer together after their ca. 1.88 Ga emplacement.

The age of the Sibasa Formation of the Soutpansberg Group (Fig. 8) is constrained by 1830 ± 15 Ma and 1832 ± 9 Ma U–Pb SHRIMP dates by the youngest zircon grains from the formation's uppermost pyroclastic units, which were interpreted as depositional ages of its ash beds (Geng et al. 2014). The lavas of the Sibasa Formation may therefore be related to the younger phase of Black Hills dyke emplacement (Geng et al. 2014). The Ngwanedzi Formation of the Soutpansberg Group is located stratigraphically above the Sibasa Formation, but is not yet dated (Fig. 8). An age on this volcanic event could potentially extend the age range of the protracted 1888-1830 Ma magmatic event, represented by early post-Waterberg and Mashonaland sills, the Black Hills dyke swarm and late Sibasa volcanism, even further, and thereby making it difficult to view this as a classical short-lived LIP. Instead, Olsson et al. (2016) view this protracted magmatic event as emplaced within a continental back-arc setting, behind a parallel and convergent plate boundary along the northwestern margin of the proto-Kalahari craton.

The Soutpansberg Group is also cut by mafic intrusions, some of which have been assigned to the 1112–1108 Ma Umkondo LIP (Jones and McElhinny 1966; Hanson et al. 2004a). However, an older ca. 1.75 Ga phase of magmatism is suggested by a highly uncertain 1749 ± 104 Ma Rb–Sr whole rock age on one mafic sill (Barton 1979) as well as Paleoproterozoic paleopoles for other intrusions (Hanson et al. 2004b). These potentially older sills are here referred to as the Soutpansberg mafic sills to distinguish these from younger (ca. 1.11 Ga) Umkondo LIP sills and dykes (Sect. 3.8), as well as the ca. 1.88 Ga post-Waterberg mafic sills; the latter of which are too old to be hosted by the Soutpansberg Group. The Soutpansberg mafic sills have been suggested to coeval with the Ngwanedzi lavas in the upper Soutpansberg Group (Dorland et al. 2006). The Bathlaros Kimberlite, the oldest preserved kimberlite cluster on the Kaapvaal craton (Fig. 8), with U–Pb perovskite ages that range from 1830 to 1650 Ma (Donnelly et al. 2011, 2012), is broadly coeval, but spatially far removed from the Soutpansberg Group.

The Richterveld terrane, in the far southwestern corner of the proto-Kalahari craton (Fig. 9) incorporates 2.0–1.73 Ga volcanic and sedimentary rocks of the Orange River Group and coeval plutonic rocks of the Vioolsdrif Suite, which likely represents an island arc terrane that accreted onto the proto-Kalahari craton during the Mesoproterozoic Namaqua-Natal orogeny (Reid 1997; Jacobs et al. 2008).

3.6 *Early Mesoproterozoic Magmatism*

Several Mesoproterozoic alkaline intrusions, including carbonatites, are known from the interior of the Kaapvaal craton, the largest of which is the Pilanesberg Complex (see Hanson et al. 2006; Verwoerd 2006 for a more detailed overview; Fig. 9). These intrusions are mostly not well-dated, but the Pilanesberg Complex recently

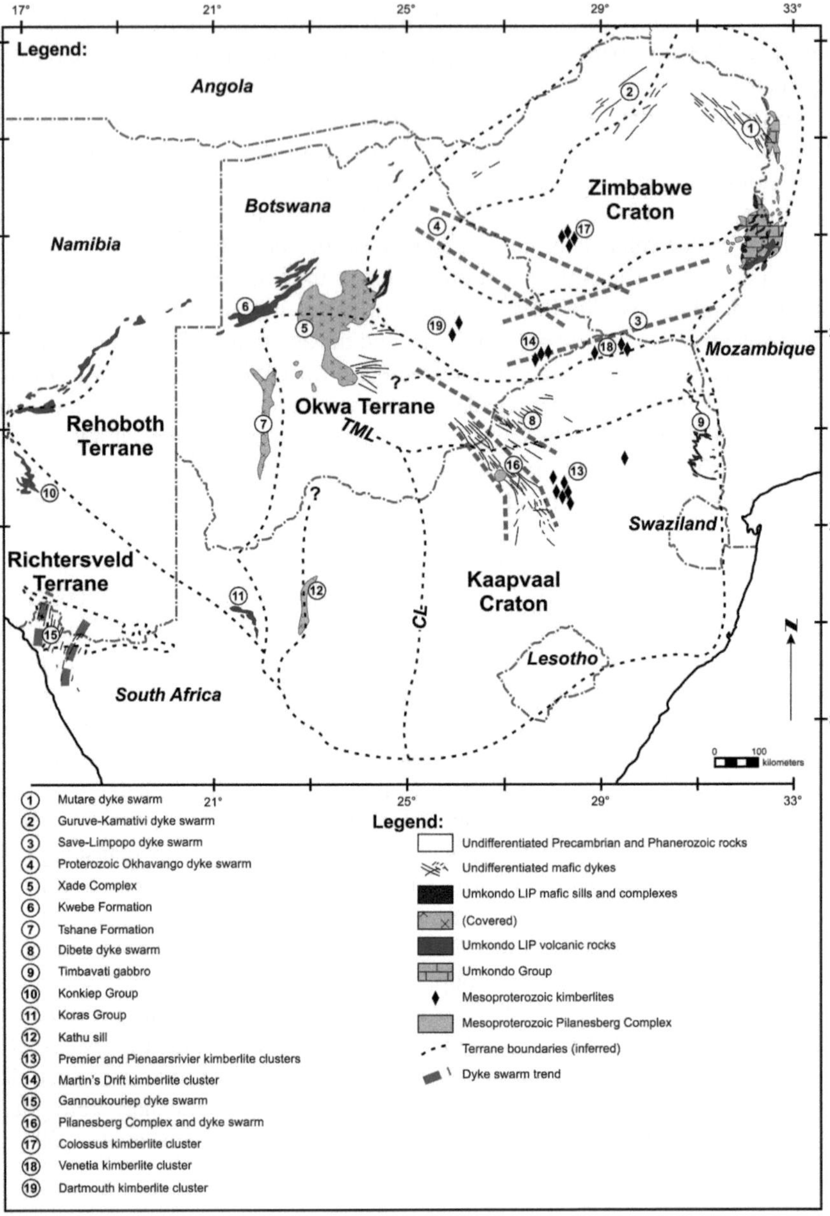

Fig. 9 Mesoproterozoic and Neoproterozoic magmatic units of the Kalahari craton. South African geology was compiled as in Fig. 2, while Botswana and Zimbabwe geology is from the 1: 1,000,000 geological maps of those respective countries

yielded a U–Pb age on titanite of 1395 +10/−11 Ma (Elburg and Cawthorn 2017).

Hanson et al. (2006) cited many well-constrained, but unpublished U–Pb zircon and titanite ages for several intrusions that range in age between 1397 and 1381 Ma. A reasonably well-constrained Rb–Sr age of ca. 1340 Ma was also reported by Harmer (1999) for the Spitskop Complex. Preliminary paleomagnetic constraints have also been reported for many of these intrusions, but paleomagnetic stability field tests are generally lacking (Gose et al. 2013). The Mesoproterozoic carbonatites and other alkaline magmatic intrusions of southern Africa appear to be spatially associated with old terrane boundaries and younger extensional faults (Friese et al. 1995). The Pilanesberg Complex is centered on a prominent, >400 km long, and south-southeast- to north-trending Pilanesberg dyke swarm (e.g., Hunter and Reid 1987). Paleomagnetic data and 1.3–1.1 Ga Rb–Sr and K–Ar ages on its many dykes suggest relatively prolonged magmatic activity, and several overlapping generations may be present within this swarm, consistent with the presence of both normal and reversely magnetized dykes (Gough and Hales 1956; Schreiner and Van Niekerk 1958; Van Niekerk 1962; McDougall 1963; Jones and McElhinny 1966; Emerman 1991). Jones and McElhinny (1966), furthermore, obtained a virtual geomagnetic pole for a ca. 1130 Ma dyke (K–Ar feldspar age; McDougall 1963) that differs significantly from other dykes in the swarm.

Numerous kimberlites on the Kaapvaal craton are also Mesoproterozoic, including the ca. 1150 Ma Premier cluster and ca. 1465 Ma Martin's Drift cluster (Wu et al. 2013; Griffin et al. 2014). The Martin's Drift cluster is associated with east-northeast-trending basement lineaments across the Kaapvaal craton, while the Premier cluster is associated with more north-trending lineaments (Jelsma et al. 2004).

3.7 *The Umkondo Large Igneous Province*

Apart from the Jurassic Karoo LIP, the most widespread magmatic event on the Kalahari craton is the ca. 1.11 Ga Umkondo LIP (Hanson et al. 2004a; Fig. 9). It was named after the extensive sills that intruded into the Umkondo Group, preserved on the eastern Zimbabwe craton along the Zimbabwe-Mozambique border (McElhinny and Opdyke 1964). The Umkondo Group was shown, on the basis of paleomagnetism and geochemistry, to host both a basaltic lava succession and coeval doleritic sills (McElhinny 1966; Munyanyiwa 1999). Since then, many mafic sills, as well as dykes across both the Zimbabwe and Kaapvaal cratons have been geochronologically and paleomagnetically linked into a remarkably this short-lived and extensive LIP event (Hanson et al. 1998; Wingate 2001; Hanson et al. 2004a; Gose et al. 2006; De Kock et al. 2014; Swanson-Hysell et al. 2015). Large subsurface Tshane, Xade and Tsetseng mafic intrusions in Botswana have also been linked to the Unkondo LIP (e.g., Hanson et al. 2006), where recent geophysical studies of the Xade Complex have dramatically increased its size to one of the largest such Mesoproterozoic mafic complexes in southern Africa (Corner et al. 2012). Intraplate 1.1 Ga volcanic rocks are also described from as far afield as the Rehoboth terrane further to the southwest (Becker et al. 2006; Miller 2008, 2012). Bimodal basaltic and rhyolitic volcanic rocks inside the northwestern Botswana rift (e.g., Key and Mapeo 1999), across the western edge of the Kalahari craton, known as the Kgwebe Formation, have also been geochronologically linked to the Umkondo LIP (Singletary et al. 2003). Granites of the same age have been revealed by geophysical data, drilling, and one poor outcrop in northwestern Botswana (Singletary et al. 2003). A small 1.1 Ga layered gabbroic to dioritic complex has also been identified within the Kwando Complex in Botswana (Singletary et al. 2003).

In Dronning Maud Land of East Antarctica the widespread 1.1 Ga Borgmassivet mafic and ultramafic intrusions of the Grunehogna terrane intrude strata of the Proterozoic Ritscherflya Supergroup as well as the Archean basement (Hanson et al. 2006). The Ritscherflya Supergroup is capped by basalts that are related to the Borgmassivet intrusions, and these are contiguous with the Umkondo Group of Zimbabwe and Mozambique in typical Gondwana reconstructions (Krynauw et al. 1991; Jones et al. 2003). Paleomagnetic data of the Borgmassivet intrusions and Ritscherflya basalts are indistinguishable from that of the Umkondo LIP (Jones et al. 2003; Gose et al. 2006; Swanson-Hysell et al. 2015).

Even before the Umkondo LIP was recognized, Wilson et al. (1987) observed that northeast-trending Guruve and Kamativi dyke swarms across Zimbabwe may on the basis of limited paleomagnetic and Rb–Sr isotopic data be coeval with the Umkondo dolerites and lavas in eastern Zimbabwe. A ca. 1.11 Ga age of some of the Guruve dykes was finally confirmed through U–Pb ID-TIMS baddeleyite age dating (De Kock et al. 2014). Other aeromagnetically pronounced dyke swarms, including the east-southeast-trending Dibete, northeast-trending Save-Limpopo and northwest-trending Okavango dyke swarms—intersecting each other across the central parts of the Kalahari craton—were all originally thought to be of Karoo age (Fig. 9). However, reconnaissance Ar–Ar age dating (Jourdan et al. 2004, 2006) revealed that these dyke swarms are made up of overlapping and apparently sub-parallel mafic dykes with both Karoo and Proterozoic ages; the latter of which were confirmed to be at least partly of Umkondo age (De Kock et al. 2014; Figs. 5 and 9). Due to the many different trending Umkondo dyke swarms, De Kock et al. (2014) also identified up to three magmatic centers for possible radiating swarms, all located along proposed passive margins that may have been conjugate to at least three other cratonic blocks, hosting coeval intrusions across a since dispersed Rodinia supercontinent. However, this proposed break-up scenario also occurred penecontemporaneous with widespread 1135–1050 Ma collisional orogenesis in many parts of the world, forming for example the Grenville orogen in Laurentia and the Namaqua-Natal orogen in southern Africa. The enigmatic tectonic regime of the relatively short-lived yet spatially extensive and voluminous Umkondo LIP event during a regional orogenic event is further accentuated by localized volcanic and sedimentary rock successions within the Koras and Konkiep groups, which both outcrop along the southwestern margin of the Kalahari craton. Here, bimodal basaltic and rhyolitic volcanic rocks of the Koras Group erupted between 1.17 and 1.09 Ga, within a post-orogenic rift that is superimposed on highly deformed metamorphic rocks in this part of the Namaqua-Natal orogen (Pettersson et al. 2007; Bailie et al. 2011; Kasbohm et al. 2016; Panzik et al. 2016), see Fig. 9. Miller (2012) introduced the Konkiep Group to include rocks previously assigned to the Sinclair Group (a term that is now discontinued) as well as other coeval units across the Rehoboth terrane. The Konkiep Group thereby includes a variety of arc-related and intraplate igneous rocks with ages between 1.38 to 1.11 Ga (Miller 2012; Cornell et al. 2015). The paleogeographic relations of the arc-related rocks to the proto-Kalahari craton is unclear, but Panzik et al. (2016) showed that the Konkiep terrane (previously the Sinclair terrane) had docked with the Kalahari craton by 1.1 Ga during the eruption of compositionally bimodal volcanic rocks of the Guperas Formation bimodal volcanic rocks and a post-Guperas dyke swarm. The paleomagnetic signature of the Guperas Formation, post-Guperas dykes and upper Koras Group are very similar to that of Umkondo LIP units, and can thereby be considered an extension of it (Kasbohm et al. 2016; Panzik et al. 2016).

3.8 Neoproterozoic Magmatic Events

The north to northeast-trending Gannakouriep dyke swarm, along the border between South Africa and Namibia, has been dated at 792–788 Ma, using more traditional U–Pb on ID-TIMS methods on both baddeleyite and zircon (Rioux et al. 2010). A slightly older preferred age of 795 ± 1 Ma was provided for one of its dykes, however, using U–Pb isotopic data on discretely digested zircons (Rioux et al. 2010). These mafic dykes intrude for over more than 300 km across several terrane boundaries within the Namaqua-Natal orogenic belt, as well as the Rehoboth terrane on the western limits of the Kalahari craton (Fig. 9). The dykes have been metamorphosed to between greenschist and amphibolite facies, and have until recently been only cursorily studied (e.g., Reid et al. 1991). The Gannakouriep dykes also cut across 890–880 Ma and 830–800 Ma granitic plutons (Frimmel et al. 2001; Miller 2008; Hanson Pers. comm.), and paleomagnetic results form both units are currently being prepared for publication (Bartholomew 2008; Hanson et al. 2011a). This magmatism is generally interpreted as heralding the breakup of Rodinia in the Kalahari craton's geological record (Frimmel et al. 2001) and are tentatively related to ca. 750 Ma bimodal volcanic rocks within the Rosh Pinah Formation, as para-autochtonous parts of the Gariep orogen farther to the west (Frimmel et al. 1996; Borg et al. 2003).

The arcuate Mutare dyke swarm trends north-northwest-trending mafic dyke swarm across the eastern Zimbabwe craton, near the border of Mozambique (Wilson et al. 1987; Ward et al. 2000), and is up to 300 km long and 100 km wide (Figs. 5 and 9). It was initially suspected to be of Umkondo-age (e.g., Hanson et al. 2006), until Mukwakwami (2005) obtained a 724 ± 4 Ma ID-TIMS U–Pb age on baddeleyites. Nevertheless, Mukwakwami (2005) thought the swarm may contain ca. 1.1 Ga parallel dykes, which could help explain the swarm's compositional heterogeneity (Ward et al. 2000).

As reviewed by Hanson (2003), there are many Neoproterozoic igneous rocks within other Pan-African orogenies that surround and may represent intraplate magmatism of the Kalahari craton. Along the northern margin of the Kalahri craton, ca. 750 Ma continental rift deposits are incorporated within the Neoproterozoic supracrustal successions of the Damara and Kaoko orogens, including extensive occurrences of tholeitic alkaline basalts within the Damara orogeny. Farther to the northeast, along in the Zambezi orogen in Zambia and Zimbabwe, there are numerous 870–730 Ma intraplate bimodal volcanic rocks and intraplate mafic and felsic intrusions. More detail and data on these occurrences are provided by Katongo et al. (2004), Johnson et al. (2005), and Johnson et al. (2007). Most of these Zambezi rocks are linked to the Congo craton by Johnson et al. (2007), but it could be debated whether some units formed along a northern rifted margin of the Kalahari craton.

A few Neoproterozoic Kimberlite occurrences from the Kalahari craton have also been U–Pb dated, using on groundmass perovskite (Griffin et al. 2014). This includes the 611 ± 5 Ma South African Dartmouth kimberlite (Griffin et al. 2014), the 572–522 Ma South African Venetia kimberlite cluster (Allsopp et al. 1995; Woodhead et al. 2009; Griffin et al. 2014), the 583 ± 64 Zimbabwean Beitbridge kimberlite (Griffin et al. 2014), and 541 ± 10 Ma Zimbabwean Colossus kimberlite cluster (Donnelly et al. 2012).

4 Discussion

4.1 A Precambrian Magmatic Barcode and Paleogeography of the Kalahari Craton

Figure 10 is a compilation of available precise U–Pb zircon and baddeleyite ages for the Precambrian mafic magmatic units, which intruded into the Kalahari craton and its constituents after stabilization (summarized in Table 1). This magmatic barcode represents a temporal fingerprint which can be compared with other Precambrian terranes globally (Bleeker and Ernst 2006), and where a sharing of at least two magmatic events can be used as a possible indication of paleogeographic continuity between the terranes. Many of the magmatic events identified in the barcode have paleomagnetic data, of variable quality, which constrain the paleolatitudinal drift of the Kaapvaal, Zimbabwe and subsequent Kalahari cratons (Fig. 11). Paleolatitudinal constraints are considered reliable if there is justifiable cause (e.g., rock magnetic investigation, field stability tests, etc.) to assign the rock age to the paleomagnetic constraint (Table 1; cases where any less reliable records are identified). In many cases, several Precambrian crustal blocks can be identified as optional nearest paleo-neighbors to the Kaapvaal, Zimbabwe or Kalahari cratons based on magmatic barcode matches, yet only in a few cases can specific paleogeographic reconstructions be proposed (Fig. 11).

From this data, the magmatic barcode of the Kaapvaal craton remains unique for Mesoarchean mafic magmatism associated with the Pongola Supergroup, Usushwana Complex and older components of the Badplaas dyke swarm, as well as the Hlagothi Complex. The Usushwana Complex has been compared paleomagnetically to the ca. 2.87 Ga Millindina Complex of the Pilbara craton of Western Australia, suggesting

Table 1 Summary of recognized magmatic events of the Kalahari craton with paleomagnetic constraints[a] listed where available

Event		Terrane	Age (Ma)	Unit	Number of units dated	Geochronology reference	Latitute	Longitude	dp	dm	Paleomagnetism reference
36	Colossos kimberlite cluster	Kalahari craton	**ca. 541**	Colossos kimberlite cluster	1	Donnelly et al. (2012)	No published data				
35	Venetia kimberlite cluster	Kalahari craton	**572–522**	Venetia kimberlite cluster	3	Allsopp et al. (1995), Woodhead et al. (2009), Griffin et al. (2014)	No published data				
34	Beitbridge kimberlite	Kalahari craton	**ca. 583**	Beitbridge kimberlite	1	Griffin et al. (2014)	No published data				
33	Dartmouth kimberlite	Kalahari craton	**ca. 611**	Dartmouth kimberlite	1	Griffin et al. (2014)	No published data				
32	Mutare dyke swarm	Kalahari craton	**ca. 724**	Mutare dyke swarm	1	Mukwakwami (2005)	No published data				
31	Rosh Pinah Formation	Kalahari craton	**ca. 750**	Rosh Pinah Formation	1	Frimmel et al. (1996)	No published data				
30	Gannakouriep dyke swarm	Kalahari craton	**795–788**	Gannakouriep dyke swarm	4	Rioux et al. (2010), Hanson et al. (2011a)	No published data				
29	Late Mesoproterozoic rifts (bimodal)	Kalahari craton	**1105–1092**	Upper Koras Group	5	Pettersson et al. (2007), Bailie et al. (2011)	**57**	**3**	**7**	**7**	Briden et al. (1979)

(continued)

Table 1 (continued)

Event		Terrane	Age (Ma)	Unit	Number of units dated	Geochronology reference	Latitute	Longitude	dp	dm	Paleomagnetism reference
				Konkiep Group	3	Panzik et al. (2016), Miller (2012), Cornell et al. (2015)	**69.8**	**4.1**	**7.4**	**7.4**	Panzik et al. (2016)
29	Umkondo LIP	Kalahari craton	**1107–1112**	Paleoproterozoic Okavango dyke swarm	2	De Kock et al. (2014)	**64**	**39**	**4**	**4**	Swanson-Hysell et al. (2015)
				Sill intruding the Kwando Complex	1	Singletary et al. (2003)					
				Xade Complex	1	Hanson et al. (2004a)					
				Umkondo Sills (Waterberg Group)	4	Hanson et al. (2004a)					
				Anna's Rust Complex	2	Hanson et al. (2004a)					
				Divuli Ranch dyke	1	De Kock et al. (2014)					
				Umkondo Sill Province	4	Hanson et al. (2004a), Wingate (2001)					

(continued)

Table 1 (continued)

Event		Terrane	Age (Ma)	Unit	Number of units dated	Geochronology reference	Latitute	Longitude	dp	dm	Paleomagnetism reference
				NE-trending intrusion (Botswana)	1	De Kock et al. (2014)					
				Dibete dyke swarm	1	De Kock et al. (2014)					
				Guruve dyke swarm	1	De Kock et al. (2014)					
				Timbavati gabbro	1	Hanson et al. (2004a)					
				Umkondo Sills (Palapye Group)	7	Hanson et al. (2004a)					
28	Late Mesoproterozoic rifts (bimodal)	Kalahari craton	**ca. 1172**	Lower Koras Group	2	Pettersson et al. (2007), Gutzmer et al. (2000)	*55*	*77*	*12.9*	*12.9*	Briden et al. (1979)
27	Premier kimberlite cluster	Kalahari craton	**ca. 1150**	Premier kimberlite	–	Wu et al. (2013)	**41**	**55**	**16**	**16**	Powell et al. (2001)
26	Pienaars River kimberlite cluster	Kalahari craton	**ca. 1332**	Pienaars River kimberlite	–	Jelsma et al (2004)	**−11.5**	**43.3**	**5.6**	**5.6**	Gose et al. (2013)
25	Martin's Drift kimberlite cluster	Kalahari craton	**ca. 1465**	Martin's Drift kimberlite	–	Griffin et al. (2014)	No published data				

(continued)

Table 1 (continued)

Event		Terrane	Age (Ma)	Unit	Number of units dated	Geochronology reference	Latitute	Longitude	dp	dm	Paleomagnetism reference
24	Pilanesberg Complex	Kalahari craton	**1397–1340**	Pilanesberg Complex	1	Elburg and Cawthorn (2017), Hanson et al. (2006), Harmer (1999)	**43.1**	**284.2**	**5.1**	**8.1**	Gose et al. (2013)
23	Kuruman kimberlite cluster	Kalahari craton	**1830–1650**	Bathlaros kimberlite	1	Donnelly et al. (2011; 2012)	**30**	**8.2**	**8.9**	**8.9**	Hargraves (1989)
22	Upper Soutpansberg lavas	Kalahari craton	**undated**	Ngwanedzi Formation	–	–	No published data				
21	Lower Soutpansberg lavas	Kalahari craton	**ca. 1830**	Sibasa Formation	2	Geng et al. (2014)	No published data				
21	Black Hills LIP	Kaapvaal craton	**1879–1839**	Black Hills dyke swarm	12	Olsson et al. (2016)	**10.1**	**18.1**	**7.9**	**7.9**	Letts et al. (2010)
				Post-Waterberg sills	3	Hanson et al. (2004b)	*8.6*	*15.4*	*17.3*	*17.3*	Hanson et al. (2004b); Recalculated by De Kock et al. (2006)
20	Mashonaland sill province	Zimbabwe craton	**1886–1872**	Mashonaland sills	5	Söderlund et al. (2010); Hanson et al. (2011b)	**6.5**	**338.5**	**5**	**5**	Hanson et al. (2011b)
19	Hartley LIP (bimodal)	Kaapvaal craton	**1928–1915**	Ttrompsburg Complex	1	Maier et al. (2003)	No published data				

(continued)

Table 1 (continued)

Event		Terrane	Age (Ma)	Unit	Number of units dated	Geochronology reference	Latitute	Longitude	dp	dm	Paleomagnetism reference
				Tsineng dyke swarm	1	Alebouyeh Semami et al. (2016)	**22.7**	**328.6**	**11.7**	**11.7**	Alebouyeh Semami et al. (2016)
				Moshaneng sill province	3	Hanson et al. (2004b)					
18	Bushveld LIP (bimodal)	Kaapvaal craton	**2058–2055**	Rustenburg Layered Suite	15	Zeh et al. (2015); Olsson et al. (2010); Mungall et al. (2016)	**19.2**	**30.8**	**5.8**	**5.8**	Letts et al. (2009)
				Molopo Farms Complex	1	De Kock et al. (2016)	No published data				
				Uitkomst Complex	2	Wabo et al. (2015a), Maier et al. (2017)	*28.7*	*58.5*	*6.2*	*9.4*	Wabo et al. (2015a)
				Moshaneng Complex	1	Mapeo et al. (2004b)	No published data				
				Marble Hall diorite	1	De Waal and Armstrong (2000)	No published data				
				Lindeques Drift intrusion	1	De Waal et al. (2006)	No published data				
				Roodekraal Complex	1	De Waal et al. (2006)	No published data				
				Bushveld Sill Province	2	Wabo et al. (2015a)	*13.1*	*44*	*14.1*	*14.1*	Wabo et al. (2015a)

(continued)

Table 1 (continued)

Event		Terrane	Age (Ma)	Unit	Number of units dated	Geochronology reference	Latitute	Longitude	dp	dm	Paleomagnetism reference
18	Phalaborwa Complex	Kaapvaal craton	**2062–2060**	Palaborwa Complex	numerous	Wu et al. (2011), Heaman and LeCheminant (1993), Reischmann (1995), French et al. (2008), Horn et al. (2000), Wingate (2000), Heaman (2009)	**27.7**	**35.8**	**6.6**	**6.6**	Letts et al. (2010)
18	Schiel Complex	Kaapvaal craton	**2054–2051**	Schiel Complex	2	Laurent and Zeh (2015)	No published data				
18	Rooiberg Group (bimodal)	Kaapvaal craton	**ca. 2061**	Rooiberg volcanics	1	Walraven (1997)	No published data				
17	Unnamed dyke	Kaapvaal craton	**ca. 2168**	Unnamed dyke	1	Larsson (2015)	No published data				
17	Machadodorp Member	Kaapvaal craton	**undated**	Machadodorp Member	–	–	No published data				
16	Hekpoort Formation	Kaapvaal craton	**ca. 2184**	Hekpoort Formation	1	Cornell et al. (1996)	*−44.1*	*40*	*9.9*	*10.4*	Humbert et al. (2017)
15	Chimbadzi Hills intrusion	Zimbabwe craton	**ca. 2262**	Chimbadzi Hills intrusion	1	Manyeruke et al. (2004)					
14	Sebanga Poort dyke swarm	Zimbabwe craton	**ca. 2408**	Sebanga Poort dyke	1	Söderlund et al. (2010)	*17.2*	*6.1*	*14.3*	*14.3*	Smirnov et al. (2013)
13	Ongeluk LIP	Kaapvaal craton	**2428–2423**	Gewonne dyke swarm	1	Gumsley et al. (2017)	**4.1**	**282.9**	**5.3**	**5.3**	Gumsley et al (2017)

(continued)

Table 1 (continued)

Event		Terrane	Age (Ma)	Unit	Number of units dated	Geochronology reference	Latitute	Longitude	dp	dm	Paleomagnetism reference
				Ongeluk volcanics	1	Gumsley et al. (2017)					
				Westerberg sill province	3	Gumsley et al. (2017), Kampmann et al (2015)					
12	Sebanga Poort dyke swarm	Zimbabwe craton	**ca. 2471**	Mtshingwe dyke	1	Söderlund et al. (2010)	No published data				
11	Sebanga Poort dyke swarm	Zimbabwe craton	**ca. 2512**	Crystal Springs dyke	1	Söderlund et al. (2010)	No published data				
10	Great Dyke and sattelites	Zimbabwe craton	**2577–2573**	Great Dyke and sattelites	5	Armstrong and Wilson (2000), Wingate (2000), Söderlund et al. (2010), Oberthür et al. (2002)	**23.6**	**57.4**	**8.9**	**8.9**	Smirnov et al. (2013)
9	Unnamed dykes	Kaapvaal craton	**2580–2574**	Unnamed dykes	2	Larsson (2015)	No published data				
8	Sand River dykes	LMC Central Zone	**2607–2602**	Stockford and Causeway dykes	2	Xie et al. (2017)	No published data				
7	White Mfolozi dyke swarm	Kaapvaal craton	**2664–2654**	White Mfolozi dyke swarm	7	Gumsley et al (2016)	No published data				
7	Proto-basinal volcanics	Kaapvaal craton	**ca. 2664**	Buffelsfontein Group	1	Barton et al (1995)	No published data				
7	Rykoppies dyke swarm	Kaapvaal craton	**2686–2659**	Rykoppies dyke swarm	5	Olsson et al (2010; 2011)	**−62.1**	**336**	**3.5**	**4.2**	Lubnina et al (2010)

(continued)

Table 1 (continued)

Event		Terrane	Age (Ma)	Unit	Number of units dated	Geochronology reference	Latitute	Longitude	dp	dm	Paleomagnetism reference
6	2.7 Ga Ventersdorp LIP (bimodal)	Kaapvaal craton	**2746–2709**	Makwassie Formation	1	Armstrong et al. (1991)	**−69.8**	**345.6**	**5.8**	**5.8**	De Kock et al. (2009)
				Makwassie Formation	1	Cornell et al. (2018)	**−69.8**	**345.6**	**5.8**	**5.8**	De Kock et al. (2009)
				Hartswater Group	2	De Kock et al. (2012)	No published data				
				Amalia Group	1	Poujol et al (2005)	No published data				
				Zoetlief Group	1	Walraven et al. (1991)	No published data				
				Goedgenoeg Formation	1	Cornell et al. (2018)	No published data				
				Unnamed dyke	1	Larsson (2015)	No published data				
				Klipriviersberg Group	1	Armstrong et al. (1991)	*−17.1*	*47.9*	*18.8*	*18.8*	Strik et al. (2007)
5	2.78 Ga Ventersdorp LIP (bimodal)	Kaapvaal craton	**2787–2781**	Derdepoort volcanics	1	Wingate (1998)	−39.6	4.7	17.5	17.5	Wingate (1998)
				Modipe gabbro	1	Denyszyn et al. (2013)	**−47.6**	**12.4**	**8.6**	**9.3**	Denyszyn et al (2013)
				Kanye Formation	3	Moore et al. (1993), Grobler and Walraven (1993)	No published data				

(continued)

Table 1 (continued)

Event		Terrane	Age (Ma)	Unit	Number of units dated	Geochronology reference	Latitute	Longitude	dp	dm	Paleomagnetism reference
				Lobatse Group	1	Walraven et al. (1996)	No published data				
				Unit previously assigned to the Goedgenoeg Formation	1	Cornell et al. (2018)	No published data				
4	Hlagothi Complex	Kaapvaal craton	**2874–2866**	Hlaghothi Complex (may include undated Thole Complex, and Post-Mozaan sills)	3	Gumsley et al. (2013); Gumsley et al. (2015)	*23.4*	*53.4*	*8.2*	*11.8*	Gumsley et al (2013)
3	Witwatersrand volcanics	Kaapvaal craton	**ca. 2914**	Crown Member	1	Armstrong et al. (1991)	No published data				
2	Mozaan volcanics	Kaapvaal craton	**ca. 2954**	Tobolsk Formation	1	Mukasa et al. (2013)	No published data				
1	Badplaas dyke swarm	Kaapvaal craton	**2980–2966**	Badplaas dyke swarm	3	Olsson et al. (2010); Gumsley et al. (2015)	**−63.6**	**285.4**	**2.3**	**4**	Lubnina et al (2010)
1	Nsuze volcanics	Kaapvaal craton	**2985–2968**	Nsuze volcanics	3	Mukasa et al. (2013); Hegner et al. (1994); Nhleko (2003)	**−67**	**285.6**	**5.3**	**9.2**	Lubnina et al (2010)
1	Usushwana Complex	Kaapvaal craton	**2990–2978**	Usushwana Complex	3	Gumsley et al. (2015)	*9.2*	*347*	*7.6*	*7.6*	Layer (1988)

[a]Paleomagnetic constraints in bold face are considered reliable, while those in regular italic face are either virtual geomagnetic poles or considered unreliable

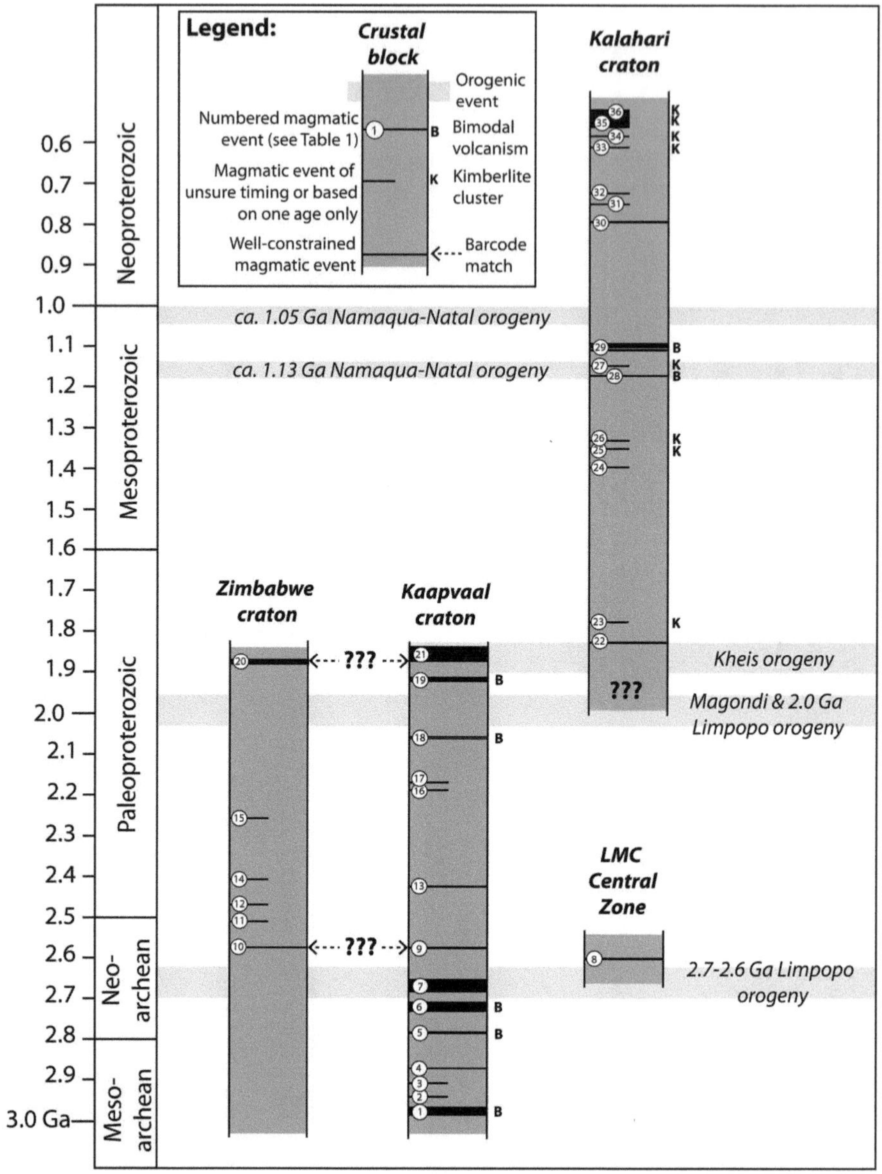

Fig. 10 Magmatic barcodes for the Kaapvaal and Zimbabwe cratons and the Central Zone of the LMC before 1.8 Ga, and a magmatic barcode for the Kalahari craton from 1.8 to 0.5 Ga

a proximity between the Kaapvaal and Pilbara cratons (Zegers et al. 1998), but a new 2.99–2.98 age from the Usushwana Complex invalidates such a paleomagnetic interpretation.

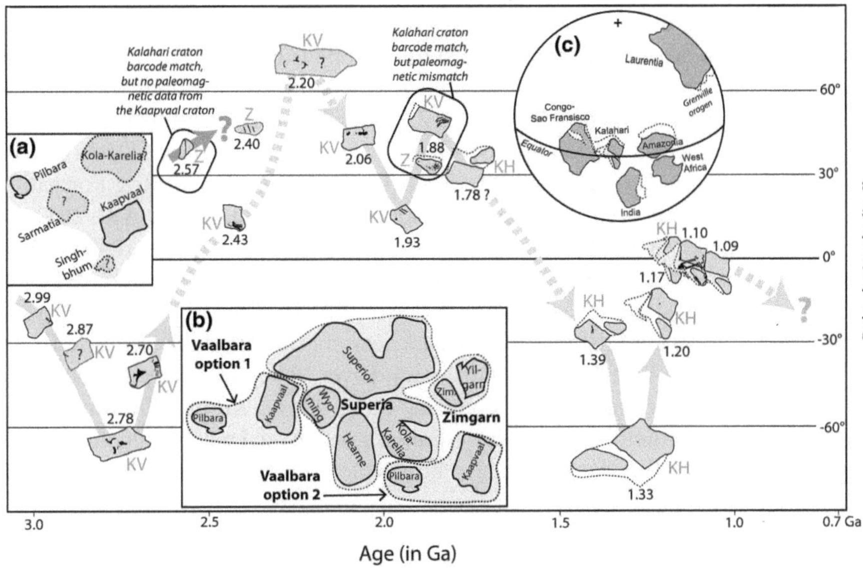

Fig. 11 Paleolatitudinal drift of the Kaapvaal (KV), Zimbabwe (Z) and Kalahari (KH) cratons between 3.0 and 0.7 Ga. Proposed relative reconstructions between the Kaapvaal, Zimbabwe and Kalahari cratons and other cratons are illustrated: **a** The Vaalbara reconstruction between 2.78 and 2.70 Ga (modified from De Kock et al. 2009)—now questionable at 2.78 Ga; **b** The Superia reconstruction, including Vaalbara and Zimgarn between 2.4 and 2.0 Ga (modified from Gumsley et al. 2017). Two possible placements of Vaalbara relative to Superia are shown; **c** Assembly of Rodinia at 1.11 Ga after De Kock et al. (2014)

In the Neoarchean, the ca. 2.78 Ga Derdepoort-Kanye-Modipe event can almost be perfectly matched with the similarly aged 2.77–2.76 Ga Black Range dykes and lower part of the Fortescue Group on the Pilbara craton (Wingate 1998; Evans et al. 2017). This match together with another approximate match between an age-bracketed 2.71–2.66 Ga Allanridge Formation and the upper part of the Fortescue Group, as well as other geological piercing points, has led to a proposal where these two cratons were nearest neighbors within a coherent crustal block named the Vaalbara "supercraton" (Cheney 1996). Based on paleomagnetic constraints a relative reconstruction of the Pilbara craton to the northwest of the present-day Kaapvaal craton was favored (De Kock et al. 2009; Fig. 11a), but better-constrained paleomagnetic results from the Kaapvaal craton's ca. 2.78 Ga Modipe gabbro has since suggest that the two blocks—although likely part of the same supercratons—were not necessarily nearest neighbors (Denyszyn et al. 2013). New paleomagnetic constraints from the 2.77–2.76 Ga Black Range dykes of the Pilbara craton, furthermore, do not support any attachment of the Kaapvaal and Pilbara cratons in the period 2.78–2.70 Ga (Evans et al. 2017; although not addressing this). During this time interval barcode matches have also been made between the Ventersdorp magmatism on the Kaapvaal craton and 2.80–2.75 Ga north-northeast-trending mafic dykes across the Singhbhum craton of India, which was attached to the south of the present-day Kaapvaal craton (Kumar et al. 2017). The Sarmartia block of eastern Europe (i.e., the combined Ukranian shield and Kursk block) has also been shown to record a very similar stratigraphic succession to the Kaapvaal craton between 2.78 and 2.04 Ga (Savko et al. 2017), thus suggesting a long-lived association between these cratons (Hunt et al. 2017). The volcanic units of the 2.78–2.60 Ga Mikhailovka Group and Krivoi Rog Formation were specifically compared with the Ventersdorp Supergroup (Savko et al. 2017).

The Great Dyke of Zimbabwe cannot be matched with any other crustal blocks at present, apart from a pair of ca. 2.57 Ga east- and southeast-trending dykes of the southeasternmost Kaapvaal craton (Larsson 2015). The ca. 2.51 Ga Crystal Springs and ca. 2.47 Ga Mtshingwe dykes are near-coeval with dykes within a Baltic LIP in the Kola-Karelia craton and Mistassini, River Valley and early Matachewan dykes in the Superior craton (e.g., Gumsley et al. 2017). Around the time of the Neoarchean-Paleoproterozoic boundary, similar geology and paleomagnetic records in supracrustal strata link the Kaapvaal craton to the Pilbara, Superior, Wyoming and Kola-Karelia cratons, although exact age barcode matches are yet to be made (Gumsley et al. 2017). The wide 2.51–2.41 Ga age range by Zimbabwean Sebanga dykes, however, overlaps with the Ringvassoy dykes (Kola-Karelia craton), Wigiemootha and Erayinia dykes (Yilgara craton), as well as the and du Chef dykes (Superior craton), suggesting that all these blocks could have been at least near-contiguous within "Superia" (Fig. 11b); i.e., the combined Superior, Kola-Karelia, Hearne and Wyoming cratons (Bleeker and Ernst 2006; Ernst and Bleeker 2010; Söderlund et al. 2010). Paleogeographic reconstructions with variable relative placements of the Zimbabwe and Yilgarn cratons have also been proposed (Smirnov et al. 2013; Pisarevsky et al. 2015). Smirnov et al. (2013) preferred a juxtaposition of the southern margin of the Zimbabwe craton to the eastern margin of the Yilgarn craton, which aligned the Sebanga Poort dyke with the Wigiemootha dyke swarm. Subsequent break-up of this configuration is envisaged to have occurred after 2.1 Ga, in order for each constituent to respectively, have collide with the Pilbara and Kaapvaal and Pilbara craton (Smirnov et al. 2013). Pisarevsky et al. (2015), however, lined up the present-day northern margin of the Zimbabwe craton with the western margin of the Yilgarn craton, in order to also align the Sebanga Poort dyke with the Wigiemootha dyke swarm (Fig. 11b). This combined "Zimgarn" block was then further restored with Superia. The discovery of a ca. 2.17 Ga dyke from the southeastern Kaapvaal craton possibly offers a further link with the Superior craton's Biscotasing dykes, as well as possibly the Dharwar craton (French and Heaman 2010). In a proposed long-lived Proterozoic reconstruction, Gumsley et al. (2017) placed Vaalbara to the northwest and Zimgarn towards the east of Superia, while Hunt et al. (2017) positioned both Vaalbara and Zimgarn along the eastern margin of Superia (Fig. 11b).

The Bushveld Complex and associated intrusions remain a globally unique mafic magmatic event. Bleeker et al. (2016) has reported near-coeval ages from the Superior craton, while Savko et al. (2017) highlighted the presence of 2.06–2.05 Ga mafic magmatic activity from eastern Sarmatia. These occurrences represent possible continuations of the Bushveld LIP that ends rather abruptly along its northern limb and far western satellite, the Molopo Farms Complex (Fig. 7).

A comparison between the Kaapvaal and Zimbabwe cratons shows firm correlations at both 1880–1830 Ma and at ca. 1110 Ma. This is consistent with these two cratons having remained together since the formation of the LMC's at either 2.70–2.60 Ga or ca. 2.05 Ga, yet do thereby not conclusively resolve this contentious issue. The ca. 1.88–1.83 Ga Mashonaland and the Waterberg sill provinces have numerous potential barcode matches from the Amazonian craton (Antonio et al. 2017), India (e.g., French et al. 2008; Belica et al. 2014), Australia (e.g., Blake et al. 1999), the North China craton (Peng et al. 2011; Peng 2015), the Baltica-Fenoscandian craton (e.g., Mertanen et al. 2006), the Superior craton (e.g., Ernst and Bleeker 2010; Ciborowski et al. 2017), the Wyoming craton (Kilian et al. 2016), and the Slave craton (e.g., Buchan et al. 2010; Mitchell et al. 2010), signifying a particularly global magmatic period. Although paleomagnetic data suggest that the Zimbabwe and Kaapvaal cratons were separated by a large distance at 1.88 Ga (Hanson et al. 2011b), and Antonio et al. (2017) proposed 1.88 and 1.83 Ga reconstructions wherein the Kalahari craton is isolated from other cratons, a later merger of these two cratons is not supported by any geological evidence. Instead, these anomalous paleomagnetic signatures may record a true polar wander event, which has been inferred elsewhere between 1.89 and 1.83 Ga (Mitchell et al. 2010; Antonio et al. 2017).

During the time interval spanning the intrusion of the ca. 1395 Ma Pilanesberg Complex up to the intrusion of the Premier kimberlite cluster at ca. 1150 Ma (Wu et al. 2013), the Kalahari craton was quite mobile (Fig. 11). It experienced latitudinal drift from a position 30° from the equator at 1.4 Ga to a high latitudinal position at 1.3 Ga, and back to a more equatorial position at 1.2 Ga (Gose et al. 2013). The Kalahari craton has also never been linked to any other cratonic unit during this time interval (e.g., 1.2 Ga reconstruction in Pesonen et al. 2003).

The voluminous Umkondo LIP event of the Kalahari craton occurred at the same time as the Keenawan LIP as well as the southwestern USA diabases of Laurentia were emplaced (Hanson et al. 2004a; Swanson-Hysell et al. 2015). Like the 1.88–1.83 Ga interval, however, 1.11 Ga is also a time of widespread global magmatism. Coeval events have been recognized from the Congo-Sao Francisco craton (Ernst et al. 2013), the Amazonian craton (Hamilton et al. 2012), and from India (Pradhan et al. 2012). Swanson-Hysell et al. (2015) preferred a reconstruction in which the Kalahari craton and Laurentia are far apart at 1110 Ma. This allowed the Namaqua-Natal orogen of the Kalahari craton to be conjugate to the Grenville margin of Laurentia, along which these two crustal units were joined to form Rodinia around 1.0 Ga. However, De Kock et al. (2014), following Ernst et al. (2013), positioned the Kalahari craton as a nearest neighbor to India, the Amazonian craton and the Congo-Sao Francisco craton at 1110 Ma (Fig. 11c), while still satisfying the paleomagnetic arguments presented by Swanson-Hysell et al. (2015).

The Gannakouriep, and perhaps also the Mutare mafic dyke swarms are near-coeval with the Gunbarrel and Franklin LIPs of Arctic Canada and the Siberia craton, respectively (e.g., MacDonnald et al. 2010; Rogers et al. 2016). These swarms are all regarded as signaling the beginning of break-up of the Rodinia supercontinent. It should be noted that only one dyke of ca. 724 Ma is dated within the Mutare dyke swarm, and it is unclear if it is representative for the whole swarm, or whether the swarm also includes some 1.11 Ga dykes. The Kalahari craton is commonly restored as conjoined with Laurentia along the Namaqua-Natal and Grenville margins, and abutting against the Congo-Sao Francisco craton in a present-day relative configuration (e.g., Dalziel 1997; Li et al. 2008). In these reconstructions the Grunehogna terrane of Dronning Maud Land (East Antarctica) is considered to be a fragment that had detached from the eastern margin of the Kalahari craton, during the breakup of Gondwanaland (Groenewald et al. 1991). The Grunehogna terrane has an established barcode that matches it firmly to the Kalahari craton at both ca. 1.11 Ga and ca. 0.18 Ga. If an age of Knoper et al. (2014) for the Grunehogna's Fingeren dyke swarm can be confirmed, this attachment is further strengthened by an additional match at ca. 0.72 Ga. Seafloor-spreading data suggests that the Grunehogna terrane is a fragment of the Kalahari craton, and serves to validate the magmatic barcode method for the to reconstruction of ancient cratons.

Several alternative reconstructions of Rodinia have, been proposed: Pisarevsky et al. (2003) presented a configuration where the eastern margin of the Kalahari-Grunehogna craton abuts against the western margin of Australia along the Maud and Pinjarra orogens, respectively, and where the Namaqua-Natal orogen remained an active continental margin against a subducting oceanic plate. Evans (2009) joined the northern margin of the Kalahari craton against both the East Antarctica craton and Australia, and thereby aligned the Albany-Fraser and the Namaqua-Natal orogens into a continuous belt. In this reconstruction, the Namaqua-Natal margin is facing the Eastern Ghats margin of cratonic India, as well as the eastern margin of the North China craton, while the Maud and Mozambique orogens of the Grunehogna terrane and the eastern Kalahari craton are interpreted as active continental margins against a subducting oceanic plate. Subsequent to this Evans (2013) revised parts of this interpretation.

4.2 Composite Nature of Dyke Swarms

Since the work of Wilson et al. (1987) and Hunter and Reid (1987), it has become apparent that many of the mafic dyke swarms across the Kalahari craton contain dykes with apparently similar trends yet distinctly different ages. This has led some authors to speculate in whether pre-existing dyke swarms form anisotropies within the crust, or the lithosphere has persisted 'weakness' zones, which are preferentially re-utilized by subsequent intrusions. Uken and Watkeys (1997) first proposed the re-utilization of pre-existing dyke trends by subsequent dyke swarms in southern Africa, but Jourdan et al. (2004) reported Ar–Ar reconnaissance work across the Kalahari craton that first showed how many Jurassic dykes appear to follow pre-existing and sub-parallel Precambrian dykes. Jourdan et al. (2004) also showed that the many northeast-trending dykes across the northeastern corner of the Kaapvaal craton did not include any Jurassic Karoo dykes, as previously inferred by, e.g., Uken and Watkeys (1997). More precise U–Pb geochronology work on this region's particularly closely-spaced SW–NE trending dykes, instead, revealed that it is made up of a Paleoproterozoic Black Hills dyke swarm (Olsson et al. 2016), that intruded sub-parallel Neoarchean dykes from the presumed northern branch of Olsson et al.'s (2011) radiating swarm. The southeast-trending branch of this proposed radiating swarm, in turn, overlaps and parallel dykes of the Mesoarchean Badplaas swarm (Olsson et al. 2010; Gumsley et al. 2015). Finally, the Sebanga dyke swarm was dated by Söderlund et al. (2010) at ca. 2512 Ma, ca. 2470 Ma and ca. 2408 Ma, indicating that the same zone of crustal weakness was (re-)activated at least three times during a remarkably long period of >100 million years.

From the many examples of overlapping or composite dyke swarms, it therefore seems relevant to note, that without absolute age constraints, any further tectonic, petrogenetic and paleomagnetic interpretations might have their limitations, and result in more speculative conclusions. An approach as demonstrated by Olsson et al. (2016) and Gumsley et al. (2015) is perhaps advisable for more complex dyke swarms, incorporating several overlapping generations, whereby only dated dykes are ideally combined with geochemical and paleomagnetic studies. However, we are also aware that interpretations have to be consistent with all available evidence, where any method has its own merit, and one must also be equally critical towards geochronological and paleomagnetic data. Thus, the implications of some dykes apparently coinciding with significantly older dykes, or even a pre-existing dyke swarm, needs to be interpreted with caution. Firstly, it is also important that dyke trends are properly documented and not just casually interpreted as overlapping and parallel, as even slightly deviating trends between two overlapping dyke generations out-rules the idea that younger dykes utilize pre-existing dyke swarms. One should also remember that several generations of cross-cutting dykes may roughly overlap other sub-parallel dykes purely by coincidence

With the above precautions in mind, it is still permissible to speculate on if/why some dykes, or even dyke swarms, appear to be overlapping in a sub-parallel fashion, and how any interpretation of such a field relationship might typically relate to magma generation in the mantle and subsequent fracture mechanics through the more brittle crust. Any intuitive thought of dykes preferentially propagating along (1) a pre-existing dyke swarm, (2) several swarms utilized a common crustal 'weakness', or (3) both scenarious, has to first and foremost be reconciled with a generally accepted fracture mechanical perception of individual dykes each generating and propagating along its own fracture, perpendicular to a least compressive stress field (e.g., Lister and Kerr 1991; Rubin 1995). This obviously applies within a more isotropic brittle crust, but where any oblique crustal weakness may intuitively offer a pathway that requires less energy for the dyke to propagate (e.g., Rivalta et al. 2015). As this is also where the topic becomes too complex to discuss further in this broad review paper, we just wish to urge readers to rigorously test any such deviations in sufficiently quantitative detail. One should, furthermore, note that it is not sufficient to quantitatively show that overlapping dyke swarms coincide with a zone of 'weaker' crust, or that it is significantly easier for a propagating dyke to follow any pre-exiting dyke swarm (or any other anisotropic basement fabric), rather than cut obliquely across, because such model interpretations also depend on whether mantle-derived mafic magmas were readily available at different times to follow such crustal anisotropies.

From the above, it is obvious that we still need to properly explain how it was possible for individual dykes, within, e.g., the Sebanga dyke swarm, to have been injected with roughly similar overlapping trends over a period of 100 Myr. This not only requires an equally permanent stress field (be it within an isotropic or anisotropic crust) but also that roughly the same underlying mantle was able to supply magmas over an equally long time interval. Our other examples of how the apparently radiating Neoarchean dyke swarm partially overlaps and parallels both an older Badplaas dyke swarm to the southeast and a younger Black Hills dyke swarm to the north-east, yet also follows 'its own' easterly Rykoppies dyke swarm trend, offer a multitude of intuitive model options, where we also note that these three 'radiating' Neoarchean branches formed over a period pf >40 Myr where each 'branch' also seem to have different geochemical signatures. While the latter discovery is interesting, it is also inconvenient that these different geochemical signatures appear to be host rock specific and conform with many older and younger dyke generations within the same basement terrain, rather than being age specific. Thus, the geochemistry has so far proved inadequate to discriminate between older and younger dyke generations within the overlapping north-eastern and south-eastern branch of the radiating Neoarchean dyke swarm. Nevertheless, whereas the south-eastern branch may have followed a pre-existing SE-trending and Mesoarchean Badplaas dyke swarm, there are no pre-existing dyke swarm to impose a similar explanation for the partially radiating north-eastern and eastern Rykoppies dyke swarm branches. Whether or not the Paleoproterozoic Black Hills dyke swarm followed the north-eastern branch of the radiating Neoarchean dyke swarm—which may also coincide with a major SW–NE trending rift zone across the Kaapvaal craton (e.g., McCarthy et al. 1990). Or simply, by coincidence, overlapped it within its own stress field, remains a speculation for further research. Some alternative model proposals are addressed by Gumsley et al (2016).

4.3 Recognition of LIPs

LIPs per definition usually cover large areas, reflecting the emplacement of large volumes of commonly mafic magma within relatively short durations (Coffin and Eldholm 1994, 2001). Ernst (2014) provided a more specific definition of LIPs as intraplate mafic magmatic provinces with or without ultramafic components, limiting both their extent (>10 Mkm^2) and volume (>0.1 Mkm^2) yet recognized that LIPs may also occur as a short magmatic pulse or multiple pulses (each pulse occurring in less than 1–5 million years) over a maximum period of <50 Myr. Ernst (2014) further noted that LIPs may be associated with silicic magmatism, carbonatites and kimberlites.

It is difficult to justify which of the above criteria and cut-off values should be used as LIP-discriminants, let alone estimate how well a heavily eroded and poorly exposed example compares to these. However, even sporadic outcrops and ages can provide some statistically valid minimum constraints that we find to be sufficient to classify the Bushveld Complex, the Great Dyke of Zimbabwe and the Umkondo as LIPs. For all of these three examples, sufficiently large areas and volumes of mafic igneous rocks were emplaced within just a few million years or less.

At the other end of the spectrum, ages derived from mafic dykes and sills associated with the Pongola, Ventersdorp and post-Waterberg/Mashonaland events indicate that these events were too long-lived to be classified as LIPs (*senso stricto)*, even if most of these candidates are sufficiently extensive and even voluminous (e.g., the thick flood basalts sequences within the Ventersdorp Supergroup). In some cases, one may argue that an age data sets records pulses within a coherent LIP event, where the proposed (1) main ca. 2.78 Ga Derdepoort-Kanye-Lobatse-Klipriviersberg pulse, (2) the ca. 2.73–2.71 Ga Platberg pulse, and (3) final emplacement of radiating ca. 2.68–2.66 Ga dyke swarms and associated proto-basinal fill volcanic rocks, pulsated for >100 Myr.

One may also argue that the ca. 1.93–1.92 Ga Hartley event is another sufficiently large and short-lived LIP event, although ages produced in the near future may link this event to a much more extended (or pulsating) event, including the subsequent ca. 1.89–1.83 Ga Black Hills dyke swarm and its associated sills and lavas, as already discussed by Olsson et al (2016). Other events, such as the Hlagothi Complex, although apparently short-lived, may not be sufficiently extensive, let alone voluminous enough to qualify as a LIP, and may also be part of a yet undisclosed longer lived, and possibly pulsating Mesoarchean LIP-event, possibly ranging from as far back as to (1) a ca. 3.07 Ga Dominion-pulse, passing through (2) a major Pongola-pulse between ca. 2.98–2.97 Ga, before terminating with (3) the ca. 2.87 Ga Hlagothi-event. Depending on whether or not the Dominon Group can be correlated to the Nsuze Group, this potential LIP could have been pulsating over a period of between 200 and 100 Myr. Alternatively, a combined Dominon-Nsuze event (including the Usushwana Complex, Badplaas dyke swarm and associated Nsuze lavas) may have been sufficiently voluminous and short-lived to be a LIP on its own.

As it is problematic to estimate magma volumes only from the dyke swarms roots of more deeply eroded Precambrian LIPs (*senso lato*), it becomes important to also correlate precisely dated mafic dykes and sills to other coeval intrusive and extrusive igneous rocks. While a lack of supracrustal successions renders this task impossible for the Zimbabwe craton, several dyke swarms across the Kaapvaal craton can now be shown to be coeval with volcanic units within the craton's better-preserved supracrustal stratigraphic record. Thus, it has become apparent that different volcanic packages in the Nsuze Group may correspond to coeval dykes present in the Badplaas mafic dyke swarm, as well as the Usushwana Complex (Gumsley et al. 2015). As mentioned, a link between the Nsuze Group and the Dominion Group requires further validation, as do potential links between the Hlagothi Complex and volcanic rocks within the Mozaan Group and Witwatersrand Supergroup. The start of Neoarchean volcanism during the so-called Ventersdorp event, is now well constrained at ca. 2.78 Ga, while its cessation at ca. 2.65 Ga coincided with the emplacements of dykes radiating across the eastern and southeastern Kaapvaal craton basement and deposition of proto-basinal fills to the Transvaal Supergroup. The compositionally bimodal volcanic rocks in the middle Ventersdorp Supergroup stratigraphy (i.e., the Platberg Group) appear to have erupted between 2730–2708 Ma (De Kock et al. 2012; Armstrong et al. 1991); whereas, more ages are needed to further resolve the older Klipriviersberg Group and younger Allanridge Formation. Regardless, U–Pb geochronology has yet to confirm the geochemical matching of Klipriviersberg Group lavas with potential dyke feeders within the Witwatersrand Supergroup (McCarthy et al. 1990). The sill-like Modipe gabbro represents the only possible feeder associated with the older flood basalt event of the Derdepoort Formation identified so far.

The Ongeluk LIP can now definitively be linked both geochronologically and paleomagnetically to the Ongeluk Formation in the Neoarchean-Paleoproterozoic Transvaal Supergroup overlying the Ventersdorp Supergroup. The volcanic rocks of the Ongeluk Formation were previously thought to be ca. 2.22 Ga (Cornell et al. 1996), and thereby correlated to the ≤2.25–2.24 Ga Hekpoort Formation volcanic rocks, further east (Schröder et al. 2016). However, a combined in situ SIMS, ID-TIMS and paleomagnetic study established a ca. 2.43 Ga age for the Ongeluk Formation and its associated LIP (Gumsley et al. 2017; Kampmann et al. 2015). The volcanic rocks of the Hekpoort Formation, Bushy Bend Member and Machadodorp Member within the Transvaal Supergroup remain undated by U–Pb, however, a ca. 2.17 Ga age established by Larsson (2015) may tentatively link to either of these unconstrained events. The middle Paleoproterozoic 2056–2055 Ma Bushveld Complex, and associated sills, complexes and volcanic rocks is now one of the most well-dated and best constrained magmatic events on the Kaapvaal craton. Evidence of its existence in Zimbabwe has never been documented, however. The same is true for the 1.93–1.92 Ga Hartley LIP event on the Kaapvaal craton. The Tsineng dyke swarm is coeval and temporally linked to the Hartley Formation within the Olifantshoek Supergroup. Dykes and sills of the 1.88–1.83 Ga post-Waterberg sills and Black Hills dyke swarm, however, are more problematic to link with volcanic units, although it has been suggested the Sibasa Formation volcanic rocks in the Soutpansberg Group are coeval (Geng et al. 2014). The 1.88–1.83 Ga magmatic events remains the first coeval match between Kaapvaal and Zimbabwe craton in the assembled Kalahari craton. This is followed by the ca. 1.11 Ga Umkondo LIP event, which is present as both dykes and sills across the whole Kalahari craton, and is coeval with volcanic rocks preserved in the Umkondo Group in Zimbabwe and Mozambique, and in the Kgwebe Formation and in subsurface in Botswana.

5 Summary

It is clear that since the pioneering work by Wilson et al. (1987), Hunter and Reid (1987), and later by Uken and Watkeys (1997), that knowledge of mafic dyke swarms and sill provinces and their relationship with volcanic successions and LIPs has advanced considerably in southern Africa. Many gaps in the knowledge of the Precambrian mafic record of the Kalahari craton remain, but with further multidisciplinary studies combining the latest developments in U–Pb geochronology and either paleomagnetism or geochemistry, with continued mapping and sampling of relevant igneous rocks, the apparent puzzle is becoming more rigorously solved. Dykes and sills can now be linked conclusively to some key magmatic units in the supracrustal successions, allowing better constraints to be placed on regional stratigraphic frameworks, although other units remain undated. Paleomagnetic clues into paleogeography of the Kalahari craton and its relationship with other cratons, as well as the petrogenesis of the magmatic events can also now be better deduced. Other events, however, require further age constraints, and the use of the geospatial database presented in this study will help to bring together all aspects of the different studies for the future development of our knowledge of the Precambrian mafic record of the Kalahari craton and its relationship with other cratons in time and space.

Acknowledgements We would like to thank Richard Hanson and Wlady Altermann for their detailed and constructive reviews. MOdK acknowledge support from the South African Department of Science and Technology and National Research Foundation (DST-NRF) Centre of Excellence for Integrated Mineral and Energy Resource Analysis (CIMERA). Additional support from the South Africa National Research Foundation through their incentive funding is acknowledged by MOdk and MBK. APG and US wish to thank both the Swedish Research Council and the Royal Physiographic Society in Lund for financial assistance. Much of the geochronology presented was funded by the Industry-Academia-Government Consortium Project "Reconstruction of Supercontinents Back to 2.7 Ga Using the Large Igneous Province (LIP) Record, with Implications for Mineral Deposit Targeting, Hydrocarbon Resource Exploration, and Earth System Evolution" (www.supercontinent.org). Here we would specifically like to thank Richard Ernst for his leading role. Many of the ages are the products of MSc and PhD projects, and we acknowledge the numerous students for their contributions.

References

Alebouyeh Semami F, De Kock MO, Söderlund U, Gumsley AP, Da Silva R, Beukes NJ, Armstrong RA (2016) New U–Pb geochronologic and palaeomagnetic constraints on the late Palaeoproterozoic Hartley magmatic event: evidence for a potential large igneous province in the Kaapvaal Craton during Kalahari assembly, South Africa. GFF https://doi.org/10.1080/11035897.2015.1124917

Allsopp HL, Smith CB, Seggie AG, Skinner EMW, Colgan EA (1995) The emplacement age and geochemical character of the Venetia kimberlite bodies, Limpopo Belt, northern Transvaal. S Afr J Geol 98(3):239–244

Altermann W, Lenhardt N (2012) The volcano-sedimentary succession of the Archean Sodium Group, Ventersdorp Supergroup, South Africa: volcanology, sedimentology and geochemistry. Precambrian Res 214–215:60–81. https://doi.org/10.1016/j.precamres.2012.02.012

Antonio PYJ, D'Agrella-Filho MS, Trindade RIF, Nédélec A, de Oliveira DC, da Silva FF, Roverato M, Lana C (2017) Turmoil before the boring billion: paleomagnetism of the 1880–1860 Ma Uatumã event in the Amazonian craton. Gondwana Res 49:106–129. https://doi.org/10.1016/j.gr.2017.05.006

Armstrong RA, Compston W, Retief EA, Williams IS, Welke HJ (1991) Zircon ion microprobe studies bearing on the age and evolution of the Witwatersrand triad. Precambrian Res 53:243–266

Armstrong RA, Wilson AH (2000) A SHRIMP U-Pb study of zircons from the layered sequence of the Great Dyke, Zimbabwe, and a granitoid anatectic dyke. Earth Planet Sci Lett 180(1–2):1–12

Bailie R, Gutzmer J, Rajesh HM, Armstrong RA (2011) Age of ferroan A-type post-tectonic granitoids of the southern part of the Keimoes Suite, Northern Cape Province, South Africa. J Afr Earth Sci 60:153–174

Barnes S-J, Maier WD, Curl EA (2010) Composition of the marginal rocks and sills of the Rustenburg Layered Suite, Bushveld Complex, South Africa: Implications for the formation of the Platinum-group element deposits. Econ Geol 105:1491–1511

Bartholomew LT (2008) Paleomagnetism of neoproterozoic intraplate igneous rocks in the southwest Kalahari craton, Namibia and South Africa. Unpublished MSc thesis, Texas Christian University, USA

Barton JM (1979) The chemical compositions, Rb-Sr isotopic systematics and tectonic setting of certain post-kinematic mafic igneous rocks, Limpopo Mobile Belt, southern Africa. Precambrian Res 9:57–80

Barton JMJ, Blignaut E, Salnikova EB, Kotov AB (1995) The stratigraphical position of the Buffelsfontein Group based on field relationships and chemical and geochronological data. S Afr J Geol 98:386–392

Barton JMJ, Barton ES, Smith CB (1996) Petrography, age and origin of the Schiel alkaline complex, northern Transvaal, South Africa. J Afr Earth Sci 22(2):133–145

Becker T, Schreiber U, Kampunzu AB, Armstrong RA (2006) Mesoproterozoic rocks of Namibia and their plate tectonic setting. J Afr Earth Sci 46:112–140

Belica ME, Piispa EJ, Meert JG, Pesonen LJ, Plado J, Pandit MK, Kamenov GD, Celestino M (2014) Paleoproterozoic mafic dyke swarms from the Dharwar craton; paleomagnetic poles for India from 2.37 to 1.88 Ga and rethinking the Columbia supercontinent. Precambrian Res 244:100–122

Beukes NJ, Cairncross B (1991) A lithostratigraphic-sedimentological reference profile for the Late Archean Mozaan Group, Pongola Sequence: application to sequence stratigraphy and correlation with the Witwatersrand Supergroup. S Afr J Geol 94(1):44–69

Blake DH, Tyler IM, Griffin TJ, Sheppard S, Thorne AM, Warren RG (1999) Geology of the Halls Creek 1:100,000 Sheet area (4461), Western Australia. Australian Geological Survey, Canberra

Bleeker W, Ernst RE (2006) Short-lived mantle generated magmatic events and theri dike swarms: the key unlocking Earth's paleogeographic record back to 2.6 Ga In: Hanski E, Mertanen S, Rämö T, Vuollo J (eds) Dike swarms: time markers of crustal evolution. Taylor & Francis/Balkema, Leiden, pp 3–26

Bleeker W, Chamberlain K, Kamo SL, Hamilton M, Kilian TM, Buchan KL (2016) Kaapvaal, Superior and Wyoming: nearest neighbours in supercraton Superia. Paper presented at the 35th international geological congress, Cape Town, South Africa

Borg G, Kärner K, Buxton M, Armstrong RA, Van der Merwe SW (2003) Geology of the Skorpion supergene zinc deposit, Southern Namibia. Econ Geol 98:749–771

Briden JC, Duff BA, Kröner A (1979) Palaeomagnetism of the Koras Group, Northern Cape province, South Africa. Precambrian Res 10:43–57

Buchan KL, Ernst RE, Bleeker W, Davies W, Villeneauve M, Van Breemen O, Hamilton M, Söderlund U (2010) Proterozoic magmatic events of the Slave craton, Wopmay orogen and environs. In: Geological Survey of Canada, Open File 5985. Geological Survey of Canada

Burke K, Kidd WSF, Kusky TM (1985) The Pongola structure of southeastern Africa: the world's oldest preserved rift? J Geodyn 2:35–49

Cawthorn RG, Davies G, Clubley-Armstrong A, McCarthy TS (1981) Sills associated with the Bushveld Complex, South Africa: an estimate of the parental magma composition. Lithos 14:1–16

Cawthorn RG, Eales HV, Walraven F, Uken R, Watkeys MK (2006) The bushveld complex. In: Johnston MR, Anhaeusser CR, Thomas RJ (eds) The geology of South Africa. The Geological Society of South Africa and Council for Geosciences, Pretoria, South Africa, pp 261–281

Cheney ES (1996) Sequence stratigraphy and plate tectonic significance of the Transvaal succession of southern Africa and its equivalent in Western Australia. Precambrian Res 79:3–24

Ciborowski T, Minifie M, Kerr AC, Baragar B, Millar IL (2017) A mantle plume origin for the Palaeoproterozoic Circum Superior Large Igneous Province. Precambrian Res 294:189–213

Clarke B, Uken R, Reinhardt J (2009) Structural and compositional constraints on the emplacement of the Bushveld Complex, South Africa. Lithos 111:21–36. https://doi.org/10.1016/j.lithos.2008.11.006

Clifford TN (1970) The structural framework of Africa. In: Clifford TN, Gass IG (eds) African magmatism and tectonics. Oliver and Boyd, Edinburgh, pp 1–26

Coffin MF, Eldholm O (1994) Large igneous provinces: crustal structure, dimensions, and external consequences. Rev Geophys 32:1–36

Coffin MF, Eldholm O (2001) Large igneous provinces: progenitors of some ophiolites? In: Ernst RE, Buchan KL (eds) Mantle plumes: their identification through time. Special Paper, vol 352. Geological Society of America, Boulder, CO, pp 59–70

Cole EG (1994) Lithostratigraphy and depositional environment of the Archaean Nsuze Group, Pongola Supergroup. Unpublished PhD thesis, Rand Afrikaans University, Johannesburg, South Africa

Cole J, Webb SJ, Finn CA (2014) Gravity models of the Bushveld Complex—have we come full circle? J Afr Earth Sci 92:97–118

Cornell DH, Schutte SS, Eglington BL (1996) The Ongeluk basaltic andesite formation in Griqualand West, South-Africa: submarine alteration in a 2222 Ma Proterozoic sea. Precambrian Res 79:101–123

Cornell DH, Armstrong RA, Walraven F (1998) Geochronology of the Proterozoic Hartley Basalt Formation, South Africa: constraints on the Kheis tectogenesis and the Kaapvaal Craton's earliest Wilson Cycle. J Afr Earth Sci 26(1):5–27

Cornell DH, Van Schjindel V, Simonsen SL, Frei D (2015) Geochronology of Mesoproterozoic hybrid intrusions in the Konkiep Terrane, Namibia, from passive to active continental margin in the Namaqua-Natal Wilson Cycle. Precambrian Res 265:166–188

Cornell DH, Zack T, Anderson T, Corfu F, Frei D, Van Schjindel V (2016) Restricted access Th-U-Pb zircon geochronology of the Palaeoproterozoic Hartley Formation porphyry by six methods, with age uncertainty approaching 1 Ma. S Afr J Geol 119:473–494. https://doi.org/10.2113/gssajg.119.3.473

Cornell DH, Meintjies PG, Van der Westerhuizen WA, Frei D (2018) Microbeam U-Pb zircon dating of the Makwassie and Goedgenoeg formations in the Ventersdorp Supergroup of South Africa. S Afr J Geol 120(4):525-540

Corner B, Verran DR, Hildebrand PR (2012) Geophysical interpretation of the nature and extent of the Xade Mafic Complex, Botswana. S Afr J Geol 115(4):485–498

Dalziel IWD (1997) Neoproterozoic-Paleozoic geography and tectonics: review, hypothesis, environmental speculation. Geol Soc Am Bull 109:16–42

De Kock MO, Evans DAD, Dorland HC, Beukes NJ, Gutzmer J (2006) Paleomagnetism of the lower two unconformity-bounded sequences of the Waterberg Group, South Africa: towards a better-defined apparent polar wander path for the Paleoproterozoic Kaapvaal Craton. S Afr J Geol 109(1):157–182

De Kock MO, Evans DAD, Beukes NJ (2009) Validating the existence of Vaalbara in the late Neoarchean. Precambrian Res 174:145–154

De Kock MO, Beukes NJ, Armstrong RA (2012) New SHRIMP U-Pb zircon ages from the Hartswater Group, South Africa: implications for correlations of the Neoarchean Ventersdorp Supergroup on the Kaapvaal craton and with the Fortesque Group on the Pilbara craton. Precambrian Res 204–205:66–74

De Kock MO, Ernst RE, Söderlund U, Jourdan F, Hofmann A, Le Gall B, Bertrand H, Chisonga BC, Beukes NJ, Rajesh HM, Moseki LM, Fuchs R (2014) Dykes of the 1.11 Ga Umkondo LIP, Southern Africa: clues to a complex plumbing system. Precambrian Res 249:129–143. https://doi.org/10.1016/j.precamres.2014.05.006

De Kock MO, Ravhura L, Vorster C, Beukes NJ, Gumsley AP (2016) Constraining the timing of the Molopo Farms Complex emplacement and provenance of its country rock. Acta Geol Sin 90:78. https://doi.org/10.1111/1755-6724.12898

De Waal SA, Armstrong RA (2000) The age of the Marble Hall diorite, its relationship to the Uitkomst Complex, and evidence for a new magma type associated with the Bushveld igneous event. S Afr J Geol 103(2):128–140
De Waal SA, Graham IT, Armstrong RA (2006) The Lindeques Drift and Heidelberg intrusions and the Roodekraal Complex, Vredefort, South Africa: comagmatic plutonic and volcanic products of a 2055 Ma ferrobasaltic magma. S Afr J Geol 109:270–300
De Waal SA, Schweitzer JK, Graham I, Gauert CDK, Ripley E (2008) A Bushveld-related high-Ti igneous suite (HITIS) derived from an alkali to transitional basaltic magma, South Africa. S Afr J Geol 111:201–224
De Wit MJ, Roering C, Hart RJ, Armstrong RA, De Ronde CEJ, Green RWE, Tredoux M, Peberdy E, Hart RA (1992) Formation of an Archaean continent. Nature 357:553–562
Denyszyn SW, Feinberg JM, Renne PR, Scott GR (2013) Revisiting the age and paleomagnetism of the Modipe Gabbro of South Africa. Precambrian Res 238:176–185. https://doi.org/10.1016/j.precamres.2013.10.002
Donnelly CL, Griffin WL, O'Reilly SY, Pearson A, Shee SR (2011) The kimberlites and related rocks of the Kuruman kimberlite province, Kaapvaal craton, South Africa. Contrib Mineral Petrol 161:351–371
Donnelly CL, Griffin WL, Yang J-H, O'Reilly SY, Li Q-L, Pearson NJ, Li X-H (2012) In situ U-Pb dating and Sr-Nd isotopic analysis of Perovskite: constraints on the age and petrogenesis of the Kuruman kimberlite province, Kaapvaal craton, South Africa. J Petrol 53(12):2497–2522
Dorland HC, Beukes NJ, Gutzmer J, Evans DAD, Armstrong RA (2006) Precise SHRIMP U-Pb zircon age constraints on the lower Waterberg and Soutpansberg Groups, South Africa. S Afr J Geol 109(1):139–156
Eglington BM, Armstrong RA (2004) The Kaapvaal Craton and adjacent orogens, southern Africa: a geochronological database and overview of the geological development of the craton. S Afr J Geol 107:13–32
Elburg M, Cawthorn RG (2017) Source and evolution of the alkaline Pilanesberg Complex, South Africa. Chem Geol 455:148–165. https://doi.org/10.1016/j.chemgeo.2016.10.007
Emerman SH (1991) Correlation of a dyke swarm in southeastern Botswana with the Pilanesberg dyke swarm, South Africa. J Afr Earth Sci 12(4):525–531
Eriksson PG, Engelbrecht JP, Res M, Harmer RE (1994) The Bushy Bend lavas, a new volcanic member of the Pretoria Group, Transvaal Sequence. S Afr J Geol 97:1–7
Eriksson PG, Altermann W, Hartzer FJ (2006) The Transvaal Supergroup. In: Johnson MR, Anhaeusser C, Thomas RJ (eds) The geology of South Africa. Geological Society of South Africa/Council for Geoscience, Johannesburg/Pretoria, pp 237–260
Erlank AJ (1984) Petrogenesis of the volcanic rocks of the Karoo Province (Special Publication No 13). Geological Society of South Africa, South Africa
Ernst RE (2014) Large igneous provinces. Cambridge University Press, Cambridge, United Kingdom
Ernst RE, Bleeker W (2010) Large igneous provinces (LIPs), giant dyke swarms, and mantle plumes: significance for breakup events within Canada and adjacent regions from 2.5 Ga to the present. Can J Earth Sci 47:695–739
Ernst RE, Head JW, Parfitt E, Grosfils EB, Wilson L (1995) Giant radiating dyke swarms on Earth and Venus. Earth Sci Rev 39:1–58
Ernst RE, Pereira E, Hamilton MA, Pisarevsky SA, Rodriques J, Tassinari CCG, Teixeira W, Van-Dunem V (2013) Mesoproterozoic intraplate magmatic 'barcode' record of the Angola portion of the Congo craton: newly dated magmatic events at 1500 and 1110 Ma and implications for Nuna (Columbia) supercontinent reconstructions. Precambrian Res 230:103–118
Evans DAD (2009) The palaeomagnetically viable, long-lived and all-inclusive Rodinia supercontinent reconstruction. In: Murphy JB, Keppie JD, Hynes AJ (eds) Ancient orogens and modern analogues, vol 327. Geological society. London, Special Publications, pp 371–404
Evans DA (2013) Reconstructing pre-Pangean supercontinents. Geol Soc Am Bull 125(11–12):1735–1751

Evans DA, Beukes NJ, Kirschvink JL (1997) Low-latitude glaciation in the Palaeoproterozoic era. Nature 386:262–266

Evans DAD, Smirnov AV, Gumsley AP (2017) Paleomagnetism and U-Pb geochronology of the Black Range dykes, Pilbara Craton, Western Australia: a Neoarchean crossing of the polar circle. Aust J Earth Sci . https://doi.org/10.1080/08120099.2017.1289981

French JE, Heaman LM (2010) Precise U-Pb dating of Paleoproterozoic mafic dyke swarms of the Dharwar craton, India: implications for the existence of the Neoarchean supercraton Sclavia. Precambrian Res 183(3):416–441

French JE, Heaman LM, Chacko T, Srivastava RK (2008) 1891–1883 Ma Southern Bastar-Cuddapah mafic igneous events, India: a newly recognized large igneous province. Precambrian Res 160(3–4):308–322. https://doi.org/10.1016/j.precamres.2007.08.005

Friese AEW, Charlesworth EG, McCarthy TS (1995) Tectonic processes within the Kaapvaal Craton during the Kimbaran (Grenville) orogeny: structural, geophyscial and isotopic constraints from the Witwatersrand Basin and environs, vol 292. Economic geology research unit information circular. University of the Witwatersrand, Johannesburg

Frimmel HE, Klötzli US, Siegfired PR (1996) New Pb-Pb single zircon age constraints on the timing of Neoproterozoic glaciation and continental break-up in Namibia. J Geol 104(4):459–469

Frimmel HE, Zartman RE, Späth A (2001) The Richtersveld Igneous Complex, South Africa: U-Pb zircon and geochemical evidence for the beginning of Neoproterozoic continental breakup. J Geol 109:493–508

Geng H, Brandl G, Sun M, Wong J, Kröner A (2014) Zircon ages defining deposition of the Palaeoproterozoic Soutpansberg Group and further evidence for Eoarchaean crust in South Africa. Precambrian Res 249:247–262. https://doi.org/10.1016/j.precamres.2014.05.020

Gold DJC (2006) The Pongola Supergroup. In: Johnson MR, Anhaeusser CR, Thomas RJ (eds) Geology of South Africa. Geological Society of South Africa/Council for Geoscience, Johannesburg/Pretoria, pp 135–147

Goldberg AS (2010) Dyke swarms as indicators of major extensional events in the 1.9–1.2 Ga Columbia supercontinent. J Geodyn 50:176–190

Gose GA, Hanson RE, Dalziel IWD, Pancake JA, Seidel EK (2006) Paleomagnetism of the 1.1 Ga Umkondo large igneous province in southern Africa. J Geophys Res 111:1–18

Gose WA, Hanson RE, Harmer RE, Seidel EK (2013) Reconnaissance paleomagnetic studies of Mesoproterozoic alkaline igneous complexes in the Kaapvaal craton, South Africa. J Afr Earth Sci 85:22–30. https://doi.org/10.1016/j.jafrearsci.2013.04.005

Gough DL, Hales AL (1956) A paleomagnetic study of the palaeomagnetism of the Pilanesberg dykes. Mon Not R Astron Soc Geophys Suppl 7:196–213

Griffin WL, Batumike JM, Greau Y, Pearson NJ, Shee SR, O'Reilly SY (2014) Emplacement ages and sources of kimberlites and related rocks in southern Africa: U-Pb ages and Sr-Nd isotopes of groundmass perovskite. Contrib Mineral Petrol 168:1032

Grobler DF, Walraven F (1993) Geochronology of Gabarone Granite Complex extensions in the area north of Mafikeng, South Africa. Chem Geol 105:319–337

Groenewald PB, Grantham GH, Watkeys MK (1991) Geological evidence for a Proterozoic to Mesozoic link between southeastern Africa and Donning Maud Land, Antarctica. J Geol Soc London 148:1115–1123

Gumsley AP, de Kock MO, Rajesh HM, Knoper MW, Söderlund U, Ernst RE (2013) The Hlagothi Complex: the identification of fragments from a Mesoarchean large igneous province on the Kaapvaal Craton. Lithos 174:333–348. https://doi.org/10.1016/j.lithos.2012.06.007

Gumsley AP, Olsson JR, Söderlund U, De Kock MO, Hofmann A, Klausen M (2015) Precise U-Pb baddeleyite age dating of the Ushushwana Complex, southern Africa—implications for the Mesoarchaean magmatic and sedimentological evolution of the Pongola Supergroup, Kaapvaal Craton. Precambrian Res 257:174–185. https://doi.org/10.1016/j.precamres.2015.06.010

Gumsley AP, Rådman J, Söderlund U, Klausen M (2016) U-Pb baddeleyite geochronology and geochemistry of the White Mfolozi Dyke Swarm: unravelling the complexities of 2.70–2.66 Ga dyke swarms across the eastern Kaapvaal Craton, South Africa. GFF 138(1):115–132

Gumsley AP, Chamberlain KR, Bleeker W, Söderlund U, De Kock MO, Larsson ER, Bekker A (2017) Timing and tempo of the Great Oxidation Evenet. Proc Natl Acad Sci USA 114:1811–1816. https://doi.org/10.1073/pnas.1608824114
Gutzmer J, Beukes NJ, Pickard A, Barley ME (2000) 1170 Ma SHRIMP age for Koras Group bimodal volcanism, Northern Cape Province. S Afr J Geol 103:32–37
Hamilton MA, Sadowski GR, Teixeira W, Ernst RE, Ruiz AS (2012) Precise, matching U-Pb ages for the Rincon del Tigre mafic layered intrusion and Huanchaca gabbro sill, Bolivia: evidence for a late Mesoproterozoic LIP in SW Amazonia? In: Geoscience at the Edge, GAC-MAC Joint Annual Meeting, St. John's
Hammerbeck ECI, Allcock RJ (1985) 1:4,000,000 Geological map of Southern Africa. The Geological Society of South Africa, Pretoria, South Africa
Hanson RE (2003) Proterozoic geochronology and tectonic evolution of southern Africa. In: Yoshida M, Windley BF, Dasgupta S (eds) Proterozoic East Gondwana: Supercontinent assembly and breakup, vol 1. Geological society special publication no 206. The Geological Society, London, pp 427–463
Hanson RE (Pers. comm.) TIMS ages for main felsic igneous rocks associated with the Gannakouriep dykes
Hanson RE, Martin MW, Bowring SA, Munyanyiwa H (1998) U-Pb zircon age for the Umkondo dolerites, eastern Zimbabwe: 1.1 Ga large igneous province in southern Africa-East Antarctica and possible Rodinia correlations. Geology 26(12):1143–1146
Hanson RE, Crowley JL, Bowring SA, Ramazani J, Gose WU, Dalziel IWD, Pancake JA, Seidel EK, Blenkinsop TG, Mukwakwami J (2004a) Coeval large-scale magmatism in the Kalahari and Laurentian cratons during Rodinia assembly. Science 304:1126–1129
Hanson RE, Gose WA, Crowley JL, Ramezani J, Bowring SA, Bullen DS, Hall RP, Pancake JA, Mukwakwami J (2004b) Paleoproterozoic intraplate magmatism and basin development on the Kaapvaal Craton: Age, paleomagnetism and geochemistry of ~1.93 to ~1.87 Ga post-Waterberg dolerites. S Afr J Geol 107:233–254
Hanson RE, Harmer RE, Blenkinsop TG, Bullen DS, Dalziel IWD, Gose WA, Hall RP, Kampunzu AB, Key RM, Mukwakwami J, Munyanyiwa H, Pancake JA, Seidel EK, Ward SE (2006) Mesoproterozoic intraplate magmatism in the Kalahari Craton: A review. J Afr Earth Sci 46:141–167
Hanson RE, Rioux M, Bowring SA, Gose GA, Bartholomew LT, Kilian TM, Evans DA, Panzik JE, Hoffmann KH, Reid DL (2011a) Constraints on Neoproterozoic intraplate magmatism in the Kalahari craton: geochronology and paleomagnetism of ~890–795 Ma extension-related igneous rocks in SW Namibia and adjacent parts of South Africa. In: Geological Society of America Abstracts with Programs, vol 5, p 371
Hanson RE, Rioux M, Gose WA, Blackburn TJ, Bowring SA, Mukwakwami J, Jones DL (2011b) Paleomagnetic and geochronological evidence for large-scale post 1.88 Ga displacement between the Zimbabwe and Kaapvaal cratons along the Limpopo belt. Geology 39(5):487–490
Hargraves RB (1989) Paleomagnetism of Mesozoic kimberlites in southern Africa and the Cretaceous apparent polar wander curve for Africa. J Geophys Res 94(B2):1851–1866
Harmer RE (1999) The petrogenetic association of carbonatite and alkaline magmatism: constraints from the Spitskop Complex, South Africa. J Petrol 40:525–548
Hart R, Moser D, Andreoli M (1999) Archean age for the granulite facies metamorphism near the center of the Vredefort structure, South Africa. Geology 27(12):1091–1094
Hartnady CJH, Joubert P, Stowe C (1985) Proterozoic crustal evolution in southwestern Africa. Episodes 8:236–244
Hatton CJ (1995) Mantle plume origin for the Bushveld and Ventersdorp magmatic provinces. J Afr Earth Sci 21(4):571–577
Heaman LM, LeCheminant AN (1993) Paragenesis and U-Pb systematics of baddeleyite (ZrO_2). Chem Geol 110:95–126
Heaman LM (2009) The application of U-Pb geochronology to mafic, ultramafic and alkaline rocks. An evaluation of three mineral standards. Chem Geol 261(1–2):43–52

Hegner E, Kröner A, Hunt P (1994) A precise U-Pb zircon age for the Archaean Pongola Supergroup volcanics in Swaziland. J Afr Earth Sci 18:339–341. https://doi.org/10.1016/0899-5362(94)90072-8

Henderson DR, Long LE, Barton JM Jr (2000) Isotopic ages and chemical and isotopic composition of the Archaean Turfloop Batholith, Pieterburg granite-greenstone terrane, Kaapvaal Craton, South Africa. S Afr J Geol 103(38):46

Hofmann A, Kröner A, Hegner E, Belyanin GA, Kramers JD, Bolhar R, Slabunov A, Reinhardt J, Horváth P (2015) The Nhlangano gneiss dome in south-west Swaziland—a record of crustal destabilization of the eastern Kaapvaal craton in the Neoarchaean. Precambrian Res 258:109–132. https://doi.org/10.1016/j.precamres.2014.12.008

Holzer L, Frei R, Barton JM Jr, Kramers JD (1998) Unraveling the record of successive high grade events in the Central Zone of the Limpopo Belt using Pb single phase dating of metamorphic minerals. Precambrian Res 87:87–115

Horn I, Rudnick RL, McDonough WF (2000) Precise elemental and isotope ratio determination by simultaneous solution nebulization and laser-ablation-ICP-M.S.: application to U-Pb geochronology. Chem Geol 167:405–425

Humbert F, Sonnette L, De Kock MO, Robion P, Horng CS, Cousture A, Wabo H (2017) Palaeomagnetism of the early Palaeoproterozoic, volcanic Hekpoort Formation (Transvaal Supergroup) of the Kaapvaal craton, South Africa. Geophys J Int 209:842–865. https://doi.org/10.1093/gji/ggx055

Hunt JP, Hatton C, De Kock MO, Bleeker W (2017) Plume activity related to the Kaapvaal craton and implications for Rhyacian plate reconstructions and ore deposits. Paper presented at the SGA Quebec 2017, Quebec City, Canada, 20–23 Aug 2017

Hunter DR, Reid DL (1987) Mafic Dyke Swarms in Southern Africa. In: Halls HC, Fahring WF (eds) Mafic Dyke Swarms. Geological Association of Canada, pp 445–456

Jacobs J, Pisarevsky S, Thomas RJ, Becker T (2008) The Kalahari Craton during the assembly and dispersal of Rodinia. Precambrian Res 160:142–158

Jelsma HA, de Wit MJ, Thiart C, Dirks PHGM, Viola G, Basson IJ, Anckar E (2004) Preferential distribution along transcontinental corridors of kimberlites and related rocks of Southern Africa. S Afr J Geol 107:301–324

Johnson SP, Rivers T, De Waele B (2005) A review of the Mesoproterozoic to early Palaeozoic magmatic and tectonothermal history of south-central Africa: implications for Rodinia and Gondwana. J Geol Soc 162(3):433–450

Johnson SP, De Waele B, Evans DM, Banda W, Tembo F, Milton JA, Tani K (2007) Geochronology of the Zambezi Supracrustal Sequence, southern Zambia: A record of Neoproterozoic divergent processes along the southern margin of the Congo Craton. J Geol 115:355–374

Jones DL, McElhinny MW (1966) Paleomagnetic correlation of basic intrusions in the Precambrian of southern Africa. J Geophys Res 71:543–552

Jones DL, Robertson IDM, McFadden PL (1974) A paleomagnetic study of the Precambrian dyke swarms associated with the great Dyke of Rhodesia. Trans Geol Soc S Afr 78:57–65

Jones DL, Bates MP, Li ZX, Corner B, Hodgkinson G (2003) Palaeomagnetic results from the ca. 1130 Ma Borgmassivet intrusions in the Ahlmannryggen region of Dronning Maud Land, Antarctica, and tectonic implications. Tectonophysics 375:247–260

Jourdan F, Féraud G, Bertrand H, Kampunzu AB, Tshoso G, Le Gall B, Tiercelin JJ, Capiez P (2004) The Karoo triple junction questioned: evidence from Jurassic and Proterozoic $^{40}Ar/^{39}Ar$ ages and geochemistry of giant Okovango dyke swarm (Botswana). Earth Planet Sci Lett 222:989–1006

Jourdan F, Féraud G, Bertrand H, Watkeys MK, Kampunzu AB, Le Gall B (2006) Basement control on dyke distribution in Large Igneous Provinces: case study of the Karoo triple junction. Earth Planet Sci Lett 241:307–322

Kampmann TC, Gumsley AP, De Kock MO, Söderlund U (2015) U-Pb geochronology and paleomagnetism of the Westerberg Sill Suite, Kaapvaal Craton - Support for a coherent Kaapvaal-Pilbara Block (Vaalbara) into the Paleoproterozoic? Precambrian Res 269:58–72

Kasbohm JJ, Evans DAD, Panzik JE, Hofmann M, Linnemann U (2016) Paleomagnetic and geochronologic data from late Mesoproterozoic red bed sedimentary rocks on the western margin of Kalahari craton. In: Li ZX, Evans DAD, Murphy JB (eds) Supercontinent cycles through earth history. Geological Society of London Special Publication, 424. Geological Society of London, London, pp 145–165

Katongo C, Koller F, Kloetzli U, Koeberl C, Tembo F, De Waele B (2004) Petrography, geochemistry, and geochronology of granitoid rocks in the Neoproterozoic-Paleozoic Lufilian-Zambezi belt, Zambia: Implications for tectonic setting and regional correlation. J Afr Earth Sci 40(5):219–244

Key R, Mapeo R (1999) The Mesoproterozoic history of Botswana and the relationship of the NW Bostwana Rift to Rodinia. Episodes 22(2):118–122

Kilian TM, Chamberlain K, Evans DAD, Bleeker W, Cousens BL (2016) Wyoming on the run—-towards final Paleoproterozoic assembly of Laurentia. Geology 44(10):863–866. https://doi.org/10.1130/G38042.1

Klausen MB, Söderlund U, Olsson JR, Ernst RE, Armoogam M, Mkhize SW, Petzer G (2010) Petrological discrimination among Precambrian dyke swarms: Eastern Kaapvaal craton (South Africa). Precambrian Res 183(3):501–522. https://doi.org/10.1016/j.precamres.2010.01.013

Knoper MW, Ernst RE, Kamo SL, Bleeker W, Söderlund U (2014) Three LIPs on two and half cratons. In: 2014 GSA Annual Meeting, Vancouver, Canada, 19–22 Oct 2014

Kramers JD, Mouri H (2011) The geochronology of the Limpopo Complex: a controversy solved. In: Van Reenen DD, Kramers JD, McCourt S, Perchuk LL (eds) Origin and evolution of precambrian high-grade Gneiss Terranes, with special emphasis on the Limpopo Complex of South Africa: geological society of America Memoir 207. The Geological Society of America, pp 85–106

Kramers JD, McCourt S, Roering C, Smit CA, Van Reenen DD (2011) Tectonic models proposed for the Limpopo Complex: mutual compatibilities and constraints. Geol Soc Am Mem 207:311–324

Krogh TE (1973) A low-contamination method for hydrothermal decomposition of zircon and extraction of U and Pb for isotopic age determinations. Geochim Cosmochim Acta 37(3):485–494

Krynauw JR, Watters BR, Hunter DR, Wilson AH (1991) A review of the field relationships, petrology and geochemistry of the Borgmassivet intrusions in the Grunehogna province, western Dronnig Maud Land, Antarctica. In: Thomson MRA, Crame JA, Thomson JW (eds) Geological evolution of Antarctica. Cambridge University Press, Cambridge, UK, pp 33–39

Kumar A, Parashuramulu V, Shankar R, Besse J (2017) Evidence for a Nearchean LIP in the Singhbhum craton, eastern India: implications to Vaalbara supercontinent. Precambrian Res 292:163–174. https://doi.org/10.1016/j.precamres.2017.01.018

Larsson ER (2015) U-Pb baddeleyite dating of intrusions in the south-easternmost Kaapvaal Craton (South Africa): revealing multiple events of dyke emplacement. MSc Thesis, Lund University, Lund, Sweden

Laurent O, Zeh A (2015) A linear Hf isotope-age array despite different granitoid sources and complex Archean geodynamics: example from the Pietersburg block (South Africa). Earth Planet Sci Lett 430:326–338. https://doi.org/10.1016/j.epsl.2015.08.082

Laurent O, Paquette J-L, Martin H, Doucelance R, Moyen J-F (2013) LA-ICP-MS dating of zircons from Meso- and Neoarchean granitoids of the Pieterburg block (South Africa): crustal evolution at the northern margin of the Kaapvaal craton. Precambrian Res 230:209–226. https://doi.org/10.1016/j.precamres.2013.02.009

Layer PW, Kröner A, McWilliams M, Burghele A (1988) Paleomagnetism and age of the Archean Usushwana Complex, Southern Africa. J Geophys Res 93(B1):449–457

Layer PW, Kröner A, McWilliams M, York D (1989) Elements of the Archean thermal history and apparent polar wander of the eastern Kaapvaal craton, Swaziland, from single grain dating and paleomagnetism. Earth Planet Sci Lett 93:23–24

Letts S, Torsvik TH, Webb SJ, Ashwal LD, Eide EA, Chunnett G (2005) Palaeomagnetism and $^{40}Ar/^{39}Ar$ geochronology of mafic dykes from the eastern Bushveld Complex (South Africa). Geophys J Int 162:36–48

Letts S, Torsvik TH, Webb SJ, Ashwal LD (2009) Palaeomagnetism of the 2054 Ma Bushveld Complex (South Africa): implications for emplacement and cooling. Geophys J Int 179:850–872. https://doi.org/10.1111/j.1365-246X.2009.04346.x
Letts S, Torsvik TH, Webb SJ, Ashwal LD (2010) New Palaeoproterozoic palaeomagnetic data from the Kaapvaal Craton, South Africa. Geol Soc Lond Spec Publ 357:9–26
Li ZX, Bogdanova SV, Collins AS, Davidson A, De Waele B, Ernst RE, Fitzsimons ICW, Fuck RA, Gladkochub DP, Jacobs J, Karlstrom KE, Lu S, Natapov LM, Pease V, Pisarevsky SA, Thrane K, Vernikovsky V (2008) Assembly, configuration, and break-up history of Rodinia: a synthesis. Precambrian Res 160:179–210
Lister JR, Kerr RC (1991) Fluidmechanical models of crack propagation and their application to magma transport in dikes. J Geophys Res 96(B6):10049–10077
Lubnina N, Ernst RE, Klausen M, Söderlund U (2010) Paleomagnetic study of Neoarchean-Paleoproterozoic dykes in the Kaapvaal Craton. Precambrian Res 183:523–552. https://doi.org/10.1016/j.precamres.2010.05.005
MacDonnald FA, Schmitz MD, Crowley JL, Roots CF, Jones DS, Maloof AC, Strauss JV, Cohen PA, Johnston DT, Schrag DP (2010) Calibrating the cryogenian. Science 327:1241–1243
Maier WD, Peltonen P, Grantham G, Mänttäri I (2003) A new 1.9 Ga age for the Trompsburg intrusion, South Africa. Earth Planet Sci Lett 212:351–360. https://doi.org/10.1016/S0012-821X(03)00281-4
Maier WD, Prevec SA, Scoates JS, Wall CJ, Barnes S-J, Gomwe T (2017) The Uitkomst intrusion and Nkomati Ni-Cu-Cr-PGE deposit, South Africa: trace element geochemistry, Nd isotopes and high-precision geochronology. Miner Depos 1–22. https://doi.org/10.1007/s00126-017-0716-x
Majaule T, Hanson RE, Key R, Singletary SJ, Martin MW, Bowring SA (2001) The Magondi Belt in northeast Botswana: regional relations and new geochronological data from the Sua Pan area. J Afr Earth Sci 32(2):257–267
Manyeruke TD, Blenkinsop TG, Buchholz P, Love D, Oberthür T, Vetter UK, Davis DW (2004) The age and petrology of the Chimbadzi Hill Intrusion, NW Zimbabwe: first evidence for early Paleoproterozoic magmatism in Zimbabwe. J Afr Earth Sc 40(5):281–292
Mapeo RBM, Armstrong RA, Kampunzu AB, Ramokate LV (2004a) SHRIMP U-Pb zircon ages of granitoids from the Western Domain of the Kaapvaal Craton, Southeastern Botswana: implications for crustal evolution. S Afr J Geol 107:159–172
Mapeo RBM, Kampunzu AB, Ramokate LV, Corfu F, Key RM (2004b) Bushveld-age magmatism in southeastern Botswana: evidence from U-Pb zircon and titanite geochronology of the Moshaneng Complex. S Afr J Geol 107:219–232
Maré LP, Fourie CJS (2012) New geochemical and palaeomagnetic results from Neoarchaean dyke swarms in the Badplaas-Barberton area, South Africa. S Afr J Geol 115(2):145–170
Marsh JS, Bowen MP, Rogers NW, Bowen TB (1989) Volcanic rocks of the Witwatersrand Triad, South Africa. II: Petrogenesis of mafic and felsic rocks of the Dominion Group. Precambrian Res 44:39–65. https://doi.org/10.1016/0301-9268(89)90075-2
Master S (1991) Stratigraphy, tectonic setting and mineralization of the Early Proterozoic Magondi Supergroup, Zimbabwe: a review, vol 238. Economic geology research institute information circular. University of the Witwatersrand, Johannesburg, South Africa
McCarthy TS, McCallum K, Myers RE, Linton P (1990) Stress states along the northern margin of the Witwatersrand Basin during Klipriviersberg Group volcanism. S Afr J Geol 93:245–260
McCourt S, Hilliard P, Armstrong RA, Munyanyiwa H (2001) SHRIMP U-Pb zircon geochronology of the Hurungwe granite northwest Zimbabwe: age constraints on the timing of the Magondi orogeny and implications for correlation between the Kheis and Magondi Belts. S Afr J Geol 104:39–46
McDougall I (1963) Potassium-argon age measurements on dolerites from Antarctica and South Africa. J Geophys Res 68:1535–1545
McElhinny MW (1966) The palaeomagnetism of the Umkondo Lavas, Eastern Southern Rhodesia. Geophys J Roy Astron Soc 10:375–381

McElhinny MW, Opdyke ND (1964) The paleomagnetism of the Precambrian dolerites of eastern Southern Rhodesia, an example of geologic correlation by rock magnetism. J Geophys Res 69:1465–1475

Meier DL, Heinrich CA, Watts MA (2009) Mafic dikes displacing Witwatersrand gold reefs: evidence against metamorphic-hydrothermal ore formation. Geology 37:607–610. https://doi.org/10.1130/G25657A

Mertanen S, Hölttä P, Pesonen LJ, Paavola L (2006) In: Hanski E, Mertanen S, Rämö T, Vuollo J (eds) Dyke Swarms—time markers of crustal evolution: selected papers of the fifth international dyke conference in Finland, Rovaniemi, Finland, 31 July–3 Aug 2005 & Fourth international dyke conference, Kwazulu-Natal, South Africa 26–29 June 2001. CRC Press, London, pp 243–256

Miller RM (2008) The geology of Namibia, vol 1, Archean to Mesoproterozoic, vol 1. Ministry of Mines and Energy, Geological Survey of Namibia, Windhoek

Miller RM (2012) Review of Mesoproterozoic magmatism, sedimentation and terrane amalgamation in southwestern Africa. S Afr J Geol 115(4):417–448

Mitchell RN, Hoffman PF, Evans DAD (2010) Coronation loop resurrected: oscillatory apparent polar wander of Orosirian (2.05–1.8 Ga) paleomagnetic poles from Slave craton. Precambrian Res 179:121–134

Moen HFG (1999) The Kheis Tectonic Subprovince, southern Africa: a lithostratigraphic perspective. S Afr J Geol 102(1):27–42

Moore M, Davis DW, Robb LJ, Jackson MC, Grobler DF (1993) Archean rapakivi granite-anorthosite-rhyolite complex in the Witwatersrand basin hinterland, southern Africa. Geology 21:1031–1034

Mukasa SB, Wilson AH, Young KR (2013) Geochronological constraints on the magmatic and tectonic development of the Pongola Supergroup (Central Region), South Africa. Precambrian Res 224:268–286

Mukwakwami J (2005) Structural geology of the Umkondo Group in eastern Zimbabwe and geochronology of associated mafic rocks and possible correlatives in Zimbabwe. Unpublished PhD thesis, University of Zimbabwe, Harare, Zimbabwe

Mungall JE, Kamo SL, McQuade S (2016) U-Pb geochronology documents out-of-sequence emplacement of ultramafic layers in the Bushveld Igneous Complex of South Africa. Nat Commun 7:13385. https://doi.org/10.1038/ncomms13385

Munyanyiwa H (1999) Geochemical study of the Umkondo dolerites and lavas in the Chimanimani and Chipinge districts (eastern Zimbabwe) and their regional implications. J Afr Earth Sci 28(2):349–365

Nhleko N (2003) The Pongola Supergroup in Swaziland. Unpublished PhD thesis, Rand Afrikaans University, Johannesburg, South Africa

Oberthür T, Davis DW, Blenkinsop TG, Höhndorf A (2002) Precise U-Pb mineral ages, Rb-Sr and Sm-Nd systematics for the Great Dyke, Zimbabwe—constraints on late Archean events in the Zimbabwe craton and Limpopo belt. Precambrian Res 113(3–4):293–305

Olsson JR, Söderlund U, Klausen MB, Ernst RE (2010) U-Pb baddeleyite ages linking major Archean dyke swarms to volcanic-rift forming events in the Kaapvaal craton (South Africa), and a precise age for the Bushveld Complex. Precambrian Res 183:490–500

Olsson JR, Söderlund U, Hamilton MA, Klausen MB, Helffrich GR (2011) A late Archean radiating dyke swarm as possible clue to the origin of the Bushveld Complex. Nat Geosci 4:865–869

Olsson JR, Klausen M, Hamilton M, März N, Söderlund U, Roberts RJ (2016) Baddeleyite U-Pb ages and geochemistry of the 1875–1835 Ma Black Hills Dyke Swarm across north-eastern South Africa: part of a trans-Kalahari Craton back-arc setting? GFF 138(1):183–202. https://doi.org/10.1080/11035897.2015.1103781

Panzik JE, Evans DAD, Kasbohm JJ, Hanson RE, Gose WA, Desormeau J (2016) Using palaeomagnetism to determine late Mesoproterozoic palaeogeographic history and tectonic relations of the Sinclair Terrane, Namaqua orogen, Namibia. In: Li ZX, Evans DAD, Murphy JB (eds) Supercontinent cycles through earth history. Geological Society of London Special Publication, 424. Geological Society of London, London, pp 119–143

Peng P (2015) Precambrian mafic dyke swarms in the North China Craton and their geological implications. Sci China: Earth Sci 58:649–675

Peng P, Guo J, Windley BF, Li X (2011) Halaqin volcano-sedimentary succession in the central-northern margin of the North China Craton: products of Late Paleoproterozoic ridge subduction. Precambrian Res 187:165–180

Pesonen LJ, Elming SÅ, Mertanen S, Pisrevsky S, D'Agrella-Filho MS, Meert JG, Schmidt PW, Abrahamsen N, Bylund G (2003) Palaeomagnetic configuration of continents during the Proterozoic. Tectonophysics 375:289–324

Pettersson Å, Cornell DH, Moen HFG, Reddy S, Evans D (2007) Ion-probe dating of 1.2 Ga collision and crustal architecture in the Namaqua-Natal Province of southern Africa. Precambrian Res 158(1–2):79–92. https://doi.org/10.1016/j.precamres.2007.04.006

Pisarevsky SA, Wingate MTD, Powell CM, Johnson S, Evans DAD (2003) Models of Rodinia assembly and fragmentation. In: Yoshida M, Windley BF, Dasgupta S (eds) Proterozoic East Gondwana: supercontinent assembly and breakup. Geological society, London, Special Publication, vol 206. Geological Society, London, pp 35–55

Pisarevsky S, De Waele B, Jones S, Söderlund U, Ernst RE (2015) Paleomagnetism and U-Pb age of the 2.4 Ga Erayinia mafic dykes on the south-western Yilgarn, Western Australia: paleogeographic and geodynamic implications. Precambrian Res 259:222–231. https://doi.org/10.1016/j.precamres.2014.05.023

Polteau S, Moore JM, Tsikos H (2006) The geology and geochemistry of the Paleoproterozoic Makganyene diamictite. Precambrian Res 148:257–274. https://doi.org/10.1016/j.precamres.2006.05.003

Poujol M, Kiefer R, Robb LJ, Anhaeusser CR, Armstrong RA (2005) New U-Pb data on zircons from the Amalia greenstone belt Southern Africa: insights into the Neoarchean evolution of the Kaapvaal Craton. S Afr J Geol 108(3):317–332

Powell C, Jones DL, Pisarevsky SA, Wingate MTD (2001) Palaeomagnetic constraints on the position of the Kalahari craton in Rodinia. Precambr Res 110 (1–4):33–46

Pradhan VR, Meert JG, Pandit MK, Kamenov G, Mondal MEA (2012) Paleomagnetic and geochronological studies of the mafic dyke swarms of Bundelkhand craton, central India: implications for the rectonic evolution and paleogeographic reconstructions. Precambrian Res 198–199:51–76

Rasmussen B, Bekker A, Fletcher IR (2013) Correlation of Paleoproterozoic glaciations based on U-Pb zircon ages for tuff beds in the Transvaal and Huronian Supergroups. Earth Planet Sci Lett 382:173–180. https://doi.org/10.1016/j.epsl.2013.08.037

Reid DL (1997) Sm-Nd age and REE geochemistry of Proterozoic arc-related igneous rocks in the Richtersveld Subprovince, Namaqua Mobile Belt, southern Africa. J Afr Earth Sci 24:621–633

Reid DL, Ransome IGD, Onstott TC, Adams CJ (1991) Time of emplacement and metamorphism of Late Precambrian mafic dykes associated with the Pan-African Gariep orogeny, Southern Africa: implications for the Nama Group. J Afr Earth Sci 13(3):531–541

Reischmann T (1995) Precise U/Pb age determination with baddeleyite (ZrO2), a case study from the Palaborwa Igneous Complex, South Africa. S Afr J Geol 98:1–4

Rioux M, Bowring SA, Dudás F, Hanson RE (2010) Characterizing the U-Pb systematics of baddeleyite through chemical abrasion: application of multi-step digestion methods to baddeleyite geochronology. Contrib Mineral Petrol 160:777–801. https://doi.org/10.1007/s00410-010-0507-1

Rivalta E, Taisne B, Bunger AP, Katz RF (2015) A review of mechanical models of dike propagation: schools of thought, results and future diections. Tectonophysics 638:1–42

Robb LJ, Armstrong RA, Waters DJ (1999) The history of granulite-facies metamorphism and crustal growth from single zircon U-Pb geochronology, Namaqualand, South Africa. J Petrol 40:1747–1770

Roering C, Van Reenen DD, Smit CA, Barton JM Jr, De Beer JH, De Wit MJ, Stettler EH, Van Schalkwyk JF, Stevens G, Pretorius S (1992) Tectonic model for the evolution of the Limpopo Belt. Precambrian Res 55:539–552

Rogers C, Mackinder A, Ernst RE, Cousens B (2016) Mafic magmatism in the Belt-Purcell Basin and Wyoming Province of western Laurentia. Geol Soc Am Spec Pap 522

Rubin AM (1995) Propagation of magma-filled cracks. Annu Rev Earth Planet Sci 23:287–336

Savko KA, Samsonov AV, Kholin VM, Bazikov NS (2017) The Sarmatia Megablock as a fragment of the Vaalbara Supercontinent: correlation of geological events at the Archean-Paleoproterozoic transition. Stratigr Geol Correl 25(2):123–145. https://doi.org/10.1134/S0869593817020058

Schaller M, Steiner O, Studer I, Holzer L, Herwegh M, Kramers JD (1999) Exhumation of Limpopo Central Zone granulites and dextral continent-scale stranscurrent movement at 2.0 Ga along the Palala Shear Zone, Northern province, South Africa. Precambrian Res 96:263–288

Schmitz MD, Bowring SA, De Wit MJ, Gartz V (2004) Subduction and terrane collosion stabilize the western Kaapvaal craton tectosphere 2.9 billion years ago. Earth Planet Sci Lett 222:363–376

Schreiner GDL, Van Niekerk CB (1958) The age of the Pilanesberg dyke from the Central Witwatersrand. Trans Proc Geol Soc S Afr 61:198–199

Schröder S, Beukes NJ, Armstrong RA (2016) Detrital zircon constraints on the tectonostratigraphy of the Paleoproterozoic Pretoria Group, South Africa. Precambrian Res 278:362–393. https://doi.org/10.1016/j.precamres.2016.03.016

Sharpe MR (1981) The chronology of magma influxes to the eastern compartment of the Bushveld Complex as exemplified by its marginal border groups. J Geol Soc Lond 138:307–326

Sharpe MR (1982) Petrology, classification and chronology of the intrusion of mafic sills beneath the eastern Bushveld Complex, vol 37. Institute for the geological research on the Bushveld Complex, Pretoria, South Africa

Singletary SJ, Hanson RE, Martin MW, Crowley JL, Bowring SA, Key RM, Ramokate LV, Direng BB, Krol MA (2003) Geochronology of basement rocks in the Kalahari Desert, Botswana, and implications for regional Proterozoic tectonics. Precambrian Res 121:47–71

Smirnov AV, Evans DAD, Ernst RE, Söderlund U, Li Z-X (2013) Trading partners: tectonic ancestry of southern Africa and western Australia, in Archean supercratons Vaalbara and Zimgarn. Precambrian Res 224:11–22. https://doi.org/10.1016/j.precamres.2012.09.020

Söderlund U, Johansson L (2002) A simple way to extract baddeleyite (ZrO_2). Geochem Geophys Geosystems 3(2). https://doi.org/10.1029/2001gc000212

Söderlund U, Hofmann A, Klausen MB, Olsson JR, Ernst RE, Persson P-O (2010) Towards a complete magmatic barcode for the Zimbabwe craton: Baddeleyite U-Pb dating of regional dolerite dyke swarms and sill complexes. Precambrian Res 183(3):388–398. https://doi.org/10.1016/j.precamres.2009.11.001

Strik G, De Wit MJ, Langereis CG (2007) Palaeomagnetism of the Neoarchaean Pongola and Ventersdorp Supergroups and an appraisal of the 3.0–1.9 Ga apparent polar wander path of the Kaapvaal Craton, Southern Africa. Precambrian Res 153:96–115

Swanson-Hysell NL, Killian TM, Hanson RE (2015) A new grand mean paleomagnetic pole for the Umkondo Igneous Province with implications for paleogeography and the geomagnetic field. Geophys J Int 203:2237–2247

Thomas RJ, Jacobs J, Eglington BM (2000) Geochemistry and isotopic evolution of the Mesoproterozoic Cape Meredith Complex. West Falkland Geol Mag 137(5):537–553

Tinker J, De Wit MJ, Grotzinger J (2002) Seismic stratigraphic constraints on Neoarchean-Paleoproterozoic evolution of the western margin of the Kaapvaal Craton, South Africa. S Afr J Geol 105:107–134

Treloar PJ (1988) The geological evolution of the Magondi Mobile Belt, Zimbabwe. Precambrian Res 38(1):55–73

Uken R, Watkeys MK (1997) An interpretation of mafic dyke swarms and their relationship with major mafic magmatic events on the Kaapvaal Craton and Limpopo Belt. S Afr J Geol 100:341–348

Van der Westhuizen WA, De Bruiyn H, Meinties PG (1991) The Ventersdorp Supergroup; an overview. J Afr Earth Sci 13(1):83–105

Van der Westhuizen WA, De Bruiyn H, Meintjies PG (2006) The Ventersdorp Supergroup. In: Johnson MR, Anhaeusser C (eds) The geology of South Africa. Geological Society of South Africa/Council for Geoscience, Johannesburg/Pretoria, pp 187–208
Van Niekerk CB (1962) The age of the Gemspost dyke from the Venterspost Gold Mine. Trans Geol Soc S Afr 65:105–111
Van Tongeren JA, Zirakparvar NA, Mathez EA (2016) Hf isotopic evidence for a cogenetic magma source for the Bushveld Complex and associated felsic magmas. Lithos 248:469–477
Verwoerd WJ (2006) The Pilanesberg Alkaline Province. In: Johnston MR, Anhaeusser C (eds) The geology of South Africa. The Geological Society of South Africa and the Council for Geoscience, Pretoria, pp 381–393
Wabo H, De Kock MO, Klausen M, Söderlund U, Beukes NJ (2015a) Paleomagnetism and chronology of B-1 marginal sills of the Bushveld Complex from the eastern Kaapvaal Craton, South Africa. GFF. https://doi.org/10.1080/11035897.2015.1099566
Wabo H, Olsson JR, De Kock MO, Humbert F, Söderlund U, Klausen M (2015b) New U-Pb age and paleomagnetic constraints from the Ultkomst Complex, South Africa: clues to the timing of intrusion. GFF. https://doi.org/10.1080/11035897.2015.1098726
Wabo H, Humbert F, De Kock MO, Söderlund U, Maré L, Beukes NJ (2019) Constraining the chronology of the Mashishing dykes from the eastern Kaapvaal craton in South Africa. In: Srivastava RK, Ernst RE, Peng P (eds) Dyke Swarms of the World: A Modern Perspective. Springer Geology. https://doi.org/10.1007/978-981-13-1666-1_6
Walraven F, Hattingh E (1993) Geochronology of the Nebo granite, Bushveld Complex. S Afr J Geol 96:31–41
Walraven F, Pape J (1994) Pb-Pb whole-rock ages for the Pongola Supergroup and the Usushwana Complex, South Africa. J Afr Earth Sci 18:297–308. https://doi.org/10.1016/0899-5362(94)90069-8
Walraven F, Smith CB, Kruger FJ (1991) Age determinations of the Zoetlief Group—a Ventersdorp Supergroup correlative. S Afr J Geol 94(2):220–227
Walraven F, Froick C, Lubala RT (1992) Pb-isotope geochronology of the Schiel Complex, northern Transvaal, South Africa. J Afr Earth Sci 15:103–110
Walraven F, Grobler DF, Key RM (1996) Age equivalents of the Plantation Porphyry and the Kanye Volcanic Formation, southeastern Botswana. S Afr J Geol 99:23–31
Walraven F (1997) Geochronology of the Rooiberg Group, Transvaal Supergroup, South Africa. EGRI Information Circular 316, University of the Witwatersrand, 21pp
Ward SE, Hall RP, Huges DJ (2000) Guruve and Mutare dykes: preliminary geochemical indication of complex Mesoproterozoic mafic magmatic systems in Zimbabwe. J Afr Earth Sci 30(3):689–701
Wilson JF, Jones DL, Kramers JD (1987) Mafic dyke swarms of Zimbabwe. In: Halls HC, Fahring AF (eds) Mafic Dyke Swarms, vol 34. Geological Association of Canada Special Papers, pp 433–444
Wilson JF, Nesbitt RW, Fanning M (1995) Zircon geochronology of Archean felsic sequences in the Zimbabwe Craton: a revision of the greenstone belt stratigraphy and model for crustal growth. In: Coward MP, Reis AC (eds) Early Precambrian Processes, Geological Society, London, Special Publication, vol 95, pp 109–126
Wingate MTD (1998) A palaeomagnetic test of the Kaapvaal-Pilbara (Vaalbara) connection at 2.78 Ga. S Afr J Geol 101(4):257–274
Wingate MTD (2000) Ion microprobe U-Pb zircon and baddeleyite ages for the Great Dyke and its satellite dykes, Zimbabwe. S Afr J Geol 103(1):74–80
Wingate MTD (2001) SHRIMP baddeleyite and zircon ages for an Umkondo dolerite sill, Nyanga Mountains, Eastern Zimbabwe. S Afr J Geol 104:13–22
Woodhead J, Hergt J, Phillps D, Paton C (2009) African kimberlites revisited: in situ Sr-isotope analysis of groundmass perovskite. Lithos 112(Supplement 1):311–317

Wu F-Y, Yang Y-H, Li Q-L, Mitchell RH, Dawson JB, Brandl G, Yuhara M (2011) In situ determinations of U-Pb ages and Sr-Nd-Hf isotopic constraints on petrogenesis of the Phalaborwa carbonatite Complex, South Africa. Lithos 127:309–322

Wu F-Y, Mitchell RH, Li Q-L, Sun J, Liu C-Z, Yang Y-H (2013) In situ U-Pb age determination and Sr-Nd isotopic analysis of perovskite from the Premier (Cullinana) kimberlite, South Africa. Chem Geol 353:83–95

Xie H, Kröner A, Brandl G, Wan Y (2017) Two orogeneic events separated by 2.6 Ga mafic dykes in the Central Zone, Limpopo Belt, southern Africa. Precambrian Res 289:129–141

Zegers TE, De Wit MJ, Dann J, White SH (1998) Vaalbara, Earth's oldest assembled continent? A combined structural, geochronological, and palaeomagnetic test. Terra Nova 10:250–259

Zeh A, Ovtcharova M, Wilson AH, Schaltegger U (2015) The Bushveld Complex was emplaced and cooled in less than one million years—results of zirconology, and geotectonic implications. Earth Planet Sci Lett 418:103–114

Constraining the Chronology of the Mashishing Dykes from the Eastern Kaapvaal Craton in South Africa

H. Wabo, F. Humbert, M. O. de Kock, G. Belyanin, U. Söderlund, L. P. Maré and N. J. Beukes

Abstract The present study focuses on NNE-trending dykes (sites LDA to LDJ) that occur near Mashishing in the eastern Kaapvaal craton. The Mashishing dykes were previously considered to be coeval and regarded as the extension of the 1.875–1.835 Ga Black Hills dyke swarm into the sedimentary rocks of the Pretoria Group. Thin sections exhibit well-preserved igneous textures, with primary minerals (e.g., hornblende, clinopyroxene, plagioclase) extensively altered to secondary minerals in most cases. Our dykes can be petrographically grouped as pyroxenite (LDB and LDC), dolerite (LDH) and diorite (remainder of samples). REE and multi-element profiles of pyroxenites and two of the diorites (LDI and LDJ)

H. Wabo (✉) · F. Humbert · M. O. de Kock · G. Belyanin · N. J. Beukes
Paleoproterozoic Mineralization Research Group (PPM Group), PO Box 524, Auckland Park 2006, South Africa
e-mail: hwabo@uj.ac.za

F. Humbert
e-mail: humbert.fabien@gmail.com

M. O. de Kock
e-mail: mdekock@uj.ac.za

G. Belyanin
e-mail: gabelyanin@uj.ac.za

N. J. Beukes
e-mail: nbeukes@uj.ac.za

H. Wabo · F. Humbert · M. O. de Kock · G. Belyanin · N. J. Beukes
Department of Geology, DST-NRF Centre of Excellence for Integrated Mineral and Energy Resource Analysis (CIMERA), University of Johannesburg, PO Box 524, Auckland Park 2006, South Africa

U. Söderlund
Department of Geology, GeoBiosphere Science Centre, Lund University, Sölvegatan 12, 223 62 Lund, Sweden
e-mail: ulf.soderlund@geol.lu.se

L. P. Maré
Council for Geoscience, Private Bag X112, Pretoria 0001, South Africa
e-mail: leoniem@geoscience.org.za

R. K. Srivastava et al. (eds.), *Dyke Swarms of the World: A Modern Perspective*, Springer Geology, https://doi.org/10.1007/978-981-13-1666-1_6

suggest a common origin of the four dykes, but show no similarity with known mafic units of the Kaapvaal craton. An age estimate between 2208 and 2276 Ma for dyke LDB, obtained from two overlapping amphibole $^{40}Ar/^{39}Ar$ plateau results, indicates that these four dykes predates the ~2.05 Ga Bushveld event. The six remaining dykes have similar chemistry to either the ~2.06 Ga Dullstroom Lavas (LDG), the 1.875–1.835 Ga Black Hills dyke swarm (LDH) or the ~1.11 Ga Umkondo dolerites (LDA, LDD, LDE and LDF). An U–Pb baddeleyite date of 1867 ± 10 Ma for dyke LDH confirms it as a member of the Black Hills dyke swarm. Demagnetization of eighty-three specimens reveals five stable magnetizations carried by titanomagnetite. Two of these are regarded as magnetic overprints while three magnetizations are likely representative of primary remanences. The corresponding virtual geomagnetic poles (Lat. −26.84°N, Long. 31.66°E; Lat. 26.07°N, Long. 11.01°E, and Lat. 55.84°N, Long. 65.02°E) resemble those from the ~2.23 Ga Hekpoort Formation, the ~1.88–1.83 Ga post-Waterberg intrusions, and the ~1.11 Ga Umkondo dolerites respectively. The above results suggest that the Mashishing dykes, despite similar trends, constitute swarms of different generations.

Keywords Kaapvaal craton · NNE-dyke · Geochemistry · Geochronology Paleomagnetism

1 Introduction

Regional mafic dyke swarms are interpreted by some authors (e.g., Ernst and Buchan 1997; Ernst 2014) to represent systems of pre-existing tensional crustal fractures along which mafic magmas emplaced. In large igneous provinces, dyke swarms constitute the link between the melt generated at depth and volcanic end products erupted in surface (e.g., Ernst and Buchan 1997).

Dyke swarms of different orientations are widespread throughout the Kaapvaal craton (KC) in South Africa (e.g. Ernst 2014). Prominent NNE-trending dykes are traceable as strong magnetic lineaments transecting the eastern KC. Some of these dykes have been studied by Olsson et al. (2016) to define the Black Hills Dyke Swarm (BHDS). Investigation of dykes can provide key information (e.g. identification of magmatic centers, occurrence of plumes, paleopoles with precise ages useful for continental reconstructions) for a better understanding of the KC evolution.

An apparent extension of the BHDS into the Transvaal Supergroup is found within the Silverton Formation shale near the town of Mashishing (formerly known as Lydenburg) (Fig. 1a). The relevant dykes (herein named the Mashishing dykes) share a NNE-trend with the BHDS as well as with the dykes that intrude the ~2.05 Ga Rustenburg Layered Suite of the Bushveld Complex (Letts et al. 2005) and the ~2.06 Ga Phalaborwa Complex (Olsson et al. 2016). The footprint of the BHDS within the Transvaal Supergroup, however, remains poorly constrained. Here, we report for the first time geochronological, geochemical and paleomagnetic data from the Mashishing dykes.

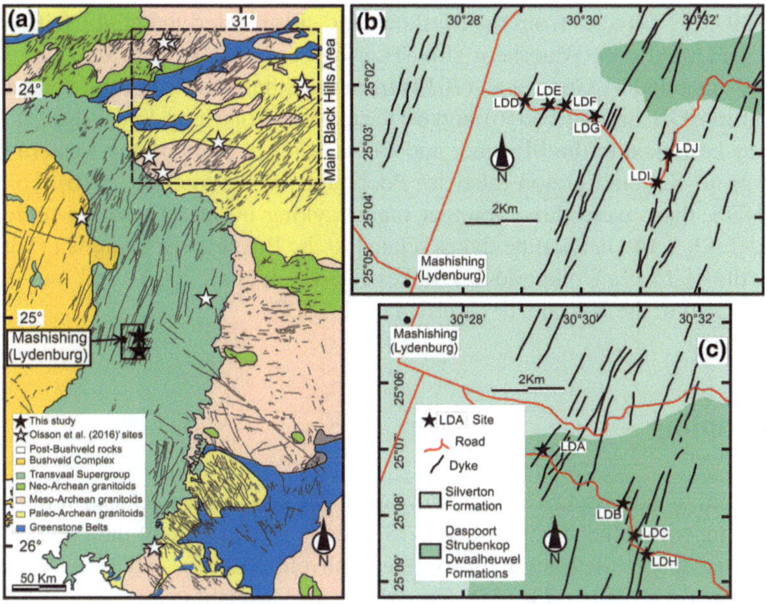

Fig. 1 **a** Generalized geological map of the eastern part of the KC. (Modified from Olsson et al. 2016), **b** and **c** more detailed maps of the sampled transects showing individual sampling sites. Dykes along northern transect are generally more altered compared than those along the southern transect. Mafic sills that occur in the area are removed for clarity purpose

2 Review of Dyking Events in The Eastern KC

Different dyke swarms have been identified by previous authors in the eastern portion of KC. The oldest of these were emplaced prior the intrusion of the ~2.05 Ga Bushveld Large Igneous Province or LIP which arguably represents the most important magmatic event of this craton (e.g., Cawthorn et al. 2006). Pre-Bushveld dyke swarms recognized so far include (i) the ~2.9 Ga SE-trending dykes (i.e. the Badplaas dyke swarm) associated with the development of the Pongola Supergroup (Olsson et al. 2010; Gumsley et al. 2013, 2015), (ii) the diversely-oriented ~2.7 to ~2.66 Ga dykes (i.e. the Rykoppies dyke swarm and the White Mfolozi dyke swarm), some of which are believed to represent feeders to the Ventersdorp volcanics and the lavas of the Lower Transvaal Supergroup (e.g., Gumsley et al. 2016), and lastly (iii) the ~2.42 Ga N-trending Gewonne dyke swarm that combines with co-magmatic sills and volcanics to form the Ongeluk LIP (Gumsley et al. 2017). The ~2.42 Ga Gewonne dyke swarm (Gumsley et al. 2017) was followed by the ~1.93 Tsineng dyke swarm (Alebouyeh Semami et al. 2016) with an intervening hiatus of ~490 Ma during which no dyke events have been recognized. After the ~1.93 Ga Tsineng dyke swarm, there was the emplacement of the Black Hill Dyke Swarm (BHDS) (Olsson et al. 2016. The 1875–1835 Ma ages obtained from the BHDS bridge a gap between the ~1.89–1.87 Ga ages from the Mashonaland sills (Söderlund et al. 2010; Hanson et al. 2011) and post-Waterberg sills (Hanson et al. 2004), and the ~1.83 Ga zircon ages obtained by Geng et al. (2014) from the Sibasa basalts of the Soutpansberg Supergroup. Olsson et al. (2016) suggest the existence of a long-lived ~1.89–1.83 LIP that may be linked to the Magondi and Okwa-Kheis arcs along the western margin of the proto-Kalahari craton (also see Alebouyeh Semami et al. 2016). Following the BHDS, there were three younger major dyking events. These are represented by the ~1.43–1.14 Ga alkaline dykes related to the Pilanesberg Complex (Verwoerd 2006), the ~1.11 Ga Umkondo LIP dykes (De Kock et al. 2014), and the ~0.18 Ga dykes related to the Karoo LIP (Jourdan et al. 2006).

3 Sampling

NNE-trending mafic dykes are generally well exposed to the northeast of the town of Mashishing within the slightly tilted (~15°) sedimentary strata of the Pretoria Group and in proximity to the eastern limb of the Bushveld Complex and its associated sills (e.g., Barnes et al. 2010) (Fig. 1a). The dykes near Mashishing intersect the Silverton Formation shale of the Pretoria Group, which lie stratigraphically underneath the Machadodorp Lavas Formation (Fig. 1).

Not all strong lineaments visible on Google Earth images correspond to dykes in the Mashishing area. Individual dykes have kilometric strike length, and range from ~20 m (dyke LDE) to ~100 m (dyke LDB) wide. Dykes further northeast of Mashishing are generally poorly exposed and more eroded compared to the ones closer to the town (Fig. 1).

The least weathered dykes were targeted for a paleomagnetic study. A total of ten dykes were sampled, four of which were near Mashishing and six of which were collected further to the northeast (Fig. 1b, c; Table 1). Knowing that mafic sills also occur in the sampling area, each dyke was rigorously identified as such using the "crosscutting" and "length versus width" criteria (e.g. Ernst 2014).

Due to the deep weathering of the Silverton Formation, particularly along contacts with the crosscutting dykes, no baked-contact tests could be performed. Paleomagnetic sampling was thus limited to the dyke outcrops, and was achieved by collecting individually oriented cores from each intrusion using a portable, hand-held petrol drill. Five to eleven oriented samples were collected from each of the dykes. Orientation of the cores was achieved by using both a magnetic compass and a sun compass. All dykes were also sampled for further investigating their magnetic mineralogy and bulk rock geochemistry. All of the dykes were also considered for the baddeleyite U–Pb dating. However, as only the dyke LDH successfully provided baddeleyite grains, the remaining sites were further processed for the $^{39}Ar/^{40}Ar$ dating, of which only the dyke LDB appeared suitable for such analyses.

Table 1 Details of individual sampling sites

Site	GPS coordinates	Thickness (m)	N	Strike of dyke
LDA	S 25°07′03.2″ E 030°29′22.3″	25	8	30
LDB*	S 25°07′52.0″ E 030°30′43.0″	100	10	20
LDC	S 25°08′23.6″ E 030°30′50.9″	50	8	25
LDD	S 25°02′13.1″ E 030°29′08.4″	70	11	25
LDE	S 25°02′19.1″ E 030°29′31.8″	20	7	35
LDF	S 25°02′21.40″ E 030°29′48.1″	50	11	30
LDG	S 25°02′33.5″ E 030°30′17.8″	80	9	30
LDH*	S 25°08′39.7″ E 030°31′05.9″	50	6	25
LDI	S 25°03′31.3″ E 030°31′18.0″	60	8	10
LDJ	S 25°03′06.8″ E 030°31′30.0″	60	5	12
Total = 10 sites (83 paleomagnetic cores)				

N = number of oriented cores drilled per site
*Indicates dykes that were dated (U/Pb method for dyke LDH and $^{40}Ar/^{39}Ar$ method for dyke LDB). Dip of country rocks is 15 WNW

4 Method

4.1 *Petrology and Geochemistry*

Mineral occurrences and textural characteristics of the studied dykes were determined by the means of an optical microscope (Fig. 2) and a CAMECA SX100 electron microprobe operating with settings 20 kV, 10 nA and WDS (Table 2). The standards used for microprobe analyses are: diopside (Si), almandine (Al), hematite (Fe), olivine (Mg), rhodochrosite (Mn), wollastonite (Ca), jadeite (Na), orthoclase (K), TiO (Ti), CrO (Cr) and NiO (Ni). A total of 376 point quantifications were used to characterize the studied samples.

All of the dyke samples were subjected to geochemical analyses at the ACME Labs, Canada (Website: http://acmelab.com/tag/bureauveritas/). Samples were crushed in a jaw crusher and agate-milled in the rock preparation laboratory at the Department of Geology of the University of Johannesburg. Major elements were analysed by X-ray fluorescent spectrometry on glass beads prepared from powdered whole-rock samples with a sample-to-flux (lithium tetraborate) ratio of 1:10. Volatiles were determined by loss on ignition. Precision for the different elements is better than ±1% of the reported values. Trace and REEs for these samples were analyzed using inductively coupled plasma mass spectrometry (ICP-MS) also at ACME Labs. The sample powders were dissolved in lithium metaborate flux fusion and the resulting molten bead was rapidly digested in weak nitric acid solution. The precision and accuracy based on replicate analysis of international rock standards are 2–5% (1σ) for most elements and ±10% for U, Sr, Nd and Ni.

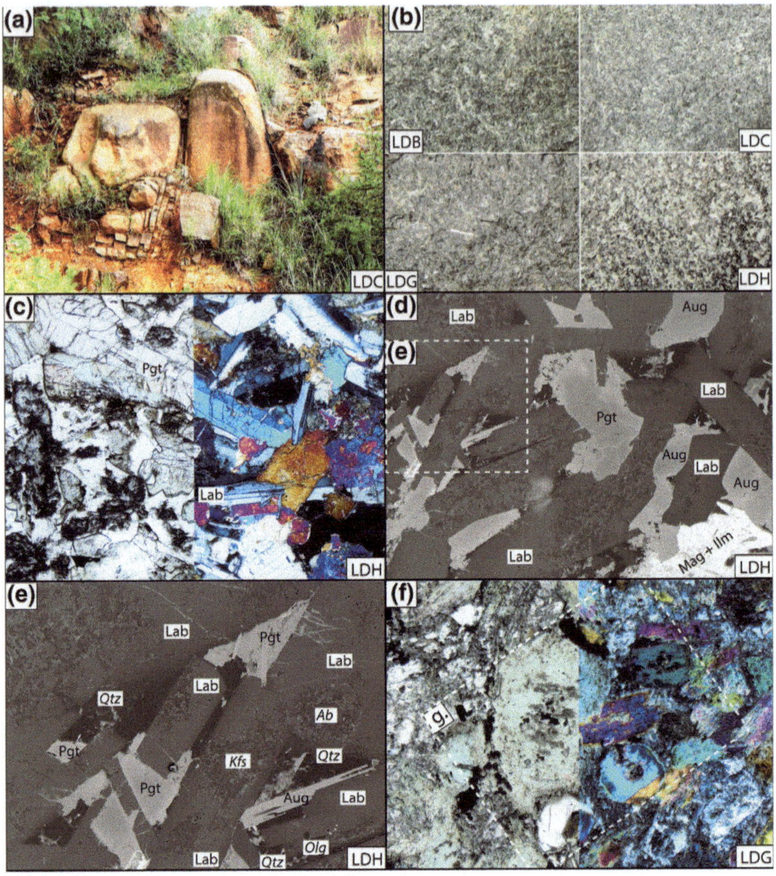

Fig. 2 **a** Typical spheroidal weathering of the Mashishing dykes. Spheroidal corestones are often undisturbed although some may have tilted or rolled, **b** typical examples of fresh cuts of the Mashishing dykes, varying from relatively preserved (LDH and LDB) to weathered (LDG) material **c** to **e** dolerite (dyke LDH) showing an assemblage pigeonite—augite—labrador—magnetite—ilmenite **f** and **g** dioritic dyke (LDG) showing phenocrysts of hornblende and andesine partly albitized **h** to **j** example of altered diorites (dykes LDA, LDD, LDE, LDF, LDI and LDJ) showing globally partly preserved hornblende crystals and totally albitized plagioclase. The primary mineral assemblage (mainly plagioclase—hornblende) has been strongly pseudomorphosed and/or recrystallized to a green-schist assemblage of actinolite—chlorite—albite—epidote **k** to **q** pyroxenite dykes LDB and LDC. These two dykes are ad-cumulates of orthopyroxenes (Opx) and show kaersutite in the intercumulus (see text for more explanations). The Opx grains are completely pseudomorphosed in actinolite, but former cleavages are preserved (**k**) and (**l**). The boundary of the Opx grains are highlighted in black dotted lines in (**k**) the same way as the boundary between kaersutite and actinolite in (**p**). *Ab* = albite; *Act* = actinolite; *Ande* = andesine; Ap = apatite; Aug = augite; *Clc* = clinochlore; *Ep* = epidote; Hbl = hornblende; Ilm = ilmenite; InCu = intercumulus; Krs = kaersutite; *Kfs* = K-feldspar; Lab = labradorite; Mag = magnetite; *Olg* = oligoclase; Pgt = pigeonite; *psd Opx* = pseudomorphosed Opx; *psd Pl* = pseudomorphosed plagioclase; *Ttn* = titanite. Secondary minerals are specified in italics. (**c**) to (**q**): thin section photomicrographs with (**k**), (**m**), (**n**) in PPL mode, (**h**), (**o**) in XPL mode, (**c**), (**f**), (**l**) in mixed PPL and XPL modes, and (**d**), (**e**), (**g**), (**i**), (**j**), (**p**), (**q**) in back scattered image

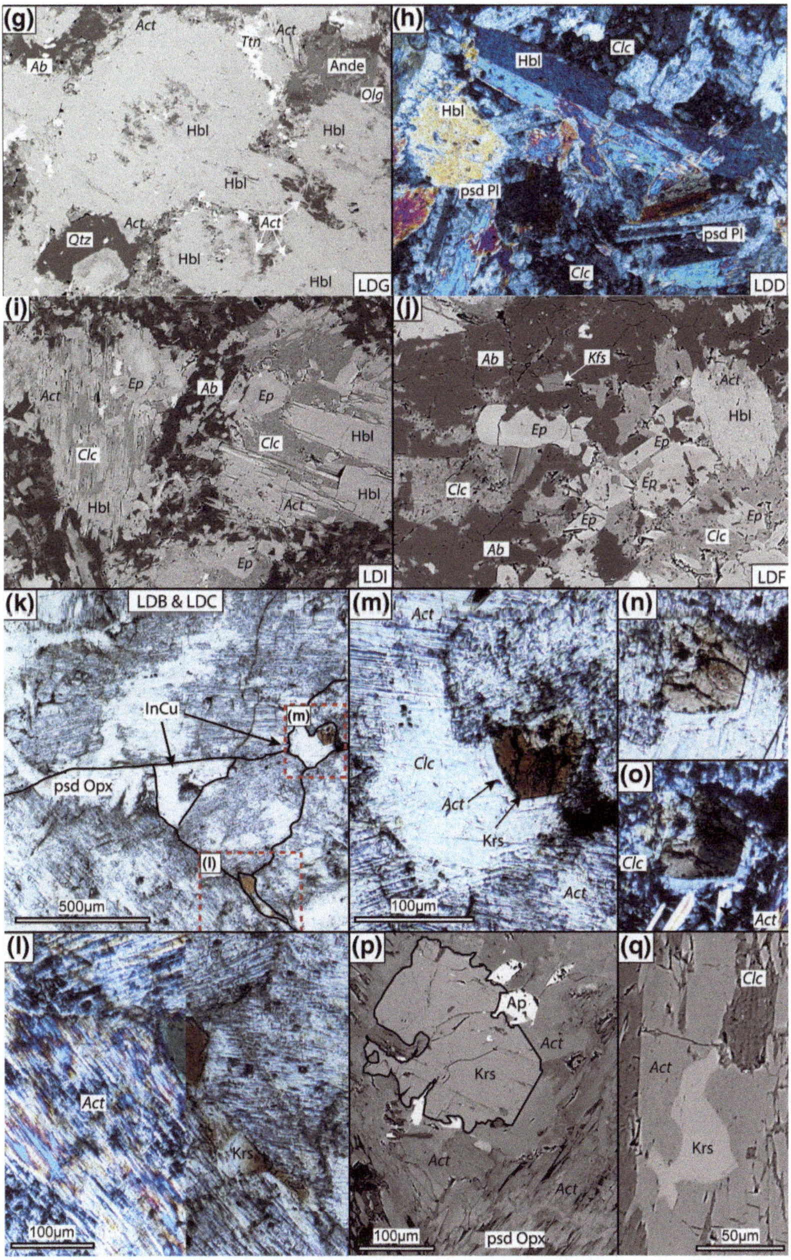

Fig. 2 (continued)

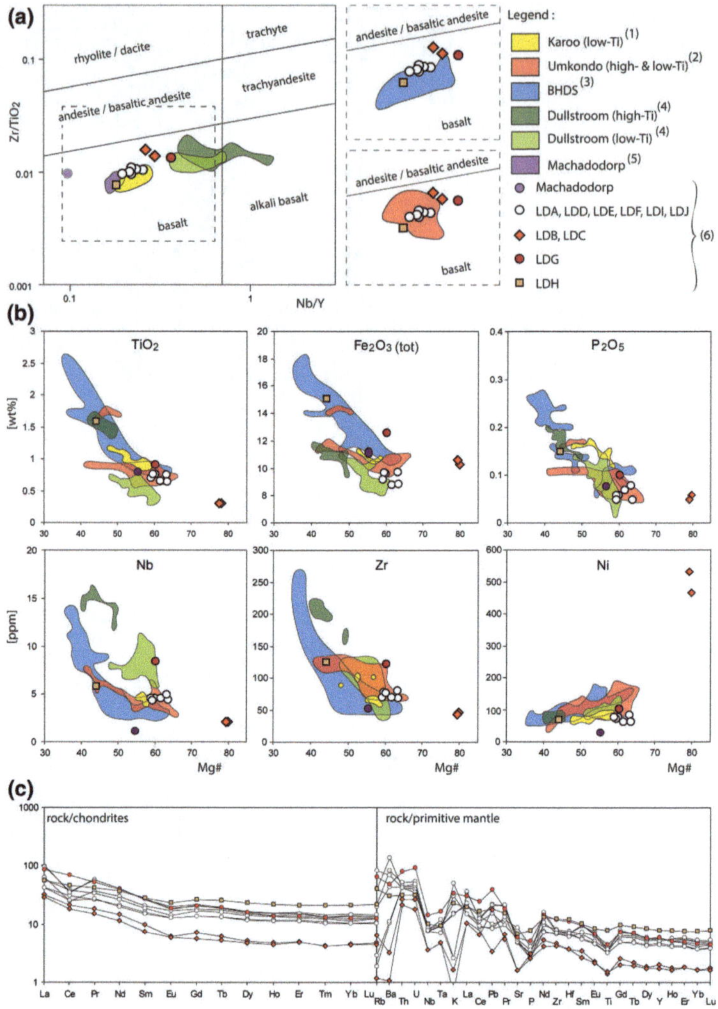

Fig. 3 **a** Nb/Y versus Zr/TiO_2 diagram comparing the Mashishing dykes (LDA to LDJ) to several igneous events chronologically between the Machadodorp Volcanic Member and the Jurassic Karoo LIP. Data are from: (1) Jourdan et al. (2007), (2) Bullen et al. (2012), (3) Klausen et al. (2010) and Olsson et al. (2016), (4) Buchanan et al. (1999), (5) Crow and Condie (1990), (6) this study, **b** same data as above represented in Harker 'like' diagrams of six representative major oxides and trace elements. We preferred P_2O_5 instead of the most commonly used CaO because it was showing a lower dispersion than the later for several data from the literature. Similarly, the Machadodorp samples (Crow and Condie 1990) are excluded due to their high scattering. For samples from the BHDS of Olsson et al. (2016), Mg# values for the 1875–1856 Ma dykes are equal or above 48, and below 48 for the 1859–1835 Ma dykes. The BHDS samples are combined with Klausen et al. (2010)'s sites from coeval dykes to provide a total of 29 samples, **c** REE patterns are normalized to chondritic values and multi-element diagrams normalized to primitive mantle after Palme and O'Neill (2014)

Table 2 Representative examples of composition of main primary and secondary minerals in Mashishing dykes, as estimated through electron microprobe analyses

Mineral	Dyke	SiO_2	TiO_2	Al_2O_3	$FeO_{(tot)}$	MnO	MgO	CaO	Na_2O	K_2O	Cr_2O_3	NiO	Total	Mg#
Kaersutite	LDB	43.3	3.52	10.27	9.9	0.09	14.77	11.55	2.52	1.08	0.12	0.09	97.22	73
Kaersutite	LDB	42.64	4.02	10.28	10.05	0.12	14.7	11.58	2.41	1.11	0.22	0.05	97.18	72
Kaersutite	LDB	42.34	4.46	10.62	10.21	0.13	14.33	11.54	2.41	1.09	0.13	0.06	97.32	71
Kaersutite	LDC	43.32	3.44	10.24	11.01	0.13	14.53	11.5	2.39	1.12	0.1	0.07	97.87	70
Kaersutite	LDC	42.92	3.73	10.5	10.63	0.11	14.38	11.41	2.33	1.11	0.08	0.08	97.27	71
Kaersutite	LDC	43.33	3.79	10.01	10.72	0.13	14.57	11.5	2.45	1.06	0.09	0.07	97.72	71
Kaersutite	LDC	42.33	3.8	10.08	10.97	0.15	14.76	11.42	2.5	1.09	0.1	0.08	97.28	71
Hornblende	LDF	51.35	0.09	5.55	11.83	0.4	14.34	12.67	0.45	0.11	0.07	0	96.86	68
Hornblende	LDF	50.54	0.13	7.05	13.07	0.38	12.96	12.61	0.57	0.15	0.05	0.02	97.52	64
Hornblende	LDF	50.83	0.11	5.96	12.73	0.39	13.53	12.79	0.52	0.12	0.37	0	97.34	65
Hornblende	LDG	49.7	0.06	6.42	15.14	0.25	13.06	12.1	0.46	0.19	0.44	0.04	97.85	61
Hornblende	LDG	49.21	0.11	7.31	15.45	0.26	11.63	12.63	0.67	0.26	0.59	0.03	98.15	57
Hornblende	LDG	47.95	0.09	8.89	15.45	0.24	11.18	12.43	0.81	0.37	0.54	0.01	97.97	56
Hornblende	LDI	50.15	0.1	6.35	12.93	0.25	13.43	12.77	0.61	0.24	0.14	0.02	96.99	65
Hornblende	LDI	51.03	0.08	5.57	13.44	0.23	12.95	12.64	0.53	0.21	0.06	0.02	96.78	63
Actinolite[a]	LDB	57.32	0.02	0.27	6.53	0.16	20.46	13.45	0.07	0.02	0.04	0.06	98.4	85
Actinolite[a]	LDB	56.85	0.03	0.38	6.73	0.15	20.14	13.3	0.1	0.01	0.01	0.05	97.77	84
Actinolite[a]	LDC	57.16	0.03	0.38	7.03	0.16	19.77	13.35	0.09	0.02	0.03	0.05	98.09	83
Actinolite[a]	LDC	56.99	0.04	0.45	6.96	0.14	19.74	13.24	0.08	0.02	0.01	0.06	97.73	83

(continued)

Table 2 (continued)

Mineral	Dyke	SiO_2	TiO_2	Al_2O_3	$FeO_{(tot)}$	MnO	MgO	CaO	Na_2O	K_2O	Cr_2O_3	NiO	Total	Mg#
Actinolite[b]	LDD	53.76	0.02	0.81	13.76	0.37	14.7	12.64	0.08	0.02	0.15	0	96.31	66
Actinolite[b]	LDG	55.05	0.1	0.89	14.73	0.3	14.26	12.84	0.14	0.04	0	0.01	98.37	63
Actinolite[b]	LDG	54.52	0.03	2.02	14.88	0.25	13.89	12.87	0.19	0.08	0.02	0	98.77	62
Actinolite[b]	LDG	53.56	0.04	2.2	15.47	0.27	13.34	12.67	0.27	0.08	0.03	0.02	97.94	61
Actinolite[b]	LDI	54.02	0.13	2.65	12.34	0.27	14.58	12.65	0.27	0.09	0.02	0.01	97.03	68
Actinolite[b]	LDI	54.52	0.04	2.21	10.95	0.25	15.78	13.14	0.2	0.08	0.04	0.02	97.21	72
Pigeonite	LDH	53	0.28	1.4	17.59	0.37	21.82	5.16	0.08	0	0.06	0.05	99.81	69
Pigeonite	LDH	51.4	0.47	1.47	23.43	0.46	17	5.76	0.1	0	0.02	0.03	100.15	56
Pigeonite	LDH	50.23	0.48	1.12	28.3	0.59	13.51	5.76	0.1	0.01	0.02	0.02	100.14	46
Pigeonite	LDH	50.26	0.49	1.08	27.26	0.55	13.95	6.14	0.08	0.01	0.02	0.01	99.84	48
Pigeonite	LDH	51.13	0.57	1.61	22.33	0.47	16.36	7.66	0.13	0	0.01	0.03	100.31	57
Augite	LDH	50.76	0.66	1.84	17.43	0.36	13.87	14.66	0.23	0	0	0.02	99.84	59
Augite	LDH	51.71	0.49	2.49	12.04	0.31	16.95	15.5	0.24	0	0.14	0.03	99.91	72
Augite	LDH	50.36	0.75	2.16	17.05	0.36	13.2	15.86	0.28	0	0.02	0.02	100.07	58
Augite	LDH	49.91	0.85	2.68	15.23	0.31	13.4	17.1	0.32	0.01	0.04	0.02	99.88	61
Augite	LDH	51.18	0.53	2.91	10.74	0.26	16.13	17.66	0.28	0.01	0.17	0.03	99.9	73
Andesine	LDG	57.62	0.06	25.87	0.62	0.03	0.05	8.75	6.35	0.6	0	0.01	99.95	
Andesine	LDG	58.81	0.05	24.79	0.55	0	0.03	7.93	6.84	0.69	0	0	99.69	
Andesine	LDG	57.47	0.04	25.37	0.56	0.01	0.05	8.73	6.43	0.6	0	0.01	99.28	
Labradorite	LDH	54.04	0.07	27.97	0.63	0.02	0.07	11.68	4.84	0.33	0.01	0	99.65	
Labradorite	LDH	52.44	0.05	28.83	0.89	0.01	0.07	12.66	4.32	0.29	0.01	0.01	99.58	

(continued)

Table 2 (continued)

Mineral	Dyke	SiO_2	TiO_2	Al_2O_3	$FeO_{(tot)}$	MnO	MgO	CaO	Na_2O	K_2O	Cr_2O_3	NiO	Total	Mg#
Labradorite	LDH	51.7	0.05	29.12	0.66	0.01	0.13	13.2	4.01	0.23	0.01	0.02	99.16	
Labradorite	LDH	53.2	0.05	28.45	0.67	0	0.1	12.31	4.41	0.3	0	0	99.5	
Labradorite	LDH	53.83	0.06	27.91	0.62	0	0.09	11.82	4.8	0.35	0.01	0.02	99.5	
Labradorite	LDH	52.76	0.06	28.72	0.67	0	0.1	12.6	4.28	0.3	0.01	0	99.5	
Labradorite	LDH	53.28	0.06	28.29	0.67	0.01	0.1	12.15	4.48	0.32	0	0.01	99.36	
Epidote	LDF	38.75	0.17	29.4	5.12	0.22	0.1	24.24	0.01	0	0.01	0	98.05	
Epidote	LDF	39.07	0.13	30.46	4.52	0.45	0.09	23.68	0.02	0.04	0.06	0	98.52	
Epidote	LDG	38.54	0.3	27.86	6.12	0.08	0.02	24.12	0.03	0.02	0.19	0	97.27	
Epidote	LDG	39.02	0.19	28.48	5.93	0.18	0.04	24.02	0.02	0.01	0.05	0.02	97.94	
Epidote	LDI	38.86	0.1	29.24	5.26	0.15	0.03	24.14	0.02	0	0.07	0.01	97.9	
Epidote	LDI	39.19	0.19	28.93	5.43	0.05	0.03	24.45	0.01	0.01	0.03	0	98.32	

Actinolite[a] = actinolites in dykes LDB and LDC; actinolite[b] = actinolites in all other dykes

4.2 U–Pb Geochronology

Baddeleyite was successfully extracted from the dyke LDH following the method of Söderlund and Johansson (2002) at Lund University (Sweden). After extraction, grains of baddeleyite were selected and combined into fractions of between 1 and 7 grains. Fractions were washed in 2–3 M HNO_3 on a hotplate and rinsed repeatedly in H_2O. A small amount of a ^{205}Pb–$^{233-236}U$ tracer solution was added before baddeleyite fractions were dissolved completely under high pressure and a temperature of ~210 °C. Dissolved fractions were dried, and redissolved in 3.1 M HCl and loaded on 50-μL columns filled with precleaned anion exchange resin (Bio-Rad 200–400 Mesh chloride). U and Pb were washed out with H_2O and collected in Tefloncapsules. A small amount of H_3PO_4 was added to each capsule before they were put on a hotplate and allowed to evaporate overnight. U and Pb were loaded on the same Re filament together with a small amount of silica gel. The mass spectrometry of sample fractions was performed using a Thermo Finnigan Triton thermal ionisation multicollector mass spectrometer at the Laboratory of Isotope Geology at the Swedish Museum of Natural History in Stockholm. The intensities of ^{208}Pb, ^{207}Pb, ^{206}Pb, ^{205}Pb were monitored during analyses with Faraday collectors. An ETP SEM equipped with a RPQ filter was used to measure the intensity of ^{204}Pb. SEM and Faraday gain was controlled by measuring a ~5–10 mV signal between runs. Measurements of Pb-isotopes were done at a filament temperature in the 1200–1320 °C range, while isotopes of U were measured in dynamic mode on the SEM with filament temperatures exceeding 1350 °C. Data reduction was performed using the Excel add-in "Isoplot" of Ludwig (2003); decay constants for ^{238}U and ^{235}U follow those of Jaffey et al. (1971). All errors in age and isotopic ratios are quoted at the 95% confidence level. Initial Pb isotope compositions were corrected for according to the global common Pb evolution model of Stacey and Kramers (1975).

4.3 $^{40}Ar/^{39}Ar$ Geochronology

Mineral aliquots of amphibole were obtained after crushing and handpicking of single grains under a binocular microscope. The minerals were repeatedly cleaned ultrasonically in distilled water and subsequently wrapped in light aluminum foils. The foils were subsequently vacuum-sealed in a ca. 15 cm long silica glass tube of 8 mm inner diameter as part of a 50 mm long pin and irradiated at the NTP radioisotopes SAFARI1 nuclear reactor at Pelindaba, South Africa, for 20 h in position B2 W (not cadmium-shielded) with the reactor running at 20 MW. Four ~2 mg aliquots of each subsample were analysed by stepwise heating, using a defocused beam from a continuous Nd-YAG 1064 nm laser and a MAP 215-50 noble gas mass spectrometer at the central analytical facility (Spectrum) of the University of Johannesburg. Signals were measured on a Johnston focused-flow electron multiplier providing excellent linearity in analogue mode. The Fish Canyon sanidine (28.201 ± 0.046 Ma) (Kuiper et al. 2008), Hb3gr amphibole (1080.40 ± 1.10 Ma) (Renne et al. 2010), as well as

McClure Mountain amphibole (523.1 ± 2.6 Ma) (Renne et al. 1998) standards, were used as monitors, yielding within uncertainty limits, identical J-values. A slight flux gradient was monitored and taken into account by placing standards at the bottom and top of the irradiation package. The value for the ^{40}K decay constant derived by Kossert and Günther (2004) was utilized. Blank measurements were obtained initially after every fourth step run. Measurement control and data reduction were carried out using an in-house software suite that includes full error propagation by Monte Carlo procedures. Age uncertainties in plateaus are shown at the 95% (2σ) confidence level. The obtained plateau ages yielded relatively large errors due to low K contents and correspondingly low ^{39}Ar signals (see Table 6). Corrections for ^{39}Ar and ^{36}Ar produced by ^{40}Ca as well as ^{40}Ar produced by ^{40}K in irradiation were made via calibrations using CaF and K_2SO_4 (see Table 6).

4.4 Paleomagnetism and Rock Magnetism

Oriented cores were cut into ~2 cm long specimens and subjected to stepwise demagnetization. All measurements were made with a vertical 2G Enterprises DC-4 K superconducting rock magnetometer with automated sample changer housed at the University of Johannesburg in South Africa. Following a measurement of natural remanent magnetization (NRM) and a low-field-strength alternating field (AF) cleaning of specimens in four steps up to 10 mT, a thermally demagnetization was performed using a ASC TD48-SC shielded furnace at decreasing intervals from 100 up to 580 °C where the specimen usually displays an erratic behaviour. Magnetic components were identified and quantified via least-squares principal component analysis (Kirschvink 1980) utilizing Paleomag 3.1.0 (Jones 2002).

Thermomagnetic experiments were performed to determine the potential carrier minerals of the magnetic remanence. The variation of magnetic susceptibility during heating was investigated on powdered dyke samples in air from ambient temperature to 700 °C using the AGICO CS-3 furnace in connection with the MFK1-FA kappabridge housed at the Council for Geoscience in Pretoria (South Africa).

5 Results

5.1 Petrology

Thin sections reveal that all dykes invariably experienced greenschist facies metamorphism (actinolite + chlorite + albite + epidote) (Fig. 2h–j). Yet, in general, magmatic textures are preserved (Fig. 2c–e), no schistosity or metamorphic foliation were observed, but most primary minerals are altered or replaced by secondary minerals (Fig. 2k–q). In several samples, we observed evidence of K-and Na-metasomatism and silicification (see next sections). When present, plagioclase is affected by albitisation which is a common alteration feature of many Paleoproterozoic-aged rocks in the Transvaal sub-basin (e.g., Humbert et al. submitted). Silicification was observed in all dykes, mixed with other alteration features in most samples, but could be quantified in the least altered sample LDH. From what precedes, it becomes apparent that the normative calculation cannot be used as reliable proxy for both the mineralogical composition and proportion for the dykes.

The dykes LDA, LDD, LDE, LDF, LDI, LDJ share a common intergranular texture as well as a similar mineralogy primarily represented by hornblende and plagioclase (Fig. 2h–j), in addition to euhedral phenocrysts of pyroxene in the case of the dyke LDA. In all the dykes, albite or albite-rich oligoclase occur as remnants of primary plagioclase. K-feldspar is commonly found associated within altered plagioclase (Fig. 2h–j), but not always. The primary ferromagnesian minerals, hornblende and pyroxenes are pseudomorphosed to actinolite (Fig. 2h–j) and clinochlore; only remnants of hornblende were observed. The hornblendes are likely primary because rocks in the Pretoria Group have only experienced lower greenschist facies of metamorphism (e.g. Eriksson et al. 2006). The secondary mineral assemblage of our samples confirms the greenschist facies. In the sample LDF, the clinochlore occurs as abundant crystals that can reach up to 80 μm (Fig. 2j). The actinolites in all six dykes (and also in dyke LDG) have roughly the same composition showing similar Mg# (with Mg# = [molar] MgO/(MgO + FeOtot)) (Table 2). Epidote is present in all of the dykes. This mineral is likely related to hydrothermal alteration, and crystallised using the calcium released from either the alteration of the hornblende (Fig. 2i) or to albitisation of plagioclase (Fig. 2j). The presence of large quantity of hornblende suggests that the six dykes are diorites although the total albitisation of plagioclase (Fig. 2i, j) prevents us to draw a clear conclusion concerning the lithology.

The dyke LDG is similar in composition and texture to the six dykes described above, but displays hornblende in higher abundance and size. These grains are frequently euhedral and relatively well preserved. Actinolite often shows an epitaxial crystallisation around it together with some occurrences of titanite (Fig. 2g). In this dyke, plagioclase is generally quite altered towards albite or oligoclase (albitisation), yet, several remnants of andesine in some crystal cores were characterized (Fig. 2g; Table 2), with an albite fraction of 58% on average. The proportions of different minerals in the dyke sample are ~65–70% of hornblende, ~30% of plagioclase and ~1–2% opaque minerals. The occurrence of andesine suggests that the dyke LDG corresponds to a hornblende-rich diorite.

Sample of dyke LDH is largely unweathered with primary igneous minerals that show no to moderate alteration (Fig. 2c–e). The texture of this dyke is subophitic with euhedral labradorite laths (An% between 51 and 64, with an average of An% = 58 on 15 crystals) that are diversely oriented and surrounded by two types of clinopyroxenes (i.e., pigeonite and augite). The pigeonite and augite, which occur in equal proportions, are mostly anhedral (Fig. 2d–e; Table 2). However, some early, subhedral pigeonite phenocrysts are also observed (Fig. 2c). The observations then suggest that crystallization of clinopyroxene (at least augite) was completed after that of the plagioclases. The alteration of pyroxenes generally gives raise to chlorite represented by chlorine. Exceptionally, we found a brunsvigite (another chloritic mineral) grain inside a magnetite-ilmenite grain in the dyke LDH (shown in Fig. 2d). The labradorites, whose presence suggests the dyke to be gabbroic in composition, show a progressive zoning with albite fraction slightly evolving from the core to the rim. Following its composition and texture, this dyke can be classified as dolerite. Large titanomagnetite crystals (i.e., up to 1 mm wide) are frequent and always contain intergrowth of ilmenite laths. Crystals of plagioclase and titanomagnetite occur in a sort of "competing" manner (complementary contacts and shapes), suggesting that the two minerals crystallized simultaneous. Some symplectites of augite and ilmenite, suggesting exsolution between the two minerals (e.g., Moseley 1984), locally occur along grain boundaries. The proportions of different minerals in the dyke LDH are ~55% of clinopyroxenes, ~40% of plagioclase and ~5% titanomagnetite + ilmenite. Quartz is present in the rock as result of localized silicification of some primary minerals, which are often ferromagnesian (Fig. 2e).

The dykes LDB and LDC, clearly differ from the other samples in terms of texture, mineral composition as well as Mg number (Mg#) (Table 3). In addition, no plagioclase was observed in these dykes, making them ultramafic. Thin sections reveal an ad-cumulate texture. The cumulate is composed by large sub-euhedral to euhedral prismatic crystals equivalent in size (between 400 μm to mm) now totally pseudomorphosed to actinolite, but still showing their characteristic primary cleavage planes (Fig. 2k–m). Based on the cleavage and the high Mg# of the samples, we interpret this/these mineral(s) to be originally pyroxene(s), probably orthopyroxene(s). Thus, the two dykes can be classified as pyroxenites. All of the actinolites in both dykes are magnesium-rich because of the high Mg# of the cumulates, and significantly differ from actinolites of other samples (Table 2).

Primary phases of the intercumulus, which represent a small volume of the rock, are intensively altered to actinolite and chlorite, but several brownish amphiboles remain preserved (Fig. 2l–q). Based on their composition (Table 2) these primary brownish amphiboles, only found in the intercumulus, are clearly kaersutites as they are rich in K (~1 wt%) and also have a significant high Ti content (3.2–4.5 wt%) comparable to that of a pure kaersutite (TiO_2 is between 5 and 10 wt%; Deer et al. 1966). Note that all other amphiboles were analyzed (152 measurements) to yield TiO_2 content lower than 0.5 wt%. Kaersutites present rhombic basal sections that show a strong pleochroism (e.g., top of Fig. 2m, n) and sometimes twinning (Fig. 2o). Kaersutite crystals are generally well preserved, although partly pseudomorphosed to actinolite (e.g., top of the kaersutites in Fig. 2m, p), and also occur as base of epitaxial crystallisations of actinolite (e.g., bottom and left parts of the kaersutite in Fig. 2m, or in Fig. 2q). Kaersutite, some rare apatite (Fig. 2p) and titanomagnetite (identified during paleomagnetic analyses) are the only remaining primary magmatic minerals in the dykes LDB and LDC.

Table 3 Major and trace element concentrations in the Mashishing dykes (LDA to LDJ) and the Machadodorp volcanics (HWC1 and HWC2)

Dyke/Sample		LDA	LDB	LDC	LDD	LDE	LDF	LDG	LDH	LDI	LDJ	HWC1	HWC2
SiO_2	(wt%)	51.58	48.56	47.08	53.2	53.44	52.45	53.84	49.56	51.66	51.73	51.46	51.8
Al_2O_3	(wt%)	14.97	7.57	8.21	14.57	13.72	15.01	9.08	13.82	14.41	14.56	13.5	13.32
Fe_2O_3	(wt%)	8.88	10.3	10.61	8.83	9.76	9.19	12.61	15.11	9.71	9.67	11.66	11.57
MgO	(wt%)	7.79	20.58	20.61	7.14	8.45	6.73	9.68	6.03	7.21	7.3	7.44	7.32
CaO	(wt%)	9.56	5.58	5.5	7.19	7.06	8.35	9.14	9.76	10.29	10.18	11.59	11.42
Na_2O	(wt%)	1.68	0.12	0.13	4.05	3.02	3.82	1.42	2.36	2.29	2.29	1.49	1.47
K_2O	(wt%)	1.56	0.02	0.05	0.93	0.47	0.45	1.04	0.71	0.08	0.08	0.22	0.22
TiO_2	(wt%)	0.65	0.29	0.31	0.66	0.75	0.71	0.91	1.59	0.77	0.77	0.85	0.85
P_2O_5	(wt%)	0.05	0.06	0.05	0.07	0.08	0.05	0.1	0.15	0.06	0.06	0.07	0.07
MnO	(wt%)	0.14	0.14	0.16	0.15	0.18	0.17	0.19	0.22	0.17	0.17	0.18	0.18
Cr_2O_3	(wt%)	0.058	0.495	0.478	0.056	0.077	0.05	0.145	0.027	0.041	0.041	0.052	0.054
Ni	(ppm)	120	855	879	93	111	96	202	70	115	119	121	117
Sc	(ppm)	34	22	22	35	38	36	27	40	38	38	37	37
LOI		2.8	5.8	6.3	2.9	2.7	2.8	1.5	0.4	3.1	2.9	1.3	1.5
Sum		99.76	99.63	99.62	99.73	99.73	99.78	99.7	99.74	99.78	99.78	99.8	99.81
Mg#		63.48	79.84	79.38	61.58	63.18	59.21	60.34	44.16	59.54	59.94	55.84	55.63
CIA		40.67	42.21	44.43	41.26	42.83	40.61	31.14	38.16	38.97	39.43	36.23	36.25
Ba	(ppm)	461	7	21	916	553	289	324	208	69	75	51	46
Co	(ppm)	32.4	89.1	87.4	28.7	26.1	38	63.6	48.3	28.1	27.6	44.4	45.4

(continued)

Table 3 (continued)

Dyke/Sample		LDA	LDB	LDC	LDD	LDE	LDF	LDG	LDH	LDI	LDJ	HWC1	HWC2
Cs	(ppm)	1.2	0.3	0.9	0.5	0.3	0.3	2	6.8	0.1	0.2	0.8	0.9
Ga	(ppm)	15.1	7.7	8.2	13.7	14.3	15.4	13.8	19.2	16.2	15.9	11.8	12.4
Hf	(ppm)	1.9	1.1	1.1	2.1	2.2	1.8	3.3	3.5	2.1	2.1	1.5	1.6
Nb	(ppm)	4.4	2.1	2.1	4.6	5	4.6	8.4	5.8	4.4	4.6	1.8	1.7
Rb	(ppm)	48.8	0.7	3.7	23.9	13.1	11.6	38.7	24.3	1.1	1.7	6.7	7
Sr	(ppm)	118	33.6	33.6	116	184.1	100.8	124.4	148.2	100.6	105.4	49	48.4
Ta	(ppm)	0.3	0.2	0.2	0.4	0.4	0.4	0.7	0.4	0.4	0.5	0.2	0.1
Th	(ppm)	3.5	1.8	2.2	3.5	4.3	3.6	6.7	2.6	3.8	3.7	0.6	0.7
U	(ppm)	0.9	0.4	0.6	0.8	1.2	1	2.1	0.7	1.1	1.1	<0.1	<0.1
V	(ppm)	233	135	147	246	268	251	221	338	259	253	274	272
W	(ppm)	0.7	<0.5	0.5	1	1.1	0.8	0.8	<0.5	0.7	0.9	0.6	<0.5
Zr	(ppm)	68.6	45.7	42.9	70.3	81.2	70.2	121.4	125.3	77.2	77.5	54.7	54.9
Y	(ppm)	19.1	8	7.1	18.1	22.8	23.6	23.3	31.1	20.3	20.9	19.3	19.5
La	(ppm)	10.1	7.6	6.9	10.1	14	15.2	20.6	13.2	23.3	24.2	4.2	4
Ce	(ppm)	19	13.2	11.2	15.9	23.8	15.4	42.8	28.7	22	21.3	9.6	9.5
Pr	(ppm)	2.5	1.68	1.4	2.52	3.16	3.48	4.85	3.96	5.35	5.48	1.31	1.33
Nd	(ppm)	10	6.9	5.3	9.5	12.6	14.1	18.9	17.5	19.4	20.7	6.3	6.4
Sm	(ppm)	2.42	1.48	1.14	2.3	2.86	3.14	4.15	4.34	4.07	4	1.84	1.95
Eu	(ppm)	0.76	0.36	0.35	0.75	0.84	1.01	1.11	1.36	1.07	1.18	0.69	0.73
Gd	(ppm)	2.8	1.48	1.16	2.8	3.49	3.71	4.23	5.47	4.39	4.46	2.69	2.85
Tb	(ppm)	0.51	0.23	0.2	0.5	0.61	0.66	0.73	0.96	0.7	0.72	0.47	0.47

(continued)

Table 3 (continued)

Dyke/Sample		LDA	LDB	LDC	LDD	LDE	LDF	LDG	LDH	LDI	LDJ	HWC1	HWC2
Dy	(ppm)	3.08	1.29	1.23	2.97	3.61	3.87	4.05	5.95	4.09	4.18	3.3	3.29
Ho	(ppm)	0.66	0.27	0.25	0.63	0.78	0.8	0.78	1.23	0.87	0.87	0.7	0.7
Er	(ppm)	1.93	0.81	0.82	1.87	2.15	2.46	2.24	3.47	2.54	2.67	2.11	2.16
Tm	(ppm)	0.29	0.11	0.11	0.27	0.32	0.35	0.33	0.55	0.35	0.38	0.28	0.28
Yb	(ppm)	1.73	0.75	0.76	1.74	1.94	2.27	2.12	3.54	2.35	2.5	1.95	2.05
Lu	(ppm)	0.27	0.11	0.12	0.26	0.31	0.32	0.31	0.54	0.34	0.36	0.28	0.29
Mo	(ppm)	3.1	1.2	1.3	3.7	3.8	2.1	3.7	0.5	3.2	3.5	4.7	4.9
Cu	(ppm)	73.7	58.1	38.5	63.2	57.3	101	112.7	192	77.5	77.9	40.8	42.3
Pb	(ppm)	2.8	0.6	3.2	1.5	1.8	1.6	7.2	3.3	4	3.9	0.7	0.6
Zn	(ppm)	29	39	40	28	43	24	28	41	46	49	19	19
Ni	(ppm)	64.4	465.8	530.7	63.7	85.6	77.2	102.5	22.4	75.8	74.7	41.8	42.5

5.2 Geochemistry

The geochemical results are presented in Table 3. The chemical index of alteration ($CIA = [Al_2O_3/(Al_2O_3 + CaO + Na_2O + K_2O)] \times 100$) (Nesbitt and Young 1982) which measures the extent to which feldspars have been converted to aluminous clays varies from ~31 to a maximum of 44 for the studied dykes. For unaltered basaltic rocks, CIA values range from 30 to 45 while clays are characterized by CIA values above 75. This may signify that the samples of the present study have not been significantly affected by weathering and can be used to investigate the magma parentage of these dykes. Due to the alkali and silica mobility observed in the least altered dykes and suspected in all other samples, the Zr/TiO_2 versus Nb/Y plot (Fig. 3a) is preferred to the more traditionally used TAS diagram (Cox et al. 1979). Although silicification will inevitably also affect concentrations of other elements, ratios such as the magnesium number (Mg#) will not be significantly influenced. We therefore used Mg# as a differentiation index in the variation diagrams (Fig. 3b).

Variation of rare earth element or REE abundances and variation of selected trace elements (i.e., multi-element diagrams) respectively normalized to chondrites and primitive mantle after Palme and O'Neill (2014) do not only show different levels of enrichment, but also dissimilar shapes (Fig. 3c).

Four of the dioritic dykes (LDA, LDD, LDE, and LDF) show similar petrography and are so similar in the bivariate diagrams and trace element plots that they could derive from a common magma. The dyke LDG shares a similar Mg# with many other sites (Table 3), but differs from those by having a relatively high Zr/TiO_2 ratio and important Fe_2O_3, Nb and Zr contents (Fig. 3a). Major and trace element plots of dyke LDG (hornblende-rich diorite; Mg# = 60) and dyke LDH (dolerite; Mg# = 44) are as dissimilar as their petrography and cannot be explained by differentiation by crystal fractionation. The two dykes are distinctly different geochemically from the other samples and it seems plausible that the magmas of these two dykes formed from distinctly different melts.

The dykes LDB and LDC are characterized by a high MgO content (~20 wt%) (Table 3) and an Mg# close to 80. The two dykes also exhibit low contents of TiO_2, Fe_2O_3, and highly incompatible elements such as Zr. In contrast, they show an enrichment in highly compatible elements (e.g., Ni, Cr or Co), which highlights an insignificant to no fractionation of olivine or spinel from the parental magma. The high LOI of the two dykes (~6 wt%) may be due to the replacement of primary ferromagnesian minerals by amphiboles and chlorites during metamorphic processes. Their low concentrations of both TiO_2 and Fe_2O_3 (Fig. 3b) can be linked to the relative high concentration of MgO. It could, however, also be attributed to an early differentiation of titanomagnetite from the parental magma, as it occurs in the arc magmatism (e.g., Elburg 2010). But such a magmatism is not expected in the central part of the KC. From what precedes, it can be said both pyroxenite dykes LDB and LDC are petrographically and geochemically similar and likely share a common magma.

The dykes LDI and LDJ were represented in Fig. 3 together with the four dykes LDA, LDD, LDE and LDF based on similar concentrations in respective elements, overlap in bivariate diagrams and comparable Mg#. Yet, an important difference between the two groups is that normalized REE ratios of the dykes LDI and LDJ are higher than those of the four other dykes (Table 4). It is interesting to see that such higher REE ratios correspond well with the values calculated for the dykes LDB and LDC (Fig. 5i; Table 4). The multi-element diagrams further confirm that the dykes LDB, LDC, LDI and LDJ share a similar pattern of trace elements, but in the exception of the Zr, Hf and P (Fig. 5j). The apparent negative anomalies in Zr and Hf elements in the dykes LDI and LDJ is certainly related to a differential enrichment relative to the REE during the crystal fractionation process. The negative anomaly in P is either related to fractionation of apatite or monazite from the parental magma, to another differential enrichment during the fractionation, or due to weathering process.

It is likely that the pyroxenite dykes LDB and LDC represent an ultramafic or undifferentiated mafic magma (20–40 times chondrites values; Fig. 5i, j) that later evolved into a differentiated liquid (50–100 times the chondrites values; Fig. 5i, j) from which the dykes LDI and LDJ crystallized. The magma differentiation likely occurred by crystal fractionation (50–100 times the chondrites values) of both olivine and pyroxene as suggested by the high differences of Cr, Ni, and Co between the primitive and the evolved two groups. The higher TiO_2 and Fe_2O_3 contents of samples LDI and LDJ relative to the pyroxenite dykes LDB and LDC argues for no fractionation of the titanomagnetite (Fig. 3b).

Given the variability in the dyke samples it is not productive to discuss positive or negative anomalies in the trace elements patterns, such as the negative anomalies in Ti (Fig. 3c), which can either be related to fractionation of titanomagnetite or ilmenite or to a specificity during the partial melting (Ti more compatible in the solid residue). A higher number of samples representing different degrees of crystal fractionation are required to interpret such anomalies correctly. The slightly negative Ce-anomaly in some of our dyke samples is certainly due to the alteration as this element is notably redox sensitive (e.g., Middelburg et al. 1988).

5.3 Geochemical Comparisons to Strata-Bounded Igneous Events in the KC

The above-defined sample groups were compared with selected geochemical data from known strata-bounded igneous events in the KC. Our sampling area is adjacent to the eastern limb of the ~2.05 Ga Rustenburg Layered Suite of the Bushveld Complex. However, major and trace elements data obtained in the present study are collectively dissimilar to the compositional range of B1, B2 and B3 Bushveld parental magmas (Barnes et al. 2010) (Fig. 4; Table 4), controverting any genetic relationship to these. We did not compare our data with the Hekpoort lavas which represent the most important magmatic unit of the Pretoria Group, as these volcanics are located

Table 4 Trace element ratios concerning the multi-element diagrams for the studied dykes and other igneous rocks in the KC

		$(La/Lu)_N$		$(La/Sm)_N$		$(Sm/Er)_N$		$(Er/Lu)_N$	
Formations or samples	n	Mean	SD	Mean	SD	Mean	SD	Mean	SD
Karoo low-Ti	7	2.50	0.31	1.77	0.21	1.38	0.10	1.02	0.09
Umkondo low-Ti	11	4.28	0.62	2.66	0.18	1.62	0.14	0.99	0.03
BHDS 1859–1835 Ma	4	2.99	0.57	1.93	0.19	1.44	0.12	1.08	0.04
BHDS 1875–1856 Ma	7	2.28	0.29	1.69	0.15	1.27	0.05	1.06	0.04
BHDS ~ 1.90 Ga	16	2.59	1.02	1.70	0.46	1.39	0.19	1.06	0.04
Bushveld RLS B3	8	3.73	1.28	2.28	0.57	1.67	0.27	0.96	0.05
Bushveld RLS B2	14	5.85	0.85	2.52	0.23	2.28	0.20	1.02	0.06
Bushveld RLS B1	10	10.97	2.84	4.18	0.40	2.89	0.37	0.90	0.09
Bushveld RLS B1 UM	7	10.68	1.56	4.11	0.34	2.87	0.30	0.91	0.04
Dullstroom low-Ti	12	7.07	1.62	3.27	0.53	–	–	–	–
Dullstroom high-Ti	9	8.20	1.62	2.67	0.23	–	–	–	–
Machadodorp Lava Member [a]	5	1.59	0.71	2.32	1.15	–	–	–	–
Machadodorp Lava Member[b]	2	1.49	0.09	1.38	0.10	0.96	0.02	1.13	0.01
Mashi. dykes LDA–LDF	4	4.38	0.50	2.91	0.22	1.37	0.05	1.10	0.05
Mashi. dykes LDH	1	2.53	–	1.94	–	1.35	–	0.97	–
Mashi. dykes LDG	1	6.89	–	3.16	–	2.00	–	1.09	–
Mashi. dykes LDB–LDC	2	6.56	0.85	3.56	0.41	1.73	0.33	1.07	0.06
Mashi. dykes LDI–LDJ	2	7.04	0.10	3.75	0.15	1.67	0.08	1.13	0.01

n = number of samples, *SD* = standard deviation, *UM* = ultramafic
[a] and [b] indicate Machadodorp results from Crow and Condie (1990) and from this study respectively

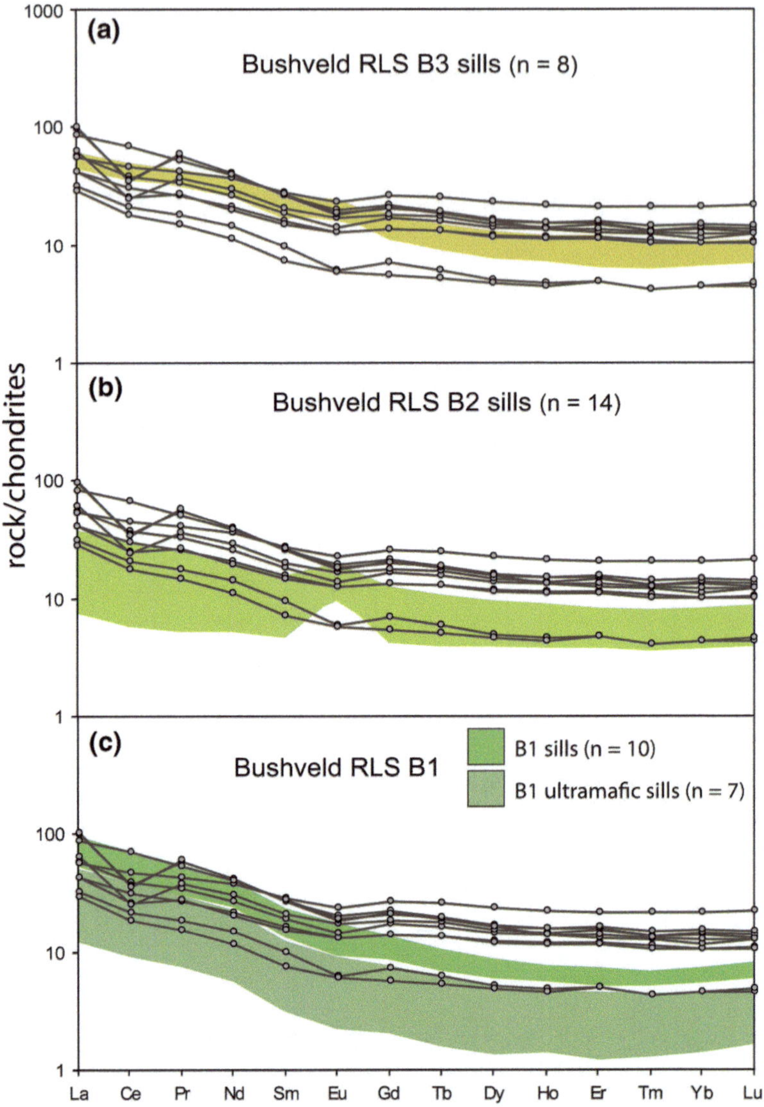

Fig. 4 REE diagrams of the Mashishing dykes (solid lines) showing no resemblance to the colored backgrounds areas representing the compositional range of B1 ultramafic, B1, B2 and B3 Bushveld magmas (Barnes et al. 2010). All data are normalized to chondrite and primitive mantle (Palme and O'Neill 2014). n = number of samples. Accordingly, trace element ratios of the two groups show significant differences (see Table 4)

stratigraphically below the Silverton Formation, thus ruling out a genetic relationship with our dykes. Figure 5 depicts the similarities revealed in trace elements patterns while Table 4 summarizes important ratios for REE profiles.

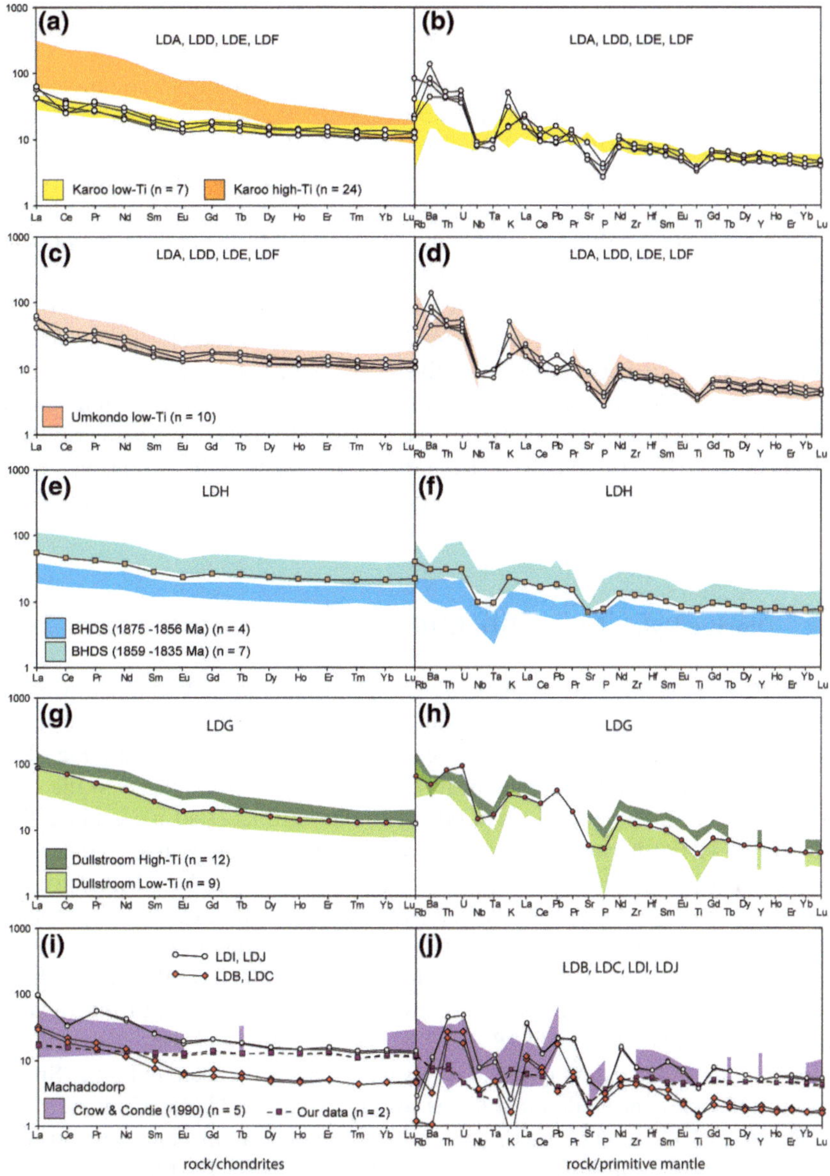

Fig. 5 REE patterns normalized to chondritic values and multi-element diagrams normalized to primitive mantle after Palme and O'Neill (2014) comparing the chemistry of Mashishing dykes (solid lines) with respect to the composition of selected magmatic units in the KC. Data used are as per Fig. 3a, b except for the BHDS that is represented by data from Olsson et al. (2016) only

5.3.1 Dolerite Dyke LDH

In the bivariate diagrams, the dyke LDH compares well with different igneous units, but the best correlation remains with the BHDS of Olsson et al. (2016) (Fig. 3a, b). The correlation of our sample with the BHDS is further confirmed by similar REE ratios (Table 4), and REE and multi-element diagrams that are sub-parallel (Fig. 5e, f). It is interesting to see that the trace element patterns of our sample plot in between the 1875–1856 Ma primitive group and the 1859–1835 Ma differentiated group of the BHDS (Fig. 5e, f). The dyke LDH shows a high TiO_2 content (1.59 wt%) compared to the values (0.29–0.91 wt%) exhibited by the other studied dykes (Table 3). Dyke LDH's TiO_2 content, however, compares well with Olsson et al. (2016)'s values and again, places our site between the primitive (TiO_2 is 0.77–1.31 wt% with mean value of 1.02 wt% $\pm$0.26) and the more evolved (TiO_2 is 1.66–3.01 wt% with mean value of 2.21 wt% $\pm$0.49) dykes of the BHDS. Note that the geochemical data provided by Klausen et al. (2010) (not showed) from dykes that enclose the two groups defined by Olsson et al. (2016) present TiO_2 content (mean value 1.41 wt% $\pm$0.44) and normalized REE ratios that are similar to dyke LDH's values (Table 4).

5.3.2 Dioritic Dykes LDA, LDD, LDE and LDF

In the bivariate diagrams, the dykes LDA, LDD, LDE and LDF compare well with both the data of Bullen et al. (2012) from the Umkondo LIP intrusions and Jourdan et al. (2007)'s data from the low-Ti rocks of the Jurassic Karoo LIP (Fig. 3a, b). The REE patterns of our samples bear some graphical resemblance to the low-Ti Karoo rocks, (Fig. 5a), but this is at odds with the distinct $(La/Lu)_N$ and $(La/Sm)_N$ ratios (Table 4). Furthermore, the respective multi-element diagrams indicate several divergences between our dykes and low-Ti Karoo rocks (Fig. 5b). This is true even if we ignore Rb, Ba, U, K, Pb, and Sr which are very fluid-mobile elements. For instance, it can be observed that our dykes' higher contents in Th, and conversely lower content in Ti are too different from Jourdan et al. (2007)'s values (Fig. 5b). The high $(La/Lu)_N$ and $(La/Sm)_N$ values (4.38 and 2.91 respectively) of our dykes are similar to the Umkondo LIP (4.28 and 2.66 respectively), and consistent with their comparable REE patterns (Fig. 5c). Our dykes are also characterized by a $(Sm/Er)_N$ ratio of 1.35 that is near identical to the average value (1.38) for Umkondo rocks. In terms of multi-element patterns, apart from minor discrepancies, all four dykes compare very well with the low-Ti group of the Umkondo LIP (Fig. 5d).

5.3.3 Hornblende-Rich Diorite LDG

Of all the studied samples, the dyke LDG displays the highest Nb/Y ratio (due to a relative high Nb content, Fig. 3b), which excludes it from being part of either the Karoo LIP, the Umkondo LIP or the BHDS. It is interesting to see that the Nb/Y ratio of the Low-Ti rocks of the Dullstroom Lavas overlaps with that of sample LDG. REE

Table 5 U–Pb baddeleyite analytical results

Analysis no (number of grains)	U/Th	Pb_c/Pb_{tot}^a	$^{206}Pb/^{204}Pb$	$^{207}Pb/^{235}U$	±2S % err	$^{206}Pb/^{238}U$	±2S % err	$^{207}Pb/^{235}U$	±2S	$^{206}Pb/^{238}U$	±2S	$^{207}Pb/^{206}Pb$	±2S	Concordance
Raw[b]				[Corr][c]				[Age, Ma]						
Bd-1 (4 grains)	8.8	0.105	623.7	5.2088	1.59	0.33049	1.57	1354.1	14	1840.7	25.2	1869	8	0.985
Bd-2 (4 grains)	10.4	0.063	1018.6	4.9416	0.83	0.31658	0.82	1309.4	7	1773	12.7	1851.6	4.5	0.958
Bd-3 (2 grains)	8.6	0.112	558.8	5.1801	1.3	0.33043	1.28	1349.4	11	1840.4	20.4	1859.4	7.4	0.99
Bd-3 (4 grains)	9.3	0.033	1907	5.2432	0.51	0.33288	0.48	1359.7	4.3	1852.3	7.7	1867.9	3.5	0.992

[a] Pb_c = common Pb, Pb_{tot} = total Pb (radiogenic + blank + initial)
[b] Measured ratio, corrected for fractionation and spike
[c] Isotopic ratios corrected for fractionation (0.1% per amu for Pb), spike contribution, blank (2 pg Pb and 0.2 pg U), and initial common Pb
Initial common Pb corrected with isotopic compositions from the model of Stacey and Kramers (1975) at the age of the sample

ratios of our sample $(La/Lu)_N$ is 6.89 and $(La/Sm)_N$ is 3.16) also compare well with the values (7.07 and 3.27 respectively) obtained by Buchanan et al. (1999) from the low-Ti Dullstroom Lavas. It is also important to note that the TiO_2 content (0.91 wt%) of the sample LDG fits in between the values for the low-Ti (0.33–0.73 wt%) and high-Ti (1.41–1.99 wt%) rocks of the Dullstroom Lavas (Buchanan et al. 1999). This is consistent with the multi-element diagram that also plots in between the two groups of Dullstroom samples. Furthermore, the Mg# (60) of the dyke LDG corresponds to the highest value reported by Buchanan et al. (1999) from the Dullstroom Lavas. The resemblances observed above are, however, regarded with caution because of dissimilar Fe_2O_3 contents of the concerned rocks (Table 3). Our sample and the low-Ti rocks of the Dullstroom Lavas further exhibit some difference in terms of trace elements even if we ignore some highly mobile large ion lithophile elements such as Rb, Ba, K, Pb, Sr, and other elements like U whose concentrations can change by secondary processes. For instance, the sample LDG is more enriched in Th and Ta relative to Buchanan et al. (1999)'s samples (Fig. 5h). Despite these discrepancies, the dyke LDG remains geochemically closer to the low-Ti rocks of the Dullstroom Lavas than to any other magmatic units in the KC.

5.3.4 Pyroxenites LDB, LDC and Diorites LDI, LDJ

In terms of REE elements, the dykes LDB, LDC, LDI and LDJ bear some resemblance to the Dullstroom Lavas (Fig. 3a); but it should be kept in mind that up to date no ultramafic rocks have been recognized for the Dullstroom volcanism. Moreover, the highest Mg# value (60) for the Dullstroom samples of Buchanan et al. (1999) is very distinct to values exhibited by the LDB and LDC dykes (80). The REE ratios of the samples LDI and LDJ are relatively similar to those of the low-Ti group of the Dullstroom Lavas (Table 4), but the Zr/TiO_2 versus Nb/Y plot (Fig. 3a) clearly highlights that the relevant rocks cannot be related.

In Fig. 5i, j, our samples are compared to the Machadodorp Lavas (Crow and Condie 1990; this study), which occur within the Silverton Formation, and constitute thus the oldest possible lavas correlative to the Mashishing dykes. It can be observed an obvious discordance that is further confirmed by different Nb/Y and Zr/TiO_2 ratios (Fig. 3a). All of these evidences are supportive of the idea that the geochemical signature of the dykes LDB, LDC, LDI and LDJ is unique so far.

5.4 Geochronology

5.4.1 U–Pb Baddeleyite Age for Dolerite Dyke LDH

U–Pb data from four baddeleyite fractions from the dyke LDH are presented in Table 5 and Fig. 6. Regression comprising the four baddeleyite fractions yields an upper intercept age of 1871 ± 5 Ma (MSWD = 1.2) and a lower intercept age of

Fig. 6 Concordia diagram for the sample LDH. All errors are reported at ±2σ

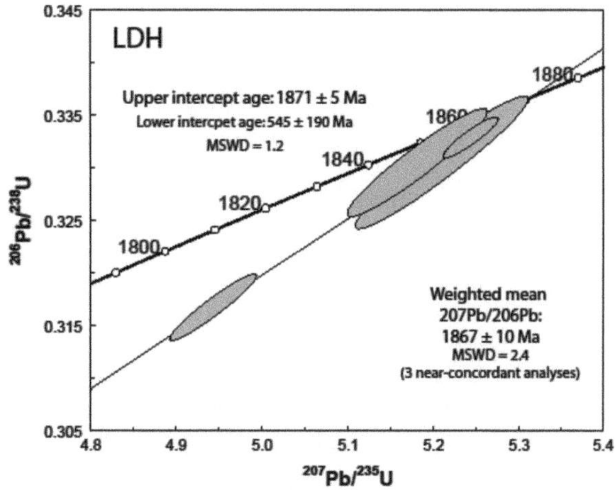

545 ± 190 Ma. A weighted mean ^{207}Pb/^{206}Pb date is calculated as 1867 ± 10 Ma (MSWD = 2.4). We conservatively consider this less precise estimate as the crystallization age of the dyke LDH.

5.4.2 ^{40}Ar/^{39}Ar Amphibole Age for Pyroxenite Dyke LDB

Due to difficulties in separating first and second generations of amphibole (kaersutite and actinolite, respectively) from each other within the sample LDB, amphibole aggregate grains containing both varieties were analyzed. Four such aggregates (labelled LDB a–d) were analyzed by step heating (Fig. 7; Table 6). All samples show very high Ca/K ratios, mostly > 200, for the significant heating steps (see Table 6), with low K-contents resulting in large uncertainties. Grains LDB-a and LDB-b yielded integrated (pooled gas) plateau ages of 2253 ± 45 Ma and 2202 ± 74 Ma with plateaus comprising 76 and 77% of ^{39}Ar released, respectively (Fig. 7a, b). Data from grains LDB-c and LDB-d yielded no plateaus as uncertainties for individual step ages do not overlap (Fig. 7c, d). It is interesting to see that for subsample LDB-c, the 95% confidence limits of the apparent integrated (pooled gas) age of 2184 ± 58 Ma overlaps those of the plateau ages of both LDB-a and LDB-b. In contrast, the temperature apparent step ages of 2377 and 2440 Ma (Fig. 7d, Table 6) probably result from excess ^{40}Ar argon inherited from the parental magma. However, for grains LDB-a and LDB-b the existence of plateaus, together with the overlap of the two ages, renders significant excess ^{40}Ar effects highly unlikely.

In order to better understand the nature of the dated amphiboles, Ca–K and Cl–K ratios were compared. For this purpose, we plotted ratios calculated from the major argon step releases (>5% of ^{39}Ar; Ca from ^{37}Ar and Cl from ^{38}Ar) and ratios obtained

Table 6 $^{40}Ar/^{39}Ar$ analytical results

Sample LDB-a (June 2015)			Relative irradiation coordinate z: 42 mm			Days after irradiation: 148	
J value: 0.008303 ± 0.000022			Heating time: 5 min				
Step, #	Laser, A	cc stp ^{39}Ar	^{40}Ar (nA)	±2SE Abs	^{39}Ar (nA)	±2SE Abs	^{38}Ar (nA)
1	9.90	7.123E−14	0.02682	4.1E−05	0.000031	2.6E−06	0.000125
2	10.14	6.685E−13	0.08962	6.2E−05	0.000298	2.9E−06	0.000339
3	10.37	7.705E−13	0.10626	6.5E−05	0.000355	3.0E−06	0.000330
4	10.61	6.936E−13	0.20332	1.1E−04	0.000307	2.5E−06	0.000315
5	10.85	8.985E−13	0.19140	7.7E−05	0.000394	3.4E−06	0.000392
6	11.08	1.528E−12	0.23285	1.8E−04	0.000654	6.6E−06	0.001149
7	11.32	1.34E−12	0.18834	8.0E−05	0.000589	3.0E−06	0.000808
8	11.55	1.252E−12	0.16551	1.4E−04	0.000572	3.9E−06	0.000794
9	11.79	1.712E−12	0.21698	1.5E−04	0.000759	6.0E−06	0.001249
10	12.03	1.309E−12	0.16596	1.0E−04	0.000574	3.4E−06	0.000847
11	12.26	1.613E−12	0.20601	9.0E−05	0.000709	6.8E−06	0.001610
12	12.50	9.56E−13	0.12028	8.7E−05	0.000425	4.6E−06	0.000546
Sample LDB−b (February 2016)			Relative irradiation coordinate z: 31 mm			Days after irradiation: 57	
J value: 0.008370 ± 0.000028			Heating time: 5 min				
Step, #	Laser, A	cc stp ^{39}Ar	^{40}Ar (nA)	±2SE Abs	^{39}Ar (nA)	±2SE Abs	^{38}Ar (nA)
1	10.00	2.14E−13	0.03231	4.8E−05	0.000091	1.9E−06	0.000135
2	10.25	1.35E−13	0.02706	3.1E−05	0.000063	3.6E−06	0.000082
3	10.49	1.19E−13	0.03878	8.8E−05	0.000051	3.5E−06	0.000046
4	10.74	2.15E−13	0.07279	5.2E−05	0.000097	3.2E−06	0.000145
5	10.98	3.69E−13	0.09069	1.1E−04	0.000165	2.8E−06	0.000160
6	11.23	4.78E−13	0.07932	8.8E−05	0.000209	3.3E−06	0.000185
7	11.47	5.56E−13	0.07341	2.1E−04	0.000248	3.1E−06	0.000158
8	11.72	3.59E−13	0.04844	3.6E−05	0.000158	3.3E−06	0.000092
9	11.96	3.62E−13	0.04400	4.0E−05	0.000167	3.2E−06	0.000087
10	12.21	6.12E−13	0.08059	4.3E−05	0.000268	4.0E−06	0.000178
11	12.45	3.71E−14	0.00624	4.3E−05	0.000020	2.9E−06	0.000023
12	12.70	5.74E−14	0.00859	2.5E−05	0.000037	5.7E−06	0.000019
Sample LDB−c (February 2016)			Relative irradiation coordinate z: 31 mm			Days after irradiation: 57	
J value: 0.008370 ± 0.000028			Heating time: 5 min				
Step, #	Laser, A	cc stp ^{39}Ar	^{40}Ar (nA)	±2SE Abs	^{39}Ar (nA)	±2SE Abs	^{38}Ar (nA)
1	9.90	5.89E−14	0.01313	6.3E−05	0.000015	7.4E−06	0.000193
2	10.14	6.45E−13	0.04346	1.0E−04	0.000281	1.3E−05	0.000476
3	10.38	5.23E−13	0.04508	2.5E−04	0.000201	1.5E−05	0.000180
4	10.62	2.05E−13	0.05023	1.5E−04	0.000109	1.1E−05	0.000091
5	10.86	6.79E−13	0.08200	5.8E−05	0.000332	1.1E−05	0.000300
6	11.10	1.03E−12	0.10870	1.6E−04	0.000447	1.3E−05	0.000481

(continued)

Table 6 (continued)

Sample LDB−c (February 2016)			Relative irradiation coordinate z: 31 mm			Days after irradiation: 57	
J value: 0.008370 ± 0.000028			Heating time: 5 min				
Step, #	Laser, A	cc stp ^{39}Ar	^{40}Ar (nA)	±2SE Abs	^{39}Ar (nA)	±2SE Abs	^{38}Ar (nA)
7	11.34	1.72E−12	0.20517	4.0E−04	0.000738	1.3E−05	0.000887
8	11.58	1.40E−12	0.14660	4.0E−04	0.000606	9.3E−06	0.000842
9	11.82	5.23E−13	0.06974	8.6E−05	0.000223	3.0E−06	0.000961
10	12.06	1.90E−12	0.23162	1.4E−04	0.000843	4.0E−06	0.001402
11	12.30	2.22E−13	0.02708	3.7E−05	0.000091	4.3E−06	0.000083
12	12.54	2.78E−13	0.03779	4.8E−05	0.000118	3.5E−06	0.000147
13	12.78	2.79E−13	0.03169	3.1E−05	0.000121	2.3E−06	0.000118
14	13.02	1.72E−13	0.03143	3.2E−05	0.000075	3.1E−06	0.000082
Sample LDB−d (June 2015)			Relative irradiation coordinate z: 42 mm			Days after irradiation: 148	
J value: 0.008303 ± 0.000022			Heating time: 5 min				
Step, #	Laser, A	cc stp ^{39}Ar	^{40}Ar (nA)	±2SE Abs	^{39}Ar (nA)	±2SE Abs	^{38}Ar (nA)
1	10.14	6.26E−15	0.00863	4.7E−05	0.000003	2.3E−06	0.000065
2	10.37	1.08E−13	0.02973	7.7E−05	0.000047	2.1E−06	0.000145
3	10.61	5.06E−13	0.06490	6.9E−05	0.000214	5.1E−06	0.000115
4	10.85	4.88E−13	0.06351	6.6E−05	0.000216	2.3E−06	0.000189
5	11.08	5.60E−13	0.13490	1.2E−04	0.000249	2.2E−06	0.000103
6	11.32	1.29E−12	0.23850	1.1E−04	0.000562	2.7E−06	0.000268
7	11.55	2.28E−12	0.30099	1.6E−04	0.001007	3.7E−06	0.001229
8	11.79	2.23E−12	0.27119	1.0E−04	0.000961	4.4E−06	0.000948
9	12.03	1.54E−12	0.19670	8.3E−05	0.000676	3.1E−06	0.000597
10	12.26	1.43E−12	0.20011	1.6E−04	0.000623	4.0E−06	0.000596
11	12.50	5.36E−12	0.76163	3.8E−04	0.002334	7.2E−06	0.003268
12	12.7	3.42E−13	0.04657	4.2E−05	0.000153	4.7E−06	0.000132
13	12.97	3.39E−13	0.04267	5.9E−05	0.000149	3.3E−06	0.000124
14	13.23	1.39E−13	0.02002	2.8E−05	0.000057	3.1E−06	0.000050
Exponential fractionation factor: −0.321							

All values regressed to time of gas inlet into mass spectrometer		
Blank measurements carried out after 3 or 4 step measurements using exactly same protocol		
Measured signals were blank corrected, using blank time functions, before regression		
Constants used		
(^{40}Ar/^{36}Ar)A	298.56 ± 0.31	Lee et al. (2006)
(^{38}Ar/^{36}Ar)A	0.1885 ± 0.0003	Lee et al. (2006)
Interfering isotope production ratios		
(^{39}Ar/^{37}Ar)Ca	(8.6 ± 0.1)E−4	Own calibrations of position B2W in Safari 1 reactor

(continued)

Table 6 (continued)

Exponential fractionation factor: −0.321		
$(^{36}Ar/^{37}Ar)Ca$	(2.8 ± 1.1)E−4	Own calibrations of position B2W in Safari 1 reactor
$(^{40}Ar/^{39}Ar)K$	0.0426 ± 0.0041	Own calibrations of position B2W in Safari 1 reactor
Decay constants		
^{40}K total	(5.541 ± 0.014)E−10 a^{-1}	Kossert and Günther (2004)
^{39}Ar	0.00258 ± 0.00003 a^{-1}	Stoenner et al. (1965)
^{37}Ar	0.01975 ± 0.0005 day^{-1}	Renne and Norman (2001)

Sample LDB−a (June 2015)	Relative irradiation coordinate z: 42 mm	Days after irradiation: 148
J value: 0.008303 ± 0.000022	Heating time: 5 min	

Step, #	±2SE Abs	^{37}Ar (nA)	±2SE Abs	^{36}Ar (nA)	±2SE Abs	$^{36}Ar/^{40}Ar$	±2SE
1	2.5E−06	0.000062	2.8E−06	0.000081	4.1E−06	0.0029232	2.8E−04
2	5.5E−06	0.000512	2.9E−06	0.000184	2.3E−06	0.0019655	5.0E−05
3	6.0E−06	0.000485	5.1E−06	0.000195	3.8E−06	0.0017553	6.7E−05
4	3.3E−06	0.000460	4.7E−06	0.000369	5.5E−06	0.0017463	5.2E−05
5	4.3E−06	0.001074	5.4E−06	0.000172	4.3E−06	0.0008446	4.6E−05
6	7.1E−06	0.004499	1.0E−05	0.000186	3.6E−06	0.0006774	4.5E−05
7	6.6E−06	0.003876	9.9E−06	0.000121	3.3E−06	0.0005196	5.1E−05
8	4.4E−06	0.004772	1.1E−05	0.000089	3.1E−06	0.000373	6.5E−05
9	8.9E−06	0.006656	1.2E−05	0.000115	3.6E−06	0.0003577	7.1E−05
10	3.1E−06	0.003423	1.1E−05	0.000075	2.9E−06	0.0003344	5.4E−05
11	1.3E−05	0.005303	1.2E−05	0.000105	4.8E−06	0.0003633	6.7E−05
12	4.2E−06	0.001846	1.0E−05	0.000042	5.5E−06	0.0002574	1.2E−04

Sample LDB−b (February 2016)	Relative irradiation coordinate z: 31 mm	Days after irradiation: 57
J value: 0.008370 ± 0.000028	Heating time: 5 min	

Step, #	±2SE Abs	^{37}Ar (nA)	±2SE Abs	^{36}Ar (nA)	±2SE Abs	$^{36}Ar/^{40}Ar$	±2SE
1	3.7E−06	0.000820	7.5E−06	0.000367	4.2E−04	0.00119	1.5E−04
2	3.3E−06	0.000855	5.3E−06	0.000388	4.5E−04	0.00176	1.7E−04
3	2.9E−06	0.000550	5.3E−06	0.000246	2.9E−04	0.00160	1.2E−04
4	3.6E−06	0.001948	6.9E−06	0.000891	1.0E−03	0.00158	7.1E−05
5	2.9E−06	0.006914	1.2E−05	0.003191	3.8E−03	0.00170	6.6E−05
6	3.9E−06	0.009995	1.6E−05	0.004619	5.4E−03	0.00095	7.4E−05
7	3.5E−06	0.010685	1.4E−05	0.004924	5.8E−03	0.00074	9.9E−05
8	3.8E−06	0.006939	1.0E−05	0.003199	3.8E−03	0.00075	1.2E−04
9	4.0E−06	0.005545	8.1E−06	0.002544	3.0E−03	0.00049	2.4E−04
10	4.6E−06	0.011380	1.9E−05	0.005247	6.2E−03	0.00036	1.1E−04
11	2.6E−06	0.000576	4.0E−06	0.000267	3.1E−04	0.00107	3.8E−02
12	3.4E−06	0.001190	7.8E−06	0.000550	6.5E−04	0.00179	2.0E−02

(continued)

Table 6 (continued)

Sample LDB−c (February 2016)			Relative irradiation coordinate z: 31 mm			Days after irradiation: 57	
J value: 0.008370 ± 0.000028			Heating time: 5 min				
Step, #	±2SE Abs	^{37}Ar (nA)	±2SE Abs	^{36}Ar (nA)	±2SE Abs	^{36}Ar/^{40}Ar	±2SE
1	1.3E−05	0.000415	1.4E−05	0.000058	6.2E−06	0.00423	1.1E−03
2	2.3E−05	0.002281	2.0E−05	0.000059	1.3E−05	0.00127	6.2E−03
3	1.6E−05	0.000720	1.5E−05	0.000076	8.2E−06	0.00161	4.2E−04
4	1.1E−05	0.001029	1.1E−05	0.000015	8.2E−06	0.00028	3.1E−03
5	1.0E−05	0.008191	2.2E−05	–	9.3E−06	–	–
6	2.1E−05	0.019039	2.9E−05	0.000002	6.9E−06	−0.00013	1.1E−04
7	1.0E−05	0.029372	2.8E−05	0.000071	9.9E−06	0.00022	1.3E−04
8	1.1E−05	0.015449	1.8E−05	0.000034	1.2E−05	0.00013	1.7E−03
9	6.5E−06	0.010608	1.4E−05	0.000015	4.0E−06	0.00008	3.7E−03
10	8.9E−06	0.011668	1.6E−05	0.000025	3.6E−06	0.00006	5.1E−05
11	5.5E−06	0.003496	8.5E−06	0.000006	3.0E−06	0.00010	1.1E−02
12	5.2E−06	0.005179	9.6E−06	0.000013	3.7E−06	0.00021	3.6E−03
13	3.1E−06	0.001905	6.5E−06	0.000018	2.4E−06	0.00050	1.8E−04
14	3.4E−06	0.002405	1.1E−05	0.000047	3.6E−06	0.00139	2.5E−04
Sample LDB−d (June 2015)			Relative irradiation coordinate z: 42 mm			Days after irradiation: 148	
J value: 0.008303 ± 0.000022			Heating time: 5 min				
Step, #	±2SE Abs	^{37}Ar (nA)	±2SE Abs	^{36}Ar (nA)	±2SE Abs	^{36}Ar/^{40}Ar	±2SE
1	5.4E−06	0.000004	2.6E−06	0.000044	3.3E−06	0.0049541	7.7E−04
2	2.8E−06	0.000057	3.7E−06	0.000076	2.8E−06	0.0024567	1.7E−04
3	1.9E−05	0.000293	6.8E−06	0.000133	1.0E−05	0.0019563	2.9E−04
4	4.7E−06	0.000165	3.5E−06	0.000105	3.1E−06	0.0015835	9.3E−05
5	4.1E−06	0.000190	2.4E−06	0.000212	3.6E−06	0.0015103	5.2E−05
6	2.0E−06	0.001068	5.5E−06	0.000301	3.1E−06	0.0011893	2.8E−05
7	6.3E−06	0.005059	1.1E−05	0.000162	4.2E−06	0.0003923	5.6E−05
8	7.3E−06	0.005369	1.2E−05	0.000119	3.3E−06	0.0002753	6.1E−05
9	3.3E−06	0.002749	8.5E−06	0.000116	2.3E−06	0.0004666	4.3E−05
10	4.9E−06	0.003441	9.6E−06	0.000107	2.4E−06	0.0003853	5.8E−05
11	8.7E−06	0.009783	1.1E−05	0.000220	2.5E−06	0.0001816	3.8E−05
12	3.8E−06	0.000498	5.5E−06	0.000010	2.5E−06	0.0001245	8.5E−03
13	3.9E−06	0.000634	4.5E−06	0.000013	3.5E−06	0.0001879	7.5E−03
14	3.5E−06	0.000315	7.2E−06	0.000006	2.9E−06	0.0001648	6.5E−03

(continued)

Table 6 (continued)

Sample LDB–a (June 2015)			Relative irradiation coordinate z: 42 mm			Days after irradiation: 148		
J value: 0.008303 ± 0.000022			Heating time: 5 min					
Step, #	%39 Ar	Age Ma	±95%	Included	Ca/K	±2SE	Cl/K	±2SE
1	0.55	1204	3101	–	75	19.5	0.46	0.116
2	5.26	1311	125	–	64	7.5	0.13	0.019
3	6.27	1437	107	–	51	6.0	0.11	0.015
4	5.42	2368	144	–	56	6.6	0.10	0.015
5	6.95	2580	82	–	104	12.1	0.12	0.017
6	11.53	2344	84	+	282	32.7	0.24	0.034
7	10.39	2271	60	+	268	30.6	0.19	0.026
8	10.10	2247	76	+	351	40.2	0.20	0.027
9	13.39	2250	74	+	372	42.8	0.24	0.033
10	10.13	2210	57	+	240	27.5	0.20	0.028
11	12.51	2239	77	+	310	35.9	0.32	0.045
12	7.50	2180	101	+	170	20.0	0.17	0.024
Sample LDB–b (February 2016)			Relative irradiation coordinate z: 31 mm			Days after irradiation: 57		
J value: 0.008370 ± 0.000028			Heating time: 5 min					
Step, #	%39 Ar	Age Ma	±95%	Included	Ca/K	±2SE	Cl/K	±2SE
1	5.77	1968	248	–	54	6.3	0.18	0.025
2	3.98	1853	737	–	83	14.4	0.15	0.030
3	3.27	2682	897	–	64	12.7	0.09	0.021
4	6.15	2725	400	–	124	16.2	0.17	0.025
5	10.45	2294	245	+	277	30.3	0.11	0.014
6	13.24	2324	143	+	321	34.9	0.12	0.015
7	15.80	2096	125	+	285	30.1	0.08	0.011
8	10.02	2138	185	+	291	33.1	0.08	0.012
9	10.59	2030	181	+	214	23.8	0.07	0.011
10	17.07	2285	110	+	280	30.2	0.09	0.012
11	1.27	1934	2262	–	181	65.6	0.14	0.066
12	2.39	1250	2644	–	207	81.1	0.06	0.034
Sample LDB–c (February 2016)			Relative irradiation coordinate z: 31 mm			Days after irradiation: 57		
J value: 0.008370 ± 0.000028			Heating time: 5 min					
Step, #	%39Ar	Age Ma	±95%	Included	Ca/K	±2SE	Cl/K	±2SE
1	0.35	0	4321	–	178	218.9	1.66	2.042
2	6.69	1089	544	–	49	7.6	0.21	0.042
3	4.79	1241	721	–	21	4.5	0.10	0.031
4	2.60	2773	638	–	56	15.1	0.10	0.042
5	7.90	2212	229	+	154	19.9	0.12	0.020

(continued)

Table 6 (continued)

Sample LDB−c (February 2016)			Relative irradiation coordinate z: 31 mm			Days after irradiation: 57		
J value: 0.008370 ± 0.000028			Heating time: 5 min					
Step, #	%39Ar	Age Ma	±95%	Included	Ca/K	±2SE	Cl/K	±2SE
6	10.65	2206	173	+	280	34.1	0.15	0.025
7	17.57	2233	136	+	260	28.6	0.17	0.020
8	14.43	2040	149	+	160	17.2	0.18	0.023
9	5.30	2474	125	+	318	33.8	0.61	0.072
10	20.08	2186	40	+	84	8.5	0.21	0.024
11	2.16	2366	316	+	250	38.2	0.13	0.028
12	2.81	2437	234	+	290	36.0	0.17	0.027
13	2.89	1958	165	+	96	10.7	0.12	0.017
14	1.78	2136	492	+	205	29.4	0.13	0.024
Sample LDB−d (June 2015)			Relative irradiation coordinate z: 42 mm			Days after irradiation: 148		
J value: 0.008303 ± 0.000022			Heating time: 5 min					
Step, #	%39 Ar	Age Ma	±95%	Included	Ca/K	±2SE	Cl/K	±2SE
1	0.04	0	6996	–	79	209.9	2.73	5.518
2	0.65	1625	1147	–	69	15.5	0.36	0.066
3	3.04	1331	492	–	78	10.8	0.06	0.023
4	2.94	1518	141	–	43	5.4	0.10	0.015
5	3.37	2278	117	+	43	5.1	0.03	0.006
6	7.73	2207	56	+	110	12.6	0.05	0.007
7	13.72	2264	54	+	314	35.7	0.17	0.024
8	13.43	2259	62	+	355	40.4	0.14	0.019
9	9.29	2163	46	+	248	28.3	0.12	0.016
10	8.60	2377	64	+	350	40.1	0.14	0.019
11	32.26	2440	36	+	256	29.1	0.19	0.027
12	2.06	2336	197	+	194	26.8	0.12	0.020
13	2.04	2265	199	+	260	32.9	0.11	0.019
14	0.84	2586	467	–	348	63.1	0.13	0.032

by electron microprobe (Fig. 8). The actinolites (Amp_2, secondary amphiboles) are characterized by very low potassium, calcium and chlorine contents and have highly variable Ca/K and Cl/K ratios. In contrast, primary amphiboles (Amp_1, kaersutites) have higher K, Ca and Cl concentrations and lower Ca/K and Cl/K. The step-heating results are much closer to kaersutite data than to actinolite.

This indicates that the argon in the plateau steps is dominated by the release from the (primary) kaersutite. The overlapping plateau estimates of 2253 ± 45 Ma and 2202 ± 74 Ma therefore most likely reflect the formation of the primary amphibole, and therefore yield the age of sample LDB. Our best estimate for this age is given by the overlap between the results for LDB a and b, i.e. between 2208 and 2276 Ma. This estimate is far removed from the 1867 ± 10 Ma crystallization age of the dyke

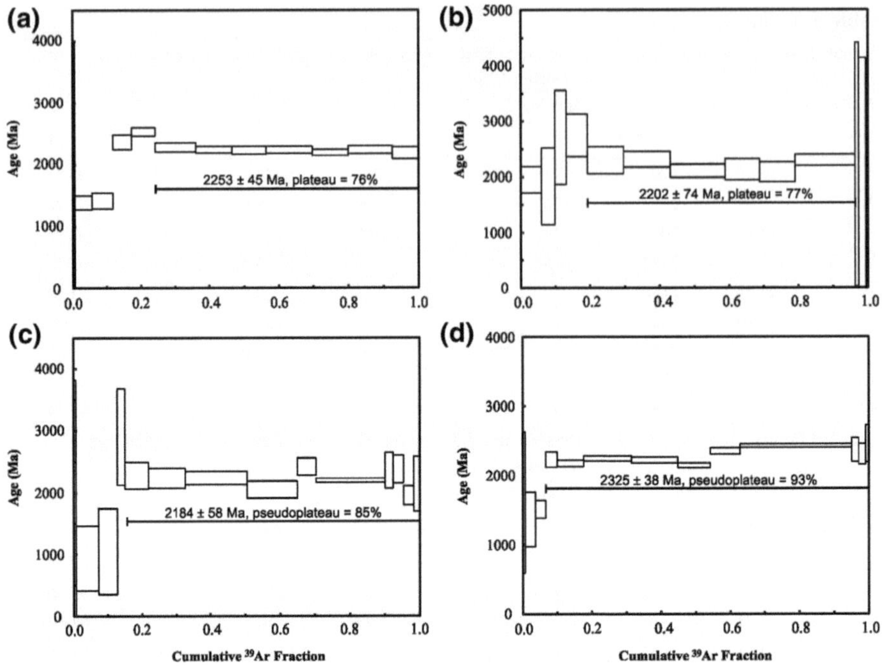

Fig. 7 $^{40}Ar/^{39}Ar$ age spectra of amphiboles from sample LDB. Uncertainty limits are 95% confidence. Note that subsamples *a* and *b* yield plateaus and *c* and *d* no plateaus

Fig. 8 Plot of Ca/K versus Cl/K ratios for amphiboles from sample LDB, incorporating results from the major argon step releases (open circles) and mineral chemistry obtained with microprobe (black circles)

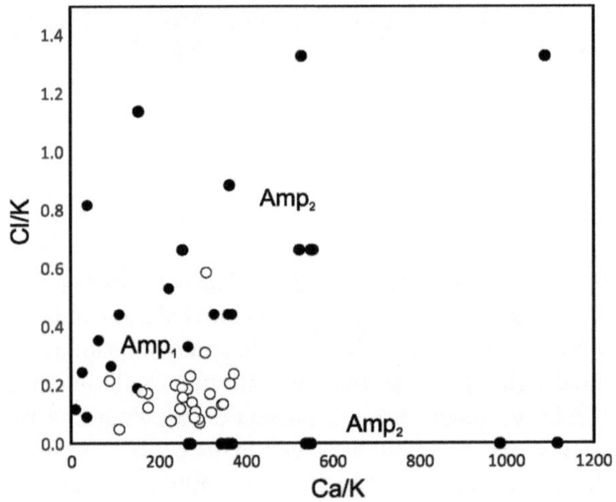

LDH (this study), and is also significantly older than the Bushveld Complex age of 2055.4 ± 0.7 Ma (Zeh et al. 2015) and differs from ages yielded by other dykes of the KC.

5.5 Paleomagnetism and Rock-Magnetism

The demagnetization behaviour of the dykes was very variable (Fig. 9). Specimens of the dykes LDD and LDE did not yield consistent magnetic components amongst samples from the same dyke, and no further description or interpretation was attempted. The remaining eight dykes did yield consistent components for samples of the same individual dyke. The dyke samples were characterized by two or three magnetic components per sample. In general, one or two components were removed by low to moderate thermal demagnetization steps in addition to a characteristic magnetization that was removed during demagnetization at high temperatures. Exceptionally, samples from the dyke LDA were characterized by a single characteristic component only, which remained stable up to high temperature demagnetization steps (Fig. 9). Individual dykes yielded very different characteristic components (Figs. 9 and 10; Table 7).

Very "soft" northeasterly and moderate downward directed components were isolated in the dyke LDG during low field-strength alternating field (AF) demagnetization steps. Poorly constrained shallow magnetizations (either northerly or southerly directed) were removed in three dykes (LDE, LDF and LDI) during the same treatment and low temperature demagnetization steps. These two low-stability components are labeled "LT". Five dykes yielded northerly and upward directed magnetizations during the same treatment. These components are similar to the Earth's present dipole field and are labeled "PF" (for present field). The removal of PF in the dyke LDB at ~250 °C and increased thermal demagnetization up to ~450 °C revealed westerly and downward components labeled MT1. The removal of PF in the dyke LDC at ~250 °C and increased thermal demagnetization up to ~350 °C and sometimes 450 °C revealed west–northwesterly and moderate upwards directed components that we labeled MT2.

Characteristic components, here labeled "HT" and consecutively numbered 1–7, were identified by demagnetization above ~400 °C (i.e., always at high temperature demagnetization steps), and after the removal of the above-mentioned components. Above 570 °C (Curie point of magnetite) the magnetic intensity of specimen either dropped sharply, or started fluctuating erratically. In the dyke LDA the characteristic magnetization is shallow and southerly directed (i.e., HT1). In the dyke LDB it is easterly, steep and downwards directed (i.e., HT2). In the dyke LDC, the characteristic components (i.e., HT3) are, northwesterly, shallow and downward directed and reminiscent of the component MT1 of the dyke LDB. The dyke LDF yielded poorly constrained shallow easterly characteristic or HT4 components. The dyke LDG yielded moderately positively inclined and northwesterly-directed HT5 components, while the dyke LDH revealed near antipodal HT6 components (i.e., moderately negatively inclined and southeasterly directed). Both dykes LDI and LDJ produced moderately positively inclined and northeasterly-directed HT7 components.

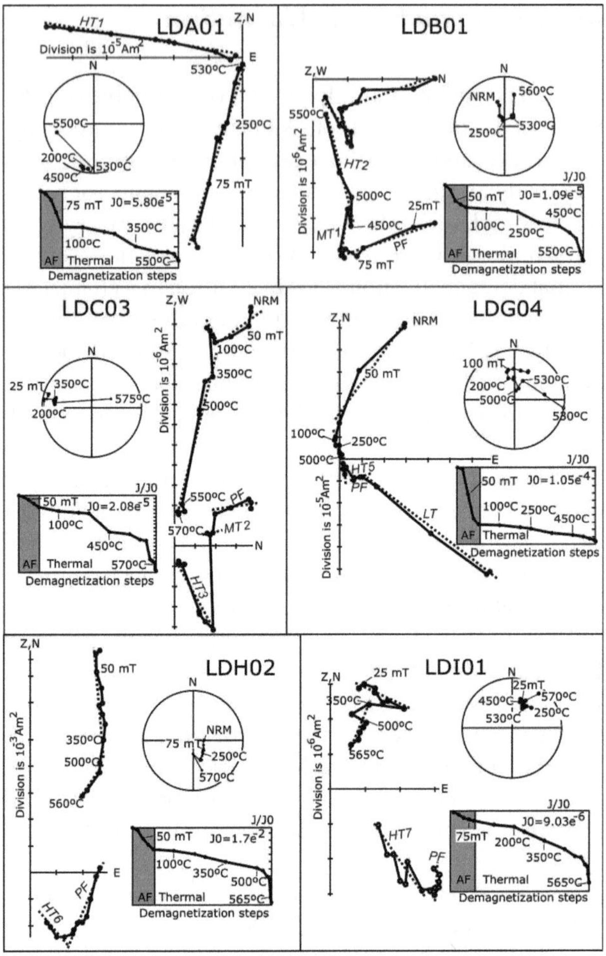

Fig. 9 Sets of vector diagrams, equal area nets and thermal decay plots from representative samples illustrating demagnetization of the studied dykes. Vector diagrams show the projection, after each demagnetization step, of the end of the magnetization vector, onto two orthogonal planes, i.e. the horizontal one (filled symbols) and the vertical one (open symbols). Dashed lines on the vector diagrams represent the least square; see the text for the names of different magnetic components. In the equal area nets, the grey dot represents the erratic behavior at higher temperature demagnetization steps. Specimen captions indicate respectively, sample number, select demagnetization steps (*NRM* = natural remanent magnetization, and *mT* = millitesla), and intensity scale on the vector diagrams

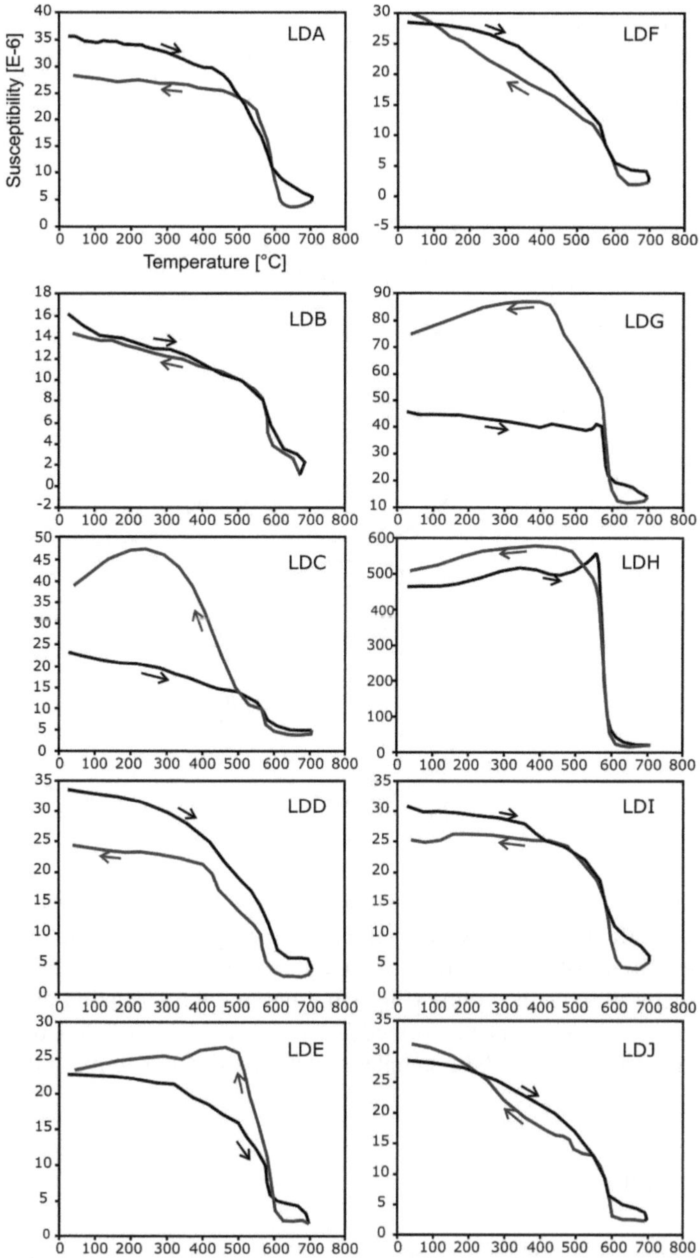

Fig. 10 Continuous heating and cooling cycles in air of the studied dykes. Black (grey) line indicates heating (cooling)

Fig. 11 Summary of site means (listed in Table 7) of various magnetic directions from the Mashishing dykes. *IS* = in situ, *TC* = tilt corrected. For LT components (see text), the α_{95} cone of confidence is represented by a solid line for the PF (present field) mean and dashed line for the other low-stability components

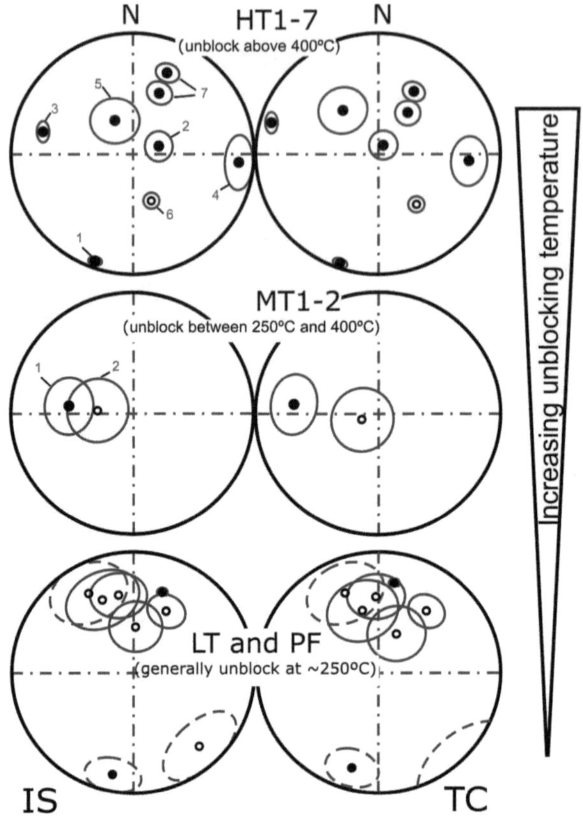

During heating all dyke samples exhibited abrupt drops in magnetic susceptibilities between 500 and 570 °C (Curie temperature range of magnetite or titanomagnetite) (Fig. 10). Furthermore, some of the dykes (i.e. LDC, LDE, LDG) displayed higher susceptibility upon cooling (Fig. 10), implying the formation of magnetized secondary minerals. These newly formed phases certainly contributed to increase the magnetic intensity of some dyke samples at demagnetization above 500–570 °C.

6 Discussion

Distinct characteristic magnetizations (i.e., HT1-7) carried by titanomagnetite were identified in addition to lower stability MT1, MT2, PF and LT components (Figs. 9 and 11; Table 7). Although we were not able to conduct paleomagnetic stability field tests, Virtual Geomagnetic Poles (VGPs) can be calculated (see Table 8). These can be compared with published paleopoles for the KC in an attempt to evaluate the relative timing of remanence acquisition (Fig. 12). A structural correction can be

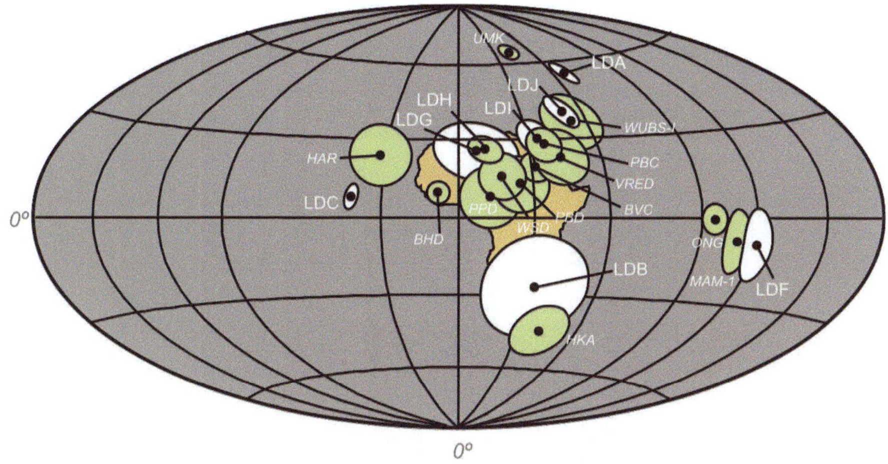

Fig. 12 VGPs from the present study (white) compared to selected published paleopoles (green) for the KC. Pole statistics and abbreviations are as per Table

applied to individual dyke VGPs in order to restore the present tilt (~15°) of the Transvaal Supergroup strata, if deemed necessary by the interpreted timing of dyke intrusion. Dykes that postdate the intrusion of the Bushveld complex do not require a correction.

6.1 *Dyking During the ~1.11 Ga Umkondo LIP*

The dykes LDA, LDD, LDE and LDF all share a similar chemistry that is consistent with the Umkondo LIP (Bullen et al. 2012) (Figs. 3a, b and 5c, d). The VGP calculated from the magnetization HT1 of the dyke LDA is comparable with the paleomagnetic pole for the Umkondo LIP (Swansen-Hysell et al. 2015) (Fig. 12) in line with this dyke's composition. The magnetization HT4 and associated VGP of the dyke LDF is not consistent with the associated chemistry, but given the highly weathered nature of this dyke, this component is regarded as spurious. Samples from both dykes LDD and LDE unfortunately did not yield consistent magnetic histories. It would be ideal to have a numerical age from at least one of the sites LDA, LDD, LDE, and LDF by which to constrain the geochemical and paleomagnetic data. Attempts to find baddeleyite grains from these dykes failed, and they were also unsuitable for $^{39}Ar/^{40}Ar$ dating. We tentatively assign a ~1.11 Ga age to dykes LDA, LDD, LDE and LDF based on their geochemical similarities with the Umkondo LIP and the paleomagnetism of dyke LDA, but this requires further testing.

Table 7 Summary of the characteristic magnetic components identified in the studied dykes

Component	Site	n/N	Present coordinates				Tilt-corrected coordinates			
			D in deg.	I in deg.	k	a_{95}	D in deg.	I in deg.	K	a_{95}
HT1	LDA	8–8	198.6	−7.7	122.31	4.70	197.1	−4.5	122.8	4.69
HT2	LDB	9–10	64.5	69.4	47.37	7.12	38.2	83.5	47.4	7.11
HT3	LDC	8–8	284.9	24.1	136.99	4.44	285.6	9.4	136	4.46
HT4	LDF	5–11	98.3	8.4	31.51	12.34	98.6	23.4	31.4	12.4
HT5	LDG	6–9	355.9	59.3	20.01	11.83	334.2	54.1	20.06	11.8
HT6	LDH	6–6	158.4	−54.9	163.17	4.80	139.9	−50.5	163.7	4.79
HT7	LDI	7–8	19.8	47.5	88.52	6.18	3.2	54.5	82.93	6.17
HT7	LDJ	5–5	23.8	30.7	139.95	5.80	15.1	39.1	139.9	5.8

D = declination, I=inclination, n=number of samples used to calculate the site mean D/I, N = number of samples drilled per site, a_{95} =semi-angle of 95% cone of confidence about mean direction, k = precision parameter of Fisher (1953)

Table 8 List of VGPs from this study (in bold) and selected published poles for the KC

Group/lithology	Label	Lat. (°N)	Long. (°E)	A95 (dp/dm in °)	Nature of pole	Rock age (Ma)	Reference for age	Reference for pole
Ongeluk formation	ONG	−0.5	100.7	5.3	Primary	2426 ± 3 Ma	Gumsley et al. (2017)	Gumsley et al. (2017)
Mamatwanore	MAM-1	−8.2	111.1	(5.6/11.1)	Primary	~2200	Evans et al. (2001)	Evans et al. (2001)
Hekpoort formation	HKA	−44.1	40	(9.9/10.4)	Primary?	2236 ± 38 Ma	Humbert et al. (2017a, b)	Humbert et al. (2017a, b)
Dyke LDB		−26.84	31.66	(20.88/20.89)	Primary	2208–2276	This study	This study
Phalaborwa complex	PBC	27.7	35.8	6.6	Primary	2060.6 ± 0.5	Wu et al. (2011)	Letts et al. (2010)
RLS (B2-magma)	BVC	19.2	30.8	5.8	Primary	2054.89 ± 0.3	Zeh et al. (2015)	Letts et al. (2009)
Dyke LDI		29.83	33.54	(6.13/8.70)	Not primary	2208–2276	This study	This study
Dyke LDJ		40.6	49.06	(4.13/6.92)	Not primary	2208–2276	This study	This study
Lower Waterberg group	WUBS-I	36.5	51.3	10.9	Primary	<2054 ± 4	Dorland et al. (2006)	De Kock et al. (2006)
Vredefort structure	VRED	22.8	41.6	10.5	Primary	2023.0 ± 4.0	Kamo et al. (1996)	Salminen et al. (2009)
NE-dykes of the BHDS	BHD	9.4	352	4.3	Primary	~1900	Lubnina et al. (2010)	Lubnina et al. (2010)
Post-Waterberg intrusions	WSD	15.6	17.1	8.9	Primary	1874.6 ± 3.9	Hanson et al. (2004)	Hanson et al. (2004)

(continued)

Table 8 (continued)

Group/lithology	Label	Lat. (°N)	Long. (°E)	A95 (dp/dm in °)	Nature of pole	Rock age (Ma)	Reference for age	Reference for pole
Post-Bushveld dykes	PBD	12.6	24.1	10.8	Primary	~1900	Letts et al. (2005)	Letts et al. (2005)
Post-Phalaborwa dykes	PPD	7.6	12.1	11.8	Primary	~1900	Letts et al. (2010)	Letts et al. (2010)
Hartley Lavas	HAR	22.7	328.6	11.7	Primary	1928 ± 4	Cornell et al. (1998)	Alebouyeh Semami et al. (2016)
Dyke LDC		7.72	318.39	(2.53/4.74)	Not primary	2208–2276	This study	This study
Dyke LDG		25.5	7.12	(11.61/16.55)	Not primary	~2061?	This study	This study
Dyke LDH		26.07	11.01	(4.81/6.80)	Primary	1867 ± 10	This study	This study
Umkondo intrusions	UMK	64	42.1	2.6	Primary	1110 ± 2	De Kock et al. (2014)	Swanson-Hysell et al. (2015)
Dyke LDA		55.84	65.02	(2.38/4.73)	Primary	~1110	This study	This study
Dyke LDF		−9.3	120.16	(6.27/12.44)	Not primary	~1110	This study	This study

6.2 *Dyking Associated with the 1875–1835 Ma BHDS*

The dyke LDH gave a U–Pb baddeleyite age of 1867 ± 10 Ma (Fig. 6), which falls within the age range 1875–1835 Ma of the BHDS (Olsson et al. 2016). The major and trace element geochemistry of the dyke LDH is also very consistent with the BHDS (Figs. 3a, b and 5e, f). Furthermore, the VGP (antipole) of characteristic magnetization HT6 agrees well with paleopoles obtained by previous authors (Letts et al. 2005, 2010; Lubnina et al. 2010) from various dual-polarity NNE-trending dykes in the eastern Kaapvaal craton (Fig. 12). It is thus clear that the dyke LDH is part of the BHDS. The associated VGP of the magnetization HT5 from the dyke LDG is indistinguishable from the magnetization HT6, and the site LDG may represent another 1875–1835 Ma generation of dyke. Such a similarity, however, cannot be reconciled with the dyke LDG's chemistry. The dyke LDG differs from the compositional range of the BHDS, but shares some compatibility with the low-Ti rocks (Buchanan et al. 1999) of the 2061 ± 2 Ma (Walvaren 1997) Dullstroom Lavas (Figs. 3a, b and 5g, h). It cannot be excluded that the magnetization HT5 represents a magnetic overprint acquired during the intrusion of the BHDS.

6.3 *Identification of New Pre-Bushveld Dykes*

The $^{40}Ar/^{39}Ar$ overlapping plateau results of 2253 ± 45 Ma and 2202 ± 74 Ma (Fig. 7) as well as the chemical composition of the dyke LDB (this study) have not been recognized before from the dykes of the KC. We have illustrated geochemical similarities between the dykes LDB, LDC, LDI and LDJ (Fig. 5i, j), suggesting that these four dykes are related to a common magmatic event. VGPs yielded by these dykes, however, are statistically distinct. VGPs for components HT3 (dyke LDC) and HT7 (dykes LDI and LDJ) respectively plot close to the ~1.9 Ga Hartley LIP pole (Alebouyeh Semami et al. 2016) and the ~2.0 Ga aged paleopoles (e.g. De Kock et al. 2006; Letts et al. 2010) (Fig. 12) and are thus interpreted as magnetic overprints. Importantly, the VGP associated with the magnetization HT2 of our dated LDB dyke is similar to a paleopole defined from the Hekpoort Formation (Humbert et al. 2017a, b) (Fig. 12). Even though the Hekpoort lavas are located stratigraphically below our sampling area and should thus be older than the studied dykes, it should be kept in mind that Humbert et al. (2017a, b) had doubt about the primary nature of their pole, and were only able to demonstrate that it is older than ~2.05 Ga. We therefore consider that the magnetization HT2 of the dyke LDB correlates well with the $^{40}Ar/^{39}Ar$ estimate obtained for this dyke. It is interesting to note here that a preliminary U–Pb baddeleyite date comparable to our $^{40}Ar/^{39}Ar$ result was obtained from a NNE trending mafic dyke that intrudes the Nelspruit granite in the south-east of our sampling area (J. Olsson, pers, comm. 2012), but this was never subject of a peer-reviewed publication. Another dolerite dyke in the southeastern part of the KC was dated at 2168 ± 5 Ma (ID-TIMS on baddeleyite) by Larsson (2015) in a MSc, but this age has not formally been published. Here, we suggest that the dykes LDB, LDC, LDJ and LDI form part of a previously unrecognized pre-Bushveld dyke swarm. A catalogue of currently known magmatic units in the KC including our new dyke swarm is available in De Kock et al. (this volume).

The Mashishing dykes occur at a comparable (slightly lower) stratigraphic level to the Machadodorp Volcanic Member. However, a geochemical comparison of our pre-Bushveld dykes with the Machadodorp volcanics (Fig. 5i, j) reveals significant differences (e.g., Nb/Yvalues). Thus, a magmatic relationship of Mashihing dykes with the Machadodorp Volcanic Member cannot be considered at this stage.

7 Conclusion

The NNE-trending dykes near Mashishing in the eastern Kaapvaal craton of South Africa are variably affected by weathering and low-grade metamorphism. The petrography and geochemistry of samples suggest four groups of dykes. One group (LDB, LDC, LDI and LDJ) is geochemically unique, while the three others have chemistry comparable to either (1) the ~2.06 Ga Dullstroom Lavas (dyke LDG), (2) the 1.875–1.835 Ga Back Hills dyke swarm (dyke LDH), and (3) the ~1.11 Ga Umkondo dolerites (dykes LDA, LDD, LDE and LDF). A magmatic relationship with the 1.875–1.835 Ga Black Hills dyke swarm is confirmed with the paleomagnetism of dyke LDH and its U–Pb baddeleyite age of 1867 ± 10 Ma. This age is significantly distinct from an $^{40}Ar/^{39}Ar$ age estimate between 2208 Ma and 2276 Ma obtained for one of the geochemically unique dykes. Our results indicate that the Mashishing dykes, despite similar trend, are not all coeval, but constitute swarms of different generations.

Acknowledgements HW received financial support from the South African National Research Foundation (NRF) and also acknowledges research funding from CIMERA as well as the PPM Research Group of the Department of Geology at the University of Johannesburg. HW thanks Prof. J. Kramers for fruitful discussions during the interpretation of $^{40}Ar/^{39}Ar$ data. The authors thank the handling editor R. Srivastava, and the two reviewers M. Klausen and R. Ernst for their constructive critics and suggestions which improve the manuscript.

References

Alebouyeh Semami F, De Kock MO, Söderlund U, Gumsley AP, Da Silva R, Beukes NJ, Armstrong R (2016) New U–Pb geochronological and paleomagnetic constraints on the late Paleoproterozoic Hartley magmatic event: evidence for a potential large igneous province in the Kaapvaal Craton during Kalahari assembly, South Africa. GFF 138:164–182

Barnes SJ, Maier WD, Curl EA (2010) Composition of the marginal rocks and sills of the Rustenburg Layered Suite, Bushveld Complex, South Africa: implications for the formation of the platinum-group element deposits. Econ Geol 105:1491–1511

Buchanan PC, Koeberl C, Reimold WU (1999) Petrogenesis of the Dullstroom formation, Bushveld magmatic province, South Africa. Contrib Miner Petrol 137:133–146

Bullen DS, Hall RP, Hanson RE (2012) Geochemistry and petrogenesis of mafic sills in the 1.1 Ga Umkondo Large Igneous province, southern Africa. Lithos 142–143:116–129

Cawthorn RG, Eales HV, Walraven F, Uken R, Watkeys MK (2006) The Bushveld complex. In: Johnston MR, Anhaeusser CR, Thomas RJ (eds) The Geology of South Africa. Geological Society of South Africa, Johannesburg, Council for Geoscience, Pretoria, pp 261–281

Cornell DH, Armstrong RA, Walraven F (1998) Geochronology of the Proterozoic Hartley formation, South Africa: constraints on the Kheis tectogenesis and the Kaapvaal Craton's earliest Wilson Cycle. J Afr Earth Sci 26:5–27

Cox KG, Bell JD, Pankhurst RJ (1979) The interpretation of igneous rocks. George Allen and Unwin, London, p 450

Crow C, Condie KC (1990) Geochemistry and origin of early Proterozoic volcanic rocks from the Transvaal and Soutpansberg successions, South Africa. Precambr Res 47:17–26

Deer WA, Howie RA, Zussman J (1966) An introduction to the rock-forming minerals. London Publisher, Longmans, Green and Co Ltd., London, p 495

De Kock MO, Ernst R, Söderlund U, Jourdan F, Hofmann A, Le Gall B, Bertrand H, Chisonga BC, Beukes N, Rajesh HM, Moseki LM, Fuchs R (2014) Dykes of the 1.11 Ga Umkondo LIP, Southern Africa: clues to a complex plumbing system. Precambr Res 249:129–143

De Kock MO, Evans DAD, Dorland HC, Beukes NJ, Gutzmer J (2006) Paleomagnetism of the lower two unconformity-bounded sequences of the Waterberg Group, South Africa: towards a better-defined apparent polar wander path for the Paleoproterozoic Kaapvaal craton. S Afr J Geol 109:157–182

Dorland HC, Beukes NJ, Gutzmer J, Evans DAD, Armstrong RA (2006) Precise SHRIMP U–Pb zircon age constraints on the lower Waterberg and Soutpansberg Groups, South Africa. S Afr J Geol 109:139–156

Elburg M (2010) Sources and processes in arc magmatism: the crucial role of water. Geologica belgica 13:121–136

Eriksson PG, Altermann W, Hartzer FJ (2006) The Transvaal Supergroup and its precursors. In: Johnson MR, Anhaeusser CR, Thomas RJ (eds) The Geology of South Africa. Geological Society of South Africa, Johannesburg, Council for Geoscience, Pretoria, pp 237–260

Ernst RE (2014) Large igneous provinces. Cambridge University Press, Cambridge, p 653

Ernst RE, Buchan KL (1997) Giant radiating dyke swarms: their use in identifying pre-Mesozoic large igneous provinces and mantle plumes. Am Geophys Union Monogr 100:297–333

Evans DAD, Gutzmer J, Beukes NJ, Kirschvink JL (2001) Paleomagnetic constraints on ages of mineralization in the Kalahari Manganese Field, South Africa. Econ Geol 96:621–631

Fisher RA (1953) Dispersion on a sphere. Proc R Soc Lond Ser A 217:295–305

Geng H, Brandl G, Sun M, Wonga J, Kröner A (2014) Zircon ages defining deposition of the Palaeoproterozoic Soutpansberg Group and further evidence for Eoarchaean crust in South Africa. Precambr Res 249:247–262

Gumsley AP, Chamberlain KR, Bleeker W, Söderlund U, De Kock MO, Larsson ER, Bekker A (2017) Timing and tempo of the Great Oxidation Event. PNAS 114:1811–1816

Gumsley AP, De Kock MO, Rajesh HM, Knoper MW, Söderlund U, Ernst RE (2013) The Hlagothi Complex: the identification of fragments from a Mesoarchaean large igneous province on the Kaapvaal Craton. Lithos 174:333–348

Gumsley A, Olsson JR, Söderlund U, De Kock MO, Hofmann A, Klausen MB (2015) Precise U–Pb baddeleyite age dating of the Usushwana Complex, Southern Africa—implications for the Mesoarchaean magmatic and sedimentological evolution of the Pongola Supergroup, Kaapvaal Craton. Precamb Res 267:174–185

Gumsley A, Rådman J, Söderlund U, Klausen MB (2016) U–Pb baddeleyite geochronology and geochemistry of the White Mfolozi Dyke Swarm: unravelling the complexities of 2.70–2.66 Ga dyke swarms on across the eastern Kaapvaal Craton. South Africa GFF 138:115–132

Hanson RE, Gose WA, Crowley JJ, Ramezani J, Bowring SA, Bullen DS, Hall RP, Pancake JA, Mukwakwami J (2004) Paleoproterozoic intraplate magmatism and basin development of the

Kaapvaal Craton: age, paleomagnetism and geochemistry of ~1.93 to ~1.87 Ga post-Waterberg dolerites. S Afr J Geol 94:233–254
Hanson RE, Riox M, Gose WA, Blackburn TJ, Bowring SA, Mukwakwami J, Jones DL (2011) Paleomagnetic and geochronological evidence for large-scale post–1.88 Ga displacement between the Zimbabwe and Kaapvaal cratons along the Limpopo belt. Geology 39:487–490
Humbert F, De Kock MO, Altermann W, Elburg MA, Lenhardt N, Masango S (2017a) Petrology, physical volcanology, and geochemistry of a Palaeoproterozoic large igneous province: the Hekpoort formation in the southern Transvaal sub-basin (Kaapvaal Craton). Precambr Res 315:232
Humbert F, Sonnette L, De Kock MO, Robion P, Horng CS, Cousture A, Wabo H (2017b) Palaeomagnetism of the early Palaeoproterozoic, volcanic Hekpoort Formation (Transvaal Supergroup) of the Kaapvaal craton, South Africa. Geophys J Int 209:842–865
Jaffey AH, Flynn KF, Glendenin LE, Bentley WC, Essling AM (1971) Precision measurement of half-lives and specific activities of ^{235}U and ^{238}U. Phys Rev 4:1889–1906
Jones CH (2002) User-driven integrated software lives: "Paleomag" paleomagnetic analysis on the Macintosh^{TM}. Comput Geosci 28:1145–1151
Jourdan F, Bertrand H, Scharer U, Blichert-Toft J, Feraud G, Kampunzu AB (2007) Major and trace elements and Sr, Nd, Hf, and Pb isotope compositions of the Karoo large igneous province, Botswana–Zimbabwe: Lithosphere versus Mantle plume contribution. J Petrol 48:1043–1077
Jourdan F, Féraud G, Bertrand H, Watkeys MK, Kampunzu AB, Le Gall B (2006) Basement control on dyke distribution in Large Igneous Provinces: case study of the Karoo triple junction. Earth Planet Sci Lett 241:307–322
Kamo SL, Reimold WU, Krogh TE, Colliston WP (1996) A 2.023 Ga age for the Vredefort impact event and a first report of shock metamorphosed zircons in pseudotachylitic breccias and Granophyre. Earth Planet Sci Lett 144:369–387
Kirschvink JL (1980) The least-squares line and plane and the analysis of palaeomagnetic data. Geophys J Roy Astron Soc 62:699–718
Klausen MB, Söderlund U, Olsson JR, Ernst RE, Armoogam M, Mkhize SW, Petzer G (2010) Petrological discrimination among Precambrian dyke swarms, eastern Kaapvaal craton (South Africa). Precambr Res 183:501–522
Kossert K, Günther E (2004) LSC measurements of the half-life of ^{40}K. Appl Radiat Isot 60:459–464
Kuiper KF, Deino A, Hilgen FJ, Krijgsman W, Renne PR, Wijbrans JR (2008) Synchronizing rock clocks of Earth history. Science 320:500–504
Larsson ER (2015) U–Pb baddeleyite dating of intrusions in the South–easternmost Kaapvaal craton (South Africa): revealing multiple events of dyke emplacement. Unpublished MSc thesis, Lund University, Lund, 30 p
Lee J-Y, Marti K, Severinghaus JP, Kawamura K, Yoo H-S, Lee J-B, Kim J-S (2006) A redetermination of the isotopic abundances of atmospheric Ar. Geochim Cosmochim Acta 70:4507–4512
Letts S, Torsvik TH, Webb SJ, Ashwal LD (2010) New palaeoproterozoic palaeomagnetic data from the Kaapvaal Craton, South Africa. Geol Soc, London, Spec Publ 357:9–26
Letts S, Torsvik TH, Webb SJ, Ashwal LD (2009) Palaeomagnetism of the 2054 Ma Bushveld complex (South Africa): implications for emplacement and cooling. Geophys J Int 179:850–872
Letts S, Torsvik TH, Ashwal L, Webb S (2005) Palaeomagnetic and $^{40}Ar/^{39}Ar$ geochronology of mafic dykes from the eastern Bushveld Complex (South Africa). Geophys J Int 162:36–48
Lubnina L, Ernst RE, Klausen MB, Söderlund U (2010) Paleomagnetic study of NeoArchean-Paleoproterozoic dykes in the Kaapvaal craton. Precambr Res 183:523–552
Ludwig KR (2003) Isoplot 3.00; a geochronological toolkit for Microsoft Excel. Berkeley Geochronological Center, Berkeley, CA, Publication No. 4, 74 p
Middelburg JJ, Van der Weijden CH, Woittiez JRW (1988) Chemical processes affecting the mobility of major, minor and trace elements during weathering of granitic rocks. Chem Geol 68:253–273
Moseley D (1984) Symplectic exsolution in olivine. Am Miner 69:139–153
Nesbitt HW, Young GM (1982) Early Proterozoic climates and plate motions inferred from major element chemistry of lutites. Nature 299:715–717

Olsson JR, Klausen MB, Hamilton MA, März N, Söderlund U, Roberts RJ (2016) Baddeleyite U-Pb ages and geochemistry of the 1875–1835 Ma Black Hills Dyke Swarm across north-eastern South Africa: part of a trans-Kalahari Craton back-arc setting? GFF 138:183–202

Olsson JR, Söderlund U, Klausen MB, Ernst RE (2010) U–Pb baddeleyite ages linking major Archean dyke swarms to volcanic rift-forming events in the Kaapvaal craton (South Africa), and a precise age of the Bushveld complex. Precambr Res 183:490–500

Palme H, O'Neill HSC (2014) Cosmochemical estimates of Mantle Composition. In: Holland HD, Turrekian KK (eds) Treatise on geochemistry. Elsevier, Amsterdam, pp 1–38

Renne PR, Swisher CC, Deino AL, Karner DB, Owens T, De Paolo DJ (1998) Intercalibration of standards, absolute ages and uncertainties in $^{40}Ar/^{39}Ar$ dating. Chemical Geology (Isotope Geoscience Section) 145:117–152

Renne PR, Mundil R, Balco G, Min K, Ludwig KR (2010) Joint determination of ^{40}K decay constants and $^{40}Ar^{*}/^{40}K$ for the Fish Canyon sanidine standard, and improved accuracy for $^{40}Ar/^{39}Ar$ geochronology. Geochim Cosmochim Acta 74:5349–5367

Renne PR, Norman EB (2001) Determination of the half-life of ^{37}Ar by mass spectrometry. Physical Review C 63(047302):3

Salminen J, Pesonen LJ, Reimold WU, Donadini D, Gibson RL (2009) Paleomagnetic and rock magnetic study of the Vredefort Impact Structure and the Johannesburg Dome, Kaapvaal craton, South Africa - Implications for the apparent polar wander path of the Kaapvaal craton during the Mesoproterozoic. Precambr Res 168:169–184

Söderlund U, Johansson L (2002) A simple way to extract baddeleyite (ZrO_2). Geochem, Geophys, Geosyst 3:10–14

Söderlund U, Hofmann A, Klausen MB, Olsson JR, Ernst RE, Persson P-O (2010) Towards a complete magmatic barcode for the Zimbabwe craton: Baddeleyite U-Pb dating of regional dolerite dyke swarms and sill complexes. Precambr Res 183:388–398

Stacey JS, Kramers JD (1975) Approximation of terrestrial lead isotope evolution by a two-stage model. Earth Planet Sci Lett 26:207–221

Stoenner RW, Schaeffer OA, Katcoff S (1965) Half-lives of argon-37, argon-39, and argon-42. Science 148:1325–1328

Swanson-Hysell NL, Kilian TM, Hanson RE (2015) A new grand mean paleomagnetic pole for the 1.11 Ga Umkondo Large Igneous Province with implications for paleogeography and the geomagnetic field. Geophys J Int 203:2237–2247

Verwoerd WJ (2006) The Pilanesberg Alkaline Province. In: Johnston MR, Anhaeusser CR, Thomas RJ (eds) The geology of South Africa. Geological Society of South Africa, Johannesburg, Council for Geoscience, Pretoria, pp 381–393

Wu F-Y, Yang Y-H, Li Q-L, Mitchell RH, Dawson JB, Brandl G, Yuhara M (2011) In situ determinations of U–Pb ages and Sr–Nd–Hf isotopic constraints on petrogenesis of the Phalaborwa carbonatite Complex, South Africa. Lithos 127:309–322

Zeh A, Ovtcharova M, Wilson AH, Schaltegger U (2015) The Bushveld Complex was emplaced and cooled in less than one million years-results of zirconology, and geotectonic implications. Earth planet Sci Lett 418:103–114

New U–Pb Baddeleyite Ages of Mafic Dyke Swarms of the West African and Amazonian Cratons: Implication for Their Configuration in Supercontinents Through Time

L. Baratoux, U. Söderlund, R. E. Ernst, E. de Roever, M. W. Jessell, S. Kamo, S. Naba, S. Perrouty, V. Metelka, D. Yatte, M. Grenholm, D. P. Diallo, P. M. Ndiaye, E. Dioh, C. Cournède, M. Benoit, D. Baratoux, N. Youbi, S. Rousse and A. Bendaoud

Abstract Eight different generations of dolerite dykes crosscutting the Paleoproterozoic basement in West Africa and one in South America were dated using the high precision U–Pb TIMS method on baddeleyite. Some of the individual dykes reach over 300 km in length and they are considered parts of much larger systems of mafic dyke swarms representing the plumbing systems for large igneous provinces (LIPs). The new U–Pb ages obtained for the investigated swarms in the southern West African Craton (WAC) are the following (oldest to youngest): 1791 ± 3 Ma for the N010° Libiri swarm, 1764 ± 4 Ma for the N035° Kédougou swarm, 1575 ± 5

L. Baratoux (✉) · M. W. Jessell · M. Benoit · D. Baratoux · S. Rousse
Laboratoire GET, Université Toulouse 3 Paul Sabatier, IRD, CNRS, UMR 5563, Observatoire Midi-Pyrénées, 31400 Toulouse, France
e-mail: lenka.baratoux@ird.fr

L. Baratoux · E. Dioh · D. Baratoux
Institut Fondamental d'Afrique Noire, Cheikh Anta Diop, Dakar, Senegal

U. Söderlund
Department of Geology, Lund University, Sölvegatan 12, 223 62 Lund, Sweden

U. Söderlund
Department of Geosciences, Swedish Museum of Natural History, Box 50 007, 104 05 Stockholm, Sweden

R. E. Ernst
Department of Earth Sciences, Carleton University, Ottawa, ON K1S 5B6, Canada

R. E. Ernst
Department of Geology and Geography, Tomsk State University, 36 Lenin Ave, Tomsk 634050, Russia

E. de Roever
Department of Geology and Geochemistry, Vrije Universiteit Amsterdam, de Boelelaan 1085, 1081 HV Amsterdam, The Netherlands

R. K. Srivastava et al. (eds.), *Dyke Swarms of the World: A Modern Perspective*, Springer Geology, https://doi.org/10.1007/978-981-13-1666-1_7

for the N100° Korsimoro swarm, ~1525–1529 Ma for the N130° Essakane swarm, 1521 ± 3 Ma for the N90° Sambarabougou swarm, 915 ± 7 Ma for the N070° Oda swarm, 867 ± 16 Ma for the N355° Manso swarm, 202 ± 5 Ma and 198 ± 16 Ma for the N040° Hounde swarm, and 200 ± 3 Ma for the sills in the Taoudeni basin. The last ones are related to the Central Atlantic Magmatic Province (CAMP) event. The Hounde swarm is oblique to the dominant radiating CAMP swarm and may be linked with the similar-trending elongate Kakoulima intrusion in Guinea. In addition, the N150° Käyser swarm (Amazonian craton, South America) is dated at 1528 ± 2 Ma, providing a robust match with the Essakane swarm in a standard Amazonia-West African craton reconstruction, and resulting in a combined linear swarm >1500 km by >1500 km in extent. The Precambrian LIP barcode ages of c. 1790, 1765–1750, 1575, 1520, 915. 870 Ma for the WAC are compared with the global LIP record to identify possible matches on other crustal blocks, with reconstruction implications. These results contribute to the refinement of the magmatic 'barcode' for the West African and Amazonian cratons, representing the first steps towards plausible global paleogeographic reconstructions involving the West African and Amazonian cratons.

1 Introduction

The history of the West African Craton (WAC) and the Guiana Shield of the Amazonian Craton spans over 3.5 Ga, with evidence of multiple mafic intrusive events, many of which in the WAC having an undetermined age (Jessell et al. 2015). The southern WAC mostly consists of a series of sub-parallel greenstone belts trending N–S

M. W. Jessell · M. Grenholm
Centre for Exploration Targeting, School of Earth Sciences, The University of Western Australia, 35 Stirling Highway, Crawley, WA 6009, Australia

S. Kamo
Jack Satterly Geochronology Laboratory, Department of Earth Sciences, University of Toronto, 22 Russell Street, Toronto, ON M5S 3B1, Canada

S. Naba
Département de Géologie, Université Joseph Ki Zerbo, Ouagadougou, Burkina Faso

S. Perrouty
Harquail School of Earth Sciences, Mineral Exploration Research Centre, Laurentian University, 935 Ramsey Lake Road, Sudbury, ON P3E 2C6, Canada

V. Metelka
CSIRO Mineral Resources, Australian Resources Research Centre, 26 Dick Perry Avenue, Kensington, WA 6151, Australia

D. Yatte · D. P. Diallo · P. M. Ndiaye
Département de Géologie, Université Cheikh Anta Diop, Dakar, Senegal

C. Cournède
Department of Earth and Planetary Sciences, University of California, 1156, High Street, Santa Cruz, CA 95064, USA

to NE–SW, separated by either tonalite–trondhjemite–granodiorite (TTG)-granite domains or by sedimentary basins. These terranes are partially overlain by Mesoproterozoic to Quaternary sediments. The crosscutting dykes have attracted attention from researchers since May (1971) interpreted some swarms in terms of the Late Triassic to Early Jurassic opening of the Atlantic Ocean. Recent dating has shown that the mafic intrusive record is not limited to this 200 Ma event and instead extends back to at least 2.73–2.68 Ga (El Bahat et al. 2013; Kouyaté et al. 2013; Tait et al. 2013; Söderlund et al. 2013a, 2014) and has continued up until at least the last 500,000 years (Crevola et al. 1994).

Mafic dyke swarms may simultaneously provide information on the geodynamic setting, geochemical characteristics of their mantle source areas, age distribution and paleomagnetic-determined positions (paleolatitude and paleo-orientations) at the time of their formation (Ernst 2014 and references therein). There have been several such studies of West African (in both the southern and northern parts of the craton) and also of Amazonian dykes (e.g. Nomade et al. 2001, 2002, 2003; de Roever et al. 2003; El Bahat et al. 2013; Kouyaté et al. 2013; Youbi et al. 2013).

Regional dyke swarms are the plumbing system for large igneous provinces (LIPs) (e.g. Ernst and Buchan 1997; Ernst 2014). They can be linear, curvilinear or have a radiating distribution with a focal point of the swarm interpreted to mark a mantle plume centre. Recently a new class of giant circumferential swarms (hundreds to nearly 2000 km in diameter) circumscribing plume centres has been recognized (Buchan and Ernst 2018a, b).

Regional-scale mapping of mafic dykes based on airborne geophysical data is the first step in the interpretation of mafic dyke swarms as it aids in regional correlation of outcrop scale observations. Mafic dykes are typically rich in magnetite giving them simultaneously a high susceptibility and remanence. Since they often cross-cut the stratigraphy, they typically show up clearly in magnetic anomaly maps. The new map of magnetic dyke swarms for the WAC based on aeromagnetic data (Jessell et al. 2015) has demonstrated the wide range of orientations of dyke swarms that occur, even at the local scale (Fig. 1a). This suggests the existence of multiple events, each of which needs to be ultimately characterized by geochemistry and paleomagnetism, but the most essential information is their age.

This paper describes the outcomes of a U–Pb geochronology study of mafic swarms from southern WAC distinguished in the Jessell et al (2015) aeromagnetic study, and which vary in length from 300 to 1000 km.

N. Youbi
Department of Geology, Faculty of Sciences-Semlalia, Cadi Ayyad University, Marrakesh, Morocco

N. Youbi
Instituto Dom Luiz, Faculdade de Ciências, Universidade de Lisboa, 1749-016 Lisbon, Portugal

A. Bendaoud
LGGIP, Faculty of Earth Sciences and Country Planning, University of Sciences and Technology, BP 32, Bab Ezzouar, Algiers, Algeria

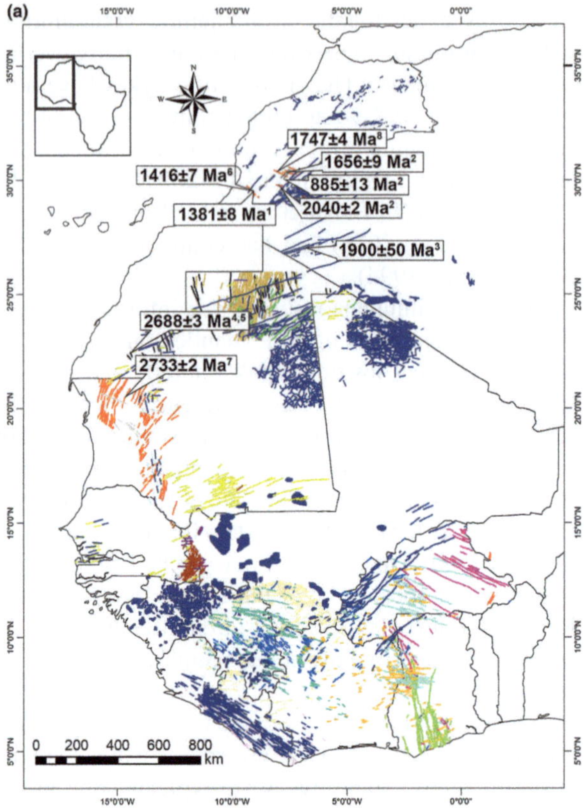

Fig. 1 **a** A map of the West African mafic dykes (after Jessell et al. 2015; and from the ArcGIS CAMP compilation of A. Marzoli and L. Parisio produced under contract to LIPs-Reconstruction Project in 2014 (cf. Ernst et al. 2013). Colour coding is based on the dyke orientation. Representative ages of all previously dated dyke swarms (only U–Pb and Pb–Pb methods on zircon and baddeleyite are shown) are indicated with the corresponding references: (1) El Bahat et al. (2013); (2) Kouyaté et al. (2013); (3) Lefort and Aïfa (2001) (4) Söderlund et al. (2013a); (5) Söderlund et al. (2014); (6) Söderlund et al. (2013b); (7) Tait et al. (2013); (8) Youbi et al. (2013). Note that for some swarms, more ages exist. See the complete list in Jessell et al. (2015). No U–Pb or Pb–Pb age exists for the southern part of the WAC. **b** Subset of the map of southern West African mafic dykes (after Jessell et al. 2015). The studied dyke swarms are colored based on their orientation and are given local names, while those in black remain undated or not dated by U–Pb methods. Dated samples are located in the map with the ages indicated (see Table 1 for details with location coordinates). **c** Simplified map of the Guiana Shield geology (northern South America) (after Kroonenberg et al. 2016) showing the dyke swarms (after Reis et al. 2013). The Käyser swarm is indicated in pink colour with the new age indicated. Im = Imataca block; Am = Amapá block; GF = Guri fault. Overall, fourteen dykes belonging to 8 swarms have been dated in this study

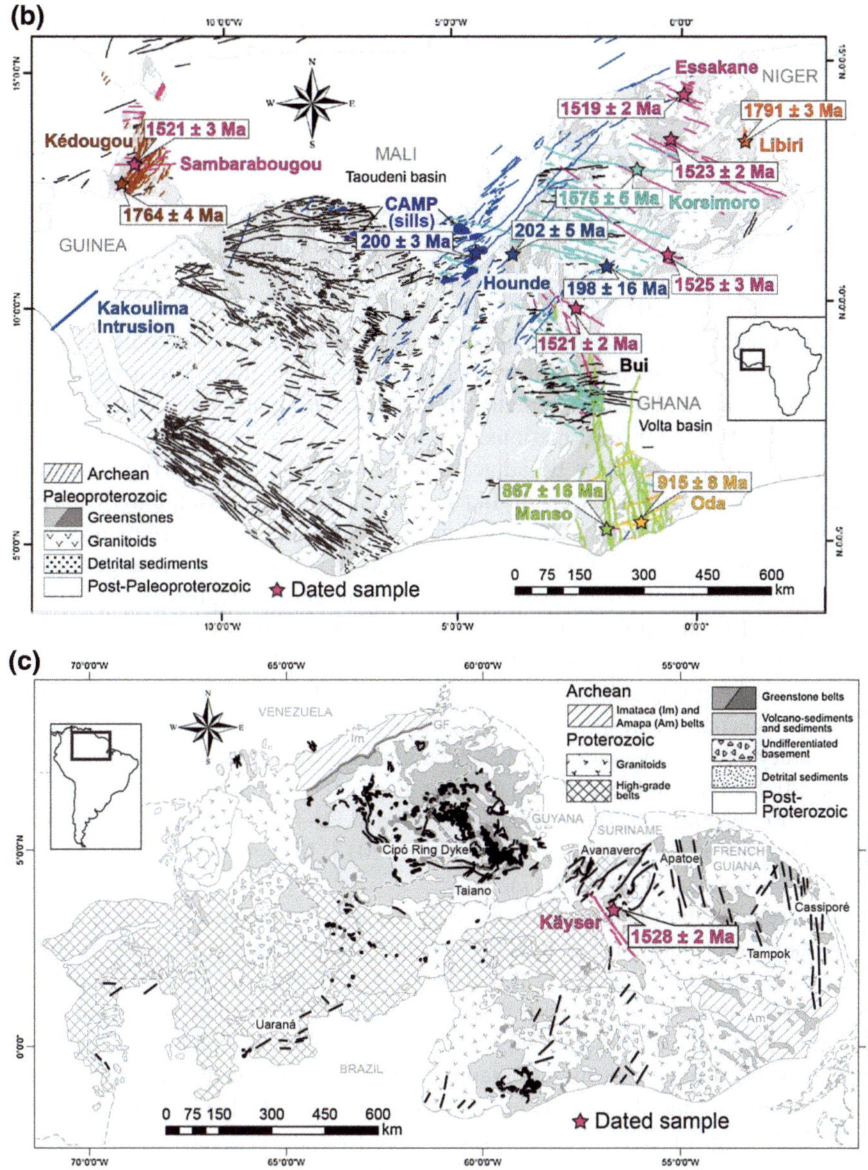

Fig. 1 (continued)

2 Regional Setting and Previous Geochronology of Dyke Swarms

2.1 West African Craton

The oldest rocks in the southern WAC (Ivory Coast, Liberia, Sierra Leone and Guinea) are Archean in age yielding magmatic and inherited ages of 3.2–3.6 Ga (Fig. 1b, Christian et al. 2004; Thiéblemont 2005; Gärtner et al. 2013; Bea et al. 2013; Montero et al. 2014; Kouamelan et al. 2015). They consist of felsic and mafic gneisses, tonalities–trondhjemites–granodiorites (TTGs) and migmatites associated with narrow greenstone belts of low to high metamorphic grade, supracrustal rocks comprising tholeiitic basalts and metasediments including turbidites, conglomerates and extensive banded iron formations.

Paleoproterozoic domains lie adjacent to the Archean domain (Fig. 1) and are characterised by the typical Archean-like greenstone-granitoid assemblages that principally consist of volcanic, volcano-sedimentary, and sedimentary sequences separated either by extensive TTG and granite provinces or by slightly younger but still Paleoproterozoic sedimentary basins. The Archean and Paleoproterozoic domains were affected by multiple tectono-metamorphic and magmatic events, referred to as Leonian (3.2–3.0 Ga, see references in Baratoux et al. (2011) for this and subsequent geochronological constraints), Liberian (2.9–2.8 Ga) and (Eo) Eburnean (2.15–2.07 Ga) orogenies. In particular, the whole complex of volcanic, volcano-sedimentary and sedimentary units has been intruded by several generations of granitoids, which were emplaced during discrete magmatic pulses from ~3200–3000 Ma, 2900–2700 Ma, and ~2250 to 2060 Ma (U–Pb and Pb–Pb zircon ages; Thiéblemont 2005; Baratoux et al. 2011; Tshibubudze et al. 2013; Kouamelan et al. 2015; Block et al. 2016; Parra-Avila et al. 2017).

The Archean and Paleoproterozoic domains are unconformably overlain by the Meso- and Neoproterozoic (up to 1200 Ma) and younger sediments of the Taoudeni, Iullemeden and Volta basins (Teal and Kah 2005; Rooney et al. 2010; Kah et al. 2012). Mafic dyke swarms (with discrete trends) and associated sills cut across the entire Proterozoic domain. Only a few swarms and associated sills have been previously dated and these studies obtained 7 clusters of ages in the Moroccan Sahara (Western Sahara) and Anti-Atlas inliers: 2688, 2040, 1380–1416, 1750, 1650, 885 and 612 Ma (Fig. 1a) (all U-Pb baddeleyite ages; El Bahat et al. 2013; Kouyaté et al. 2013; Söderlund et al. 2013a, b, 2014; Youbi et al. 2013; Aït Lahna et al. 2016; El Bahat et al. 2017; Ikenne et al. 2017). Some dykes in the WAC belong to the widely distributed 201 Ma Central Atlantic Magmatic Province (CAMP, May 1971; Marzoli et al. 1999, 2018; Bertrand et al. 2014; Davies et al. 2017). Others are overlain by the Taoudeni Basin, and hence are pre-CAMP, whereas volcanic rocks in Senegal can be as young as 500,000 years (Crevola et al. 1994).

The CAMP (LIP) was emplaced at 201 Ma, in part coincident with the Triassic–Jurassic boundary (e.g. Schoene et al. 2010; Blackburn et al. 2013; Davies et al. 2017), during the early stages of the break-up of the supercontinent of Pangaea that led to the opening of the Central Atlantic Ocean. CAMP magmatism is nowadays represented by remnants of intrusive (layered intrusions, sills, dykes) and extrusive rocks that occur in once-contiguous parts of northwestern Africa, southwestern Europe, and North and South America (e.g. McHone and Puffer 2003; Youbi et al. 2003; Knight et al. 2004; Marzoli et al. 2018; Bertrand et al. 2014). It may have covered over ~10^7 km^2. More recently, Davies et al. (2017) obtained high precision U–Pb zircon ages for sills, dykes and layered intrusion from all over the CAMP. Their data suggest that onset of CAMP started at 201.6 Ma and preceded the 201.564 Ma end-Triassic mass extinction event, contributing to an ongoing discussion about the casual relationship between the CAMP volcanism and mass extinctions (Tanner et al. 2004; Deenen et al. 2010).

One of the main characteristics of the WAC is the absence of Mesoproterozoic events or rocks suggesting a quiescent period between 1.7 and 1.0 Ga (e.g. Ennih and Liégeois 2008 and references therein). The lack of detrital zircons of that age can be considered characteristic of sediments sourced from the WAC (e.g. Abati et al. 2010; Gärtner et al. 2013). However, this conventional viewpoint has been challenged by recent U–Pb results (El Bahat et al. 2013; Kouyaté et al. 2013; Söderlund et al. 2013b; Youbi et al. 2013; Aït Lahna et al. 2016; Ikenne et al. 2017; Tapsoba et al. 2018) which demonstrate important intraplate magmatic events at 1.28–1.24, 1.38–1.41, 1.65 and 1.75 Ga based on dating of mafic dykes by the U–Pb method on baddeleyites. Furthermore, rhenium–osmium (Re–Os) geochronology from the sedimentary succession of the Atar Group in the Taoudeni basin (Mauritania) indicate emplacement ages ranging from 1109 ± 22 to 1105 ± 37 Ma (Rooney et al. 2010). This Mesoproterozoic age is confirmed by chemostratigraphy (Kah et al. 2012) and microfossils (Beghin et al. 2017). Our study identifies four new events for the southern WAC that add supporting evidence for Mesoproterozoic mafic intrusive activity.

In the southern part of the WAC, prior to the present study all of the existing ages were obtained by K–Ar and Ar–Ar method on bulk rock or biotite (Table 1). The dyke set (now named the Libiri swarm) was dated by Ama Salah (1991) and the ages range from 896 ± 25 to 1378 ± 36 Ma (K–Ar on bulk rock). The K–Ar bulk rock ages for the Kédougou dyke swarm in Senegal range between 853 ± 11 and 1791 ± 24 Ma (Saraya-Kédougou and Kossanto-Dalafi map sheets; Delor et al. 2010; Goujou et al. 2010). The Essakane swarm (Burkina Faso) was dated at 1280 ± 26 and 1236 ± 20 Ma (Ar/Ar on whole rock, Tapsoba et al. 2018) and at 1814 ± 21 Ma (K–Ar on biotite; Pissila map sheet, Kagone et al. 2003; Castaing et al. 2003). K–Ar bulk rock ages of 1165 ± 15, 945 ± 13, and 806 ± 11 Ma were obtained for the Sambarabougou swarm in Senegal (Kossanto-Dalafi and Bakel-Semme map sheets; Lahondère et al. 2010; Goujou et al. 2010). The CAMP ages in West Africa cluster around 201 Ma as elsewhere in the CAMP event (Youbi et al. 2003; Davies et al. 2017). The now named Korsimoro, Oda, and Manso swarms were not previously dated.

Table 1 Location, orientation and ages (literature and this work) of the studied dykes and sills

	Swarm	Swarm label in Jessell et al. (2015)	Dyke trend	Sample	Latitude	Longitude	Age (Ma) in literature	Method	Reference	Age (Ma) (this paper)
West Africa	Libiri	U	N005-010°	DO109	13.398	1.25478	896 ± 25 to 1378 ± 36	K–Ar on bulk rock	Ama Salah (1991)	1791 ± 3
	Kédougou	L	N030-040°	BK031	12.68266	−12.192	853 ± 11 to 1791 ± 24	K–Ar on bulk rock	Delor et al. (2010), Goujou et al. (2010)	1764 ± 4
	Korsimoro	S	N100°	DO112	12.88598	−1.07054	Undated			1575 ± 5
	Essakane	T	N130°	DO115	14.42709	−0.01882	1814 ± 26	K–Ar on biotite	Castaing et al. (2003)	1519 ± 2
	Essakane	T	N130°	YU14	11.07739	−0.48632	1814 ± 26	K–Ar on biotite	Castaing et al. (2003)	1525 ± 3
	Essakane	T	N130°	DO09	13.49736	−0.32885	1814 ± 26	K–Ar on biotite	Castaing et al. (2003)	1523 ± 2
	Essakane	T	N130°	BN29	9.9983	−2.43392	1814 ± 26	K–Ar on biotite	Castaing et al. (2003)	1521 ± 2
	Essakane (Sam-barabougou)	M	N090°	BK001	13.10357	−11.924	1165 ± 15; 945 ± 13 and 806 ± 11	K–Ar on bulk rock	Lahondère et al. (2010), Goujou et al. (2010)	1521 ± 3
	Oda	W	N070°	MG-8	5.433776	−1.15767	Undated			915 ± 7
	Manso	X	N355°	GH370	5.308322	−1.87074	Undated			855 ± 10
	Hounde	R	N030-045°	BN24	10.86821	−1.77979	185–197	Ar–Ar on bulk rock	Mauche et al. (1989), Lanphere and Dalrymple (1971, 1976), Dalrymple et al. (1975) and Wahl et al. (2007)	198 ± 16

(continued)

Table 1 (continued)

	Swarm	Swarm label in Jessell et al. (2015)	Dyke trend	Sample	Latitude	Longitude	Age (Ma) in literature	Method	Reference	Age (Ma) (this paper)
	Hounde	R	N030-045°	HO136	11.16114	−3.76703	185–197	Ar–Ar on bulk rock	Mauche et al. (1989), Lanphere and Dalrymple (1971, 1976), Dalrymple et al. (1975) and Wahl et al. (2007)	202 ± 5
	CAMP	Sills are not included in the paper	Sill	HO427	11.32165	−4.36509	185–197	Ar–Ar on bulk rock	Mauche et al. (1989), Lanphere and Dalrymple (1971, 1976), Dalrymple et al. (1975) and Wahl et al. (2007)	200 ± 3
South America	Käyser	Not included in the paper	N150°	69Sur74	3.488373	−56.733	1501 ± 5	Ar–Ar on biotite	de Roever et al. (2003)	1528 ± 2

2.2 The Guiana Shield, Amazonian Craton

The Guiana Shield forms the northern part of the Amazonian Craton and is composed of two Archean nuclei located at the east and west of the Shield, respectively, and Paleoproterozoic greenstone-granitoid domain in the central part (Fig. 1c) (Gibbs and Barron 1993; Delor et al. 2003; Kroonenberg and de Roever 2010; Kroonenberg et al. 2016; Santos et al. 2000; Tassinari et al. 2000). The Venezuelan Imataca block in the west is separated from the Paleoproterozoic domain by the Guri fault and is presumably a counterpart of the West African Man craton delimited by the Sassandra fault (Nomade et al. 2003; Bispo-Santos et al. 2014a). The Imataca block and the eastern Amapá block in Brazil were dated between 3.3 and 2.6 Ga (Montgomery and Hurley 1978; Montgomery 1979; Tassinari et al. 2001; Avelar et al. 2003).

The Paleoproterozoic domain is built of a long TTG-greenstone belt in the north (2.26–2.07 Ga) along with a high-grade metamorphic belt and a belt of felsic volcanics and granites in the central part (2.00–1.96 Ga) (Delor et al. 2003). The Guiana Shield was assembled mainly during the Trans-Amazonian Orogeny, which was polyphase and occurred between 2.18 and 1.95 Ga (Ledru et al. 1994; Delor et al. 2003; Vanderhaeghe et al. 1998). Younger phases of this orogeny may be correlated with the Eburnean Orogeny in the WAC. The Trans-Amazonian Orogeny resulted in deformation and metamorphism of the volcanic and sedimentary units in a collisional setting with metamorphic conditions attaining granulite facies in the central part of the Guiana Shield in the Bakhuis and Cauarane-Coeroeni belts (Delor et al. 2003; Kroonenberg et al. 2016; Vanderhaeghe et al. 1998).

The southernmost part of the Guiana Shield is formed of felsic volcanics and granitic intrusions, dated at 1.89–1.81 Ga (Reis et al. 2003). This consists of (1) the large Uatumã SLIP (Silicic LIP) dated between 1.89 and 1.86 Ga (U–Pb zircon and baddeleyite) observed also in Central-Brazil Shield (Antonio et al. 2017; Klein et al. 2012; Teixeira et al. 2016) and (2) a minor intrusion of the 1.82 Ga SLIP of the Madeira Suite located only in northern craton (Bastos Neto et al. 2014; Siachoque et al. 2017). These felsic events are associated with well-dated mafic dykes (Teixeira et al. 2016). The Río Negro belt in the southwest yields even younger ages of 1.86–1.72 Ga and is intruded by Mesoproterozoic granitoids (1.59–1.51 Ga) (Tassinari et al. 1996; Tassinari and Macambira 1999; Santos et al. 2000). Most of the ages on basement rocks were obtained by conventional U–Pb or Pb–Pb methods on zircon; however, U–Pb SHRIMP ages on baddeleyites were also effectuated and they yield an age of 1527 ± 7 Ma for the Repartimento anorthosite of the Mucajaí complex (Santos et al. 1999). Scarce sediments of the younger platform cover (1.3–1.2 Ga) were found on the Río Negro belt (Kroonenberg et al. 2016). Around 1200 Ma the Nickerie Metamorphic Episode caused the development of large SW–NE shear zones and K–Ar and Rb–Sr resetting in the western half of the Shield. The Nickerie Metamorphic Episode has been attributed to the Grenvillian Laurentia-Amazonia collision along the western border of the Shield (e.g., Priem et al. 1971; Cordani et al. 2010; Kroonenberg and de Roever 2010).

Three prominent dyke swarms (Avanavero, Apatoe and Käyser) have been identified in the northern part of the Guiana Shield (de Roever et al. 2003). The N45° Avanavero Suite which also includes sills, yields U–Pb zircon and baddeleyite ages of 1794 ± 4 Ma (Norcross et al. 2000), 1782 ± 3 Ma (Santos et al. 2001, 2002, 2003). U–Pb ages of 1795 ± 2 and 1793 ± 1 Ma from sills and dykes in the Pakaraima Block in the north of the Guiana Shield and Urupi Block in the south, respectively, confirm that they belong to the Avanavero event (Reis et al. 2013). A cooling age supposed to approximate the emplacement age of 1505 ± 5 Ma was obtained for the N150° Käyser swarm by the Ar–Ar method on biotite (de Roever et al. 2003). No U–Pb age were previously published for this swarm. The N160° Apatoe dykes belong to the CAMP LIP suite dated on the basis of approximate 190–200 Ma ages obtained by the Ar–Ar method on biotite and amphibole (Deckart 1996; Deckart et al. 1997). However, no precise U–Pb ages exist for the Apatoe dykes, but their link with CAMP has also confirmed by paleomagnetic data (Nomade et al. 2000).

This study focuses on nine different dyke swarms, 8 from the southern part of the WAC and 1 dyke swarm in the Guiana Shield, Amazonian craton (Fig. 1; Table 1). The swarms were named according to a village located nearby the dated outcrop or using the existing nomenclature. From the oldest to the youngest in the southern WAC (according to the dating results in this paper), the swarms are: Libiri swarm in Niger (N005-010°), Kédougou swarm in Senegal (N030-040°), Korsimoro swarm in Burkina Faso (N100°), Essakane swarm which is widespread in Burkina Faso and extends to Ghana (N130°), Sambarabougou swarm in Senegal (N090°), Oda (N070°) and Manso (N355°) swarms in southern Ghana, and CAMP dolerites, which are represented by the Houndé swarm (N030-045°) and flat-lying sills in the Taoudeni basin. The Käyser swarm (N150°) is present in Suriname (Amazonian craton, South America).

3 Analytical Procedures for Geochronology: Thermal Ionization Mass Spectrometry (TIMS)

Mineral separation on 13 samples comprising 8 dyke swarms from southern WAC and one swarm from the Amazonian Craton (Fig. 1) was performed at Lund University, Sweden, following the extraction technique of Söderlund and Johansson (2002). The yield was typically between 20 and 60 baddeleyite grains after processing approximately 100–200 g of crushed and milled sample. Baddeleyite grains are dark to moderately brown for the Precambrian samples, whereas the younger samples (ca. 200 Ma) are faintly brown to almost colourless, indicating that the shade of colouring is related to the degree of accumulated radiation damage over time. Grains from all samples are fresh without any trace of alteration (Fig. 2).

For this sample set, 1–7 grains per fractions were combined and transferred to clean Teflon® capsules. Grains were washed in numerous steps using ~3M HNO_3, including one step on a hotplate (15–30 min). A small amount of a ^{205}Pb–$^{233-236}U$

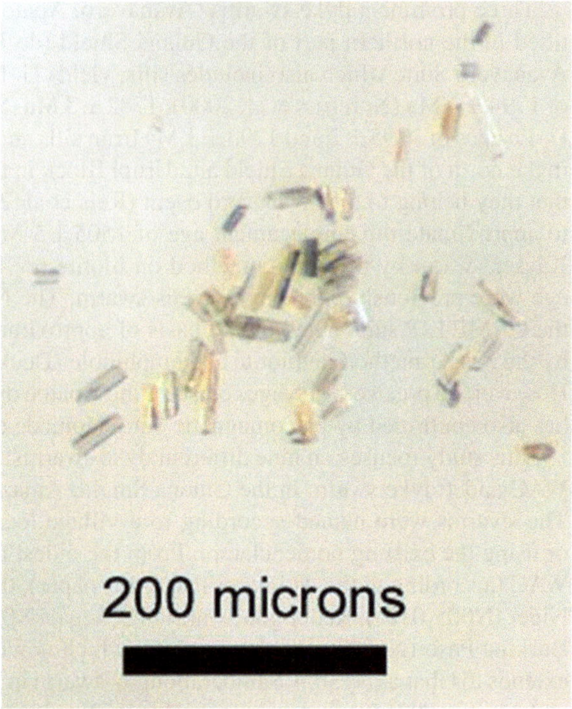

Fig. 2 Representative image of separated baddeleyites from the sample HO136. The grains are faintly brown to colourless, fresh, without any trace of alteration

tracer solution and 10 drops of HF–HNO_3 (10:1) were added to each capsule. Baddeleyite grains were dissolved completely after 3 days in oven under high pressure at a temperature of ~190 °C. The samples were evaporated on hotplate and then dissolved in 10 drops of ~3M HCl.1 drop of 0.25M H_3PO_4 was added to each capsule before they were put on a hotplate and evaporated. U and Pb were loaded on outgassed Re filament together with a small amount of silica gel.

Thermal Ionization Mass Spectrometry (TIMS) on all samples except HO136 was performed at the Laboratory of Isotope Geology at the Museum of Natural History in Stockholm using a Thermo Finnigan Triton thermal ionisation mass spectrometer. For the larger samples, the intensities of 208^{Pb}, ^{207}Pb, ^{206}Pb and ^{205}Pb were measured in static mode using the Faraday detectors. An ETP SEM equipped with a RPQ filter was used to measure the intensity of ^{204}Pb. The SEM-Faraday gain was controlled by measuring a ~5–10 mV signal alternating between Faraday and SEM detectors prior to each Pb analysis. Smaller samples were analyzed in dynamic (peak-switching) mode using the SEM detector. Measurements of Pb-isotopes were done at a filament temperature in the 1220–1260 °C range, while isotopes of U were measured in dynamic mode on the SEM with filament temperatures exceeding 1350 °C. Data reduction was performed using the Excel add-in "Isoplot" of Ludwig (2003); decay constants for ^{238}U and ^{235}U follow those of Jaffey et al. (1971). All errors in age and isotopic ratios are quoted at 2σ. Initial Pb isotope compositions were corrected

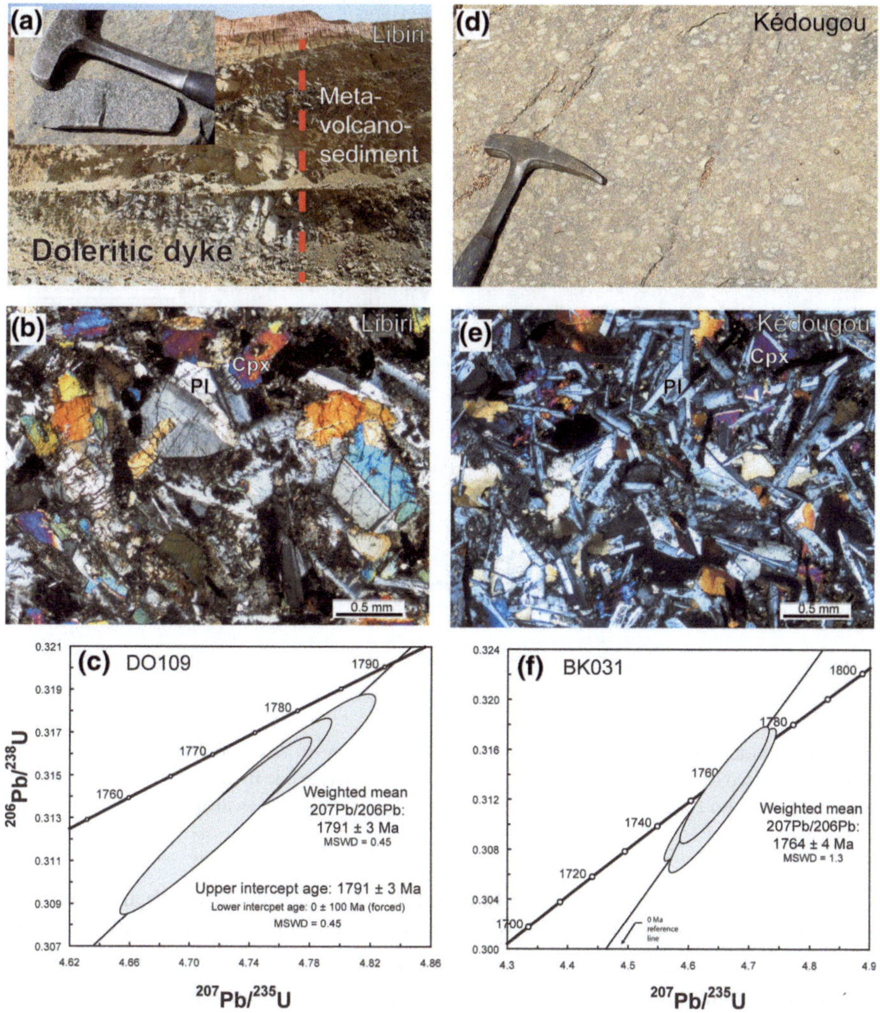

Fig. 3 **a** The Libiri dyke swarm in the Libiri pit in Niger showing the contact between the dyke and the meta-volcano-sedimentary host rock. The small inset in the upper left corner shows the homogeneous doleritic texture. **b** Doleritic texture with predominant clinopyroxene and plagioclase in the Libiri dyke swarm. **c** U–Pb Concordia diagram for the sample DO109 (1791 ± 3 Ma, Libiri swarm); **d** Fine-grained dolerite with large plagioclase phenocrysts of the Kédougou swarm. **e** Fine grained matrix of the Kédougou swarm with abundant plagioclase laths and interstitial clinopyroxene. **f** U–Pb Concordia diagram for the sample BK031 (1764 ± 4 Ma, Kédougou swarm)

according to the global common Pb evolution model of Stacey and Kramers (1975). U–Pb data are presented in Tables 1 and 2 and the fractions are plotted in the concordia diagrams in Figs. 3, 4, 6, 7 and 9.

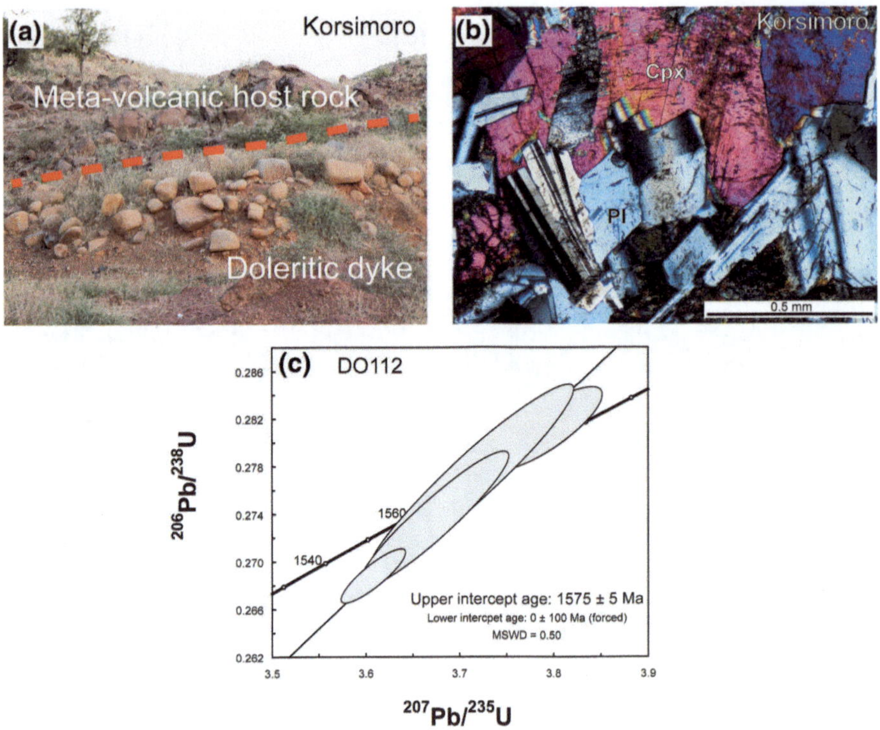

Fig. 4 **a** Contact between the Korsimoro swarm and the host volcanic rock. **b** Fresh doleritic texture composed of sub-automorphic plagioclase and clinopyroxene in the sample DO112. **c** U–Pb Concordia diagram for the sample DO112 (1575 ± 5 Ma, Korsimoro swarm)

Sample HO136 was dated at the Jack Satterly Geochronology Lab at the University of Toronto, Canada. U–Pb analysis was effectuated by isotope dilution thermal ionization mass spectrometry methods (ID-TIMS). Prior to dissolution, baddeleyite crystals were cleaned in 8N HNO_3 at room temperature. For dissolution, the separates were placed in Teflon capsules with a known quantity of a mixed ^{205}Pb–$^{233-235}$U spike (from the EARTHTIME Project; http://www.earth-time.org/u-pb.html), ~0.10 ml of concentrated HF acid and ~0.02 ml of 7N HNO_3 and left at 195 °C for ~3 days (Krogh 1973). Samples were dried to a precipitate and re-dissolved in ~0.15 ml of 3N HCl overnight. U and Pb were isolated from the baddeleyite solution using anion exchange columns and HCl (Krogh 1973), deposited onto outgassed rhenium filaments with silica gel (Gerstenberger and Haase 1997), and analyzed in a VG354 mass spectrometer using a Daly detector in pulse counting mode. Corrections of the ^{206}Pb–^{238}U ages for initial ^{230}T^{h} disequilibrium in the baddeleyite have been made assuming a Th/U ratio in the magma of 4.2. All common Pb was assigned to procedural Pb blank. Dead time of the measuring system for Pb and U was 16 and 14 ns, respectively. The mass discrimination correction for the Daly detector is constant at 0.05% per atomic mass unit. Amplifier gains and Daly characteristics were monitored using the SRM 982 Pb standard. Thermal mass discrimination corrections are 0.10% per atomic mass unit for Pb and were corrected internally (cycle by cycle) for U. Decay constants are those of Jaffey et al. (1971). All age errors quoted in the text and table, and error ellipses in the Concordia diagrams are given at the 95% confidence interval. VG Sector software was used for data acquisition. In-house data reduction software in Visual Basic by D. W. Davis was used. Plotting and age calculations were done using Isoplot 3.00 (Ludwig 2003).

4 Field Description, Petrography, and U–Pb Geochronology of the Studied Dykes

As discussed above, local names were given to individual dyke swarms that had not been previously distinguished or named, otherwise the existing names were kept (Sambarabougou, Kédougou swarms). The dated swarms are related to the swarm labels used by Jessell et al. (2015) in Table 1. Most of the swarms are represented by one sample. However, the Essakane swarm was dated on five samples from West Africa. This swarm extends from Burkina Faso through Ghana (Essakane swarm) to eastern Senegal, where the local name for these dykes is the Sambarabougou swarm. Three samples were dated from the CAMP in Burkina Faso: two of them from the NE Hounde swarm and one from a sill in the Taoudeni basin sediments. All dykes correspond to tholeiitic basalts except for the Käyser swarm, which is alkaline, and the Manso swarm, which is transitional (unpublished data; detailed geochemical analysis being prepared as a separate publication). One sample from the Käyser swarm was dated in northern South America. The studied samples are discussed below in the order of decreasing U–Pb age obtained in this study.

4.1 The 1790–1750 Ma Events

4.1.1 Libiri Swarm

Sample DO109

Sample DO109 was taken in the Samira Mine, Libiri pit in Niger (Fig. 3a). The dyke consists of dark grey to black, medium to fine-grained dolerite. Chilled margins are preserved. The sampled dyke is approximately 20 m wide. It trends ~N005-010°, and is mapped as part of a NNE trending swarm extending to the north from the sampled site (Fig. 1). Host rocks are greenschist facies Birimian volcano-sedimentary rocks, (graphitic) shales and argillites of the Samira Formation, which are part of the volcano-sedimentary sequences of the Birimian basement. The sample is composed of 0.5–1 mm large ophitic clinopyroxene and plagioclase, with minor Fe and Ti oxides. The sample exhibits only slight sericitization and some pyroxenes show overgrowths by amphibole (Fig. 3b).

The best-quality baddeleyite grains were divided into three fractions. As shown in the Concordia diagram (Fig. 3c) the fractions overlap, plotting discordant by 1–2%. The weighted mean $^{207}Pb/^{206}Pb$ date is identical with the upper intercept of 1791 ± 3 Ma, when using a forced lower intercept through the origin. This result is considered as the best age estimate for the emplacement age of this Libiri dyke.

4.1.2 Kédougou Swarm

Sample BK031

Sample BK031 was obtained from a dolerite dyke in eastern Senegal. This N030-040° trending sub-vertical 30–100 m wide mafic dyke is located ~30 km SW of Mako and it intrudes the Diale-Dalema sedimentary basin. Large plagioclase phenocrysts (up to 3 cm in size), often sericitized, are randomly distributed in medium-grained (0.5–1 mm) ophitic matrix composed of olivine, plagioclase, clinopyroxene, minor Fe and Ti oxides and secondary magnesiohornblende and chlorite (Fig. 3d, e). Some samples of this swarm contain large plagioclase phenocrysts (up to 4 cm, Fig. 3d).

The three analysed baddeleyite fractions plot less than ca. 1% discordant, and are all concordant within error. The weighted mean of 1764 ± 4 Ma is considered the best age estimate of this dyke (Fig. 3f; Table 2).

4.2 The 1575 Event

4.2.1 Korsimoro Swarm

Sample DO112

This sample originates from a subvertical, 30 m wide dyke that belongs to a N100° trending swarm. It was sampled in a road cut 10 km SW of Korsimoro, on the road to Kaya. The sampled dyke is well-preserved, massive but fractured by later faults/joints. Chilled margins are narrow. The host rocks are Birimian greenschist facies andesites of the Goren-Kaya belt (Fig. 4a). This dyke is composed of medium-grained dolerite. The sample contains clinopyroxene (2–5 mm), automorphic plagioclase laths (1–3 mm), and minor Fe and Ti oxides (Fig. 4b). Pyroxene is overgrown by amphibole and plagioclase is affected by weak sericitization.

Three of four baddeleyite fractions plot concordant within error and the weighted mean of 1575 ± 5 Ma is interpreted to be the best age estimate of this sample (Fig. 4c; Table 2).

Table 2 U-Pb ID-TIMS data

Analysis no. (number of grains)	U/Th	Pbc/Pbtot[a]	^{206}Pb/^{204}Pb raw[b]	^{207}Pb/^{235}U (corr)[c]	±2s % err	^{206}Pb/^{238}U	± 2s% err	^{207}Pb/^{235}U (age, Ma)	^{206}Pb/^{238}U	^{207}Pb/^{206}Pb	±2s	Concordance
DO 109												
Bd-1 (2 grains)	8.9	0.063	998.9	4.7785	0.77	0.31604	0.71	1781.1	1770.4	1793.7	6.1	0.987
Bd-2 (2 grains)	5.4	0.074	841.3	4.7411	0.92	0.31417	0.90	1774.5	1761.2	1790.3	5.3	0.984
Bd-3 (4 grains)	6.7	0.068	933.6	4.7174	1.11	0.31257	1.09	1770.3	1753.3	1790.4	5.5	0.979
BK 031												
Bd-1 (7 grains)	11.7	0.093	673.4	4.6133	0.95	0.31051	0.92	1751.7	1743.2	1761.8	5.7	0.989
Bd-2 (4 grains)	9.5	0.178	337.3	4.6559	1.56	0.31189	1.53	1759.4	1750.0	1770.5	9.3	0.988
Bd-3 (1 grain)	18.5	0.095	678.1	4.6593	1.29	0.31312	1.23	1760.0	1756.1	1764.6	8.6	0.995
DO 112												
Bd-1 (4 grains)	4.2	0.361	123.3	3.7923	1.14	0.28125	0.83	1591.1	1597.7	1582.5	13.6	1.010
Bd-2 (3 grains)	18.6	0.142	435.2	3.7146	1.39	0.27716	1.32	1574.5	1577.0	1571.1	9.3	1.004
Bd-3 (3 grains)	9.2	0.121	490.2	3.6071	0.79	0.26875	0.70	1551.1	1534.5	1573.8	7.3	0.975

(continued)

Table 2 (continued)

Analysis no. (number of grains)	U/Th	Pbc/Pbtot[a]	$^{206}Pb/^{204}Pb$ raw[b]	$^{207}Pb/^{235}U$ (corr)[c]	±2s % err	$^{206}Pb/^{238}U$	± 2s% err	$^{207}Pb/^{235}U$ (age, Ma)	$^{206}Pb/^{238}U$	$^{207}Pb/^{206}Pb$	±2s	Concordance
Bd-4 (7 grains)	11.4	0.230	254.9	3.6716	1.78	0.27350	1.72	1565.2	1558.6	1574.2	12.0	0.990
YU 14												
Bd-1 (4 grains)	20.9	0.041	1494.9	3.4327	0.35	0.26230	0.26	1511.9	1501.6	1526.4	4.1	0.984
Bd-2 (4 grains)	9.2	0.043	1498.1	3.4239	0.56	0.26191	0.51	1509.9	1499.6	1524.4	4.7	0.984
Bd-3 (7 grains)	12.9	0.065	940.0	3.4054	0.64	0.26043	0.61	1505.6	1492.0	1524.8	4.0	0.979
DO 09												
Bd-1 (3 grains)	11.0	0.042	1531.4	3.4448	0.51	0.26396	0.46	1514.7	1510.1	1521.1	4.5	0.993
Bd-2 (3 grains)	6.5	0.025	2569.5	3.4194	0.40	0.26201	0.35	1508.9	1500.1	1523.7	3.9	0.985
Bd-3 (4 grains)	17.2	0.058	1090.7	3.4354	0.51	0.26294	0.48	1512.5	1504.8	1523.3	4.0	0.988
BK 001												
Bd-1 (4 grains)	19.6	0.041	1604.2	3.4287	0.52	0.26308	0.50	1511.0	1505.6	1518.6	3.5	0.991
Bd-2 (4 grains)	24.0	0.067	975.7	3.4231	0.77	0.26199	0.75	1509.7	1500.0	1523.4	4.9	0.985

(continued)

Table 2 (continued)

Analysis no. (number of grains)	U/Th	Pbc/Pbtot[a]	$^{206}Pb/^{204}Pb$ raw[b]	$^{207}Pb/^{235}U$ (corr)[c]	±2s % err	$^{206}Pb/^{238}U$	± 2s% err	$^{207}Pb/^{235}U$ (age, Ma)	$^{206}Pb/^{238}U$	$^{207}Pb/^{206}Pb$	±2s	Concordance
Bd-3 (1 grain)	22.6	0.107	575.6	3.4735	0.81	0.26569	0.68	1521.2	1518.9	1524.5	8.5	0.996
BN 29												
Bd-1 (5 grains)	8.5	0.017	3109.0	3.4517	0.35	0.26459	0.29	1516.3	1513.3	1520.4	3.3	0.995
Bd-2 (3 grains)	12.3	0.019	2820.4	3.4523	0.25	0.26456	0.21	1516.4	1513.1	1521.0	2.6	0.995
Bd-3 (3 grains)	11.2	0.067	772.8	3.4364	0.62	0.26372	0.55	1512.8	1508.8	1518.3	5.4	0.994
DO 115												
Bd-1 (4 grains)	11.4	0.019	3499.8	3.3221	0.34	0.25478	0.30	1486.3	1463.1	1519.5	3.0	0.963
Bd-2 (3 grains)	8.9	0.035	1871.7	3.4276	0.53	0.26277	0.51	1510.8	1504.0	1520.2	3.2	0.989
Bd-3 (5 grains)	19.6	0.084	744.6	3.3869	0.74	0.26033	0.69	1501.4	1491.5	1515.3	5.6	0.984
Bd-4 (4 grains)	11.6	0.097	639.2	3.5282	0.78	0.27093	0.76	1533.6	1545.5	1517.1	4.9	1.019
69Sur74												
Bd-1 (1 grain)	27.8	0.042	1599.0	3.4763	0.67	0.26520	0.63	1521.8	1516.4	1529.4	5.1	0.991

(continued)

Table 2 (continued)

Analysis no. (number of grains)	U/Th	Pbc/Pbtot[a]	^{206}Pb/^{204}Pb raw[b]	^{207}Pb/^{235}U (corr)[c]	±2s % err	^{206}Pb/^{238}U	± 2s% err	^{207}Pb/^{235}U (age, Ma)	^{206}Pb/^{238}U	^{207}Pb/^{206}Pb	±2s	Concordance
Bd-2 (2 grains)	52.3	0.131	453.5	3.4787	0.62	0.26539	0.53	1522.4	1517.4	1529.4	6.2	0.992
Bd-3 (4 grains)	50.6	0.037	1797.9	3.4576	0.55	0.26425	0.51	1517.6	1511.6	1526.0	4.3	0.991
Bd-4 (3 grains)	30.9	0.029	2271.1	3.4515	0.42	0.26347	0.39	1516.2	1507.6	1528.3	3.5	0.986
MG 8												
Bd-1 (3 grains)	17.3	0.233	297.6	1.4485	2.97	0.15210	2.81	909.3	912.7	900.9	25.9	1.013
Bd-2 (4 grains)	12.5	0.215	315.2	1.4765	2.64	0.15370	2.49	920.8	921.7	918.7	23.1	1.003
Bd-3 (4 grains)	6.2	0.143	468.0	1.4646	1.90	0.15286	1.82	915.9	917.0	913.3	15.5	1.004
GH 370												
Bd-a (5 grains)	3.6	0.160	395.1	1.3133	1.88	0.14012	1.75	851.6	845.4	867.8	17.1	0.974
Bd-b (4 grains)	4.1	0.597	100.5	1.3271	6.87	0.14228	6.50	857.6	857.5	857.8	61.5	1.000

(continued)

Table 2 (continued)

Analysis no. (number of grains)	U/Th	Pbc/Pbtot[a]	$^{206}Pb/^{204}Pb$ raw[b]	$^{207}Pb/^{235}U$ (corr)[c]	±2s % err	$^{206}Pb/^{238}U$	± 2s% err	$^{207}Pb/^{235}U$ (age, Ma)	$^{206}Pb/^{238}U$	$^{207}Pb/^{206}Pb$	±2s	Concordance
Bd-c (6 grains)	2.7	0.843	57.0	1.3416	9.16	0.14404	8.61	863.9	867.4	854.9	90.0	1.015
BN 24												
Bd-1 (4 grains)	0.8	0.462	128.6	0.2153	9.10	0.03114	8.38	198.0	197.7	201.5	101.2	0.981
HO 427												
Bd-1 (4 grains)	20.9	0.361	162.7	0.2190	2.31	0.03164	1.91	201.1	200.8	204.6	32.6	0.981
Bd-2 (4 grains)	15.8	0.265	250.5	0.2172	2.81	0.03159	2.36	199.6	200.5	189.1	38.3	1.060
Bd-3 (2 grains)	26.7	0.664	100.8	0.2138	8.28	0.03090	7.28	196.8	196.2	203.3	104.8	0.965
HO 136												
Bd-1 (15 grains)	5.4	0.750	49.3	0.2	6.60	0.03185	0.50	192.8	201.6	83.5	170	−136

[a]Pbc = common Pb; Pbtot = total Pb (radiogenic + blank + initial)
[b]Measured ratio, corrected for fractionation and spike
[c]Isotopic ratios corrected for fractionation (0.1% per amu for Pb), spike contribution, blank (0.6 pg Pb and 0.06 pg U), and initial common Pb. Initial common Pb corrected with isotopic compositions from the model of Stacey and Kramers (1975) at the age of the sample

4.3 The 1520 Event

4.3.1 Essakane Swarm

Sample DO09
The dyke was sampled in the open pit of the Taparko gold mine. This subvertical dyke has a trend of N130° and has a minimum width of 20 m. The host rocks are greenschist facies Birimian volcanoclastic rocks. The dyke is located close to a tectonically modified intrusive contact with a tonalite, which is undated, but the estimated age is between 2200 and 2100 Ma. The dyke is composed of medium-grained dolerite in the central part with the preserved chilled margin at the rims (Fig. 5a). The dyke postdates the gold mineralization in the mine. Sample DO09 exhibits medium-grained ophitic texture (0.5–1 mm) with minor overgrowth of clinopyroxene by amphibole and weak sericitization of plagioclase. Fe and Ti oxides are present.

Baddeleyite grains were combined into fractions comprising 3–4 grains in each. The three fractions plot about 1% discordant forming a cluster with largely overlapping ellipses. Their weighted mean is 1523 ± 2 Ma, and is interpreted to best reflect the crystallisation age of the dyke (Fig. 6a; Table 2).

Sample DO115
Sample DO115 was obtained from the interior of a N130° trending 50 m wide, sub-vertical dyke located 5 km west of the Essakane gold mine (Fig. 5b). The dyke post-dates gold mineralization. Host rocks are greenschist facies sandstones and arkoses interpreted as early Birimian sediments by Tshibubudze and Hein (2013). The intrusive dyke is a dark grey to black, medium to coarse-grained dolerite, at places greenish due to hornblende secondary alteration (Fig. 5c). Automorphic phenocrysts of clinopyroxene (1–5 mm) are overgrown by magnesio-hornblende. Randomly distributed plagioclase grains (1–5 mm) are sericitized. Some of them form granophyric intergrowths with quartz. Small grains of Fe and Ti oxides are randomly distributed in the rock.

The four baddeleyite fractions analysed are linearly distributed and regression yields an upper intercept age of 1519 ± 2 Ma with a lower forced intercept at 0 Ma (Fig. 6b; Table 2).

Sample YU14
This sample is from a N130° trending subvertical dyke, which has a width of ~30 m. It was sampled in a large outcrop near the road to the Youga gold mine. Chilled margins are not exposed in the outcrop, consisting of disturbed boulders and blocks. The dyke post-dates gold mineralization. Host rocks are greenschist facies andesites that belong to the Birimian units in the Youga/Zabre region. The dolerite has a relatively coarse-grained texture (1–5 mm) with local overgrowths of pyroxene by amphibole. Plagioclase and quartz form granophyric intergrowths with typical sericitization. K-feldspar and Fe and Ti oxides are present in a small amount.

The optically best baddeleyite grains were combined into fractions comprising 3–7 grains in each. The fractions of baddeleyite from sample YU14 largely overlap,

Fig. 5 **a** Subvertical NW-trending ~40 m wide dyke of the Essakane swarm. **b** Contact between the Essakane subvertical dolerite dyke and basement volcanic rocks in the gold mine of Taparko. **c** Coarse-grained dolerite of the Essakane dyke sample. **d** Large boulders of the Sambarabougou swarm. **e** Coarse-grained dolerite/gabbro of the Sambarabougou swarm. Note the high magnetic susceptibility (10.8). **f** Coarse-grained texture of the Essakane swarm composed of olivine, pyroxene and plagioclase. **g** Coarse-grained texture of altered sample BK001 composed of plagioclase, clinopyroxene, magnetite and secondary amphibole. **h** Fresh medium-grained doleritic texture of the Käyser swarm composed of plagioclase, olivine, clinopyroxene, and Fe and Ti oxides

plotting 1.6% discordant. Using a forced lower intercept of 0 ± 100 Ma yields an upper intercept age of 1525 ± 3 Ma, taken as the best age estimate of this sample (Fig. 6c; Table 2).

Sample BN029

The subvertical dyke has a trend of N130°, a minimum width of 10 m, based on

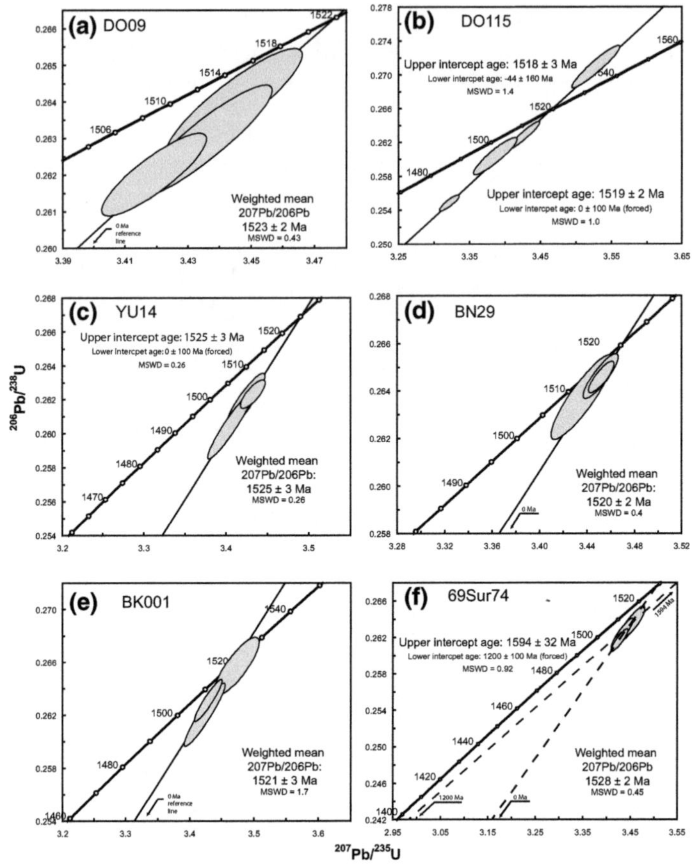

Fig. 6 U–Pb Concordia diagram for the 4 samples of Essakane swarm, 1 of Sambarabougou swamr and 1 for Käyser swarm: Essakane swarm: DO09 (1523 ± 2 Ma), DO115 (1519 ± 2 Ma), YU14 (1525 ± 3 Ma), BN29 (1520 ± 2 Ma), Sambarabougou swarm: BK001 (1521 ± 3 Ma), and Kayser swarm in Amazonia 69Sur74 (1528 ± 2 Ma)

field observation, and a maximum width up to 200 m, based on aeromagnetic data interpretation (depending on whether the thermally affected host rocks in the contact zone are contributing to the aeromagnetically determined width; Schwarz et al. 1987; Ernst 1991).

The dyke was sampled from an outcrop along the side of the road from Wa to Loggo in northern Ghana. The host rocks are Paleoproterozoic granitoids. The dyke is located near the contact zone between Koudougou-Tumu granitoid domain and Wa-Lawra greenstone belt. The automorphic feldspar laths (0.5–1 mm) form intergrowths with coarse-grained and poikilitic clinopyroxene (1–5 mm). Olivine (0.5–1 mm) is abundantly present in this sample and has sub-automorphic to rounded shapes (Fig. 5f). Biotite forms intergrowths with Fe and Ti oxides is randomly distributed.

The optically best baddeleyite grains were combined into fractions, each comprising 3–5 grains. The three fractions analysed form a cluster just below the Concordia, all about 1% discordant (Fig. 6d; Table 2). An age of 1520 ± 2 Ma was obtained by applying a forced regression through 0 Ma. This age is a reliable minimum age of this sample, and due to the only 1% discordance we consider the true emplacement age is within the error interval (forcing through 300 Ma yields an upper intercept of 1523 ± 3 Ma).

4.3.2 Sambarabougou Swarm

Sample BK001

The Sambarabougou dyke is subvertical with a trend of N090° and has an average width of 30–50 m (Fig. 5d). It can be traced for 100 km and has a very strong signal in airborne magnetic data (Fig. 5e). This dyke is a plagioclase-phyric gabbro that intrudes into host rocks which are greenschist grade sediments of the Diale-Dalema Supergroup. The coarse-grained clinopyroxene and plagioclase (3–7 mm) are massively replaced by hornblende and sericite, respectively (Fig. 5g). Biotite is present in the form of intergrowths with Fe and Ti oxides (0.5–1 mm). Abundant automorphic apatite grains suggest a secondary hydrothermal alteration of the rock.

All three baddeleyite fractions plot close to the Concordia curve (less than 1.5% discordant) with the single grain fraction being concordant within error (Fig. 6e; Table 2). The weighted mean is 1521 ± 3 Ma, and is considered the best age estimate of this dyke.

4.3.3 Käyser Swarm

Sample 69Sur74

The Käyser swarm in Suriname has a trend of N150° and can be traced for ca. 300 km. A member of the swarm was sampled along the Westrivier River, and is <50 m wide and can be traced for 20 km. The host rocks of the Käyser swarm are coarse-grained biotite granite, leucogranite and felsic metavolcanics. The sample, a medium-grained olivine dolerite, consists of equal amounts of plagioclase (labradorite) and mafic minerals, olivine and pinkish clinopyroxene. The clinopyroxene, a titaniferous augite, occurs as interstitial masses between euhedral plagioclase crystals and subhedral olivine grains (Fig. 5h). Fe and Ti oxides are common and usually overgrown by broad rims of reddish-brown biotite. Some reddish-brown kaersutite-like amphibole is also present. The sample looks fairly fresh, but shows local alteration to chlorite, green biotite, actinolite and sericite.

Four fractions of 1–4 baddeleyite grains each were analysed (Fig. 6f; Table 2). They vary from 0.8–1.4% discordant, with a weighted mean $^{207}Pb/^{206}Pb$ age of 1528 ± 2 Ma. Taking into account the ca. 1200 Ma age of the Nickerie tectonothermal event, we explore a Pb loss line to 1200 Ma in Fig. 3, but consider an upper intercept of ca. 1569 Ma less likely as it would not explain the tight clustering of the fractions, each consisting of only one or only a couple of grains. Given the near concordant nature of the analyses we consider the weighted mean $^{207}Pb/^{206}Pb$ age as the best estimate of the crystallization age at this stage.

4.4 The Neoproterozoic Events

4.4.1 Oda Swarm

Sample MG-8

The Oda swarm in southern Ghana can be traced for ca. 200 km with a trend of N070°. The sampled dyke was obtained from near Oda village (Fig. 7a). It is 50–100 m wide. The doleritic dyke intrudes the granitoid-greenstone Paleoproterozoic basement rocks. Phenocrysts of clinopyroxene and rare orthopyroxene of 1–2 mm in size are surrounded by finer-grained laths of plagioclase (Fig. 7b). Large phenocrysts of plagioclase of up to 1 cm are occasionally present. Fe and Ti oxides are randomly distributed in the rock.

Three baddeleyite fractions of 3–5 each plot concordant. As recorded by the $^{206}Pb/^{205}Pb$ ratio the U and Pb content of the baddeleyite grains of this sample seems lower compared to the other samples, which explains the relative large error ellipses (Fig. 7c; Table 2). The Concordia age is 915 ± 7 Ma (MSWD = 0.38) which is taken as the age of the MG-8 dyke.

4.4.2 Manso Swarm

Sample GH370

A prominent N355° trending dyke swarm in Ghana is about 400 km long, with subvertical dykes. The sampled dyke is approximately 200 m wide, based on the geophysical data (though the true width could be less if contact metamorphosed host rocks contribute to anomaly width; Schwarz et al. 1987; Ernst 1991). It is a coarse-grained dolerite/gabbro sampled from outcrop in a small river along the road from Abosso to Wassa Nkran near the village of Amanfro (Fig. 7d). The host rocks consist of metamorphosed volcanoclastic rocks, phyllites, basalts, gabbros and TTGs belonging to the 2200–2100 Ma Birimian series of the Ashanti greenstone belt. The medium to coarse-grained dolerite is composed of clinopyroxene and plagioclase, frequently replaced by amphibole and sericite, respectively (Fig. 7e). Biotite forms intergrowths with ilmenite. Fe and Ti oxides are present as a minor phase.

Baddeleyite grains are fresh with no zircon rims observed. The optically best grains were handpicked and combined into three fractions, each comprising 4–6 grains.

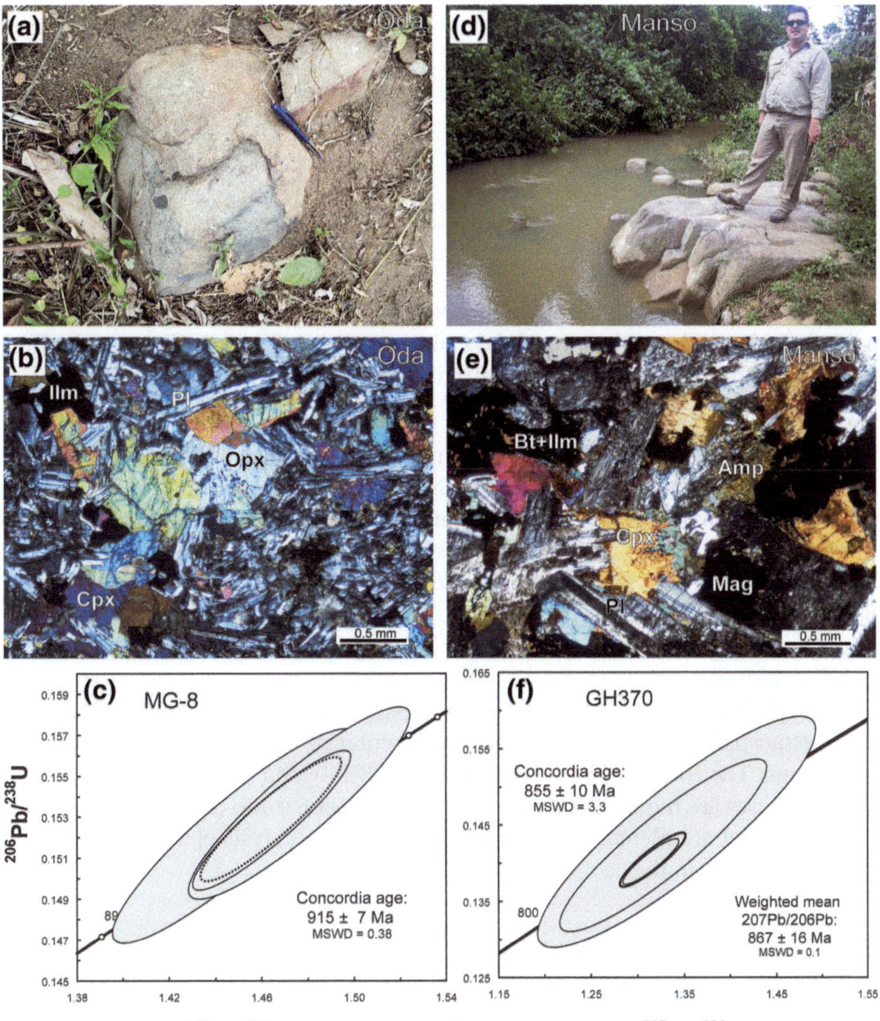

Fig. 7 **a** A small outcrop of Oda swarm. **b** Phenocrysts of clinopyroxene and rare orthopyroxene of 1–2 mm in size are surrounded by finer-grained to larger laths of plagioclase and Fe–Ti oxides. **c** U–Pb Concordia diagram for the sample MG-8 (915 ± 7 Ma, Oda swarm). **d** Outcrop of Manso swarm in the river. **e** The medium to coarse-grained dolerite composed of clinopyroxene and plagioclase, frequently replaced by amphibole and sericite, respectively. Biotite forms intergrowths with ilmenite. **f** U–Pb Concordia diagram for the sample GH370 (867 ± 16 Ma, Manso swarm)

The three baddeleyite fractions overlap with a mean centred just below the Concordia yielding a Concordia age of 855 ± 10 Ma (Fig. 7f; Table 2). However, the MSWD value of 3.2 indicates the fractions are not truly concordant but suffer from minor loss of Pb. On the assumption that Pb-loss is recent (0 Ma) the weighted mean result of 7/6 ages is 867 ± 16 Ma, which we consider to be the best age estimate of this sample.

4.5 The CAMP Events

4.5.1 Hounde Swarm

Two U–Pb ages were obtained from the NE-trending Hounde swarm: sample BN24 from an outlier dyke to the east of the main part of the swarm in Ghana and sample HO136 from the main part of the swarm in Burkina Faso.

Sample BN24

Sample BN24 was collected from an outcrop aside the road from Navrongo to Tumu in northwestern Ghana. Dyke sample BN24 is part of a swarm of N45° trending dykes of the Hounde swarm, found also in Burkina Faso, where this regional swarm is most prominent (Fig. 1). The host rocks are granitoids of the Koudougou-Tumu domain. This NE-trending swarm is at least 700 km long and 400 km wide. The sampled dyke is a typical basaltic dolerite with fine-grained ophitic texture (Fig. 8a) composed of clinopyroxene, orthopyroxene, plagioclase (0.5–1 mm) and minor Fe and Ti oxides.

Ten light brown baddeleyite grains were extracted. The largest and optically best grains (transparent with no fractures) were combined into one fraction comprising four grains. The fraction plots concordant at 198 ± 16 Ma (Fig. 9a; Table 2). Additional analyses are required for improving the precision (time-consuming due to low abundance, but are planned), nevertheless this single analysis clearly indicates that the dyke belongs to the ca. 201 Ma CAMP event.

Sample HO136

Sample HO136 was collected from an outcrop aside the road close to the village of Kongolikan, 45 km SW of Hounde. Dyke sample HO136 is part of a swarm of N45° trending dyke swarm prominent in Burkina Faso (Fig. 1). The host rocks are metavolcanic rocks of the Hounde belt. This NE-trending Hounde swarm is at least 700 km long and 400 km wide. The sampled dyke is a typical basaltic dolerite with fine-grained ophitic texture composed of clinopyroxene, orthopyroxene, plagioclase (0.5–1 mm) and minor Fe and Ti oxides (Fig. 8c).

Baddeleyite was recovered using the method of Söderlund and Johansson (2002) at the University of Lund and shipped to the Jack Satterly Geochronology Laboratory at the University of Toronto for U–Pb analysis. Baddeleyite grains were typical pale brown, unaltered, flat, blade-like crystals that were highly translucent and exhibited

Fig. 8 **a** Medium to fine-grained doleritic texture of the Hounde swarm. **b** Doleritic sill overlying the sediments of the Taoudeni basin. **c** Fine grained dolerite composed of orthopyroxene phenocrysts in plagioclase and clinopyroxene matrix. **d** Coarse-grained texture of CAMP sill including plagioclase, orthopyroxene and clinopyroxene replaced by amphibole and biotite. **e** Granophyric texture in the CAMP sill. **f** Titano-magnetite with ilmenite core and rare chalcopyrite in the CAMP sill

sharp crystal edges. The total amount of baddeleyite recovered was consumed in a single U–Pb analysis (~15 grains containing a total of 5 picograms of Pb). The one result is concordant with a $^{206}Pb/^{238}U$ age of 201.8 ± 1.0 Ma (Fig. 9b).

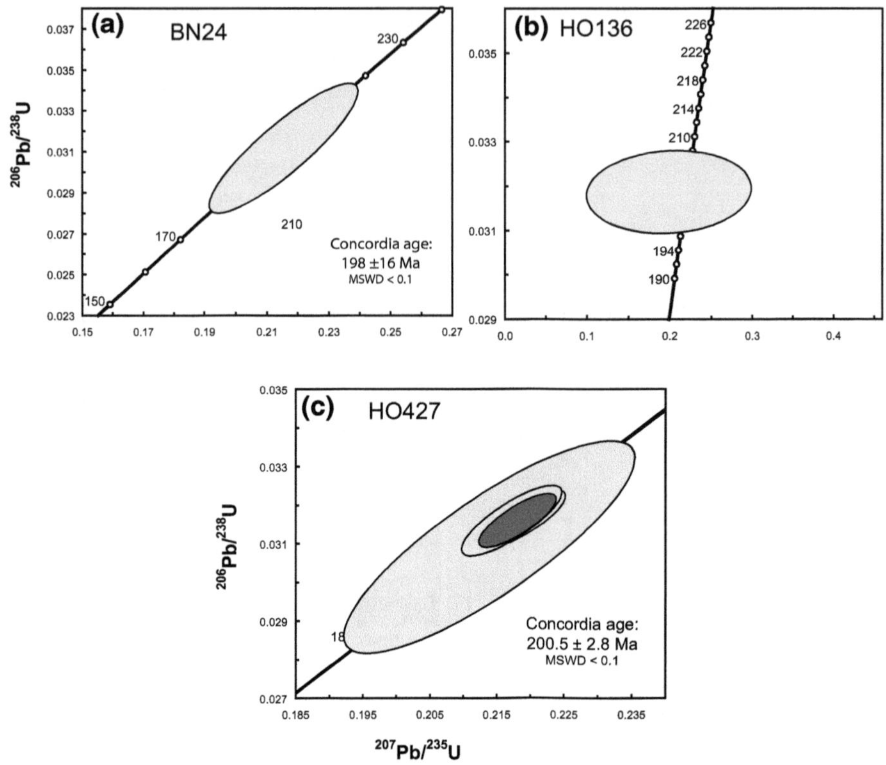

Fig. 9 U–Pb Concordia diagram for the CAMP events: BN24 (198 ± 16 Ma, Hounde swarm) HO136 (202 ± 5 Ma, Hounde swarm) and HO427 (200 ± 3 Ma, Taoudeni sill)

4.5.2 Sills in the Taoudeni Basin

Sample HO427

Dolerite sample HO427 was obtained from a flat lying post-Taoudeni sill, at least 30 m thick near the abandoned sandstone quarry 20 km west of Bobo-Dioulasso (Burkina Faso) (Fig. 1). The host rock is sandstone of the Taoudeni basin (post-1000 Ma) (Fig. 8b). The doleritic basalt is medium to coarse-grained and consists of subautomorphic orthopyroxene, clinopyroxene and automorphic laths of plagioclase (Fig. 8d). Plagioclase forms with quartz abundant granophyres (Fig. 8e). The sample exhibits some overgrowth of pyroxene by amphibole. Minor Fe and Ti oxides are present (Fig. 8f).

About 30 light brown baddeleyite grains up to ca. 80 μm in the longest dimension were extracted. Three fractions of 2–4 grains in each fraction were analysed. The fractions are all concordant and yield a Concordia age of 200 ± 3 Ma (Fig. 9c; Table 2).

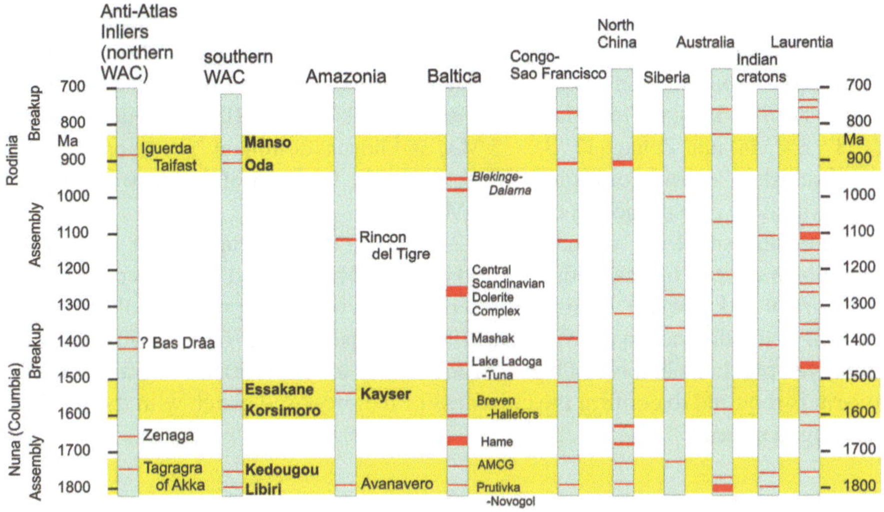

Fig. 10 LIP Barcode diagram that compares the LIP record of the WAC (both northern and southern portions) with that of Amazonia, Baltica, Congo (SF), Siberia, Australia, India and Laurentia. Sources of information for the northern WAC are from Youbi et al. (2013), and for Amazonia are from Reis et al. (2013) as discussed in the text. The LIP barcode information for the other crustal blocks is from Ernst (2014) and references therein. Yellow bands correspond to the ages (bolded) of units of the southern WAC and Amazonia dated herein

5 Discussion

5.1 Barcode Significance

The new ages along with the previous ages from the Anti-Atlas inliers provide an updated more robust LIP record for the West African craton which can be summarized as a LIP 'barcode' (Bleeker and Ernst 2006; Ernst and Bleeker 2010) (Fig. 10).

5.1.1 The Libiri (1790 Ma), Kédougou (1765 Ma), and Tagragra of Akka (ca. 1750 Ma) Event(s)

Two U–Pb ages obtained in this study, 1791 ± 3 Ma for the Libiri swarm of Niger and the 1764 ± 4 Ma age for the Kédougou swarm in eastern Senegal (Figs. 3 and 10; Table 2) represent new barcode lines for the West African craton.

Other units of each age would be expected to occur across the southern WAC which requires testing by further U–Pb dating of the many remaining undated swarms (Jessell et al. 2015).

Further insight into the nature of this 1765 Ma magmatism arises from integration with the previous U-Pb geochronology of the Anti-Atlas Inliers in the northern West

African craton. In the Anti-Atlas Inliers in northern part of the West African craton ca. 1750 Ma dykes have been dated by the U-Pb method on baddeleyite and zircon (Youbi et al. 2013; Söderlund et al. 2013b). These dykes have been dated in four inliers, Tagragra of Akka (ca. 1758 Ma), Tafeltast-Kerdous (1741 ± 10 Ma), Iguerda-Taïfast (1747 ± 4 Ma) and Zenaga (1734 ± 5 Ma) and have a dominant NW trend. The data are discordant in some cases and the age for this Tagragra of Akka swarm is only approximately constrained as ca. 1750 Ma.

Given the small age difference (4–24 myr, given uncertainties, it is therefore possible that the 1765 Ma Kédougou dykes should be grouped with this ca. 1750 Ma Tagragra of Akka swarm. If this link is correct, then we would provisionally recognize two LIP barcode lines in the WAC, a 1790 Ma pulse and a 1765 (−1750) Ma pulse. Improved precision in the U–Pb geochronology of the ca. 1750 Ma Tagragra of Akka swarm is required to confirm the closeness of the proposed match with the 1765 Ma Kédougou swarm.

5.1.2 The Korsimoro Swarm—1575 Ma Event

This swarm can be traced across Burkina Faso for a distance of about 400 km (on the basis of aeromagnetic maps; Fig. 1) and obviously must extend into the adjacent countries (Mali to the west and Ghana, Togo, Benin and Niger to the east).

5.1.3 The Essakane, Sambarabougou, and Käyser Swarms—1520–1528 Ma Event

With the six matching ages now obtained for the Essakane, Sambarabougou and Käyser swarms (from N–S across the swarm, and from the east to the west: 1519 ± 2, 1523 ± 2 1526 ± 3, 1521 ± 2; 1521 ± 3; 1528 ± 2 Ma), the swarm has a minimum width of 1500 km and a minimum length of 1500 km (Fig. 1). The swarm can be traced across Burkina Faso and continues to Senegal and Suriname. It is probable that the dykes labelled P and/or Q in Jessell et al. (2015) (Fig. 1) located in Mali also belong to this swarm. The swarm probably extends further to the east as well. Experience with other LIPs would suggest that coeval erosional remnants of volcanic rocks, sills and perhaps silicic rocks remain to be discovered and linked to the Essakane LIP. Until we have more information on the overall trend and distribution (radiating swarm?) we can only speculate on whether it can be linked with breakup (or attempted breakup) along the SE margin of the WAC. As noted below the 1520 Ma age for the Käyser swarm of Suriname, northern South America, extends the Essakane swarm into formerly adjacent Amazonian craton of South America.

The age difference between the 1520 Ma Essakane and 1575 Ma Korsimoro swarm is sufficiently large, 55 Ma, that these two events are unlikely to belong to the same LIP event (cf. Chap. 1 in Ernst 2014).

5.1.4 The Oda—915 Ma Event

The N070° Oda swarm of southern Ghana is at least 200 km long and 200 km wide, which represents a minimum surface of 400,000 km^2, and has now been dated at 915 ± 7 Ma. The Bui swarm, further to the north (Fig. 1), has a similar trend (N100° vs. N070° for the Oda swarm) and pending further U–Pb dating may extend the width of the Oda swarm. Small dykes to the west in Ivory Coast might also belong to this swarm. If the Bui dykes are part of the Oda swarm then the combined extent is >300 km long and >450 km wide (Fig. 1).

5.1.5 The Manso Swarm—867 Ma Event

The Manso swarm has an average trend of about 355° and extends over an area of about 400 km long and 200 km wide (Fig. 1). To the north it dies out and does not show up on the aeromagnetic maps in Burkina Faso (Fig. 1). On the north side of the WAC in the Iguerda-Taïfast and Zenaga inliers of the Anti Atlas mountains two dykes yielded discordant approximate ages of ca. 885 Ma by the U–Pb method (Kouyaté et al. 2013; Söderlund et al. 2013b). These dykes termed the Iguerda-Taïfast swarm have a NNE trend and have been linked with a widespread swarm of N-trending dykes in the northern WAC with an age of ca. 850 Ma (Kouyaté et al. 2013; swarm E38 in Ernst et al., 1996). On the basis of the similar trend and age it would seem possible that a single ca. 850–880 Ma swarm extends more than 1500 km between the two areas (in southern and northern parts of the WAC). However, the aeromagnetic maps do not show the Manso swarm continuing northward into Burkina Faso and thus the two similarly aged swarms in northern and southern WAC are likely separate swarms, and it remains unknown whether there is any genetic or tectonic significance to the similarity of their ages.

5.1.6 The CAMP—201 Ma Event

Two U–Pb ages were acquired for the Hounde swarm (202 ± 5 and 198 ± 16 Ma) and one age for a Taoudeni sill (202 ± 3 Ma) confirming a distribution for the swarm of >700 by 400 km. The ages for the Hounde swarm confirm a link with CAMP (particularly given the higher resolution 202 ± 5 Ma age), and allow us to more robustly discuss this Hounde swarm as a major, but previously unknown part of CAMP, whose trend is distinct from the main radiating pattern (Fig. 1). In Sect. 5.2.6 we further explore the implications of the recognition of Hounde swarm as part of the CAMP event.

The 202 ± 2 Ma age for a sill in the Taoudeni basin confirms these also as a part of the CAMP LIP and provides an additional support for existing U–Pb dates which indicate a short duration of magmatism at 201 Ma for this huge (10 Mkm^2) LIP (e.g. Davies et al. 2017; see also Sect. 2.1)

5.2 Reconstruction Implications

Here we consider regional and global continental reconstructions based on improved LIP barcode obtained for the WAC through this study. The basic idea is that matching LIP records (matching barcodes) for different crustal blocks represents a tool for identifying crustal blocks which were nearest neighbours (Bleeker and Ernst 2006; Ernst and Bleeker 2010; Ernst et al. 2013) and complements the constraints provided by paleomagnetic studies.

The LIP record of the WAC is compared with that of Amazonia, Baltica, Congo, North China, Siberia, Australia, India and Laurentia (Fig. 10).

5.2.1 The 1790–1750 Ma Event(s)

As summarized in the previous section, this portion of the WAC barcode consists of two LIPs: 1790 Ma from the southern WAC and the 1765–1750 Ma from southwestern and northern WAC. Specifically, the latter comprises a precise 1765 Ma age from the southern WAC and more approximate ages of 1750 Ma from the northern WAC (in the Anti-Atlas inliers). Below these timings are compared with the global LIP barcode record (Fig. 4; Ernst 2014).

1790 Ma Pulse

The precise U–Pb 1790 Ma age for the Libiri dyke matches directly with the age of the Avanavero LIP of Amazonia (Reis et al. 2013), the Florida (Uruguayan) LIP of Rio de la Plata Craton (Halls et al. 2001; Teixeira et al. 2013) and the Prutivka-Novogol LIP of Baltica (Sarmatia portion) (Bogdanova et al. 2013), the Hart dolerites of the North Australian craton (e.g. Li 2000), the Taihang swarm of the North China craton (Peng 2010, 2015), and a pulse within the Para de Minas swarm of the São Francisco craton (Cederberg et al. 2016).

The Avanavero LIP has an overall distribution (including dykes, sills and volcanic rocks) that likely originally extended over an area of at least 300,000 km^2 (e.g., Reis et al. 2013). The associated dykes show some variation in trend but the main dykes trend NNE-NE (Gibbs 1987; Gibbs and Barron 1993; Delor et al. 2003; de Roever et al. 2003; Reis et al. 2013).

The 1790 Ma Prutivka-Novogol LIP of the Samartian portion of Baltica is present as mafic-ultramafic intrusions widely distributed in the Voronezh massif and also in the Ukrainian shield (Shumlyanskyy et al. 2015, 2016, 2017). Dykes are observed only in the Ukrainian shield and their trends are variable, but are mainly NW trending (Fig. 1 in Bogdanova et al. 2013). The Ukrainian shield also hosts anorthosite-mangerite-charnockite granite (AMCG) complexes of two generations. The older 1800–1780 Ma matches the Prutivka-Novogol LIP and there is a younger pulse at 1760–1750 Ma (see below), which might be correlated with the U–Pb 1765 Ma age for the Kédougou dykes in Senegal.

The 1790 Ma ENE–WSW trending Florida dyke swarm of the Rio de la Plata craton is also related to the emplacement of within-plate rapakivi granites that crop out in the neighbouring Nico-Pérez Neoproterozoic terrane (Teixeira et al. 2013).

The Hart dolerite rocks in the North Australia craton consist of a series of massive sills, up to 1800-m thick. The Hart intrusions are exposed along the southeastern and southwestern margins of the Speewah and Kimberley Basins. One dolerite sill was dated with the SHRIMP U–Pb zircon method and gave an age of 1790 ± 4 Ma (reported in Li 2000).

The Xiong'er-Taihang LIP of the North China craton includes NNW–SSE-trending dykes that are evenly distributed throughout the central North China Craton, as well as subordinate NE–SW and E–W trending dykes and coeval Xiong'er volcanism at the southern side of the North China craton. The Taihang dykes are up to 100 m in width and can be traced for up to 60 km. One of the dykes has been dated to 1789 ± 28 Ma (U–Pb TIMs on baddeleyite, Peng et al. 2006), and many more dolerites have yielded similar (ca. 1.78 Ga) U–Pb zircon and baddeleyite ages (Peng et al. 2006; Peng 2010, 2015).

U–Pb baddeleyite ages demonstrate the presence of three dyke generations within the NW-trending Para de Minas swarm (São Francisco craton, South America), the oldest of which is dated at ca. 1795 Ma (1798 ± 4, 1791 ± 7 and 1793 ± 18 Ma) (Cederberg et al. 2016), The other two age pulses are ca. 1710 and ca. 770 Ma.

1765–1755 Ma Pulse

The second barcode pulse for the WAC consists of the precisely dated 1765 Ma Kédougou swarm of Senegal, and the more approximately dated 1750 Ma Tagragra of Akka swarm of the Anti-Atlas inliers (Fig. 3). More precise dating of the latter will be required to determine whether the Tagragra of Akka swarm represents a discrete pulse that can be separated from the more precisely determined 1765 Ma Kédougou swarm. In any case, the closeness in the ages as currently known (1765 Ma vs. ca. 1750 Ma; only about 15 my apart), suggests that they both belong to the same magmatic event.

Global age matches for 1765 Ma include: the 1765 Ma Pipilia swarm of India (Singhbhum craton) (Shankar et al. 2014; Srivastava et al. 2017), and the younger AMCG pulse in the Ukrainian shield of Sarmatia (Bogdanova et al. 2013; Shumlyanskyy et al. 2015, 2017). The slightly younger timing of ca. 1750 Ma would have matches with the 1750 Ma Timpton event of Siberia (Gladkochub et al. 2010), and the 1750 Ma Kivalliq event of northern Laurentia (Ernst et al. 2016; Peterson et al. 2015).

The Newer dolerite dykes of the Singhbhum craton are dated at 1766 ± 1 and 1765 ± 1 Ma (Shankar et al. 2014) and their name has been changed to the Pipilia swarm (Srivastava et al. 2017).

On the other hand, the Sarmatia magmatism shows some intermediate ages between the two main pulses at ca. 1790 and ca. 1760 Ma (e.g. Bogdanova et al. 2013; Shumlyanskyy et al. 2015, 2016, 2017) suggesting that the 1800–1750 Ma magmatism should be considered as a single event with two main pulses at ca. 1790 and ca. 1760 Ma. If the latter is true then the 1790 Ma (Libiri) and 1765–1750 Ma pulses in the WAC could also be viewed as pulses of the same event. Some LIPs have an age span of 10s of myr (Ernst 2014) and so it is plausible that these two pulses, 25 myr apart at 1790 and 1765 Ma, could represent a single LIP.

1790–1750 Ma Reconstruction
With this in mind we consider the overall distribution of 1790 and 1765–1750 Ma magmatism in the WAC and notice that the three widely separated dyke sets converge to the eastern margin of the WAC. Such radiating swarms are typically interpreted as the location of a mantle plume centre (e.g. Ernst and Buchan 1997; Ernst 2014). However, the broad age range would not be typical for a single radiating swarm as the plate would typically be moving significantly over such a time period and therefore not remain stationary above a single underlying plume. Nevertheless, it is noted that depending on subcontinental lithospheric topography, plumes can be channelled sideways for considerable distance into lithospheric thin spots (e.g. Bright et al. 2014; Sleep, 2006), and so a repeated location for two different plume centres even in the presence of plate movement may not be unusual. Also a plume head can become detached from the plume tail and remain attached to a plate when the plate is moving (e.g. pp. 492–493 in Ernst 2014).

Provisionally we carry this radiating swarm analysis further (while cognizant of the unusual large age range). The position of Amazonia with respect to WAC is well constrained with evidence for an attachment since approximately 2.0 Ga until their final breakup in association with opening of the Atlantic Ocean at ca. 120 Ma with some periods, which are less well constrained at ca. 1400 Ma and between 900 and 550 Ma (e.g. Trompette 1994; Nomade et al. 2003; Bispo-Santos et al. 2014a; Johansson 2009; D'Agrella-Filho et al. 2016b; Matthews et al. 2016). Therefore, it is noteworthy that in this standard Amazonia—West African craton reconstruction the dykes associated with the Avanavero LIP have trends which are consistent with the radiating swarm pattern mentioned above, and would in fact be aligned with (and belonging to the same sub-swarm as) the 1790 Ma Libiri swarm.

In this reconstruction of Amazonia with West African and with Baltica added in the SAMBA (South America—Baltica) reconstruction (Johansson 2009, 2014 (Fig. 11), it is notable that the dykes of the 1790–1750 Ma Sarmatia event trend also have a dominant regional trend which is consistent with the overall radiating pattern (cf. Bogdanova et al. 2013). The precise ages for most of these Sarmatia dykes are unknown. Many belong to the 1790 Ma pulse but some may belong to the younger 1765–1750 Ma pulse. We can also see that in this SAMBA reconstruction, the 1790–1750 Ma Sarmatia magmatism appears adjacent to the 1765 Ma Kédougou dyke of Senegal. Overall, this story is consistent with a radiating pattern consisting of two pulses, a western to central 1790 Ma pulse consisting of the Amazonia and Libiri dykes and some of the Sarmatia dykes and a central to eastern pulse consisting of the 1765 Ma Kédougou dykes of Senegal, the younger pulse of Sarmatia and the 1750 Ma Anti-Atlas dykes. New paleomagnetic results for Amazonia at 1790 Ma (Bispo-Santos et al. 2014a, b) for Avanavero sills supports a SAMBA connection with Baltica.

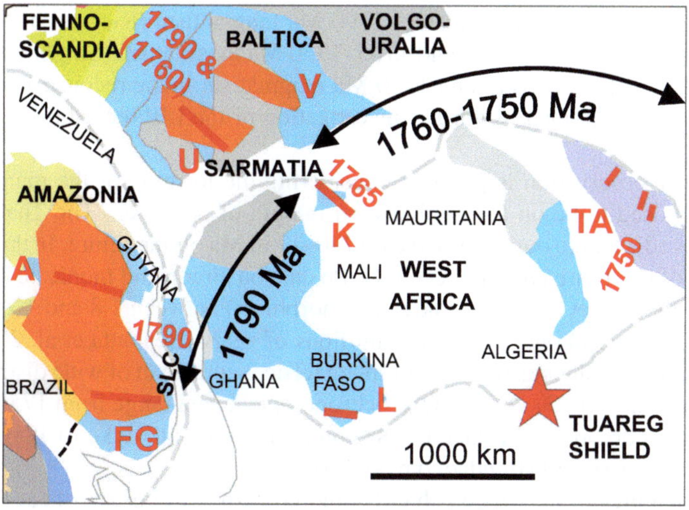

Fig. 11 Distribution of 1790–1750 Ma dykes and associated units dated in West Africa, Amazonia and Baltica using the SAMBA reconstruction of Johansson (2009, 2014). L = 1791 Ma Libiri dykes (dated herein), and K = 1762 Ma Kédougou swarm, TA = c. 1750 Ma Tagragra of Akka (Youbi et al. 2013; Ernst 2014). A = Avanavero dykes (see Reis et al. 2013), FG = French Guiana of ca: 1784 Ma age (see Théveniaut et al. 2015), SLC = São Luis Craton. In the Sarmatian portion of Baltica there is the U = Ukrainian shield and V = Voronezh magmatism of approximately ca. 1790–1750 Ma age (e.g. Shumlaynskyy et al. 2017). The dominant trend of associated dykes is NW and these dykes are mainly 1790 Ma (Fig. 1 in Bogdanova et al. 2013). Basement units are from Johansson (2009): Grey = Archean nuclei. Blue = areas with 1.95–2.22 Ga orogenic belts. Yellow = 1.8–1.95 Ga orogenic belts, light purple equals region of Paleozoic with Proterozoic Inliers in the northern WAC. The data fall into two pulses: a ca. 1790 Ma pulse and a ca. 1765–1750 Ma pulse, and the two fanning patterns converge to the same region in south-eastern Algeria which represents a potential plume centre (marked by red star). The ca. 1800–1750 Ma magmatism is widespread in the Tuareg shield which is adjacent to the WAC in the vicinity of the proposed plume centre region. See text for further discussion

The SAMBA reconstruction has been criticized recently on two grounds: that the Mesoproterozoic belt structure in Fennoscandia on the side adjacent to Amazonia in this reconstruction is orthogonal to the trend of the same age belts in western Amazonia, (Pisarevsky et al. 2014; Bogdanova and Pisarevsky 2015) and a paleomagnetic argument that at 1440 Ma Baltica and Amazonia were not connected, but allowing for a possible connection at 1790 Ma (Pisarevsky et al. 2014). Concerns about the reconstruction of West Amazonia with Baltica at ca. 1400 Ma are also expressed in D'Agrella-Filho et al. (2016a). There is also paleomagnetic data indicating that the components of Baltica (Sarmatia-Volga Uralia and Fennoscandia) were not assembled at 1790 and only finished by 1750 Ma (Bogdanova et al. 2013; Pisarevsky et al. 2014). However, our reconstruction only requires Sarmatia to have been adjacent to Amazonia and the WAC, and the difference in belt trends in Fennoscandia is not relevant.

The location of the other areas of 1790 Ma magmatism (Rio de La Plata, SFC, North Australia, North China), and 1765–1750 Ma magmatism (Singhbhum craton, northern Laurentia, Siberia), and 1750 Ma magmatism in the reconstruction would remain to be determined by further investigation of the LIP record integrated with paleomagnetism.

A further observation is possible considering the location of the mantle plume centre proposed to be on the northeastern margin of the WAC. The Tuareg shield (Fig. 11) is adjacent to our proposed ca. 1790–1750 Ma plume centre. In the westernmost part of the Tuareg Shield, recent U–Pb data on zircon and monazite suggest an important thermal event (high grade metamorphism) between 1.8 and 1.75 Ga, that appears to be spread out over tens of millions of years (Benbatta et al. 2017). This event is contemporaneous with an intraplate felsic magmatism of wide distribution in this region (Caby 2003). Available geochronological data range from 1754 ± 10 Ma (Caby and Andréopoulos-Renaud 1983) on a meta-rhyolite dyke to 1810 ± 38 Ma on an orthogneissic sill (Bendaoud et al. 2017).

These features in the Tuareg shield could correspond to greater heat flux above the proposed plume centre leading to partial melting of lower crust. However, it has been generally interpreted that the Tuareg shield region only became attached to the WAC along the Trans-Saharan orogenic during the Pan-African orogeny (e.g. Caby 2003).

We may speculate that the presence of the impressive ca. 1800–1750 Ma felsic province in the Tuareg adjacent to our interpreted 1800–1750 Ma plume centre would be consistent with an interpretation that the Tuareg shield block was approximately proximal to this NE part of the WAC at this time.

5.2.2 The 1575 Ma Event

With respect to the 1575 Ma Korsimoro swarm, the closest ages on other blocks are slightly older: 1590–1599 Ma LIP magmatism is found in several blocks including the Gawler craton (bimodal magmatism and the Olympic Dam iron oxide copper gold deposit), NW Laurentia (diabase and the Wernecke Breccias), and southern Rio de la Plata craton (Tandilia dolerite dykes) (Hamilton and Buchan 2010 and references therein; Teixeira et al. 2013; Furlanetto et al. 2013; Rogers et al. 2018).

In addition, Baltica has intraplate magmatism with a range of ages that overlap with the 1575 Ma Korsimoro swarm. Specifically, the Breven-Hällefors dykes have yielded U–Pb ages of 1595 ± 3.5 and 1602 ± 2 Ma (Söderlund et al. 2005). The Värmland and Åland-Åboland dykes yield slightly younger U–Pb ages that range from 1577 to 1540 Ma with the most precise age for a Värmland dyke being 1569 ± 3 Ma (Söderlund et al. 2005; #128 in Ernst and Buchan 2001). These younger ages in Baltica are the most direct match with the 1575 ± 5 Ma Korsimoro pulse in Burkina Faso. Based on the closest age correlation, we are investigating the possibility of a nearest-neighbor relationship between the WAC and Baltica.

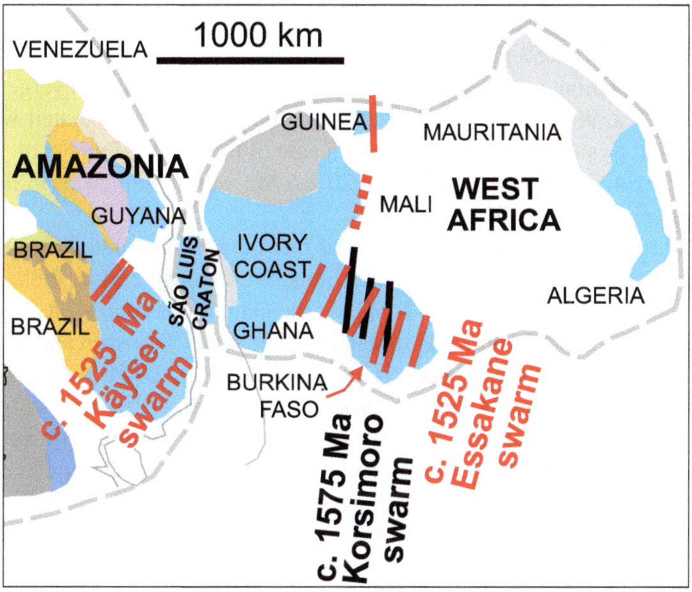

Fig. 12 Distribution of 1520 and 1575 Ma dykes in West Africa and Amazonia using part of the SAMBA reconstruction of Johansson (2009, 2014). Grey = Archean nuclei. Blue = areas with 1.95–2.22 Ga orogenic belts. Yellow, pink, light brown and green = 1.8–1.95 Ga and younger orogenic belts

5.2.3 The 1520–1528 Ma Event

The discovery of the huge ca. 1519–1528 Ma Esskakane-Sambarabougou-Käyser swarm provides an important LIP barcode lines for the Leo Man shield of the SW West African craton (WAC; Fig. 1) and formerly adjacent Amazonian craton. Reconstructing the West African craton and Amazonia (in South America) in the Gondwana fit of West Africa and South America, restores the Käyser, Sambarabougou, and Essakane swarms into an approximately parallel dyke swarm nearly 1500 km across (Fig. 12).

This 1519–1528 Ma swarm of the reconstructed WAC and Amazonia can be used as a tool for paleocontinental reconstruction with other crustal blocks. No precise age matches are yet recognized on other blocks (Table 1.2 in Ernst 2014). The closest match is with the slightly younger 1505 Ma LIP magmatism from the Siberian and the São-Francisco-Congo cratons, (Ernst et al. 2016; Silveira et al. 2013).

5.2.4 The 915 Ma Event

The 915 ± 7 Ma age for the NE trending Oda swarm is relatively unique in the global LIP record (Ernst 2014). However, there are three approximate matches of significance: in the North China craton, (Peng et al. 2011a), in the Congo Craton (Ernst et al. 2008) and São Francisco Craton (Chaves et al. 2018; Evans et al. 2016).

In the North China craton a ca. 920 Ma fanning dyke swarm converges toward a sill complex (ca. 900 Ma; Peng et al. 2011b) and is further associated with rifting (Xu–Huai Rift System) on the southeastern side of the craton (Peng et al. 2011a, b). Ages of 924 ± 4, 922 ± 3, 926 ± 2 and 920 ± 6 Ma were obtained from the fanning dyke swarm (Peng et al. 2011a).

The 920 Ma magmatism is also reported from the reconstructed Congo and São Francisco cratons (e.g. Evans et al. 2016). In the West Congo craton, there are significant volcanics and sills of ca. 920 Ma, e.g., the Gangila basalts, typical continental flood basalts, and subsequent Mayumbian rhyolitic lavas (Franssen and André 1988; Tack et al. 2001). In the São Francisco craton (Brazil), U–Pb dating indicates a ca. 920 Ma age for the Bahia fanning swarm (Evans et al. 2016) that was previously thought to be 1.02–1.01 Ga based on Ar–Ar dating (e.g. D'Agrella-Filho et al. 2004). Correa-Gomes and Oliveira (2000) correlated the radiating Bahia dykes with those in Congo craton, also suggesting a paleogeographic link.

Geochemical and paleomagnetic study of the ca. 915 Ma Oda swarm of the West African craton will help test the reconstruction of the West African versus São Francisco-Congo craton at this time and confirm whether the Oda swarm is part of the ca. 920–920 Ma Bahia—Gangilia-Mayumbian LIP of the combined São Francisco-Congo craton.

5.2.5 The 867 Ma Event

The Manso swarm provides an important barcode line at ca. 870 Ma in the southern WAC, which can be matched with the ca. 885 Ma Iguerda-Taïfast swarm from the northern WAC (Anti Atlas Inliers). Whether these represent a single swarm is not known (see discussion in Sect. 5.1.5) but together these form a major barcode line.

The southern margin of the WAC was connected to Amazonia prior to Gondwana breakup. Indeed these areas were probably connected since about 2.0 Ga (Ledru et al. 1994; Nomade et al. 2003), and only broke apart during the Atlantic Ocean opening at ca. 120 Ma. At 870 Ma Ghana would have been adjacent to the São Luis craton of Amazonia and a key prediction is therefore that the Manso swarm should continue into this portion of formerly adjacent Amazonia. However, there is also some uncertainty about the reconstruction location of the São Luis craton at this time (e.g. Onstott and Hargraves 1981).

5.2.6 The Hounde Swarm of the 201 Ma CAMP Event

As presented above, two U–Pb ages were acquired for the Hounde swarm (202 ± 5 and 198 ± 16 Ma) and one age for a Taoudeni sill (202 ± 3 Ma). Here we focus on the implications of the discovery that the NE-trending >700 by 400 km Hounde swarm is part of CAMP LIP.

The most dramatic characteristic of the CAMP event is its giant radiating dyke swarm covering over 7 million km^2 and converging (prior to opening of the Atlantic) (Fig. 13). However, the trend of the N030-040° trending Hounde swarm does not fit with the overall radiating pattern, and indicates emplacement under a different stress regime. Note the contrast in Fig. 1 between the trend of coast- parallel Liberian dykes, which are part of the radiating pattern, and the Hounde dykes. An interpretation is that this >700 km by >400 km Hounde swarm has been emplaced along a regional zone of weakness (e.g., as argued by Jourdan et al. 2006, for the Okavango swarm of the 180 Ma Karoo event of southern Africa), but the trends of regional swarms appear, in general, to be independent of local host rock grain (Ernst 2014). This leads us to consider a link between the Hounde swarm and the Kakoulima intrusion.

The Kakoulima layered mafic-ultramafic intrusion of Guinea—Conakry has been dated at ca. 200 Ma by the Ar–Ar method (Deckart et al. 1997; Nomade et al. 2007). It has been dated also by the high-precision U–Pb method on zircon and yielded an age of 201.635 ± 0.029 Myr (Davies et al. 2017). This is the oldest age acquired for CAMP related magmatism. It is around 150 ± 38 kyr older than the North Mountain basalt in Canada and 71 ± 61 kyr older than the Amelal sill in Argana basin of Morocco both of which were previously the oldest dated CAMP units (Schoene et al. 2010; Blackburn et al. 2013). The Kakoulima massif has been explored for Ni–Cu, PGEs, Co, and Cr (e.g., Konaté and Pan 2013); however, it remains poorly studied, particularly its geometry. It is described as a laccolithic intrusion, which is elongated in a NE–SW trend, for a distance of 50 km up to perhaps 100 km (Deckart et al. 1997; Konaté and Pan 2013). Its vertical extent is at least 1 km. Given the distinct elongation (in plan view) of this layered intrusion, it has greater similarity to the geometry of the Great Dyke of Zimbabwe (dyke-like layered intrusion) than to that of the Bushveld (sill-like layered intrusion). As shown by gravity data (Podmore and Wilson 1987) the Great Dyke of Zimbabwe has a Y-shaped geometry in which surface widths of up to 11 km typically contract to less than 1 km within less than 3 km of the surface. A similar trough-shaped geometry is inferred for the Muskox layered intrusion (which narrows from >11 km to the feeder dyke width of 500 m; Francis 1994).

Provisionally we consider that the elongate Kakoulima intrusion has similarities to other dyke-like layered intrusions such as the Great Dyke and therefore its direction of elongation would have tectonic significance. It is possible that the NE trend of this layered intrusion simply represents magma filling along a local plane of weakness. However, we interpret that the NE elongation trend of the Kakoulima intrusion has regional significance, and note a similar NE-trend shown by the 700×400 km Hounde dyke swarm which may suggest that the two (Hounde swarm and Kakoulima intrusion) have a common origin, distinct from the overall radiating pattern associated with the main CAMP magmatic pulse at 201 Ma. Two swarms in South America the Taiano and Uaraná have Mesozoic ages (Fig. 6 in Reis et al. 2013) and are on the same trend as the Hounde swarm and therefore may represent the continuation of the Hounde swarm into South America.

One possible explanation for their unusual NE trend is that both units, the Hounde dykes and the Kakoulima intrusion, are similarly emplaced within a regional zone

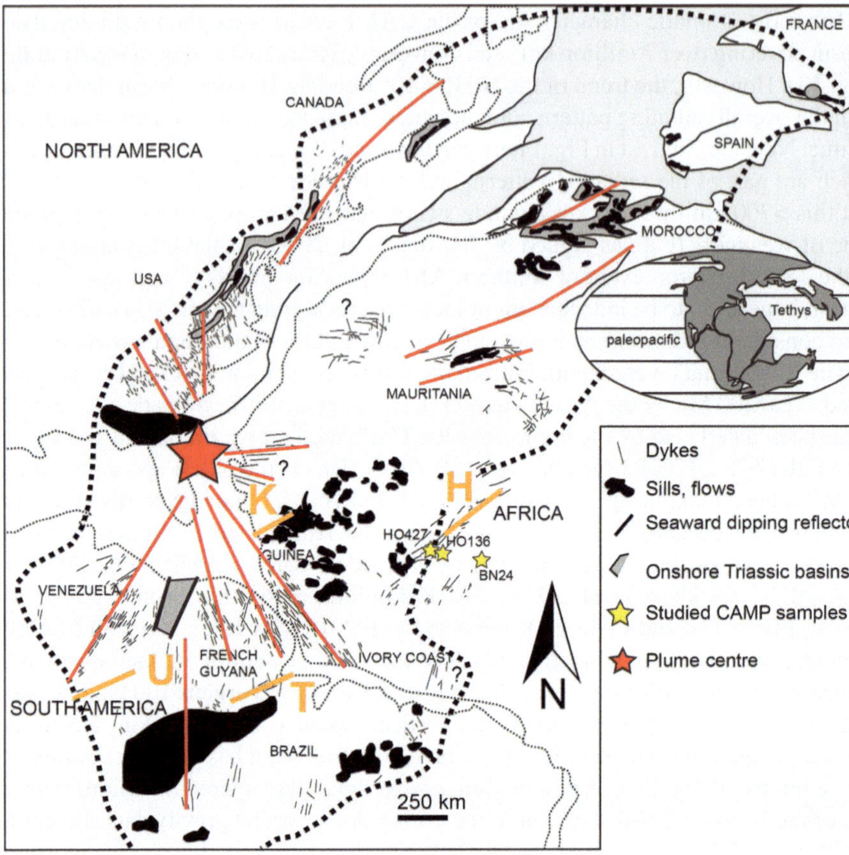

Fig. 13 Distribution of the CAMP event modified after Jourdan et al. (2009) with superimposed locations for the samples dated herein, two from the Hounde swarm (HO136 and BN24) and one from a sill in the Taoudeni basin (HO427). Superimposed red lines show generalized distribution of overall radiating swarm with red star marking inferred plume centre. Orange line marks general trend of Hounde (H) which is oblique to radiating pattern. Kakoulima (K) intrusion which has similar trend (see discussion in text). The same trend is also observed for Taiano (T) and Uaraná (U) swarms in Brazil with Mesozoic ages (Fig. 6 in Reis et al. 2013)

of weakness of the same trend. However, no such zone is apparent on geological or geophysical maps around the Kakoulima intrusion.

We suggest that the shift in trend between the prominent NE orientation of the Hounde swarm and Kakoulima intrusion to the radiating pattern (for the main 201 Ma CAMP pulse) reflects a change in stress regime. According to the mantle plume models (e.g., Campbell 2007), rapid changes in stress regime can occur and so the age difference between the radiating (domal uplift) and the NE-trending stress pattern implied by the linking of the Hounde swarm and Kakoulima intrusion, need only be a few Ma. The postulated link of the Hounde swarm with the precisely dated

201.635 ± 0.029 Ma Kakoulima intrusion (oldest dated unit so far in CAMP) would suggest emplacement of these NE trending units (Hounde swarm and Kakoulima intrusion) occurred at the beginning stages of the CAMP event likely prior to the radiating swarm.

6 Conclusions

Fourteen new ages spreading between 1790 and 200 Ma were obtained by high precision U–Pb TIMS dating of baddeleyites from doleritic dykes in the southern part of the West African Craton. Eight individual dyke swarms (including CAMP sills), which all fulfil the size requirements of a LIP, were documented. Most of the LIP ages were not previously reported from the WAC before and represent therefore significant new LIP barcode constraints for the paleogeographic reconstruction of Proterozoic supercontinents. The ages confirm significant Mesoproterozoic magmatic activity across the WAC, already documented for the northern part of the WAC in the Anti-Atlas inliers by El Bahat et al. (2013), Kouyaté et al. (2013), Söderlund et al. (2013b) and Youbi et al. (2013), contrastingly to the previous hypothesis of tectonic quiescence between 1.7 and 1.0 Ga (e.g. Ennih and Liégeois 2008 and references therein). Interestingly, two swarms may extend under the Taoudeni Basin between the northern and southern WAC: the 1765–1750 Ma and the 867–855 Ma events. Specifically, the 1765 Ma from the Kédougou swarm can be potentially linked with the ca. 1750 Ma Tagragra of Akka swarm in the Anti-Atlas inlier.

The ~1525 Ma Essakane and Sambarabougou swarms of the southern WAC can be traced into Amazonia as the Käyser swarm dated herein yielding an event that is 1500 by 1500 km in extent.

The 1790–1750 Ma dykes of the WAC can define an overall fanning pattern to which the dykes of the 1790 Ma Avanavero swarm of Amazonia and 1790–1750 Ma dykes of Sarmatia can be added.

The 915 Ma Oda swarm matches with the LIPs in the North China craton and the São Francisco craton.

The large 201 Ma Hounde swarm of the southern WAC has trend oblique to the main CAMP radiating swarm and may be linked to the elongate Kakoulima intrusion that has similar trend.

Acknowledgements We would like to acknowledge the mining companies who provided the necessary on-site logistics and let us acquire samples from their sites: SEMAFO Inc. for access to the Libiri pit, IAMGOLD Corp. for access to the Essakane mine, High River Gold Mines Ltd. for access to the Taparko mine, Etruscan Mining Corporation for access to Youga mine. We would like to thank for the logistical support to the IRD center in Ouagadougou and to Moumouni Kone. The research was co-financed by the West African Exploration Initiative (WAXI). We wish to acknowledge AMIRA International and the industry sponsors, including AusAID and the ARC Linkage Project LP110100667, for their support of the WAXI project (P934A). The field work was co-financed by Centre National de la Recherche Scientifique—Institut National des Sciences de

l'Univers SYSTER research grant. Special thanks to Laurent Aillères for posing as a scale. Sergei Pisarevsky is also appreciated for discussion of some paleomagnetic implications.
This is publication no. 71 of the Large Igneous Provinces—Supercontinent Reconstruction—Resource Exploration Project (CAMIRO Project 08E03, and NSERC CRDPJ 419503-11) (www.supercontinent.org, www.camiro.org/exploration/ongoing-projects). REE was partially supported from Mega-Grant 14.Y26.31.0012 of the Russian Federation. This is publication no. 2 of the Swedish Research Council project: The Precambrian Large Igneous Provinces (LIPs). Distribution, origin, and links to ore deposits and climate change (2015-05875) (grant to Ulf Söderlund, Nasrrddine Youbi and Richard E. Ernst). We appreciate the comments from Reviewers 1 and 2 which led to important improvements in the manuscript. In particular we thank Reviewer 1 for the suggestion that the Mesozoic Taiano and Uaraná dyke swarms in South America (Brazil) are potentially matching in age to the CAMP age Hounde swarm.

References

Ama Salah I (1991) Pétrographie et relations structurales des formations métavolcaniques et sédimentaires du Birimien du Niger occidental. Problème de l'accrétion crustale au Protérozoïque inferieur. Unpublished PhD thesis, Orleans University, p 254

Antonio PYJ, D'Agrella-Filho MS, Trindade RIF, Nédélec A, de Oliveira DC, da Silva FF, Roverato M, Lana C (2017) Turmoil before the boring billion: paleomagnetism of the 1880–1860 Ma Uatumã event in the Amazonian craton. Gondwana Res 49:106–129

Abati J, Aghzer A, Gerdes A et al (2010) Detrital zircon ages of Neoproterozoic sequences of the Moroccan Anti-Atlas belt. Precambr Res 181(1–4):115–128

Aït Lahna A, Tassinari CCG, Youbi N et al (2016) Refining the stratigraphy of the Taghdout Group by using the U-Pb geochronology of the Taghdout Sill (Zenaga inlier, Anti-Atlas, Morocco). In: Seventh international dyke conference (IDC7), Beijing, China, 18–20 Aug 2016. http://idc7.csp.escience.cn/dct/page/1

Avelar VG, Lafon JM, Delor C et al (2003) Archean crustal remnants in the easternmost part of the Guiana Shield: Pb-Pb and Sm-Nd geochronological evidence for Mesoarchean versus Neoarchean signatures. Géol de la France 2-3-4:83–99

Baratoux L, Metelka V, Naba S et al (2011) Juvenile Paleoproterozoic crust evolution during the Eburnean orogeny (~2.2–2.0 Ga), western Burkina Faso. Precambr Res 191:18–45

Bastos Neto AC, Ferron JTMM, Chauvet A, Chemale F, de Lima EF, Barbanson L, Costa CFM (2014) U-Pb dating of the Madeira Suite and structural control of the albite-enriched granite at Pitinga (Amazonia, Brazil): evolution of the A-type magmatism and implications for the genesis of the Madeira Sn–Ta–Nb (REE, cryolite) world-class deposit. Precambr Res 243:181–196

Bea F, Montero P, Haissen F et al (2013) 2.46 Ga kalsilite and nepheline syenites from the Awsard pluton, Reguibat Rise of the West African Craton, Morocco. Generation of extremely K-rich magmas at the Archean-Proterozoic transition. Precambr Res 224:242–254

Beghin J, Storme J-Y, Blanpied, C et al (2017) Microfossils from the late Mesoproterozoic – early Neoproterozoic Atar/El Mreïti Group, Taoudeni Basin, Mauritania, northwestern Africa. Precambr Res 291:63–82. https://doi.org/10.1016/j.precamres.2017.01.009

Benbatta A, Bendaoud A, Garrido C et al (2017) A complex Paleoproterozoic evolution of the In Ouzzal Terrane (Hoggar, Algeria) revealed by zircon and monazite geochronology and P-T path of Al-Fe granulites. In: 2nd Symposium IGCP 638, Casablanca, Marocco, 7–12 Nov 2017

Bendaoud A, Caby R, Djemaï S et al (2017) Precambrian evolution of the Tirek terrane (Hoggar, Algeria): evidence for the existence of an Archeo-Paleoproterozoïc continent in the western part of the Tuareg shield. In: 2nd Symposium IGCP 638, Casablanca, Marocco, 7–12 Nov 2017

Bertrand H, Fornari M, Marzoli A et al (2014) The Central Atlantic Magmatic Province extends into Bolivia. Lithos 188:33–43

Bispo-Santos F, D'Agrella-Filho MS, Janikian L et al (2014a) Towards Columbia: Paleomagnetism of 1980–1960 Ma Surumu volcanic rocks, Northern Amazonian Craton. Precambr Res 244:123–138

Bispo-Santos F, D'Agrella-Filho MS, Trindade RIF, Janikian L, Reis NJ (2014b) Was there SAMBA in Columbia? Paleomagnetic evidence from 1790 Ma Avanavero mafic sills (Northern Amazonian craton). Precambr Res 244:139–155. https://doi.org/10.1016/j.precamres.2013.11.002

Blackburn TJ, Olsen PE, Bowring SA et al (2013) Zircon U-Pb geochronology links the end-Triassic extinction with the Central Atlantic Magmatic Province. Science 340:941–945

Block S, Baratoux L, Zeh A et al (2016) Paleoproterozoic juvenile crust formation and stabilisation in the south-eastern West African Craton (Ghana); New insights from U-Pb-Hf zircon data and geochemistry. Precambr Res 287:1–30. https://doi.org/10.1016/j.precamres.2016.10.011

Bleeker W, Ernst R (2006) Short-lived mantle generated magmatic events and their dyke swarms: the key unlocking Earth's paleogeographic record back to 2.6 Ga. In: Hanski E, Mertanen S, Rämö T, Vuollo J (eds) Dyke swarms – time markers of crustal evolution. Taylor and Francis/Balkema, London, pp 3–26

Bogdanova SV, Pisarevsky SA (2015) Were Amazonia and Baltica Connected in Nuna and Rodinia? AGU Fall Meeting. Am Geol Soc San Francisco 2015:T13A–2966

Bogdanova SV, Gintov OB, Kurlovich DM et al (2013) Late Palaeoproterozoic mafic dyking in the Ukrainian Shield of Volgo-Sarmatia caused by rotation during the assembly of supercontinent Columbia (Nuna). Lithos 174:196–216

Bright RM, Amato JM, Denyszyn SW et al (2014) U-Pb geochronology of 1.1 Ga diabase in the southwestern United States: testing models for the origin of a post-Grenville Large Igneous Province. Lithosphere 6:135–156

Buchan KL, Ernst RE (2018a) A giant circumferential dyke swarm associated with the High Arctic Large Igneous Province (HALIP). Gondwana Res (GR Focus Rev) 58:39–57

Buchan KL, Ernst RE (2018b) Giant circumferential dyke swarms: catalogue and characteristics. In: Srivastava RK, Ernst RE, Peng P (eds) Dyke swarms of the world – a modern perspective. Springer book volume (accepted)

Caby R (2003) Terrane assembly and geodynamic evolution of Central-Western Hoggar: a synthesis. J Afr Earth Sci 37:133–159

Caby R, Andréopoulos-Renaud U (1983) Age à 1800 Ma du magmatisme sub-alcalin associé aux métasédiments monocycliques dans la chaîne pan-Africaine du Sahara central. J Afr Earth Sci 1:193–197

Campbell IH (2007) Testing the plume theory. Chem Geol 241:153–176

Castaing C, Billa M, Milesi JP et al (2003) Notice explicative de la carte géologique et minière au 1/1,000,000 du Burkina Faso. Edition du BRGM, p 147

Cederberg J, Söderlund U, Oliveira EP et al (2016) U-Pb baddeleyite dating of the Proterozoic Pará de Minas dyke swarm in the São Francisco craton (Brazil) – implications for tectonic correlation with Siberian, Congo and North China cratons. GFF 138:219–240. https://doi.org/10.1080/11035897.2015.1093543

Chaves AO, Ernst RE, Söderlund U et al (2018) The 920–900 Ma Bahia-Gangila LIP of the São Francisco and Congo cratons and link with Dashigou-Chulan LIP of North China craton: new insights from U-Pb geochronology and geochemistry. Precambr Res. https://doi.org/10.1016/j.precamres.2018.08.023

Christian J, Egal E, Cocherie A et al (2004) Archean evolution of the Leo Rise and its Eburnean reworking. J Afr Earth Sci 39:97–104. https://doi.org/10.1016/j.jafrearsci.2004.07.059

Cordani UG, Fraga LM, Reis N et al (2010) On the origin and tectonic significance of the intra-plate events of Grenvillian-type age in South America: a discussion. J South Am Earth Sci 29:143–159. https://doi.org/10.1016/j.jsames.2009.07.002

Correa-Gomes LC, Oliveira EP (2000) Radiating 1.0 Ga mafic dyke swarms of eastern Brazil and western Africa: evidence of post-assembly extension in the rodinia supercontinent? Gondwana Res. https://doi.org/10.1016/s1342-937x(05)70291-4

Crevola G, Cantagrel JM, Moreau C (1994) Le volcanisme cénozoïque de la presqu'île du Cap-Vert (Sénégal) cadre chronologique et géodynamique Bull Soc Géol France 165(5):437–446
D'Agrella-Filho MS, Pacca IIG, Trindade RIF et al (2004) Paleomagnetism and 40Ar/39Ar ages of mafic dikes from Salvador (Brazil): new constraints on the São Francisco craton APW path between 1080 and 1010 Ma. Precambr Res 132:55–77
Davies JHFL, Marzoli A, Bertrand H et al (2017) End-Triassic mass extinction started by intrusive CAMP activity. Nat Commun. https://doi.org/10.1038/ncomms15596
Dalrymple GB, Gromme CS, White RW (1975) Potassium-argon age and paleomagnetism of diabase dikes in Liberia: initiation of central Atlantic rifting. Geol Soc Am Bull 86:399–411
D'Agrella-Filho MS, Trindade RIF, Queiroz MVB, Meira VT, Janikian L, Ruiz AS, Bispo-Santos F (2016a) Reassessment of Aguapeí (Salto do Céu) Paleomagnetic pole of the Amazonian Craton and implications for Proterozoic supercontinents. Precambr Res 272:1–17
D'Agrella-Filho MS, Bispo-Santos F, Trindade RIF, Antonio PYJ (2016b) Paleomagnetism of the Amazonian Craton and its role in paleocontinents. Braz J Geol 46:275–299
de Roever EWF, Kroonenberg SB, Delor C et al (2003) The Käyser dolerite, a Mesoproterozoic alkaline dyke suite from Suriname. Géol de la France 2-3-4:161–174
Deckart K (1996) Etude du magmatisme associé au rifting de l'Atlantique Central et Sud: géochronologie 40Ar/39Ar et géochimie sur les intrusions jurassiques de Guinée et de Guyane française/Suriname, et crétacées du Brésil. Université de Nice Sophia-Antipolis. Unpublished PhD thesis, p 221
Deckart K, Féraud G, Bertrand H (1997) Age of Jurassic continental tholeiites of French Guyana, Surinam and Guinea: implications for the initial opening of the Central Atlantic Ocean. Earth Planet Sci Lett 150:205–220
Deenen MHL, Ruhl M, Bonis NR et al (2010) A new chronology for the end-Triassic mass extinction. Earth Planet Sci Lett. https://doi.org/10.1016/j.epsl.2010.01.003
Delor C, de Roever EWF, Lafon JM et al (2003) The Bakhuis ultra-high-temperature granulite belt (Suriname): II. Implications for the late Transamazonian crustal stretching in a revised Guiana Shield framework. Géol de la France 2-3-4:207–230
Delor C, Couëffé R, Goujou JC et al (2010) Notice explicative de la carte géologique à 1/200 000 du Sénégal, feuille Saraya-Kédougou Est. Ministère des Mines, de l'Industrie, de l'Agro-Industrie et des PME, Direction des Mines et de la Géologie, Dakar
El Bahat A, Ikenne M, Söderlund U et al (2013) U-Pb ages and geochemistry of dolerite dykes in the Bas Drâa Inlier of the Anti-Atlas of Morocco: newly identified 1380 Ma event in the West African Craton. Lithos 174:85–98
El Bahat A, Ikenne M, Cousens B et al (2017) New constraints on the geochronology and Sm-Nd isotopic characteristics of Bas-Drâa mafic dykes, Anti-Atlas of Morocco. J Afr Earth Sci 127:77–87
Ennih N, Liégeois JP (2008) The boundaries of the West African craton, with a special reference to the basement of the Moroccan metacratonic Anti-Atlas belt. In: Ennih N, Liégeois JP (eds) The boundaries of the West African craton, vol 297. Geological Society, London, Spec Pub., pp 1–17
Ernst RE (1991) The near-surface geometry of the Great Abitibi Dyke, Superior Province, Canada: Geological Society of America Map & Chart Series, MCH-071
Ernst RE (2014) Large Igneous Provinces. Cambridge University Press, p 653
Ernst RE, Bleeker W (2010) Large Igneous Provinces (LIPs), giant dyke swarms, and mantle plumes: significance for breakup events within Canada (and selected adjacent regions) from 2.5 Ga to present. Can J Earth Sci 47:695–739
Ernst RE, Buchan KL (1997) Giant radiating dyke swarms: their use in identifying pre-Mesozoic large igneous provinces and mantle plumes. In: Mahoney J, Coffin M (eds) Large Igneous Provinces: continental, oceanic, and planetary volcanism. AGU geophysical monograph series, vol 100, pp 297–333
Ernst RE, Buchan KL (2001) Large mafic magmatic events through time and links to mantle-plume heads. In: Ernst RE, Buchan KL (eds) Mantle Plumes: their identification through time. Geological Society of America, Boulder, CO, Special Paper 352, pp 483–575

Ernst RE, Buchan KL, West TD et al (1996) Diabase (dolerite) dyke swarms of the world: first edition. Geological Survey of Canada Open File 3241, Scale 1:35,000,000 map and a report, p 104

Ernst RE, Bleeker W, Söderlund U et al (2013) Large Igneous Provinces and supercontinents: toward completing the plate tectonic revolution. Lithos 174:1–14

Ernst RE, Hamilton MA, Söderlund U et al (2016) Long-lived connection between southern Siberia and northern Laurentia in the Proterozoic. Nat Geosci 9:464–469. https://doi.org/10.1038/NGEO2700

Ernst REE, Wingate MTD, Buchan KL et al (2008) Global record of 1600–700 Ma Large Igneous Provinces (LIPs): implications for the reconstruction of the proposed Nuna (Columbia) and Rodinia supercontinents. Precambr Res 160:158–178. https://doi.org/10.1016/j.precamres.2007.04.019

Evans DAD, Trindade RIF, Catelani EL et al (2016) Return to Rodinia? Moderate to high paleolatitude of São Francisco/Congo craton at 920 Ma. In: Li ZX, Evans DAD, Murphy JB (eds) Supercontinent cycles through earth history. Geological Society of London Spec Pub. 424, pp 167–190. https://doi.org/10.1144/sp424.1

Francis D (1994) Chemical interaction between picritic magmas and upper crust along margins of the Muskox Intrusion, Northwest Territories: Geological Survey of Canada Paper 92-12, p 94

Franssen L, André L (1988) The Zadinian group (Late Proterozoic, Zaire) and its bearing on the origin of the West-Congo orogenic belt. Precambr Res. https://doi.org/10.1016/0301-9268(88)90003-4

Furlanetto F, Thorkelson DJ, Daniel Gibson H et al (2013) Late Paleoproterozoic terrane accretion in northwestern Canada and the case for circum-Columbian orogenesis. Precambr Res 224:512–528

Gärtner A, Villeneuve M, Linnemann U et al (2013) An exotic terrane of Laurussian affinity in the Mauritanides and Souttouffides (Moroccan Sahara). Gondwana Res 24:687–699

Gerstenberger H, Haase G (1997) A highly effective emitter substance for mass spectrometric Pb isotope ratio determinations. Chem Geol 136:309–312

Gibbs AK (1987) Contrasting styles of continental mafic intrusions in the Guiana Shield. In: Halls HC, Fahrig WF (eds) Mafic dyke swarms. Geological Association of Canada Special Paper 34, pp 457–465

Gibbs AK, Barron CN (1993) Geology of the Guiana Shield. Oxford University Press, p 246

Gladkochub DP, Pisarevsky SA, Donskaya TV et al (2010) Proterozoic mafic magmatism in Siberian craton: an overview and implications for paleocontinental reconstruction. Precambr Res 183:660–668

Goujou JC, Buscail F, Théveniaut H et al (2010) Notice explicative de la carte géologique à 1/200 000 du Sénégal, feuille Kossanto-Dalafi Est, 2010, Ministère des Mines, de l'Industrie, de l'Agro-Industrie et des PME. Direction des Mines et de la Géologie, Dakar

Halls HC, Campal N, Davis DW et al (2001) Magnetic studies andU–Pb geochronology of the Uruguayan dyke swarm, Rio de la Plata craton, Uruguay: paleomagnetic and economic implications. J South Am Earth Sci 14:349–361

Hamilton MA, Buchan KL (2010) U-Pb geochronology of the Western Channel Diabase, northwestern Laurentia: Implications for a large 1.59 Ga magmatic province, Laurentia" APWP and paleocontinental reconstructions of Laurentia, Baltica and Gawler craton of southern Australia. Precambr Res 183:463–473

Ikenne M, Söderlund U, Ernst RE et al (2017) A c.1710 Ma mafic sill emplaced into a quartzite and calcareous series from Ighrem, Anti-Atlas Morocco: Evidence that the Taghdout passive margin sedimentary group is nearly 1 Ga older than previously thought. J Af Earth Sci 127:62–76. In: Magmatism, metamorphism and associated mineralization in North Africa and related areas edited by Moha Ikenne, Mustapha Souhassou and Shoji Arai. https://doi.org/10.1016/j.jafrearsci.2016.08.020

Jaffey AH, Flynn KF, Glendenin LE et al (1971) Precision measurement of half-lives and specific activities of 235U and 238U. Phys Rev 4:1889–1906

Jessell MW, Santoul J, Baratoux L et al (2015) An updated map of West African mafic dykes. J Afr Earth Sci 112(B):440–450. https://doi.org/10.1016/j.jafrearsci.2015.01.007

Johansson A (2009) Baltica, Amazonia and the SAMBA connection—1000 million years of neighbourhood during the Proterozoic? Precambr Res 175:221–234

Johansson Å (2014) From Rodinia to Gondwana with the "SAMBA" model: a distant view from Baltica towards Amazonia and beyond. Precambr Res 244:226–235

Jourdan F, Féraud G, Bertrand H et al (2006) Basement control on dyke distribution in Large Igneous Provinces: case study of the Karoo triple junction. Earth Planet Sci Letters. 241:307–322

Jourdan F, Marzoli A, Bertrand H et al (2009) 40Ar/39Ar ages of CAMP in North America: implications for the Triassic-Jurassic boundary and the 40 K decay constant bias. Lithos 110:167–180

Kagone O, Koala F, Napon S et al (2003) Notice explicative de la Carte géologique du Burkina Faso à 1/200 000; Feuille Pissila

Kah LC, Bartley JK, Teal DA (2012) Chemostratigraphy of the Late Mesoproterozoic Atar Group, Taoudeni Basin, Mauritania: Muted isotopic variability, facies correlation, and global isotopic trends. Precambr Res 200–203:82–103

Klein EL, Rodrigues JB, Lopes ECS et al (2012) Diversity of Rhyacian granitoids in the basement of the Neoproterozoic-Early Cambrian Gurupi Belt, northern Brazil: geochemistry, U–Pb zircon geochronology, and Nd isotope constraints on the Paleoproterozoic magmatic and crustal evolution. Precambr Res 220–221:192–216. https://dx.doi.org/10.1016/j.precamres.2012.08.007

Knight KB, Nomade S, Renne PR et al (2004) The Central Atlantic magmatic province at the Triassic-Jurassic boundary: paleomagnetic and 40Ar/30Ar evidence from Morocco for brief, episodic volcanism. Earth Planet Sci Lett 228:143–160

Konaté AA, Pan H (2013) Mount Kakoulima: An overview and analysis. Earth Sci Res 2(2):23–33. https://doi.org/10.5539/esr.v2n2p23

Kouamelan AN, Djro SC, Allialy ME et al (2015) The oldest rock of Ivory Coast. J Afr Earth Sci. https://doi.org/10.1016/j.jafrearsci.2014.12.004

Kouyaté D, Söderlund U, Youbi N et al (2013) U-Pb baddeleyite and zircon ages of 2040 Ma, 1650 Ma and 885 Ma on dolerites in the West African Craton (Anti-Atlas inliers): Possible links to break-up of Precambrian supercontinents. Lithos 174:71–84. https://doi.org/10.1016/j.lithos.2012.04.028

Kroonenberg SB, de Roever EWF (2010) Geological evolution of the Amazonian Craton. In: Hoorn C, Wesselingh FP (eds) Amazonia, landscape and species evolution. Blackwell Publishing Chichester, pp 7–28

Krogh TE (1973) A low contamination method for hydrothermal decomposition of zircon and extraction of U and Pb for isotopic age determinations. Geochim Cosmochim Acta 37:485–494

Kroonenberg SB, de Roever EWF, Fraga LM et al (2016) Paleoproterozoic evolution of the Guiana Shield in Suriname: a revised model. J Geosci 95(4):491–522

Lahondère D, Théveniaut H, Ndiaye PM et al (2010) Notice explicative de la carte géologique à 1/200,000 du Sénégal, feuille Bakel-Semme Sud-Ouest Est, Ministère des Mines, de l'Industrie, de l'Agro-Industrie et des PME. Direction des Mines et de la Géologie, Dakar

Lanphere MA, Dalrymple GB (1971) A test of the 40Ar/39Ar age spectrum technique on some terrestrial materials. Earth Planet Sci Lett 12:359–372

Lanphere MA, Dalrymple GB (1976) Identafication of excess 40Ar by the 40Ar/ 39Ar age spectrum technique. Earth Planet Sci Lett 32(2):141–148

Ledru P, Johan V, Milési JP et al (1994) Markers of the last stages of the Paleoproterozoic collision: evidence for a 2 Ga continent involving Circum-South Atlantic provinces. Precambr Res 69:169–191

Lefort JP, Aïfa T (2001) Paleomagnetic and Geochronological Evidence for a Large Rotation of the West African Craton between 1.9 and 1.4 Ga: a kinematic model for the fragmentation of Columbia and the accretion of Rodinia. Gondwana Res 4:678–679

Li ZX (2000) Palaeomagnetic evidence for unification of the North and West Australian craton by ca. 1.7 Ga: new results from the Kimberley Basin of northwestern Australia. Geophys J Int 142:173–180

Ludwig KR (2003) User's manual for Isoplot 3.00: a geochronological toolkit for Microsoft Excel: Berkeley Geochronology Center, Spec Pub No. 4, p 71

Matthews KJ, Maloney KT, Zahirovic S, Williams SI, Seton M, Müller D (2016) Global plate boundary evolution and kinematics since the late Paleozoic. Global Planet Change 146:226–250

Mauche R, Faure G, Jones LM et al (1989) Anomalous isotopic compositions of Sr, Ar, and O in the Mesozoic diabase dikes of Liberia, West Africa. Contrib Miner Petrol 10:12–18

Marzoli A, Renne PR, Piccirillo EM et al (1999) Extensive 200 million-year-old continental flood basalts of the Central Atlantic Magmatic Province. Science 284:616–618

Marzoli A, Callegaro S, Dal Corso J, Davies JHFL, Chiaradia M, Youbi N, Bertrand H, Reisberg L, Merle R, Jourdan F (2018) The Central Atlantic Magmatic Provinces (CAMP): a review. In: Tanner LH (ed) The late triassic world, topics in geobiology, vol 46, Springer, Berlin, pp 91–125

May PR (1971) Pattern of Triassic-Jurassic diabase dykes around the North Atlantic in context of predrift position of the continents. Geol Soc Am Bull 82:1285–1292

McHone JG, Puffer J (2003) Flood Basalt Provinces of the Pangaean Atlantic Rift: Regional Extent and Environmental Signaficance. In: LeTourneau PM, Olsen PE (eds) The Great Rift Valleys of Pangea in Eastern North America, vol 1. Columbia University Press, pp 141–154

Montero P, Haissen F, El Archi A et al (2014) Timing of Archean crust formation and cratonization in the Awsard-Tichla zone of the NW Reguibat Rise, West African Craton. A SHRIMP, Nd–Sr isotopes, and geochemical reconnaissance study. Precambr Res 242:112–137

Montgomery CW (1979) Uranium-lead geochronology of the Archean Imataca Series, Venezuelan Guyana shield. Contrib Miner Petrol 69:167–176

Montgomery CW, Hurley PM (1978) Total rock U-Pb and Rb-Sr systematics in the Imataca Series, Guyana Shield, Venezuela. Earth Planet Sci Lett 39:281–290

Nomade S, Théveniaut H, Chen Y et al (2000) Paleomagnetic study of French Guyana Early Jurassic dolerites: hypothesis of a multistage magmatic event. Earth and Planet Sci Lett 184:155–168

Nomade S, Chen Y, Feraud G et al (2001) First paleomagnetic and Ar-40/Ar-39 study of Paleoproterozoic rocks from the French Guyana (Camopi and Oyapok rivers), northeastern Guyana Shield. Precambr Res. https://doi.org/10.1016/S0301-9268(01)00149-8

Nomade S, Féraud G, Renne P et al (2002) New 40Ar/39Ar ages for Central Atlantic Magmatic Province in French Guyana: a younger volcanism? Goldschmidt Conference Abstracts, 2002, p A559

Nomade S, Chen Y, Pouclet A et al (2003) The Guiana and West-African Shield Paleoproterozoic grouping: new paleomagnetic data for French Guiana and Ivory Coast. Geophys J Int 154:677–694

Nomade S, Knight KB, Beutel E et al (2007) Chronology of the Central Atlantic Magmatic Province: implications for the Central Atlantic rifting processes and the Triassic–Jurassic biotic crisis. Palaeogeogr Palaeoclimatol Palaeoecol 244(1–4):326–344

Norcross C, Davis DW, Spooner ETC et al (2000) U-Pb and Pb-Pb age constraints on Paleoproterozoic magmatism, deformation and gold mineralization in the Omai area, Guyana Shield. Precambr Res 102:69–86

Onstott TC, Hargraves RB (1981) Proterozoic transcurrent tectonics: palaeomagnetic evidence from Venezuela and Africa. Nature 289:131–136

Parra-Avila LA, Kemp AIS, Fiorentini ML et al (2017) The geochronological evolution of the Paleoproterozoic Baoulé-Mossi domain of the Southern West African Craton. Precambr Res. https://doi.org/10.1016/j.precamres.2017.07.036

Peng P (2010) Reconstruction and interpretation of giant mafic dyke swarms: a case study of 1.78 Ga magmatism in the North China craton. In Kusky TM, Zhai M-G, Xiao W (eds) The evolving continents: understanding processes of continental growth, vol 338. Geol Soc London, Spec Pub, pp 163–178

Peng P (2015) Precambrian mafic dyke swarms in the North China Craton and their geological implications. Sci China Earth Sci 58:649–675. https://doi.org/10.1007/s11430-014-5026-x

Peng P, Zhai MG, Guo JH (2006) 1.80-1.75 Ga mafic dyke swarms in the central North China craton: implications for a plume-related break-up event. In: Hanski E, Mertanen S, Ramo, T, Vuollo J (eds) Dyke swarms – time makers of crustal evolution. Taylor and Francis Ltd., pp 99–112

Peng P, Bleeker W, Ernst RE et al (2011a) U-Pb baddeleyite ages, distribution and geochemistry of 925 Ma mafic dykes and 900 Ma sills in the North China craton: evidence for a Neoproterozoic mantle plume. Lithos 127:210–221
Peng P, Zhai M, Li Q et al (2011b) Neoproterozoic (~900 Ma) Sariwon sills in North Korea: geochronology, geochemistry and implications for the evolution of the south-eastern margin of the North China Craton. Gondwana Res 20:243–254
Peterson TD, Scott JMJ, LeCheminant AN et al (2015) The Kivalliq Igneous Suite: anorogenic bimodal magmatism at 1.75 Gain the western Churchill Province, Canada. Precambr Res 262:101–119
Pisarevsky SA, Elming SÅ, Pesonen LJ et al (2014) Mesoproterozoic paleogeography: supercontinent and beyond. Precambr Res 244:207–225
Podmore F, Wilson AH (1987) A reappraisal of the structure, geology and emplacement of the Great Dyke, Zimbabwe. In: Halls HC, Fahrig WF (eds) Mafic dyke swarms. Geological Association of Canada Special Paper 34, pp 317–330
Priem HA, Boelrijk NM, Hebeda E, Verdurmen ET, Verschure R (1971) Isotopic ages of the Trans-Amazonian acidic magmatism and the Nickerie metamorphic episode in the Precambrian basement of Suriname, South America. Geol Soc Am Bull 82:1667–1680
Reis NJ, Fraga LM, de Faria MSG et al (2003) Geologia do Estado de Roraima, Brasil. Géol de la France 2-3-4:121–134
Reis NJ, Teixeira W, Hamilton M et al (2013) Avanavero mafic magmatism, a late Paleoproterozoic LIP in the Guiana Shield, Amazonian Craton: U-Pb ID-TIMS baddeleyite, geochemical and paleomagnetic evidence. Lithos 174:175–195
Rogers CC, Kamo S, Soderlund U et al (2018) Geochemistry and U-Pb geochronology of 1590 and 1550 Ma mafic dyke swarms of western Laurentia: Mantle plume magmatism shared with Australia. Lithos 314–315:216–235. https://doi.org/10.1016/j.lithos.2018.06.002
Rooney AD, Selby D, Houzay JP et al (2010) Re–Os geochronology of a Mesoproterozoic sedimentary succession, Taoudeni basin, Mauritania: Implications for basin-wide correlations and Re–Os organic-rich sediments systematics. Earth Planet Sci Lett 289:486–496
Santos J, Reis N, Hartmann L, McNaughton N, Fletcher I (1999) Associação anortosito-charnockito-granito rapakivi no calimiano do norte do cráton Amazônico, Estado de Roraima: evidências obtidas por geocronologia U–Pb (SHRIMP) em zircão e baddeleyíta. 6t^{h} Simpósio de Geologia da Amazônia (Manaus), pp 503–506
Santos JOS, Hartmann LA, Gaudette HE et al (2000) A new understanding of the provinces of the Amazon Craton based on integration of field mapping and U-Pb and Sm-Nd geochronology. Gondwana Res 3:453–488
Santos JOS, Potter PE, Reis NJ et al (2001) Age and regional stratigraphy of the Roraima Supergroup and Roraima-like outliers in northern South America based on U-Pb geochronology. In: Simpósio de Geologia da Amazônia, 7, Resumos expandidos Belém, 2001
Santos JOS, Hartmann LA, McNaughton NJ et al (2002) Timing of mafic magmatism in the Tapajos Province (Brazil) and implications for the evolution of the Amazonian Craton: evidence from baddeleyite and zircon U-Pb SHRIMP geochronology. J South Am Earth Sci 15:409–429
Santos JOS, Potter PE, Reis NJ et al (2003) Age, source and regional stratigraphy of the Roraima Supergroup and Roraima-like outliers in northern South America based on U-Pb geochronology. Geol Soc Am Bull 115:331–348
Schoene B, Guex J, Bartolini A et al (2010) Correlating the end-Triassic mass extinction and flood basalt volcanism at the 100 ka level. Geology 38:387–390
Shankar R, Vijayagopal B, Kumar A (2014) Precise Pb-Pb baddeleyite ages of 1765 Ma for a Singhbhum “ewer dolerite”dyke swarm. Curr Sci 106:1306–1310
Shumlyanskyy L, Ernst RE, Söderlund U et al (2016) New U–Pb ages for mafic dykes in the Northwestern region of the Ukrainian shield: coeval tholeiitic and jotunitic magmatism. GFF 138:79–85

Shumlyanskyy L, Mitrokhin O, Billström K et al (2015) The ca. 1.8 Ga mantle plume related magmatism of the central part of the Ukrainian shield. GFF 138(1):86–101. https://doi.org/10.1080/11035897.2015.1067253

Shumlyanskyy L, Hawkesworth C, Billström K et al (2017) The origin of the Palaeoproterozoic AMCG complexes in the Ukrainian shield: new U-Pb ages and Hf isotopes in zircon. Precambr Res 292:216–239

Siachoque A, Salazar CA, Trindade R (2017) Emplacement and deformation of the A-type Madeira granite (Amazonian Craton, Brazil). Lithos 277:284–301

Silveira EM, Soderlund U, Oliveira EP et al (2013) First precise U-Pb baddeleyite ages of 1500 Ma mafic dykes from the São Francisco Craton, Brazil, and tectonic implications. Lithos. https://doi.org/10.1016/j.lithos.2012.06.004

Sleep NH (2006) Mantle plumes from top to bottom. Earth-Sci Rev 77:231–271

Söderlund U, Johansson L (2002) A simple way to extract baddeleyite (ZrO_2). Geochem Geophys Geosyst. https://doi.org/10.1029/2001GC000212

Söderlund U, Ernst RE, Youbi N et al (2013a) A major Archean (2688 Ma) dyke swarm discovered in the western Reguibat Shield, West African craton, Morocco. Unpublished Report #A131 (20 February 2013). Consortium Project for Reconstruction of Supercontinents Supercontinents Back Back to 2.7 Ga Using the Large Igneous Province (LIP) Record (www. supercontinent.org), CAMIRO (Canadian Mining Industry Research Organization) and Ernst Geosciences, p 6

Söderlund U, Ernst RE, Youbi N et al (2014) A major Archean (2688 Ma) dyke swarm discovered in the western Reguibat Shield, West African craton, Morocco. In: 25t[h] Colloquium of African Geology (CAG 25). Dar es Salaam, Tanzania, 14–16 Aug 2014

Söderlund U, Ibanez-Mejia M, El Bahat A et al (2013b) Reply to Comment on "U–Pb baddeleyite ages and geochemistry of dolerite dykes in the Bas-Drâa inlier of the Anti-Atlas of Morocco: Newly identified 1380 Ma event in the West African Craton" by André Michard and Dominique Gasquet. Lithos 174:101–108. https://doi.org/10.1016/j.lithos.2013.04.003

Söderlund U, Isachsen CE, Bylund G et al (2005) U–Pb baddeleyite ages and Hf, Nd isotope chemistry constraining repeated mafic magmatism in the Fennoscandian Shield from 1.6 to 0.9 Ga. Contrib Min Pet 150:174–194

Srivastava RK, Söderlund U, Ernst RE et al (2017) Precambrian mafic dyke swarms in the Singhbhum craton (eastern India) and their links with dyke swarms of the eastern Dharwar craton (southern India). Precambr Res, Special Issue (submitted)

Schwarz EJ, Hood PJ, Teskey DJ (1987) Magnetic expression of Canadian diabase dykes and downward modelling. In: Halls HC, Fahrig WF (eds) Mafic dyke swarms. Geol Assoc of Canada Spec Pub. 34, pp 153–162

Stacey JS, Kramers JD (1975) Approximation of terrestrial lead isotope evolution by a 2-stage model. Earth Planet Sci Lett 26:207–221

Tack L, Wingate MTD, Liegeois JP et al (2001) Early Neoproterozoic magmatism (1000-910 Ma) of the Zadinian and Mayumbian Groups (Bas-Congo): onset of Rodinia rifting at the western edge of the Congo craton. Precambr Res 110:277–306

Tait J, Straathof G, Söderlund U et al (2013) The Ahmeyim Great Dyke of Mauritania: a newly dated Archaean intrusion. Lithos 174:323–332. https://doi.org/10.1016/j.lithos.2012.09.014

Tanner LH, Lucas SG, Chapman MG (2004) Assessing the record and causes of Late Triassic extinctions. Earth-Sci Rev 65:103–139. https://doi.org/10.1016/S0012-8252(03)00082-5

Tapsoba B, Lo CH, Wenmenga U et al (2018) Chemical and Sr-Nd compositions and 40Ar/39Ar ages of NW-trending dolerite dikes of Burkina Faso: evidence for a Mesoproterozoic magmatism in the West African Craton. Geosci Front (accepted)

Tassinari CCG, Macambira MJB (1999) Geochronological provinces of the Amazonian Craton. Episodes 1999:174–182

Tassinari CCG, Cordani UG, Nutman AP et al (1996) Geochronological systematics on basement rocks from the Rio Negro-Juruena Province Amazonian Craton, and tectonic implications. Int Geol Rev 38:161–175

Tassinari CCG, Bettencourt JS, Geraldes MC et al (2000) The Amazonian Craton. In: Cordani UG, Milani EJ, Thomas Filho A, Campos DA (eds) Tectonic evolution of South America. In: 31s[t] international geological congress, Rio de Janeiro, Brazil, pp 41–95

Tassinari CCG, Teixeira W, Nutman AP et al (2001) Archean crustal evolution of the Imataca Complex, Amazonian Craton: Sm-Nd, Rb-Sr e U-Pb (SHRIMP) evidences. In: Simpósio de Geologia da Amazônia, vol 7, Belém

Teal DJ, Kah LC (2005) Using C-isotopes to constrain interbasinal stratigraphic correlations, Mesoproterozoic Atar Group, Mauritania. In: Geological Society of America, abstracts with programs, vol 37, p 45

Théveniaut H, Söderlund U, Bleeker W, Hamilton M, Delor C, Roig JY (2015) The ca. 1784 Ma Oyapock dyke, an eastern part of the feeder system for the Avanavero LIP, Amazonia. Unpublished report A215 (25 February, 2015) of the Large Igneous Provinces, Supercontinent Reconstruction, Resource Exploration Project. www.largeigneousprovinces.org, p 8

Thiéblemont D (2005) Géologie et pétrologie de l' Archéen de Guinée: une contribution régionale à la formation de la croûte continentale. Unpublished PhD thesis, University of Brest, p 149

Teixeira W, D'Agrella-Filho MS, Ernst RE et al (2013) U-Pb (ID-TIMS) baddeleyite ages and paleomagnetism of 1.79 and 1.50 Ga tholeiitic dyke swarms, and position of the Rio de la Plata craton within the Columbia supercontinent. Lithos 174:157–174

Teixeira W, Hamilton MA, Girardi VAV, Faleiros FM (2016) Key Dolerite Dyke Swarms of Amazonia: U-Pb Constraints on Supercontinent Cycles and Geodynamic Connections with Global LIP Events Through Time. Acta Geol Sin Engl Ed 90:84–85

Trompette R (1994) Geology of Western Gondwana (2000–500 Ma). A.A. Balkema, Rotterdam, Brookfield, 350 p

Tshibubudze A, Hein KAA (2013) Structural setting of gold deposits in the Oudalan-Gorouol volcano-sedimentary belt east of the Markoye Shear Zone, West African Craton. J Arf Earth Sci 80:31–47. https://doi.org/10.1016/j.jafrearsci.2012.11.010

Tshibubudze A, Hein KAA, Peters LFH et al (2013) Oldest U-Pb crystallisation age for the West African Craton From the Oudalan-Gorouol Belt of Burkina Faso. South Afr J Geol 116:169–181. https://doi.org/10.2113/gssajg.116.1.169

Vanderhaeghe O, Ledru P, Thiéblemont D et al (1998) Contrasting mechanism of crustal growth Geodynamic evolution of the Paleoproterozoic granite–greenstone belts of French Guiana. Precambr Res 92:165–193

Wahl RR, Thorman CH, Tysdal RG et al (2007) Geologic, geophysical and mineral localities map of Liberia, a digital compilation. USGS Open File Report, pp 2007–1258

Youbi N, Martins LT, Munhá JM et al (2003) The Late Triassic-Early Jurassic volcanism of Morocco and Portugal in the framework of the Central Atlantic Magmatic province: an overview. In: Hames WE, McHone JG, Renne PR, Ruppel C (eds) The Central Atlantic Magmatic Province: Insights from Fragments of Pangea. AGU Geophysical Mono-graph Series, vol 136, pp 179–207

Youbi N, Kouyaté D, Söderlund U et al (2013) The 1750 Ma Magmatic Event of the West African Craton (Anti-Atlas, Morocco). Precambr Res 236:106–123

Spatial and Temporal Distribution Patterns of Mafic Dyke Swarms in Central Asia: Results from Remote-Sensing Interpretation and Regional Geology

Qianwen Feng, Jinyi Li, Jianfeng Liu and Kefa Zhou

Abstract Mafic dykes in continents are a special type of intrusive rock formed by different tectonic events including earlier fracturing deformation in host rocks and successive magma intrusions originating at depth. Dyke swarms indicate regional crustal deformation related to continental formation and evolution, and play the role of messenger regarding magma activities from the deeper crust or even mantle. The large number of mafic dykes in Central Asia hold the keys to resolve some debated tectonic problems such as subduction-accretion-collision processes during the formation of Central Asia (the western part of the Central Asian Orogenic Belt). However, spatial-temporal distribution patterns of mafic dykes in this vast area have not been comprehensively described and discussed to date. This study carried out a fundamental work on the spatial-temporal distribution patterns of dykes in Central Asia, and is intended to provide basic preparation for more in-depth studies in the future. In Enhanced Thematic Mapper Plus (bands 7, 4, and 2) images, major mafic dykes (greater than 5 m wide) displayed as dark-colored linear objects against their host rocks, and can be distinguished and plotted one-by-one and further analyzed using Geographic Information System software. The results indicate that more than 99% of the mafic dykes were emplaced in the Eastern Tianshan and Beishan, Western Mongolian-Altai, Eastern Junggar, North and West bank of Balkhash, Western

Q. Feng · K. Zhou
Xinjiang Research Center for Mineral and Geology, Xinjiang Institute of Ecology and Geography, Chinese Academy of Sciences, 830011 Urumqi, China
e-mail: fengqianwen@ms.xjb.ac.cn

K. Zhou
e-mail: zhoukf@ms.xjb.ac.cn

Q. Feng · J. Li (✉) · J. Liu · K. Zhou
Institute of Geology, Chinese Academy of Geological Sciences, 100037 Beijing, China
e-mail: jyli@cags.ac.cn

J. Liu
e-mail: wenjv@aliyun.com

Q. Feng · K. Zhou
Xinjiang Key Laboratory of Mineral Resources and Digital Geology, 830011 Urumqi, China

R. K. Srivastava et al. (eds.), *Dyke Swarms of the World: A Modern Perspective*, Springer Geology, https://doi.org/10.1007/978-981-13-1666-1_8

Junggar, and Chingis-Taerbahatai. Most dykes formed during the Late Paleozoic, and the occurrence of these dykes is a result of various regional fracturing processes along different tectonic boundaries as well as immense magmatic intrusions (related to subduction, post-collisional processes, or large igneous provinces/giant mantle plumes). Other much older dykes were emplaced in the marginal areas of cratons, such as the Neoproterozoic dykes that intruded into the metamorphosed basement of the Tarim Craton (Aksu blueschist complex) and dykes emplaced in the Kuruktag area, which record the break-off history of the cratons. This study provides a general framework of mafic dykes in Central Asia on a large scale, and more intensive studies of mafic dykes at medium to small scales in different areas need a combined application of different observation methods, which will provide a better understanding of the continental evolution of Central Asia.

Keywords Dyke swarms · Central Asia · Satellite images · Visual interpretation Tectonic evolution

1 Introduction

As a special type of intrusive rock, mafic dykes are the consequence of earlier fracturing within the upper crust overprinted by successive magma intrusion from the deeper crust or even mantle. Compared to dykes within oceanic crust (mafic sheeted dyke complex in ophiolite suite; Moores 1982), continental mafic dykes are much more complicated; nevertheless, they are significant for re-constructing tectonic histories and geodynamical settings at different scales. At a continental scale, dyke swarms are considered to represent the break-off and destruction of cratons (e.g. Morgan 1971; Campbell and Griffiths 1990; Goldberg and Butler 1990; Zhao and McCulloch 1993; Kamo and Gower 1994; Zhao and McCulloch 1994; Kamo et al. 1995; Li et al. 2012), such as the largest dyke on Earth (the Giant Dyke) in Zimbabwe (Wilson 1982, 1996; Mukasa et al. 1998), dyke swarms in the North China craton (Li et al. 1997; Hou et al. 1998; Li et al. 2001; Shao and Zhang 2002; Hou et al. 2003; Peng et al. 2004; Peng 2010; Liu et al. 2017; Wang et al. 2017; Zhang and Cheng 2017), and huge radial dyke swarms that are indicative of mantle plumes (e.g. LeCheminant and Heaman 1991; Mertanen et al. 1996; Ma et al. 2000; Lu and Jiang 2003; Xu et al. 2007). On a regional scale, dykes of different strike and width elaborate spatial distribution patterns of dyke-filled fractures and intensities of mafic magmatic intrusions. The dykes play an important role in the formation of new crust in some orogenic zones (e.g. Li et al. 2005a; Chen et al. 2013). Moreover, some mineral resources are closely linked to dyke swarms (e.g. Oberthur et al. 1997; Schoenberg et al. 2003; Luo et al. 2008, 2012; Liu et al. 2014).

The significance of dyke swarms is increasingly highlighted by geological societies. It can be concluded from the seven International Dyke Conferences (IDCs) held in recent decades and numerous published papers that studies of dyke swarms can be divided into the following aspects: (1) Geochemical information of dyke rocks, which is very important to understand the processes of continental break-up, giant mantle plumes, and the interactions between mantle and crust; (2) Transportation processes and mechanisms during intrusion of dyke magma within the fractures in crust, as well as their contributions to continental growth and mineral enrichment; (3) Later deformation acting on previous dykes, for instance, some dykes are cut-off or curved; and (4) Formation mechanisms of dyke-filled fractures in host rocks, and the driving paleo-stress conditions.Central Asia is a major part of the Asian-European continent, and occupies most of the western Central Asian Orogenic Belt (CAOB; Jahn et al. 2000; Jahn 2004; Windley et al. 2007; Xiao et al. 2008). The crust of this vast area was mainly created by subduction and closure of the paleo-Asian Ocean (PAO) and a collage of continental slices, islands, marginal accretionary complexes, and sporadic oceanic crust fragments (Şengör et al. 1993; Windley et al. 2007; Li et al. 2009b; Xiao et al. 2009; Xiao et al. 2010a, b). To date, some crucial issues related to accretion and collision evolution histories in different parts of the CAOB, such as the timing of the closure of the PAO, are still hotly debated (Li et al. 2002b; Shu et al. 2004; Gao et al. 2006; Xiao et al. 2006b; Zhang et al. 2007a). Meanwhile, the existence of Permian large igneous provinces (LIPs) or giant mantle plumes in the Tarim and Eastern Tianshan in Xinjiang (Northwest China) also needs comprehensive discussion (Xia et al. 2004, 2006, 2008; Zhang and Zou 2013). Previous studies have mostly focused on accretionary complexes, granitic rocks, and volcanic and sedimentary rocks. However, the large number of mafic dykes in the CAOB have not paid sufficient attention. Investigations of the spatial-temporal distribution patterns of dykes on a large scale of this area have not been carried out.

Considering this situation, it is crucial to accomplish a survey of the spatial-temporal distribution patterns of dykes in Central Asia to provide fundamental data and instructions for further specific studies, especially discussing the role of mafic dyke swarms in the evolutionary history of the CAOB. Our investigations of the mafic dyke swarms in Central Asia started in 2009, and detailed studies have been carried out in some dyke-concentrated areas (such as the Western Junggar and Eastern Tianshan). This study introduces the methods applied to distinguish and extract dykes from host rocks, and reports the spatial-temporal distribution patterns of dyke-swarms on a large scale in Central Asia.

2 Tectonic Background and Dyke-Related Studies in Central Asia

2.1 Tectonic Background of Central Asia (Western Part of the CAOB)

The CAOB, alternatively known as the Altaids (Şengör et al. 1993) or the North Asian Orogenic Region (NAOR; Li et al. 2006a, 2009b), is situated between the Siberian craton to the north and the Tarim–North China craton to the south (Fig. 1a). It extends

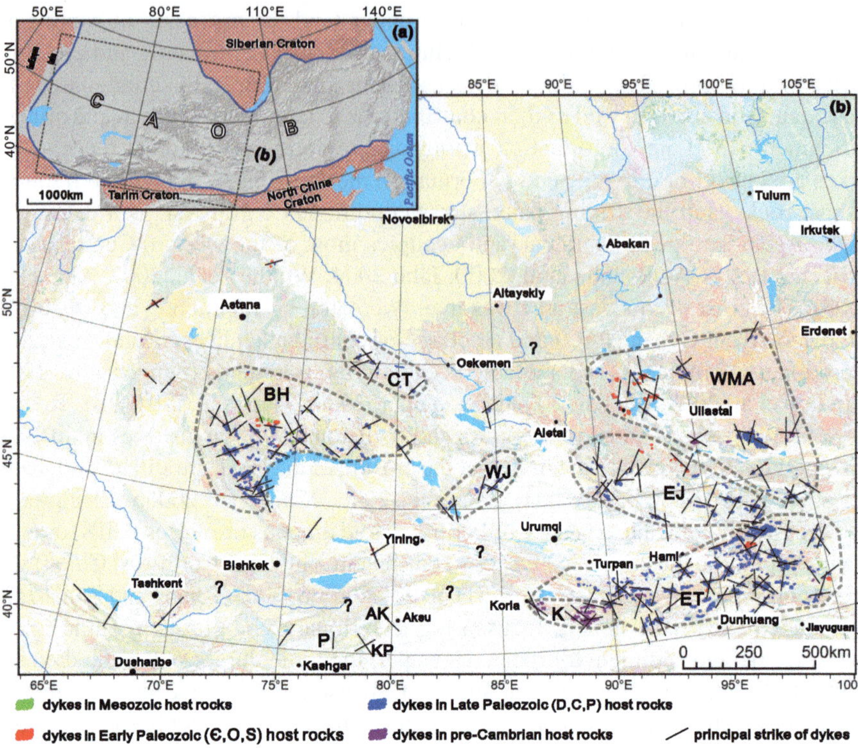

Fig. 1 Location of study area and distribution of mafic dykes in central Asia. **a** Tectonic location of the Central Asian Orogenic Belt. **b** Distribution patterns of dykes in central Asia, resulted from visual interpretation of ETM+ (bands 7, 4, and 2) images. ET: Eastern Tianshan and Beishan; WMA: Western Mongolian-Altai; BH: North and West bank of Balkhash; EJ: Eastern Junggar; WJ: Western Junggar; CT: Chingis-Taerbahatai; K: Kuruktag; AK: Aksu blueschist; KP: Keping; P: Piqiang area. (The dykes are too dense to be displayed as linear features in such a small-scale map. They are displayed as dyke swarms)

from the Ural Mountains in the west to the Pacific Ocean in the east. It is one of the largest Phanerozoic orogens on Earth, and exposes important geological units that can broaden our understanding of the subduction–accretion–collision processes that were active during the assembly of the Asian continent.

In recent years, investigations of key lithological and structural features within the CAOB have provided great insight regarding its general tectonic framework and Phanerozoic evolution, which was dominated by the marginal accretions and collisions between continental blocks during the Late Paleozoic, and the intracontinental orogenic processes that transpired during the Mesozoic and Cenozoic (e.g. Li 1980; Li et al. 2006a; Windley et al. 2007; Xiao et al. 2008; Wilhem et al. 2012). However, some significant issues regarding the evolution of the CAOB are still hotly debated, particularly the timing of closure of the PAO and the geodynamic setting during the Carboniferous-Permian. Xia et al. (2004, 2006, 2008) hypothesized that the Paleozoic ocean in the region closed in the Late Devonian and that during the Carboniferous and Permian both an intracontinental rift setting and a large igneous province occurred. Li (2004), Li et al. (2006a), Shu et al. (2004), Gao et al. (2006) suggested that the Paleozoic oceans closed in the Late Carboniferous, and the subsequent geodynamic setting was marked by the collision and amalgamation of continental blocks. Meanwhile, Late Carboniferous-Permian magmatism occurred within post-collisional settings (Han et al. 1999; Chen and Jahn 2004; Han et al. 2006). However, other studies argued that the Paleozoic oceans remained until the end of the Permian, and possibly persisted until the beginning of the Triassic (e.g. Li et al. 2002b, 2005b; Zhang et al. 2005; Xiao et al. 2006b; Zhang et al. 2007a).

2.2 *Previous Studies of Dykes in Central Asia*

Mafic dyke swarms in different parts of Central Asia have attracted some attention over the last 10 years (e.g. Zhang et al. 2007b, 2008a; Yin et al. 2009; Luo et al. 2012; Tang et al. 2012; Wang et al. 2015; Yang et al. 2015). For instance, the dykes in Western Junggar have been dated via different methods (Li et al. 2004; Xu et al. 2008; Zhou et al. 2008; Feng et al. 2012a, b; Yin et al. 2012; Zhang and Zou 2013), and the occurrence of diorite dyke swarms have been explained as the result of ridge subduction (Ma et al. 2012) or vertical magmatic intrusion and regional extension during post-collision evolution (Li et al. 2005a). Some curved dykes in Beishan (Sun et al. 2010) and truncated dykes in Eastern Tianshan (Feng et al. 2012c) indicate the shearing sense along local large faults. The correlation of distribution patterns between dykes and mineral deposits is also discussed in some areas (Qi 1993; Luo et al. 2008, 2012).

According to the previous aforementioned studies and observations of this work in different areas, the distribution patterns of mafic dykes are characterized by the following: (1) Most dykes are linear geological bodies with a certain width and length (aspect ratio greater than 10:1). Most dykes in Central Asia are less than 10 m wide, and smaller dykes less than 1 m are also widespread; (2) Dykes are usually intensively exhibited as dyke swarms, in some dyke-concentrated areas, the quantity of dykes is approximately 2–3 orders of magnitude greater than that of major faults; (3) Dykes of different strikes are usually crosscut and displaced among each other, and some displacement along the dykes is too small to be observed in satellite images; and (4) Dykes are usually accompanied by several small faults, some are fault-cut dykes and others are filled by dykes.

3 Data and Methods

3.1 *Difficulties of Field Observation on a Large Scale*

Field observations are indispensable to investigate individual dyke-segments and dyke swarms in a small area. To describe distribution patterns in a large region such as the whole Central Asia, field observation results of all dyke-emplaced areas (usually with different levels of detail) over the entirety of Central Asia must be collected and summarized. However, in the situation that the dyke-swarms have not been thoroughly investigated at small scales in previous studies, it is impossible to finish this immense task in a quite limited time. As a compromised step, alternatively efficient methods must be tested and applied, to estimate the distribution patterns of dykes on a large scale for the first time.

3.2 *Visual Interpretation Method and Employed Data*

Through experience of dyke-related surveys in recent years, we have found that a remote-sensing interpretation method is reliable for regional investigations of mafic dykes. Fortunately, most parts of Central Asia are arid and outcrops of dykes are rarely covered by vegetation. In different types of satellite images of different spatial resolutions, mafic dykes are displayed as dark-colored linear objects against lighter-colored host rocks (Fig. 2), and can be easily distinguished using visual interpretation method (Zhang et al. 2007b, 2008a, Feng et al. 2012a, b, c).

Fig. 2 Mafic dykes displayed as dark-colored linear objects in different satellite images

Numerous ETM+ images were collected covering the whole study area (bands 7, 4, and 2 of the Landsat 7 data are synthesized as red, green, and blue channels, respectively; the product images were freely downloaded from https://wist.echo.nasa.gov/api/ in 2010). The spatial resolution of the ETM+ images is approximately 14.25 m, which allows one to distinguish major dykes. The images can be imported into ArcGIS software, and each individual dyke segment is visually distinguished and plotted one-by-one and stored as vector-formatted lines in an ArcGIS database. The lengths and strikes of the dykes are calculated by the coordinates of their endpoints automatically. Concentrations of dykes in different areas are analyzed and compared based on an analysis of individual dykes.

3.3 Corroboration and Evaluation of Results

The existence and accuracy of distinguished dykes are corroborated and evaluated in intentionally arranged field surveys. The results show that only major dykes greater than 5 m wide can be correctly distinguished; smaller dykes cannot be seen in ETM+ images.

4 Spatial and Temporal Distributions of Dyke Swarms

4.1 Spatial Distribution Patterns of Dykes on a Large Scale

Visual interpretation using ETM+ images distinguished approximately 30,830 major dykes (more than 5 m wide) emplaced in Central Asia (Fig. 1b).

On a large scale, these dykes are not evenly distributed (Fig. 3), and the optimal directions of the dykes vary among different areas in Central Asia (Fig. 4). More than 99% of the dykes are concentrated in the following areas (from the most to the least): (1) Eastern Tianshan and Beishan (ET); (2) Western Mongolian-Altai (WMA); (3) North and West of Balkhash (BH); (4) Eastern Junggar (EJ); (5) Western Junggar (WJ); and (6) Chingis-Taerbahatai (CT). Eastern Tianshan and Beishan is the most dyke-concentrated area in Central Asia, accommodating more than 60% dykes (Figs. 1b, 3).

Other areas such as the Western Tianshan and Altai in Xinjiang are covered by vegetation or glaciers. The existence and number of mafic dykes in these areas is unknown (see question mark labels in Fig. 1b), and have rarely been reported in previous studies (e.g. the NE–SW trending dykes in the Southwest Altai; Cai et al. 2010).

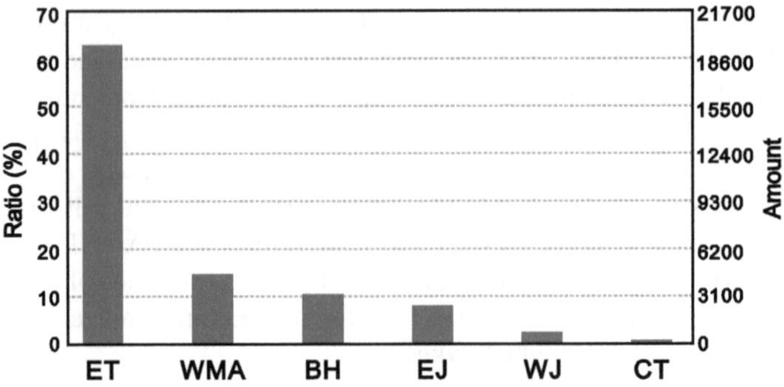

Fig. 3 The amount of dykes in dyke-concentrated areas. ET, WMA, BH, EJ, WJ, and CT stand for the same areas with Fig. 1(b). The ratio refers to the amount of dykes in these areas compare with the whole study area

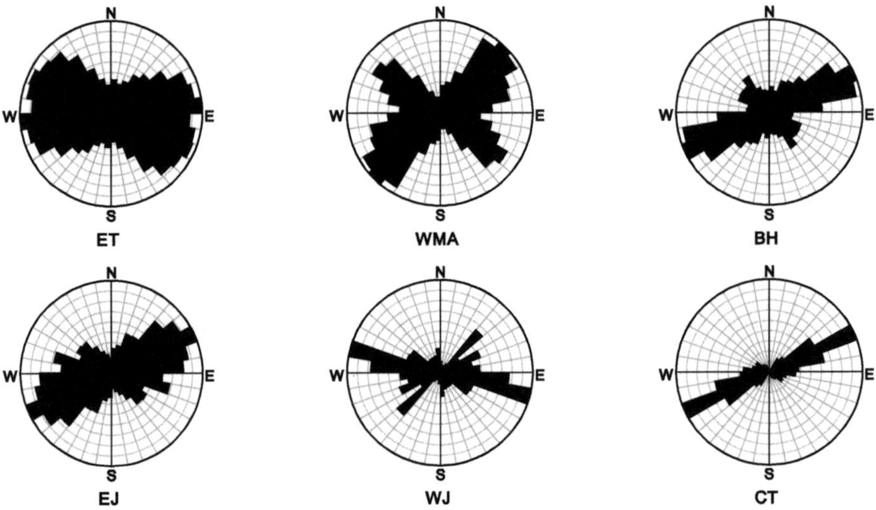

Fig. 4 Rose diagrams of dykes in dyke-concentrated areas. ET, WMA, BH, EJ, WJ, and CT stand for the same areas with Fig. 1(b)

4.2 *Temporal Distribution of Dykes*

Chronological studies of dykes have not been thoroughly completed in available previous studies thus far. This study discusses the ages of dykes indirectly by: (1) Ages of the host rocks (dykes must be younger than their host rocks); (2) Nearby mafic magmatic events; and (3) regional fracturing events.

According to regional geological maps and published geochronological results, most host rocks of the dykes in Central Asia formed during the Paleozoic, except for the Archean and Proterozoic host rocks intruded by Neoproterozoic dykes in

Kuluketage and the western marginal area of the Tarim Basin. Further classification of host rocks reveals that among the Paleozoic host rocks, the majority formed during the Late Paleozoic.

5 Primary Understanding of Some Mafic Dyke Swarms in Central Asia

5.1 Multi-Period Dykes in the Eastern Tianshan and Beishan

The dykes in these areas are the result of multi-period tectonic events. The most widely distributed dykes intruded during the Late Carboniferous to Early Permian (Liu et al. 1999; Li et al. 2006b; Xiao et al. 2006a; Pirajno et al. 2008; Qin et al. 2011; Su et al. 2011). Some studies have considered that the origination of these dykes is related to LIPs or mantle plumes during that period (Pirajno et al. 2008; Qin et al. 2011; Su et al. 2011); however, in some outcrops the dyke-filled fractures are obviously controlled by regional shearing displacement (Fig. 5a, b). Other dykes sparsely emplaced during intrusion of Mesozoic round-shaped plutons (Li et al. 2002a; Zhang et al. 2006; Wang et al. 2008; Li et al. 2010, 2014; Zhang et al. 2016) are much younger (Fig. 5c), and resulted from intra-continental evolution with weak magmatic activities.

5.2 Permian and Neoproterozoic Dykes in Western Tarim and Kuruktag

A dyke swarm in Keping in the western Tarim (a segment of the passive margin of the Tarim during the Paleozoic), constrained as occurring from 270–290 Ma by volcanic-sedimentary formations (Li et al. 2017a) cut by the dykes, is also related to the Permian Tarim LIP. Besides, some Cenozoic mafic dykes (48–46 Ma) in Piqiang basin, western Tarim, and the dyke-filled fractions are caused by S–N compression (Li et al. 2009a).

The protolith of the Aksu blueschist is a part of the oceanic crust accreted to Tarim as part of an accretionary wedge during the Neoproterozoic. It experienced high-pressure metamorphism (860–870 Ma) during the processes of deep subduction and exhumation (Zhang et al. 2008b; Zheng et al. 2008; Shen and Geng 2012; Zhang et al. 2014). Dozens of NW-SE trending diabase dykes emplaced in the Aksu blueschist, and the widths of the dykes range from several centimeters up to 20 m. The strikes of the dykes are nearly orthogonal to the metamorphic foliations (Fig. 6a, b). Previous chronological work revealed that these dykes occurred during 759–807 Ma (Chen et al. 2004; Zhan et al. 2007; Zhang et al. 2009a). In the Kuruktae area, the dykes are ultra-high-densely emplaced (the highest density of dykes in Central Asia, Fig. 6c,

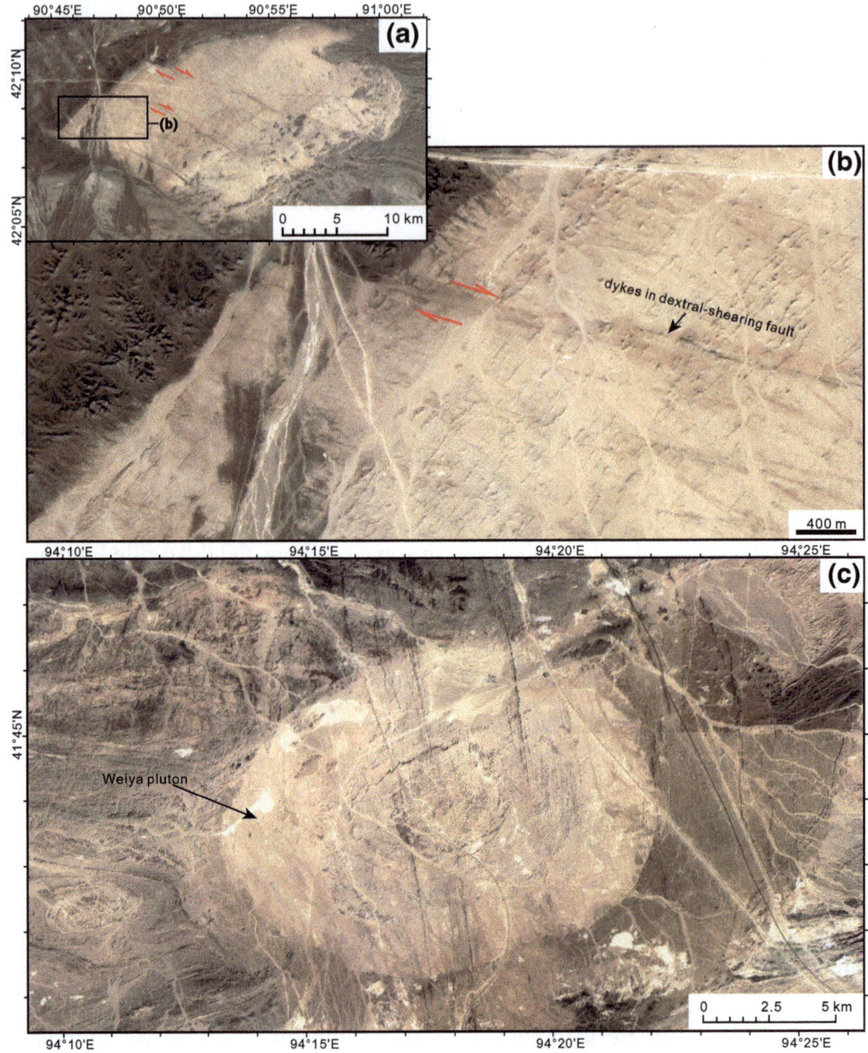

Fig. 5 Some mafic dykes emplaced in Eastern Tianshan **a**, **b** Dykes intruded in dextral-shearing faults, in a granite pluton. **c** Several NNW-SSE trending dykes emplaced in Mesozoic round-shaped pluton in Weiya area (The images are downloaded from Google Earth)

d), and a notable zebra-like scene occurs in the field (Zhang et al. 2009b). The dykes in Kuruktag were intruded 823 ± 8.7 Ma and 776.8 ± 8.9 Ma, which may record the influence of mantle-originating magma during the break-off of Rodinia (Zhang et al. 2009b).

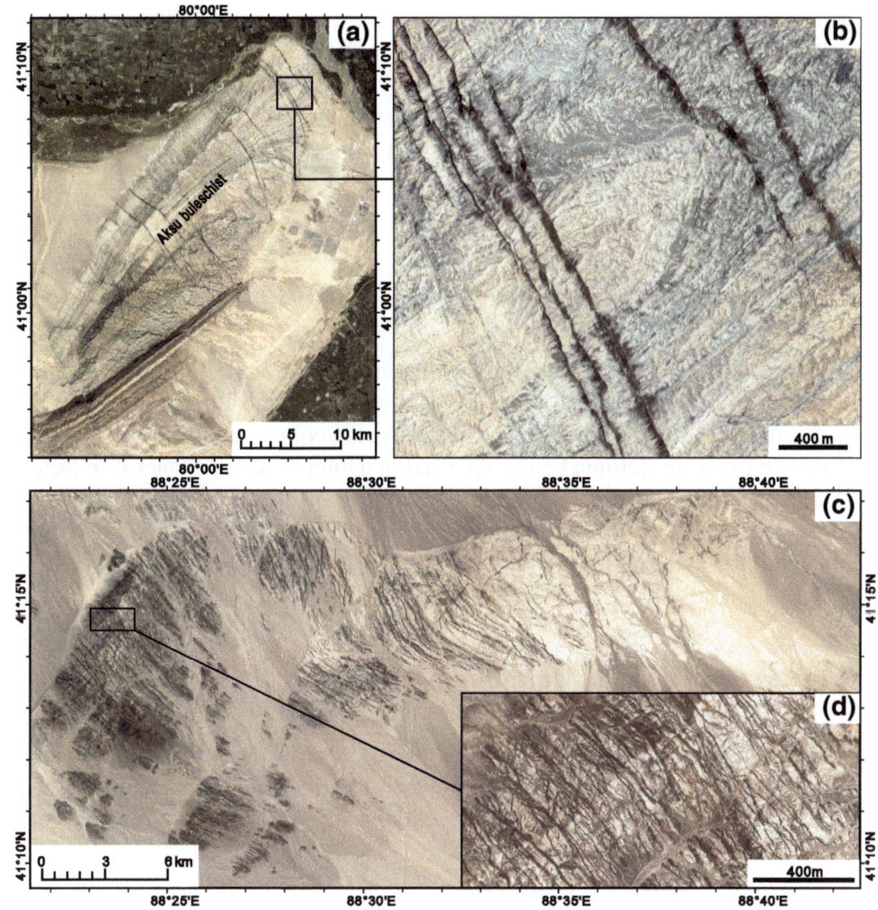

Fig. 6 Mafic dykes emplaced in Aksu buleschist and Kuruktag area **a**, **b** Dykes emplaced in Aksu buleschist (Western margin of the Tarim) **c**, **d** Kuruktag (Eastern Tianshan) is the most dyke-concentrated area among the Central Asia.(The images are downloaded from Google Earth)

5.3 Late Paleozoic Dykes in Western Junggar, Eastern Junggar and Chingis-Taerbahatai

Dykes in these areas mostly emplaced during the Late Carboniferous to Early Permian, based on previous chronological studies of dykes in Western Junggar (Li et al. 2004; Xu et al. 2008; Zhou et al. 2008; Feng et al. 2012a, b; Yin et al. 2012; Zhang and Zou 2013), constrained by the cutting relationship between the dykes and the host rocks (plutons and volcanic-sedimentary formations) in Eastern Junggar (Feng et al. 2015) and Chingis-Taerbahatai.

The spatial distribution patterns of dykes are obviously influenced by the NE–SW–trending Darabut fault/suture line in Western Junggar, two NW–SE-trending suture lines (Karamaili and Aermantai) in Eastern Junggar, and the NW–SE-trending suture line in Chingis-Taerbahatai (Han et al. 2010), respectively. Dykes in these areas play an important role in providing more insights regarding subductions and collisions during the Late Paleozoic.

5.4 Paleozoic Dykes in the North and West Bank of Balkhash

The host rocks intruded by the dyke swarms mainly formed during the Paleozoic (more so during the Late Paleozoic and less so during the Early Paleozoic). The spatial and temporal distribution of dyke swarms is coincidentally overlapped by a well-known orocline tectonic belt (composed of a huge C-shaped Paleozoic tectonic belt) in Kazakhstan (Van der Voo et al. 2007; Abrajevitch et al. 2008; Xiao et al. 2010a; Li et al. 2017b). More detailed studies of dykes may more convincingly elaborate on the mechanisms and processes of this orocline (Fig. 7).

5.5 Dykes in Western Mongolian-Altai

The host rocks of the dykes in this area formed during both the Early and Late Paleozoic. The optimal directions of NW–SE and NE–SW in this area are very clear; however, whether this assemblage is representative of conjugate fractures needs more kinematic evidence on smaller scales.

6 Discussions and Conclusions

6.1 Observational Methods of Dyke Swarms at Different Scales

This study only reports the distribution patterns of dyke swarms on a large scale in Central Asia, and provides a framework for more comprehensive dyke-related studies on a regional scale in the future. However, detailed studies in smaller areas (medium scale) and higher resolution images (such as Google Earth images) must be applied. For anatomical study (small to microscopic scale), observation in outcrops and under a microscope must be carried out.

In recent years, the popularity of unmanned aerial vehicles (UAVs) has made low-cost aerial photography a reality. Aerial photos with ultra-high resolution provide a new visual aspect of dykes, which can greatly enhance work efficiency.

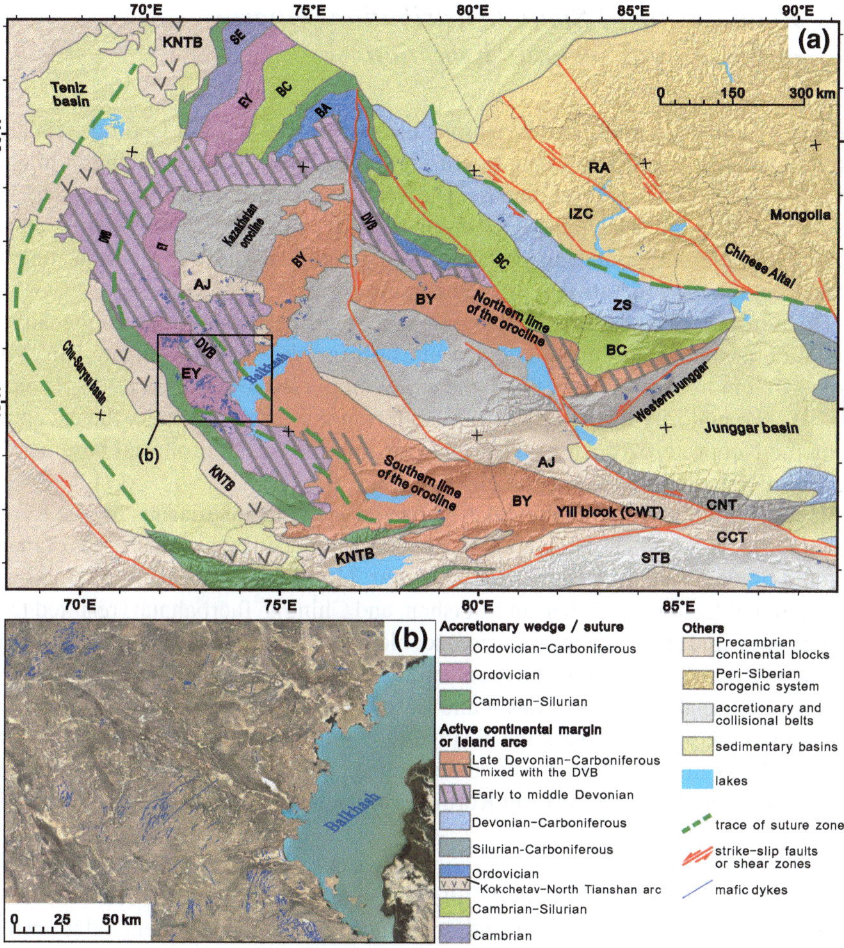

Fig. 7 Spatial relationship of mafic dykes and Kazakhstan orocline. **a** The Kazakhstan orocline and mafic dykes emplaced in Balkhash area. **b** Some dykes emplaced in Western Balkhash (the image is downloaded from Google Earth). The base map of the Kazakhstan orocline is re-made according to Li et al. 2017b. RA: Rudny Altai; IZC:Irtysh-Zaisan complex; ZS: Zharma-Saur arc; BC: Boshchekul-Chingiz arc; BA: Baydaulet-Akbastau arc; EY: Erementau-Yili belt; SE: Selety arc; DA: Dulate arc; YA: Yemaquan arc; BY:Balkhash-Yili arc; DVB: Early to Middle Devonian volcanic belt; KNTB: Kokchetav-North Tianshan belt; IMT: Ishim-Middle Tianshan block; AJ: Aktau-Junggar block; CCT: Chinese central Tianshan block; CNT: Chinese North Tianshan belt; STB: South Tianshan belt; CWT: Chinese western Tianshan

6.2 Relationship of Spatial Distribution Patterns and Regional Tectonic Evolution

The large number of mafic dykes exhibited in the present crust of Central Asia are a result of tectonic events including fracturing in the upper crust and rising of magma from the lower crust or even mantle. The spatial-temporal distribution patterns of the mafic dykes are multipurpose records to understand the continental evolution including angles of fracturing processes and related crustal deformation, magma sequences, interaction between crust and mantle, etc.

The density variation of the dykes in space and period reflect the intensity of crustal deformation and magmatic activities during their emplacement. This work reveals that more than 99% of the mafic dykes emplaced in six areas as displayed in Fig. 1b. The majority of the dykes formed during the Late Paleozoic, the exception being the Neoproterozoic dykes that intruded into the metamorphosed basement of the Tarim craton and Kuruktag block.

In the distribution map of the dykes, it can be seen that the reasons for the formation of these mafic dykes are diverse. Most dykes emplaced in marginal areas of major tectonic units (West Junggar, Eastern Junggar, North and West bank of Balkhash, some parts of Eastern Tianshan and Beishan, and Chingis-Taerbahatai) resulted from compression and shearing between contact blocks or extensional fractures caused by subduction-related magma or post-collision-related magma. The Permian dykes in the west marginal area of the Tarim and Eastern Tianshan may be related to an LIP or giant mantle plume beneath the Tarim craton during the Permian. Some Neoproterozoic dykes intruded into the Aksu blueschist complex (exposed metamorphosed basement of the Tarim craton) and Kuruktag block record the break-off of the Tarim craton and Kuruktag block.

For more comprehensive studies of mafic dykes in the future, the geometric-kinematic characteristics of the dyke-filled fractures in the host rocks can be revealed showing evidence of movement and crosscutting and displacement relationships between differently oriented dykes (widespread dextral/sinistral displacement along the strike of dykes hints that most fractures in nature are not caused by extensional stress, but shear or even compressional stress conditions), and boundary surfaces between the dykes and their host rocks. The significance of dyke-related magma could be revealed by testing the dykes using chronological and geochemical methods, as well as correlation of the dykes and host rocks.

Acknowledgements This study was co-supported by the National Natural Science Foundation of China (project number 41502201) and the "Western Light" project of the Chinese Academy of Sciences (project number XBBS201301). We are grateful to Prof. Rajesh Srivastava and Prof. Peng for their thoughtful arrangements. We also thank two reviewers for constructive comments which led to major improvements in the manuscript.

References

Abrajevitch A, Van der Voo R, Bazhenov ML et al (2008) The role of the Kazakhstan orocline in the late Paleozoic amalgamation of Eurasia. Tectonophysics 455(1–4):61–76
Cai KD, Sun M, Yuan C et al (2010) Geochronological and geochemical study of mafic dykes from the northwest Chinese Altai: implications for petrogenesis and tectonic evolution. Gondwana Res 18:638–652
Campbell IH, Griffiths RW (1990) Im plication of mantle plume structure for the evolution of flood basalts. Earth Planet Sci Lett 99:79–93
Chen B, Jahn BM (2004) Genesis of post-collisional granitoids and basement nature of the Junggar Terrane, NW China: Nd–Sr isotope and trace element evidence. J Asian Earth Sci 23(5):691–703
Chen NH, Dong JJ, Li ZL et al (2013) Permian crustal extension of Beishan area in Xinjiang, NW China: estimation from the statistical thickness of exposed mafic dyke swarms. Acta Petrol Sinica 29(10):3540–3546
Chen Y, Xu B, Zhang S et al (2004) First Mid-Neoproterozoic paleomagnetic results from the Tarim Basin (NW China) and their geodynamic implications. Precambr Res 133(3):271–281
Feng QW, Li JY, Liu JF et al (2012a) Ages and geological significance of the dark dykes emplaced in the Karamay pluton and adjacent area, in Western Junggar, Xinjiang, NW China: Evidence form LA-ICP-MS zircon chronology and Ar-Ar Amphibole chronology. Acta Petrol Sinica 28(7):2158–2170
Feng QW, Li JY, Liu JF et al (2012b) Ages of the Hongshan granite and intruding dioritic dyke swarms, in western Junggar, Xinjiang, NW China: Evidence from LA-ICP-MS zircon chronology. Acta Petrol Sinica 28(9):2935–2949
Feng QW, Li JY, Zhang J et al (2012c) About 9 km sinistral stike-slip displacement of the East segment of Xindi fracture zone. Ecidences from satellite remote sensing images. Xinjiang Geology, Xinjiang, NW China, pp 210–213
Feng QW, Zhou KF, Wang JL (2015) The Late Paleozoic geodynamic environment of Qiongheba arc, eastern Junggar: evidence from spatial and temporal distribution patterns of dark dykes in Hersai pluton. Chinese J Geol 54(4):1068–1082
Gao J, Long LL, Qian Q et al (2006) South Tianshan: a Late Paleozoic or a Triassic orogeny? Acta Petrol Sinica 22(5):1049–1061
Goldberg SA, Butler JR (1990) Late Proterozoic rift-related dykes of the southern and central Appalachians, eastern USA. In: Parker (ed) Mafic dykes and emplacement mechanisms, pp 131–144
Han BF, He GQ, Wang SG (1999) Postcollisional mantle-derived magmatism, underplating and implications for basement of the Junggar Basin. Science China (Series D) 42(2):113–119
Han BF, Ji JQ, Song B et al (2006) Late Paleozoic vertical growth of continental crust around the Junggar Basin, Xinjiang, China (Part I): Timing of post-collisional plutonism. Acta Petrol Sinica 22(5):1077–1086
Han BF, Guo ZJ, Zhang ZC et al (2010) Age, geochemistry, and tectonic implications of a Late Paleozoic stitching pluton in the North Tian Shan suture zone, western China. Geol Soc Am Bull 122:627–640
Hou GT, Zhang C, Qian XL et al (1998) The formation mechanism and tectonic stress field of the Mesoporoterozoic mafic dike swarms in the North China Craton. Geol Rev 44(3):309–314
Hou GT, Li JH, Halls HC et al (2003) The flow structures and mechanics of Late Precambrian mafic dyke swarms in North China Craton. Acta Geol Sin 77(2):210–216
Jahn BM, Griffin WL, Windley BF (2000) Continental growth in the Phanerozoic: evidence from Central Asia. Tectonophysics 328:vii–x
Jahn BM (2004) Phanerozoic continental growth in Central Asia. J Asian Earth Sci 23:599–603
Kamo SL, Gower CF (1994) U–Pb baddeleyite dating clarifies age of characteristic paleomagnitic remanence of Long Range dykes, southeastern Labrador. Atl Geol 30:259–262
Kamo SL, Krpgh TE, Kumarapeli PS (1995) Age of the Grenville dyke swarm, Ontario-Quebec: implication for the time of lapetan rifting. Can J Earth Sci 32:273–280

LeCheminant AN, Heaman LM (1991) U–Pb ages for the 1.27 Ga Mackenzie igneous events, Canada: support for a plume initiation model (abstract). Program with anstracts, A73, Geological Association on Canada: Waterloo, p 16

Li CY (1980) A preliminary study of plate tectonics of China. Bull Chin Acad Sci 2(1):11–22

Li DD, Luo ZH, Huang JX et al (2009a) The chronological and geological implications of the Cenozoic basic magmatic activities in Piqiang Basin, Xinjiang China. Earth Sci Front 16(3):270–281

Li DD, Wang YW, Wang JB et al (2014) New evidence of the mafic-ultramafic pluton emplacement age in Weiya, Eastern Tianshan Xinjiang. Xinjiang Geol 32(1):1–5

Li HB, Zhang ZC, Lv LS et al (2012) Petrological, geochemical and geometric characteristics of dyke swarms and their tectonic significance. Acta Petrol ET Mineral 31(1):91–103

Li JH, He WY, Qian XL (1997) Genetic mechanism and tectonic setting of Proterozoic mafic dyke swarm: it implication for paleoplate reconstruction. Geol J China Univ 3(3):272–281

Li JH, Hou GT, Qian XL et al (2001) Single-zircon U–Pb age of the initial Mesoproterozoic basic dike swarms in Hengshan mountain and its implication for the tectonic evolution of the North China Craton. Geol Rev 17(3):234–238

Li JY (2004) Structural characteristics of crustal "mosaicking and superimposition" of the continent of China and its evolution. Geol Bull China 23:986–1004

Li JY, He GQ, Xu X et al (2006a) Crustal tectonic framework of Northern Xinjiang and adjacent regions and its formation. Acta Geol Sin 80(1):148–468

Li JY, Song B, Wang KZ et al (2006b) Perimian mafic-ultramafic complexes on the southern margin of the Tu–Ha Basin, East Tianshan Mountains: Geological records of vertical crustal growth in Central Asia. Acta Geoscientica Sinica 27(5):424–446

Li JY, Zhang J, Yang TN et al (2009b) Crustal tectonics division and evolution of the Southern part of the North Asian orogenic region and its adjacent areas. J Jilin Univ (Earth Sci Edition) 39(4):584–605

Li JY, Liu JF, Zheng RG et al (2017a) The eruptive timing of Permian basalts in Keping area, The northwest margin of Tarim Basin, China: evidence from detrital zircons. J Earth Sci Environ 39(3):301–325

Li M, Wang YW, Wang JB et al (2010) Zircon U–Pb chronologic evidence for magma mixing in the Weiya ore district, Xinjiang. Geol China 37(1)

Li P, Sun M, Rosenbaum G et al (2017b) Geometry, kinematics and tectonic models of the Kazakhstan Orocline, Central Asian Orogenic Belt. J Asian Earth Sci

Li SL, Li WQ, Feng XC et al (2002a) Age of formation of Weiya composite stocks in Eastern Tianshan mountains. Xinjiang Geol 20(4):357–359

Li XZ, Han BF, Ji JQ et al (2004) Geology, geochemistry and K–Ar ages Karamay basic-intermediate dyke swarm from Xinjiang, China. Geochimica 33(6):574–584

Li XZ, Han BF, Li ZH et al (2005a) Mechanism of the Karamay basic-intermediate dyke swarm from Xinjiang and tectonic implications. Geol Rev 51(5):517

Li YJ, Wang ZM, Wu HR et al (2002b) Discovery of radiolarian fossils from the Aiketik group at the Western end of the South Tianshan mountains of China and its implications. Acta Geol Sin 76:146–153

Li YJ, Sun LD, Wu HR et al (2005b) Permo-Carboniferous radiolarians from the Wupata'erkan group, Western South Tianshan, Xinjiang, China. Acta Geol Sin 79:16–23

Liu J, Jin SY, Wang YJ (2014) The summarization of platinum group metals deposits in the Great Dyke, Zimbabwe. Precious Met 35(3):79–82

Liu S, Feng CX, Feng GY et al (2017) Zircon U–Pb age, geochemical and Sr–Nd–Pb isotopic data: Constraints on the genetic model of the mafic dykes from the North China Craton. Acta Petrol Sinica 33(6):1667–1685

Liu YL, Zhang ZC, Guo ZJ et al (1999) K–Ar isochron dating of Kuluktag mafic dykes, Xinjiang autonomous region, and discussion on some related questions. Geol J China Univ 5(1):54–58

Lu SN, Jiang MM (2003) Mantle plumes and giant radiating dyke swarms. Geol Surv Res 26(3):136–144

Luo ZH, Lu XX, Wang BZ et al (2008) Post-orogenic dike complexes and implications for metallogenesis. Earth Sci Front 15(4):1–12

Luo ZH, Chen BH, Jiang XM et al (2012) A preliminary attempt for targeting prospecting districts using the wide composition-spectrum dike swarms: an example of the South Alatao Mountains, Xinjiang, China. Acta Petrol Sinica 28(7):1949–1965

Ma C, Xiao W, Windley BF et al (2012) Tracing a subducted ridge–transform system in a late Carboniferous accretionary prism of the southern Altaids: Orthogonal sanukitoid dyke swarms in Western Junggar, NW China. Lithos 140–141:152–165

Ma F, Mu ZG, Li JH (2000) Geochemistry and petrogenesis of Precambrian mafic dyke swarms. Geol-Geochem 28(4):58–64

Mertanen S, Personen LJ, Huhma H (1996) Paleomagnetism and Sm–Nd ages of the Neoproterozoic diabase in Laanila and Kautokeino, northern Fennoscandia, in Precambrian crustal evolution in the North Atlantic region. Spec Publ 112:331–358

Moores EM (1982) Origin and emplacement of ophiolites. Rev Geophys Space Phys 20(4):735–760

Morgan WJ (1971) Convection plumes in the lower mantle. Nature 230:42–43

Mukasa SB, Wilson AH, Carlson RW (1998) A multielement geochronologic study of the Great Dyke, Zimbabwe: significance of the robust and reset ages. Earth Planet Sci Lett 164(1–2):353–359

Oberthur T, Cabri LJ, Weisner TW et al (1997) Pt, Pd and other trace elements in sulfides of the Main Sulfide Zone, Great Dyke, Zimbabwe: a reconnaissance study. Can Mineral 35(35):597–609

Peng P, Zhai MG, Zhang HF et al (2004) Geochemistry and geological significance of the 1.8 Ga mafic dyke swarms in the North China Craton: an example from the juncture of Shanxi, Hebei and Inner Mongolia. Acta Petrol Sinica 20(3):439–456

Peng P (2010) Reconstruction and interpretation of giant mafic dyke swarms: a case study of 1.78 Ga magmatism in the North China craton. Geol Soc, London, Spec Publ 338:163–178

Pirajno F, Mao J, Zhang Z et al (2008) The association of mafic–ultramafic intrusions and A-type magmatism in the Tian Shan and Altay orogens, NW China: implications for geodynamic evolution and potential for the discovery of new ore deposits. J Asian Earth Sci 32(2–4):165–183

Qi JY (1993) Geology and genesis of vein rock group in Western Zhunggar, Xinjiang. Acta Petrol Sinica 9(3):288–299

Qin KZ, Su BX, Sakyi PA et al (2011) SIMS zircon U–Pb geochronology and Sr–Nd isotopes of Ni–Cu–Bearing Mafic-Ultramafic Intrusions in Eastern Tianshan and Beishan in correlation with flood basalts in Tarim Basin (NW China): constraints on a ca. 280 Ma mantle plume. Am J Sci 311(3):237–260

Schoenberg R, Nägler TF, Gnos E et al (2003) The source of the great dyke, Zimbabwe, and Its tectonic significance: evidence from Re–Os isotopes. J Geol 111(5):565–578

Şengör AMC, Natal'in BA, Burtman VS (1993) Evolution of the Altaid tectonic collage and Paleozoic crustal growth in Eurasia. Nature 364:299–307

Shao JA, Zhang LQ (2002) Mesozoic dyke swarms in the north of North China. Acta Petrol Sinica 18(3):312–318

Shen QH, Geng YS (2012) The tempo-spatial distribution, geological characteristics and genesis of blueschist belts in China. Acta Geol Sin 86(9):1407–1446

Shu LS, Guo ZJ, Zhu WB et al (2004) Post-collision tectonism and basin-range evolution in the Tianshan belt. Geol J China Univ 10(3):393–404

Su BX, Qin KZ, Sakyi PA et al (2011) U–Pb ages and Hf–O isotopes of zircons from late Paleozoic mafic–ultramafic units in the southern Central Asian Orogenic Belt: tectonic implications and evidence for an early-permian mantle plume. Gondwana Res 20(2–3):516–531

Sun B, Shan YH, Qiao YB (2010) Curved mafic dike swarm in a Late-Paleozoic granitoid in the Northern Mountain, Gansu, northwestern China: genesis and implications. Chinese J Geol 45(1):80–91

Tang GJ, Wang Q, Wyman DA et al (2012) Late carboniferous high εNd(t)–εHf(t) granitoids, enclaves and dikes in western Junggar, NW China: Ridge-subduction-related magmatism and crustal growth. Lithos 140(141):86–102

Van der Voo R, Abrajevitch S, Bazhenov ML et al. (2007) Late Paleozoic History of the Kazakhstan orocline. American Geophysical Union

Wang L, Wang GH, Lei SB et al (2017) The Early Carboniferous Xiaomiaogou granite porphyry dykes in the northern margin of the North China Craton: implication for craton-mantle interaction and intraplate magmatism. Geol J 52:489–509

Wang LM, Chao WD, Li YJ et al (2015) Zircon LA–ICP–MS U–Pb dating and tectonic settings implication of the early permian intermediate to Mafic Dyke swarms in Urho Area, West Junggar. Xinjiang Geol 33(3):297–304

Wang YW, Wang JB, Wang LJ et al (2008) Zorcon U–Pb age, Sr–Nd isotope geochemistry and geological siginificances of the weiya mafic-ultramafic complex, xinjiang. Acta Petrol Sinica 24(4):781–792

Wilhem C, Windley BF, Stampfli GM (2012) The Altaids of central Asia: a tectonic and evolutionary innovative review. Earth Sci Rev 113:303–341

Wilson AH (1982) The geology of the Great 'Dyke', Zimbabwe: the ultramafic rocks. J Petrol 23(2):240–292

Wilson AH (1996) The Great Dyke of Zimbabwe. Dev Petrol 15(96):365–402

Windley BF, Alexeiev D, Xiao WJ et al (2007) Tectonic models accretion of the central Asian Orogenic Belt. J Geol Soc 164:31–47

Xia LQ, Xia ZC, Xu XY et al (2004) Carboniferous Tianshan igneous megaprovince and mantle plume. Geol Bull China 23(9–10):903–910

Xia LQ, Li XM, Xia ZC et al (2006) Carboniferous-Permian rift-related volcanism and mantle plume in the Tianshan, Northwestern China. Northwest Geol 39(1):1–49

Xia LQ, Xia ZC, Xu XY et al (2008) Petrogenesis of Carboniferous-Early Permian rift-related volcanic rocks in the Tianshan and its neighboring areas, Northwestern China. Northwest Geol 41(4):1–68

Xiao PX, Huang YH, Wang YX et al (2006a) Geochemical characteristics and isotope dating of moyite at the southeastern margin of the Kuruktag block, Xinjiang, China. Geol Bull China 25(6):725–729

Xiao W, Kröner A, Windley B (2009) Geodynamical evolution of central Asia in the Paleozoic and Mesozoic. Int J Earth Sci 98(6):1185–1188

Xiao WJ, Han CM, Yuan C et al (2006b) Unique Carboniferous-Permian tectonic-metallogenic framework of Northern Xinjiang (NW China): constraints for the tectonics of the Southern Paleoasian Domain. Acta Petrol Sinica 22(5):1062–1076

Xiao WJ, Shu LS, Gao J et al (2008) Continental dynamics of the central Asian Orogenic Belt and its metallogeny. Xinjiang Geol 26(1):4–8

Xiao WJ, Huang BC, Han CM et al (2010a) A review of the western part of the Altaids: a key to understanding the architecture of accretionary orogens. Gondwana Res 18(2–3):253–273

Xiao WJ, Mao QG, Windley BF et al (2010b) Paleozoic multiple accretionary and collisional processes of the Beishan orogenic collage. Am J Sci 310(10):1153–1594

Xu QQ, Ji JQ, Han BF et al (2008) Petrology, geochemistry and geochronology of the intermediate to mafic dykes in Northern Xinjiang since late Paleozoic. Acta Petrol Sinica 24(5):977–996

Xu YG, He B, Huang XL et al (2007) The dabate over mantle plumes and how to test the plume hypothesis. Earth Sci Front 14(2):1–9

Yang CX, Wang QH, Gao X et al (2015) Geochemical characteristics and tectonic setting of diabase in Beishan Gansu. Gansu Geol 24(1):19–23

Yin JY, Yuan C, Sun M et al (2009) Sanukitic dykes in west Junggar, Xinjiang: geochemical features, petrogenensis and links to Cu–Au mineralization. Geochimica 38(5):413–423

Yin JY, Yuan C, Sun M et al (2012) Age, geochemical features and possible petrogenesis mechanism of early Permian magnesian diorite in Hatu Xinjiang. Acta Petrol Sinica 28(7):2171–2182

Zhan S, Chen Y, Xu B et al (2007) Late Neoproterozoic paleomagnetic results from the Sugetbrak Formation of the Aksu area, Tarim basin (NW China) and their implications to paleogeographic reconstructions and the snowball earth hypothesis. Precambr Res 154(3):143–158

Zhang CL, Li ZX, Li XH et al (2009a) Neoproterozoic mafic dyke swarms at the northern margin of the Tarim Block, NW China: age, geochemistry, petrogenesis and tectonic implications. J Asian Earth Sci 35(2):167–179

Zhang CL, Zou HB (2013) Comparison between the Permian mafic dykes in Tarim and the western part of central Asian Orogenic Belt (CAOB), NW China: implications for two mantle domains of the Permian Tarim Large igneous province. Lithos 174:15–27

Zhang J, Zhang CL, Li HK et al (2014) Revisit to time and tectonic environment of the Aksu blueschist terrane in northern Tarim, NW China: new evidence from zircon U–Pb age Hf isotope. Acta Petrol Sinica 30(11):3357–3365

Zhang J, Cheng SH (2017) Are those mafic dikes in Hengshan-Xuanhua-Xinghe area of North China sanukitoids? J Earth Sci Environ 39(1):65–82

Zhang LF, Ai YL, Li Q et al (2005) The formation and tectonic evolution of UHP metamorphic belt in southwestern Tianshan, Xinjiang. Acta Petrol Sinica 21(4):1029–1038

Zhang LF, Ai YL, Li XP et al (2007a) Triassic collision of western Tianshan orogenic belt, China: evidence from SHRIMP U–Pb dating of zircon from HP/UHP eclogitic rocks. Lithos 96:266–280

Zhang XR, Zhao GC, Eizenhöfer PR et al (2016) Tectonic transition from Late Carboniferous subduction to early Permian post-collisional extension in the Eastern Tianshan, NW China: insights from geochronology and geochemistry of mafic–intermediate intrusions. Lithos 256–257:269–281

Zhang Z, Zhu W, Shu L et al (2009b) Neoproterozoic ages of the Kuluketage diabase dyke swarm in Tarim, NW China, and its relationship to the breakup of rodinia. Geol Megazine 146(1):150–154

Zhang ZL, Qin QM, Cao B et al (2007b) Application of high-resolution satellite images to extraction of dyke attributes. Geogr Geo-Inf Sci 23(3):15–18

Zhang ZL, Qin QM, Tian W et al (2008a) Emplacement characteristics and spatial distribution of Permian Mazhartager basic dike swarms in Bachu area, Tarim basin. Acta Petrol Sinica 24(10):2273–2280

Zhang ZY, Zhu WB, Shu LS et al (2008b) Thermo-tectonic evolution of Precambrian blueschists in Aksu, Northwest Xinjiang, China. Acta Petrol Sinica 24(12):2849–2856

Zhang ZZ, Gu LX, Wu CZ et al (2006) Weiya quartz syenite in early Indosinina from eastern Tianshan mountains: Petrogenesis and tectonic implication. Acta Petrol Sinica 22(5):1135–1149

Zhao JX, McCulloch MT (1993) Sm-Nd mineral isochron ages of Late Proterozoic Dyke swarms in Australia: evidence for two distintive events of mafic magmatism and crustal extension. Chem Geol 109:341–354

Zhao JX, McCulloch MT (1994) Characterization of a plume-related ~800 Ma magmatic event and its implication for basin information in central-southern Australia. Earth Planet Sci Lett 121:349–367

Zheng BH, Zhu WB, Shu LS et al (2008) The protolith of the Aksu Precambrian blueschist and its tectonic setting. Acta Petrol Sinica 24(12):2839–2848

Zhou J, Ji JQ, Han BF et al (2008) $^{40}Ar/^{39}Ar$ Geochronology of mafic dykes in North Xinjiang. Acta Petrol Sinica 24(5):997–1010

Neoarchean-Mesoproterozoic Mafic Dyke Swarms of the Indian Shield Mapped Using Google Earth™ Images and ArcGIS™, and Links with Large Igneous Provinces

Amiya K. Samal, Rajesh K. Srivastava, Richard E. Ernst and Ulf Söderlund

Abstract We present dyke swarm maps generated using Google Earth™ images, ArcGIS™, field data, and available geochronological ages of Neoarchean-Mesoproterozoic (ranging in age from ~2.80 to ~1.10 Ga) mafic dyke swarms and associated magmatic units of the different Archean cratons of the Indian shield which represent the plumbing system of Large Igneous Provinces (LIPs). The spatial and temporal distributions together with the trends of the dyke swarms provide important informations about geodynamics. Twenty four dyke swarms (17 have been precisely dated), mostly mafic in nature, have been mapped from the different cratons and named/re-named to best reflect their location, trend, distribution and distinction from other swarms. We have identified 14 distinct magmatic events during the Neoarchean-Mesoproterozoic in the Indian shield. These intraplate magmatic events (many of LIP scale) of the Indian shield and their matches with coeval LIPs on other crustal blocks suggest connections of the Indian shield within known supercontinents, such as Kenorland/Superia (~2.75–2.07 Ga), Columbia/Nuna (1.90–1.38 Ga), and Rodinia (1.20–0.72 Ga). However, further detailed U–Pb geochronology and associated paleomagnetism are required to come to any definite constraints on the position of the Indian cratons within these supercontinents.

Keywords Dyke swarm · Neoarchean-Mesoproterozoic · Google earth™ image ArcGIS™ · LIP · Indian shield

---

A. K. Samal · R. K. Srivastava (✉)
Centre of Advanced Study in Geology, Institute of Science,
Banaras Hindu University, Varanasi 221005, India
e-mail: rajeshgeolbhu@gmail.com

R. E. Ernst
Department of Earth Sciences, Carleton University, Ottawa ON, Canada

R. E. Ernst
Department of Geology and Geography, Tomsk State University,
36 Lenin Ave, Tomsk 634050, Russia

U. Söderlund
Department of Geology, Lund University, Sölvegatan 12, 223 62 Lund, Sweden

R. K. Srivastava et al. (eds.), *Dyke Swarms of the World: A Modern Perspective*,
Springer Geology, https://doi.org/10.1007/978-981-13-1666-1_9

1 Introduction

Identification of Large Igneous Provinces (LIPs) has significance to many aspects of geology including constraining continental reconstructions, providing exploration targeting strategies for a range of ore deposit types, and having implications for climate change including mass extinction events through Earth history (Bleeker and Ernst 2006; Ernst and Bleeker 2010; Ernst et al. 2010, 2013; Ernst 2014; Ernst and Jowitt 2013, 2017; Ernst and Youbi 2017; and references therein). In general, a LIP may comprises flood basalts, or their erosional/deformational remnants and a plumbing system of dykes, sills, and layered intrusions (Ernst and Buchan 2001a, b; Ernst and Srivastava 2008; Ernst 2014). LIPs may also be associated with felsic magmatism, carbonatites and in some cases lamprophyres, lamproites and kimberlites (Bryan and Ernst 2008; Ernst and Bell 2010). Recognition of a LIP emplaced during the Mesozoic-Cenozoic is easier than a LIP preserved during the Paleozoic-Precambrian. This is because flood basalts are well preserved in the Mesozoic-Cenozoic, whereas in Paleozoic-Precambrian time, flood basalts are typically more deeply eroded thus exposing the plumbing system, in the form of giant dyke swarms, sill provinces, and layered intrusions. Therefore, in an ancient shield terrain, mafic dykes/dyke swarms are a reliable indicator of ancient LIPs. Specifically, mafic dyke swarms with an average dyke width of $\geq$10 m are recognized as a proxy criteria for LIPs and associated rifting (Ernst et al. 1995; Ernst 2014). Such widths are typical of swarms in shield areas including those in India (Murthy 1987; Ramachandra et al. 1995; Samal et al. 2015). In contrast, mafic dyke swarms with a smaller average widths of only a few metres belong to other settings (collisional zones, arcs, small intrusions, spreading ridges and ophiolites). Once a Precambrian LIP has been identified on the basis of a dyke swarm, further research typically identifies additional magmatic units which can be linked, such as remnant volcanics and sills and layered intrusions, and also silicic magmatism, carbonatites and kimberlites (Ernst and Buchan 2001a, b; Ernst 2014).

Dykes/dyke swarms (linked to LIPs, as per the rationale above) have been mapped from cratonic blocks around the world. These include Australia (Pirajno and Hoatson 2012; Thorne et al. 2014; Wingate and Martin 2016), North China craton (Peng 2015, 2016), Canada and adjacent regions of the United States and Greenland (Buchan and Ernst 2004, 2013), West Africa (Jessell et al. 2015), Russia (Ernst et al. 2016), Egypt (Hamimi et al. 2016), North Africa (Youbi et al. 2016), and Southwestern Algeria (Amara et al. 2016). Murthy (1995) prepared a mafic dyke map of India at 1:2,000,000 scale, but given many new U–Pb ages (Halls et al. 2007; French et al. 2008; French and Heaman 2010; Ratre et al. 2010; Srivastava et al. 2011a, 2018; Kumar et al. 2012b, 2014, 2015, 2017; Pradhan et al. 2012; Pisarevsky et al. 2013; Shankar et al. 2014; Nagaraju et al. 2018; Söderlund et al. 2018; Pivarunas et al. 2018), and the opportunity for remapping of swarms using Google Earth™ images, the record of dyke swarms in India needs to be updated.

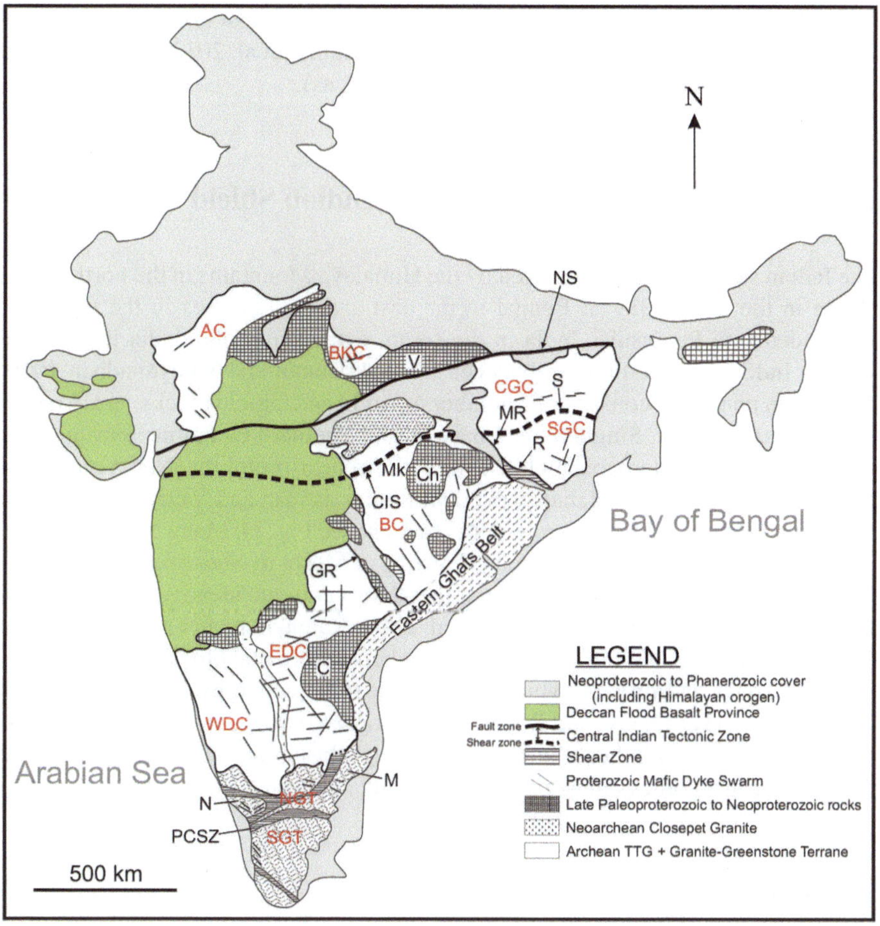

Fig. 1 Generalized geological and tectonic map of the Indian shield (modified after French et al. 2008). Archean cratons: Southern Granulite Terrain (SGT); Dharwar craton includes Eastern Dharwar craton (EDC), Western Dharwar craton (WDC) and Northern Granulite Terrain (NGT); Bastar craton (BC); Singhbhum craton includes Singhbhum Granite Complex (SGC) and Chhotanagpur Gneissic Complex (CGC); Bundelkhand craton (BKC); Aravalli craton (AC). Other major geological features: Ch, Chattisgarh Basin; CIS, Central Indian Shear Zone; GR, Godavari Rift; M, Madras Block; Mk, Malanjkhand; MR, Mahanadi Rift; N, Nilgiri Block; NS, Narmada-Son Fault Zone; PCSZ, Palghat-Cauvery Shear Zone; R, Rengali Province and Kerajang Shear Zone; S, Singhbhum Shear Zone; V, Vindhyan Basin

The aim of this communication is to present series of updated maps of Neoarchean-Mesoproterozoic dyke swarms and their related magmatic units (sills, volcanic rocks, etc.), emplaced within different Archean cratons of the Indian shield (Fig. 1). These maps are prepared using Google Earth™ images, field surveys, and previously published geological maps (Halls 1982; Roy et al. 1995; Ramachandra et al. 1995;

Mandal et al. 2006; French et al. 2008; French and Heaman 2010; Pradhan et al. 2012; Samal et al. 2015; Srivastava et al. 2016a; Kumar et al. 2017) (see Table 1 for complete list of identified swarms and key references).

2 A Brief Outline of Geology of the Indian Shield

The Indian subcontinent is bordered by the Himalaya Mountains in the north, Indian Ocean in the south, Bay of Bengal in the east and Arabian Sea in the west. It is subdivided into Peninsular India in the south, extra-Peninsular India in the north and the Indo-Gangetic plain in between (Ramakrishnan and Vaidyanadhan 2010). The Indian peninsula consists of a number of Archean cratonic blocks, viz. Aravalli, Bundelkhand, Bastar, Singhbhum, Dharwar, and Southern Granulite Terrain (SGT) and Meso- to Neoproterozoic mobile belts with Archean protoliths (Naqvi et al. 1974; Naqvi and Rogers 1987; Mahadevan 2008; Ramakrishnan and Vaidyanadhan 2010; French and Heaman 2010 and references therein; see Fig. 1). Many have considered the SGT as a Proterozoic mobile belt consisting of Meso- to Neo-Archean protoliths, identified as the Pandyan mobile belt, rather than a cratonic identity (Ramakrishnan 1988, 1993; Ghosh et al. 2004; Clark et al. 2009; Ramakrishnan and Vaidyanadhan 2010 and references therein). The boundaries between crustal blocks are represented by thrust zones and/or rifts, which are thought to be active by at least ~1500 Ma and perhaps earlier (Naqvi and Rogers 1987; Rogers 1996). Evidence from the dyke swarms (Sect. 5.1.8) indicate in some cases (e.g. Bastar and Dharwar cratons) that the Indian crustal blocks must have been already connected by 1.9 Ga.

3 Methodology

Indirect methods of geological mapping using remote sensing have been applied to extract the general geometry and distribution of dyke swarms. These methods include both aerial photography and satellite-based passive (e.g. Google Earth™, Landsat, SPOT, WorldView) and active (e.g. Synthetic Aperture Radar—SAR) imaging systems. In addition, in recent years, airborne and terrestrial light detection and ranging (LiDAR), unmanned aerial vehicle (UAV) photogrammetry have also become efficient ways to extract the detailed geometry of dyke swarms and their cross-cutting relationships (Ni et al. 2018; Cruden et al. 2016; Dering et al. 2016). During the present work, Google Earth™ images have been used to trace mafic dykes. Published geological maps have also been taken into consideration in both mapping dykes and grouping them into swarms. The line work was imported into ArcGIS™ advanced version 10.0 (ArcInfo) in order to prepare the final dyke maps for the different Indian cratons.

Table 1a Precisely dated Precambrian mafic dykes and other associated magmatic units of the Indian shield

S. No.[a]	Name (location)	Dyke trend	Age and method	Reference(s)
Dharwar Craton				
A. NE–SW to ESE–WNW trending ~2.37 Ga Bangalore-Karimnagar swarm				
1.	Yeragumballi (near Henur)	WNW–ESE	2366.7 ± 1 Ma; U–Pb baddeleyite	Halls et al. (2007)
2.	Harohalli	E–W	2365.4 ± 1 Ma; U–Pb baddeleyite	French and Heaman (2010)
3.	Penukonda	E–W	2365.9 ± 1.5 Ma; U–Pb baddeleyite	French and Heaman (2010)
4.	Chennekottapalle	E–W	2368.6 ± 1.3 Ma; U–Pb baddeleyite	French and Heaman (2010)
5.	Hyderabad	NE–SW	2367.1 ± 3.1 Ma; U–Pb baddeleyite	Kumar et al. (2012b)
6.	Karimnagar	NE–SW	2368.5 ± 2.6 Ma; U–Pb baddeleyite	Kumar et al. (2012b)
7.	Horsley Hill	NE–SW	2368 ± 2 Ma; U–Pb baddeleyite	Söderlund et al. (2018)
8.	Madenapalle	NE–SW	2368 ± 2 Ma; U–Pb baddeleyite	Söderlund et al. (2018)
9.	Bhongir	NE–SW	2368 ± 2 Ma; U–Pb baddeleyite	Söderlund et al. (2018)
9A.	Krishnagiri	WNW–ESE	2363 ± 6.6 Ma; U–Pb zircon	Pivarunas et al. (2018)
B. N–S to NNE–SSW trending ~2.25–2.26 Ga Ippaguda-Dhiburahalli swarm				
10.	Chennur	N–S	2257 ± 4 Ma; U–Pb baddeleyite	Söderlund et al. (2018)
11.	Yenugonda	N–S	2256 ± 4 Ma; U–Pb baddeleyite	Söderlund et al. (2018)
12.	Dhiburahalli	NNE–SSW	2257 ± 2 Ma; U–Pb baddeleyite	Söderlund et al. (2018)
13.	Ippaguda	N–S	2251 ± 3 Ma; U–Pb baddeleyite	Söderlund et al. (2018)
C. N–S to NNW–SSE trending ~2.22 Ga Kandlamadugu swarm				
14.	Kandlamadugu	N–S	2220.5 ± 4.9 Ma; U–Pb baddeleyite	French and Heaman (2010)
15.	Nelahalu	N–S	2215.2 ± 2 Ma; U–Pb baddeleyite	Srivastava et al. (2011a, 2014b)
16.	Haniyur	N–S	2211.7 ± 0.9 Ma; U–Pb baddeleyite	Srivastava et al. (2011a, 2014b)
17.	Thammadihalli	NNW–SSE	2215.9 ± 0.3 Ma; U–Pb baddeleyite	Kumar et al. (2014)

(continued)

Table 1a (continued)

S. No.[a]	Name (location)	Dyke trend	Age and method	Reference(s)
18.	Nelahalu	N–S	2220 ± 2 Ma; U–Pb baddeleyite	Söderlund et al. (2018)
19.	Kushtagi	NNW–SSE	2216.6 ± 0.7 Ma; U–Pb baddeleyite	Nagaraju et al. (2018)
20.	Tippanapalle	NW–SE	≥2210 ± 2 Ma; U–Pb baddeleyite	Söderlund et al. (2018)
D. NW–SE to WNW–ESE trending ~2.21 Ga Anantapur-Kunigal swarm				
21.	Somala	NW–SE	2209.3 ± 2.8 Ma; U–Pb baddeleyite	French and Heaman (2010)
22.	Kandukur	WNW–ESE	2208 ± 15 Ma; U–Pb baddeleyite	Söderlund et al. (2018)
23.	Kunigal	NW–SE	2206 ± 4 Ma; U–Pb baddeleyite	Söderlund et al. (2018)
24.	Konapuram	NW–SE	2209 ± 3 Ma; U–Pb baddeleyite	Söderlund et al. (2018)
25.	Narapally	WNW–ESE	2208 ± 5 Ma; U–Pb baddeleyite	Söderlund et al. (2018)
E. NW–SE to WNW–ESE trending ~2.18 Ga Mahbubnagar-Dandeli swarm				
26.	Bandepalem	WNW–ESE	2176.5 ± 3.7 Ma; U–Pb baddeleyite & Zircon	French and Heaman (2010)
27.	Dandeli	NW–SE	2180.8 ± 0.9 Ma; U–Pb baddeleyite	French and Heaman (2010)
F. NW–SE to NE–SW trending ~2.08 Ga Devarabanda swarm				
28.	Neredugommu	N–S	2081.8 ± 0.7 Ma; U–Pb baddeleyite	Kumar et al. (2015)
29.	Puttamgandi	N–S	2081.1 ± 0.7 Ma; U–Pb baddeleyite	Kumar et al. (2015)
30.	Malyala	NW–SE	2081.8 ± 1.1 Ma; U–Pb baddeleyite	Kumar et al. (2015)
31.	Mukundapuram	NNE–SSW	2082.8 ± 0.9 Ma; U–Pb baddeleyite	Kumar et al. (2015)
32.	Ramannapeta	NNE–SSW	2081 ± 3 Ma; U–Pb baddeleyite	Söderlund et al. (2018)
33.	Marakuntapalle	NE–SW	2087 ± 5 Ma; U–Pb baddeleyite	Söderlund et al. (2018)
34.	Devarabanda	NW–SE	2080 ± 6 Ma; U–Pb baddeleyite	Söderlund et al. (2018)
35.	Pyapili	NW–SE	2083 ± 2 Ma; U–Pb baddeleyite	Söderlund et al. (2018)

(continued)

Table 1a (continued)

S. No.[a]	Name (location)	Dyke trend	Age and method	Reference(s)
G. E–W to NW–SE trending ~1.88–1.89 Ga Hampi swarm				
36.	Hampi	E–W	1894 Ma; U–Pb baddeleyite	Halls et al. (2007)
37.	Near Anantpur	E–W	1879 ± 5 Ma; $^{40}Ar/^{39}Ar$ whole-rock	Chatterjee and Bhattacharji (2001)
38.	Pulivendla (Cuddapah basin)	Mafic sill	1885.4 ± 3.1 Ma; U–Pb baddeleyite	French et al. (2008)
39.	Pulivendla (Cuddapah basin)	Mafic sill	1899 ± 20 Ma; $^{40}Ar/^{39}Ar$ phlogopite	Anand et al. (2003)
H. NW–SE trending ~1.79 Ga Pebbair Swarm				
40.	Pebbair	NW–SE	1794 ± 7 Ma; U–Pb baddeleyite	Söderlund et al. (2018)
I. ~1.42 Ga lamproite intrusion				
41.	Chelima (Cuddapah basin)	Lamproite intrusion	1417.8 ± 8.2 Ma; $^{40}Ar/^{39}Ar$ phlogopite	Chalapathi Rao et al. (1999)
J. ~1.10–1.12 Ga Kimberlite intrusions				
42.	Narayanpet	Kimberlite intrusion	1123 ± 17 Ma to 1129 ± 12 Ma; U–Pb perovskite	Chalapathi Rao et al. (2013)
43.	Raichur	Kimberlite intrusion	1093 ± 18 Ma; U–Pb perovskite	Chalapathi Rao et al. (2013)
44.	Wajrakarur	Kimberlite intrusion	1099 ± 12 Ma to 1129 ± 12 Ma; U–Pb perovskite	Chalapathi Rao et al. (2013)
Bastar Craton				
A. ~2.25 Ga Syenite intrusion				
45A.	Nuapada	Syenite intrusion	2473 ± 8 Ma; U–Pb zircon	Santosh et al. (2018)
A. NW–SE trending ~1.88–1.89 Ga Bastanar swarm				
45.	Dhurli (near Bacheli)	NW–SE	1891.1 ± 0.9 Ma; U–Pb baddeleyite	French et al. (2008)
46.	Bastanar	NW–SE	1883 ± 1.4 Ma; U–Pb baddeleyite & Zircon	French et al. (2008)
47.	Bhanupratappur	NW–SE	1882.4 ± 1.5 Ma; Pb–Pb baddeleyite	Shellnutt et al. (2018)
B. N–S to NNE–SSW trending ~1.44–1.46 Ga Lakhna swarm				
48.	Lakhna	N–S; Rhyolitie	1466 ± 3 Ma; U–Pb Zircon	Pisarevsky et al. (2013)

(continued)

Table 1a (continued)

S. No.[a]	Name (location)	Dyke trend	Age and method	Reference(s)
49.	West of Chhindekelai	N–S; Rhyolite	1450 ± 22 Ma; U–Pb Zircon	Ratre et al. (2010)
50.	Lakhna	N–S; Trachyte	1453 ± 19 Ma; U–Pb Zircon	Ratre et al. (2010)
51.	South of Chhindekelai	N–S; Alkali Gabbro	1442 ± 30 Ma; U–Pb Zircon	Ratre et al. (2010)
C. ENE–WSW trending ~1.42 Ga Bandalimal swarm				
52.	Bandalimal	ENE–WSW	1421 ± 23 Ma; Sm–Nd mineral and whole-rock	Das et al. (2011)
D. ~1.05 Ga Lamproite intrusions				
53.	Nuapada	Lamproite intrusion	1055 ± 10 Ma; $^{40}Ar/^{39}Ar$ whole-rock	Sahu et al. (2013)
54.	Nuapada	Lamproite intrusion	1045 ± 9 Ma; $^{40}Ar/^{39}Ar$ whole rock	Chalapathi Rao et al. (2016)
Singhbhum Craton				
A. NE–SW trending ~2.80 Ga Keshargaria swarm				
55.	Keshargaria	NE–SW	2800.2 ± 0.7 Ma; Pb–Pb baddeleyite	Kumar et al. (2017)
B. NNE–SSW to NE–SW trending ~2.75–2.76 Ga Ghatgaon swarm				
56.	Ghatgaon	NNE–SSW	2763.7 ± 0.8 Ma; Pb–Pb baddeleyite	Kumar et al. (2017)
57.	Dumuria	NNE–SSW	2763.5 ± 0.8 Ma; Pb–Pb baddeleyite	Kumar et al. (2017)
58.	Ghatgaon	NNE–SSW	2764.4 ± 0.8 Ma; Pb–Pb baddeleyite	Kumar et al. (2017)
59.	Nuapada	NNE–SSW	2763.2 ± 0.9 Ma; Pb–Pb baddeleyite	Kumar et al. (2017)
60.	Jhumpura	NNE–SSW	2760 ± 0.6 Ma; Pb–Pb baddeleyite	Kumar et al. (2017)
61.	Khairpal	NNE–SSW	2761 ± 1 Ma; Pb–Pb baddeleyite	Kumar et al. (2017)
62.	Hakai	NE–SW	2752 ± 0.9 Ma; Pb–Pb baddeleyite	Kumar et al. (2017)
C. NE–SW to ENE–WSW trending ~2.26 Ma Kaptipada swarm				
63.	Kaptipada	NE–SW	2256 ± 6 Ma; U–Pb baddeleyite	Srivastava et al. (2018)

(continued)

Table 1a (continued)

S. No.[a]	Name (location)	Dyke trend	Age and method	Reference(s)
D. WNW–ESE trending ~1.77 Ga Pipilia swarm				
64.	Pipilia	WNW–ESE	1766.2 ± 1.1 Ma; Pb–Pb baddeleyite	Shankar et al. (2014)
65.	Pipilia	WNW–ESE	1764.5 ± 0.9 Ma; Pb–Pb baddeleyite	Shankar et al. (2014)
Bundelkhand Craton				
A. ~2.10 Ga Gabbro sill				
66.	Gwalior (Gwalior basin)	Gabbro sill	2104 ± 23 Ma; Sm–Nd mineral-whole rock	Samom et al. (2017)
B. NW–SE trending ~1.98 Ga Jhansi Swarm				
67.	Chhatarpur	NW–SE	1979 ± 8 Ma, U–Pb Zircon	Pradhan et al. (2012)
C. NE–SW to E–W trending ~1.11 Ga Mahoba swarm				
68.	Mahoba	ENE–WSW	1113 ± 7 Ma, U–Pb Zircon	Pradhan et al. (2012)
Vindhyan Basin				
A. ~1.10 Ga kimberlite intrusions				
69.	Majhgaon	Kimberlite intrusion	1073.5 ± 13.7 Ma; $^{40}Ar/^{39}Ar$ phlogopite	Gregory et al. (2006)

Ernst and Buchan (2016) have provided comprehensive guidelines for preparing a regional mafic dyke swarm map for any area. The Indian mafic dykes can easily be distinguished from surrounding country rocks comprised of Archaean/Proterozoic granitoids by tonal differences (mafic dykes are darker than their host granitoids). In general, dykes can appear as long serrated ridges, such as linearly arranged hills and as stony blocks, which can easily be distinguished in Google Earth™ Images. Due to limited outcropping, most geological maps underestimate the continuity of dykes.

The grouping of dykes into a particular swarm is prerequisite for establishing the swarm geometry (such as linear, radiating, arcuate or circumferential) (Ernst and Buchan 1997; Buchan and Ernst 2018a, b). Generally, dykes having similar trends are thought to be emplaced at the same time and would therefore represent a single swarm. However, there are examples from India (and elsewhere) that suggest swarms of different ages can have similar trends (e.g. Srivastava 2006a, b; Srivastava and Gautam 2008, 2009; Srivastava et al. 2014a; Samal et al. 2015; Samal and Srivastava 2016). Therefore, during the present study, we used multiple criteria (such as geochemistry, geochronology, and dyke thickness and spacing) in addition to trend, for grouping of dykes into different swarms.

Table 1b Other available radiometric and estimated ages (based on cross-cutting relationships with precisely dated mafic dykes and country rocks) of Precambrian mafic dykes and other associated magmatic units of the Indian shield

S. No.	Name (location)	Dyke trend	Age and method	Reference(s)
Bastar Craton				
A. WNW–ESE to NW–SE trending ~2.7 Ga Keshkal swarm				
i.	Southern Bastar	WNW to NW	Overlapping field relationships and Sm–Nd isochron estimate	Srivastava et al. (2009a)
B. NW–SE to NE–SW trending ~2.4–2.5 Ga Dantewara swarm				
ii.	Gatam-Kateklyan	NW–SE	Overlapping field relationships and U–Pb 2118 ± 2 Ma metamorphic rutile age of a sample from the BN swarm	Srivastava et al. (2011b)
C. ~2.18 Ga silicic magmatic rock				
iii.	Bijli	Rhyolite intrusion	2180 ± 25 Ma; Rb–Sr whole rock	Divakara Rao et al. (2000)
D. ~1.10 Ga Kimberlite intrusions				
iv.	Tokpal	Kimberlite intrusion	~1100 Ma	Chalapathi Rao et al. (2014)
Singhbhum Craton				
A. Early-Paleoproterozoic Keonjhar swarm				
v.		E–W to ENE–WSW	Based on cross cutting field relationship	Srivastava et al. (2018)
B. Middle-Paleoproterozoic Bhagamunda swarm				
vi.		NW–SE to NNW–SSE	Based on cross cutting field relationship	Srivastava et al. (2018)
C. Late-Paleoproterzoic Barigaon swarm				
vii.		N–S to NNE–SSW	Based on cross cutting field relationship	Srivastava et al. (2018)
Bundelkhand Craton				
A. ~1.98 Ga Gabbro sill				
viii.	Dargawan (Bijawar basin)	Gabbro sill	1967 ± 140 Ma; Rb–Sr whole rock	Pandey et al. (2012)
Aravalli Craton				
A. 2.8 Ga Jagat Swarm				
ix.	Jagat	NW–SE	2828 ± 46 Ma, Sm–Nd whole rock	Gopalan et al. (1990)

(continued)

Table 1b (continued)

S. No.	Name (location)	Dyke trend	Age and method	Reference(s)
B. 2.2 Ga (???) Amet swarm				
x.	Amet	NE–SW	~2.2 Ga; based on overlapping field relationship	Mondal et al. (2008b)
Himalayan Mountain Range				
A. ~1.80–1.90 Ga mafic magmatic intrusions in NW Himalaya				
xi.	South of Budhakedar	Gabbro intrusive	1907 ± 91 Ma; Rb–Sr whole rock	Ahmad et al. (1999)
xii.	Rampur	Metabasalts	1800 ± 13 Ma; $^{207}Pb/^{206}Pb$ single zircon	Miller et al. (2000)

[a]See respective geological maps (Figs. 2, 3, 4, 5 and 6) for location of dated sample

4 Distribution of Mafic Dykes in the Indian Shield

The Indian shield comprises numerous crosscutting and multiple generations of Precambrian mafic dyke swarms and associated sills, mafic-ultramafic intrusions, and volcanic supracrustal sequences (Murthy 1987, 1995; Devaraju 1995; Srivastava et al. 2008; Srivastava and Ahmad 2008, 2009; Srivastava 2011). The following sections present dyke maps and descriptions of all the currently identified mafic dyke swarms of each of the Indian cratons. Where known, we also identify additional units related to the swarm (in a LIP context). Most of the available radiometric ages on the Precambrian dykes of the Indian shield are tabulated in Table 1.

4.1 Dharwar Craton (Both Eastern and Western)

The Dharwar craton consists of (i) eastern Dharwar craton, (ii) western Dharwar craton, and (iii) Northern Granulite Terrain (high-grade crustal block) in the southern part (Drury et al. 1984; Naqvi and Rogers 1987; Ramakrishnan 1988, 1993; Chetty and Bhaskar Rao 2006; Santosh et al. 2009; Clark et al. 2009; Sharma 2009; Ramakrishnan and Vaidyanadhan 2010; see Fig. 2). The major E–W trending Palghat-Cauvery shear zone (PCSZ) separates the Southern Granulite Terrain (SGT; also known as the Pandyan Mobile Belt) from the Northern Granulite Terrain (NGT) (see Fig. 2; Drury et al. 1984; Ramakrishnan 1988, 1993; Ramakrishnan and Vaidyanadhan 2010). The NGT of the Dharwar craton, the region between the Fermor Line (Fermor 1936) and PCSZ, comprises Archean and Neoproterozoic lithologies such as orthogneiss, charnockite, mafic granulite and ultramafic intrusive in association with metasedimentary units (Chetty and Bhaskar Rao 2006; Santosh et al. 2009; Ramakrishnan and Vaidyanadhan 2010 and references therein). The Fermor Line denotes the orthopyroxene isograd and separates charnockitic (granulitic) and non-charnockitic (gneiss-schist) terrains (Pichamuthu 1965; Naqvi and Rogers 1987). In other words, the Fermor line locates a regional amphibolite–granulite facies transition zone in the southern Indian Precambrian shield. The SGT is devoid of any Paleoproterozoic mafic dykes; however, a number of Cretaceous mafic dykes are encountered from this region (Radhakrishna et al. 1999; Radhakrishna and Joseph 2012; Buchan and Ernst 2018a).

The Chitradurga fault, exists along the eastern boundary of the Chitradurga greenstone belt, separates the EDC and the WDC; both have distinct geological features (see Fig. 2; Swami Nath and Ramakrishnan 1981; Rogers 1986; Naqvi and Rogers 1987; Jayananda et al. 2006, 2013a, b; Mahadevan 2008; Chardon et al. 2008 and references therein). It is considered that the EDC, the WDC and the high-grade crustal block in the southern part share common magmatic activity during the Proterozoic, particularly Paleoproterozoic mafic dykes (Halls 1982; Drury 1984; Murthy 1987; Devaraju 1995; Murthy 1995; Ernst and Srivastava 2008; Srivastava et al. 2008; Srivastava 2011; Dash et al. 2013); therefore, the mafic dyke swarms emplaced in these three regions are discussed together.

Although mafic dykes are present in all three portions of the southern Indian shield, they are more conspicuous in the eastern Dharwar craton (Chatterjee and Bhattacharji 2001; Halls et al. 2007; Radhakrishana et al. 2007; Jayananda et al. 2008; French and Heaman 2010; Piispa et al. 2011; Kumar et al. 2012a, b; Srivastava et al. 2014a, b, 2015; Söderlund et al. 2018 and references therein) than in the western Dharwar craton (Naqvi et al. 1972; Devaraju et al. 2008; Radhakrishna 2009; Radhakrishna et al. 2004; Halls et al. 2007; Ramakrishan 2009; French and Heaman 2010), or in the NGT in the southern part of the Dharwar craton (Radhakrishna 2009; Dash et al. 2013) (see Fig. 2). On the basis of available radiometric ages, geochemistry, and dyke distribution and trend (Halls et al. 2007; French and Heaman 2010; Piispa et al. 2011; Kumar et al. 2012a, b; Srivastava et al. 2014a, b, 2015; Söderlund et al. 2018 and references therein), the Paleoproterozoic mafic dykes of the southern Indian shield are classified into eight different dyke swarms, viz. (i) NE–SW to ESE–WNW trending ~2.37 Ga Bangalore-Karimnagar swarm, (ii) N–S to NNE–SSW trending ~2.25–2.26 Ga Ippaguda-Dhiburahalli swarm (iii) N–S to NNW–SSE trending ~2.22 Ga Kandlamadugu swarm, (iv) NW–SE to WNW–ESE trending ~2.21 Ga Anantapur-Kunigal swarm (v) NW–SE to WNW–ESE trending ~2.18 Ga Mahbubnagar-Dandeli swarm, (vi) NW–SE to NE–SW trending ~2.08 Ga Devarabanda swarm, (vii) E–W to NW–SE trending ~1.88–1.89 Ga Hampi swarm and (viii) NW–SE trending ~1.79 Ga Pebbair swarm.

4.1.1 NE–SW to ESE–WNW Trending ~2.37 Ga Bangalore-Karimnagar Swarm

This is a fanning dyke swarm with trends ranging from NE–SW to ESE–WNW. It is concentrated in the EDC. ENE–WSW trending dykes are exposed around southern part of the EDC and NE–SW trending dykes around Karimnagar and Hyderabad in the north. The ENE–WSW trending dykes are also present in the WDC, whereas

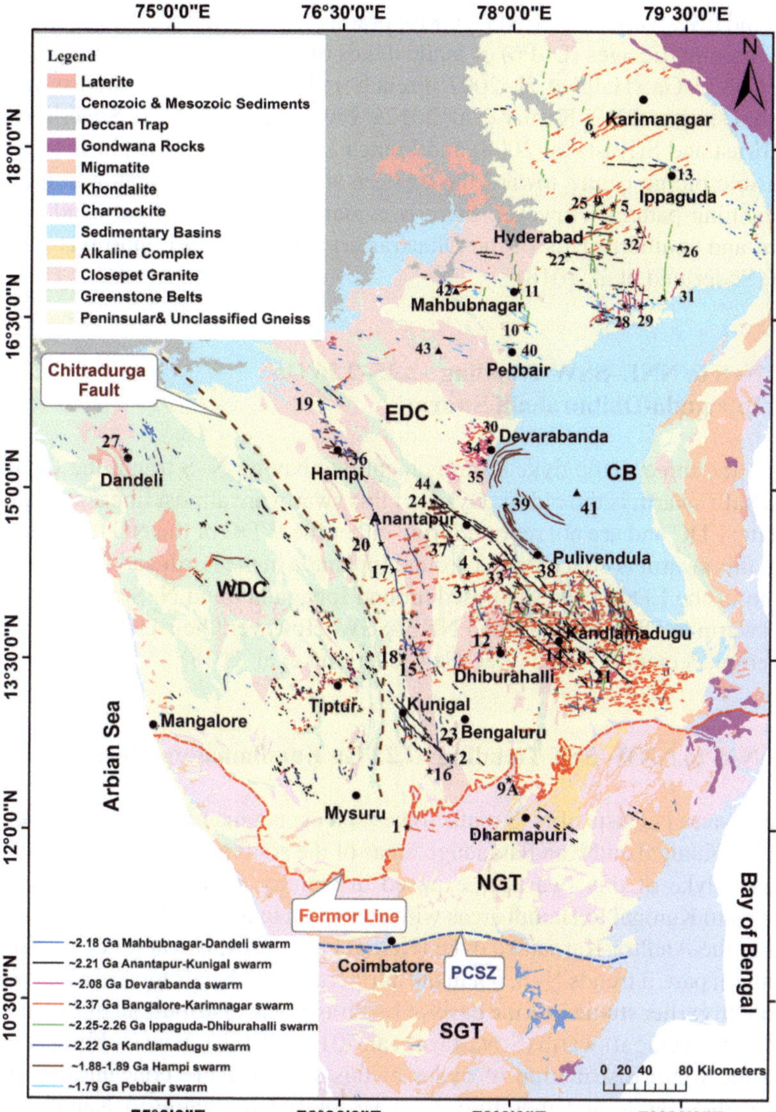

Fig. 2 Geological map showing distribution of distinct Paleoproterozoic mafic dyke swarms emplaced within the Dharwar craton (based on Drury et al. 1984; French et al. 2008; Mahadevan 2008; French and Heaman 2010; Ramakrishnan and Vaidyanadhan 2010; Söderlund et al. 2018 and references therein). EDC: Eastern Dharwar Craton; WDC: Western Dharwar Craton; NGT: North Granulite Terrain; SGT: South Granulite Terrain; PCSZ: Palghat-Cauvery Shear Zone; CB: Cuddapah Basin. Locations of dated dykes (*) and kimberlite/lamproite intrusions (▲) (see Table 1; S. Nos. 1–44)

dykes trend ESE–WNW in the NGT, corroborating their overall radiating pattern. The precise radiometric ages (U–Pb) of mafic dykes of this swarm date its emplacement as ~2.36–2.37 Ga (Halls et al. 2007; French and Heaman 2010; Srivastava et al. 2011a, b; Demirer 2012; Kumar et al. 2012a; Pivarunas et al. 2018; Söderlund et al. 2018). Ernst and Srivastava (2008) and French and Heaman (2010) have suggested the possible plume centre toward the western side of Dharwar craton at the focus of the fanning pattern. However, there is evidence for an oroclinal bend between northern and southern parts of the Dharwar craton that would undo this radiating pattern (Söderlund et al. 2018).

4.1.2 N–S to NNE–SSW Trending ~2.25–2.26 Ga Ippaguda-Dhiburahalli Swarm

Amongst the three mafic dyke swarms identified having N–S trend, the Ippaguda-Dhiburahalli swarm is the oldest. Dykes of this swarm are almost linear and exposed only in the EDC and are not reported from either the WDC or the NGT. As the name reflects, this swarm is concentrated around Ippaguda in the north and Dhiburahalli in the south of the EDC. Dykes exposed around Ippaguda trend N–S, whereas, dykes of this swarm at Dhiburahalli trend NNE–SSW. However, their U–Pb ages confirm coeval emplacement at ~2.25–2.26 Ga (Söderlund et al. 2018).

4.1.3 N–S to NNW–SSE Trending ~2.22 Ga Kandlamadugu Swarm

Another swarm consists of N–S mafic dykes and occurs mainly in EDC, particularly around the Kunigal and Kandlamadugu areas of the southern EDC. One particularly significant dyke of this swarm is exposed in vicinity of the Closepet granite but extends from Kunigal to Hampi areas with an overall length of 500 km; this has been termed as the Andhra-Karnataka dyke (Kumar et al. 2014; Nagaraju et al. 2018). In the southern part, it trends N–S, but in the northern part, it bends toward a NNW–SSE direction. In earlier studies, on the basis of U–Pb ages and distribution, this swarm was termed as the Kunigal swarm (Srivastava et al. 2011a, 2014b). New U–Pb ages clearly suggest the wider distribution of dykes of this swarm, more concentrated around Kandlamadugu area, and therefore we have revised the name to Kandlamadugu swarm (Söderlund et al. 2018). A few N–S dykes exposed in the WDC are also thought to be part of this swarm. No dyke of this swarm is known from the NGT. The change in trend from N–S, in the south, to NNW–SSE in the north is consistent with the oroclinal bend identified in Söderlund et al. (2018).

4.1.4 NW–SE to WNW–ESE Trending ~2.21 Ma Anantapur-Kunigal Swarm

NW–SE to WNW–ESE trending dykes are also significant throughout the EDC, the WDC and the NGT. French and Heaman (2010) was the first to date (using the U–Pb method) one of the NW–SE trending dykes from the Somala area located on the southern part of the EDC and reported an emplacement age of 2209 ± 3 Ma. Similar U–Pb ages of NW–SE trending mafic dykes from Kunigal and Anantapur areas and WNW–ESE trending mafic dykes from Hyderabad area (see Table 1) suggest the swarm is widespread in the Dharwar craton (Söderlund et al. 2018) and also present in the NGT. On the basis of its distribution and available ages, this swarm is distinguished as the Anantapur-Kunigal dyke swarm (Söderlund et al. 2018). The swarm has a radiating distribution with NW–SE trends in the southern part (EDC, WDC and NGT) and WNW–ESE dykes in the northern part of the EDC with an interpreted plume centre (at the focus of the radiating pattern) located in the NW of the Dharwar craton, beneath the Deccan province. However, the oroclinal bend in the Dharwar craton may undo this radiating pattern (Söderlund et al. 2018).

4.1.5 NW–SE to WNW–ESE Trending ~2.18 Ga Mahbubnagar-Dandeli Swarm

On the basis of available ca. 2.18 Ga ages (see Table 1) on mafic dykes (Sm–Nd and preliminary U–Pb; Pandey et al. 1997; French et al. 2004, respectively), Ernst and Srivastava (2008) identified a ~2.18 Ga WNW–ESE to NW–SE trending Mahbubnagar swarm. Later, French and Heaman (2010) provided two U–Pb precise ages, i.e. one from a WNW–ESE trending mafic dyke from the northern part of the EDC, near Bandepalem, and another from a NW–SE trending mafic dyke from the northern part of the WDC, near Dandeli, which yield almost identical ages (2176.5 ± 3.7 Ma and 2180.8 ± 0.9 Ma, respectively). This leads to the identification of a NW–SE to WNW–ESE trending dyke swarm across the Dharwar craton, identified as the Northern Dharwar dyke swarm (French and Heaman 2010). Considering its distribution, trends, and locations of dated samples, Söderlund et al. (2018) re-named it as Mahbubnagar-Dandeli dyke swarm. Mafic dykes of this group are mainly exposed in the entire EDC and northern parts of the WDC. No dyke of this swarm is identified from the NGT. The Mahbubnagar-Dandeli swarm has a radiating pattern which has been interpreted to mark a plume centre on the NW margin of the Dhawar craton (French and Heaman 2010; Ernst and Srivastava 2008), but this radiating pattern may also be undone by correcting for the oroclinal bend recently recognized in the Dharwar craton (Söderlund et al. 2018).

4.1.6 N–S to NE–SW Trending ~2.08 Ga Devarabanda Swarm

This is the third dyke swarm from the southern part of Indian shield that includes N–S trending mafic dykes. Dykes of this swarm trend N–S to NE–SW and are exposed all along the margin of the Cuddapah basin showing an overall radiating pattern that consists of N–S trending dykes in the northern parts of the EDC, NW–SE trending dykes near Devarabanda, and NE–SW trending dykes in the southern parts of the EDC that also extend up to the southern region of the WDC (see Fig. 2). The grouping of these dykes into a single swarm is based on new and existing U–Pb ages (Demirer 2012; Söderlund et al. 2018) and Pb–Pb (Kumar et al. 2015) ages.

4.1.7 E–W to NW–SE Trending ~1.88–1.89 Ga Hampi Swarm

This is another radiating swarm identified from the Dharwar craton, with expression in the Cuddapah basin and extension also into the Bastar craton (Sect. 4.2.3) (French et al. 2008; Ernst and Srivastava 2008). Dykes of this swarm trend E–W in the central region of the Dharwar craton (mostly in the EDC with couple of dykes in the WDC), and NW–SE between Hyderabad and Ippaguda in the northern EDC. A U–Pb baddeleyite age of 1894 Ma applies to the E–W trending dykes from Hampi area (Halls et al. 2007) and an Ar–Ar age of ~1879 Ma applies to ENE–WSW trending dykes from the Kalyandurg area (Chatterjee and Bhattacharji 2001). Geochemical and paleomagnetic studies also help to define the extent of this swarm (French and Heaman 2010; Piispa et al. 2011; Belica et al. 2014; Srivastava et al. 2014a). Besides mafic dykes, mafic sills of the same age (~1.88–1.89 Ga) are also emplaced within the Cuddapah basin (Anand et al. 2003; French et al. 2008). Similar age NW–SE trending dykes termed as the Bastanar swarm (formerly BD2) are also present in the Bastar craton (Sect. 4.2.3), and define a fanning pattern with the E–W to NW–SE trending Hampi dykes of the EDC (French et al. 2008; see also discussion in Ernst and Srivastava 2008).

4.1.8 NW–SE Trending ~1.79 Ga Pebbair Swarm

This youngest swarm has recently been identified in the EDC. A NW–SE trending dyke, exposed at Pebbair area, yielded a U–Pb age of 1794 ± 7 Ma (Söderlund et al. 2018). Dykes of this swarm are currently recognized mostly around the Pebbair area of the EDC.

4.2 Bastar Craton

The rectangular-shaped Bastar craton is bounded by major structural elements viz. Godavari rift in the SW, Mahanadi rift in the NE, Narmada-Son fault zone (Deccan Traps age rift system) in the NW, and Eastern Ghat Mobile Belt (EGMB) in the SE; and also Deccan LIP in the west (see Fig. 1; after Naqvi and Rogers 1987; French et al. 2008). The Bastar craton comprises granitoids of different generations, including 3.5–3.6 Ga TTG basement genisses and ~2.5 Ga un-deformed and un-metamorphosed granites, supracrustal rocks, and a number of intracratonic Proterozoic sedimentary basins (Crookshank 1963; Ramakrishnan 1990; Naqvi and Rogers 1987; Chaudhuri et al. 2002; French et al. 2008; Srivastava and Gautam 2009, 2015; Rajesh et al. 2009; Ramakrishnan and Vaidyanadhan 2010 and references therein).

Mafic dykes of different generations are well preserved in the Bastar craton. Most of them intrude into the Archean granitoids and supracrustals (Crookshank 1963; Ramachandra et al. 1995; Srivastava and Gautam 2009, 2015 and references therein). Crookshank (1963) presented a dyke and volcanics map of the southern Bastar craton, but Ramachandra et al. (1995) were first to publish a dyke map of the entire Bastar craton. However, subsequent U–Pb dating requires a new detailed dyke map with updated identification of the different generations of dyke swarms. With the help of field mapping and geochemistry (Ramakrishnan 1990; Srivastava et al. 1996, 2011b; Srivastava 2006a, b; Srivastava and Singh 2004; Srivastava and Gautam 2008, 2009, 2012) and U–Pb geochronology (French et al. 2008), three sets of mafic dyke, with similar trends (i.e. NW–SE to WNW–ESE) but different ages, were identified in the southern Bastar craton (Srivastava and Singh 2004; Srivastava and Gautam 2008, 2009, 2012; Srivastava et al. 2011b). These are the NW–SE to WNW–ESE trending ~2.7 Ga sub-alkaline metabasite (BD1), NW–SE trending ~2.4–2.5 Ga boninite-norite (BN) and NW–SE trending ~1.88–1.89 Ga sub-alkaline dolerite (BD2) swarms. In the northern Bastar craton, Srivastava and Gautam (2015) identified four sets, i.e. undated NW–SE trending North Bastar dykes (NBD), undated ENE–WSW trending Dongargarh-Chhura dykes (DCD), ~1.42 Ga ENE–WSW trending Bandalimal dykes (BDD) and ~1.44–1.46 Ga N–S trending Lakhna dykes (LKD). During the present study, a careful examination of available field (Crookshank 1963; Ramakrishnan 1990; Ramachandra et al. 1995; Srivastava et al. 1996), geochemical (Srivastava 2006a, b; Srivastava and Gautam 2008, 2009, 2012, 2015; Srivastava et al. 2011b) and geochronological data (French et al. 2008; Ratre et al. 2010; Das et al. 2011; Pisarevsky et al. 2013; Shellnutt et al. 2018), allowed five distinct mafic dyke swarms to be identified for the Bastar craton as a whole. In some cases the swarms were re-named (on the basis of their distribution and because previously used names such as BD1, BD2 and BN caused confusion with similar named swarms in other parts of the world). These five swarms are: (i) WNW–ESE to NW–SE trending ~2.7 Ga Keshkal swarm (formerly BD1) (ii) NW–SE to NE–SW trending ~2.4–2.5 Ga Dantewara swarm (formerly BN) (iii) NW–SE trending ~1.88–1.89 Ga Bastanar swarm (formerly BD2) (iv) N–S trending ~1.44–1.46 Ga Lakhna swarm, and (v) ENE–WSW trending ~1.42 Ga Bandalimal swarm. Based on this new classification, a dyke swarm map of the Bastar craton has been prepared (see Fig. 3) and is discussed in subsequent sections.

The Bastar craton also comprises intrusions of ~1.05 Ga lamproites (Sahu et al. 2013; Chalapathi Rao et al. 2016) and ~1.10 Ga kimberlites (Chalapathi Rao et al. 2014), which might belong to the widespread ~1.10–1.12 Ga magmatic event of

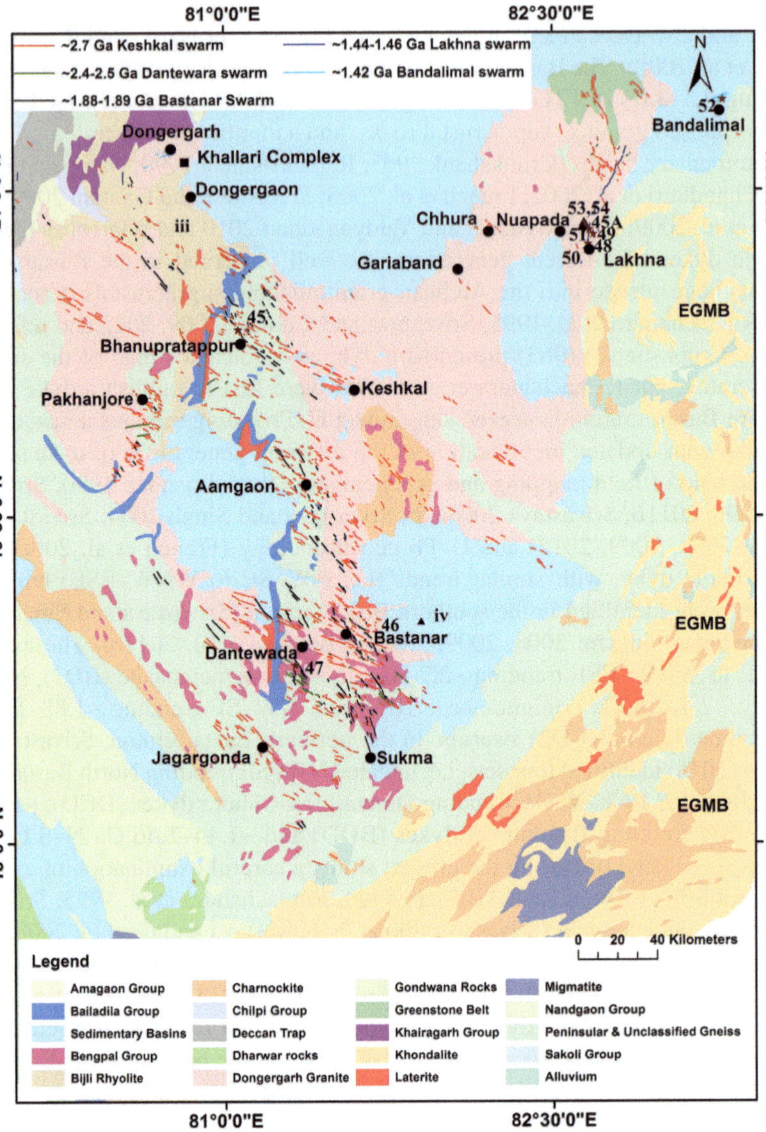

Fig. 3 Geological map showing distribution of distinct Neoarchean-Mesoproterozoic mafic dyke swarms emplaced within the Bastar craton (based on Crookshank 1963; Ramachandra et al. 1995; French et al. 2008; Srivastava and Gautam 2015; Srivastava et al. 2016a, b and references therein). Locations of dated dykes (*), kimberlite/lamproite intrusions (▲), syenite intrusion (◆), and Bijli rhyolite (see Table 1; S. Nos. 45A to 54 and iii, iv). EGMB—Eastern Ghat Mobile Belt

the Bundelkhand craton (Sect. 4.4.2). In addition, Bijli rhyolites, exposed in the northern part of the Bastar craton, have a Rb–Sr age of 2180 ± 25 Ma (Divakara Rao et al. 2000) and are possibly related to the ~2.18 Ga event of the Dharwar craton (Sect. 4.1.5; also see Ernst and Srivastava 2008), although there is no mafic dyke of this age yet reported from the Bastar craton.

However, recently Santosh et al. (2018) have reported U–Pb zircon age for a syenitie body exposed near the Nuapada and placed it at 2473 ± 8 Ma. Furthermore, they also presented U–Pb zircons (extracted from the Nuapada lamproites) which yield three Paleoproterozoic ages ca. 2.4, 2.2, and 2.0 Ga. They interpreted the 2.4 Ga age for xenocrysts derived from the syenite or basement rocks during the lamproite intrusion, the 2.2 Ga age for emplacement of the lamproites, and the 2.0 Ga age for a thermal event associated with major mafic dyke emplacement in southern Bastar and the adjacent Dharwar Craton (Santosh et al. 2018). However, we favour the interpretation of a ~1.05 Ga emplacement age for the Nuapada lamroites and that all zircons extracted from the Nuapada lamproites are xenocrystic and probably represent thermal events associated with the 2.4–2.5 Ga Dantewara swarm (the Bastar craton; Sect. 4.2.2), the ~2.18 Ga Mahbubnagar-Dandeli swarm (the Dharwar craton; Sect. 4.1.5) and Bijli rhyolites (the Bastar craton), and the ~1.88–1.89 Ga Bastanar swarm (the Bastar craton; Sect. 4.2.3) and Hampi swarm (Dharwar craton; Sect. 4.1.7) events. Nevertheless, more detailed work is required to assess which of these two interpretations is correct.

4.2.1 WNW–ESE to NW–SE Trending ~2.7 Ga Keshkal Swarm

WNW–ESE to NW–SE trending mafic dykes are distributed throughout the Bastar craton. Although there is no precise geochronological data available for these dykes, on the basis of Nd-isotope data, Srivastava et al. (2009a) suggested an emplacement age of ~2.7 Ga. This is supported by field relationships with country rocks including dykes being cut by ~2.3–2.5 Ga granitic veins (Ramakrishnan 1990; Srivastava et al. 1996). Dykes of this swarm are more conspicuous around the Keshkal, Bhanupratappur, Aamgaon, and Bastanar regions. This swarm includes the earlier reported ~2.7 Ga BD1 swarm from the southern Bastar craton (Srivastava et al. 1996, 2009a; Srivastava 2006a, b; Srivastava and Gautam 2008), Keshkal (Srivastava and Gautam 2012), and North Bastar dykes (NBD) from the northern Bastar craton (Srivastava and Gautam 2015). These are also reported as high temperature metabasite dykes as they show metamorphism under amphibolite facies P–T conditions (Srivastava et al. 2009a).

4.2.2 NW–SE to NE–SW Trending ~2.4–2.5 Ga Dantewara Swarm

Another set of NW–SE trending dykes, mostly boninitic (-noritic) in nature, has also been discovered in the southern Bastar craton, particularly around Dantewara, and identified as the BN swarm (Srivastava and Singh 2003; Srivastava 2006b,

2008), herein renamed as the Dantewara swarm. In addition, boninitic volcanic rocks associated with the BN (now Dantewara) dyke swarm are reported from the southern Bastar craton (Srivastava et al. 2004). Boninite and high-Mg norite dykes are also reported from central and northern parts of the Bastar craton (Subba Rao et al. 2008; Chalapathi Rao and Srivastava 2009; Srivastava and Gautam 2012). Srivastava and Gautam (2015) have noticed boninitic geochemical characteristics of the Dongergarh-Chhura dykes (DCD) from the northern Bastar craton. During the present study, we recorded a few NE–SW to ENE–WSW trending dykes near Pakhanjore and Bhanupratappur, and north of Lakhna areas, which may also belong to the BN (now Dantewara) swarm. However, further data, particularly geochronological, is required to corroborate the proposed correlation of all these dyke sets. If established by this exercise, the overall swarm shows a radiating pattern. Although no geochronological data are available for this swarm, a 2.4–2.5 Ga age is predicted based on field relationships, particularly the observation of a Dantewara dyke cutting a 2.7 Ga Keshkal dyke and that none of the Dantewara dykes is reported to cut ~2.3 Ga granites (Ramakrishnan 1990, Srivastava 2006b). The interpreted age is further supported by a U–Pb metamorphic rutile age (2118 ± 2 Ma; Srivastava et al. 2011a) and available ages on boninitie-noritic magmatism elsewhere in the world, which is thought to be broadly age correlated (Srivastava 2008; Srivastava and Ernst 2013; Srivastava et al. 2016b). The 2.47 Ga syenite intrusion at the Nuapada, northern Bastar craton (Santosh et al. 2018) may be part of the Dantewara magmatic event.

4.2.3 NW–SE Trending ~1.88–1.89 Ga Bastanar Swarm

This is another prominent NW–SE trending mafic dyke swarm, reported from the Bastar craton and concentrated in southern and central regions around Bastanar, Sukma, Aamgaon, Bhanupratappur areas. Dykes of this swarm are also sporadically present in northern parts of the craton. Baddeleyite/zircon grains separated from two dykes of this BD2 swarm have been dated by the high precision U–Pb ID-TIMS method at 1891.1 ± 0.9 Ma and 1883.0 ± 1.4 Ma (French et al. 2008), and a third dyke from the Bhanupratappur area has been dated by the Pb–Pb ID-TIMS method at 1882.4 ± 1.5 Ma (Shellnutt et al. (2018). This Bastanar swarm (new name herein) includes the BD2 swarm from the southern and central Bastar (Srivastava and Singh 2004; Srivastava and Gautam 2008, 2012 and references therein) and the NBD swarm from the northern Bastar regions (Srivastava et al. 2015). Overall, this swarm shows a linear NW–SE trend, but has been correlated with the coeval E–W to NW–SE trending Hampi swarm of the Dharwar craton (Sect. 4.1.7).

4.2.4 N–S to NNE–SSW Trending ~1.44–1.46 Ga Lakhna Swarm

A group of N–S trending mafic dykes is present in the Lakhna area, in close proximity to Eastern Ghat Mobile Belt (EGMB). During the present study, some longer NNE–SSW trending dykes were identified at the northern end of the Lakhna region, which we suggest may also belong to the Lakhna swarm (see Fig. 3). A few N–S trending mafic dykes are also encountered in the western side of the northern Bastar craton. This extensive Lakhna swarm comprises a variety of compositional types including dolerite, alkali gabbro, rhyolite, and trachyte (Ratre et al. 2010; Pisarevsky et al. 2013; Srivastava and Gautam 2015). The U–Pb SHRIMP zircon ages of associated rhyolite, trachyte and alkali gabbro units suggest emplacement at ~1442–1453 Ma (Ratre et al. 2010). A U–Pb zircon age of 1466.4 ± 2.6 Ma was obtained for a N–S trending rhyolite dyke intruded at Lakhna (Pisarevsky et al. 2013). Geochemistry on mafic members of this swarm suggests its alkaline nature and that it was derived from melts generated within the garnet stability field (Srivastava and Gautam 2015).

4.2.5 ENE–WSW Trending ~1.42 Ga Bandalimal Swarm

A few ENE–WSW trending mafic dykes are also reported from northern region of the Bastar craton. Those in the Bandalimal area intruded within the Proterozoic Chhattisgarh sedimentary basins, and Sm–Nd mineral and whole rock data suggest emplacement at ~1.42 Ga (Das et al. 2011; Sinha et al. 2011). This age is close to that of the N–S trending Lakhna swarm, but the ENE–WSW trending Bandalimal dykes are provisionally considered as a separate swarm because of their different trend. However, it remains possible that the Lakhna and the Bandalimal dykes belong to a single protracted magmatic event (~1.42–1.46 Ga). A few dykes of similar trend i.e. ENE–WSW, are also encountered in the Dongergarh area and may also belong to the Bandalimal swarm.

4.3 Singhbhum Craton

The triangular shaped Singhbhum craton is surrounded its three sides by the Mahanadi rift (and Sukinda thrust) on the SW, the Narmada–Son rift (together with Gangetic alluvium and Mahakoshal–Satpura belt) on the NNW, and the Indo-Gangetic Bengal basin (and the ~118 Ma Rajmahal traps) on the SE (Naqvi and Rogers 1987; Sharma 2009; see Fig. 1). The Singhbhum craton consists of two major geological domains viz. the Singhbhum Granite Complex (SGC) and the Chotanagpur Gneissic Complex (CGC), and these domains are separated by the Singhbhum Mobile Belt (SMB) (cf. Naqvi and Rogers 1987; Sharma 2009; Srivastava et al. 2009b, 2012, 2014c). The geology of the Singhbhum craton, particularly the SGC part, has been presented in a number of publications (Moorbath and Taylor 1988; Basu et al. 1993; Goswami et al. 1995; Misra et al. 1999; Acharyya et al. 2010; Roy and Bhattacharya 2012; Prabhakar and Bhattacharya 2013). Recent work (Upadhyay et al. 2014; Nelson et al. 2014; Dey et al. 2017) suggests that the SGC comprises (i) two pulses of granitoid suites at 3.45–3.44 and 3.35–3.32 Ga, (ii) tonalities, trondhjemites of the Older Metamorphic Tonalitic Gneisses (OMTG) together with the Singhbhum granites at 3.45–3.44 Ga, and (iii) the OMTG, with slightly younger phases (~3.35–3.32 Ga) of Singhbhum granite. An early phase of relatively high-grade metamorphism at 3.30–3.28 Ga followed by extensive fluid-induced alteration during low-grade metamorphism at 3.19–3.12 and 3.02–2.96 Ga has also been described. Episodic Archean plume-related mafic-ultramafic (basaltic and komatiitic) magmatic underplating and intraplating in an oceanic plateau setting is also recorded (Sharma et al. 1994; Chaudhuri et al. 2015; Dey et al. 2017). Distinct phases of Precambrian mafic magmatic activity in the form of volcanics, dyke swarms and layered intrusions are reported from the Singhbhum craton, and these are more conspicuous in the SGC portion (Dunn and Dey 1942; Saha et al. 1973; Mallik and Sarkar 1994; Saha 1994; Mahadevan 2002; Kumar and Ahmad 2007; Bose 2008, 2009; Srivastava et al. 2012, 2018).

Volcanic rocks of ~2.2–2.8 Ga ages are displayed in the periphery of the Singhbhum granite in several Archean to early Proterozoic volcanic basins viz. Similipal, Dhanjori, Dalma, Jagannathpur, Malangtoli (Bose et al. 1989; Saha 1994; Roy et al. 2002; Misra and Johnson 2005; Misra 2006; Bose 2008, 2009). Mafic dykes of Neoarchean-Paleoproterozoic age, popularly labelled as "Newer Dolerite", are mostly concentrated in the Singhbhum granite complex (Saha 1994; Saha et al. 1973; Mallik and Sarkar 1994; Shankar et al. 2014; Srivastava et al. 2016a, 2018; Kumar et al. 2017). There is limited precise age data available for the mafic dykes of the Singhbhum craton (see Table 1; Shankar et al. 2014; Srivastava et al 2016a, 2018; Kumar et al. 2017). Shankar et al. (2014) have provided a ~1765–1766 Ma Pb–Pb baddeleyite ages for a set of 'Newer dolerite' trending WNW–ESE, whereas Kumar et al. (2017) have investigated another other set of mafic dykes ('Newer dolerite') trending NNE–SSW to NE–SW and obtained precise Pb–Pb baddeleyite TE-TIMS dates of 2800.2 ± 0.7, 2762.4 ± 2.0 and 2752.0 ± 0.9 Ma, which they interpreted to represent emplacement ages. Also, the WNW–ESE trending Pipilia swarm (Srivastava et al. 2016a) has been precisely dated as 1765.3 ± 1 Ma by the Pb–Pb baddeleyite TE-TIMS method (Shankar et al. 2014).

Recently, Srivastava et al. (2018) have mapped and re-classified the mafic dykes of Singhbhum Granite Complex (SGC) into seven swarms on the basis of available precise geochronological data and cross cutting field relationships. These include (i) the ~2.80 Ga NE–SW trending Keshargaria swarm, (ii) the ~2.75–2.76 Ga NNE–SSW to NE–SW trending Ghatgaon swarm, (iii) the ~2.26 Ma NE–SW to ENE–WSW trending Kaptipada swarm, (iv) the ~1.77 Ga WNW–ESE trending Pipilia swarm, (v) the early-Paleoproterozoic E–W to ENE–WSW trending Keonjhar swarm, (vi) the middle-Paleoproterozoic NW–SE to NNW–SSE trending Bhagamunda swarm, and (vii) the late-Paleoproterzoic N–S to NNE–SSW trending Barigaon swarm. Based on this classification scheme, a dyke swarm map for the distinct mafic dyke swarms of the Singhbhum craton has been prepared (Fig. 4) and a summary on each swarm is presented below.

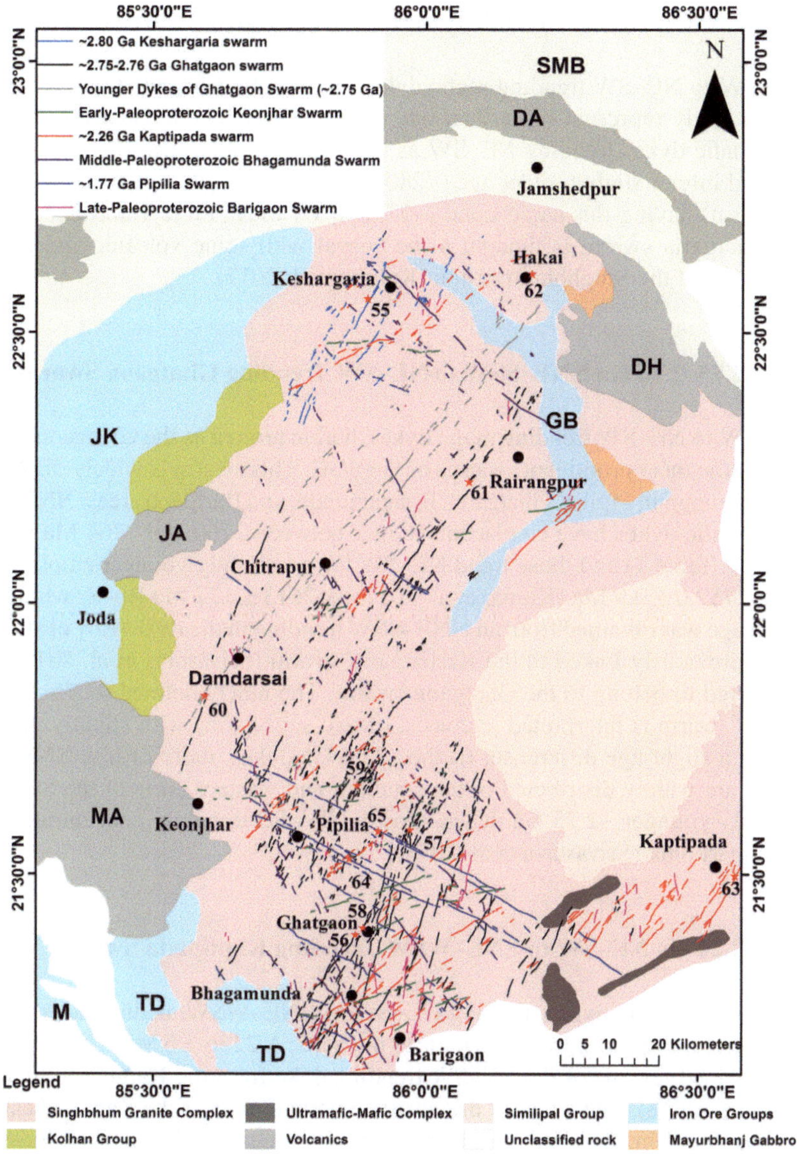

Fig. 4 Geological map showing distribution of distinct Neoarchean-Paleoproterozoic mafic dyke swarms emplaced within the Singhbhum Granite Complex (SGC), Singhbhum craton (based on Saha 1994; Misra 2006; Srivastava et al. 2018 and references therein). CGC, Chhotanagpur Gneissic Complex; DA, Dalma; DH, Dhanjori; MA, Malangtoli; JA, Jagannathpur; GB, Garumahishani-Badampahar; JK, Jamda-Koira; TD, Tomka-Daiteri; M, Malayagiri. (*) Locations of dated dykes (see Table 1; S. Nos. 55–65)

4.3.1 ~2.80 Ga NE–SW Trending Keshargaria Swarm

NNE–SSW to NE–SW trending mafic dykes are widely distributed throughout the SGC and likely represent more than one swarm (Table 1). Herein we distinguish a set of mafic dykes trending NE–SW as the Keshargaria swarm (Srivastava et al. 2018), and infer a slightly older age (~2800 Ma) than the other mafic dykes (Ghatgaon swarm) having this same trend (~2752–2764 Ma) (Table 1 and Sect. 4.3.2). The Keshargaria swarm is thought to be coeval with some volcanic and granitic magamatism of the Singhbhum craton (Kumar et al. 2017).

4.3.2 ~2.75–2.76 Ga NNE–SSW to NE–SW Trending Ghatgaon Swarm

NNE–SSW to NE–SW trending mafic dykes, herein termed as the Ghatgaon swarm, are one of the most prominent swarms of the SGC. This swarm is widely distributed around Rairangpur, Pipilia, Ghatgaon, Bhagamunda, and Barigaon areas. NNE–SSW trending mafic dykes have emplacement ages between 2760 and 2764 Ma (Kumar et al. 2017; Table 1) and those trend NE–SW have a slightly younger emplacement age i.e. (2752.0 $\pm$ 0.9 Ma (Kumar et al. 2017). A 2613 $\pm$ 177 Ma Rb–Sr whole rock isochron age was obtained from an NNE–SSW trending mafic dyke (Roy et al. 2004) that was previously linked to the Keshargaria swarm (Srivastava et al. 2016a), but now inferred to belong to the Ghatgaon swarm. The newly defined ~2.75–2.76 Ga Ghatgaon swarm is interpreted to consist of two subswarms with slightly different trend and a slight age difference: (i) the ~2.76 Ga dykes, more or less NNW–SSE trending, are widely distributed in the southern and western parts of the SGC, and (ii) slightly younger ~2.75 Ga dykes, mostly NW–SE trending, concentrate in the northern part only (Srivastava et al. 2018).

4.3.3 ~2.26 Ga NE–SW to ENE–WSW Trending Kaptipada Swarm

Another set of mafic dykes trending NE–SW to ENE–WSW, is distributed around the southern, northern and western parts of the SGC. A 2256 $\pm$ 6 Ma U–Pb ID-TIMS age is obtained for one of the NE–SW trending dyke from the Kaptipada area (Srivastava et al. 2018). The dykes of this swarm, herein termed as the Kaptipada swarm are encountered in the southern (Kaptipada-Bhagamunda-Ghatgaon regions) and western (Piplia-Keonjhar regions) parts of the SGC and mostly NE–SW trending. However, the portion of the Kaptipada swarm in the northern region (around Keshargaria) shows ENE–WSW trends, probably suggesting an overall radiating geometry for the Kaptipada swarm.

4.3.4 ~1.77 Ga WNW–ESE Trending Pipilia Swarm

A set of WNW–ESE trending mafic dykes is emplaced in and around the southern part of the SGC. This set of dykes is mostly concentrated around Pipilia, in the southern part of the SGC, hence herein named as the Pipilia swarm. A few dykes of this swarm are also distributed around central (Damdarsai) and northern (Keshargaria) parts of the SGC. Dyke orientations suggest a linear geometry. Baddeleyite Pb–Pb ages of two dykes are 1766.2 ± 1.1 Ma and 1764.5 ± 0.9 Ma (Shankar et al. 2014).

4.3.5 Early-Paleoproterozoic E–W to ENE–WSW Trending Keonjhar Swarm

A few ENE–WSW trending mafic dykes, mostly distributed around the southern part of the SGC, are also reported, but no geochronological data are available for this set of dykes. However, careful examinations of Google Earth™ images and observed crosscutting relationships indicate that this set of dykes is younger than the ~2.75–2.76 Ga Ghatgaon swarm, as the former offset the latter at a number of places. These dykes are also reported to be cross-cut by dykes of the ~2.26 Kaptipada swarm the Late Paleoproterozoic N–S to NNE–SSW trending Barigaon swarm and the Middle Paleoproterozoic NW–SE to NNW–SSE trending Bhagamunda swarm. These cross-cutting relationships reveal emplacement of ENE–WSW trending dykes in the early-Paleoproterozoic (between 2.76 and 2.26 Ga) and these dykes are herein identified as the Keonjhar swarm. A set of E–W trending mafic dykes in the northern region also shows similar field relationships and tentatively grouped with the Keonjhar swarm. If these two dyke sets are of the same age, it would suggest a radiating geometry for the overall swarm.

4.3.6 Middle-Paleoproterozoic NW–SE to NNW–SSE Trending Bhagamunda Swarm

NW–SE to NNW–SSE trending mafic dykes are distributed throughout the entire SGC, except in the southeastern region, and are concentrated around the Bhagamonda, Pipilia, Rairangpur and Keshargaria areas. There are no geochronological data available for this group of dykes. However, mafic dykes of this swarm are cross-cut by dykes of the ~1.76 Ga WNW–ESE trending Pipilia as well as the N–S to NNE–SSW trending Barigaon swarms, whereas they crosscut mafic dykes of the ~2.75–2.76 Ga NNE–SSW to NE–SW trending Ghatgaon and ~2.26 Ga NE–SW to ENE–WSW trending Kaptipada swarms; hence a middle-Paleoproterozoic age (2.26–1.76 Ga) is assigned, and these dykes are identified as a separate swarm and termed as the Bhagamunda swarm (Srivastava et al. 2018). Dykes in the southern parts of SGC that trend in a NNW–SSE direction also show similar field relationships as those of the Bhagamunda swarm, and are therefore provisionally included as a part of this swarm. Overall, the distribution pattern suggests a radiating geometry for this Bhagamunda swarm.

4.3.7 Late-Paleoproterzoic N–S to NNE–SSW Trending Barigaon Swarm

N–S to NNE–SSW trending mafic dykes exposed around the Barigaon area crosscut almost all the identified mafic dyke swarms in the SGC and are termed the Barigaon swarm (Srivastava et al. 2018). This suggests that these are the youngest swarm of the SGC with probable emplacement in the late-Paleoproterozoic or later.

4.4 Bundelkhand Craton

The triangular-shaped Bundelkhand craton, covering an area of 26,000 km^2, is separated from the Satpura mobile belt in the south and Aravalli craton in the west by the Proterozoic Vindhyan basin. The Bundelkhand craton consists of three distinct litho-tectonic units: (i) an Archean enclave suite consisting of highly deformed older gneisses–greenstone rocks, (ii) a granite suite with undeformed multiphase granitoid plutons and associated quartz reefs, and (iii) an intrusive suite with mafic dykes and other intrusions (Sharma and Rahman 2000; Meert et al. 2010). The oldest component of Bundelkhand craton is manifested by the TTG magmatism at ca 3.27 Ga and granitoid emplacement at ca 2.55–2.52 Ga indicating the stabilization of the craton (Mondal et al. 2002, 2008a). Singh and Slabunov (2014) were the first to discuss the greenstone belts in the Bundelkhand craton and classified them into two complexes viz. Central Bundelkhand (Babina and Mauranipur belts) and South Bundelkhand (Girar and Madaura belts) complex. Quartz veins/reefs of estimated Paleoproterozoic age, representing polyphase tectonically-controlled hydrothermal activity, show spectacular distribution throughout the craton and beyond and trending NE–SW to NNE–SSW (Pati et al. 2007). Proterozoic rift basins also exist at the northern and southern periphery of the Bundelkhand craton (Chakraborty et al. 2016; see Fig. 5).

In the Bundhelkhand craton, mafic dykes crosscut all lithologies suggesting they are the youngest magmatic activity (Basu 1986). They mostly trend NW–SE, but also trend ENE–WSW and NE–SW (Basu 1986; Mondal and Zainuddin 1996; Mondal et al. 2002; Rao 2004; Pati et al. 2008; Pradhan et al. 2012; Ernst 2014). Earlier workers dated these mafic dykes by Ar–Ar and K–Ar methods and placed them broadly between ~1560 and ~2150 Ma (Sarkar et al. 1997; Rao 2004; Rao et al. 1995). However, precise U–Pb geochronology, presented by Pradhan et al. (2012) suggests two phases of dyke emplacement; one at ~1.98 Ga and the other at ~1.11 Ga. Based on these ages and trends, the mafic dykes of the Bundelkhand craton have been classified and re-named into two distinct mafic dyke swarms viz. (i) ~1.98 Ga NW–SE trending Jhansi swarm, and (ii) ~1.11 Ga ENE–WSW trending Mahoba swarm. Figure 5 shows their distribution. Gabbro sills are reported to intrude Proterozoic rift basins of the Bundelkhand craton (Fig. 5). For instance, in the north, the Gwalior basin is intruded by a 2104 $\pm$ 23 Ma (Sm–Nd isochron age) gabbro sill (Samom et al. 2017), whereas in the south, a ~1.98 Ga gabbro sill, likely linked to the coeval Jhansi swarm, is intruded within the Bijawar basin (Pandey et al. 2012).

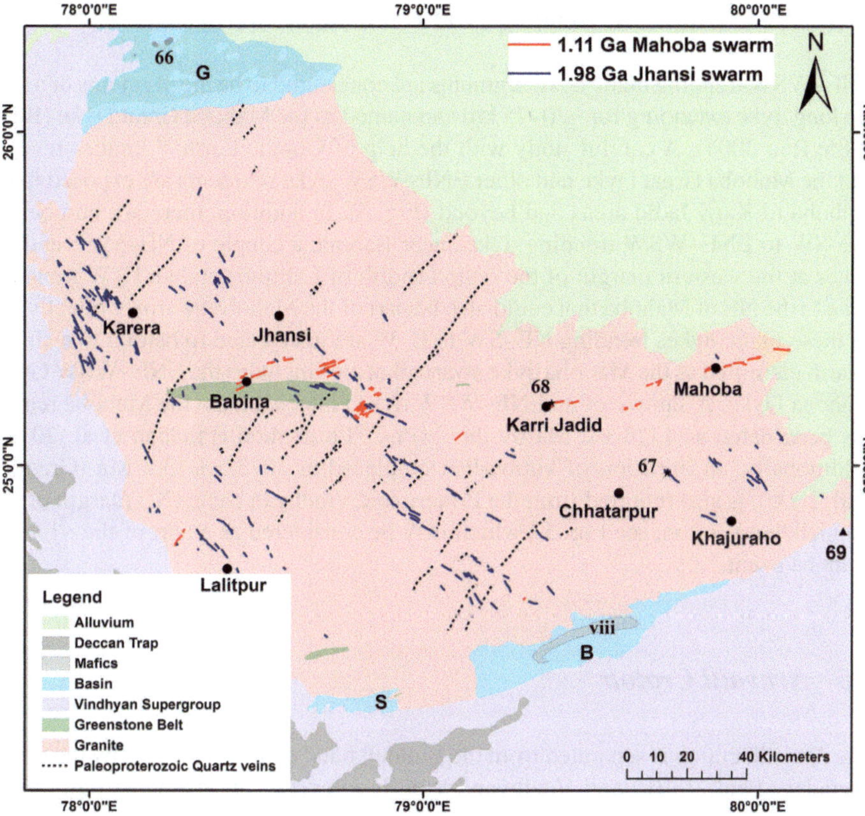

Fig. 5 Geological map showing distribution of distinct Paleo-Mesoproterozoic mafic dyke swarms emplaced within the Bundelkhand craton (based on Basu 1986; Mondal et al. 2008a, b; Pati et al. 2008; Pradhan et al. 2012 and references therein). Locations of dated dykes (*), gabbro sills, and approximate location of Majhgaon kimberlite intrusion (▲) (see Table 1; S. Nos. 66 to 69 and viii). Pre-Vindhyan sedimentary basins: *G* Gwalior basin; *S* Sonrai basin; *B* Bijawar basin (cf. Chakraborty et al. 2016). NE–SW trending dotted lines represent occurrence of giant quartz veins/reefs (cf. Pati et al. 2007)

4.4.1 NW–SE Trending ~1.98 Ga Jhansi Swarm

NW–SE trending mafic dykes are most prominent in Khajuraho, Lalitpur, Babina, Jhansi and Karera areas. Rao (2004) dated a NW–SE mafic dyke by $^{40}Ar/^{39}Ar$ method at ~2.0 Ga and identified a Bundelkhand dyke swarm. Later, a sample of this swarm was dated at 1979 ± 8 Ma (U–Pb zircon/baddeleyite concordia age; Pradhan et al. 2012), which gives a precise constraint on the emplacement of this swarm. Ernst (2014) renamed this swarm as the Jhansi swarm (based on consultation with J. Meert) since these dykes are concentrated in the Jhansi region and because it is better to use a local name rather than a craton name for naming any dyke swarm. The Jhansi dyke swarm has a linear geometry.

4.4.2 NE–SW to E–W Trending ~1.11 Ga Mahoba Swarm

ENE–WSW trending mafic dyke segments are considered to be the segments of a single long dyke extending for ~70–75 km and named as the Mahoba Great Dyke (Basu 1986; Rao 2004). A careful study with the help of Google Earth™ images reveals that the Mahoba Great Dyke and other ENE–WSW dyke segments are exposed from Mahoba to Karri Jadid areas and beyond (Fig. 5). In addition, there are number of NE–SW to ENE–WSW trending dykes near Babina, a couple of NE–SW trending dykes at the western margin of the craton (north of Lalitpur) and an E–W trending dyke to the NE of Mahoba that could also be part of the Mahoba swarm. Collectively, all these mafic dykes trending NE–SW to E–W are interpreted to belong to a single swarm identified as the Mahoba dyke swarm that also includes the ENE–WSW Great Mahoba Dyke. A sample of an ENE–WSW dyke collected from the Mahoba region has been dated as 1113 ± 7 Ma by the $^{207}Pb/^{206}Pb$ method (Pradhan et al. 2012). Additionally, an intrusion of kimberlite, emplaced at 1073.5 ± 13.7 Ma (Gregory et al. 2006), is also reported from the Proterozoic Vindhyan basin (NE margin of the Bundelkhand craton; see Fig. 5), which may be considered as a part of the ~1.1 Ga Mahoba event.

4.5 *Aravalli Craton*

The Aravalli craton is separated from the Bundelkhand craton in the east by the Great Boundary Fault (GBF) and Vindhyan basin. It comprises six tectono-stratigraphic units viz. Mesoarchean Mewar gneissic complex, Neoarchean Mangalwar complex with Bhilwara group, Paleoproterozoic Hindoli group and Aravalli supergroup along with mineralized supracrustal belts, Mesoproterozoic Delhi supergroup and Neoproterozoic Marwar supergroup, 750 Ma Malani igneous suite (part of a Silicic LIP; e.g. Ernst 2014) and Neoarchean (Cryogenian) Erinpura granite (Heron 1953; Ramakrishnan and Vaidyanadhan 2010). Based on cover relationships and grade of metamorphism, the Banded Gneissic Complex (BGC) is classified into an older BGC-I, which occurs to the east and south of Nathdwara, and BGC-II, which occurs to the south of Nathdwara (Heron 1953; Ahmad et al. 2017). The BGC serves as a basement upon which two sedimentary sequences viz. Aravalli supergroup and Delhi supergroup have been deposited (Heron 1953; Gupta et al. 1980). The absence of greenstone belts and very limited occurrence of mafic dykes in any of the lithological units demarcates the uniqueness of the Aravalli craton with respect to other cratons of the Indian shield.

Ahmad et al. (2008) reviewed the status of mafic volcanic rocks of the Aravalli craton and discussed the geochemistry of Proterozoic metavolcanic rocks from the Bhilwara, Aravalli, Jharol and Delhi belts. Besides these metavolcanic rocks, a number of metabasic dykes within the basement rock are also reported (Roy et al. 1995; Shekhawat et al. 2000, 2001, 2007); however, their distribution is poorly known. The mafic dykes, mostly metadolerite and amphibolite in nature, are sporadically distributed in and around Mavli, Jagat, Nathdwara and Amet areas of Udaipur. Upadhyaya et al. (1992) has mapped the amphibolite dykes of the Jagat group and studied their geochemical characteristics. Amphibolite dykes are also reported from the Malvi area which are thought to be equivalent to the Jagat amphibolite dykes (Gopalan et al. 1990). The Sm–Nd whole-rock isochron age of Mavli amphibolite dykes suggests an emplacement age of 2828 ± 46 Ma (Gopalan et al. 1990). Another set of metadolerite dykes, with probable emplacement after 2.5 Ga, has also been reported from the Mavli region (Gopalan et al. 1990). Based on trend and cross-cutting field relationships with other lithological units, Mondal et al. (2008b) have classified these metadolerite/metabasite dykes of the Aravalli craton, mostly exposed in and around the Jagat, Mavli, Amet, and Nathdwara areas, into three broad groups: (i) NW–SE trending ~2.8 Ga (ii) NE–SW trending ~2.2 Ga, and (iii) N–S trending ~1.8 Ga mafic dykes. Our map (Fig. 6) focuses on those dykes that can be readily traced from the Google™ Earth images viz. (i) the NW–SE trending ~2.8 Ga Jagat swarm swarm, and (ii) the NE–SW trending ~2.2 Ga Amet swarm (see Fig. 6). The N–S trending set is more difficult to recognize in Google Earth images and is not included on our map. Precise geochronological data on all these mafic dykes is required for better correlation and understanding.

5 Discussion

5.1 Correlation of Identified Magmatic Events in the Indian Shield and Recognition of LIPs

As discussed above, the Google Earth™ images, field surveys and previously published geological maps have been used to prepare dyke swarm maps for each of the Indian cratons. We have identified fourteen distinct mafic dyke swarms during the Neoarchean-Mesoproterozoic (~2.8 and ~1.1 Ga) (Sect. 4; Table 2). All of these swarms are thought to have an intraplate setting and many extend into adjoining cratonic blocks and are considered to be parts of the plumbing system of LIPs. A LIP is defined as a large volume (>0.1 Mkm^3), mainly mafic (+ultramafic) magmatic event of intraplate affinity, characteristically of short duration (<1 myr) or consisting of multiple short pulses over a maximum of a few 10 s of myr (Bryan and Ferrari 2013; Ernst 2014; Ernst and Youbi 2017, and references therein). However, as noted in the introduction (Sect. 1), a swarm (independent of size) in a craton exhibiting an average dyke width of >10 m can be inferred to also identify a LIP. In the discussion below, we focus on eleven LIP events identified from the Indian shield (Table 2), which can be correlated with similar events identified from other parts of the globe.

Table 2 Identified Neoarchean-Mesoproterozoic mafic dyke swarms and associated magmatic units of different Archean cratons of the Indian shield and their possible correlation

Identified Magmatic events	Possible LIP	Dharwar craton	Bastar craton	Singhbhum craton	Bundelkhand craton	Aravalli craton
~2.80 Ga	2.75–2.80 LIP			~2.80 Ga Keshargaria swarm		*~2.80 Ga Jagat swarm*
~2.75–2.76 Ga			*~2.70 Ga Keshkal swarm*	~2.75–2.76 Ga Ghatgaon swarm		
Neoarchean-Paleoproterozoic boundary (2.45–2.50 Ga)						
~2.37 Ga	2.36–2.50 Ga LIP	~2.37 Ga Bangalore-Karimnagar swarm	*~2.4–2.5 Ga Dantewara swarm* ~2.5 Ga syenite intrusion			
~2.25–2.26 Ga	2.25–2.26 Ga LIP	~2.25–2.26 Ga Ippaguda-Dhiburahalli swarm		~2.26 Ga Kaptipada swarm		
~2.22 Ga	2.21–2.22 Ga LIP	~2.22 Ga Kandlamadugu swarm				
~2.21 Ga		~2.21 Ga Anantapur-Kunigal swarm				
~2.18 Ga	2.18 Ga LIP	~2.18 Ga Mahbubnagar-Dandeli swarm	*~2.18 Ga Bijli rhyolites*			

(continued)

Table 2 (continued)

Identified Magmatic events	Possible LIP	Dharwar craton	Bastar craton	Singhbhum craton	Bundelkhand craton	Aravalli craton
~2.08 Ga	2.08 Ga LIP	~2.08 Ga Devarabanda swarm			~2.10 Ga gabbro sill in Gwalior basin	
~1.98 Ga	1.98 Ga LIP				~1.98 Jhansi swarm + *~1.98 Ga gabbro sill in Bijawar basin*	
~1.88–1.90 Ga	1.88–1.90 Ga LIP	~1.88–1.89 Ga Hampi swarm (includes Pulivendla sills)	~1.88–1.89 Ga Bastanar swarm			
~1.77–1.79 Ga	1.77–1.79 Ga LIP	~1.79 Ga Pebbair swarm		~1.77 Ga Pipilia swarm		
Paleoproterozoic-Mesoproterozoic boundary (1.6 Ga)						
~1.44–1.46 Ga	1.42–1.46 Ga LIP		~1.44 Ga Lakhna swarm			
~1.42 Ga		~1.42 Ga lamproites	*~1.42 Ga Bandlimal swarm*			
~1.10–1.12 Ga	1.10–1.12 Ga LIP	~1.10 Ga kimberlites	~1.05–1.10 Ga lamproites and kimberlites		~1.11 Ga Mahoba swarm	

Ages in italics are not precise. For source of these ages, please see text. A widespread ~1.80–1.90 Ga mafic magmatic event is reported from all along the Himalayan range. ~1.10 Ga kimberlites, intruded in the Vindhyan rocks, are also reported. See Table 1 for source of all these age data

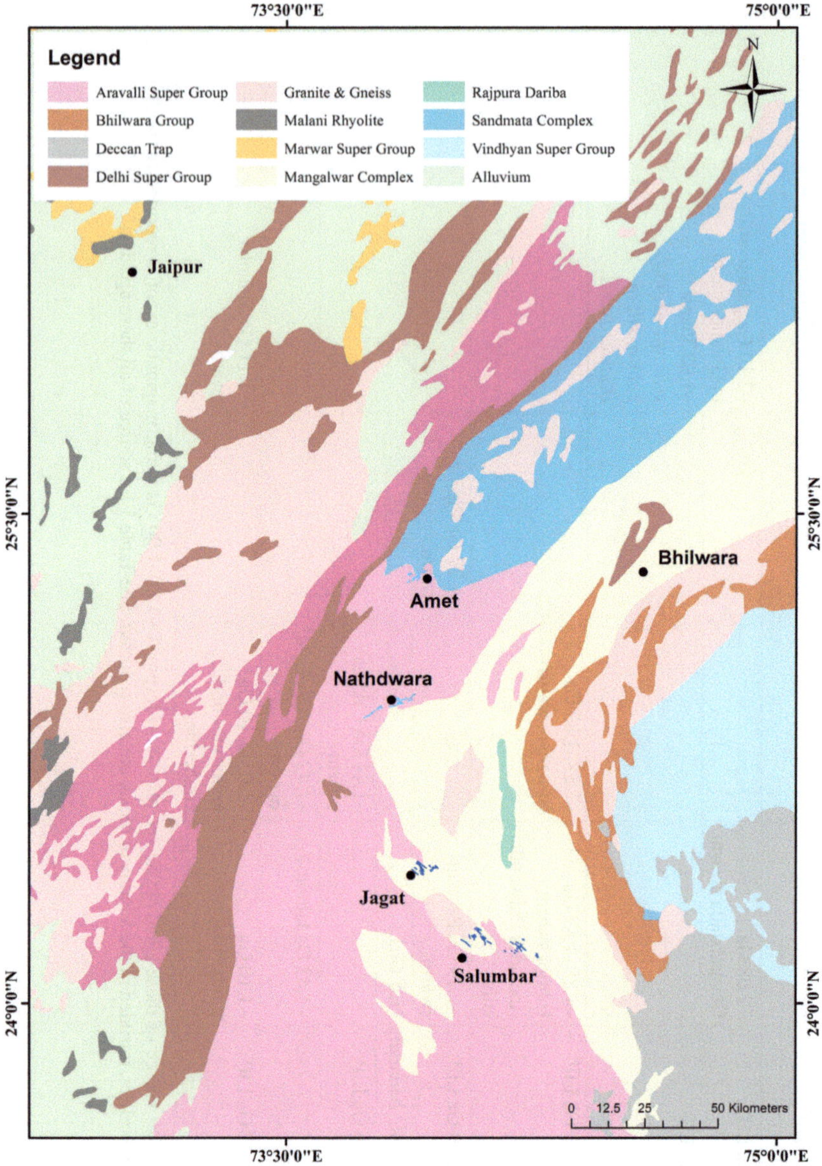

Fig. 6 Geological map showing distribution of distinct Precambrian mafic dyke swarms emplaced within the Aravalli craton (based on Gupta et al. 1980; Gopalan et al. 1990; Upadhyaya et al. 1992; Roy et al. 1995; Shekhawat et al. 2007; Ahmad et al. 2017 and references therein)

5.1.1 ~2.75–2.80 Ga LIP

The ~2.8 Ga mafic dyke swarm is probably the oldest identified mafic dyke swarm in the Indian shield. The northern part of the Singhbhum Granite Complex is transected by the ~2.80 Ga Keshargaria swarm (see Fig. 4) (Sect. 4.3.1). Although there is no other exact match reported from the Indian shield, the poorly dated ~2.80 Ga Jagat swarm (Sect. 4.5) is known from the Aravalli craton. A slightly younger ~2.75–2.76 Ga mafic dyke event is recorded from the Singbhum craton as well (Kumar et al. 2017). It is identified as the Ghatgaon swarm (Srivastava et al. 2018) (Sect. 4.3.2). Söderlund et al. (2018) have correlated the Ghatgaon swarm with the poorly dated ~2.70 Ga Keshkal swarm of the Bastar craton (Sect. 4.2.1). All together, these ages suggests a LIP event(s) ca. 2.75–2.80 Ga on several cratonic blocks of India.

The ~2.75–2.80 Ga event(s) of the Indian shield is an analogue for other roughly coeval events around the globe (Srivastava et al. 2018). Ernst (2014 and references therein) has documented ca. 2.70–2.78 Ga LIP fragments from crustal blocks of the Pilbara, Kaapvaal, Superior, Baltica, and others. For example, (i) ~2.70–2.73 Ga LIPs: the Zimbabwe craton (Bulawayan-Zeederbergs-Belingwe; Prendergast 2004), Wyoming craton (Stillwater; Wall et al. 2012), Slave craton (Kam group; Bleeker and Hall 2007), and Superior craton (Abitibi belt; e.g. Ayer et al. 2002), (ii) ~2.73–2.75 Ga LIPs: the West African craton (the Ahmeyim Great Dyke of Mauritania event of the Reguibat shield; Tait et al. 2013), the Superior craton (the Ring of Fire event; Mungall et al. 2010 and Bird River sill; Wang 1993 in Mealin 2006), Yilgarn craton (mafic-ultramafic intrusives and volcanics in the Eastern Goldfields terrane; Ivanic et al. 2010; Barnes et al. 2012), and (iii) ~2.78 Ga LIPs: the Kaapvaal craton (Derdepoort-Gaberone pulse of the Ventersdorp group of events, Pirajno and Hoatson 2012), the Pilbara craton (Mount Roe-Black Range pulse of the Fortescue group of events; Pirajno and Hoatson 2012). The ca. 2.80 Ga Keshargaria swarm could be matched with the ~2.8 Ga magmatism in the Pilbara craton and the 2.81 Ga intrusions of the Eastern Goldfields superterrane (Ivanic et al. 2010; Kumar et al. 2017; Srivastava et al. 2018).

5.1.2 ~2.36–2.50 Ga LIPs

The ~2.36–2.37 Ga Bangalore-Karimnagar mafic dyke swarm represents a prominent magmatic event in the Dharwar craton (Sect. 4.1.1). There is no similar event recorded from any other cratons of the Indian shield, however the poorly dated ~2.4–2.5 Ga Dantewara swarm and ~2.47 Ga Nuapada syenite intrusion of the Bastar craton (Sect. 4.2.2) is close in age to the Bangalore-Karimnagar swarm. The ~2.36–2.37 Ga Bangalore-Karimnagar swarm of the Dharwar craton is roughly coeval with the ~2.38–2.41 Ga Graedefjord-Scourie LIP(s) from the North Atlantic Craton (Nilsson et al. 2013; Davies and Heaman 2014; Hughes et al. 2014). Some other reported ~2.41 Ga LIP events are the Sebanga Poort event from the Zimbabwe craton (Söderlund et al. 2010), Ringvassoy event from the Karelia-Kola craton (Kullerud et al.

2006), Du Chef swarm of the eastern Superior craton (Ernst and Bleeker 2010), and the Widgiemooltha-Erayinia LIP from the Yilgarn craton (Smirnov et al. 2013; Pisarevsky et al. 2015).

5.1.3 ~2.25–2.26 Ga LIP

This LIP event is well represented by the ~2.25–2.26 Ga Ippaguda-Dhiburahalli swarm of the Dharwar craton (Sect. 4.1.2) and ~2.26 Ga Kaptipada swarm of the Singhbhum craton (Sect. 4.3.3). Based on the presence of these matching ~2.25–2.26 Ga swarms in the Dharwar and Singhbhum cratons, Srivastava et al. (2018) has predicted that the intervening Bastar craton must also host ~2.25–2.26 Ga dykes. Globally, 2.24 Ga dykes are known from the Vestfold Hills of the Mawson craton of East Antarctica (Ernst and Buchan 2001a, b). The Hekpoort lavas of the Kaapvaal craton also have an age constrained to ~2.23–2.25 Ga (Gumsley et al. 2017; Humbert et al. 2017). Another similar age is from a 2262 ± 2 Ma troctolite intrusion near Chimbadzi, in the Zimbabwe craton (Manyeruke et al. 2004).

5.1.4 ~2.21–2.22 Ga LIP

There are two mafic dyke swarms, with slightly different ages that represent this LIP event: the ~2.22 Ga Kandlamadugu swarm and ~2.21 Ga Anantapur-Kunigal swarm both from the Dharwar craton (Sects. 4.1.3 and 4.1.4). 2.21–2.22 Ga LIP events are known from different localities around the globe including (i) ~2.22 Ga BN1 event from the North Atlantic Craton (Ernst and Bleeker 2010), (ii) 2.23–2.21 Ga MacKay-Malley event(s) from the Slave craton (Buchan et al. 2010; Ernst and Bleeker 2010), (iii) 2.21 Ga Koli (Karjalitic) event of the Karelian craton (Vuollo and Huhma 2005; see also Huhma et al. 2018), (iv) 2.21–2.22 Ga Ungava-Nipissing event of the Superior craton (Buchan et al. 1998; Ernst and Bleeker 2010), and (v) 2.21 Ga Turee Creek-Cheela Springs event of the Pilbara craton (Müller et al. 2005; Martin and Morris 2010).

5.1.5 ~2.18 Ga LIP

The ~2.18 Ga Mahbubnagar-Dandeli LIP of the Dharwar craton (Sect. 4.1.5) is not recognized in other Indian cratons except for the ~2.18 Ga Bijli rhyolites in the Bastar craton. Globally LIPs of approximately this age are prominent on other crustal blocks: the ~2.17 Ga Biscotasing LIP from the Superior craton, the South Pass–Powder River (formerly Wind River) event from the Wyoming craton (Ernst and Bleeker 2010; Killian et al. 2016). Slightly older LIP events (~2.19 Ga), such as Dogrib (Slave craton) and Tulemolu-MacQuoid (Rae craton), are also identified (Buchan et al. 2010; Ernst and Bleeker 2010), and slight younger events include the

2.16–2.15 Ga Rabbit Creek event of the Wyoming craton (Killian et al. 2016) and the 2.15 Ga Riviere du Gue of the Superior craton (e.g. Ernst and Bleeker 2010), and the 2.15 Ga Hengling event of the North China craton (Peng et al. 2005; Peng 2015)

5.1.6 ~2.08 LIP

The ~2.08 Ga Devarabanda LIP of the Dharwar craton (Sect. 4.1.6) has no recognized age matches in the other Indian cratons except for the ~2.10 Ga gabbro sill within the Gwalior basin of the Bundelkhand craton (see Fig. 5). However, it is known from other crustal blocks, such as the ~2.07 Ga Fort Frances, Lac Esprit and Cauchon dyke swarms of the Superior craton (Ernst and Bleeker 2010) and units of similar age in the Karalia-Kola craton (Huhma et al. 2018).

5.1.7 ~1.98 Ga LIP

The only known ~1.98 Ga magmatic activity in the Indian shield is the Jhansi swarm of the Bundelkhand craton (Sect. 4.4.1) and a poorly dated ~1.98 Ga gabbro sill within the Bijawar basin. The ~1.96–1.98 Ga Pechenga-Onega LIP from the Karelia-Kola craton is close in age to the Jhansi swarm (Ernst and Bleeker 2010) as the ~1.97 Ga Xiwangshan LIP from the North China craton (Peng et al. 2005; Peng 2015). The Povungnituk volcanics from the Superior craton were thought to be ca. 1.97 Ga (Ernst and Bleeker 2010), but are now well dated as 2.0 Ga as part of the Minto-Povungnituk LIP (Kastek et al. 2016) and is, therefore, not coeval with the Jhansi event.

5.1.8 ~1.88–1.90 Ga LIP

~1.88–1.89 Ga magmatic activity is widespread in the Indian shield (French et al. 2008; Ernst and Srivastava 2008; Srivastava and Samal 2018; Shellnutt et al. 2018). Most prominently, this event comprises a radiating dyke swarm consisting of NW–SE trending Bastanar dykes (Bastar craton) and E–W to NW–SE trending Hampi dykes (Dharwar craton) (Sects. 4.2.3 and 4.1.7, respectively). This swarm converges at the eastern side of the Dharwar craton, marking the location of an 1890 Ma mantle plume centre (French et al. 2008; Ernst and Srivastava 2008; Belica et al. 2014; Shellnutt et al. 2018). The importance of this event is further substantiated by the presence of a sill (i.e. the dated Pulivendla sill; French et al. 2008) and a geophysical anomaly marking a major layered intrusion in the Cuddapah basin (Radhakrishana et al. 2007; Ernst and Srivastava 2008). Additionally, a number of mafic intrusive rocks within the Himalayan Mountain Range are reported (Ahmad et al. 1999; Miller et al. 2000), which are also thought to be part of the 1.88–1.90 Ga event (Srivastava and Samal 2018). Collectively, all these mafic occurrences indicate the presence of a LIP during ~1.88–1.90 Ga that is distributed over most parts of the Indian shield (Srivastava and Samal 2018).

Ernst and Srivastava (2008) have pointed out that 1880–1900 Ma LIPs are found on many different blocks and probably have more than one centre of mantle upwelling (e.g. plume) of this age. The ~1880–1900 Ma LIPs are found in the Superior, Slave, Rae, Kaapvaal, Zimbabwe, Amazonian, Siberian, and East European cratons (French et al. 2008; Ernst and Srivastava 2008; Ernst 2014; Teixeira et al. 2018a, b; and references therein). Dykes of same age has also been reported from the Yilgarn Craton, Western Australia, which suggests that the southern India and the Western Australia were attached during the Paleoproterozoic and separated after ~1.88 Ga (Stark et al. 2018; Shellnutt et al. 2018). A widespread mainly Silicic LIP, the 1.89–1.86 Ga Uatumã plutonic-volcanic LIP event, is present in the Amazonian craton (Teixeira et al. 2018b).

5.1.9 ~1.77–1.79 LIP

This event is mainly recorded in the eastern Dharwar craton and the Singhbhum Granite Complex of the Singhbhum craton (Sects. 4.1.8 and 4.3.4, respectively). Söderlund et al. (2018) reported the ~1.79 Ga Pebbair swarm in the eastern Dharwar craton, whereas Shankar et al. (2014) documented ~1.77 Ga mafic dykes, which have been termed as the Pipilia swarm by Srivastava et al. (2018). No other craton of the Indian shield has evidence for magmatism of this age, however Srivastava et al. (2018) have projected that the Bastar craton should also host this event. Their argument is based on the presence of ~1.77–1.79 Ga magmatic event in the Dharwar and Singhbhum cratons which are on opposite sides of the Bastar craton.

Globally, ~1.77–1.79 Ga LIPs are known from the Xiong'er-Taihang event of the North China craton (Halls et al. 2000; Wang et al. 2004; Peng et al. 2005, 2006; Peng 2010, 2015), the Hart-Carson magmatism in the Kimberley and Halls Creek Element of the North Australian craton (Page et al. 2000; Thorne et al. 2014; Pirajno and Hoatson 2012), the Avanavero LIP of the Amazonian craton (Norcross et al. 2000; Reis et al. 2013), the Florida (Uruguayan) swarm from the Rio de Plata craton (Halls et al. 2001; Teixeira et al. 2013), and mafic/ultramafic dyke swarms and associated AMCG magmatism from the Ukrainian shield (Amelin et al. 1994; Shumlyanskyy et al. 2012, 2015, 2016; Bogdanova et al. 2013). Most of these cratonic areas are potentially linked to the supercontinent Columbia/Nuna (e.g. Shankar et al. 2014).

5.1.10 ~1.42–1.46 LIP

This is the first Mesoproterozoic LIP event(s) identified in the Indian shield. It is mainly represented by ~1.42 Ga Bandalimal swarm (Das et al. 2011) and ~1.44–1.46 Ga Lakhna swarm (Ratre et al. 2010; Pisarevsky et al. 2013) of the Bastar craton (Sects. 4.2.4 and 4.2.5). A contemporary event is also observed from the Dharwar craton where ~1.42 Ga lamproites are intruded within the Cuddapah basin (Chalapathi Rao et al. 1999). However, this event(s) is not reported from any of the other Indian cratons. Possible global matches include the 1.42 Ga Highland-

Purcell event of western Laurentia (Rogers et al. 2018) and the ca. 1.44–1.42 Ga units from Amazonia (Teixeira et al. 2016). There are also some slightly older events that may be linked; ~1.43–1.46 Ga Michael-Shabogamo LIP of eastern Laurentia (Ernst and Buchan 2001a; Gower et al. 2002), ~1.46 Ga Tuna-Lake Ladoga LIP of Baltica (Ernst et al. 2008), and ~1.46 Ga West Bangemall-Edmund LIP of the North Australian craton (Pirajno and Hoatson 2012).

5.1.11 ~1.10–1.12 Ga LIP

The ~1.11 Ga Mahoba mafic dyke swarm is known from the Bundelkhand craton (Sect. 4.4.2; Pradhan et al. 2012; Ernst 2014). Mafic dykes of this age are not reported from any other Indian craton; however possibly related ~1.05–1.12 Ga kimberlite/lamproite intrusions are recorded from the Dharwar and Bastar cratons and in the Proterozoic Vindhyan basin; these include ~1.10–1.12 Ga kimberlites of the Dharwar craton (Chalapathi Rao et al. 2013), ~1.05 Ga lamproites from the Bastar craton (Sahu et al. 2013; Chalapathi Rao et al. 2016), and ~1.10 Ga kimberlite intrusion within the Vindhyan basin (Gregory et al. 2006).

The Umkondo LIP of the Kalahari craton (Bullen et al. 2012; de Kock et al. 2014), the GN-Huila LIP of Congo craton (Ernst et al. 2013), and the Rincon del Tigre-Huanchaca LIP of the Amazonian craton (Hamilton et al. 2012; Teixeira et al. 2018a) show exact age matches with the Mahoba swarm and provisional reconstructions have been suggested (Ernst et al. 2013; de Kock et al. 2014). Some other similar age events of uncertain relationship with the 1.11 Ga LIPs are (i) ~1.09–1.12 Ga Keweenawan LIP of Laurentia (Heaman et al. 2007; Miller and Nicholson 2013), (ii) ~1.09 Ga SW Laurentia LIP of Laurentia (Bright et al. 2014), and (iii) ~1.08 Ga Warakuna LIP of Australia (Pirajno and Hoatson 2012).

5.2 *Possible Paleoreconstructions*

There are many proposals for configurations of supercontinents that existed during the Precambrian (e.g. Rogers and Santosh 2002; Bleeker 2003; Li et al. 2008; Meert 2012; Ernst et al. 2013; Evans 2013; Pisarevsky et al. 2014; Nance et al. 2014). These include the 0.6 Ga Gondwana (Pannotia), 1.20–0.72 Ga Rodinia, 1.80–1.38 Ga Columbia/Nuna, and 2.75–2.07 Ga Kenorland (Superia + Sclavia). In addition, a supercraton Vaalbara that combined Kaapvaal craton (southern Africa) and Pilbara craton (Western Australia) existed during 3.1–2.8 Ga (e.g. Zegers et al. 1998; Bleeker 2003; de Kock et al. 2009; Smirnov et al. 2013). The supercontinent Ur is also proposed for Western Dharwar, Singhbhum, Kaapvaal, and Pilbara cratons at ~3 Ga (Rogers 1996; Rogers and Santosh 2003). The supercontinent cycle in Earth's history is marked by the episodic tectonic processes of assembly and breakup of continents (Worsley et al. 1982, 1984; Li et al. 2008; Nance et al. 2014). Rift-related mafic dyke swarms are one of the best types of evidence to mark the breakup of supercontinents (Worsley et al. 1982, 1984; Windley 1984; Condie 1989). This is further emphasized by Ernst (2014) who indicates that every major Precambrian breakup and formation of a new ocean should be linked to a LIP (and its associated dyke swarms) based on the evidence from the Gondwana supercontinent breakup record (e.g. Storey 1995).

All the identified mafic dyke swarms and their LIPs from the Indian shield and their correlation with possible matches from the other parts of the globe (see Sect. 5.1; Table 3) provide a basis for a robust evaluation of the position of the various Indian cratons within the Neoarchean-Mesoproterozoic supercontinents, through both using the LIP barcode matching method (Bleeker and Ernst 2006; Ernst et al. 2013) and paleomagnetic studies which rely greatly on mafic dyke swarms (e.g. Evans 2013; Buchan 2013; Pisarevsky et al. 2015). These topics are outside the scope of the present manuscript, but are the focus of a separate research, in progress.

5.3 Further U–Pb Geochronology Targets

Here, we identify some additional targets for U–Pb geochronology from among the many remaining undated swarms of the Indian cratons, which will provide further constraints for supercontinent reconstructions.

i. A number of precise U–Pb ages are available for swarms of the eastern Dharwar craton, but only one precise age is available for the western Dharwar craton. For instance, we predict that ~2.70–2.80 Ga dykes should be present in the western Dharwar craton.
ii. Additional events are predicted for the Bastar craton. Srivastava et al. (2018) pointed out that since the Singhbhum and Dharwar cratons share at least two mafic dyke events ~2.25–2.26 and ~1.76 Ga, then the intervening Bastar craton should host both ages of swarms. In addition, other dyke events such as those at 2.36–2.37, 2.21–2.22, 2.18, and 2.08 Ga, which are observed in the eastern Dharwar craton, may be present in the Bastar and Singhbhum cratons.
iii. Three distinct dyke swarms, emplaced within the Singhbhum Granite Complex and are currently undated, must be precisely dated. Dykes of the Chhotanagpur Gneissic Complex of the Singhbhum craton should also be precisely dated. This would help to establish a proper correlation between these two portions of the Singhbhum craton, i.e. the Singhbhum Granite Complex and Chhotanagpur Gneissic Complex.
iv. Geochronological data is also needed for the undated swarms of the Aravalli craton which currently lacks precise ages.
v. In addition, there are abundant undated mafic intrusive bodies (mainly sills) reported from Proterozoic sedimentary rift basins of the Indian shield, and we wish to test their links with already recognized dyke swarms as another part of LIP plumbing systems. Obtaining precise ages on these mafic sills is also important for better understanding of basin evolution.

Table 3 Global correlation of identified LIP events of the Indian shield (see text for more details)

2.75–2.80	2.36–2.50	2.25–2.26	2.21–2.22	2.08–2.18	1.98	1.88–1.90	1.77–1.79	1.42–1.46	1.10–1.12
Kenorland/Superia					*Columbia/Nuna*				*Rodinia*
Indian Shield									
	Dharwar	Dharwar	Dharwar	Dharwar		Dharwar	Dharwar	Dharwar	Dharwar
Bastar	Bastar	Bastar (?)		Bastar (?)		Bastar	Bastar (?)	Bastar	Bastar
Singhbhum		Singhbhum					Singhbhum		
				Bundelkhand	Bundelkhand				Bundelkhand
Aravalli									
									Vindhyans
						Himalayan Mt. Range			
North American shield									
Wyoming				Wyoming					
Superior	Superior		Superior	Superior	Superior	Superior			
Slave			Slave	Slave		Slave			
				Rae		Rae			
	North Atlantic		North Atlantic						
								Laurentia	Laurentia
South America shield									
							Rio de Plata		
						Amazonian	Amazonian		Amazonian
African shield									

(continued)

Table 3 (continued)

2.75–2.80	2.36–2.50	2.25–2.26	2.21–2.22	2.08–2.18	1.98	1.88–1.90	1.77–1.79	1.42–1.46	1.10–1.12
West African									
Zimbabwe	Zimbabwe	Zimbabwe				Zimbabwe			
Kaapvaal		Kaapvaal				Kaapvaal			
									Kalahari
									Congo
							African		
Australian shield									
Yilgarn	Yilgarn					Yilgarn			Yilgarn
Pilbara			Pilbara						
							North Australia	North Australia	
European shield									
	Karelia-Kola		Karelia-Kola		Karelia-Kola				
						Siberian			
							Ukrainian		
								Baltica	
Other shield areas									
		East Antarctica							
					North China		North China		
Kenorland/Superia					*Columbia/Nuna*				*Rodinia*

Full characterization of the LIP records of the Indian cratons (both their dyke swarms and associated intrusions) would help to constrain the timing and pattern of assembly of the various Indian cratons and any post-assembly rotations between the various cratons. The LIP records can also be used in targeting for new metallogenic systems. It is now well established that LIPs can also provide significant input of energy and metals and contribute to the genesis of various types of mineralization (Ernst and Jowitt 2013, 2017).

6 Conclusions

Based on Google Earth™ images, field data, and available geochronological ages, 24 distinct Neoarchean-Mesoproterozoic mafic dyke swarms have been identified from different Archean cratons of the Indian shield. These dyke swarms range in ages from ~2.80 Ga to ~1.10 Ga.

- In the Dharwar craton (consisting of eastern Dharwar craton, western Dharwar craton and North Granulite Terrain), these include eight swarms: (i) NE–SW to ESE–WNW trending ~2.37 Ga Bangalore-Karimnagar swarm, (ii) N–S to NNE–SSW trending ~2.25–2.26 Ga Ippaguda-Dhiburahalli swarm, (iii) N–S to NNW–SSE trending ~2.22 Ga Kandlamadugu swarm, (iv) NW–SE to WNW–ESE trending ~2.21 Ga Anantapur-Kunigal swarm, (v) NW–SE to WNW–ESE trending ~2.18 Ga Mahbubnagar-Dandeli swarm, (vi) N–S, NW–SE and ENE–WSW trending ~2.08 Ga Devarabanda swarm, (vii) E–W to NW–SE trending ~1.88–1.89 Ga Hampi swarm, and (viii) NW–SE ~1.79 Ga Pebbair swarm.
- In the Bastar craton, these include five swarms: (i) WNW–ESE to NW–SE trending ~2.7 Ga Keshkal swarm, (ii) NW–SE to NE–SW trending ~2.4–2.5 Ga Dantewara swarm, (iii) NW–SE trending ~1.88–1.89 Ga Bastanar swarm (formerly BD2), (iv) N–S trending ~1.44–1.46 Ga Lakhna swarm, and (v) ENE–WSW trending ~1.42 Ga Bandalimal swarm.
- In the Singhbhum Granite Complex (a major southern part of the Singhbhum craton), these include seven swarms: (i) NE–SW trending ~2.80 Ga Keshargaria swarm, (ii) NNE–SSW to NE–SW trending ~2.75–2.76 Ga Ghatgaon swarm, (iii) NE–SW to ENE–WSW trending ~2.26 Ma Kaptipada swarm, (iv) WNW–ESE trending ~1.77 Ga Pipilia swarm, (v) E–W to ENE–WSW trending early-Paleoproterozoic Keonjhar swarm, (vi) NW–SE to NNW–SSE trending middle-Paleoproterozoic Bhagamunda swarm, and (vii) N–S to NNE–SSW trending late-Paleoproterzoic Barigaon swarm.
- In the Bundelkhand craton, these include two swarms: (i) NW–SE trending ~1–98 Ga Jhansi swarm, and (ii) ENE–WSW trending ~1.11 Ga Mahoba swarm.
- In the Aravalli craton, these includes two swarms: (i) NW–SE trending ~2.8 Ga Jagat swarm, and (ii) NE–SW trending ~2.2 Ga Amet swarm. An additional N–S swarm is indicated in the literature but not visible on Google Earth images and was not mapped.

- Overall, we have identified 14 distinct mafic magmatic events in the Indian shield during the Neoarchean-Mesoproterozoic. These comprise ~2.80, ~2.75–2.76, ~2.37, ~2.25–2.26, ~2.22, ~2.21, ~2.18, ~2.08, ~1.98, ~1.88–1.90, ~1.77–1.79, ~1.44–1.46, ~1.42, and ~1.10–1.12 Ga dyke swarms. All of these swarms represent the exposed plumbing system for large igneous provinces (LIPs).
- This expanded record of dated LIPs/dyke swarms from the Indian shield have been correlated with matching LIPs on other crustal blocks and clearly indicate that the Indian shield was an integral part of all the known supercontinents during Neoarchean-Mesoproterozoic. These data along with new paleomagnetic studies will provide robust constraints for the setting of the Indian cratons within these supercontinents.

Acknowledgements This work is part of a number of projects sanctioned to RKS and he is thankfully acknowledged the Department of Science and Technology, Government of India, New Delhi (Research Scheme number SR/S4/ES-590/2011), the Ministry of Earth Sciences, Government of India, New Delhi (Research Scheme number MoES/16/10/11-RDEAS), and the Council of Scientific and Industrial Research, New Delhi (Research Scheme number 24 (0348)/17/EMR-II) for financial supports. REE was partially supported from Russian Mega-Grant 14.Y26.31.0012. RKS and AKS are also thankful to the Head of the Department of Geology, Banaras Hindu University, for extending all necessary facilities developed with DST-PURSE grant (Scheme 5050) and UGC-CAS-II grant (Scheme 5055) during this work. The authors would like to thank Keneth L. Buchan and K. R. Hari, the reviewers of the MS, and the handling Editor Peng Peng for extremely constructive reviews which have improved the MS significantly.

References

Acharyya SK, Gupta A, Orihashi Y (2010) New U-Pb zircon ages from Paleo-Mesoarchean TTG gneisses of the Singhbhum Craton, eastern India. Geochem J 44:81–88

Ahmad T, Mukherjee PK, Trivedi JR (1999) Geochemistry of Precambrian mafic magmatic rocks of the Western Himalaya, India: petrogenetic and tectonic implications. Chem Geol 160:103–119

Ahmad T, Deb M, Raza M (2008) Proterozoic mafic volcanism in the Aravalli-Delhi orogen, northwestern India: geochemistry and tectonic framework. J Geol Soc India 72:93–111

Ahmad I, Mondal MEA, Bhutani R, Satyanarayanan M (2017) Geochemical evolution of the Mangalwar Complex, Aravalli Craton, NW India: Insights from elemental and Nd-isotope geochemistry of the basement gneisses. Geosci Front. https://doi.org/10.1016/j.gsf.2017.07.003. (in press)

Amara M, Benmammar A, Ouadahi S, Ernst RE, Bendaoud A, Jessell M, Djemaï S, Hamoudi M (2016) Mapping the Dyke Swarms of the Eglab-Yetti Region, Southwestern Algeria. Acta Geol Sin (Engl Ed) 90(supp. 1):51

Amelin YV, Heaman LM, Verchogliad VM, Skobelev VM (1994) Geochronological constraints on the emplacement history of an anorthosite-rapakivi granite suite: U-Pb zircon and baddeleyite study of the Korosten complex, Ukraine. Contrib Mineral Petrol 116:411–419

Anand M, Gibson SA, Subbarao KV, Kelly SP, Dickin AP (2003) Early Proterozoic melt generation processes beneath the intracratonic Cuddapah Basin, southern India. J Petrol 44:2139–2171

Ayer J, Amelin Y, Corfu F, Kamo S, Ketchum J, Kwok K, Trowell N (2002) Evolution of the southern Abitibi greenstone belt based on U-Pb geochronology: autochthonous volcanic constructions followed by plutonism, regional deformation and sedimentation. Precambrian Res 115:63–95

Barnes S-J, Van Kranendonk MJ, Sonntag I (2012) Geochemistry and tectonic setting of basalts from the Eastern Goldfields Superterrane. Aust J Earth Sci 59:707–735

Basu AK (1986) Geology of parts of Bundelkhand and Granite Massif Central India. Geol Surv India Rec 117:61–124

Basu AR, Sharma M, Premo WR (1993) U–Pb age of an older metamorphic group micaschist: earliest terrain of the eastern Indian craton. Recent Res Geol Geophys Precambrians. Recent Res Geol 16:93–102. (In: Saha AK (ed))

Belica ME, Piispa EJ, Meert JG, Pesonen LJ, Plado J, Pandit MK, Kamenov GD, Celestino M (2014) Paleoproterozoic mafic dyke swarms from the Dharwar craton; paleomagnetic poles for India from 2.37 to 1.88 Ga and rethinking the Columbia supercontinent. Precambrian Res 244:100–122

Bleeker W (2003) The Late Archean record: a puzzle in ca. 35 pieces. Lithos 71:99–134

Bleeker W, Ernst RE (2006) Short-lived mantle generated magmatic events and their dyke swarms: the key unlocking Earth's paleogeographic record back to 2.6 Ga. In: Hanski E, Mertanen S, Rämö T, Vuollo J (eds) Dyke Swarms—time markers of crustal evolution: London. Francis, Taylor, pp 3–26

Bleeker W, Hall B (2007) The Slave craton: geology and metallogenic evolution. In: Goodfellow WD (ed) Mineral deposits of Canada: a synthesis of major deposit types, district metallogeny, the evolution of geological provinces, and exploration methods, vol 5. Geological Association of Canada, Mineral Deposits Division, Special Publication, pp 849–879

Bogdanova SV, Gintov OB, Kurlovich D, Lubnina NV, Nilsson M, Orlyuk MI, Pashkevich IK, Shumlyanskyy LV, Starostenko VI (2013) Late Palaeoproterozoic mafic dyking in the Ukrainian Shield (Volgo-Sarmatia) caused by rotations during the assembly of supercontinent Columbia. Lithos 174:196–216

Bose MK (2008) Petrology and geochemistry of Proterozoic 'Newer Dolerite' and associated ultramafic dykes within Singhbhum granite pluton, eastern India. In: Srivastava RK, Sivaji C, Chalapathi Rao NV (eds) Indian Dykes: geochemistry, geophysics and geochronology. Narosa Publishing House Pvt. Ltd., New Delhi, pp 413–445

Bose MK (2009) Precambrian mafic magmatism in the Singhbhum Craton, Eastern India. J Geol Soc India 73:13–35

Bose MK, Chakraborti MK, Saunders AD (1989) Geochemistry of the lavas from Proterozoic Dalma volcanic belt, Singhbhum, eastern India. Geol Rundchau 70:504–518

Bright RM, Amato JM, Denyszyn SW, Ernst RE (2014) U-Pb geochronology of 1.1 Ga diabase in the southwestern United States: testing models for the origin of a post-Grenville Large Igneous Province. Lithosphere 6:135–156

Bryan SE, Ernst RE (2008) Revised definition of Large Igneous Provinces (LIPs). Earth Sci Rev 86:175–202

Bryan SE, Ferrari L (2013) Large Igneous Provinces and silicic large igneous provinces: progress in our understanding over the last 25 year. Geol Soc Am Bull 125:1053–1078

Buchan KL (2013) Key paleomagnetic poles and their use in Proterozoic continent and supercontinent reconstructions: a review. Precambrian Res 238:93–110

Buchan KL, Ernst RE (2004) Dyke swarms and related units in Canada and adjacent regions. Geological Survey of Canada Map 2022A (scale 1:5,000,000) and accompanying booklet

Buchan KL, Ernst RE (2013) Diabase dyke swarms of Nunavut, Northwest Territories, and Yukon, Canada. Geological Survey of Canada, Open File 7464

Buchan KL, Ernst RE (2018a) Giant circumferential dyke swarms: catalogue and characteristics. In: Srivastava RK, Ernst RE, Peng P (eds) Dyke Swarms of the world—a modern perspective. Springer (this volume)

Buchan KL, Ernst RE (2018b) A giant circumferential dyke swarm associated with the High Arctic Large Igneous Province (HALIP). Gond Res. https://doi.org/10.1016/j.gr.2018.02.006. (in press)

Buchan KL, Mortensen JK, Card KD, Percival JA (1998) Palaeomagnetism and U-Pb geochronology of diabase dyke swarms of Minto Block Superior Province, Quebec, Canada. Can J Earth Sci 35:1054–1069

Buchan KL, Ernst RE, Bleeker W, Davis WJ, Villeneuve M, van Breemen O, Hamilton M, Söderlund U (2010) Proterozoic magmatic events of the Slave Craton, Wopmay Orogen and Environs. Geological Survey of Canada Open File 5985

Bullen DS, Hall RP, Hanson RE (2012) Geochemistry and petrogenesis of mafic sills in the 1.1 Ga Umkondo large igneous province, southern Africa. Lithos 142–143:116–129

Chakraborty PP, Pant NC, Paul PP (2016) Control on sedimentation in Indian Paleoproterozoic basins: clues from the Gwalior and Bijawar basins, central India. In: Mazumder R, Eriksson PG (eds) Precambrian Basins of India: stratigraphic and tectonic context, vol 43. Geological Society of London, Memoirs, pp 67–83

Chalapathi Rao NV, Srivastava RK (2009) A new find of boninite dyke from the Palaeoproterozoic Dongargarh Super group: Inference for a fossil subduction zone in the Archaean of the Bastar craton, central India: Neues Jahrbuch für Mineralogie—Abhandlungen, vol 186, pp 271–282

Chalapathi Rao NV, Miller JA, Gibson SA, Pyle DM, Madhavan V (1999) Precise $^{40}Ar/^{39}Ar$ dating of Kotakonda kimberlite andChelima lamproite, India: implication to the timing of mafic dykes warm activity in the Eastern Dhawar craton. J Geol Soc India 53:425–432

Chalapathi Rao NV, Wu FY, Mitchell RH, Li LQ, Lehmann B (2013) Mesoproterozoic U-Pb ages, trace element and Sr–Nd isotopic composition of perovskite from kimberlites of the Eastern Dharwar craton, southern India: Distinct mantle sources and a widespread 1.1 Ga tectonomagmaticevent. Chem Geol 353:48–64

Chalapathi Rao NV, Lehmann B, Panwar B, Kumar A, Mainkar D (2014) Petrogenesis of the crater-facies Tokapalkimberlite pipe, Bastar craton, Central India. Geosci Front 5:81–790

Chalapathi Rao NV, Atiullah, Burgess R, Nanda P, Choudhary AK, Sahoo S, Lehmann B, Chahong N (2016) Petrology, $^{40}Ar/^{39}Ar$ age, Sr-Nd isotope systematics, and geodynamic significance of an ultrapotassic (lamproitic) dyke with affinities to kamafugite from the eastern-most margin of the Bastar Craton, India. Miner Pet 110:269–293

Chardon D, Jayananda M, Chetty TRK, Peucat J-J (2008) Precambrian continental strain and shear zone patterns: South Indian case. J Geophys Res 113:B08402. https://doi.org/10.1029/2007JB005299

Chatterjee N, Bhattacharji S (2001) Origin of the felsic dykes and basaltic dykes and flow in the Rajula-Palitana-Sihor area of the Deccan Traps, Saurashtra, India: a geochemical and geochronological study. Int Geol Rev 43:1094–1116

Chaudhuri AK, Saha D, Deb GK, Deb SP, Mukherjee MK, Ghosh G (2002) The Purana basins of southern cratonic province of India – a case for Mesoproterozoic fossil rifts. Gondwana Res 5:23–33

Chaudhuri T, Mazumder R, Arima M (2015) Petrography and geochemistry of Mesoarchaean komatiites from the eastern iron ore belt, Singhbhum craton, India, and its similarity with Barberton type komatiite. J Afr Earth Sci 101:135–147

Chetty TRK, Bhaskar Rao YJ (2006) The Cauvery shear zone, southern granulite terrain, India: a crustal-scale flower structure. Gondwana Res 10:77–85

Clark C, Collins AS, Timms NE, Kinny PD, Chetty TRK, Santosh M (2009) SHRIMP U-Pb age constraints on magmatism and high-grade metamorphism in the Salem Block, southern India. Gondwana Res 16:27–36

Condie KC (1989) Plate tectonics and crustal evolution, 3rd edn. Pergamon Press, New York, p 488

Crookshank H (1963) Geology of Southern Bastar and Jeypore from the Bailadila range to Eastern Ghats. Mem Geol Sur India 87:150

Cruden A, Vollgger S, Dering G, Micklethwaite S (2016) High spatial resolution mapping of dykes using unmanned aerial vehicle (UAV) photogrammetry: new insights on emplacement processes. Acta Geol Sin (Engl Ed) 90(supp. 1):52–53

Das P, Das K, Chakraborty PP, Balakrishnan S (2011) 1420 Ma diabasic intrusives from the Mesoproterozoic Singhora Group, Chhattisgarh Supergroup, India: implications towards non-plume intrusive activity. J Earth Syst Sci 120:223–236

Dash JK, Pradhan SK, Bhutani R, Balakrishnan S, Chandrasekaran G, Basavaiah N (2013) Paleomagnetism of ca. 2.3 Ga mafic dyke swarms in the northeastern Southern Granulite Terrain, India: constraints on the position and extent of Dharwar craton in the Paleoproterozoic. Precambrian Res 228:164–176

Davies JHFL, Heaman LM (2014) New U-Pb baddeleyite and zircon ages for the Scourie dyke swarm: a ling-lived large igneous province with implications for the Paleoproterozoic evolution of NW Scotland. Precambrian Res 249:180–198
de Kock MO, Evans DAD, Beukes NJ (2009) Validating the existence of Vaalbara in the Neoarchean. Precambr Res 174:145–154
de Kock MO, Ernst RE, Söderlund U, Jourdan F, Hofmann A, Gall BL, Bertrand H, Chisonga BC, Beukes N, Rajesh HM, Moseki LM, Fuchs R (2014) Dykes of the 1.11 Ga Umkondo LIP, Southern Africa: clues to a complex plumbing system. Precam Res 249:129–143
Demirer K (2012) U–Pb Baddeleyite Ages from Mafic Dyke Swarms in Dharwar Craton, India-Links to an Ancient Supercontinent. Dissertations in Geology at Lund University (Master's thesis), 308 p
Dering G, Micklethwaite S, Barnes SJ, Fiorentini M, Cruden A, Tohver E (2016) An elevated perspective: dyke-related fracture networks analysed with Uav photogrammetry. Acta Geol Sin (Engl Ed) 90(supp. 1):54–55
Devaraju TC (1995) Dyke Swarms of Peninsular India, vol 33. Geological Society of London, Memoirs, 451 p
Devaraju TC, Alapieti TT, Sudhakara, Kaukonen R (2008) Calc-Alkaline Mafic Dykes Swarms of Volcanic Arc, Ocean Floor and N-MOR Basalt Affinity with Features of Destructive Plate Margin Emplacement in the Northern Segment of Western Dharwar Craton. In: Srivastava RK, Sivaji C, Chalapathi Rao NV (eds) Indian dyke: geochemistry, geophysics and geochronology. Narosa Publishing House Pvt. Ltd., New Delhi, pp 215–237
Dey S, Topno A, Liu Y, Zong K (2017) Generation and evolution of Palaeoarchaean continental crust in the central part of the Singhbhum craton, eastern India. Precambrian Res 298:268–291
Divakara Rao V, Narayana BL, Rama Rao P, Murthy NN, Subba Rao MV, Rao JM, Reddy GLN (2000) Precambrian acid volcanism in central India—geochemistry and origin. Gondwana Res 3:215–226
Drury SA (1984) Proterozoic dyke swarms and thermal evolution in south India. J Geol Soc India 25:437–444
Drury SA, Harris NB, Holt RW, Reeves-Smith GJ, Wightman RT (1984) Precambrian tectonics and crustal evolution in South India. J Geol 92:3–20
Dunn JD, Dey AK (1942) The geology and petrology of Eastern Singhbhum and surrounding areas. Mem Geol Surv 69:281–456
Ernst RE (2014) Large igneous provinces. Cambridge University Press, 653 p
Ernst RE, Bell K (2010) Large Igneous Provinces (LIPs) and carbonatites. Miner Pet 98:55–76
Ernst RE, Bleeker W (2010) Large Igneous Provinces (LIPs), giant dyke swarms, and mantle plumes: Significance for breakup events within Canada and adjacent regions from 2.5 Ga to the Present. Can J Earth Sci 47:695–739
Ernst RE, Buchan KL (2001a) The use of mafic dyke swarms in identifying and locating mantle plumes. In: Ernst RE, Buchan KL (eds) Mantle plumes: their identification through time, vol 352. Geological Society of America Special Paper, pp 247–265
Ernst RE, Buchan KL (2001b) Large mafic magmatic events through time and links to mantle-plume heads. In: Ernst RE, Buchan KL (eds) Mantle plumes: their identification through time, vol 352. Geological Society of America Special Paper, pp 483–566
Ernst RE, Buchan KL (2016) Guidelines for preparing comprehensive regional mafic dyke swarm maps. Acta Geol Sin (Engl Ed) 90(supp. 1):20–21
Ernst RE, Jowitt SM (2013) Large Igneous Provinces (LIPs) and metallogeny. In: Colpron M, Bissig T, Rusk BG, Thompson JFH (eds) Tectonics, metallogeny, and discovery. The North American Cordillera and similar accretionary settings, vol 17. Society of Economic Geologists Special Publication, pp 17–51
Ernst RE, Jowitt SM (2017) Multi-commodity, multi-scale exploration targeting using the Large Igneous Province record. Geological Survey of Western Australia Record 2017/6, pp 41–44
Ernst RE, Srivastava RK (2008) India's place in the Proterozoic world: constraints from the Large Igneous Province (LIP) record. In: Srivastava RK, Sivaji C, Chalapathi Rao NV (eds) Indian

Dyke: geochemistry, geophysics and geochronology. Narosa Publ. House Pvt. Ltd., New Delhi, pp 413–445
Ernst RE, Youbi N (2017) How Large Igneous Provinces affect global climate, sometimes cause mass extinctions, and represent natural markers in the geological record. Palaeogeogr Palaeoclim Palaeoecol 478:30–52
Ernst RE, Head JW, Parfitt E, Grosfils EB, Wilson L (1995) Giant radiating dyke swarms on Earth and Venus. Earth Sci Rev 39:1–58
Ernst RE, Buchan KL (1997) Giant radiating dyke swarms: their use in identifying pre-Mesozoic large igneous provinces and mantle plumes. Geophysical Monogr Am Geophys Union 100:297–334
Ernst RE, Wingate MTD, Buchan KL, Li Z (2008) Global record of 1600–700 MaLarge Igneous Province (LIPs): implications for the reconstruction of the pro-posed Nuna (Columbia) and Rodinia supercontinents. Precambrian Res 160:159–178
Ernst RE, Srivastava RK, Bleeker W, Hamilton M (2010) Precambrian Large Igneous Provinces (LIPs) and their dyke swarms: new insights from high-precision geochronology integrated with paleomagnetism and geochemistry. Precambrian Res 183:vi–xi
Ernst RE, Bleeker W, Söderlund U, Kerr AC (2013) Large Igneous Provinces and supercontinents: toward completing the plate tectonic revolution. Lithos 174:1–14
Ernst RE, Buchan KL, Botsyun S (2016) Map of mafic dyke swarms and related units of russia and adjacent regions. Acta Geol Sin (Engl Ed) 90(supp. 1):22–23
Evans DAD (2013) Reconstructing pre-Pangean supercontinents. Geol Soc Am Bull 125:1735–1751
Fermor LL (1936) An attempt at the correlation of the ancient schistose formations of Peninsular India, vol 70, pt 2, no 1. Memoirs of the Geological Survey of India
French JE, Heaman LM (2010) Precise U-Pb dating of Paleoproterozoic mafic dyke swarms of the Dharwar craton, India: implications for the existence of the Neoarchean supercraton Sclavia. Precambrian Res 183:416–441
French JE, Heaman LM, Chacko T, Rivard B (2004) Global mafic magmatism and continental breakup at 2.2 Ga: evidence from the Dharwar craton, India. Geol Soc Am Abstr Program 36(5):340
French JE, Heaman LM, Chacko T, Srivastava RK (2008) 1891–1883 Ma Southern Bastar Cuddapah mafic igneous events, India: a newly recognized large igneous province. Precambrian Res 160:308–322
Ghosh JG, de Wit MJ, Zartman RE (2004) Age and tectonic evolution of Neoproterozoic ductile shear zones in the Southern Granulite Terrain of India, with implications for Gondwana studies. Tectonics 23(3):TC3006. https://doi.org/10.1029/2002TC001444
Gopalan K, Macdaugall JD, Roy AB, Murali AV (1990) Sm-Nd evidence for 3.3 Ga old rocks in Rajasthan, northwestern India. Precambrian Res 48:287–297
Goswami JN, Misra S, Wiedenbeck M, Ray SL, Saha AK (1995) 3.55 Ga old zircon from Singhbhum-Orissa iron ore craton, Eastern India. Curr Sci 69:1008–1011
Gower CF, Rivers T, Krogh TE (2002) A U-Pb geochronological review of the Proterozoic history of the eastern Grenville Province. Can J Earth Sci 39:795–829
Gregory LC, Meert JG, Tamrat E, Malone S, Pandit MK, Pradhan V (2006) A paleomagnetic and geochronologic study of the Majhgawan kimberlite, India: implications for the age of the Upper Vindhyan supergroup. Precambrian Res 149:69–75
Gumsley AP, Chamberlain KR, Bleeker W, Söderlund U, de Kock MO, Larsson ER, Bekker A (2017) Timing and tempo of the Great Oxidation Event. Proc Natl Acad Sci USA 114:1811–1816
Gupta SN, Arora YK, Mathur RK, Iqbaluddin, Prasad B, Sahai TN, Sharma SB (1980) Lithostratigraphic map of Aravalli region. Geological Survey of India (Hyderabad)
Halls HC (1982) The importance and potential of mafic dyke swarms in studies of geodynamic processes. Geosci Can 9:145–154
Halls HC, Li J-H, Davis D, Hou G-T, Zhang B-X, Qian X-L (2000) A precisely dated Proterozoic paleomagnetic pole from the North China Craton, and its relevance to paleocontinental construction. Geophys J Int 143:185–203

Halls HC, Campal N, Davis DW, Bossi J (2001) Magnetic studies and U-Pb geochronology of the Uruguayan dyke swarm, Rio de la Plata craton, Uruguay: paleomagnetic and economic implications. J S Amn Earth Sci 14:349–361

Halls HC, Kumar A, Srinivasan R, Hamilton MA (2007) Paleomagnetism and U-Pb geochronology of eastern trending dykes in the Dharwar craton, India: feldspar clouding, radiating dyke swarms and the position of India at 2.37 Ga. Precambrian Res 155:47–68

Hamilton MA, Sadowski GR, Teixeira W, Ernst RE, Ruiz AS (2012) Precise, matching U–Pb ages for the Rincon del Tigre mafic layered intrusion and Huanchaca gabbro sill, Bolivia: evidence for a late Mesoproterozoic LIP in SW Amazonia? GAC MAC Joint Annual Meeting, St. John's 2012 Geoscience at the Edge, vol 35

Hamimi Z, Zoheir B, Hasan SM, Ernst RE (2016) Mapping the Dyke Swarms of the Eastern Desert, Egypt. Acta Geol Sin (Engl Ed) 90(supp. 1):28

Heaman LM, Easton RM, Hart TR, Hollings P, MacDonald CA, Smyk MC (2007) Further refinement to the timing of Mesoproterozoic magmatism, Lake Nipigon region, Ontario. Can J Earth Sci 44:1055–1086

Heron AM (1953) The geology of central Rajputana. Mem Geol Surv India 79:1–389

Hughes HSR, McDonald I, Goodenough KM, Ciborowski TJR, Kerr AC, Davies JHFL, Selby D (2014) Enriched lithospheric mantle keel below the Scottish margin of the North Atlantic Craton: evidence from the Paleoproterozoic Scourie dyke swarm and mantle xenoliths. Precambrian Res 250:97–126

Huhma H, Hanski E, Kontinen A, Vuollo J, Mänttäri I, Lahaye Y (2018) Sm-Nd and U-Pb isotope geochemistry of the Palaeoproterozoic mafic magmatism in eastern and northern Finland, vol 405. Geological Survey of Finland, Bulletin, 150 p

Humbert F, Sonnette L, de Kock MO, Robion P, Horng CS, Cousture A, Wabo H (2017) Palaeomagnetism of the early Palaeoproterozoic, volcanic Hekpoort Formation (Transvaal Supergroup) of the Kaapvaal craton, South Africa. Geophys J Int 209:842–865

Ivanic TJ, Wingate MTD, Kirkland CL, van Kranendonk MJ, Wyche S (2010) Age and significance of voluminous mafic—ultramafic magmatic events in the Murchison Domain, Yilgarn Craton. Aust J Earth Sci 57:597–614

Jayananda M, Chardon D, Peucat J-J, Capdevila R, Martin H (2006) 2.61 Ga potassic granites and crustal reworking, western Dharwar craton (India): tectonic, geochronologic and geochemical constraints. Precambrian Res 150:1–26

Jayananda M, Mahesha N, Srivastava RK, Mahabaleshwar B, Blais S (2008) Petrology and geochemistry of Paleoproterozoic high-magnesian norite and dolerite dyke swarms from the Halagur-Satnur areas, eastern Darwar craton, southern India. In: Srivastava RK, Sivaji C, Chalapathi Rao NV (eds) Indian dyke: geochemistry, geophysics and geochronology. Narosa Publ. House Pvt. Ltd., New Delhi, pp 239–260

Jayananda M, Peucat J-J, Chardon D, Krishna Rao B, Corfu F (2013a) Neoarchean greenstone volcanism, Dharwar craton, Southern India: Constraints from SIMS zircon geochronology and Nd isotopes. Precambrian Res 227:55–76

Jayananda M, Tsutsumi Y, Miyazaki T, Gireesh RV, Kapfo K-U, Tushipokla, Hidaka H, Kano T (2013b) Geochronologic constraints on Meso and Neoarchean regional metamorphism and magmatism in the Dharwar craton, southern India. J Asian Earth Sci 78:18–38

Jessell MW, Santoul J, Baratoux L, Youbi N, Ernst RE, Metelka V, Miller J, Perrouty S (2015) An updated map of West African mafic dykes. J Afr Earth Sci 112:440–450

Kastek N, Ernst RE, Baragar WRA, Söderlund U, Kamo SL, Bleeker W, Sylvester P (2016) U-Pb Geochronology and geochemistry of the povungnituk group of the cape smith belt: a part of a craton-scale circa 2.0 Ga large igneous province (LIP), northern Superior craton. GAC-MAC Annual Meeting 2016 June 2, Whitehorse, Yukon

Killian TM, Bleeker W, Chamberlain K, Evans DAD, Cousens B (2016) Palaeomagnetism, geochronology and geochemistry of the Palaeoproterozoic Rabbit Creek and Powder River dyke swarms: implications for Wyoming in supercraton Superia. In: Li ZX, Evans DAD, Murphy J

(eds) Supercontinent cycles through earth history, vol 424. Geological Society London Special Publications, pp 15–45
Kullerud K, Skjerlie KP, Corfu F, de la Rosa J (2006) The 2.40 Ringvassøy mafic dykes, West Troms Basement Complex, Norway: the concluding act of early Palaeoproterozoic continental breakup. Precambrian Res 150:183–200
Kumar A, Ahmad T (2007) Geochemistry of the mafic dykes in parts of Chotanagpur gneissic complex: petrogenetic and tectonic implications. Geochem J 41:173–186
Kumar A, Hamilton MA, Halls HC (2012a) A Paleoproterozoic giant radiating dyke swarm in the Dharwar Craton, southern India. Geochem Geophys Geosystem 7:Q02011. https://doi.org/10.1029/2011GC003926
Kumar A, Nagaraju E, Besse J, Bhaskar Rao YJ (2012b) New age, geochemical and paleomagnetic data on a 2.21 Ga dyke swarm from south India: Constraints on Paleoproterozoic reconstruction. Precambrian Res 220–221:123–138
Kumar A, Nagaraju E, Srinivasa Sarma D, Davis DW (2014) Precise Pb baddeleyite geochronology by the thermal extraction-thermal ionization mass spectrometry method. Chem Geol 372:72–79
Kumar A, Parashuramulu V, Nagaraju E (2015) A 2082 Ma radiating dyke swarm in the Eastern Dharwar Craton, southern India and its implications to Cuddapah basin formation. Precambrian Res 266:490–505
Kumar A, Parashuramulu V, Shankar R, Besse J (2017) Evidence for a Neoarchean LIP in the Singhbhum craton, eastern India: implications to Vaalbara supercontinent. Precambrian Res 292:163–174
Li ZX, Bogdanova SV, Collins AS, Davidson A, De Waele B, Ernst RE, Fitzsi-mons ICW, Fuck RA, Gladkochub DP, Jacobs J, Karlstrom KE, Lu S, Natapov LM, Pease V, Pisarevsky SA, Thrane K, Vernikovsky V (2008) Assembly, configuration, and break-up history of Rodinia: a synthesis. Precambrian Res 160:179–210
Mahadevan TM (2002) Geology of Bihar and Jharkhand. GSI Publications, 2(1)
Mahadevan TM (2008) Precambrian geological and structural features of the Indian Peninsula. J Geol Soc India 72:35–55
Mallik AK, Sarkar A (1994) Geochronology and geochemistry of mafic dykes from the Precambrians of Keonjhar, Orissa. Indian Miner 48:13–24
Mandal N, Mitra AK, Misra S, Chakraborty C (2006) Is the outcrop topology of dolerite dykes of the Precambrian Singhbhum Craton fractal? J Earth Syst Sci 115:643–660
Manyeruke TD, Blenkinsop TG, Buchholz P, Love D, Oberthür T, Vetter UK, Davis DW (2004) The age and petrology of the Chimbadzi Hill Intrusion, NW Zimbabwe: first evidence for early Paleoproterozoic magmatism in Zimbabwe. J Afr Earth Sci 40:281–292
Martin D McB, Morris PA (2010) Tectonic setting and regional implications of ca. 2.2 Ga mafic magmatism in the southern Hamersley Province, Western Australia. Aust J Earth Sci 57:911–931
Mealin CA (2006) Geological investigations in the Bird River Sill, southeastern Manitoba (part of NTS 52L5): geology and preliminary geochemical results. Report of Activities 2006, Manitoba Science, Technology, Energy and Mines, Manitoba Geological Survey, pp 214–225
Meert JG (2012) What's in a name? The Columbia (Paleopangaea/Nuna) supercontinent. Gondwana Res 21:987–993
Meert JG, Pandit MK, Pradhan VR, Banks JC, Sirianni R, Stroud M, Newstead B, Gifford J (2010) The Precambrian tectonic evolution of India: a 3.0 billion year odyssey. J Asian Earth Sci 39:483–515
Miller JD, Nicholson SW (2013) Geology and mineral deposits of the Mid-continent rift—an overview. In: Miller JD (ed) Field guide to the Cu-Ni-PGE deposits of the Lake Superior Region, Precambrian Research Center Guidebook Series 13–1. University of Minnesota Press, Duluth
Miller C, Klotzli U, Fran W, Thoni M, Grasemann B (2000) Proterozoic crustal evolution in the NW Himalaya (India) as recorded by circa 1.80 Ga mafic and 1.84 Ga granitic magmatism. Precambrian Res 103:191–206
Misra S (2006) Geochronological constraints on evolution of Singhbhum mobile belt and associated basic volcanics of eastern Indian Shield. Gondwana Res 9:543–544

Misra S, Johnson PT (2005) Geochronological constraints on evolution of Singhb-hum mobile belt and associated basic volcanics of eastern Indian shield. Gondwana Res 8:129–142

Misra S, Deomurari MP, Wiedenbeck M, Goswami JN, Ray S, Saha AK (1999) ^{207}Pb/^{206}Pb zircon ages and the evolution of the Singhbhum Craton, eastern India: an ion microprobe study. Precambrian Res 93:139–151

Mondal MEA, Zainuddin SM (1996) Evolution of Archean-Palaeoproterozoic Bundelkhand massif, Central India—evidence from granitoid geochemistry. Terra Nova 8:532–539

Mondal MEA, Goswami JN, Deomurari MP, Sharma KK (2002) Ion microprobe 207Pb/206Pb ages of zircons from the Bundelkhand Massif, northern India: implications for crustal evolution of the Bundelkhand-Aravalli supercontinent. Precambrian Res 117:85–100

Mondal MEA, Chandra R, Ahmad T (2008a) Precambrian mafic magmatism in Bundelkhand craton. J Geol Soc India 72:113–122

Mondal MEA, Raza M, Ahmad T (2008b) Geochemistry of the mafic dykes of the Aravalli-Bundelkhand proto-continent: implications for sub-continental lithosphere evolution of north Indian shield. In: Srivastava RK, Sivaji C, Chalapathi Rao NV (eds) Indian dyke: geochemistry, geophysics and geochronology. Narosa Publ. House Pvt. Ltd., New Delhi, pp 527–545

Moorbath S, Taylor PN (1988) Early Precambrian crustal evolution in eastern India: the ages of the Singhbhum Granite and included remnants of older gneiss. J Geol Soc India 31:82–84

Müller SG, Krapež B, Barley ME, Fletcher IR (2005) Giant iron-ore deposits of the Hamersley province related to the breakup of Paleoproterozoic Australia: new insights from in situ SHRIMP dating of baddeleyite from mafic intrusions. Geology 33:577–580

Mungall JE, Harvey JD, Balch SJ, Azar B, Atkinson J, Hamilton MA (2010) Eagle's Nest: a magmatic Ni-Cu-PGE deposit in the James Bay Lowlands, Ontario, Canada. Soc Econ Geol Spec Publ 15:539–557

Murthy NGK (1987) Mafic dyke swarms of the Indian shield. In: Halls HC, Fahriig WF (eds) Mafic Dyke Swarms, vol 34. Geological Association of Canada Special Paper, pp 393–400

Murthy NGK (1995) Proterozoic mafic dykes in southern peninsular India: a review: In: Dyke swarms of peninsular India, vol 33. Geological Society of India, Memoirs, pp 81–98

Nagaraju E, Parashuramulu V, Kumar A, Srinivas Sarma D (2018) Paleomagnetism and geochronological studies on a 450 km long 2216 Ma dyke from the Dharwar craton, southern India. Phys Earth Planet Inter 274:222–231

Nance RD, Murphy JB, Santosh M (2014) The supercontinent cycle: a retrospective essay. Gondwana Res 25:4–29

Naqvi SM, Rogers JJW (1987) Precambrian geology of India. Oxford Monographs on Geology and Geophysics No 6. Oxford University Press, New York, 233 p

Naqvi SM, Divakara Rao V, Satyanarayana K, Hussain SM (1972) Petrochemistry of dolerite dykes from Shimoga and Chitaldrug schist belts, Mysore. Geophys Res Bull (NGRI) 10:109–123

Naqvi SM, Divakara Rao V, Narain Hari (1974) Archean protocontinental growth of the Indian Shield and antiquity of its rift valleys. Precambrian Res 1:345–398

Nelson DR, Bhattacharya HR, Thern ER, Altermann W (2014) Geochemical and ion-microprobe U-Pb zircon constraints on the Archaean evolution of Singhbhum Craton, eastern India. Precambrian Res 255:412–432

Nilsson MKM, Klausen MB, Söderlund U, Ernst RE (2013) Precise U-Pb ages and geochemistry of Paleoproterozoic mafic dykes from southern West Greenland: Linking the North Atlantic and the Dharwar Cratons. Lithos 174:255–270

Ni N, Chen N, Ernst RE, Yang S, Chen J (2018) Semi-automatic extraction and mapping of dyke swarms based on multi-resolution remote sensing images: Applied to the dykes in the Kuluketage region in the northeastern Tarim Block. Precambrian Res

Norcross C, Davis DW, Spooner ET, Rust A (2000) U-Pb and Pb–Pb age constraints on Paleoproterozoic magmatism, deformation and gold mineralization in the Omai area, Guyana Shield. Precambrian Res 102:69–86

Page RW, Jackson MJ, Krassay AA (2000) Constraining sequence stratigraphy in north Australian basins: SHRIMP U-Pb zircon geochronology between Mt Isa and McArthur River. Aust J Earth Sci 47:431–461

Pandey BK, Gupta JN, Sarma KJ, Sastry CA (1997) Sm-Nd, Pb-Pb and Rb-Sr geochronology and petrogenesis of the mafic dyke swarm of Mahbubnagar, South India: implications for Paleoproterozoic crustal evolution of the Eastern Dharwar Craton. Precambrian Res 84:181–196

Pandey UK, Sastry DFVLN, Pandey BK, Roy M, Rawat TPS, Ranjan Rajeeva, Shrivastava VK (2012) Geochronological (Rb-Sr and Sm-Nd) studies on intrusive gabbros and dolerite dykes from parts of Northern and Central Indian cratons: implications for the age of onset of sedimentation in Bijawar and Chattisgarh basins and uranium mineralisation. J Geol Soc India 79:30–40

Pati JK, Patel SC, Pruseth KL, Malviya VP, Arima M, Raju S, Pati P, Prakash K (2007) Geochemistry of giant quartz veins from the Bundelkhand craton, Central India and its implications. J Earth Syst Sci 116:510–697

Pati JK, Raju S, Malviya VP, Bhushan R, Prakash K, Patel SC (2008) Mafic dykes of Bundelkhand craton, Central India: field, petrological and geochemical characteristics. In: Srivastava RK, Sivaji C, Chalapathi Rao NV (eds) Indian dyke: geochemistry, geophysics and geochronology. Narosa Publ. House Pvt. Ltd., New Delhi, pp 547–569

Peng P (2010) Reconstruction and interpretation of giant mafic dyke swarms: a case study of 1.78 Ga magmatism in the North China craton. In: Kusky TM, Zhai MG, Xiao W (eds) The evolving continents: understanding processes of continental growth, vol 338. Geological Society London Special Publications, pp 163–178

Peng P (2015) Precambrian mafic dyke swarms in the North China craton and their geological implications. Sci China Earth Sci 58:649–675

Peng P (2016) 1:2,500,000 Map of Precambrian Dyke Swarms and Related Units in North China. Acta Geol Sin (Engl Ed) 90(supp. 1):16

Peng P, Zhai MG, Zhang HF, Guo JH (2005) Geochronological constraints on the Paleoproterozoic evolution of the North China Craton: SHRIMP zircon ages of different types of mafic dykes. Int Geol Rev 47:492–508

Peng P, Zhai MG, Guo JH (2006) 1.80–1.75 Ga mafic dyke swarms in the central North China Craton implications for a plumerelated break-up event. In: Hanski E, Mertanen S, Ramö T, Vuollo J (eds) Dyke swarms—time markers of crustal evolution. Taylor & Francis, London, pp 99–112

Pichamuthu CS (1965) Regional metamorphism and charnockitization in Mysore State, India. Indian Miner 6:46–49

Piispa EJ, Smirnov AV, Pesonen LJ, Lingdevaru M, Murthu KSA, Devraju TC (2011) An integrated study of Proterozoic dykes, Dharwar craton, southern India. In: Srivastava RK (ed) Dyke swarms: keys for geodynamic interpretation. Springer, Berlin, pp 33–45

Pirajno F, Hoatson DM (2012) A review of Australia's large igneous provinces and associated mineral systems: implications for mantle dynamics through geological time. Ore Geol Rev 48:2–54

Pisarevsky SA, Biswal TK, Wang X-C, De Waele B, Ernst R, Söderlund U, Tait JA, Ratre K, Singh YK, Cleve M (2013) Palaeomagnetic, geochronological and geochemical study of Mesoproterozoic Lakhna Dykes in the Bastar Craton, India: implications for the Mesoproterozoic supercontinent. Lithos 174:125–143

Pisarevsky SA, Elming S-Å, Pesonen LJ, Li Z-X (2014) Mesoproterozoic paleogeography: supercontinent and beyond. Precambrian Res 244:207–225

Pisarevsky SA, De Waele B, Jones S, Söderlund U, Ernst RE (2015) Paleomagnetism and U-Pb age of the 2.4 Ga Erayinia mafic dykes in the south-western Yilgarn, Western Australia: paleogeographic and geodynamic implications. Precambrian Res 259:222–231

Pivarunas PF, Meert JG, Pandit MK, Sinha A (2018) Paleomagnetism and geochronology of mafic dykes from the Southern Granulite Terrane, India: expanding the Dharwar craton southward. Tectonopys. https://doi.org/10.1016/j.tecto.2018.01.024. (in press)

Prabhakar N, Bhattacharya A (2013) Paleoarchean partial convective overturn in the Singhbhum Craton, Eastern India. Precambrian Res 231:106–121

Pradhan VR, Meert JG, Pandit MK, Kamenov G, Mondal MEA (2012) Paleomagnetic and geochronological studies of the mafic dyke swarms of Bundelkhand craton, central India: implications for the tectonic evolution and paleogeographic reconstructions. Precambrian Res 198–199:51–76

Prendergast MD (2004) The Bulawayan Supergroup: a late Archean passive margin-related large igneous province in the Zimbabwe craton. J Geol Soc 161:431–445

Radhakrishana T, Krishnendu NR, Balasubramonian G (2007) Mafic magmatism around Cuddapah Basin: age constraints, petrological characteristics and geochemical inference for a possible magma chamber on the South western margin of the basin. J Geol Soc India 70:194–206

Radhakrishna T (2009) Precambrian mafic magmatism in South Indian granulite terrain. J Geol Soc India 73:131–142

Radhakrishna T, Joseph M (2012) Geochemistry and paleomagnetism of Late Cretaceous mafic dykes in Kerala, southwest coast of India in relation to large igneous provinces and mantle plumes in the Indian Ocean region. Bull Geol Soc Am 124:240–255

Radhakrishna T, Maluski H, Mitchell JG, Joseph M (1999) $^{40}Ar/^{39}Ar$ and K/Ar geochronology of the dykes from the south Indian. Tectonophy 304:109–129

Radhakrishna T, Balasubramonian G, Joseph M, Krishnendu NR (2004) Mantle processes and geodynamics: inferences from mafic dykes of south India. In: Ravindra Kumar GR, Subhash N (eds) Earth system science and natural resource management (Silver Jubilee Compendium). Centre for Earth Science Studies, Trivandrum, pp 3–25

Rajesh HM, Mukhopadhyay J, Beukes NJ, Gutzmer J, Belyanin GA, Armstrong RA (2009) Evidence for an early Archaean granite from Bastar craton, India. J Geol Soc 166:193–196

Ramachandra HM, Mishra VP, Deshmukh SS (1995) Mafic dykes in the Bastar Precambrian: study of the Bhanupratappur–Keskal mafic dyke swarm. In: Devaraju TC (ed) Mafic Dyke Swarms of Peninsular India, vol 33. Geological Society of India Memoirs, pp 183–207

Ramakrishan M (2009) Precambrian mafic magmatism in the Western Dharwar Craton, southern India. J Geol Soc India 73:101–116

Ramakrishnan M (1988) Tectonic evolution of the Archean high grade terrain of south India. J Geol Soc India 31:118–119

Ramakrishnan M (1990) Crustal development in Southern Bastar Central Indian craton. Geol Surv India Spec Publ 28:44–66

Ramakrishnan M (1993) Tectonic evolution of granulite terrains of southern India. Geol Soc India Mem 25:35–44

Ramakrishnan M, Vaidyanadhan R (2010) Geology of India. Geological Society of India, Bangalore, 994 p

Rao JM (2004) The wide-spread 2 Ga dyke activity in the Indian shield—evidences from Bundelkhand mafic dyke swarm, Central India and their tectonic implications. Gondwana Res 7:1219–1228

Rao JM, Bhattacharji S, Rao MN, Hermes OD (1995) $^{40}Ar/^{39}Ar$ ages and geochemical characteristics of dolerite dykes around the Proterozoic Cuddapah Basin, South India. In: Devaraju TC (ed) Dyke swarms of Peninsular India, vol 33. Geological Society of India, Memoirs, pp 307–328

Ratre K, De Waele B, Biswal TK, Sinha S (2010) SHRIMP geochronology for the 1450 Ma Lakhna dyke swarm: Its implication for the presence of Eoarchaean crust in the Bastar Craton and 1450–517 Ma depositional age for Purana basin (Khariar), Eastern Indian Peninsula. J Asian Earth Sci 39:565–577

Reis NJ, Teixeira W, Hamilton MA, Santos FB, Almeida ME, Filho MSD (2013) Avanavero mafic magmatism, a late Paleoproterozoic LIP in the Guiana Shield, Amazonian Craton: U-Pb ID-TIMS baddeleyite, geochemical and paleomagnetic evidence. Lithos 174:175–195

Rogers JJW (1986) Dharwar craton and the assembly of Peninsular India. J Geol 94:129–143

Rogers JJW (1996) A history of continents in the past three billion years. J Geol 104:91–107

Rogers JJW, Santosh M (2002) Configuration of Columbia, a Mesoproterozoic supercontinent. Gondwana Res 5:5–22

Rogers JJW, Santosh M (2003) Supercontinents in Earth history. Gondwana Res 6:357–368

Rogers C, Söderlund U, Ernst RE, Cousens B (2018) The 1418 Ma Highland-Purcell event of western Laurentia: distinguished from the 1460–1470 Ma Moyie LIP. (Abstr) Resources for Future Generations (http://www.rfg2018.org/) conference, Vancouver, Canada, 16–21 June 2018
Roy AB, Bhattacharya HN (2012) Tectonic and stratigraphic reappraisal constrain-ing the growth and evolution of the Singhbhum Craton, eastern India. J Geol Soc India 80:455–469
Roy AB, Kataria P, Upadhyaya R, Sharma BL (1995) Dyke rocks in the Precambrian crust of the Aravalli Mountain, Rajasthan. In: Devaraju TC (ed) Dyke swarms of Peninsular India, vol 33. Geological Society of India, Memoirs, pp 169–182
Roy A, Sarkar A, Jeyakumar S, Aggrawal SK, Ebihara M (2002) Sm–Nd age and mantle source characteristics of the Dhanjori volcanic rocks, Eastern India. Geochem J 36:503–518
Roy A, Sarkar A, Jeyakumar S, Aggarwal SK, Ebihara M, Satoh H (2004) Late Archaean mantle metasomatism below eastern Indian craton: Evidence from trace elements, REE geochemistry and Sr-Nd-O isotope systematics of ultramafic dykes, vol 113. Proceedings of the Indian Academy of Science (Earth and Planetary Sciences), pp 649–665
Saha AK (1994) Crustal evolution of Singhbhum-North Orissa, eastern India, vol 27. Geological Society of India, Memoirs, 341 p
Saha AK, Sankaran AV, Bhattacharyya TK (1973) Geochemistry of the newer dolerite suite of intrusions within the Singhbum granite—a preliminary study. J Geol Soc India 14:229–346
Sahu N, Gupta T, Patel SC, Khuntia DBK, Behra D, Pande K, Das SK (2013) Petrology of lamproites from the Nuapada lamproite field, Bastar craton, India. In: Pearson DG, Grutter HS, Harris JW, Kjarsgaard BA, O'Brien H, Chalapathi Rao NV, Sparks RSJ (eds) Proceedings of X International Kimberlite Conference on, Special Issue Journal of Geological Society of India, vol 1, 137–166
Samal AK, Srivatava RK, Sinha LK (2015) ArcGIS studies and field relationships of Paleoproterozoic mafic dyke swarms from the south of Devarakonda area, Eastern Dharwar Craton, southern India: implications for their relative ages. J Earth Syst Sci 124:1075–1084
Samal AK, Srivastava RK (2016) Does geochronology of few dykes of a swarm are true representative of all dykes of the same magmatic event?: constraints from the geochemistry and Google™ Earth Image–ArcGIS™ studies of the Paleoproterozoic mafic dyke swarms of the eastern Dharwar craton, Southern India. Acta Geol Sinica 90:2–3
Samom JD, Ahmad T, Choudhary AK (2017) Geochemical and Sm–Nd isotopic constraints on the petrogenesis and tectonic setting of the Proterozoicmafic magmatism of the GwaliorBasin, central India: the influence of Large Igneous Provinces on Proterozoic crustal evolution, vol 463. Geological Society, London, Special Publications. https://doi.org/10.1144/SP463.10
Santosh M, Maruyama S, Sato K (2009) Anatomy of a Cambrian suture in Gondwana: Pacific-type orogeny in southern India? Gondwana Res 16:321–341
Santosh M, Hari KR, He X-F, Han Y-S, Manu Prasanth MP (2018) Oldest lamproites from peninsular India track the onset of Paleoproterozoic plume-induced rifting and the birth of large igneous province. Gondwana Res 55:1–20
Sarkar A, Ghosh S, Singhai RK, Gupta SN (1997) Rb–Sr geochronology of the Dargawan sill: constraint on the age of the type Bijawar sequence of Central India. International Conference on Isotopes in Solar System, 11–14 Nov, vol 5, pp 100–101
Shankar R, Vijayagopal B, Kumar A (2014) Precise Pb-Pb baddeleyite ages of 1765 Ma for a Singhbhum 'newer dolerite' dyke swarm. Curr Sci 106:1306–1310
Sharma RS (2009) Cratons and fold belts of India. Springer, Heidelburg, p 304
Sharma KK, Rahman A (2000) The early Archaean-Paleoproterozoic crustal growth of the Bundelkhand craton northern Indian shield. In: Deb M (ed) Crustal evolution and metallogeny in the northwestern Indian Shield. Narosa Publishing House, New Delhi, pp 51–72
Sharma M, Basu AR, Ray SL (1994) Sm-Nd isotopic and geochemical study of the archean tonalite-amphibolite association from the eastern Indian craton. Contrib Miner Petrol 117(1):45–55
Shekhawat LS, Joshi DW, Pandit M (2000) Stratigraphic status of the conglomerate occurring north and northwest of Salumber in western Indian craton: implications on geology of Palaeoproterozoic Aravalli Supergroup. Gondwana Res 3:245–249

Shekhawat LS, Joshi DW, Pandit MK (2001) A relook into the status of granitoids and conglomerate in Salumber-Jaisamand area, southern Rajasthan: implications for the stratigraphy of the Palaeoproterozoic Aravalli Fold Belt. J Geol Soc India 58:53–63
Shekhawat LS, Pandit MK, Joshi DW (2007) Geology and geochemistry of palaeoproterozoic low-grade metabasic volcanic rocks from Salumber area, Aravalli Supergroup, NW India. J Earth Syst Sci 116:511–524
Shellnutt JG, Hari KR, Liao AC, Denyszyn SW, Vishwakarma N (2018) A 1.88 Ga giant radiating mafic dyke swarm across Southern India and Western Australia. Precambrian Res. https://doi.org/10.1016/j.precamres.2018.01.021. (in press)
Shumlyanskyy L, Billström K, Hawkesworth C, Elming S-Å (2012) U-Pb age and Hf isotope compositions of zircons from the north-western region of the Ukrainian shield: mantle melting in response to post-collision extension. Terra Nova 24:373–379
Shumlyanskyy L, Mitrokhin O, Billström K, Ernst R, Vishnevska E, Tsymbal S, Cuney M, Soesoo A (2015) The ca. 1.8 Ga mantle plume related magmatism of the central part of the Ukrainian shield. GFF 138:86–101
Shumlyanskyy L, Ernst RE, Söderlund U, Billström K, Mitrokhin O, Tsymbal S (2016) New U-Pb ages for mafic dykes in the Northwestern region of the Ukrainian shield: coeval tholeiitic and jotunitic magmatism. GFF 138:79–85
Singh VK, Slabunov A (2014) The Central Bundelkhand Archean greenstone complex, Bundelkhand craton, central India: geology, composition, and geochronology of supracrustal rocks. Int Geol Rev 57:1349–1364
Sinha DK, Jain SK, Naganath KP (2011) Tectonic significance and age of doleritic sill near Bandhalimal in the Singhora Protobasin of Chhattisgarh Basin, Central India. In: Srivastava RK (ed) Dyke Swarms: keys for geodynamic interpretation. Springer, Heidelberg, pp 167–187
Smirnov AV, Evans DAD, Ernst RE, Söderlund U, Li Z-X (2013) Trading partners: tectonic ancestry of Southern Africa and Western Australia, in Archean supercratons Vaalbara and Zimgarn. Precambrian Res 224:11–22
Söderlund U, Hofmann A, Klausen MB, Olsson JR, Ernst RE, Persson P-O (2010) Towards a complete magmatic barcode for the Zimbabwe craton: baddeleyite U-Pb dating of regional dolerite dyke swarms and sill provinces. Precambrian Res 183:388–398
Söderlund U, Bleeker W, Demirer K, Srivastava RK, Hamilton MA, Nilsson M, Pesonen L, Samal AK, Jayananda M, Ernst RE, Srinivas M (2018) Emplacement ages of Paleoproterozoic mafic dyke swarms in eastern Dharwar craton, India: implications for paleoreconstructions and support for a ~30° change in dyke trends from south to north. Precambrian Res (revised version submitted)
Srivastava RK (2006a) Precambrian mafic dyke swarms from the Central Indian Bastar craton: temporal evolution of the subcontinental mantle. In: Hanski E, Mertanen S, Ramo T, Vuollo J (eds) Dyke swarms: time markers of crustal evolution. Taylor & Francis, London, pp 147–159
Srivastava RK (2006b) Geochemistry and petrogenesis of Neoarchaean high-Mg low-Ti mafic igneous rocks in an intracratonic setting, central India craton: evidence for boninite magmatism. Geochem J 40:15–31
Srivastava RK (2008) Global intracratonic boninite-norite magmatism during the Neoarchean Paleoproterozoic: evidence from the central Indian Bastar craton. Int Geol Rev 50:61–74
Srivastava RK (2011) Dyke swarms: keys for geodynamic interpretation. Springer, Heidelburg, 605 p
Srivastava RK, Ahmad T (2008) Precambrian Mafic Magmatism in the Indian Shield—Part I. J Geol Soc India 72:1–140
Srivastava RK, Ahmad T (2009) Precambrian Mafic Magmatism in the Indian Shield—Part II. J Geol Soc India 73:1–152
Srivastava RK, Ernst RE (2013) Global intracratonic boninite-norite magmatism during the Neoarchean-Paleoproterozoic—Revisited. Publication Large Igneous Provinces Commission: http://www.largeigneousprovinces.org, LIP of the Month, pp 1–13
Srivastava RK, Gautam GC (2008) Precambrian Mafic Dyke Swarms from the Southern Bastar Central India Craton: present and future perspectives. In: Srivastava RK, Sivaji C, Chalapathi

Rao NV (eds) Indian dykes: geochemistry, geophysics and geochronology. Narosa Publishing House Pvt. Ltd., New Delhi, pp 367–376
Srivastava RK, Gautam GC (2009) Precambrian mafic magmatism in the Bastar craton, central India. J Geol Soc India 73:52–72
Srivastava RK, Gautam GC (2012) Early Precambrian mafic dyke swarms from the Central Archean Bastar craton, India: geochemistry, petrogenesis and tectonic implications. Geol J 47:144–160
Srivastava RK, Gautam GC (2015) Geochemistry and petrogenesis of Paleo-Mesoproterozoic mafic dyke swarms from northern Bastar craton, central India: geodynamic implications in reference to Columbia supercontinent. Gondwana Res 28:1061–1078
Srivastava RK, Samal AK (2018) Geochemical characterization, petrogenesis, and emplacement tectonics of Paleoproterozoic high-Ti and low-Ti mafic intrusive rocks from the western Arunachal Himalaya, northeastern India and their possible relation to the ~1.9 Ga LIP event of the Indian shield. Geol J 1–21. https://doi.org/10.1002/gj.3172
Srivastava RK, Singh RK (2003) Geochemistry of high-Mg mafic dikes from the Bastar craton: evidence of Late Archean boninite-like rocks in an intracratonic setting. Curr Sci 85:808–812
Srivastava RK, Singh RK (2004) Trace element geochemistry and genesis of Precambrian sub-alkaline mafic dykes from the central Indian craton: evidence for mantle metasomatism. J Asian Earth Sci 23:373–389
Srivastava RK, Hall RP, Verma R, Singh RK (1996) Contrasting Precambrian mafic dykes of the Bastar craton, central India: petrological and geochemical characteristics. J Geol Soc India 48:537–546
Srivastava RK, Singh RK, Verma SP (2004) Neoarchaean mafic volcanic rocks from the southern Bastar greenstone belt, Central India: petrological and tectonic significance. Precambrian Res 131:305–322
Srivastava RK, Sivaji C, Chalapathi Rao NV (2008) Indian dyke: geochemistry, geophysics and geochronology. Narosa Publishing House Pvt. Ltd., New Delhi, p 626
Srivastava RK, Ellam RM, Gautam GC (2009a) Sr-Nd isotope geochemistry of the early Precambrian sub-alkaline mafic igneous rocks from the southern Bastar craton, central India. Miner Pet 96:71–79
Srivastava RK, Chalapathi Rao NV, Sinha AK (2009b) Cretaceous alkalineintrusives with affinities to aillikites from the Jharia area: magmatic expression of metasomatically veined and thinned lithospheric mantle beneath the Singhbhum Craton, Eastern India. Lithos 112:407–418
Srivastava RK, Hamilton MA, Jayananda M (2011a) 2.21 Ga large igneous province in the Dharwar Craton, India. In: International symposium on Large Igneous Provinces of Asia, Mantle Plumes and Metallogeny, Irkutsk, Russia, Extended Abstract, pp 263–266
Srivastava RK, Heaman LM, French JE, Filho CFF (2011b) Evidence for a Paleoproterozoic event of metamorphism in the Bastar Craton, Central India: P–T–t constraints from mineral chemistry and U–Pb geochronology of mafic dykes. Episodes 34:13–24
Srivastava RK, Sinha AK, Kumar S (2012) Geochemical characteristics of Mesoproterozoic metabasite dykes from the Chotanagpur Gneissic Terrain, eastern India: implications for their emplacement in a plate margin tectonic environment. J Earth Syst Sci 121:509–523
Srivastava RK, Jayananda M, Gautam GC, Gireesh V, Samal AK (2014a) Geochemistry of an ENE–WSW to NE–SW trending ~2.37 Ga mafic dyke swarm of the Eastern Dharwar Craton, India: does it represent a single magmatic event? Chemie Erde – Geochem 74:251–265
Srivastava RK, Jayananda M, Gautam GC, Samal AK (2014b) ~2.21–2.22 Ga N-S to NNW–SSE trending Kunigal mafic dyke swarm from Eastern Dharwar Craton, India: implications for Paleoproterozoic large igneous provinces and supercraton Superia. Miner Pet 109:695–711
Srivastava RK, Kumar S, Sinha AK, Chalapathi Rao NV (2014c) Petrology and geochemistry of high-titanium and low-titanium mafic dykes from the Damodar valley, Chhotanagpur Gneissic Terrain, eastern India and their relation to Cretaceous mantle plume(s). J Asian Earth Sci 84:34–50
Srivastava RK, Samal AK, Gautam GC (2015) Geochemical characteristics and petrogenesis of distinct Paleoproterozoic mafic dyke swarms in space and time and associated large igneous provinces from the Eastern Dharwar Craton, India. Int Geol Rev 57:1462–1484

Srivastava RK, Söderlund U, Ernst RE, Mondal SK, Samal AK (2016a) Neoarchaean-Palaeoproterozoic Mafic Dyke Swarms from the Singhbhum Granite Complex, Singhbhum Craton, Eastern India: implications for identification of Large Igneous Provinces and their possible continuation on other formerly adjacent crustal blocks. Acta Geol Sin (Engl Ed) 90(supp. 1):17–18

Srivastava RK, Pimentel MM, Gautam GC (2016b) Nd isotope and geochemistry of an early Palaeoproterozoic high-Si high-Mg boninite–norite suite of rocks in the southern Bastar craton, central India: petrogenesis and tectonic significance. Int Geol Rev 58:1596–1615

Srivastava RK, Söderlund U, Ernst RE, Mondal SK, Samal AK (2018) Precambrian mafic dyke swarms in the Singhbhum craton (eastern India) and their links with dyke swarms of the eastern Dharwar craton (southern India). Precambrian Res. https://doi.org/10.1016/j.precamres.2018.08.001. (in press)

Stark JC, Wang X, Denyszyn SW, Li Z-X, Rasmussen B, Zi J-W, Sheppard S, Liu Y (2018) Newly identified 1.89 Ga mafic dyke swarm in the Archean Yilgarn Craton, Western Australia suggests a connection with India. Precambrian Res. https://doi.org/10.1016/j.precamres.2017.12.036. (in press)

Storey BC (1995) The role of mantle plumes in continental breakup: case histories from Gondwanaland. Nature 377:301–308

Subba Rao DV, Sridhar DN, Balaram V, Nagaraju K, Rao TG, Keshavakrishna A, Singh UP (2008) Proterozoic mafic-ultramafic dyke swarms in the vicinity of Chhattisgarh Khariar-Singhora basins in northern Bastar craton, central India. In: Srivastava RK, Sivaji C, Chalapathi Rao NV (eds) Indian dykes: geochemistry, geophysics and geochronology. Narosa Publishing House Pvt. Ltd., NewDelhi, pp 377–396

Swami Nath J, Ramakrishnan M (1981) The early Precambrian supracrustals of southern Karnataka, vol 112. Geological Survey of India, Memoirs, 350 p

Tait J, Straathof G, Söderlund U, Ernst RE, Key R, Jowitt SM, Lo K, Dahmada MEM, N'Diaye O (2013) The ahmeyim great dyke of Mauritania: a newly dated Archaean intrusion. Lithos 174:323–332

Teixeira W, D'Agrella-Filho MS, Hamilton MA, Ernst RE, Girardi VAV, Mazzucchelli M, Bettencourt JS (2013) U-Pb (ID-TIMS) baddeleyite ages and paleomagnetism of 1.79 and 1.59 Ga tholeiitic dyke swarms, and position of the Rio de la Plata Craton within the Columbia supercontinent. Lithos 174:157–174

Teixeira W, Ernst RE, Hamilton MA, Lima G, Ruiz A, Geraldes MC (2016) Widespread ca. 1.4 Ga intraplate magmatism and tectonics in a growing Amazonia. GFF 138:241–254

Teixeira W, Hamilton MA, Girardi VAV, Faleiros FM, Ernst RE (2018a) U-Pb baddeleyite ages of key dyke swarms in the Amazonian Craton (Carajás/Rio Maria and Rio Apa areas): tectonic implications for events at 1880, 1110 Ma, 535 Ma and 200 Ma. Precambrian Res. https://doi.org/10.1016/j.precamres.2018.02.008. (in press)

Teixeira W, Reis NJ, Bettencourt JS, Klein EF, Oliveira DC (2018b) Intraplate prpterozoic magmatism in the Amazonian craton reviewed: geochronology, crustal tectonics and global barcode matches. In: Srivastava RK, Ernst RE, Peng P (eds), Dyke Swarms of the World—a modern perspective, Springer (this volume)

Thorne JP, Highet LM, Cooper M, Claoue Long JC, Hoatson DM, Jaireth S, Huston DL, Gallagher RG (2014) Australian Mafic-Ultramafic Magmatic Events GIS Dataset, 1:5,000,000 scale [Digital Dataset]. Geoscience Australia, Commonwealth of Australia, Canberra; www.ga.gov.au

Upadhyay D, Chattopadhyay S, Kooijman E, Mezger K, Berndt J (2014) Magmatic and metamorphic history of Paleoarcheantonalite–trondhjemite–granodiorite (TTG) suite from the Singhbhum craton, eastern India. Precambrian Res 252:180–190

Upadhyaya R, Sharma BL Jr, Sharma BL, Roy AB (1992) Remnants of greenstone sequence from the Archaean rocks of Rajasthan. Curr Sci 63:87–92

Vuollo J, Huhma H (2005) Paleoproterozoic mafic dykes in NE Finland. In: Lehtinen M, Nurmi PA, Rämö OT (eds) Precambrian Geology of Finland—key to the evolution of the Fennoscandian Shield. Elsevier, Amsterdam, pp 195–236

Wall CJ, Scoates JS, Friedman RM, Meurer WP (2012) Age of the Stillwater Complex. Age of the Bushveld Complex. Abstract, 22nd V.M. Goldschmidt conference "Earth in Evolution", 24–29 June, Montreal, Quebec, Canada

Wang YJ, Fan WM, Zhang YH, Guo F, Zhang HF, Peng P (2004) Geochemical, ^{40}Ar-^{39}Ar geochronological and Sr–Nd isotopic constraints on the origin of Paleoproterozoic mafic dykes from the southern Taihang Mountains and implications for the ca. 1800 Ma event of the North China Craton. Precambrian Res 135:55–77

Windley BF (1984) The evolving continents, 2nd edn. Wiley, New York, p 399

Wingate MTD, Martin DM (2016) Updated digital map of mafic dyke swarms and large igneous provinces in Western Australia. Acta Geol Sin (Engl Ed) 90(supp. 1):13–14

Worsley TR, Nance RD, Moody JB (1982) Plate tectonic episodicity: a deterministic model for periodic "Pangeas", vol 65. Eos, Transactions of the American Geophysical Union, p 1104

Worsley TR, Nance RD, Moody JB (1984) Global tectonics and eustasy for the past 2 billion years. Mar Geol 58:373–400

Youbi N, Ernst RE, Söderlund U, Boumehdi MA, Bensalah MK, Aarab EM (2016) Morocco, North Africa: a Dyke Swarm Bonanza. Acta Geol Sin (Engl Ed) 90(supp. 1):15

Zegers TE, de Wit MJ, Dann J, White SH (1998) Vaalbara, Earth's oldest assembled continent? A combined structural, geochronological, and palaeomagnetic test. Terra Nova 10:250–259

Petrology and Mineral Chemistry of a Porphyritic Mafic Dyke, Jonnagiri Schist Belt, Eastern Dharwar Craton, India: Implications for Its Magmatic Origin

V. V. Sesha Sai, S. N. Mahapatro, Santanu Bhattacharjee, Tarun C. Khanna and M. M. Korakoppa

Abstract We present field, petrology and mineralogy of a porphyritic mafic dyke that traverses the granite-greenstone terrain of the Neoarchean Jonnagiri schist belt, eastern Dharwar craton, India. The undeformed porphyritic dyke is characterised by the presence of euhedral plagioclase megacrysts (0.5–3.5 cm) exhibiting primary magmatic alignment. At places, partial resorption is noticed in the plagioclase phenocrysts indicating crystal-melt interaction. The groundmass consists of andesine and titan augite. Sub-ophitic/ophitic textures are conspicuously noticed. Ilmenite, titano-magnetite, apatite and baddeleyite are the accessory phases. Exsolved ilmenite and titano-magnetite along with euhedral apatite is also observed in the porphyritic dyke. Mineral chemistry of plagioclase shows variation between $Or_{3.50–6.44\%}$, $Ab_{43.18–62.98\%}$ and $An_{30.57–52.46\%}$. Clustering of biotite, is noticed at places. EPMA analyses of plagioclase reveals the presence of normal zoning; $Ab_{45.73}$ in the core to $Ab_{49.61}$ in the rim. The predominantly andesine composition of plagioclase, position of clinopyroxene in the Ca + Na versus Ti binary mineral chemistry diagram, and the micron size euhedral baddeleyite indicate transitional nature of the Jonnagiri porphyritic

V. V. Sesha Sai (✉) · S. Bhattacharjee · M. M. Korakoppa
Geological Survey of India, Bandlaguda, Hyderabad, India
e-mail: seshu1967@gmail.com

S. Bhattacharjee
e-mail: bhattacharjeesantanu21@gmail.com

M. M. Korakoppa
e-mail: korakoppam@gmail.com

S. N. Mahapatro
Geological Survey of India, Central Region, Raipur, India
e-mail: snmahagsi@gmail.com

T. C. Khanna
CSIR-National Geophysical Research Institute, Habsiguda, Hyderabad, India
e-mail: khannangri@ngri.res.in

R. K. Srivastava et al. (eds.), *Dyke Swarms of the World: A Modern Perspective*, Springer Geology, https://doi.org/10.1007/978-981-13-1666-1_10

dyke. The clinopyroxene is compositionally a Ti-augite ($Wo_{40}En_{34}Fs_{26}$). The temperature and oxygen fugacity estimates for the coexisting magnetite-ilmenite solid solution pair yielded an equilibration temperature of ~756 °C and $10^{-15.6}$atmf_{O2} for the Jonnagiri porphyritic dyke.

Keywords Mafic dyke · Porphyritic · Magmatic origin · Jonnagiri belt
Eastern Dharwar craton · India

1 Introduction

Emplacement of dyke swarms in a shield area provide critical information on the palaeocontinental reconstruction (e.g. Hou et al. 2008; Zhao et al. 2004; Santosh 2012), and, therefore, a key for geodynamic interpretation (Srivastava 2011). The Indian shield has recorded extensive activity of mafic magmatism (e.g. Naqvi and Rogers 1987; Srivastava et al. 2008; Srivastava and Ahmed 2009; Srivastava 2011; Srivastava et al. 2015). The granite-greenstone terrain of the Dharwar craton, the Proterozoic Cuddapah basin, and the high-grade granulite belt in the southern part of Indian shield, witnessed extensive dyke and sill intrusions, particularly during Paleoproterozoic (e.g. Radhakrishna and Joseph 1996; Halls et al. 2007; Radhakrishna 2008; Jayananda et al. 2008; Srivastava 2011; Kumar et al. 2012, 2015; Khanna et al. 2013; Belica et al. 2013; Srivastava et al. 2014a, b, 2015; Sesha Sai et al. 2017). Chronologically, two distinct phases (~2370 and ~1890 Ma) of dyke swarms in the Dharwar craton, which predominantly trend in east-west direction, have been identified (Piispa et al. 2011; Srivastava et al. 2014a, c, 2015). Additionally, two other dyke swarms, emplaced at ~2.21 and ~2.18 Ga, have also recorded from the eastern Dharwar craton (Srivastava et al. 2015). Recently, Kumar et al. (2015) have reported ~2.08 Ga mafic dyke swarm in the eastern Dharwar craton (EDC). Although, plagioclase megacryst bearing dykes are pervasively noticed as large swarms in many granite-greenstone terrains in the Archean shield areas (e.g. Phinney and Morrison 1986; Morrison et al. 1988), mafic dyke(s) with plagioclase megacrysts are less known from the Dharwar craton. Plagioclase megacrysts bearing leuco-gabbronorite dyke have been reported from the Bastar craton (Dora et al. 2016).

2 Geological Setting

The Jonnagiri schist belt (JSB) is located close to the vicinity of the western margin of the Proterozoic Cuddapah basin in the EDC, Southern India. The northern part of the schist belt is of 11 km long and up to 5 km wide and exposed in sigmoid shape trending WNW–ESE. In the southeastern part, it abruptly changes into a linear band of about 0.8 km width, trending NNW–ESE to NNE–SSW for a length of 16 km from Aminabad to Gooty (Sreeramachandra Rao et al. 2001). WNW–ESE trending mafic

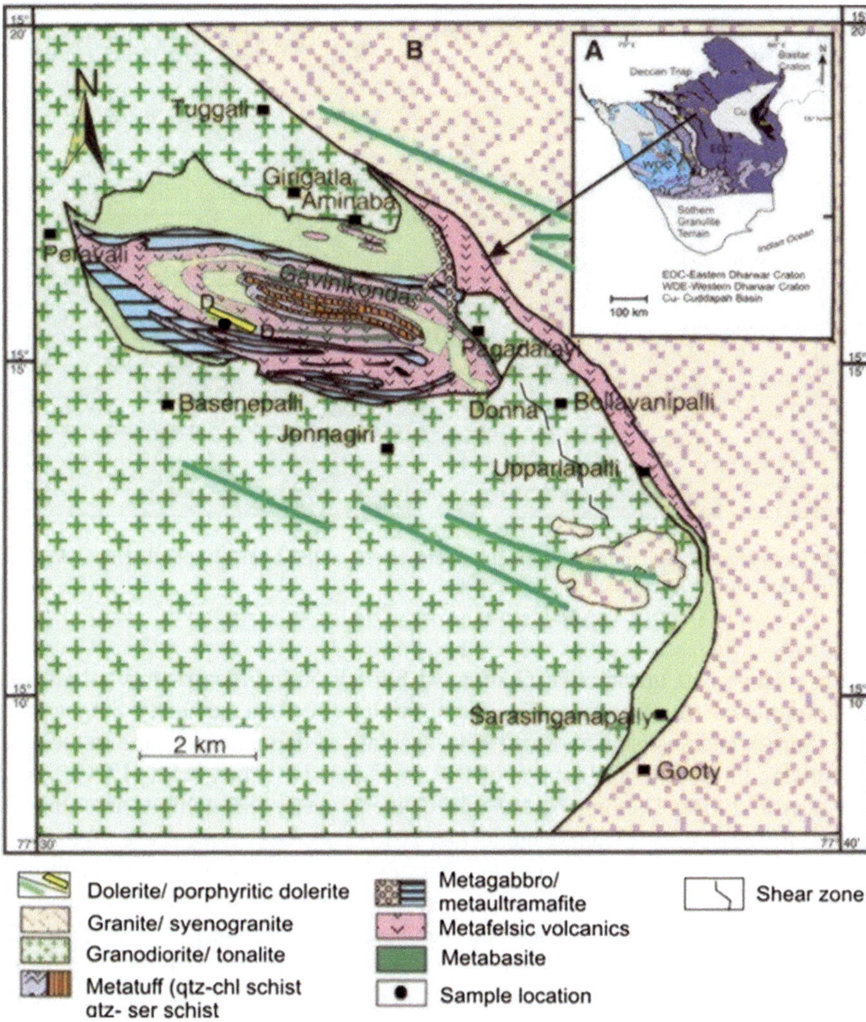

Fig. 1 Geological map of Jonnagiri and adjoining areas, Eastern Dharwar Craton, India (Modified after Jairam et al. 2001). Inset map showing the location of the JSB close the western margin of the Proterozoic Cuddapah basin

dykes of Proterozoic age have been recorded within the granite-greenstone terrain around Jonnagiri area in the EDC (e.g. Jairam et al. 2001). This belt represents the Jonnagiri shear zone which is traceable for 25 km further south-southwest up to Julakalva schist belt near the border of Proterozoic Cuddapah Basin. The belt lies between 15°05′ and 15°20′ latitudes and 77°30′ and 77°40′ longitudes.

Lithologically, the JSB is comprised of metamorphosed basic and acid volcanic together with tuffaceous rocks in the form of quartz sericite schist and sericite-

Fig. 2 Field photograph showing plagioclase phenocrysts in the Jonnagiri porphyritic dyke. Note the alignment in larger plagioclase phenocrysts

chlorite schist and ultramafite. Geological Survey of India carried out studies on the granitoids of the Peninsular Gneissic Complex (PGC) in the Gooty-Singanamala area (Suresh and Viswanatha Rao 1994); specialised thematic mapping in the areas comprising the southern extension of JSB and Julakalva schist belt in EDC (Suresh and Jaiswal 2003). The metabasalts of the Neoarchaean JSB exhibit 'arc—nascent back—arc signatures' (Manikyamba et al. 2015).

The JSB has been subjected to a number of phases of deformation and is intruded by younger granitoids and dolerite dykes, while undeformed dolerite dykes (porphyritic) of Proterozoic age are noticed in and around the JSB (e.g., Sreeramachandra Rao et al. 2001; Jairam et al. 2001). The present paper deals with the field, petrological and mineral chemistry studies of a major WNW–ESE trending porphyritic mafic dyke emplaced in the granite-greenstone terrain of the JSB, EDC, India; located within the folded metavolcanic sequence of the JSB and lies between Pagadarayi and Peravali (Fig. 1). Field observations indicate that the dyke is intermittently exposed over a stretch of>2 km as low-lying outcrops striking in WNW–ESE direction. The size of the plagioclase megacrysts ranges from 0.5 cm to a maximum of 7 cm in diameter. The dyke is undeformed and is characterised by the presence of aligned large megacrysts of plagioclase (Fig. 2). The euhedral nature of the phenocrysts and preservation of terminal ends in some of the plagioclase phenocrysts indicate that the alignment of plagioclase phenocrysts is magmatic. Phenocryst groundmass ration varies from 15 (phenocryst): 85 (groundmass) in the peripheral part to 30 (phenocryst): 70 (groundmass) in the central part of the porphyritic dyke. Partial resorption is some plagioclase phenocryst indicate crystal-melt interaction.

3 Methodology

Samples were collected during the course of field studies. Based on megascopic observations, 15 representative samples have been selected from the Jonnagiri porphyritic dyke for preparation of thin cum polished sections. Mineral chemistry was determined by EPMA, Petrology Division, GSI, SR, Hyderabad by CAMECA SX 100. Analyses conditions: Accelerating voltage: 15 kV, current: 15 nA. Beam size: 1 μm. Signal used Na Ka, Mg Ka, Al Ka, K Ka, Cr Ka, Mn Ka, Fe Ka, P Ka, Ni Ka, Zr La, Ca Ka, Ti K. All natural standards have been used except for Mn and Ti for which synthetic standards have been used.

4 Petrography and Mineral Chemistry

Petrographic studies reveal that the dyke exhibits porphyritic texture and essentially composed of plagioclase and clinopyroxene, while apatite is noticed as a conspicuous accessory mineral. Large euhedral phenocrysts of plagioclase (Fig. 3a) are embedded in subhedral grains of clinopyroxene and groundmass plagioclase. Sub-ophitic texture is conspicuously noticed in these samples, wherein plagioclase is partially enclosed in clinopyroxene (Fig. 3b).

At places, ophitic texture is also noticed (Fig. 3c). Opaque phases are mostly confined to the clinopyroxene (Fig. 3d). At places, it is observed that euhedral to subhedral opaques are partially enclosed in clinopyroxene (Fig. 3e). It is also noticed that partially enclosed opaque in ophitic plagioclase within the clinopyroxene (Fig. 3f). Megascopic studies show that the relatively large plagioclase phenocrysts show alignment at places. Microscopic studies of the plagioclase in the groundmass of Jonnagiri porphyritic dyke indicate the presence of randomly oriented plagioclase laths. At places, it is noticed that the plagioclase is altered along the margins. Mineral chemistry studies indicate the presence of euhedral baddeleyite. EPMA analyses have been carried out by selecting areas in plagioclase that are fresh and unaltered.

Presence of greenish amphibole and minor amounts of biotite, after alteration of the clinopyroxene is also noticed in the Jonnagiri porphyritic dyke. Opaques are represented by Fe–Ti oxides i.e. ilmenite and titano-magnetite. Jonnagiri porphyritic dyke is characterized by the presence of both apatite and Fe–Ti oxides. Both ilmenite and magnetite together constitute about ~10% in the rock are noticed as subhedral to euhedral grains that are more or less uniformly distributed in the rock. Apatite is noticed as minute euhedral inclusion in clinopyroxene. The petrographic obser-

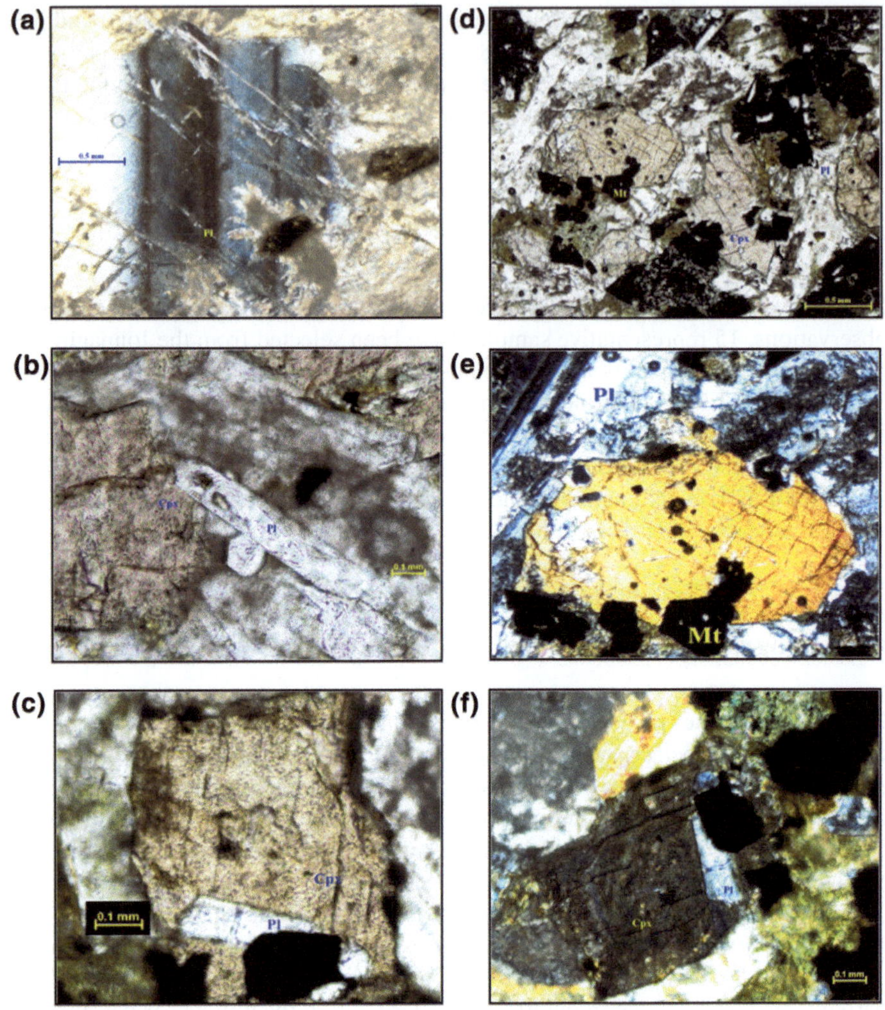

Fig. 3 **a** Plagioclase phenocryst in Xed. **b** Sub-ophitic texture in PPL. **c** Ophitic texture in PPL. **d** Opaques in clinopyroxene (Cpx) PPL. **e** Opaques in clinopyroxene Xed. Note the lamellar twining in plagioclase (Pl), and **f** Photomicrograph in Xed showing partially enclosed opaque in ophitic plagioclase within the clinopyroxene

vations are substantiated by SEM images (Fig. 4a–c), while the EPMA analyses of the plagioclase and pyroxene have been plotted in the respective mineral chemistry diagrams (Figs. 5 and 6). The EPMA analyses of the pyroxene from the Jonnagiri porphyritic dyke has been plotted in the Ca + Na versus Ti binary mineral chemistry diagram for pyroxene (Fig. 7; after Leterrier et al. 1982) which indicate the transitional nature of the Jonnagiri porphyritic dyke. Small grains of euhedral baddeleyite are also observed (Fig. 8).

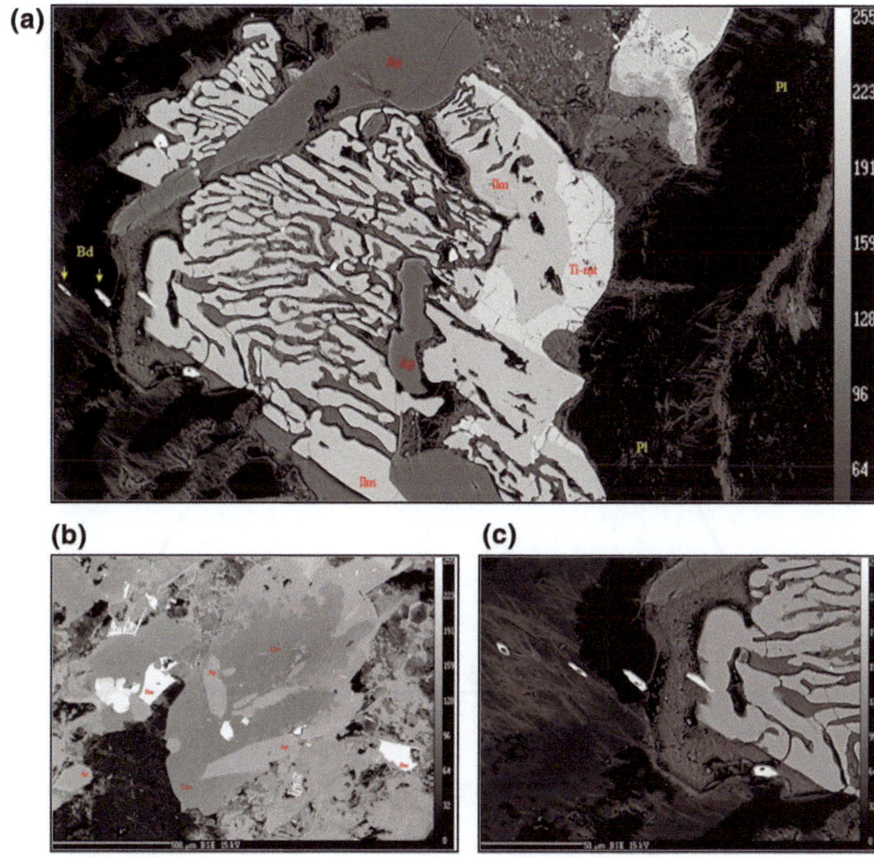

Fig. 4 **a** BSE image showing exsolved ilmenite and titano-magnetite along with apatite in the porphyritic dyke. Note that the titano-magnetite occurs at the rim and ilmenite in the core, **b** BSE image showing euhedral apatite inclusions within the clinopyroxene in Jonnagiri porphyritic dyke, and **c** BSE image showing the oriented baddeleyite grains ranging in size from 10 to 12 μm. Note that one of the baddeleyite grain partially transects the ilmenite

4.1 Plagioclase

Plagioclase occurs large size phenocrysts as well as relatively fine-grained laths in groundmass. Under crossed-nicols, it exhibits characteristic lamellar twining (Figs. 3a, e). EPMA analyses of plagioclase grains indicate that SiO_2 range from 53.90 to 58.34%, Al_2O_3 from 23.92 to 27.24%, CaO contents range from 7.49 to 10.04% and Na_2O contents from 4.82 to 7.09%. K_2O content of 0.61 to 1.24%. Zoning is noticed in plagioclase.

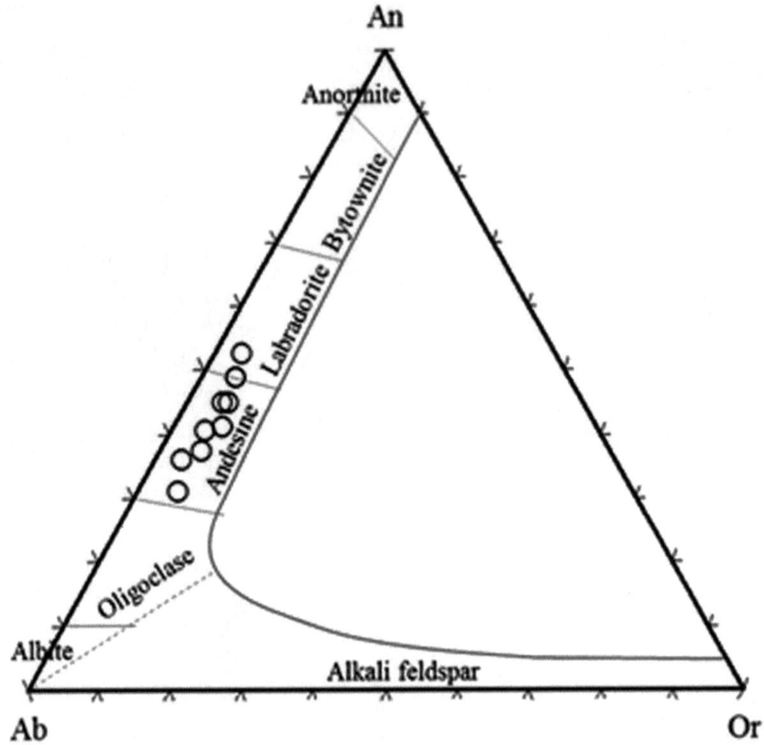

Fig. 5 Orthoclase (Or)—Albite (Ab)—Anorthite (An) ternary mineral chemistry diagram showing the position of plagioclases

End member calculation from the EPMA analytical data of plagioclases suggest their compositional variation from orthoclase$_{3.50–6.44\%}$, albite$_{43.18–62.98\%}$ to anorthite$_{30.57–52.46\%}$. On the Orthoclase (Or)—Albite (Ab)—Anorthite (An) ternary mineral chemistry diagram, analyses of the plagioclase core in the phenocrysts indicate labradorite composition, while the analyses of plagioclase rim in phenocrysts as well as groundmass plagioclase of the Jonnagiri porphyritic dyke falls in the field of andesine (Fig. 5). The andesine to labradorite composition of the plagioclase in Jonnagiri dyke, perhaps indicate relatively less density of the plagioclase than the basaltic magma from which it crystallised.

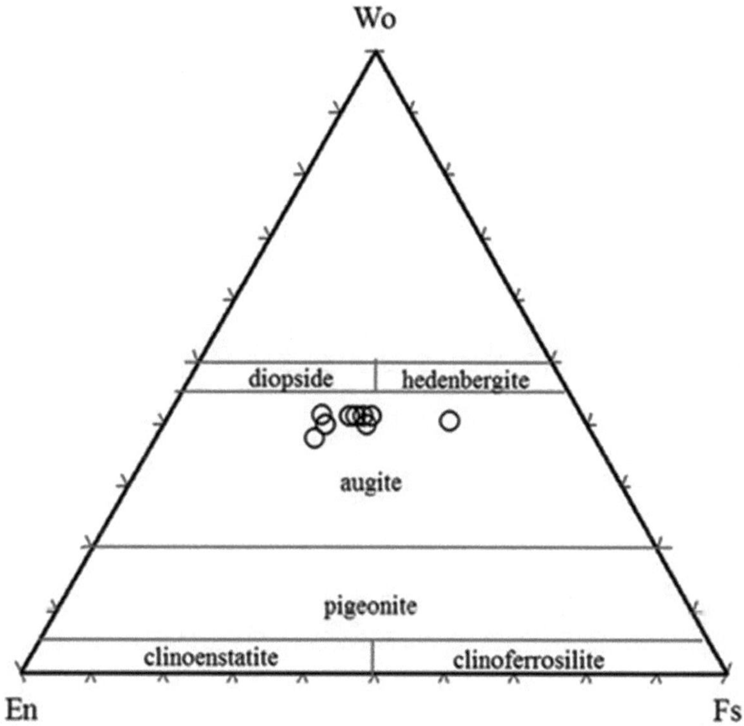

Fig. 6 Wollastonite (Wo)—Enstatite (En)—Ferrosilite (Fs) mineral chemistry ternary diagram showing position of pyroxenes

Further, predominant andesine composition indicates its relatively alkaline nature suggesting emplacement in a crustally thinned tectonic domain (e.g. Khanna et al. 2013). The mineral chemistry of the plagioclase feldspar from the Jonnagiri porphyritic dyke is presented in Table 1.

4.2 *Clinopyroxene*

Clinopyroxene occurs as euhedral to subhedral grains showing moderate relief and well-developed cleavage in plane polarised light (Fig. 3d), while in crossed nicols it is anisotropic, exhibits inclined extinction and second order interference colours (Fig. 3e). It is usually noticed as an intergranular phase in between lath shaped plagioclase.

EPMA analyses of the clinopyroxene indicate a relatively lower SiO_2 content (<50%). SiO_2 contents in the clinopyroxene range from 48.08 to 49.82%. TiO_2 contents in clinopyroxene range from 0.74 to 1.33%, while Al_2O_3 ranges from 0.35 to

Table 1 Mineral chemistry of plagioclase from the porphyritic dyke, Jonnagiri schist belt, EDC, India

	1	2	17	18	19	20	31	32	33	34
SiO_2	55.400	56.828	53.904	57.032	54.410	57.014	57.678	58.348	55.941	59.241
TiO_2	0.156	0.004	0.104	0.096	0.089	0.063	0.108	0.168	0.094	0.081
Al_2O_3	27.246	26.443	27.234	25.469	26.863	25.648	25.031	25.040	26.219	23.925
Cr_2O_3	0.000	0.000	0.015	0.000	0.003	0.001	0.010	0.013	0.001	0.000
FeOT	0.580	0.533	0.429	0.414	0.511	0.520	0.446	0.681	0.537	0.475
MnO	0.000	0.030	0.045	0.099	0.028	0.000	0.041	0.000	0.000	0.031
MgO	0.063	0.047	0.066	0.054	0.061	0.056	0.027	0.014	0.053	0.027
CaO	10.012	9.328	10.602	8.276	10.046	8.471	7.498	7.542	9.308	6.227
Na_2O	5.162	5.642	4.823	6.193	5.289	5.873	6.310	6.887	5.594	7.090
K_2O	0.900	0.868	0.738	0.921	0.864	1.249	1.090	0.611	1.041	1.102
NiO	0.082	0.000	0.148	0.088	0.031	0.061	0.000	0.005	0.000	0.000
Total	99.601	99.723	98.107	98.642	98.212	98.960	98.249	99.309	98.820	98.199
Fe_2O_3	0.269	0.000	0.334	0.414	0.511	0.520	0.000	0.681	0.537	0.132
FeO	0.338	0.533	0.128	0.000	0.000	0.000	0.446	0.000	0.000	0.356
Total	99.628	99.723	98.141	98.642	98.195	98.956	98.239	99.309	98.788	98.212
Cations:	8(O)	8(O)	8(O)	8(O)	8(O)	8(O)	8(O)	8(O)	8(O)	8(O)
Si	2.514	2.569	2.487	2.601	2.507	2.595	2.635	2.635	2.555	2.698
Ti	0.005	0.000	0.004	0.003	0.003	0.002	0.004	0.006	0.003	0.003

(continued)

Table 1 (continued)

	1	2	17	18	19	20	31	32	33	34
Al	1.458	1.409	1.481	1.369	1.459	1.376	1.348	1.333	1.412	1.284
Fe^{+3}	0.009	0.000	0.012	0.014	0.018	0.018	0.000	0.023	0.018	0.005
Fe^{+2}	0.013	0.020	0.005	0.000	0.000	0.000	0.017	0.000	0.000	0.014
Mn	0.000	0.001	0.002	0.004	0.001	0.000	0.002	0.000	0.000	0.001
Mg	0.004	0.003	0.005	0.004	0.004	0.004	0.002	0.001	0.004	0.002
Ca	0.487	0.452	0.524	0.404	0.496	0.413	0.367	0.365	0.456	0.304
Na	0.454	0.495	0.432	0.548	0.473	0.518	0.559	0.603	0.495	0.626
K	0.052	0.050	0.043	0.054	0.051	0.073	0.064	0.035	0.061	0.064
Ni	0.003	0.000	0.005	0.003	0.001	0.002	0.000	0.000	0.000	0.000
Total	5.000	4.999	5.000	5.004	5.013	5.001	4.998	5.001	5.004	5.000
Orthoclase	5.247	5.024	4.348	5.329	4.983	7.224	6.420	3.509	5.996	6.442
Albite	45.735	49.631	43.189	54.457	46.358	51.527	56.488	60.113	48.973	62.988
Anorthite	49.019	45.345	52.463	40.215	48.659	41.149	37.092	36.378	45.030	30.571

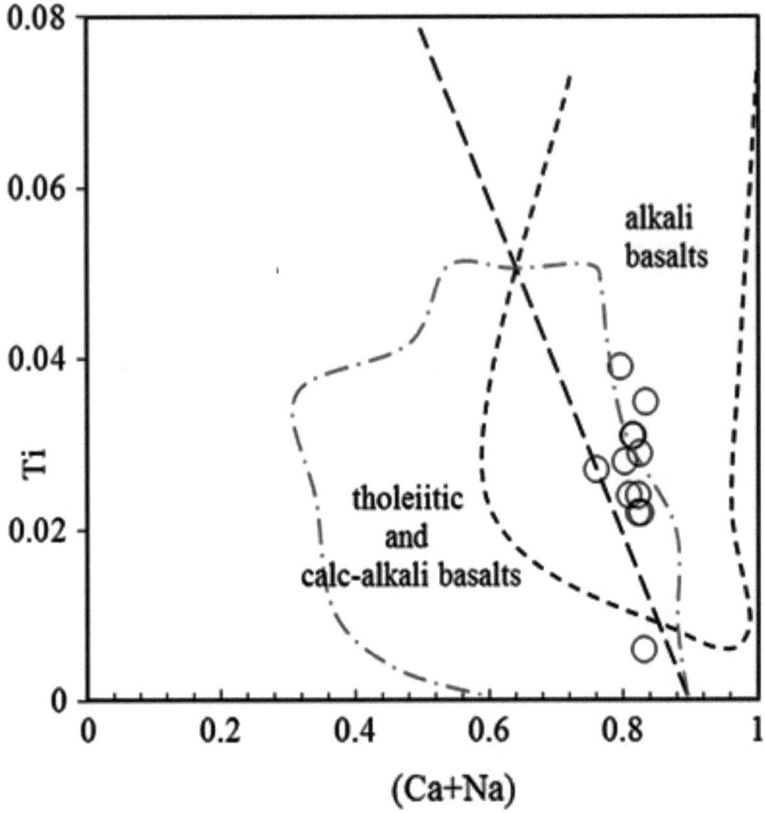

Fig. 7 Ca + Na versus Ti binary mineral chemistry diagram for pyroxene (after Leterrier et al. 1982), showing the alkaline nature of the clinopyroxenes

2.95%, FeO^T contents range from 13.13 to 17.14% and MnO contents range from 0.16 to 0.49%. The clinopyroxene is compositionally a Ti-augite ($Wo_{40}En_{34}Fs_{26}$). MgO contents vary from 9.93 to 13.53%, CaO from 17.92 to 19.90%, Na_2O from 0.21 to 0.32%, maximum K_2O content up to 0.14%, NiO up to 0.13% and Cr_2O_3 content up to 0.02%. End member calculation from the EPMA analytical data indicate variation in composition from wollastonite$_{31.60–41.61\%}$, enstatite$_{818.62–31.69\%}$ to ferrosilite$_{022.71–040.76\%}$. On the Wollastonite (Wo)-Enstatite (En)-Ferrosilite (Fs) ternary diagram, pyroxenes of the Jonnagiri porphyritic dyke fall in the field of augite, but close to the boundary of augite and diopside (Fig. 6).

The titanium concentrations in these clinopyroxenes slightly overlap with basalts derived from tholeiitic and calc-alkaline lineage (Fig. 7; Leterrier et al. 1982). However, it is observed that the pyroxenes of the Jonnagiri porphyritic dyke indicate the alkaline nature when plotted in the Ca + Na versus Ti binary mineral chemistry diagram (after Leterrier et al. 1982). Plagioclase with andesine composition (Fig. 5) supports this observation. The mineral chemistry of the pyroxene of the Jonnagiri porphyritic dyke is presented in Table 2.

4.3 Fe–Ti Oxides

Notable amounts of titano-magnetite and ilmenite are noticed as euhedral to subhedral grains at the vicinity of the clinopyroxene (Figs. 3d, e, f). Both Ti-oxide and T–Fe oxide are more or less uniformly distributed in the rock. At places, Fe–Ti oxides are partially enclosed in the plagioclase laths that exhibit ophitic relation with the clinopyroxene. Exsolution intergrowth texture is conspicuously noticed in the Fe–Ti oxides. Presence of exsolved phases of Fe–Ti oxides in magmatic rocks are often associated both with the mafic intrusions and in the anorthosite massifs of Proterozoic age (Lister 1966; Haggerty and Rumble 1976; Ashwal 1982; Frost and Lindsley 1991). Titano-magnetite indicates the physical and chemical conditions of the mafic intrusions during their crystallization (e.g. Tan et al. 2016). The mineral chemistry of the ilmenite and titano-magnetite from the Jonnagiri porphyritic dyke is presented in Table 3.

EPMA data of ilmenite (FeO TiO_2) indicate that TiO_2 contents range from 46.07 to 49.71%, and FeO^T contents range from 46.17 to 46.44%. Cr_2O_3 content up to 0.53% has been recorded, while a maximum NiO content up to 0.11% has been observed. End member calculation from the EPMA analytical data indicate variation between ilmenite$_{87.64–92.46\%}$, geikelite$_{0.008–0.075\%}$, pyrophanite$_{3.82–4.40\%}$, and hematite$_{3.60–7.93\%}$.

EPMA composition of titano-magnetite (FeO Fe_2O_3 TiO_2) indicate that the FeO^T contents range from 79.39 to 79.62% and TiO_2 from 11.53 to 12.1%. MnO content range from 0.42 to 0.50%, while Cr_2O_3 content up to 0.15% and NiO content up to 0.04%. BSE images indicate that at places both ilmenite and titano-magnetite are noticed in continuity; and in such places ilmenite is noticed in the core, while titano-magnetite occurs at the peripheral parts of the ilmenite (Fig. 4a). End member calculation from EPMA analytical data indicate that their composition from ilmenite$_{21.54–22.69\%}$, geikelite up to$_{0.13\ \%}$, pyrophanite$_{0.95–1.23\%}$, to hematite$_{76.05–77.35\ \%}$.

4.4 Apatite

Apatite is a conspicuous accessory phase and found to be associated with clinopyroxene. Euhedral apatite is often noticed as inclusions within the clinopyroxene (Fig. 4b). EPMA data indicate CaO content up to 53.94% and P_2O_5 content up to 40.31 for the apatite inclusions in clinopyroxenes.

Table 2 Mineral chemistry of clinopyroxene from the porphyritic dyke, Jonnagiri schist belt, EDC, India

Point no.	3	4	8	9	11	12	15	16	23	24	25	26
SiO_2	49.724	49.488	49.713	49.764	49.821	49.194	49.629	49.542	49.488	49.680	48.087	49.127
TiO_2	0.744	1.229	0.963	0.743	0.938	1.065	1.015	0.836	1.054	0.812	0.183	1.336
Al_2O_3	1.396	2.158	1.831	1.328	1.904	2.004	1.957	1.599	1.970	1.523	0.354	2.950
Cr_2O_3	0.000	0.014	0.020	0.006	0.000	0.017	0.009	0.007	0.000	0.000	0.019	0.019
FeOT	16.551	13.224	13.765	17.090	13.611	15.570	15.288	17.141	15.983	17.144	23.556	13.132
MnO	0.423	0.162	0.343	0.496	0.359	0.366	0.427	0.328	0.376	0.498	0.475	0.401
MgO	10.399	12.561	12.417	9.993	13.539	11.173	11.073	10.412	10.768	10.251	6.163	12.346
CaO	19.255	19.906	18.913	19.266	17.929	19.143	19.451	18.950	19.046	19.334	18.690	18.787
Na_2O	0.220	0.254	0.294	0.308	0.324	0.322	0.296	0.253	0.302	0.296	0.216	0.290
K2O	0.007	0.006	0.000	0.000	0.012	0.000	0.006	0.000	0.000	0.010	0.014	0.000
NiO	0.000	0.044	0.000	0.132	0.000	0.069	0.000	0.016	0.000	0.014	0.000	0.000
Total	98.722	99.085	98.259	99.145	98.442	98.987	99.163	99.084	98.987	99.562	97.758	98.389
Fe_2O_3	1.224	2.490	1.702	1.636	2.447	2.403	1.683	1.678	1.376	2.194	2.611	1.129
FeO	15.449	10.983	12.234	15.618	11.409	13.408	13.773	15.631	14.745	15.169	21.206	12.116
Total	98.842	99.295	98.429	99.290	98.682	99.164	99.320	99.252	99.125	99.782	98.019	98.501
Cations:	6(O)	6(O)	6(O)	6(O)	6(O)	6(O)	6(O)	6(O)	6(O)	6(O)	6(O)	6(O)
Si	1.937	1.890	1.916	1.935	1.907	1.900	1.913	1.924	1.916	1.921	1.955	1.889
Ti	0.022	0.035	0.028	0.022	0.027	0.031	0.029	0.024	0.031	0.024	0.006	0.039

(continued)

Table 2 (continued)

Point no.	3	4	8	9	11	12	15	16	23	24	25	26
Al	0.064	0.097	0.083	0.061	0.086	0.091	0.089	0.073	0.090	0.069	0.017	0.134
Cr	0.000	0.000	0.001	0.000	0.000	0.001	0.000	0.000	0.000	0.000	0.001	0.001
Fe^{+3}	0.036	0.072	0.049	0.048	0.070	0.070	0.049	0.049	0.040	0.064	0.080	0.033
Fe^{+2}	0.503	0.351	0.394	0.508	0.365	0.433	0.444	0.508	0.477	0.491	0.721	0.390
Mn	0.014	0.005	0.011	0.016	0.012	0.012	0.014	0.011	0.012	0.016	0.016	0.013
Mg	0.604	0.715	0.714	0.579	0.773	0.643	0.636	0.603	0.621	0.591	0.373	0.708
Ca	0.804	0.814	0.781	0.803	0.735	0.792	0.803	0.789	0.790	0.801	0.814	0.774
Na	0.017	0.019	0.022	0.023	0.024	0.024	0.022	0.019	0.023	0.022	0.017	0.022
K	0.000	0.000	0.000	0.000	0.001	0.000	0.000	0.000	0.000	0.000	0.001	0.000
Ni	0.000	0.001	0.000	0.004	0.000	0.002	0.000	0.000	0.000	0.000	0.000	0.000
Total	4.000	4.000	4.000	4.000	4.000	4.000	4.000	4.000	4.000	4.000	4.000	4.000
Wollastonite	40.990	41.618	40.068	41.079	37.609	40.618	41.274	40.256	40.698	40.815	40.608	40.376
Enstatite	30.798	36.535	36.597	29.643	39.511	32.982	32.689	30.772	32.011	30.106	18.629	36.914
Ferrosilite	28.212	21.847	23.335	29.278	22.880	26.400	26.037	28.972	27.292	29.079	40.763	22.710

Table 3 Mineral chemistry of ilmenite and titano-magnetite from the porphyritic dyke, Jonnagiri schist belt, EDC, India

	EPMA analyses of ilmenite					EPMA analyses of titano-magnetite		
	14/1	19/1	25/1	29/1	30/1	15/1	24/1	31/1
SiO_2	0	0.016	0.049	0.029	0.02	0.207	0.046	0.051
TiO_2	46.207	49.716	46.368	46.076	46.241	12.169	11.615	11.538
Al_2O_3	0.005	0.024	0.005	0	0.002	0.036	0.682	0.03
Cr_2O_3	0.027	0	0.053	0.037	0.012	0.076	0.106	0.151
FeOT	46.314	46.173	46.448	46.208	46.357	79.629	78.394	79.48
MnO	1.817	1.75	1.925	1.943	1.889	0.505	0.421	0.494
MgO	0.002	0.026	0.019	0.005	0.003	0.034	0.008	0
CaO	0.012	0	0.028	0.133	0.086	0.144	0.027	0.016
Na_2O	0.004	0	0.002	0	0	0.011	0.027	0.031
K_2O	0.013	0.005	0.001	0.029	0.035	0.001	0	0.019
NiO	0	0.065	0.014	0.111	0.072	0.044	0	0
Total	94.401	97.775	94.912	94.571	94.717	92.856	91.326	91.81
Fe_2O_3	7.423	3.715	7.488	7.870	7.730	77.004	76.118	77.531
FeO	39.634	42.830	39.710	39.126	39.401	10.338	9.899	9.714
Total	95.144	98.147	95.662	95.359	95.491	100.568	98.949	99.575
Cations:	3(O)	3(O)	3(O)	3(O)	3(O)	3(O)	3(O)	3(O)
Si	0.000	0.000	0.001	0.001	0.001	0.005	0.001	0.001

(continued)

Table 3 (continued)

	EPMA analyses of ilmenite					EPMA analyses of titano-magnetite		
	14/1	19/1	25/1	29/1	30/1	15/1	24/1	31/1
Ti	0.926	0.963	0.924	0.921	0.923	0.238	0.231	0.229
Al	0.000	0.001	0.000	0.000	0.000	0.001	0.021	0.001
Cr	0.001	0.000	0.001	0.001	0.000	0.002	0.002	0.003
Fe^{+3}	0.149	0.072	0.149	0.157	0.154	1.510	1.514	1.538
Fe^{+2}	0.883	0.923	0.880	0.869	0.874	0.225	0.219	0.214
Mn	0.041	0.038	0.043	0.044	0.042	0.011	0.009	0.011
Mg	0.000	0.001	0.001	0.000	0.000	0.001	0.000	0.000
Ca	0.000	0.000	0.001	0.004	0.002	0.004	0.001	0.000
Na	0.000	0.000	0.000	0.000	0.000	0.001	0.001	0.002
K	0.000	0.000	0.000	0.001	0.001	0.000	0.000	0.001
Ni	0.000	0.001	0.000	0.002	0.002	0.001	0.000	0.000
Ilmenite	88.434	92.465	88.122	87.641	87.954	22.693	22.203	21.541
Geikelite	0.008	0.100	0.075	0.020	0.012	0.133	0.032	0.000
Pyrophanite	4.106	3.827	4.327	4.408	4.271	1.123	0.956	1.109
Hematite	7.452	3.609	7.476	7.931	7.763	76.051	76.809	77.350

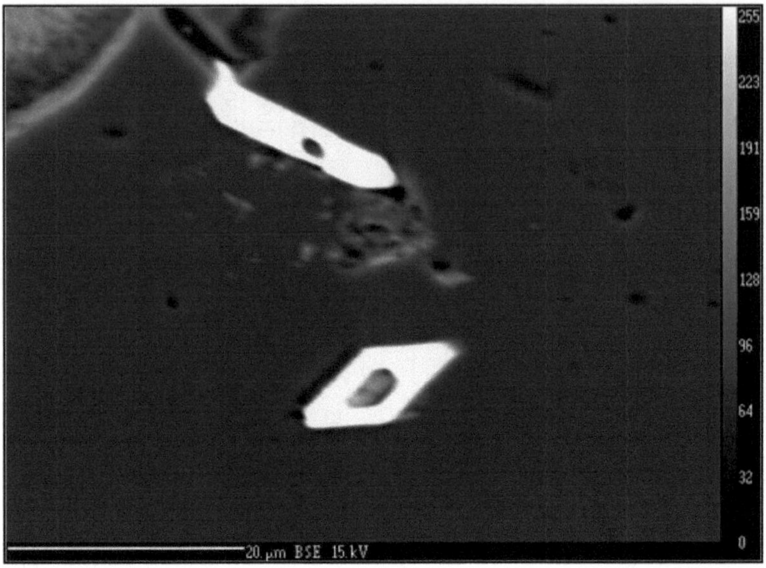

Fig. 8 BSE image showing the euhedral nature of the micron size baddeleyite grains in the Jonnagiri porphyritic dyke

4.5 *Baddeleyite*

The baddeleyite grains show alignment and at places partially transect ilmenite grains (Fig. 4c). BSE image shows euhedral nature of baddeleyite (Fig. 8). Due to the high uranium and minimal initial lead contents, baddeleyite (ZrO_2) occur as accessory phase in mafic rocks and is considered to be a reliable geochronometer (Smith 2010). However, baddeleyite in Jonnagiri porphyritic dyke occur as minute grains ranging in size from 10 to 12 μm.

5 Discussions

The present study deals with plagioclase megacrysts bearing porphyritic mafic dyke from the Jonnagiri greenstone belt, in EDC, Southern India. The plagioclase phenocrysts in the Jonnagiri porphyritic dyke range in size from 0.5 to 3.2 cm. Field and megascopic studies indicate that the euhedral plagioclase phenocrysts with well preserve terminal ends indicate the undeformed nature and presence of primary magmatic alignment in the Jonnagiri porphyritic dyke. Partial resorption in some plagioclase phenocryst indicates crystal—melt interaction. Alignment of plagioclase phenocrysts in mafic dykes has been described as a feature resulting due to flow differentiation (e.g. Ross 1986). Increase in phenocryst groundmass from the centre

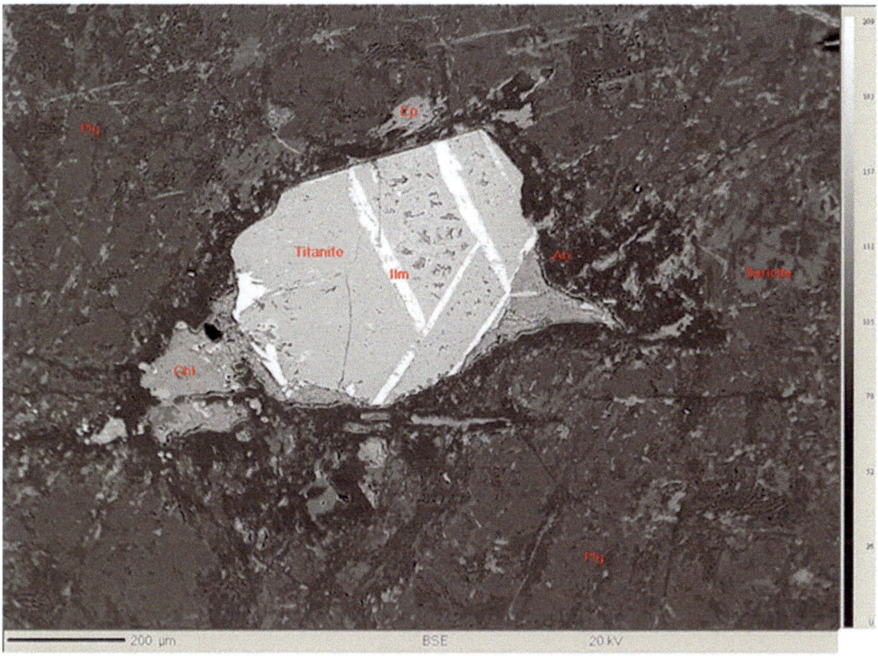

Fig. 9 BSE image showing exsolved ilmenite in titanite inclusion in plagioclase megacryst in the Jonnagiri porphyritic dyke

to the peripheral parts in porphyritic dyke has been explained in terms of increase in the velocity gradient and the fluid viscosity (e.g. Komar 1972). Euhedral plagioclase megacrysts of 2–5 cm size have been recorded in the Tarssartoq dyke swarm to the north of Isua Greenstone belt of Greenland (White et al. 2000). The larger plagioclase phenocrysts confined to the central part of the dyke crystallize when the magma resides at relatively deeper level and the groundmass phases crystallize when the magma is emplaced to sub crustal levels. Presence of sub-ophitic/ophitic textures in Jonnagiri porphyritic dyke (Figs. 3b, c) substantiate the magmatic origin. Ophitic/sub-ophitic textures in coarse grained dykes contain magmatic pyroxene (e.g. White et al. 2000). Exsolution textures in the form of ilmenite and titanomagnetite intergrowth conspicuously noticed in the Jonnagiri porphyritic dyke also support the magmatic origin. Fe and Ti enriched melts crystallize abundant titanomagnetite at an early stage (e.g. Pang et al. 2008). Partially enclosed euhedral to subhedral opaques in the ophitic plagioclase lath (Fig. 3c) and clinopyroxene (Fig. 3e) indicate early crystallisation of the Fe–Ti oxides in the Jonnagiri porphyritic dyke. Notable variation both in the abundance and size of plagioclase phenocrysts can be noticed in outcrop scale in the megacrystic plagioclase bearing amphibolites in Flekkefjord area, South Norway (Falkum and Grundvig 2001). The size of the plagioclase megacrysts from gabbroic dykes in the Gardar Province vary from <1 cm to 1 m (Halama et al. 2002). Low degree alkaline nature is evidenced by the presence of plagioclase phenocryst

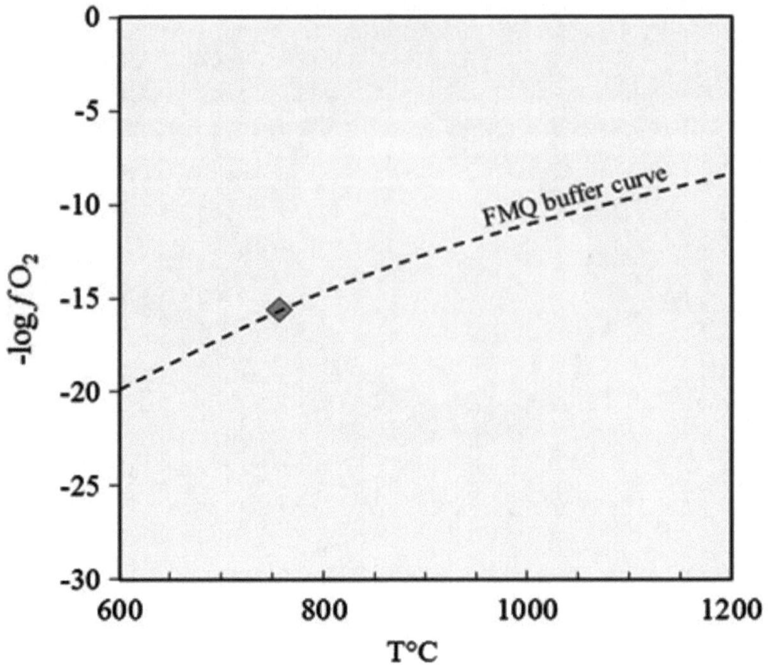

Fig. 10 Binary plot showing temperature and oxygen fugacity estimates for the coexisting magnetite-ilmenite solid solution pairs Lepage (2014). FMQ curve in the figure is after Frost (1991)

with composition varying from andesine to labradorite composition in the porphyritic mafic dyke from the Jonnagiri greenstone belt. Mafic dykes with low degree alkaline, transitional to tholeiitic character have been reported in a continental tectonic from the Izera Complex, West Sudetes, Poland (Ilnicki 2010).

Microprobe analyses of plagioclase grains in the Jonnagiri porphyritic dyke indicate the presence of zoning phenomenon. The core analysed Na_2O content of 5.16% and CaO content of 10.01%, while the rim analysed Na_2O content of 5.64% and CaO content of 9.32% (Table 2). Increase in Na ($Ab_{45.73}$ to $Ab_{49.61}$) and decrease in Ca ($An_{49.01}$ to $An_{45.34}$) from core to rim indicate normal zoning in the plagioclase, a characteristic feature indicating relatively high temperature in core and lower temperature in the rim resulted from crystallisation of magma. Further presence of exsolved ilmenite within the titanite inclusion in the plagioclase megacryst (Fig. 9) affirm the magmatic origin for the dyke.

Alteration studies on granites indicated replacement of titanite by ilmenite (e.g. Broska et al. 2007), however, implications of exsolved ilmenite in titanite inclusion in the plagioclase megacrysts in mafic dykes is less understood. It is notable that composition of most of the plagioclase grains in Jonnagiri porphyritic dyke are predominantly andesine. Wyers and Barton (1986) have reported transitional lavas of alkaline—sub-alkaline composition from the Patmos province in Greece. The predominantly andesine composition of plagioclase grains, position of the clinopyroxene in the Ca + Na versus Ti binary mineral chemistry diagram (after Leterrier et al. 1982), and micro size euhedral baddeleyite grains indicate the transitional nature of the Jonnagiri porphyritic dyke that is presumably emplaced in a rift related setting. However, detailed geochemical and geochronological studies are the gap areas that can be taken up for further research.

The temperature and oxygen fugacity estimates for the coexisting magnetite-ilmenite solid solution pairs using the calculation software given by Lepage (2014) yielded an equilibration temperature of ~756 °C and $10^{-15.6}$atmf_{O2}. The equilibrium conditions are consistent with an oxide equilibration path defined by a fayalite-magnetite-quartz (FMQ) buffer curve (Fig. 10; Frost 1991). The basaltic rocks generated in arc settings are commonly characterized by oxygen fugacities necessarily more oxidizing than the FMQ buffer (10^{-8}–10^{-3}) at a given temperature (1200 °C; cf. Frost 1991). The estimate for the Jonnagiri dyke is consistent with a relatively non-oxidizing i.e. reducing environment.

6 Conclusions

The porphyritic mafic dyke from the Jonnagiri greenstone belt described in the present study is characterised by presence of euhedral plagioclase megacrysts showing magmatic alignment. Partially resorbed plagioclase phenocryst indicates crystal-melt interaction. Petrographically Jonnagiri dyke exhibits porphyritic texture with magmatically aligned plagioclase phenocrysts set in groundmass andesine and titan augite. Fe–Ti oxides; ilmenite and titano-magnetite exhibiting exsolved textures are notable, while apatite and baddeleyite are the accessory phases. The predominant andesine compositions of the plagioclase grains indicate transitional nature of the Jonnagiri porphyritic dyke. The temperature and oxygen fugacity estimates for the coexisting magnetite-ilmenite solid solution pairs yielded an equilibration temperature of ~756 °C and $10^{-15.6}$atmf_{O2}.

Acknowledgements The Director General, Geological Survey of India, is thankfully acknowledged, for providing an opportunity to work in the field area as part of the GSI-Geoscience Australia field workshop in March 2017. Additional Director General and HOD, Geological Survey of India, Southern Region, Hyderabad is thanked for his kind support. Michael Doublier, Geoscience Australia, is profusely thanked for helpful discussions during a field Workshop in Jonnagiri schist belt. M. L. Dora and Jeff Karson are profusely thanked for critical and constructive review. Editorial comments of Rajesh K. Srivastava have also helped to improve the MS. VVSS and SB thanks Pravir Pankaj, GSI, for the support in the field.

References

Ashwal LD (1982) Mineralogy of mafic and Fe–Ti oxide rich differentiates of the Marcy anorthosite massif, Adirondacks, New York. Am Mineral 67:14–27

Belica ME, Piispa EJ, Meert JG, Pesonen LJ, Plado J, Pandit MK, Kamenov GD, Celestino M (2013) Paleoproterozoic mafic dyke swarms from the Dharwar craton; palaeomagnetic poles for India from 2.37 to 1.88 Ga and rethinking the Columbia supercontinent. Precambr Res 244:100–122

Broska I, Harlov D, Tropper P, Siman P (2007) Formation of magmatic titanite and titanite-ilmenite phase relations during granite alteration in the Tribec Mountains, western carpathians, Slovakia. Lithos 95:58–71

Dora ML, Mahapatro SN, Singh H, Malaviya VP, Shareef M, Wankhade SH, Ranadive K (2016) Plagioclase Megacrysts in Proterozoic Leucogabbronorite Dyke from Bastar Craton, Central India: Melt-crystal interaction at crustal level. Paper presented at the 35th international geological congress, Cape Town, South Africa, 27th August–4th September 2016

Falkum T, Grundvig S (2001) Origin of plagioclase-megacrystic, orthopyroxene amphibolites within a Precambrian banded gneiss suite, Flekkefjord area Vest-Agder, South Norway. NGU Bull 438:5

Frost BR (1991) Magnetic petrology: factors that control the occurrence of magnetite in crustal rocks. Rev Miner Geochem 25:489–509

Frost BR, Lindsley DH (1991) Occurrence of iron-titanium oxides in igneous rocks. oxide minerals: petrologic and magnetic significance. Miner Soc Am, Rev Miner 25:433–468

Haggerty SE, Rumble D (1976) Opaque mineral oxides in terrestrial igneous rocks. Oxide Miner: Miner Soc Am, Rev Miner 3:Hg101–Hg300

Halama R, Waight T, Markl G (2002) Geochemical and isotopic zoning patterns of plagioclase megacrysts in gabbroic dykes from the Gardar Province, South Greenland: implications for crystallisation processes in anorthositic magmas. Contrib Mineral Petrol 144(1):109–127

Halls HC, Kumar A, Srinivasan R, Hamilton MA (2007) Palaeomagnetism and U–Pb geochronology of eastern trending dykes in the Dharwar craton, India: feldspar clouding, radiating dyke swarms and position of India at 2.37 Ga. Precambr Res 155:47–68

Hou GT, Santosh M, Xianglin Qian GS, Li JH (2008) Configuration of the Late Paleoproterozoic supercontinent Columbia: insights from radiating mafic dyke swarms. Gondwana Res 14:395–409

Ilnicki S (2010) Petrogenesis of continental mafic dykes from the Izera Complex, Karkonosze-Izera Block (West Sudetes, SW Poland). Int J Earth Sci 99(4):745–773

Jairam MS, Roop Kumar D, Srinivasan KN (2001) Classification of greenstones and associated granitoids of Jonnagiri schist belt, Kurnool District, Andhra Pradesh. Geol Surv India Spec Pub 55:59–66

Jayananda M, Mahesha N, Srivastava RK, Blais S (2008) Petrology and geochemistry of Paleoproterozoic high-magnesian norite and dolerite dyke swarms from the Halagur-Satnur areas, Eastern Dharwar Craton, Southern India. In: Srivastava RK, Sivaji C, Chalapathi Rao NV (eds) Indian Dykes: Geochemistry, Geophysics and Geochronology. Narosa Publishing House Pvt. Ltd., New Delhi, pp 239–260

Khanna TC, Sesha Sai VV, Zhao GC, Subba Rao DV, Keshav Krishna A, Sawant SS, Charan SN (2013) Petrogenesis of mafic alkaline dikes from the ~2.18 Ga Mahbubnagar Large Igneous Province, Eastern Dharwar Craton, India: geochemical evidence for uncontaminated intra continental mantle derived magmatism. Lithos 179:84–98

Komar PD (1972) Mechanical interactions of phenocrysts and flow differentiation of igneous dikes and sills. Geol Soc Am Bull 83(4):973–988

Kumar A, Nagaraju E, Besse J, Bhaskar Rao YJ (2012) New age, geochemical and palaeomagnetic data on a 2.21 Ga dyke swarm from south India: constraints on Paleoproterozoic reconstruction. Precambr Res 220:123–138

Kumar A, Parashuramulu V, Nagaraju E (2015) A 2082 Ma radiating dyke swarm in the Eastern Dharwar Craton, southern India and its implications to Cuddapah basin formation. Precambr Res 266:490–505

Lepage DL (2014) ILMAT: A magnetite-Ilmenite Geothermobarometry Program (version 1.20c)

Leterrier J, Maury RC, Thonon P, Girard D, Marchal M (1982) Clinopyroxene composition as a method of identification of the magmatic affinities of paleo-volcanic series. Earth Planet Sci Lett 59:139–154
Lister GF (1966) The composition and origin of selected iron–titanium deposits. Econ Geol 61:275–310
Manikyamba C, Ganguly S, Santosh M, Singh RM, Saha A (2015) Arc-nascent back-arc signatures in metabasalts from the neoarchaean Jonnagiri greenstone terrane, Eastern Dharwar Craton, India. Geol J 50:651–669
Morrison DA, Phinney WC, Maczuga DE (1988) The Petrogenetic significance of plagioclase megacrysts in Archean rocks. Lunar and Planetary Inst., Workshop on the Deep Continental Crust of South India, pp 112–114
Naqvi SM, Rogers JJW (1987) Precambrian geology of India, Oxford Monographs on Geology and Geophysics. Oxford University Press, Oxford, p 223
Pang KN, Zhou MF, Lindsley D, Zhao D, Malpas J (2008) Origin of Fe–Ti oxide ores in mafic intrusions: evidence from the Panzhihua Intrusion, SW China. J Petrol 49(2):295–313
Phinney WC, Morrison DA (1986) Petrogenesis of calcic plagioclase megacrysts in Archean rocks. Lunar and Planetary Inst. Workshop on Early Crustal Genesis: The World's Oldest Rocks p 90–91
Piispa EJ, Smirnov AV, Pesonen LJ, Lingadevaru M, Anantha Murthy KS, Devaraju TC (2011) An integrated study of proterozoic dykes, Dharwar Craton, Southern India. In: Srivastava RK (ed) Dyke Swarms: Keys for Geodynamic Interpretation. Springer, New York, pp 33–45
Radhakrishna BP (2008) Foreword—Precambrian mafic magmatism in Indian shield. J Geol Soc India 72:5
Radhakrishna T, Joseph M (1996) Proterozoic palaeomagnetism of the mafic dyke swarms in the high grade region of south India. Precambr Res 76:31–46
Ross ME (1986) Flow differentiation, phenocryst alignment, and compositional trends within a dolerite dike at rockport, Massachusetts. Bull Geol Soc Am 97(2):232–240
Santosh M (2012) India's Paleoproterozoic legacy. In: Mazumder R, Saha D (eds) Paleoproterozoic of India: geological society, Special Publications, London, vol. 365, pp 263–288
Sesha Sai VV, Tripathy V, Bhattacharjee S, Khanna TC (2017) Paleoproterozoic magmatism in the Cuddapah basin India. J Indian Geophys Union 21(6):516–525
Smith KF (2010) Potential for baddeleyite in U–Pb dating of mafic alkaline rocks from the Balcones Igneous Province, Texas. In Session No. 277. Recent Advances in Mineralogy and Petrology. Presented at the annual meeting of the Geological Society of America, Denver, vol 42(5), p 647
Sreeramachandra Rao K, Roop Kumar D, Jairam MS, Bhattacharjee S, Ananda Murthy S, Krishna Rao PV (2001) Interpretation of geological characteristic of Dona gold prospect, Jonnagiri schist belt, Kurnool district, Andhra Pradesh. Geol Surv India Spec Pub 58:217–231
Srivastava RK (ed) (2011) Dyke Swarms: keys for geodynamic interpretations, Springer, ISBN 978-3-642-12496-9
Srivastava RK, Ahmed T (2009) Precambrian mafic magmatism in the Indian shield: retrospect and prospect. J Geol Soc India 73(1):7–11
Srivastava RK, Sivaji C, Chalapathi Rao NV (eds) (2008) Indian dykes through space and time: retrospect and prospect. Geochemistry, Geophysics and Geochronology, Narosa Publishing House Pvt. Ltd., New Delhi, Indian Dykes, pp 1–18
Srivastava RK, Kumar S, Sinha AK, Chalapathi Rao NV (2014a) Petrology and geochemistry of high titanium and low titanium mafic dyes from the Damodar Valley, Chottanagpur Gneissic terrain, eastern India and their relation to Cretaceous mantle plumes (s). J Asian Earth Sci 84:34–50
Srivastava RK, Jayananda M, Gautam GC, Gireesh V, Samal AK (2014b) Geochemistry of an ENE–WSW to NE–SW trending ~2.37 Ga mafic dyke swarm of the eastern Dharwar craton, India: does it represent a single magmatic event? Chemie der Erde Geochemistry 74:251–265
Srivastava RK, Jayananda M, Gautam GC, Samal AK (2014c) Geochemical studies and petrogenesis of ~2.21–2.22 Ga Kunigal mafic dyke swarm (trending N–S to NNW–SSE) from eastern Dharwar craton, India: implications for Paleoproterozoic large igneous provinces and supercraton superia. Miner Pet 108:695–711

Srivastava RK, Samal A, Gautam GC (2015) Geochemical characteristics and petrogenesis of four Palaeoproterozoic mafic dike swarms and associated large igneous provinces from the eastern Dharwar craton, India. Int Geol Rev 57(11–12):1462–1484

Suresh G, Jaiswal N (2003) Specialised thematic Mapping of the southern extension of Jonnagiri schist belt, Julakalva schist belt and adjoining granitoids of Peninsular Gneissic Complex. Rec Geol Surv India 136(5):11–12

Suresh G, Viswanatha Rao N (1994) Study of granitoids of Gooty—Singanamala area, Anantapur district. Rec Geol Surv India 128(5):376–377

Tan W, Peng L, He H, Wang YC, Liang X (2016) Mineralogy and origin of exsolution in Ti-rich magnetite from different magmatic Fe–Ti oxide-bearing intrusions. Can Miner 54(3):539–553

White RV, Crowley JL, Myers JS (2000) Earth's oldest well-preserved mafic dyke swarms in the vicinity of the Isua greenstone belt, southern West Greenland. Geol Greenl Surv Bull 186:65–72

Wyers PG, Barton M (1986) Petrology and evolution of transitional alkaline - sub alkaline lavas from Patmos, Dodecanesos, Greece: evidence for fractional crystallization, magma mixing and assimilation. Contrib Miner Petrol 93(3):297–311

Zhao G, Sun M, Wilde SA, Li S (2004) A Paleo-Mesoproterozoic supercontinent: assembly, growth and breakup. Earth Sci Rev 67:91–123

Geochemistry, Petrogenesis and Tectonic Significance of the Proterozoic Mafic Dykes from the Bomdila Area, NE Lesser Himalaya, India

Shaik A. Rashid, Javid A. Ganai, Irfan Maqbool Bhat and Naqeebul Islam

Abstract The present study documents a new set of comprehensive whole rock geochemical data of metabasic rocks from Lesser Himalayan sequence, western Arunachal Pradesh, NE India. They occur as sills and dykes within Paleoproterozoic Bomdila and Rupa Group of rocks and are metamorphosed to amphibolite grade metamorphism. Hornblende and plagioclase are the dominating minerals with minor constituents of Fe-oxides, quartz, biotite and apatite in these rocks. They are sub-alkaline tholeiitic in nature, having composition of basalt to basaltic andesite. They are enriched in LREE (light rare earth elements) and LILE (large ion lithophile elements) with depleted HFSE (high field strength elements) characteristics. The geochemical signatures e.g., Fe-enrichment, high Ti/Zr ratio (>52) but low Th/Ta ratio (0.23–0.71) and large variation in La/Yb (5.43–10.21), Zr/Y (0.15–15.14), Ti/Y (261–1133) and Zr/Nb (0.15–16.42) ratios indicate that these rocks were probably derived from an enriched lithospheric mantle source due to varying degrees of partial melting and subsequently the melt was modified by fractional crystallization of olivine, pyroxene and plagioclase rather than crustal contamination. Originally these rocks were emplaced in a continental rift tectonic environment and witnessed amphibolite grade metamorphism during Himalayan orogeny. The geochemical characteristics of the studied metabasic rocks are consistent with the Paleoproterozoic mafic rocks reported from other parts of Lesser Himalaya like Chail and Juthogh amphibolites, Garhwal flows and dykes, Mandi-Darla-Rampur volcanics indicating a same magmatic event along the Lesser Himalaya around Paleoproterozoic.

S. A. Rashid (✉) · N. Islam
Department of Geology, Aligarh Muslim University, Aligarh, India
e-mail: rashidamu@hotmail.com; shaik.rashidamu@gmail.com

N. Islam
e-mail: nislam47@gmail.com

J. A. Ganai · I. M. Bhat
Department of Earth Sciences, University of Kashmir, Srinagar, India
e-mail: jganai.ganai9@gmail.com

I. M. Bhat
e-mail: imbhat89@gmail.com

R. K. Srivastava et al. (eds.), *Dyke Swarms of the World: A Modern Perspective*, Springer Geology, https://doi.org/10.1007/978-981-13-1666-1_11

Keywords Mafic dykes · Proterozoic geochemistry · Petrogenesis · Bomdila NE Himalaya · India

1 Introduction

Mafic dykes constitute a common expression of crustal extension and represent a major prospect by which basaltic magma is transferred from the mantle to the upper crust, a process which has operated periodically throughout the last 3.0 Ga of the earth's history. In addition, mafic dykes provide a window into the composition and evolution of the sub-continental mantle and, hence, the tectonic evolutions of the overlying continental crust (Tarney 1992). Mafic rocks of Precambrian to Eocene age are well-preserved, mainly as volcanic flows, dykes and sills, throughout the Himalayan orogenic belt (Ahmad and Bhat 1987; Ahmad et al. 1999). Ahmad et al. (1999) suggested that the age concordance of Precambrian mafic magmatism covering the northern margin of Indian plate imply that 1800–2000 Ma was a period of major crustal evolution in a rift tectonic environment. The western Himalayan magmatism is generally restricted to the Lesser Himalaya and has been the subject of extensive geological and geochemical investigations (Bhat and Ahmad 1990; Ahmad and Tarney 1991; Sahai and Srivastava 1997; Ahmad et al. 1999; Srivastava and Sahai 2001; Ahmad 2008; Srivastava et al. 2009; Bikramaditya Singh and Singh 2013; Srivastava and Samal 2018). However, in the eastern Himalaya, studies have been confined to voluminous mafic magmatism of the Siang window (Jain and Thakur 1978; Bhat 1984; Roychowdhury 1984; Bhat and Ahmad 1990; Singh 1993; Sengupta et al. 1996; Singh 2006; Singh and Bikramaditya Singh 2012). Besides, only few occurrences of mafic magmatic rocks of different geochemical characteristics are reported from western Arunachal Himalaya. Geochemical studies on Sela group mafic intrusives from Central Crystallines of western Arunachal have depleted geochemical characteristics (Srivastava et al. 2009) similar to Vaikrita amphibolites of western Garhwal Himalaya (Ahmad et al. 1999). However, metabasic rocks associated with Lesser Himalayan Sequence of western Arunachal have enriched geochemical characteristics (Bikramaditya Singh and Singh 2013; Rashid and Islam 2016) consistent with Chial and Jutogh amphibolites of western Himalaya (Ahmad et al. 1999; Ahmad 2008) and Proterozoic Basal Aravalli volcanics (Ahmad and Tarney 1994). Recently, Srivastava and Samal (2018) studied Paleoproterozoic mafic intrusive rocks from the western Arunachal Himalaya for their geochemical characteristics and petrogenesis and suggested presence of a widespread ~1.9 Ga large mafic magmatic event in the Indian shield. Not much detailed geochemical studies of metabasic dykes and sills from the Bomdila and Rupa Groups of the Lesser Himalayan sequence have been carried out earlier. Rashid and Islam (2016) have conducted preliminary geochemical characters of these mafic rocks. The aim of this paper is to study major and trace element geochemical characteristics of metabasic rocks from the Lesser Himalayan sequence and to decipher their petrogenetic processes and possible emplacement environment.

2 Geology of the Study Area

The geology of the western Arunachal Himalaya has been discussed by previous workers (Verma and Tandon 1976; Bhushan et al. 1991; Kumar 1997). In western Arunachal Pradesh, Himalaya has been divided into three major tectonic units from south to north, as the Sub-Himalaya, Lesser Himalaya and Higher Himalaya (Thakur 1986). The Sub-Himalaya or outer Himalaya consists of a thick sequence of Siwalik sedimentary rocks, and thrusts with Brahmaputra alluvium in the south along north dipping Himalayan Frontal Thrust (HFT) and towards north is in thrust contact with Lesser Himalaya along the Main Boundary Thrust (MBT). The Lesser Himalaya is represented by low-medium grade metamorphic rocks known as Lesser Himalayan Crystallines (LHC) with a minor portion of the Lesser Himalayan Sedimentary Sequence (LHS) and Gondwana Group. The LHC overlies the Gondwana Group along the Bomdila Thrust, further north the LHC is in thrust contact with the Higher Himalayan Crystallines (HHC) along the Main Central Thrust (MCT).

The study area is situated in the Lesser Himalaya, West Kameng district of Arunachal Pradesh, NE India and is bounded by many important thrust faults on either side (Fig. 1). Although many workers have studied the area and proposed different stratigraphic names (Acharyya 1971; Acharyya et al. 1975; Verma and Tandon 1976; Kumar 1997), however, on the basis of lithostratigraphy, grade of regional metamorphism and associated igneous intrusives, the Proterozoic rocks of Arunachal Himalaya have been grouped into three major tectono-stratigraphic units by Srinivasan (2001); these include Sela Group, Rupa Group and Bomdila Group. The Sela Group which is considered to be the oldest sequence (Paleoproterozoic) among the three is best exposed around Sela Pass along Bomdila-Tawang road in western Arunachal Himalaya near Bhutan border. It consists of calc-silicates, marble, kyanite-sillimanite $\pm$ staurolite polyphase deformed schists, migmatites, high-grade ortho-augen gneisses and amphibolites etc. with younger intrusions of hornblende granite (481 ± 23 Ma, Dikshitulu et al. 1995), tourmaline granite (29 ± 7 Ma, Bhalla and Bishui 1989), pegmatites and aplites. The Rupa Group of rocks belong to Mesoproterozoic and constitutes thick sequence of low to medium grade garnetiferous-biotite-muscovite schists, phyllites, sericite quartzites, calc-silicates and tremolite - actinolite marbles. Main Central Thrust (MCT) separates the Sela Group of Higher Himalayan Crystallines from Rupa Group which in turn unconformably overlies the Bomdila Group in the Lesser Himalaya. The Bomdila Group essentially comprises low to medium grade metasedimentary rocks (mainly phyllites, garnetiferous mica schists and quartzites) intruded by Paleoproterozoic Bomdila augen gneisses and metabasic dykes. The Bomdila gneiss, which has been characterized as orthogneiss on the basis of textural properties, is a batholithic dimension body occupying ~500 km^2 area in the western Arunachal Pradesh, NE India. It is characterized by medium to coarse grained, well defined porphyritic augen gneisses wherein the quartz/albite augens, measuring 1–10 cm, are wrapped with biotite and muscovite. The augen gneisses exposed around Bomdila town have been dated as 1914 ± 23 Ma by Rb/Sr technique by Dikshitulu et al. (1995) and hence are considered to be Paleoproterozoic in age. Various metabasic rocks occur within Rupa and Bomdila Group rocks have discordant field relations and reaction zones noticed at number of places. These field characteristics clearly demonstrate that these metabasics are younger (Paleoproterozoic) than host gneisses and might have intruded into the basement rocks as well.

3 Field and Petrographic Studies

The sampling locations of metabasic dykes and sills intruded in the Rupa Group (i.e., granitoids and metasediments) and Bomdila Group rocks (i.e., metasediments) of Lesser Himalaya are shown in the Fig. 1. These mafic rocks are metamorphosed to amphibolites, commonly foliated in nature and occur as intrusives in gneisses (Fig. 2a) and metasediments (Fig. 2b, c), representing the younger magmatic phase. Besides aplitic and pegmatitic dykes that intrude all phases of the granitoid plutons, two types of mafic dykes were observed in the field. First type includes circular to ovoid shaped basic rocks, ranging in size from 20 to 1 m probably representing the older basement components. Sporadic occurrences of sulphide mineral specks are observed in almost all these mafic enclaves. The other type includes the mafic intrusive rocks which occur in the form of dykes and sills within granitoid gneisses. These mafic intrusive rocks range in thickness from few meters to 10 m. Grain size variation from margin to core is conspicuous with and without amygdales. The contact between the mafic rocks and host granitoids is sharp; however, mafic rocks

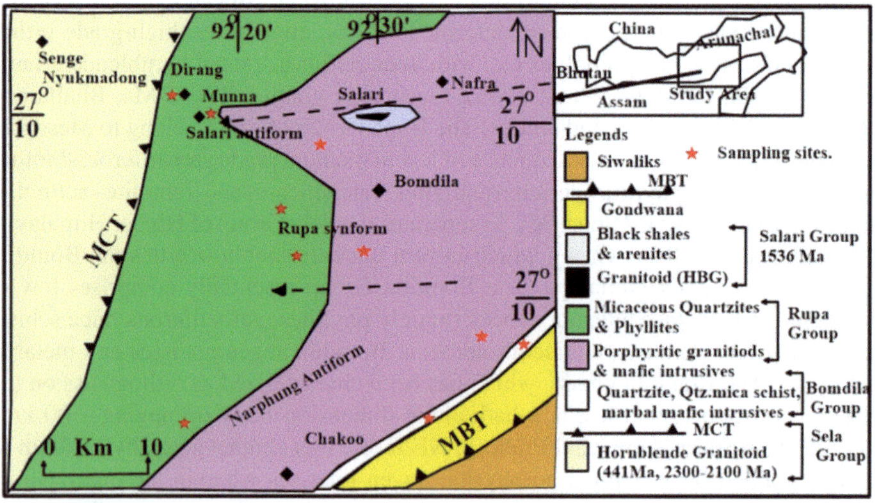

Fig. 1 Simplified geological map of the study area, western Arunachal Himalaya (after Srinivasan 2001)

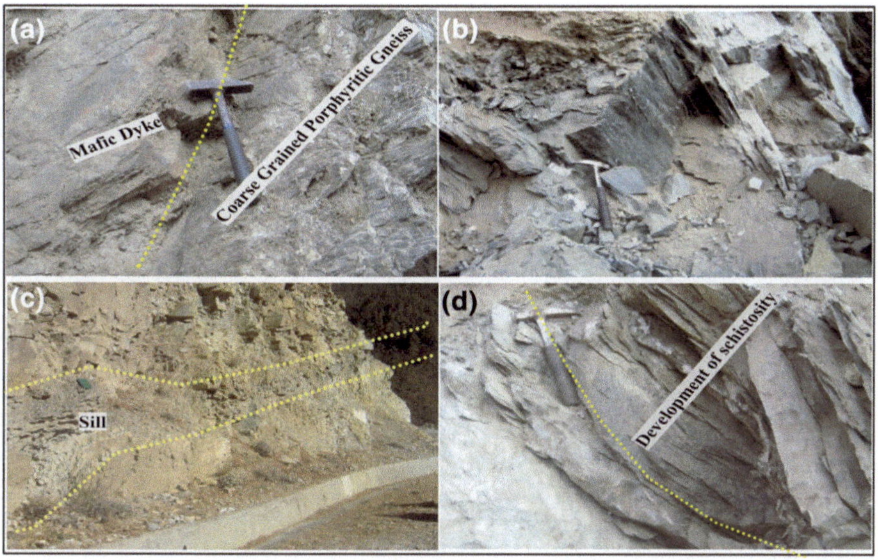

Fig. 2 Showing field relationship of Bomdila and Rupametabasic rocks within Lesser Himalaya **a** and **b** mafic dyke intruding into coarse grained porphyritic gneiss at Bomdila town, **c** mafic sill in porphyritic gneiss of Bomdila Group and **d** development of schistosity in the metabasics

have become almost schistose along the contact (Fig. 2d). At number of places, it has been observed that a marginal zone of about 1 m between granitoids and mafic rocks has been developed where considerable reaction between the two rock types has taken place. Least weathered samples of metabasic rocks from the study area were collected from the central part of the exposure, with minimum possibilities of contamination.

Photomicrographs of the studied thin sections of metabasics reveal sparsely distributed Plagioclase and hornblende phenocrysts. The chief mineral constituents are hornblende and plagioclase (oligoclase-andesine) with minor constituents of quartz and biotite. Opaques (Fe–Ti oxides), chlorite, epidote and sphene constitute minor accessory mineral phases. No garnet was observed in the studied rocks. These metabasics have pronounced schistosity which is marked by parallel orientation of hornblende and plagioclase crystals (Fig. 3a, b). At few places, recrystallization of quartz crystals with well-defined sutured margins are developed (Fig. 3b). In some thin sections, the plagioclase crystals with well-preserved twinning are encountered with biotite. Large euhedral crystals of plagioclase abut against foliated biotite; this indicates post-tectonic crystallization of plagioclase. Biotite flakes, relatively less abundant than hornblende, are aligned parallel to hornblende crystals. Subhedral epidotes are present in association with amphiboles; the epidote is secondary in nature that is formed by the breakdown of amphiboles due to late stage hydrothermal alteration. Calcite occurs as subhedral crystals and at places is associated with quartz.

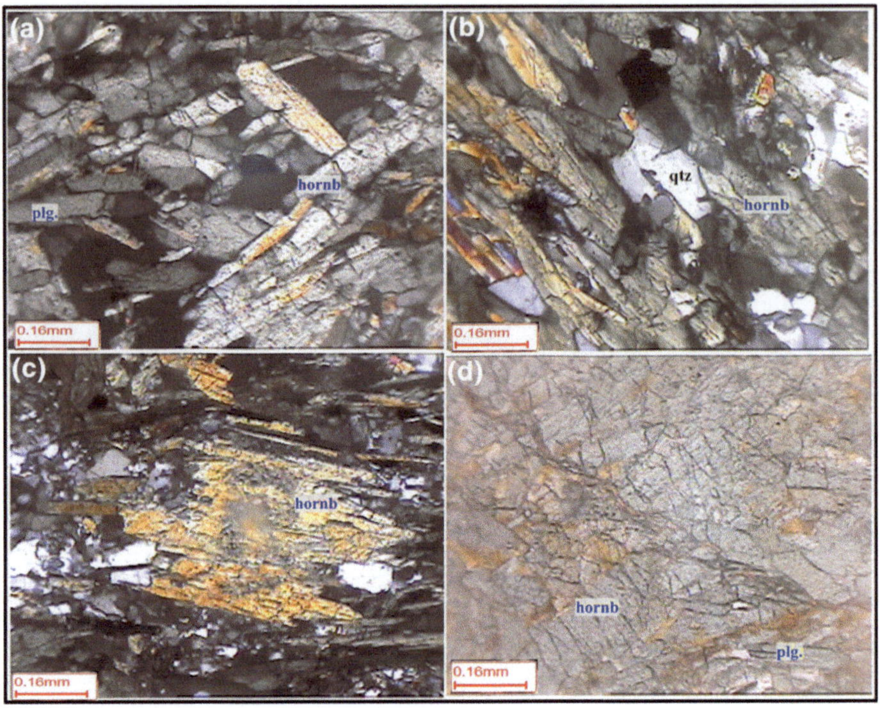

Fig. 3 Photomicrographs of Bomdilametabasic rocks from Lesser Himalaya with 5× magnification; amphibolites showing **a** and **b** the weak alignment of hornblende (hornb) and plagioclase (plg.) minerals, development of abundant hornblende and few quartz grains, **c** prismatic crystal of hornblende with two sets of cleavage, surrounded by plagioclase and opaque grains and **d** well-developed grains of hornblende and plagioclase

4 Geochemistry

4.1 Analytical Methods

After petrographic study of collected rock samples, eleven least weathered samples were selected for whole rock major and trace element analysis. All the major oxides were determined by X-ray fluorescence (XRF) spectrometry (Siemens SRS-3000 Sequential X-ray Spectrometer) at the Wadia Institute of Himalayan Geology, Dehradun, India by using pressed pellets. Accuracy and precision were obtained by repeated analysis of geostandards and evaluated by the laboratory. For XRF analyses, the accuracy at 95% confident was well within 5%while the precision was better than 3% RSD for major elements. Trace elements together with rare earth elements, were analyzed after digestion of samples with HF–HNO_3 (7:3, acid mixture) in Savillex screw-top vessels. Solutions were analyzed at National Geophysical Research Institute (NGRI), Hyderabad, by high-resolution inductively coupled mass spectrometer

(HR-ICP-MS; Nu Instruments Attom, Wrexham, United Kingdom) in jump-wiggle mode at a moderate resolution of 300, which permits all the analytes of interest to be measured accurately. Certified reference materials G-2 (USGS) and JG-2 (Japan) along with couple of procedural blanks were also prepared with the sample batch by adopting the same protocol described above to negate errors due to reagent and handling. For trace elements the procedure, precision and detection limits are the same as given by Balaram et al. (1996). The analytical results demonstrate a high degree of machine accuracy and precision better than an RSD of 3% for the majority of trace elements. The whole-rock geochemical data of studied metabasic rocks are presented in Table 1.

4.2 Post Crystallization Alteration Effects on Whole Rock Geochemistry

Prior to any interpretation of representative geochemical data, it is necessary to identify the effect of post-emplacement alteration processes on the whole-rock chemistry. Alterations like saussuritization, epidotization, silicification and chloritization occur purely as a response to progressive low-grade regional metamorphism (Condie et al. 1977). The studied rocks have undergone amphibolite grade metamorphism, which may have mobilized some of the elements. Alkali elements are the prime suspect of alteration. To elucidate the nature and the extent of migration of these alkali elements, samples were plotted in the Na_2O/K_2O versus $Na_2O + K_2O$ diagram (Fig. 4a) of Miyashiro (1975) and CaO/Al_2O_3–$MgO/10$-$SiO_2/100$ ternary diagram (Fig. 4b) of Schweitzer and Kroner (1985). It is evident from these plots that all the samples fall mainly in "not altered" field. However, slight increase in the values of Na_2O may be attributed to the albitization or spilitisation process which might have replaced Ca by Na to a varying degree. The consistent variations of majority of the major and trace elements denote that the present rocks have preserved much of their primary igneous chemistry, giving rise to acceptable intra-suite correlations for the immobile elements. Large ion lithophile elements (LILE) such as Ba, Rb, Sr etc. except Th generally show mobile nature during secondary alteration effects (Cann 1970; Pearce and Cann 1973; Condie and Sinha 1996). This effect on LILE concentrations can be revealed by Rb/Sr ratio which is very low (0.007) in least altered basaltic rocks and very high (8) in highly altered mafic rocks (Lafleche et al. 1992). The present studied metabasic samples have lower ratios of Rb/Sr (0.02–0.42) thus indicates that the secondary processes have not altered primary concentrations of LILEs. In addition, normalized REE and multielement patterns (Fig. 8) are regular and consistent suggesting that these incompatible trace element abundances and their ratios appear to reflect primary magmatic characteristics. It can be inferred from the above observations that though most of the elements in mafic rocks show their primary igneous signatures, some nonsystematic variations, however, warrant a caution against the relying too much on major element data while using for petrogenetic interpretation.

Table 1 Major oxides and trace element concentrations of metabasic rocks from the Bomdila and Rupa Group rocks of Lesser Himalaya, western Arunachal Himalaya

Major oxide (wt%)	BG-57	BG-5A	BG-58	BG-58A	BG-60	BG-61	BG-59	BG-79	DG-8(AR)	DG-11	RG-33C
SiO_2	53.52	51.01	44.82	52.92	46.64	49.09	52.54	47.91	46.3	46.46	50.17
TiO_2	1.72	1.42	1.32	2.49	2.17	2.27	2.41	3.15	3.6	2.07	2.73
Al_2O_3	12.55	12.05	11.76	12.15	12.22	11.44	13.67	11.92	12.28	12.66	13.66
$Fe_2O_3^t$	14.55	13.96	16.13	15.72	15.64	15.13	10.94	16.28	15.93	15.64	14.6
MnO	0.24	0.23	0.24	0.21	0.21	0.18	0.14	0.24	0.23	0.23	0.21
MgO	4.1	7.23	7.91	7.66	6.72	10.5	4.8	5.94	6.26	7.73	5.68
CaO	5.67	6.79	7.62	2.59	5.88	2.62	10.49	9.49	9.7	11.36	7.81
Na_2O	3.45	3.37	2.64	2.65	4.37	2.76	3.21	2.24	1.91	1.21	2.43
K_2O	2.72	2.11	1.84	0.24	1.07	3.05	0.47	1.37	1.12	0.93	1.22
P_2O_5	0.24	0.18	0.18	0.23	0.2	0.16	0.41	0.75	0.41	0.25	0.54
LOI	0.94	1.4	1.95	4.47	2.15	3.63	1.52	0.88	0.44	0.77	1.19
Sum	99.71	99.75	96.41	101.33	97.27	100.83	99.08	99.29	97.74	98.54	99.5
Mg#	37.30	52.23	50.87	50.71	47.57	59.44	48.09	43.51	45.35	51.06	45.10
CaO/Al_2O_3	0.45	0.56	0.65	0.21	0.48	0.23	0.77	0.80	0.79	0.90	0.57
$Fe_2O_3^t/MgO$	3.55	1.93	2.04	2.05	2.33	1.44	2.28	2.74	2.54	2.02	2.57

(continued)

Table 1 (continued)

Major oxide (wt%)	BG-57	BG-5A	BG-58	BG-58A	BG-60	BG-61	BG-59	BG-79	DG-8(AR)	DG-11	RG-33C
Trace elements (in ppm)											
V	295	329	334	322	298	363	353.95	429.99	396.30	344.10	354.23
Sc	41	43	42	39	30	31	26.25	51.14	42.36	39.26	35.92
Co	38	40	44	45	38	39	31.28	61.32	76.16	69.40	56.60
Cr	67	79	95	42	83	88	71.01	87.83	76.79	69.34	82.37
Rb	40	24	23	22	26	41	11.03	64.98	58.02	24.35	68.23
Ba	515	314	277	35	183	240	79.35	323.31	257.53	91.41	328.19
Sr	107	97	81	nd	81	12	518.81	232.24	349.18	141.85	262.49
Zr	197	133	143	212	141	147	5.94	12.73	8.22	6.79	18.18
Y	27	19	20	14	12	12	39.92	72.09	43.31	28.87	55.71
Nb	12	9	10	22	12	12	40.04	32.66	46.20	20.36	30.90
Ni	nd	nd	nd	nd	nd	nd	14.57	23.37	32.89	36.42	20.73
Cu	nd	nd	nd	nd	nd	nd	38.43	94.56	198.55	108.10	76.17
Zn	nd	nd	nd	nd	nd	nd	73.28	127.78	121.52	87.79	134.13
Ga	nd	nd	nd	nd	nd	nd	23.27	27.63	24.28	20.50	26.51
Cs	nd	nd	nd	nd	nd	nd	0.45	3.95	2.91	0.49	9.77
Hf	nd	nd	nd	nd	nd	nd	0.33	0.87	0.57	0.47	0.41
Ta	nd	nd	nd	nd	nd	nd	9.86	7.26	10.59	7.96	8.71
Pb	nd	nd	nd	nd	nd	nd	8.94	8.03	6.83	7.62	11.20
Th	nd	nd	nd	nd	nd	nd	2.26	3.82	4.16	2.02	6.15

(continued)

Table 1 (continued)

Major oxide (wt%)	BG-57	BG-5A	BG-58	BG-58A	BG-60	BG-61	BG-59	BG-79	DG-8(AR)	DG-11	RG-33C
Rare-earth elements (in ppm)											
La	30	20.1	20	nd	24.3	17.6	21.60	34.46	28.07	15.96	36.56
Ce	66.3	42.4	45.2	nd	47.9	42.4	48.64	78.35	63.58	36.13	76.70
Pr	nd	nd	nd	nd	nd	nd	5.42	8.83	7.08	4.07	8.45
Nd	36	23.8	23.2	nd	23.7	26.5	30.18	50.44	39.32	22.82	45.46
Sm	6.61	4.79	4.66	nd	5.44	5.43	6.35	11.78	8.61	5.02	9.92
Eu	1.5	1.24	1.24	nd	1.4	1.61	2.27	3.72	2.80	1.75	2.83
Gd	6.1	4.53	4.69	nd	4.61	4.87	7.69	15.04	10.23	6.17	12.36
Tb	nd	nd	nd	nd	nd	nd	1.20	2.35	1.55	0.96	1.86
Dy	6.05	4.67	4.83	nd	4.49	4.32	6.15	11.85	7.45	4.77	9.16
Ho	nd	nd	nd	nd	nd	nd	1.31	2.46	1.50	0.99	1.85
Er	2.87	2.82	2.75	nd	2.5	2.68	4.15	7.55	4.59	3.08	5.66
Tm	nd	nd	nd	nd	nd	nd	0.65	1.22	0.73	0.50	0.87
Yb	3.84	3.13	3.36	nd	2.38	2.6	3.19	6.35	3.75	2.68	4.44
Lu	0.25	0.24	0.25	nd	0.14	0.17	0.43	0.93	0.57	0.42	0.60

nd represents not determined

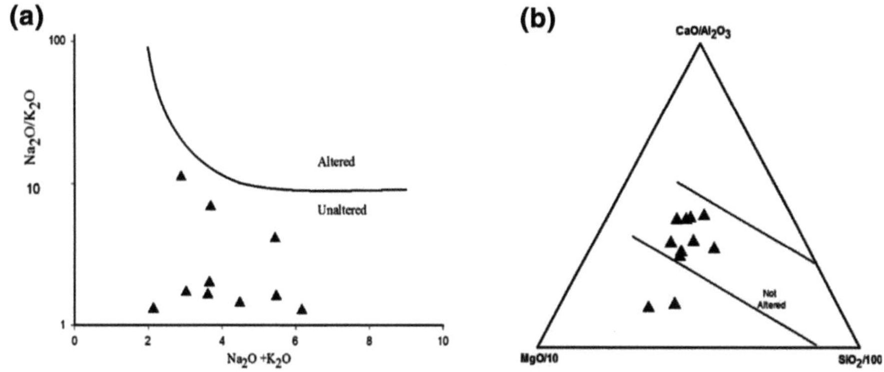

Fig. 4 **a** Na_2O/K_2O versus Na_2O+K_2O binary diagram (after Miyashiro 1975), **b** CaO/Al_2O_3–$MgO/10$-$SiO_2/100$ ternary plot (Schweitzer and Kroner 1985) for the Bomdilametabasic rocks of Lesser Himalaya

For this reason, the elements which are known to be immobile or less mobile during secondary alteration processes such as rare earth elements (REE), HFSE as well as Ti, P, Cr, Ni can be used for petrogenetic interpretation (Winchester and Floyd 1977; Saunders et al. 1980; Rollinson 1993).

4.3 Geochemical Characteristics

The studied metabasic rocks have restricted range of SiO_2 (45–53wt%). The concentration of other major elements such as; CaO, Al_2O_3, TiO_2, P_2O_5 and MnO ranges from (7.18–11.36 wt%), (11.2–13.01), (2.5–4.0 wt%), (0.16–0.8 wt%) and (0.15–0.25 wt%) respectively. In total alkali silica (TAS) classification diagram after (Le Maitre 2002) the studied Bomdila and Rupa metabasic rocks plot mainly in the field of basalt to basaltic-andesite (Fig. 5). Most of the samples have Nb/Y ratios <0.7 indicating their sub alkaline affinity (Pearce and Gale 1977).

In YTC ($Y = Y + Zr$, $Ti = TiO_2$ wt% $\times$ 100, $C = Cr$) ternary diagram (Davies et al. (1979), most of the studied samples plot along and above the tholeiitic trend (Fig. 6a). This tholeiitic nature is very clearly observed on Jensen's cation triangular plot (Al–(Fe + Ti)–Mg) (Jensen 1976), in which studied samples plot in Fe-tholeiitic field (Fig. 6b).

Various binary variation diagrams have been plotted to understand the crystallization behavior of studied rock samples (Fig. 7). In these binary plots, we use some known mobile elements e.g., CaO, Na_2O, K_2O and Sr and less mobile or immobile elements e.g., SiO_2, Al_2O_3, TiO_2, P_2O_5, Y, Yb, La, Ce and Nb (Pearce and Cann 1973; Flyod and Winchester 1978; Rollinson 1993) against MgO. These plots show overall magmatic trends, suggesting that these elements are immobile. The studied samples show systematic increase in SiO_2, Al_2O_3, CaO, Na_2O and P_2O_5 with

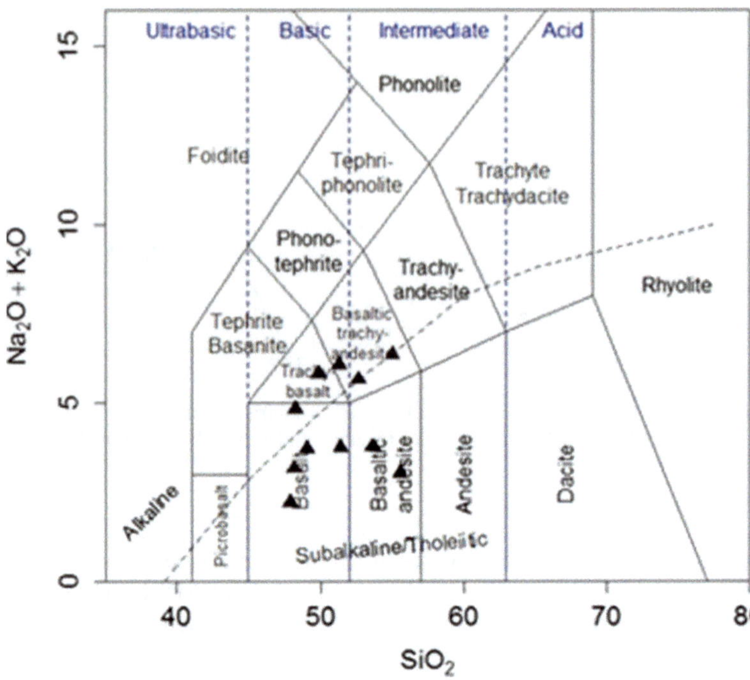

Fig. 5 Total Alkali Silica (TAS) classification diagram (after Le Maitre 2002) for the classification of the metabasic rocks of the Lesser Himalaya from Bomdila area

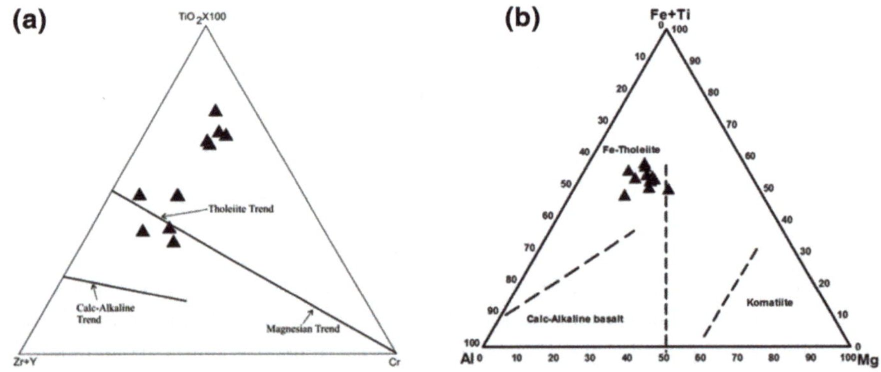

Fig. 6 **a** Y (Zr + Y)–T (TiO_2 × 100)–C (Cr) ternary plot (after Davies et al. 1979) showing the tholeiitic affinity of the basic rocks and **b** Ternary cation% Al–(Fe + Ti)–Mg plot (after Jensen 1976) for Bomdilametabasic rocks

decreasing MgO, whereas TiO_2 show increasing trend with MgO increase. These trends suggest fractionation of mafic magma wherein minerals like olivine, pyroxene and feldspar have fractionated earlier followed by Fe–Ti oxides. Also, in Fig. 7,

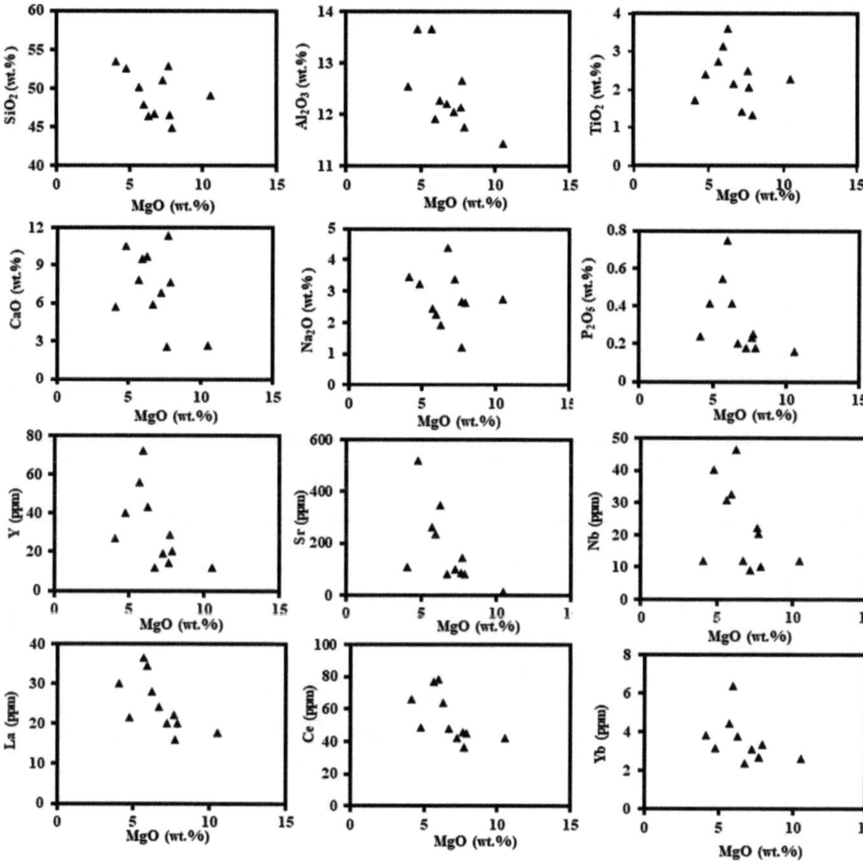

Fig. 7 Harker variation diagrams for Bomdilametabasic rocks showing major and trace element variation against MgO

trace elements such as Y, La, Ce, Nb, Yb and Sr show negative correlation with MgO which is consistent with crystallization of minor minerals at later stage (Srivastava 2012). Negative correlation of Sr with MgO is consistent with high level plagioclase fractionation. The higher concentration of $Fe_2O_3^t$ (10–16 wt%) and TiO_2 (1.32–3.6 wt%) in the studied samples indicates fractional crystallization of Mg and Ca—rich minerals at depth, facilitating late stage crystallization of Fe–Ti oxides at higher levels (Ahmad and Tarney 1991).

The chondrite normalized REE and primitive mantle-normalized incompatible multi-element patterns for the present metabasic rocks are shown in Fig. 8. The studied samples have sub-parallel REE patterns with enriched LREE [$(La/Yb)_N$ ratio 3.89–7.3] and slightly depleted HREE patterns with low Eu negative anomaly ($Eu/Eu^* = 0.72$–0.96) (Fig. 8a), reflecting their generation from similar mantle sources. Also the enrichment of large ion lithophile elements (LILE) is evident in

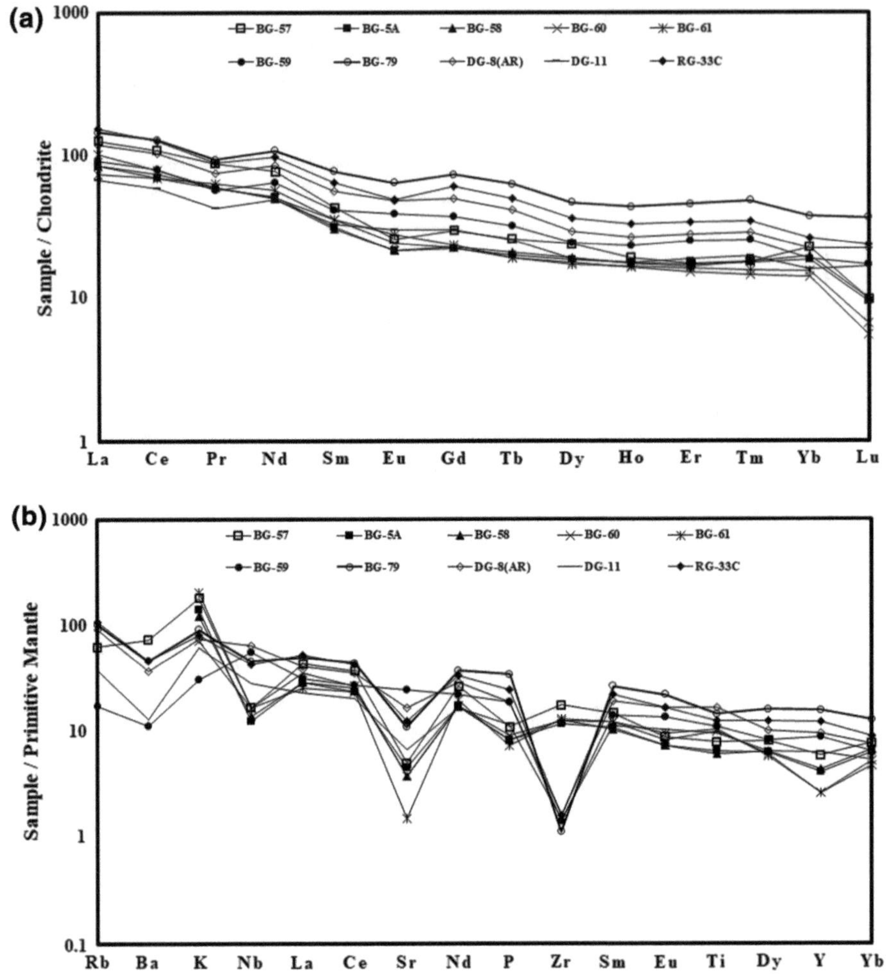

Fig. 8 **a** Chondrite-normalized REE patterns and **b** primitive mantle normalized multi- element patterns for Bomdilametabasic samples. Chondrite and primitive mantle normalized values are after Sun and McDonough (1989)

multi-element patterns (Fig. 8b) with distinct negative Sr anomaly in all the samples but selective depletion of HFSE. Some samples are depleted in Nb, P and Y whereas few samples depict Zr negative anomaly. This variable depletion of HFSE in the studied rocks is likely due to differences in depth and degrees of partial melting (Ahmad et al. 1999). Observed prominent negative Sr anomaly in studied samples indicates shallow level fractionation of plagioclase feldspar during evolution of the melt (Tarney and Jones 1994) which is also corroborated by the low Eu negative anomaly in few samples.

The enriched LREE-LILE and depleted HFSE patterns in the studied metabasic rock (Fig. 8b) may be attributed to either enriched mantle source or crustal contamination (Ahmad and Tarney 1991). Various geochemical criteria are often used to monitor the possibility of crustal contamination (Rollinson 1993). The incompatible element ratios such as Nb/La and Nb/Ce are proxy indicators used to constrain the influence of crustal inputs on the mafic rock compositions (Li et al. 2006; Sandeman et al. 2006). The Nb/La values of 0.40–1.85 (average 0.83) and Nb/Ce values of 0.18–0.82 (average 0.37) of studied metabasic samples are higher than average bulk crustal composition of 0.69 and 0.33 respectively of Taylor and McLennan (1985) reflecting crustal contamination. However, their Fe-enrichment, lower Th/Ta ratio (<0.71) and higher Ti/Zr ratios (>52) does not support the above interpretation (Arndt and Jenner 1986). Furthermore, their regular REE patterns are against crustal contamination which would otherwise exhibit steeper slope for LREE compared to HREE. Therefore, the observed trace element characteristics of the studied metabasic rocks are primarily inherited from a LREE-LILE enriched mantle source depleted in HFSE (Ahmad and Tarney 1991).

5 Discussion

5.1 Petrogenesis

Due to the immobile and incompatible behavior of trace elements during post-crystallization alteration processes, the incompatible trace element ratios and multi-element spidergrams are commonly useful in magma identification and source characterization of any magma suite (Pearce 1982, 2008; Rollinson 1993). Ratios of highly incompatible trace elements such as Nb/La, Ce/Nd, Zr/Nb, Ta/Th etc. do not change significantly due to moderate degrees of melting and fractional crystallization so are expected to reflect the source characteristics (Saunders et al. 1988; Ahmad and Tarney 1991; Petterson and Windley 1992; Condie 1997; Srivastava et al. 2008; French and Heaman 2010). Some of the important characters like low magnesium numbers (mostly < 60) and negative Sr as well as Eu anomalies of the studied metabasic rocks suggest that fractional crystallization has played a role in the generation of these rocks. The positive correlation of CaO/Al_2O_3 versus $Fe_2O_3^t/MgO$ (Table 1) further indicates that olivine, pyroxene and plagioclase were the main fractionating phases. However, a large variation in incompatible element ratios (e.g., La/Yb, Zr/Y, Ti/Y, Zr/Nb) in the studied rocks cannot be explained by crystal fractionation alone. In Zr/Y versus Zr plot (Fig. 9), large variation in Zr/Y ratio is depicted with increasing Zr abundance. Generally, pyroxene and amphibole fractionation is believed to slightly increase this ratio in basaltic rocks (Floyd 1993) however, this ratio is more sensitive to variable degrees of partial melting of similar mantle sources (Ahmad et al. 1999). Therefore, such large variation of Zr/Y and other incompatible element ratios in studied rocks depicts different degrees of partial melting of a similar mantle

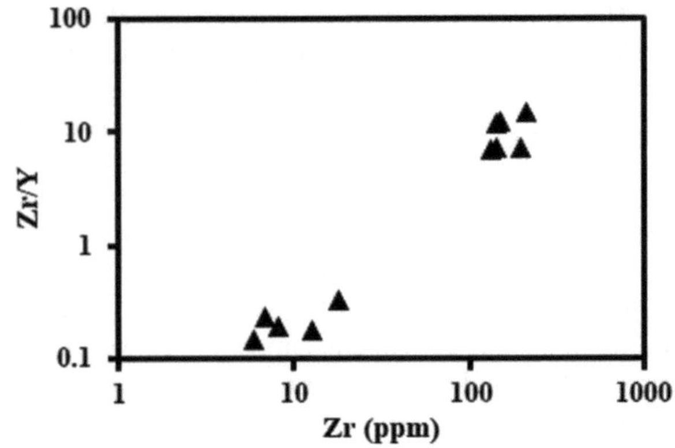

Fig. 9 Binary plot of Zr/Y versus Zr (after Sun and Nesbitt 1977) for the Bomdilametabasics from Lesser Himalaya

source. Similar observations were observed in Garhwal dykes of Lesser Himalaya by Ahmad et al. (1999). It may be concluded from the above discussion that the studied metabasic rocks were derived due to different degrees of partial melting of an enriched mantle source subsequently followed by fractionation.

The studied samples plotted on $(La/Yb)_N$ versus $(La)_N$ plot (Fig. 10) that further supports the inference that these rocks were generated by different degrees of partial melting of a common mantle source. In general, on this type of plot, a linear relationship will be defined by the samples if they are derived by different degrees of partial melting of a common source however, the samples may shift horizontally to higher X-axis values because of crystal fractionation (Feigenson et al. 1983; Bradshaw et al. 1993; Vimal et al. 2012; Hirahara et al. 2015). A broad positive linear pattern is exhibited by the studied rocks in $(La/Yb)_N$ versus $(La)_N$ diagram, indicating that they were formed by different degrees of melting of a source with similar $(La/Yb)_N$. Lower degrees of melting resulted the samples with higher La and La/Yb whereas samples with lower La and La/Yb have formed by higher degree of melting.

To know the depth of mantle melting incompatible element ratio Sm/Yb can trace the presence or absence of residual garnet because fractional crystallization has no effect on it (McKenzie and O'Nions 1991; Zi et al. 2008). Yb is compatible in garnet as compared to Sm and La which are incompatible so are enriched in melts during low degrees of partial melting. The high La/Sm and Sm/Yb ratios greater than 1 in studied samples indicate a magma origin involving low degrees of melting of a garnet bearing mantle source.

The studied metabasic rocks of Lesser Himalayan Sequence are compared for their geochemical characters with two other occurrences from the western Arunachal Himalaya (Fig. 11). Strikingly similar REE patterns were observed for studied metabasics and metabasic of Bikramaditya Singh and Singh (2013) from Lesser Himalaya, however, both are quite distinct with Sela group metabasics of Higher Himalayan Crystallines. This observation suggests that the metabasic rocks reported from the Lesser Himalaya and Higher Himalaya have different petrogenetic history. Srivastava et al. (2009) suggested that the Sela group metabasics of Higher

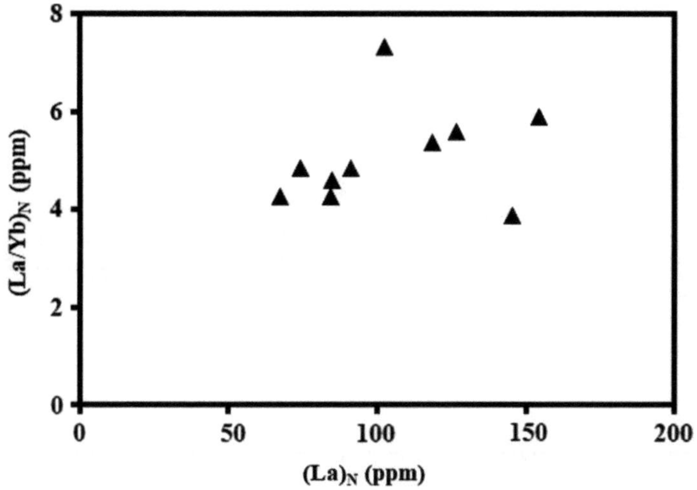

Fig. 10 $(La/Yb)_N$ versus $(La)_N$ for Bomdilametabasic rocks from Lesser Himalaya showing the positive relationship (normalized to primitive mantle after Sun and McDonough 1989)

Himalaya are derived from a depleted asthenospheric mantle and are related to Western Himalayan metabasic rocks such as Vaikrita amphibolites (Ahmad et al. 1999). Srivastava and Sahai (2001) suggested the same for the mafic magmatic rocks reported from the Central Crystallines of the Bhagirath–Yamuna valleys. Recently, Srivastava and Samal (2018) reported high Ti mafic intrusives from the western Arunachal Himalaya with inclined REE patterns similar to the present studied metabasic rocks indicating enriched mantle sources.

The studied metabasic rocks have striking similarity with other Proterozoic metabasic rocks of Lesser Himalaya like Chail and Jutogh amphibolites, Garhwal volcanics and dykes of Ahmad (2008) and metabasics of Bikramaditya Singh and Singh (2013) in terms of normalized REE patterns (Fig. 12), multi-element patterns and some important incompatible element ratios. Also they show similar geochemical characteristics to that of high Ti mafic intrusives of Srivastava and Samal (2018). All these rocks show tholeiitic nature, overall low Mg-numbers (reflecting their evolved nature), enriched LREE-LILE and depleted HFSE characteristics. These geochemical signatures are similar with most of the continental rift volcanics and Proterozoic dyke swarms (Thompson et al. 1983; Weaver and Tarney 1984; Tarney 1992; Hawkesworth et al. 1993).

5.2 Tectonic Setting

An attempt was made to elucidate the tectonic setting of emplacement of Bomdila metabasic rocks. It has been suggested that the basalts from different tectonic set-

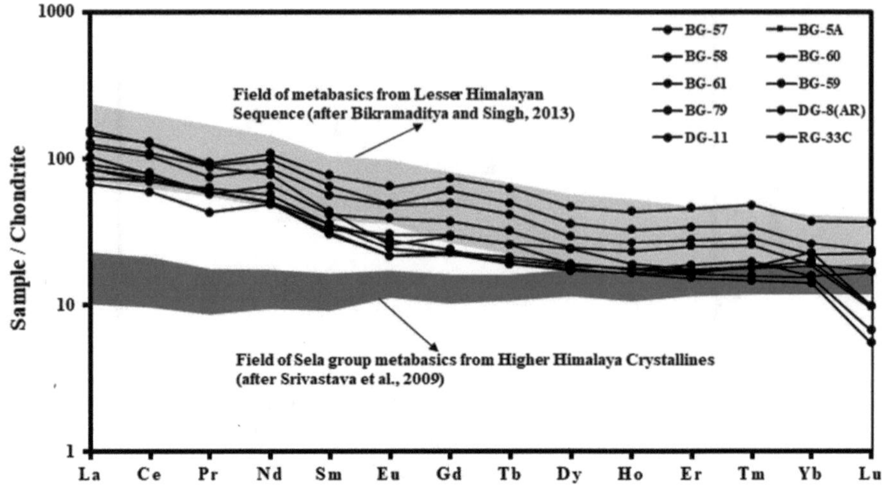

Fig. 11 Comparison of rare-earth element patterns of metabasic rocks (present study) from Lesser Himalayan sequences with Sela group metabasic rocks of Higher Himalayan Crystallines (Srivastava et al. 2009) and metabasic rocks of Lesser Himalayan sequences (after Bikramaditya Singh and Singh 2013)

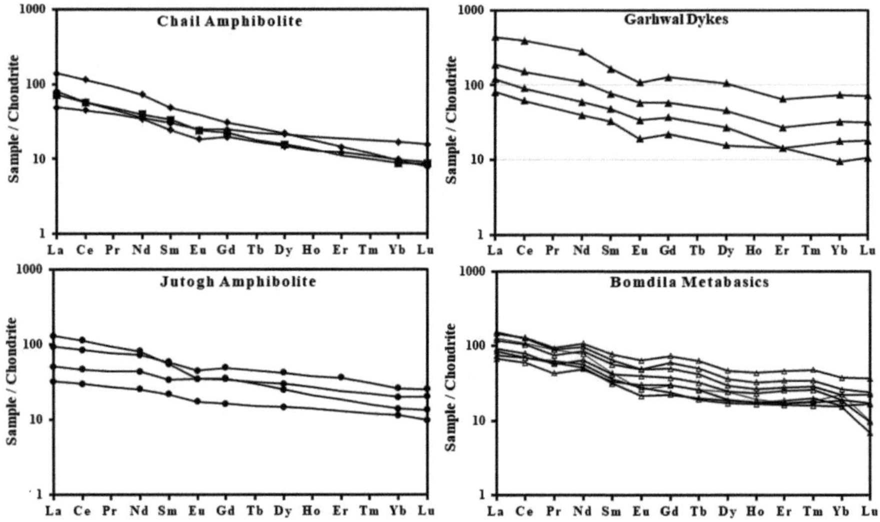

Fig. 12 Comparison of rare-earth element patterns of Bomdilametabasic rocks (present study) with Chail amphibolites, Garhwal dykes and Jutoghamphibolites of Lesser Himalaya (dataafter Ahmad 2008)

tings may often be distinguished by the relative abundances of 'immobile' elements such as Ti, P, Zr, Nb, Y, Cr and REE (Pearce and Cann 1973; Pearce 1975). Saunders et al. (1980) have observed that the basalts from active and remnant arcs are

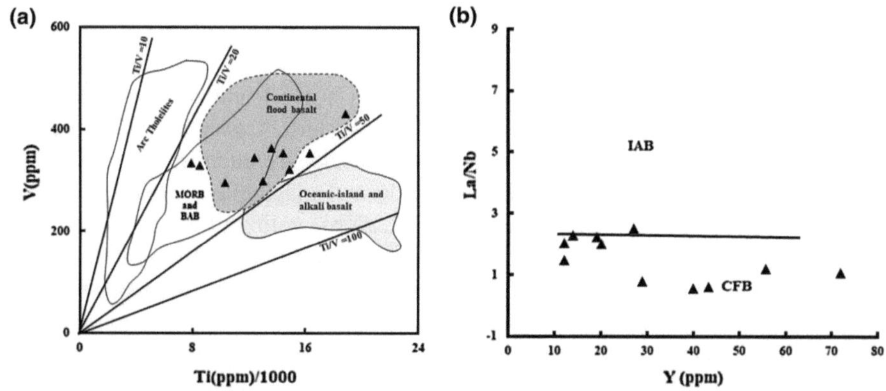

Fig. 13 **a** Ti/1000 versus V diagram (after Shervais 1982) and **b** La/Nb versus Y diagram (after Floyd et al. 1991) distinguishing Bomdilametabasics as continental flood basalts. CFB = continental flood basalts; IAB = island arc basalts

characterized by higher abundance of LILE (Cs, Rb, K, Ba, Sr, Th, U) relative to HFSE (Ti, P, Zr, Hf, Nb and Ta) than ocean ridge basalts. Wang and Glover III (1992) suggested that particularly the continental rift basalts invariably plot outside the within-plate field. The present studied samples are characterized by high V concentration (295–429 ppm), Ti/V > 20, lower La/Nb (0.54–2.50), Zr (<250 ppm) and Ti/Y < 410 all reflect continental Parana flood basalt geochemical characteristics (Erlank et al. 1988). Incompatible immobile element concentrations in basalts are proxy indicators for determining their tectonic environment of eruption especially ratios of La/Nb, Ti/V and Y, V concentrations (Winchester and Floyd 1976; Shervais 1982; Floyd et al. 1991; Kuzmichev et al. 2005). In Ti/1000 versus V tectonic discrimination diagram of Shervais, (1982), the Ti and V are relatively immobile up to high grade granulite facies of metamorphism (Rollinson 1993). In studied metabasic rocks, the Ti/V ratio ranges from 24 to 46 and plots within continental flood basalt field (Fig. 13a). In La/Nb versus Y plot of Floyd et al. (1991) the Bomdila metabasics also plot in the field of continental flood basalts (Fig. 13b). As discussed earlier the enrichment of LREE-LILE and depletion in HFSE also indicates rift tectonic environment. Also the low Nb/La ratio (<1) of studied samples is a characteristic feature of many continental flood basalts like Parana, Karoo and Deccan.

The continental rift tectonic scenario for studied metabasic rocks is corroborated by their remarkable similarities in trace element characteristics with Lesser Himalayan Mandi-Darla-Rampur volcanics of Himachal Pradesh (Ahmad and Bhat 1987; Bhat and Le Fort 1992), Garhwal volcanics (flows and dykes), Chail and Jutogh amphibolites of Utrakhand (Ahmad and Tarney 1991, Ahmad et al. 1999), Proterozoic Basal Aravalli volcanics of NW Indian shield (Ahmad and Rajamani 1991; Ahmad and Tarney 1994) and also with high Ti mafic intrusives (HTMI) from western Arunachal Himalaya (Srivastava and Samal 2018).

6 Conclusions

The Bomdila metabasics occur as sills and dykes in the Rupa and Bomdila group of rocks in Lesser Himalayan Sequence of the western Arunachal Himalaya, and show amphibolite grade metamorphism. These rocks are chemically classified as sub-alkaline tholeiites and contain hornblende and plagioclase feldspar in major proportion. The studied rocks have similar $(La/Yb)_N$ ratios (4.27–5.91) but varying $(La)_N$ (67–154) values indicating that they were formed by different degrees of melting of a source with similar $(La/Yb)_N$ ratios. Presence of enriched LREE and LILE and depleted HFSE concentrations suggest that these rocks were generated from enriched mantle source comparable to sub-lithospheric mantle in rift tectonic environment. The studied metabasic rocks have striking similarity with other Proterozoic metabasic rocks of the Lesser Himalaya like Chail and Jutogh amphibolites, Garhwal volcanics and dykes in terms of normalized REE patterns, multi-element patterns and some important incompatible element ratios, thus pointing towards a strong magmatic event in rift tectonic environment during the Proterozoic time along the Lesser Himalayas, as also suggested by Srivastava and Samal (2018).

Acknowledgements We are thankful to the Chairpersons, Department of Geology, AMU, Aligarh and Department of Earth Sciences, University of Kashmir, Srinagar for providing necessary facilities to carry out this work. Dr. N. K. Saini and Dr. A. K. Singh, WIHG, Dehradun and Dr. V. Balaram, NGRI, Hyderabad are thankfully acknowledged for analytical facility. SAR is thankful to the DST, New Delhi for financial support in the form of a major project. The authors are also thankful to anonymous reviewers for their constructive comments that helped us to improve the quality of the manuscript. We are thankful to Profs. R. K. Srivastava, Richard and Peng for editorial handling.

References

Acharyya SK (1971) Structure and stratigraphy of the Darjeeling frontal zone, Eastern Himalaya. Misc Publ Geol Survey of India 24:71–90

Acharyya SK, Ghosh SC, Ghosh RN, Shah SC (1975) The continental Gondwana Group and associated marine sequences of Arunachal Pradesh (NEFA), Eastern Himalaya. Himalayan Geol 5:60–82

Ahmad T, Bhat MI (1987) Geochemistry and petrogenesis of the Mandi-Darla volcanics, north western Himalayas. Precambr Res 37:231–256

Ahmad T, Tarney J (1991) Geochemistry and petrogenesis of Garhwal volcanics: implications for evolution of the north Indian lithosphere. Precambr Res 50:69–88

Ahmad T, Rajamani V (1991) Geochemistry and petrogenesis of the basal Aravalli volcanics near Nathdwara, Rajasthan. Precambr Res 49:327–340

Ahmad T, Tarney J (1994) Geochemistry and petrogenesis of late Archean Aravalli volcanics, basement enclaves and granitoids, Rajasthan. Precambr Res 65:1–23

Ahmad T, Mukherjee PK, Trivedi JR (1999) Geochemistry of Precambrian mafic magmatic rocks of the Western Himalaya, India: petrogenetic and tectonic implications. Chem Geol 160:103–119

Ahmad T (2008) Precambrian mafic magmatism in the Himalayan mountain range. J Geol Soc India 72:85–92

Arndt NT, Jenner GA (1986) Crustally contaminated komatiites and basalts from Kambalda, Western Australia. Chem Geol 56:229–255

Balaram V, Ramesh SL, Anjaiah KV (1996) New trace element and REE data in thirteen GSF reference samples by ICP-MS. Geostand Geoanal Res 20:71–78

Bhalla JK, Bishui PK (1989) Geochronology and geochemistry of granite emplacement and metamorphism in northeastern Himalaya. Rec Geol Surv India 122:18–20

Bhat MI (1984) Abor volcanics: further evidence for the birth of the Tethys Ocean in the Himalyan segment. J Geological Society of London 141:763–775

Bhat MI, Ahmad T (1990) Petrogenesis and the mantle source characteristics of the Abor volcanic rocks, eastern Himalayas. Geological Society of India 36:227–246

Bhat MI, Le Fort P (1992) Sm-Nd age and Petrogenesis of Rampur meta-volcanic rocks, NW-Himalayas: Late Archaean relicts in the Himalayan belt. Precambr Res 56:191–210

Bikramaditya Singh RK, Singh AK (2013) Geochemistry and petrogenesis of metabasic rocks from the Lesser Himalayan Crystallines, western Arunachal Himalaya, northeast India. Geosci J 17:27–41

Bradshaw TK, Hawkesworth CJ, Gallagher K (1993) Basaltic volcanism in the Southern Basin and Range: no role for a mantle plume. Earth and Planetary Science Letters 116:45–62

Bhushan SK, Bindal CM, Aggarwal RK (1991) Geology of Bomdila group in Arunachal Pradesh. Himalayan Geol 2:207–214

Cann JR (1970) Rb, Sr, Y, Zr and Nb in some ocean floor basaltic rocks. Earth Planet Sci Lett 10:7–11

Condie KC, Viljoen MJ, Kable EJD (1977) Effects of alteration on element distributions in Archean tholeiites from the Barberton greenstone belt, South Africa. Contrib Miner Petrol 64:75–89

Condie KC, Sinha AK (1996) Rare earth and other trace element mobility during mylonitization: a comparison of the Brevard and Hope Valley shear zones in the Appalachian Mountains, USA. J Met Geol 14:213–226

Condie KC (1997) Sources of Proterozoic mafic dyke swarms: constraints from Th/Ta and La/Yb ratios. Precambr Res 81:3–14

Davies JF, Grant RWE, Whitehead RES (1979) Immobile trace elements and Archean volcanic stratigraphy in the Timmins mining area, Ontario. Can J Earth Sci 16:305–311

Dikshitulu GR, Pandey BK, Krishna V, Raju RD (1995) Rb-Sr systematics of granitoids of the central gneissic complex, Arunachal Himalaya: implications on tectonism, stratigraphy and source. Geological Society of India 45:51–56

Erlank AJ, Duncan AR, Marsh JS (1988) A laterally extensive geochemical discontinuity in subcontinental Gondwana lithosphere. In: Proceedings of the V Conference on Geochemical Evolution of the Continental Crust, Brazil, pp 1–10

Feigenson MD, Hofmann AW, Spera FJ (1983) Case studies on the origin of basalt. Contrib Miner Petrol 84:390–405

Floyd PA, Winchester JA (1978) Identification and discrimination of altered and metamorphosed volcanic rocks using immobile elements. Chem Geol 21:291–306

Floyd PA, Kelling SL, Gokcen SL, Gokcen N (1991) Geochemistry and tectonic environment of basaltic rocks from the Misis ophiolitic melange, south Turkey. Chem Geol 89:263–280

Floyd PA (1993) Geochemical discrimination and petrogenesis of alkalic basalt sequences in part of the Ankara mélange, central Turkey. Geol Soc London 150:541–550

French JE, Heaman LM (2010) Precise U–Pb dating of Paleoproterozoic mafic dyke swarms of the Dharwar craton, India: implications for the existence of the Neoarchean supercraton Sclavia. Precambr Res 183:416–441

Hirahara Y, Kimura J-I, Senda R, Miyazaki T, Kawabata H, Takahashi T, Chang Q, Vaglarov BS, Sato T, Kodaira S (2015) Geochemical variations in Japan Sea back-arc basin basalts formed by high-temperature adiabatic melting of mantle metasomatized by sediment subduction components. Geochem Geophy Geosys 16:1324–1347

Hawkesworth CJ, Gallagher K, Pearson G, Turner SP, Calsteren V (1993) The continental lithosphere: a geochemical perspective. An Acad Bras Ci 65:199–225

Jain AK, Thakur VC (1978) Abor volcanics of the Arunachal Himalaya. Geol Soc India 19:335–349

Jensen LS (1976) A new cation plot for classifying subalkalic volcanic rocks. Min Nat Res Ontario Div Min Misc paper 66:20

Kumar G (1997) Geology of Arunachal Pradesh. Geological Society of India, Bangalore, p 217

Kuzmichev A, Kroner A, Hegner E, Dunyi L, Yusheng W (2005) The Sheshkhid ophiolite, northern Mongolia: a key to the reconstruction of a Neoproterozoic island-arc system in central Asia. Precambr Res 138:125–150

Lafleche MR, Dupuy C, Bougault H (1992) Geochemistry and petrogenesis of Archean volcanic rocks of the southern Abitibi Belt, Quebec. Precambr Res 57:207–241

Le Maitre RW (2002) Igneous rocks: a classification and glossary of terms, 2nd edn. Cambridge University Press, Cambridge, p 236

Li XH, Li ZX, Wingate MTD, Chung SL, Liu Y, Lin GC, Li WX (2006) Geochemistry of the 755 Ma Mundine well dyke swarm, North Western Australia: part of a Neoproterozoic mantle superplume beneath Rodinia. Precambr Res 146:1–15

McKenzie D, O'Nions RK (1991) Partial melt distributions from inversion of rare earth element concentrations. J Petrol 32:1021–1091

Miyashiro A (1975) Classification, characteristics, and origin of ophiolites. J Geol 83:249–281

Pearce JA, Cann JR (1973) Tectonic setting of basic volcanic rocks determined using trace element analyses. Earth Planet Sci Lett 19:290–300

Pearce JA (1975) Basalt geochemistry used to investigate past tectonic environments on Cyprus. Tectonophysics 25:41–67

Pearce JA, Gale DH (1977) Identification of ore deposition environment from trace element geochemistry. Geol Soc Lond Spec Publ 7:14–24

Pearce JA (1982) Trace element characteristics of lavas from destructive plate boundaries. In: Thorpe RS (ed) Orogenic Andesites. Wiley, Chichester, UK, pp 528–548

Pearce JA (2008) Geochemical fingerprinting of oceanic basalts with applications to ophiolite classification and the search for Archean oceanic crust. Lithos 100:14–48

Petterson MG, Windley BF (1992) Field relations, geochemistry and petrogenesis of the Cretaceous basaltic Jutal dykes, Kohistan, northern Pakistan. J Geol Soc Lond 149:107–114

Rashid SA, Islam N (2016) Geochemical characteristics of proterozoic mafic dykes from the Bomdila Group of rocks, NE Lesser Himalaya, India. Acta Geologica Sinica 90:122

Rollinson HR (1993) Using geochemical data: evaluation, presentation, interpretation. Longman Scientific Technical, Essex, UK, p p344

Roychowdhury J (1984) The Abor group of rocks in Arunachal Pradesh. Record Geol Surv India 113:48–57

Sahai A, Srivastava RK (1997) Structural and geochemical characteristics of amphibolites from the Bhagirathi and Yamuna valleys of Main Central Thrust Zone, Garhwal Himalaya. J Himalayan Geol 18:191–201

Sandeman HA, Hanmer S, Tella S, Arimetage AA, Davis WJ, Ryan JJ (2006) Petrogenesis of Neoarchaean rocks of the Mc Quoid supracrustal belt: a back-arc setting for the northwestern Hearne Subdomain, western Churchill Province, Canada. Precambr Res 144:140–165

Saunders AD, Tarney J, Weaver SD (1980) Transverse geochemical variations across the Antarctic Peninsula: implications for the genesis of calc-alkaline magmas. Earth Planet Sci Lett 46:344–360

Saunders AD, Norry MJ, Tarney J (1988) Origin of MORB and chemically depleted mantle reservoirs: trace element constraints. In: Menzies MA, Cox KG (eds) Oceanic and continental lithosphere: similarities and differences. J Petrol, pp 415–455

Schweitzer J, Kroner A (1985) Geochemistry and petrogenesis of early Proterozoic intra-cratonic volcanic rocks of the Ventersdorp Supergroup, South Africa. Chem Geol 51:265–288

Sengupta S, Acharyya SK, De Smeth JB (1996) Geochemical characteristics of the Abor volcanics, NE Himalaya, India: nature and early Eocene magmatism. J Geol Soc London 153:695–704

Shervais JW (1982) Ti–V plots and petrogenesis of modern and ophiolitic lavas. Earth Planet Sci Lett 59:108–118

Singh AK (2006) Petrography, geochemistry and petrogenesis of Abor Volcanics, Eastern Himalayan Syntaxial Bend. Himalayan Geol 27:163–181

Singh AK, Bikramaditya Singh RK (2012) Petrogenetic evolution of the felsic and mafic volcanic suite in the Siang window of Eastern Himalaya, Northeast India. Geoscience Frontier 3:613–634
Singh S (1993) Geology and tectonics of the eastern syntaxial bend, Arunachal Himalaya. J Himalayan Geol 4:149–163
Srinivasan A (2001) Stratigraphy and structure of low grade meta-sedimentaries in eastern Bhutan and western Arunachal Pradesh. J Himalayan Geol 22:83–98
Srivastava RK, Sahai A (2001) High-Field Strength element geochemistry of mafic intrusive rocks from the Bhagirathi and Yamuna valleys, Garhwal Himalaya, India. Gondwana Res 4:455–463
Srivastava RK (2012) Petrological and geochemical studies of paleoproterozoic mafic dykes from the Chitrangi Region, Mahakoshal Supracrustal Belt, Central Indian Tectonic Zone: Petrogenetic and tectonic significance. J Geol Soc India 80:369–381
Srivastava RK, Samal AK (2018) Geochemical characterization, petrogenesis and emplacement tectonics of Paleoproterozoic high-Ti and low-Ti mafic intrusive rocks from the western Arunachal Himalaya, Northeastern India and their possible relation to the ~1.9 Ga LIP event of the Indian shield. Geological Journal in press. https://doi.org/10.1002/gj.3172
Srivastava RK, Sivaji C, Chalapathi Rao NV (2008) Indian Dyke: geochemistry geophysics and geochronology. Narosa Publishing House Ltd., New Delhi, p p626
Srivastava RK, Srivastava HB, Srivastava V (2009) Petrology and geochemistry of Proterozoic olivine tholeiite intrusives from the Central Crystallines of the western Arunachal Himalaya, India: evidence for a depleted mantle. Curr Sci 97:1355–1361
Sun SS, Nesbitt RW (1977) Chemical heterogeneity of the Archean mantle composition of the bulk earth and mantle evolution. Earth Planet Sci Lett 35:429–448
Sun SS, McDonough WF (1989) Chemical and isotopic systematics of oceanic basalts: implications for mantle composition and processes. Magmatism in the ocean basins 42:313–345
Tarney J (1992) Geochemistry and significance of mafic dyke swarms in the Proterozoic. In: Condie KC (ed) Proterozoic crustal evolution. Elsevier, Amsterdam, pp 151–179
Tarney J, Jones CE (1994) Trace element geochemistry of orogenic igneous rocks and crustal growth models. J Geol Soc London 151:855–868
Taylor SR, McLennan SM (1985) The continental crust: its composition and evolution. Blackwell Publishers, Oxford, p p312
Thakur VC (1986) Tectonic zonation and tectonic framework of Eastern Himalaya. Science de la Terre Memoir 47:347–360
Thompson RN, Morrison MA, Dickin AP, Henery GL (1983) Continental flood basalts—Arachnids rule OK. In: Hawkesworth CJ, Norry MJ (eds) Continental Basalts and Mantle Xenoliths. Shiva Publication Limited, Cheshire, pp 58–85
Verma PK, Tandon SK (1976) Geological observations in a part of the Kameng district, Arunachal Pradesh (NEFA). J Himalayan Geol 6:259–286
Vimal R, Banerjee R, Gupta S, Krishna V, Achar KK, Babu PR, Parihar PS, Maithani PB (2012) Geochemistry and Sr and Pb isotope systematics of basement granitoids from north and west of Palnad Sub-basin, Guntur and Nalgonda districts, Andhra Pradesh. J Appl Geochem 14:295–315
Wang P, Glover L (1992) A tectonics test of the most commonly used geochemical discriminant diagrams and patterns. Earth-Sci Rev 33:111–131
Weaver BL, Tarney J (1984) Empirical approach to estimating the composition of the continental crust. Nature 310:575–577
Winchester JA, Floyd PA (1977) Geochemical discrimination of different magma series and their differentiation products using immobile elements. Chem Geol 20:325–344
Zi J, Fan W, Wang Y, Peng T, Guo F (2008) Geochemistry and petrogenesis of the Permian mafic dykes in the Panxi region, SW China. Gondwana Research 14(3):368–382

Petrology and Tectonic Setting of Dyke Swarms Emplaced in the Upper Jurassic Qorveh Granitoid Complex (Majidabad and Kangareh), Kurdistan Province, Iran

Ashraf Torkian

Abstract The studied area is located in the south to east of the Qorveh (Majidabad and Kangareh; Kurdistan Province) in the Sanandaj-Sirjan zone. Based on field survey, the dykes are gabrroic to gabbro-norite rocks in composition, and followed by granites dykes with low abundances and they have intruded in the granitoid rocks. The most abundance dykes are diorites (monzodioritic and quartz-monzodioritic rocks). They consist of plagioclase ($An_{54.87–45.95}$), pyroxene ($Wo_{32.92–50.77}$ $En_{36.28–40.55}$ $Fs_{8.67–27.89}$), hornblende and sometimes K-feldspar and quartz. The mafic rocks include plagioclase ($An_{66.28–58.72}$), clinopyroxene, orthopyroxene and hornblende which based on mineralogy and geochemical classification are divided to two groups: gabbroic and gabbro-noritic rocks. Biotite, apatite, titanite, zircon and opaque minerals are their accessory minerals in both of group. Microscopic studies and microprobe analyses data show amphiboles and biotites are Mg-hornblende and annite, in composition, respectively. The mafic rocks have tholeiitic-calc-alkaline affinity. Spider diagrams imply enrichment in LILE (such as Cs, Rb, U and Pb), depletion in HFSE (Nb and Ba) and high LILE/HFSE ratio. The enrichment in LILE and Pb suggest the crustal contamination. The tectonic setting discrimination diagrams reveal all of rocks belong to volcanic arc related to an active continental margin setting.

Keywords Diorite · Gabbro · Dyke · Calc-alkaline · Volcanic arc · Contamination Iran

1 Introduction

The studied area is located in nearby the Qorveh (Kurdistan Province) in the Sanandaj-Sirjan zone (SaSZ) (Fig. 1a) between 47°33′- 47°48′E-Longitude and 35°10′- 35°11′N-Latitude. The 148–157 Ma Qorveh Granitoid Complex intruded into low to medium metamorphic rocks such as schist, amphibolites and marble

A. Torkian (✉)
Department of Geology, Faculty of Science, Bu–Ali Sina University, Hamedan, Iran
e-mail: a-torkian@basu.ac.ir

R. K. Srivastava et al. (eds.), *Dyke Swarms of the World: A Modern Perspective*, Springer Geology, https://doi.org/10.1007/978-981-13-1666-1_12

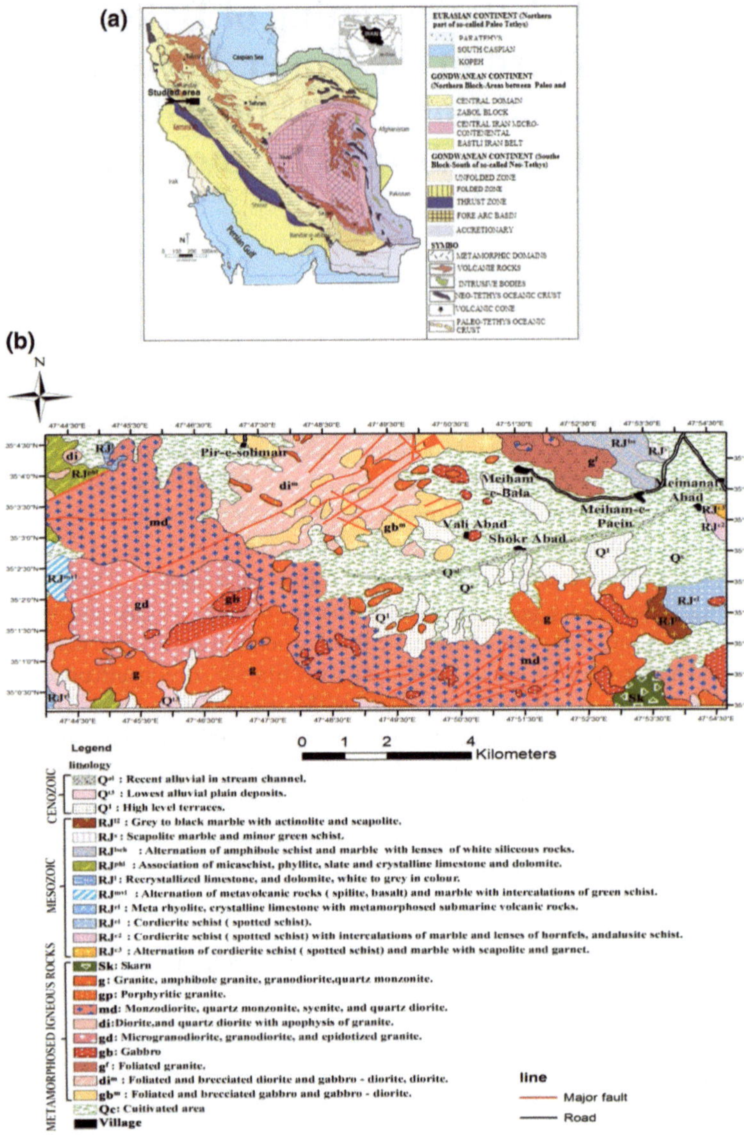

Fig. 1 **a** Tectonic zones of the Zagros Orogen in western Iran (after Alavi 1994) and location of the study area (polygonal inset) the Qorveh town in the Sanandaj–Sirjan Zone. Simplified geological maps of **b** the Majidabad and **c** the Kangareh in south and east of the Qorveh, respectively (modified from Hosseini 1997)

(Azizi et al. 2015; Mahmoudi et al. 2011). The Qorveh Granitoid Complex (QGC), consists of garnitoid and intermediate-mafic plutons (Torkian and Khalili 2008), mafic- intermediate dyke swarms (Torkian 2011). This area covers more than 35 km^2

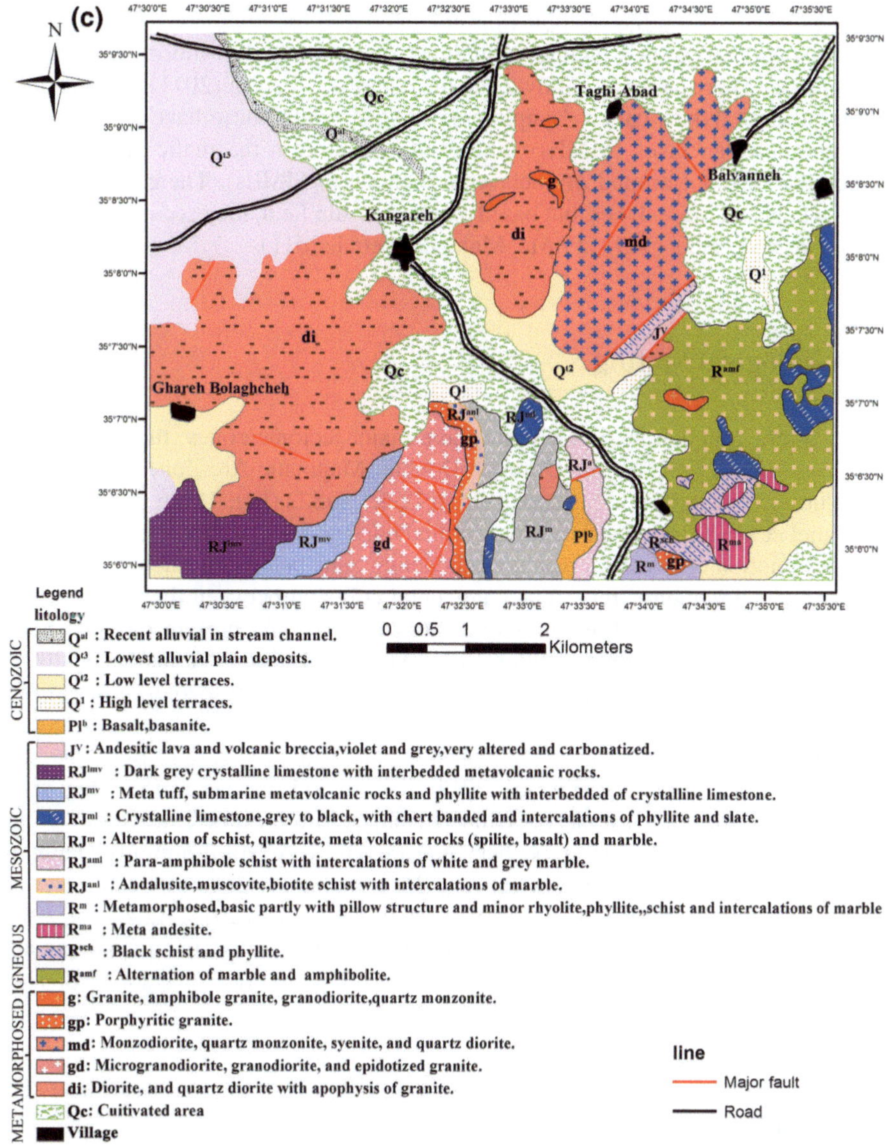

Fig. 1 (continued)

in south to east Qorveh (Fig. 1a), in approximately 85 km northwest of Hamedan city. They are ideal for studying the linkage tectonic setting and evolution of igneous rocks between host rocks and dykes. This study focuses on the petrology and tectonic setting of the dioritic–gabbroic dykes in the Majidabad (Fig. 1b) and the Kangareh (Fig. 1c) regions, in south and east the Qorveh, respectively.

The Qorveh Granitoid Complex covers an area of approximately 70 km^2 and it show a multi-intrusion structure, containing unites of gabbro-diorite-granodiorite and granite. There are many of dykes in the complex. Torkian (2011) have reported some of these dykes, called "synplutonic dykes", which these have created mafic microgranular enclaves (MMEs). This study focuses on the mafic-dioritic dykes which they are not around of enclaves (unrelated to MMEs). There are not repot about of ages of these dykes, but based on relationship field, they seem younger than their own host rocks (149–157 Ma, Mahmoudi et al. 2011).

2 Geological Setting

The western Iran can be divided into three major tectonic zones: the Zagros Fold and Thrust Zone (ZFTZ), the Urumieh-Dokhtar Magmatic Arc Zone (UDMZ) and the plutonic-metamorphic Sanandaj-Sirjan Zone (SaSZ) (Fig. 1a) (e.g. Alavi 1994; Mohajjel et al. 2003). The SaSZ was formed during closure of the Neo-Tethys, i.e. by consumption of oceanic crust at a subduction zone below the stable Central Iran block, and subsequent Cretaceous continental collision between the Afro-Arabian and Iranian continental fragments (e.g. Şengör 1984). At present day, the SaSZ move northwards 14 mm/year with respect to (stable) Eurasia (Vernant et al. 2004). It has experienced major tectono-magmatic events during Mesozoic-Tertiary times (Ghasemi and Talbot 2006).

Numerous intrusive rocks of various size intrude the SaSZ (e.g. Berberian and Berberian 1981; Torkian and Rezaii 2013; Torkian and Furman 2015). They show large range of variation in rock types (most belong to Jurassic time), most of the garanitoids in this zone are S-and I-type, but some are typical A-type granitoids. Dating of individual granitoid complexes in SaSZ by Rb-Sr and U-Pb methods yielded crystallization ages ranging from 131 to 185 Ma (Azizi and Asahara 2013; Azizi et al. 2011; Yajam et al. 2015). They have dykes intrusive, granitic to mafic in composition and the temporal relationships between the various dykes are not yet established. They are important for geologic studies.

3 Petrography

The intermediate–mafic dykes are located in the QGC and are mainly distributed in the host granitic to granodioritic, with lesser gabbroic (in Kangareh area) rocks (Fig. 1) Detailed petrographic studies of unaltered intermediate-mafic dykes show that the dykes are gabbroic to dioritic in composition, and followed by granitic dykes (aplites, most in Majidabad area, Torkian and Khalili 2008) with low abundances.

3.1 Dioritic Dykes

The most abundant dykes are diorites including monzodioritic and quartz-monzodioritic rocks. They are fresh at their center and show slight alteration towards their margins. They exhibit intergranular and granular to subophitic and even pegmatoidic textures. The intergranular and subophitic textures are very characteristic of many dioritic samples. The diorites with pegmatitic texture are occurred in the Kangareh and the Majidabad, which includes coarse-grained hornblende and plagioclase crystals. The pegmetitic dykes are dikes is lower than other dykes. These rocks consist mainly of plagioclase (45–55%; $An_{54.87-45.95}$), pyroxene (10–15%; $Wo_{32.92-50.77}$ $En_{36.28-40.55}$ $Fs_{8.67-27.89}$) (Table 1), Mg-hornblende (5–10%), quartz (5–13%), alkali-feldspar (0–5%), with accessory apatite, zircon and opaque minerals (Fig. 3) and

Table 1 Data of microprobe electronic analyses of clinopyroxene minerals in dioritic rocks

Sample	B2.K32	B2.K32	B2.K32
Lithology	Diorite	Diorite	Diorite
SiO_2	49.280	55.050	53.510
TiO_2	2.960	1.320	1.150
Al_2O_3	8.040	2.930	3.090
FeO	0.040	0.000	0.000
MnO	8.200	2.200	6.100
MgO	0.150	0.200	0.200
CaO	12.800	11.860	12.100
Na_2O	14.940	20.660	20.190
K_2O	2.270	0.970	1.580
Total	107.60	97.77	104.50
Si	1.723	2.027	1.910
Ti	0.078	0.037	0.031
Al	0.331	0.127	0.130
Fe^{3+}	0.235	0.071	0.177
Fe^{2+}	0.240	0.068	0.182
Mn	0.004	0.006	0.006
Mg	0.667	0.651	0.644
Ca	0.560	0.815	0.772
Na	0.154	0.069	0.109
K	0.00	0.00	0.00
Cation	3.993	3.872	3.961
$Mg/(Mg+Fe^{2+})$	0.735	0.905	0.779
Wo	32.90	50.77	43.50
En	39.22	40.55	36.28
Fs	27.89	8.67	20.22

Fig. 2 Hand specimen of pegmetitic diorite, including white-colored plagioclases and amphiboles (Majidbad)

biotite as secondary mineral. The grain size is largely invariable, showing textures ranging from fine-grained to coarse-grained, with relatively large plagioclase and amphibole crystals (3–5 mm) in pegmatoidic diorites (Fig. 2). Plagioclase occurs as euhedral to subhedral laths that are sometimes altered to sericite, clinozoisite, epidote and calcite. Pyroxene and hornblende occurs as euhedral to subhedral laths, 0.2–1.5 mm in length. Quartz and alkali-feldspar occur interstitially.

3.2 Gabbroic Dykes

The mafic rocks include plagioclase (50–65%; $An_{66.28-58.72}$) (Table 2), clinopyroxene and orthopyroxene (15–25%) and hornblende (0–13%) (Fig. 3) which based on mineralogy and geochemical classification are divided to two groups: gabbroic and gabbro-noritic rocks. The gabbroic rocks have hypidiomorphic granular and ophitic textures (Fig. 3) similar to the diorites. Titanite and opaque minerals are their accessory minerals in both of groups (Fig. 4).

4 Geochemistry

4.1 Sampling and Analytical Methods

Rock-forming minerals (amphibole, plagioclase and pyroxene) were analyzed using a XGT 7200 Electron Probe Micro Analyzer (EPMA) at Kansaran Binalud Co.and Zarazma Minaral Studies Co. (Tehran, Iran). Analytical conditions were optimized

Table 2 Representative plagioclase crystals analysis from the gabbroic dykes

Sample	B2K.32	B2K.40	B2K.40	BK.40	B2K.40	B2K.40	B2K.40	B2K.16	B2K.16	B2K.42	B2K.42
Lithology	Gabbro	Gabbro	Gabbro	Gabbro	Gabbro	Gabbro	Gabbro	Gabbro	Gabbro	Gabbro	Gabbro
SiO_2	59.35	57.18	57.88	56.01	61.08	54.02	56.95	55.61	55.29	60.53	59.98
TiO_2	0.02	0.14	0.11	0.11	0.22	0.04	0.37	0.04	0	0.14	0.09
Al_2O_3	26.61	26.47	26.61	26.61	18.06	26.09	23.73	27.35	28.62	23.98	26.14
FeO	0.16	0.93	0.49	0.50	3.29	5.22	2.85	2.85	0.17	0.22	0.22
MnO	0.08	0.01	0.01	0.0	0.07	0.09	0.08	0.08	0	0	0.21
MgO	0	0.38	0.06	0.06	2.77	1.92	1.64	1.64	0	0	0.14
CaO	11.45	10.87	11.47	11.47	7.64	8.42	10.53	10.53	13.87	9.24	9.85
Na_2O	3.14	2.75	3.54	3.54	5.78	3.27	2.74	4.04	2.08	5.96	4.43
K_2O	0.12	1.13	0.83	0.83	0.82	1.35	0.77	0.06	0.11	0.07	0.07
Total	100.11	100.90	100.45	98	99.63	99.42	99.66	99.98	100.05	99.61	99.94
Si	2.741	2.641	2.660	2.679	2.774	2.489	2.638	2.538	2.558	2.728	2.743
Ti	0.001	0.005	0.004	0.004	0.008	0.001	0.013	0.001	0	0.005	0.003
Al	1.394	1.441	1.382	1.325	0.967	1.383	1.298	1.471	1.555	1.283	1.355
Fe^{2+}	0.007	0.036	0.021	0.020	0.139	0.224	0.123	0.012	0.008	0.008	0.009
Mn	0.003	0	0	0	0.003	0.004	0.003	0	0	0	0
Mg	0	0.026	0.004	0.004	0.188	0.133	0.113	0	0	0	0.001
Ca	0.567	0.538	0.565	0.588	0.367	0.416	0.523	0.617	0.688	0.449	0.483
Na	0.281	0.246	0.315	0.328	0.509	0.292	0.246	0.357	0.185	0.525	0.393
K	0.007	0.067	0.049	0.051	0.048	0.079	0.046	0.003	0.006	0.004	0.004
An%	66.28	63.23	60.80	39.73	52.80	64.19	63.07	78.23	45.95	54.87	
Ab%	32.89	28.95	33.95	55.12	37.11	30.22	36.57	21.03	53.64	44.66	
Or%	0.827	7.83	5.24	5.15	10.81	5.59	0.36	0.74	0.41	0.46	

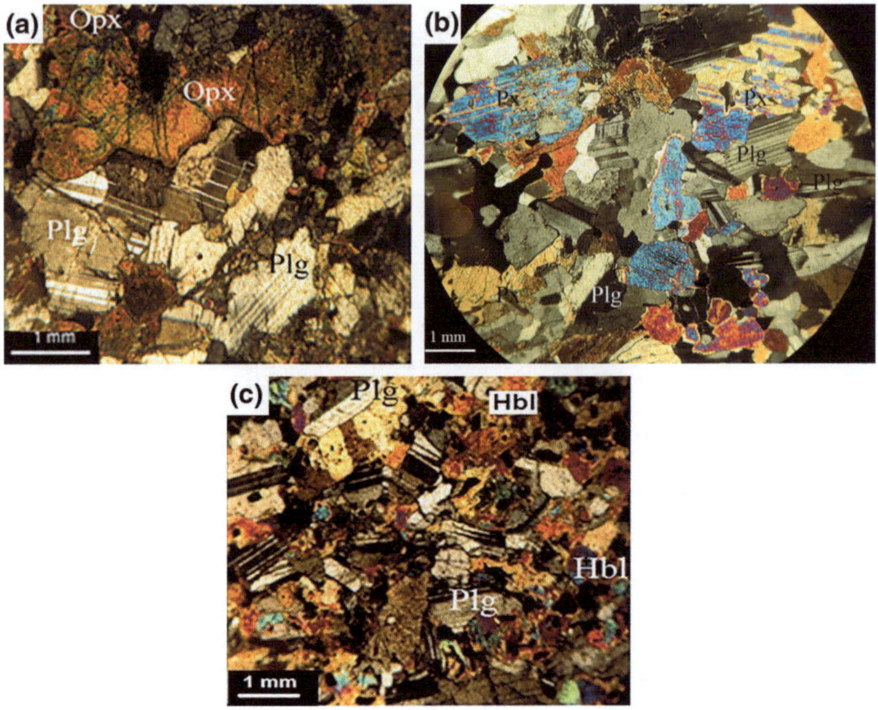

Fig. 3 Photomicrographs of (**a**, **b**) gabbro and (**c**) dioritic rocks. Plg: Plagioclase; Hbl: Hornblende; Opx: Orthopyroxene

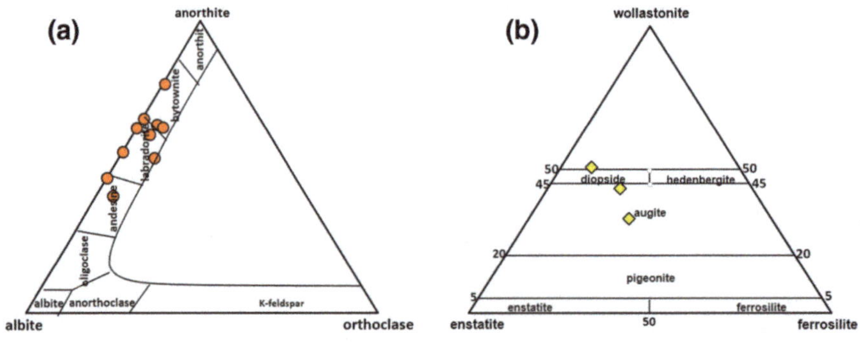

Fig. 4 Plot of plagioclases composition in Or-Ab-An diagram (**a**) pyroxene composition into the Wo-En-Fs diagram (**b**), pyroxenes exhibit augite-diopsidic composition

using standard silicates and oxides at 10 kV accelerating voltage, 10 nA beam current, and integration time of 10 s on peak and background (Tables 1 and 2).

Nineteenth fresh samples from different units were selected for geochemical analyses. The collected rocks were crushed in a hardened jaw crusher and then powdered in an agate mill to <75 μm. Whole-rock major and trace element analyses were performed by X-ray fluorescence and ICP-OES techniques in the Geological Survey of Iran. Data processing was by means of the program Minpet (version 2.02; Richard 1995). Major, trace and REE element compositions of the analyzed rocks are given in Table 3.

4.2 Results

The dioritic rocks have SiO_2 = 61.8–50.70 wt%, TiO_2 = 1–2 wt%, Al_2O_3 = 14.8–17.70 wt%, MgO = 1.80–6 wt%, Fe_2O_3 = 11.30–6.90 wt%, Na_2O = 0.60–4.10 wt%, K_2O = 0.2–1.5 wt% (Table 3). Chemically, they belong to the sub-alkaline series (Fig. 5a; Irvine and Baragar 1971) and metaluminous. The gabbros have SiO_2 = 44.88–50.4 wt%, TiO_2 = 1.1–3.55 wt%, Al_2O_3 = 15.19–17.5 wt%, MgO = 4.84–10.01 wt%, Fe_2O_3 = 0.4–9.9 wt%, Na_2O = 1.61–3.35 wt%, K_2O = 0.17–0.57 wt% (Table 3). The mafic–intermediates dykes plot in the tholeiitic to calc-alkaline series field on an AFM diagram (Miyashiro 1978) (Fig. 5b). The mafic and intermediate dykes can be clearly distinguished in this diagram.

Chemically, they fall in the sub-alkaline series (Fig. 5a; Irvine and Baragar 1971). The mafic rocks have tholeiitic affinity, but dioritic dykes belong to the calc-alkaline series (Fig. 5b). The results of representative rock analyses are plotted on Harker variation diagrams (Fig. 6). The Harker diagrams suggest magma differentiation in the rocks analyzed. Positive trends of incompatible elements against each other, suggests that role of fractionation crystallization in the production of the dioritic—gabbroic dykes (Fig. 7).

Spider diagrams imply enrichment in LILE (such as Cs, Rb, U and Pb), depletion in HFSE (Nb and Ba) and high LILE/HFSE ratio (Fig. 8). The enrichment in LILE and

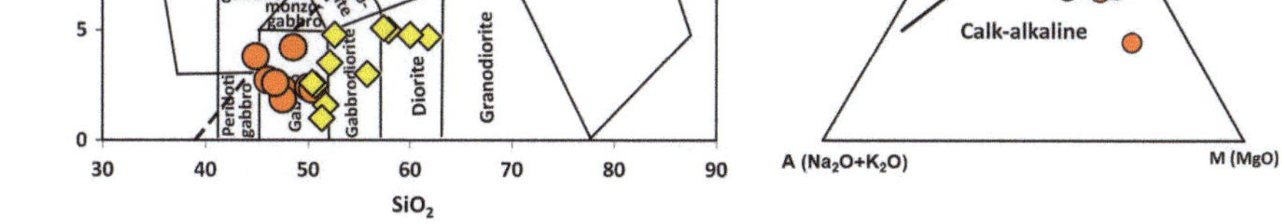

Fig. 5 **a** Total alkalies—SiO_2 and **b** plot of samples from the mafic-intermediate dykes in the AFM diagram

Table 3 Chemical compositions (major, trace and REEs) of the studied rocks. L.O.I. = loss of ignition

	TDP1-2	TD2-2	TD4-1	TDA1-2	TDA2-1	TD5-1	TDA1-1	TDV1-1	TDM1-1	TD3-1
	Diorite	Diorite	Diorite	Diorite	Diorite	Diorite	Diorite	Monzodiorite	Monzodiorite	Quartzmonzo diorite
N	35°05′19″	35°05′29″	35°07′37″	35°05′10″	35°02′47″	35°01′01″	35°02′34″	34°39′37″	34°59′43″	34°59′43″
E	47°47′54″	47°47′46″	47°47′38″	47°47′54″	47°48′08″	47°48′20″	47°48′43″	47°51′55″	47°51′01″	47°50′47″
SiO_2	51.7	50.7	51.3	61.8	55.8	52.1	60	57.9	52.6	57.4
TiO_2	1.1	1.4	1.2	1	0.9	1	1.1	1.5	1.9	2
Al_2O_3	17.7	16.8	17.1	16.6	15.2	16.3	16.9	16.2	15.1	14.8
Fe_2O_3	10.6	10.4	10.6	6.9	8.3	10.8	7.9	9.6	11.3	11.1
MgO	6	5.9	5.4	1.8	6.5	6.1	1.8	2.4	2.6	2.2
MnO	0.2	0.2	0.2	0.1	0.1	0.2	0.1	0.2	0.2	0.2
CaO	9.6	10.1	9.1	5.6	8.7	7.3	6	5.9	6	5.4
Na_2O	1.4	2.1	0.6	4.1	2.8	3	4.1	3.8	3.9	3.6
K_2O	0.2	0.4	0.4	0.6	0.2	0.5	0.7	1.2	0.9	1.5
P_2O_5	0.1	0.3	0.2	0.3	0.2	0.2	0.4	0.4	0.7	0.5
L.O.I.*	1.25	1.45	1.48	0.75	1.05	2.27	0.58	0.5	0.54	1.02
Sum	99.85	99.75	99.58	99.55	99.75	99.77	99.58	99.6	99.74	99.72
Ba	77	103	129	168	34.9	68.9	177	260	197	235
Ce	5.8	3.4	<1	52.7	19.3	17.4	27.2	32.4	60.7	30.8
Co	14.3	17.4	26.7	15.4	3.5	24.7	6.1	10	8.1	15.4
Cr	229	282	197	56	173	241	71	51	71	29
Cs	5.1	10	6	7.3	6.7	5.3	5.7	4.5	4	5.2
Cu	20.4	58.1	84.2	7.4	14.9	19	3.4	81.9	13.4	18.8

(continued)

Table 3 (continued)

	TDP1-2	TD2-2	TD4-1	TDA1-2	TDA2-1	TD5-1	TDA1-1	TDV1-1	TDM1-1	TD3-1
	Diorite	Diorite	Diorite	Diorite	Diorite	Diorite	Diorite	Monzodiorite	Monzodiorite	Quartzmonzo diorite
N	35°05′19″	35°05′29″	35°07′37″	35°05′10″	35°02′47″	35°01′01″	35°02′34″	34°39′37″	34°59′43″	34°59′43″
E	47°47′54″	47°47′46″	47°47′38″	47°47′54″	47°48′08″	47°48′20″	47°48′43″	47°51′55″	47°51′01″	47°50′47″
Dy	3.5	3.7	4.6	5.8	6.1	4.8	7.4	6.2	8.7	6.7
Er	1.1	1.3	0.9	5.3	1.4	2	1.8	1.5	2.2	1.8
Eu	1.4	0.6	1.4	0.8	1.6	1.4	1.4	1.3	1.5	1.7
Ga	14.4	22.6	18.6	25.2	21.2	20.6	24.5	32.7	32.2	35.1
Gd	1.6	2.6	2.2	3.1	1.8	4.4	2.6	4.4	5.9	5
Ge	2.1	1.7	2.2	1.7	1.8	1.8	1.4	1.7	2.2	1.7
Hf	9.6	11.5	11.6	10.8	16.7	9.8	8.9	10.2	10.6	9.9
Ho	1.3	1.3	1.1	1.1	1	1.3	1.2	1.3	1.5	1.1
La	7.8	5.5	6.5	32.7	11.2	14.7	25.6	30.8	32.9	20.4
Li	18.4	20.4	17.6	16	17.4	34.9	15.7	18.5	17.4	22.8
Lu	0.2	0.2	0.2	0.3	0.1	0.4	0.2	0.4	0.4	0.4
Mo	4.8	2.3	4.7	0.2	3	0.3	4.9	2.9	0.2	4.5
Nb	16.6	21.5	17	24.1	18.5	17.8	21.7	25.2	40.4	33.8
Nd	12.3	1.4	9.3	22.1	16.1	16.1	22.5	19.1	28.2	22.2
Ni	79.2	54.8	79.8	5.4	92.6	67.8	9.2	15.8	10.4	12.1
Pb	35.7	33.4	45.5	28.2	19.5	39.8	21.7	22.9	32.2	32.9
Pr	3.1	3.4	3.5	4.5	3.1	4.8	4.2	6.2	8	6
Rb	72.5	81.3	86.7	97.3	74.1	86.9	85.9	82.1	84.9	84

(continued)

Table 3 (continued)

	TDP1-2	TD2-2	TD4-1	TDA1-2	TDA2-1	TD5-1	TDA1-1	TDV1-1	TDM1-1	TD3-1
	Diorite	Diorite	Diorite	Diorite	Diorite	Diorite	Diorite	Monzodiorite	Monzodiorite	Quartzmonzo diorite
N	35°05′19″	35°05′29″	35°07′37″	35°05′10″	35°02′47″	35°01′01″	35°02′34″	34°39′37″	34°59′43″	34°59′43″
E	47°47′54″	47°47′46″	47°47′38″	47°47′54″	47°48′08″	47°48′20″	47°48′43″	47°51′55″	47°51′01″	47°50′47″
Sb	0.2	0.4	0.2	0.2	0.2	0.4	0.3	0.3	0.4	0.1
Sc	39.7	43.9	33.5	18.6	22.8	36.3	16.9	24.2	29.9	30.9
Se	0.1	0.1	0.1	0.05	0.1	0.1	0.1	0.1	0.1	0.1
Sm	4.9	3.1	2.7	4.8	4.6	4	4.5	5.3	8.5	5.6
Sn	4.9	4.9	4.4	3.7	3.9	5.1	3.8	5	6.4	5.6
Sr	274	339	273	375	296	279	360	224	230	225
Ta	4.2	4.4	5.3	4.4	4	5.9	4.5	4.2	5.9	6.4
Tb	1.3	1.4	1	1.2	1	1.1	1.1	1.5	2.4	2.3
Te	0.2	0.1	0.1	0.1	0.1	0.1	0.1	0.2	0.2	0.2
Th	38.7	38.3	34.9	30.4	30.5	39.2	28.9	38.2	51	44.9
Tl	1.1	1.1	0.8	0.5	0.9	1.1	0.6	1.1	0.8	0.8
Tm	0.1	0.2	0.1	0.2	0.1	0.3	0.2	0.3	0.4	0.3
U	2.1	2.5	2	2.5	1.1	4.7	1.9	4.1	4.7	4.5
V	190	204	170	128	179	186	117	147	189	169
Y	17.7	17.5	23.3	27.9	37.4	24.3	36.5	28.2	41.9	34
Yb	38.2	2.7	3.1	3.2	5.2	3.6	4.6	38.3	5.8	39.2
Zn	41.3	53	72	29.8	24	33.9	25.8	84.6	81	122
Zr	86.8	113.1	119.9	95.3	253.3	112.5	105	96	102	95

(continued)

Table 3 (continued)

	TD2-1	B2K32	B2K33	B2K40	B2K34	B1K40	TDP1-1	TD 2-3	B1K42
	Gabbro diorite	Gabbro	Gabbro	Gabbro	Gabbro	Gabbro	Gabbro	Gabbro	Gabbro
N	34°59′42″	35°07′37″	35°08′49″	35°07′47″	35°07′47″	35°07′34″	35°00′43″	35°00′26″	35°07′44″
E	47°50′26″	47°31′16″	47°32′51″	47°32′17″	47°32′17″	47°31′23″	47°51′32″	47°51′46″	47°31′40″
SiO_2	50.4	46	44.88	47.58	48.51	47.48	50	50.4	46.79
TiO_2	1.2	1.44	3.55	0.39	1.57	1.65	1.1	1.1	2
Al_2O_3	17.5	16.34	15.19	17.02	15.66	17.3	17.2	17.2	15.9
Fe_2O_3	9.6	0.77	0.04	0.45	0.8	0.76	9.4	9.9	0.8
MgO	5.9	8.6	4.84	10.04	7.21	8.92	6.4	6.3	8.1
MnO	0.2	0.17	0.24	0.1	0.16	0.17	0.2	0.2	0.16
CaO	10.4	11.54	10.01	13.63	9.42	10.45	10.5	9.6	11.62
Na_2O	2.2	2.51	3.35	1.74	3.7	1.61	1.9	1.9	2.45
K_2O	0.4	0.24	0.46	0.57	0.54	0.26	0.5	0.4	0.17
P_2O_5	0.2	0.07	1.63	0.01	0.22	0.13	0.2	0.2	0.02
L.O.I.*	1.71	1.93	0.9	2.35	1.55	1.01	2.18	2.37	1.24
Sum	99.71	96.5	100.4	99.09	99.17	100.1	99.58	99.57	99.58
Ba	108	75	80	40	72	52	65.9	97	65
Ce	1	18	89	12	27	19	1	1	13
Co	14.7	39.2	33	29.8	36.1	39	19.1	20.8	40.4
Cr	402	214	10	415	136	203	234	273	250
Cs	6.2	0.5	0.7	0.7	0.6	0.5	1	9.5	0.5

(continued)

Table 3 (continued)

	TD2-1	B2K32	B2K33	B2K40	B2K34	B1K40	TDP1-1	TD 2-3	B1K42
	Gabbro diorite	Gabbro	Gabbro	Gabbro	Gabbro	Gabbro	Gabbro	Gabbro	Gabbro
N	34°59′42″	35°07′37″	35°08′49″	35°07′47″	35°07′47″	35°07′34″	35°00′43″	35°00′26″	35°07′44″
E	47°50′26″	47°31′16″	47°32′51″	47°32′17″	47°32′17″	47°31′23″	47°51′32″	47°51′46″	47°31′40″
Cu	67.3	65	68	59	53	58	41.5	114.4	88
Dy	2.9	2.83	10.97	1.32	4.36	2.65	3	3.3	2.28
Er	0.6	1.54	5.82	0.69	2.59	1.48	0.7	0.8	1.24
Eu	0.9	1.26	3.47	0.53	1.36	1.12	1.4	1.3	0.98
Ga	27.2	–	–	–	–	–	15.3	14.8	
Gd	2.8	3.15	13.63	1.34	4.54	2.95	2.1	2.8	2.37
Ge	2	–	–	–	–	–	1.6	1.9	
Hf	13.1	1.02	1.2	0.56	1.79	1.16	9.2	12	0.9
Ho	1.3	–	–	–	–	–	1	1.3	–
La	3.2	7	43	4	12	6	2.1	7.7	4
Li	24	–	–	–	–	–	21.4	21.8	
Lu	0.3	0.18	0.56	0.1	0.31	0.18	0.2	0.2	0.16
Mo	0.6	–	–	–	–	–	1.5	3.3	–
Nb	20.6	1.6	29	1.1	8.2	4.2	16.5	16.6	2.8
Nd	12.2	9.6	5.2	3.8	13.6	8.3	8.2	2.9	6.2
Ni	49	76	11	97	38	80	67.3	63.5	51
Pb	19.2	3	4	4	4	3	34.8	30.6	34
Pr	3.9	1.67	10.4	0.69	2.64	1.47	3.5	3.6	1.05
Rb	87.9	2	6	9	4	4	77.7	80.9	1

(continued)

Table 3 (continued)

	TD2-1	B2K32	B2K33	B2K40	B2K34	B1K40	TDP1-1	TD 2-3	B1K42
	Gabbro diorite	Gabbro	Gabbro	Gabbro	Gabbro	Gabbro	Gabbro	Gabbro	Gabbro
N	34°59′42″	35°07′37″	35°08′49″	35°07′47″	35°07′47″	35°07′34″	35°00′43″	35°00′26″	35°07′44″
E	47°50′26″	47°31′16″	47°32′51″	47°32′17″	47°32′17″	47°31′23″	47°51′32″	47°51′46″	47°31′40″
Sb	0.3	–	–	–	–	–	0.3	0.4	–
Sc	41.3	22.6	20.3	30	25.2	23.6	30	40.4	28.2
Se	0.1	0.5	0.8	0.5	0.5	0.5	0.1	0.05	0.5
Sm	2.6	2.65	11.91	1.03	3.56	2.22	3.1	2	1.84
Sn	4.9	0.7	3.6	1'	1.3	1.6	3.8	4.8	0.7
Sr	415	335	329	261	273	316	246	359	294
Ta	6.8	0.33	1.47	0.31	0.63	0.54	5	5.6	0.33
Tb	1.3	0.45	1.74	0.21	0.66	0.41	1	1.1	0.36
Te	0.1	0.2	0.1	0.14	0.12	0.5	0.1	0.2	0.21
Th	30.9	0.3	1.29	0.78	1.07	0.33	27.2	37.6	0.27
Tl	0.6	0.1	0.11	0.1	0.1	0.1	0.7	0.9	0.1
Tm	0.2	0.2	0.7	0.1	0.35	0.19	0.2	0.2	0.16
U	2.8	0.12	0.6	0.2	0.3	0.1	2.2	2.9	0.1
V	197	121	144	67	127	117	161	199	155
Y	14.7	12.1	45.1	6.5	20.4	13.1	13.2	14.9	10
Yb	2.3	1.3	4.3	0.6	2.3	1.2	1.9	2.2	1.1
Zn	55.4	46	50	30	59	52	46.6	67.7	51
Zr	117	37	44	18	71	52	139	94.5	32

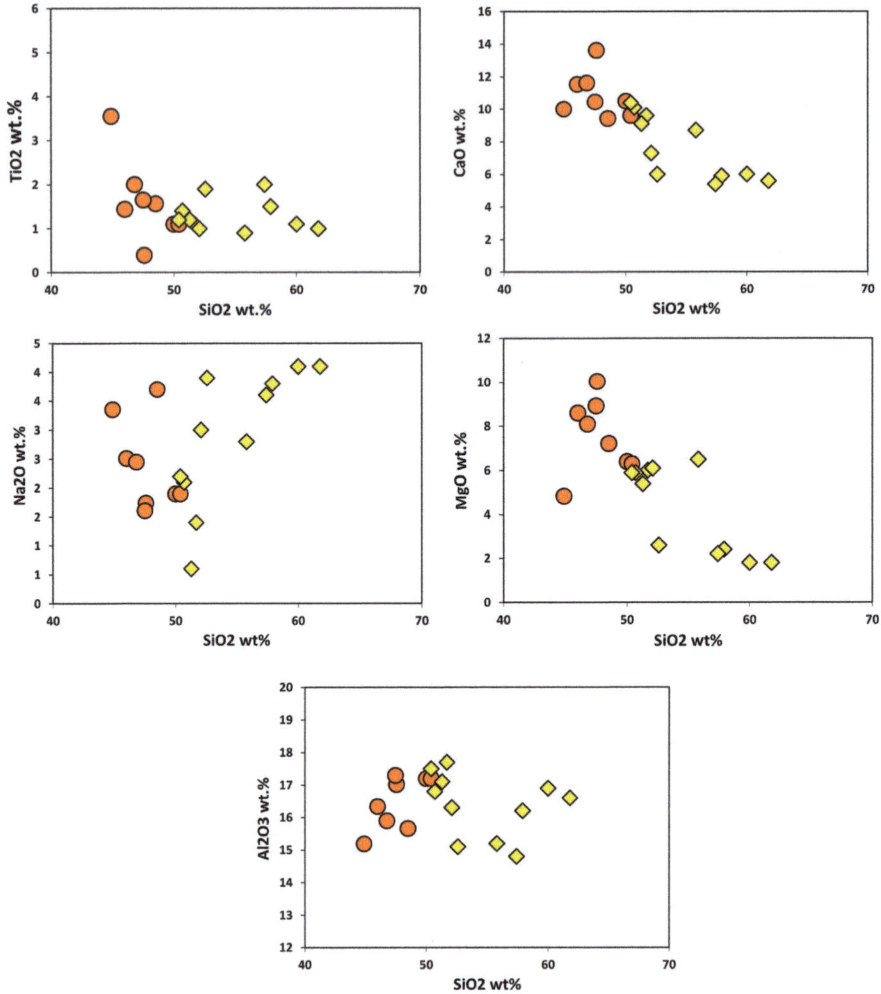

Fig. 6 Harker variation diagrams for some of major elements from the mafic-dioritic rocks. Symbols as in Fig. 5

Pb suggest the crustal contamination (Arslan and Aslan 2006). The tectonic setting discrimination diagrams reveal all of rocks belong to volcanic arc related to an active continental margin setting (Fig. 9). The geochemical features indicate generation of the parent melts by slab dehydration, wedge melting processes, and limited degrees of crustal contamination and/or melting of subducted sediment (Fig. 10). Concentration La/Nb and La/Ta rations as well as the enrichment of LREE and LILE reveal that dykes were derived from enriched lithospheric mantle. In addition, enrichment of LREE elements relative to HREE represent that there are garnet phase or amphibole in source.

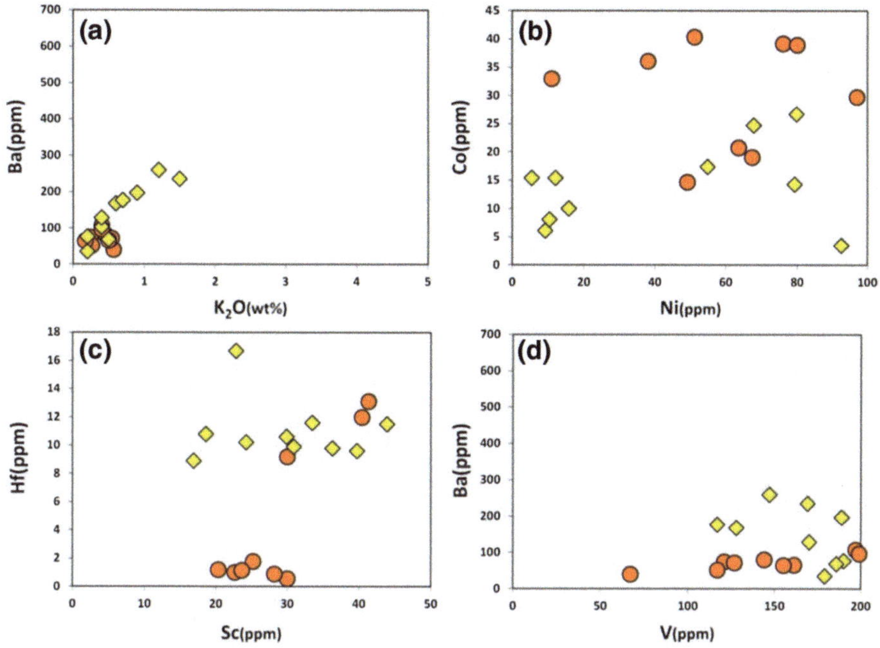

Fig. 7 Linear and positive trends of incompatible elements against each other, the observed chemical variations could be explained by fractionation crystallization for the diorites-gabbros. Symbols as in Fig. 5

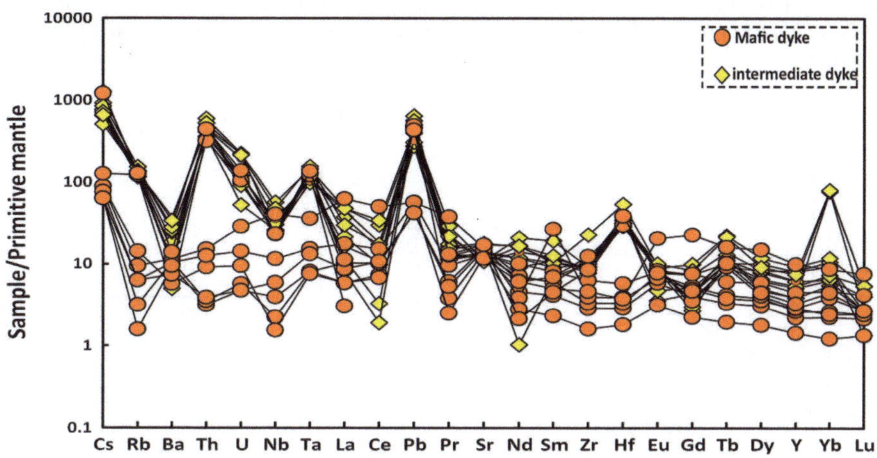

Fig. 8 Spider diagram normalized by primitive mantle (Sun and McDonough 1989)

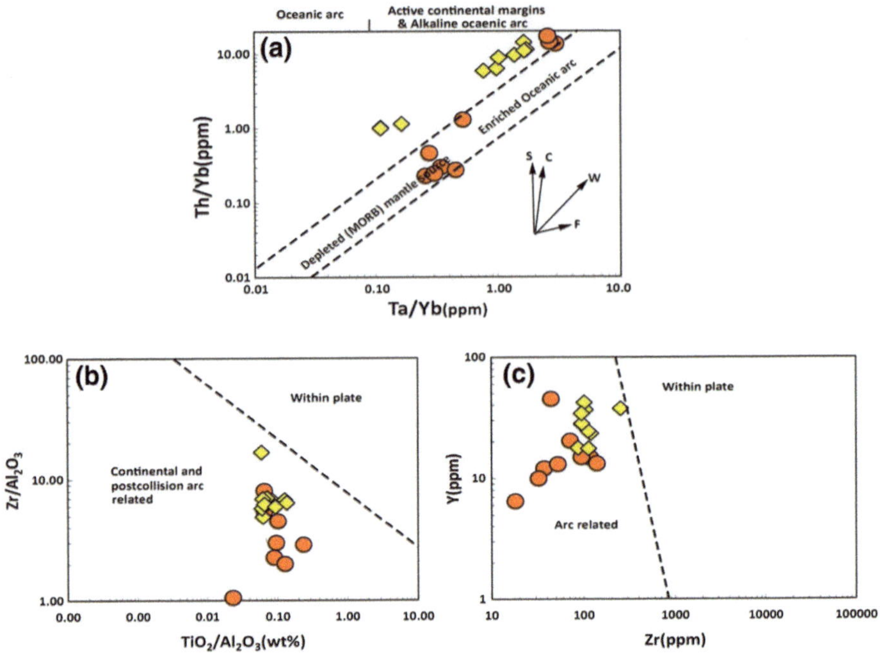

Fig. 9 **a** Plot of intermediate-mafic dykes in Th/Yb-Ta/Yb diagram (Pearce 1982), vectors show trends produced by subduction zone enrichment (S), crustal contamination (C), within-plate enrichment (w) and fractional crystallization (F) showing the roles of fractional crystallization for dykes. **b, c** Diagrams (after Jiang et al. 2006) for determination of tectonic setting of rocks of studied area. These reveal all of rocks belong to volcanic arc related to an active continental margin. Symbols as in Fig. 8

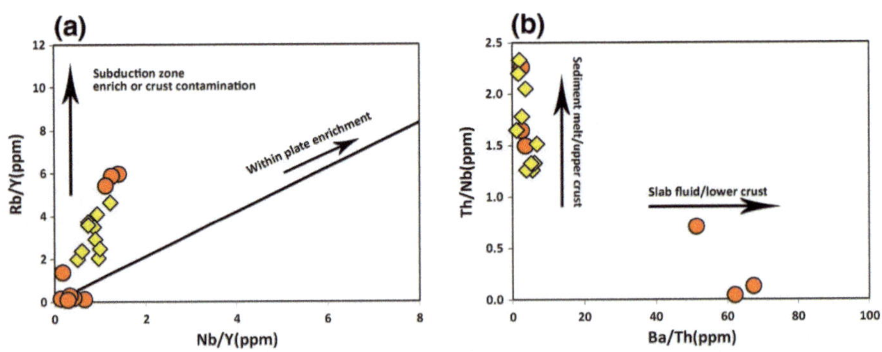

Fig. 10 **a** Rb/Y versus Nb/Y diagram illustrating subduction zone enriched or/and the crustal contamination, **b** Th/Nb-Ba/Th diagram (see Fig. 9 in Niu et al. 2016) suggesting role of sediment melting or/and contamination with upper crust. Symbols as in Fig. 8

5 Conclusions

The mafic-intermediate dykes intruded in the Qorveh Granitoid Complex, in the Sanandaj-Sirjan subduction zone, formed during subduction beneath continental crust in the Jurassic. These dykes are dioritic (monzodioritic and quartz-monzodioritic) and gabrroic (gabbroic and gabbro-noritic) in composition with sub-alkaline affinities. The dioritic rocks are more abundance than the gabbros and they include plagioclase + pyroxene + Mg-hornblende ± K-feldspar ± quartz. The mafic dykes consist of plagioclase orthopyroxene and clinopyroxene, hornblende. Geochemical features such as enrichment of LILIE and LREE relative to HREE, positive Pb anomalies show that dykes are most likely the result of magmatism in an active continental margin arc setting.

Acknowledgements Some of this work was supported by the Bu-Ali Sina University (Hamedan, Iran). I express my appreciation to organizers of 7th International Dyke Conference and to Profs. Rajesh K. Srivastava, Richard Ernst and Peng Peng (IDC7 volume Book Springer editors) for encouraging me to contribute to the IDC-7 volume. Also, I thanks from Prof. N. Youbi, and the other reviewer for notes and suggestions.

References

Alavi M (1994) Tectonics of the Zagros Orogenic belt of Iran: new data and interpretations. Tectonophysics 229:211–238

Arsalan M, Aslan Z (2006) Mineralogy, petrography and whole-rock geochemistry of the Tertiary granitic intrusions. J Asian Earth Sci 27:177–193

Azizi H, Asahara Y (2013) Juvenile granite in the Sanandaj-Sirjan Zone, NW Iran: Late Jurassic–Early Cretaceous arc–continent collision. Int Geol Rev 55:1523–1540

Azizi H, Asahara Y, Mehrabi B, Chung SL (2011) Geochronological and geochemical constraints on the petrogenesis of high-K granite from the Suffiabad area, Sanandaj-Sirjan Zone, NW Iran. Chem Erde 71:363–376

Azizi H, Najari M, Asahara Y, Catlos EJ (2015) U–Pb zircon ages and geochemistry of Kangareh and Taghiabad mafic bodies in northern Sanandaj–Sirjan Zone, Iran: evidence for intra-oceanic arc and back-arc tectonic regime in Late Jurassic. Tectonophysics 660:47–64

Berberian F, Berberian M (1981) Tectono-plutonic episodes in Iran. Geol SurvIran 52:565–592

Ghasemi A, Talbot CJ (2006) A new tectonic scenario for the Sanandaj-Sirjan Zone (Iran). J Asian Earth Sci 26:683–693

Hosseini M (1997) Geological Quadrangle Map of Qorveh (No. 5560, scale 1:100,000). Geol SurvIran

Irvine TN, Baragar WRA (1971) Guide to the chemical classification of the common volcanic rocks. Can J Earth Sci 8:523–484

Jiang YH, Jiang SY, Zhao KD, Lingp HF (2006) Petrogenesis of Late Jurassic Qianlishan granites and mafic dykes, Southeast China: implications for a back-arc extension setting, Geol Mag 457–474

Mahmoudi S, Corfu F, Masoudi F, Mehrabi B, Mohajjel M (2011) U–Pb dating and emplacement history of granitoid plutons (northern Sanandaj–Sirjan Zone). J Asian Earth Sci 410:238–249

Miyashiro A (1978) Nature of alkalic volcanic rock series. Contrib Mineral Petrol 66:91–104

Mohajjel M, Fergusson CL, Sahandi MR (2003) Cretaceous-tertiary convergence and continental collision, Sanandaj- Sirjan Zone, western Iran. J Asian Earth Sci 21:397–412

Niu X, Chen B, Feng G, Liu F, Yang J (2016) Origin of lamprophyres from the northern margin of the North China Craton: implications for mantle metasomatism. https://doi.org/10.1144/jgs2016-044

Pearce JA (1982) Trace element characteristics of lavas from destructive plate boundaries. In: Thorp RS (ed) Andesites: Orogenic Andesites and Related Rocks. Wiley, New York

Richard LR (1995) MinPet: mineralogical and petrological data processing system, version 2.02. MinPet Geological Software, Québec, Canada

Şengör AMC (1984) The Cimmeride orogenic system and tectonics of Eurasia. Geol Soc Am Spec paper 195

Sun SS, McDonough WF (1989) Chemical and isotopic systematic of oceanic basalts: implications for mantle composition and processes (Saunders AD, Norry MJ (eds) Magmatism in ocean basins). Geol Soc London Spec Pub 42:313–345

Torkian A (2011) The Parishan Pluton in Qorveh Area, an example for Magma Mingling process, Southeastern Sanandaj, Iran. In: Srivastava RK (ed) Dyke swarms: keys for geodynamic interpretation. Springer, Berlin, Heidelberg

Torkian A, Furman T (2015) The significance of mafic microgranular enclaves in the petrogenesis of the Qorveh Granitoid Complex, northern Sanandaj-Sirjan Zone, Iran. Neu Jb Mineral Abh 192:117–133

Torkian A, Rezaii M (2013) The study the granitoid rocks in shear zone in SE-Qorveh (Kurdistan, Iran): with emphasis on geochemical behavior of whole-rock and mineral chemistry of Biotite and Feldspar (abstract). In: Goldschmidt conference, Florence, Italy

Torkian A, Khalili M, Sepahi AA (2008) Petrology and geochemistry of the I-type calc-alkaline Qorveh Granitoid Complex, Sanandaj-Sirjan Zone, western Iran. Neu Jb Mineral Abh 185:131–142

Vernant P, Nilforoushan F, Hatzfeld D, Abbassi M, Vigny C, Masson F, Nankali H, Martinod J, Ashtiani A, Bayer R, Tavakoli F, Chery J (2004) Contemporary crustal deformation and plate kinematics in Middle East constrained by GPS measurements in Iran and Northern Oman. Geophys J Int 157:381–398

Yajam S, Montero P, Scarrow JH, Ghalamghash J, Razavi SMH, Bea F (2015) The spatial and compositional evolution of the Late Jurassic Ghorveh-Dehgolan plutons of the Zagros Orogen, Iran: SHRIMP zircon U-Pb and Sr and Nd isotope evidence. Geochim Cosmochim Acta 13:25–43

From Ophiolites to Oceanic Crust: Sheeted Dike Complexes and Seafloor Spreading

Jeffrey A. Karson

Abstract Persistent, coordinated plate separation and dike intrusion generate sheeted dike complexes in oceanic crust at mid-ocean ridge spreading centers and other magmatic rifts. Although sheeted dike complexes were first described in ophiolite complexes, investigations of dikes, dike intrusion events and sheeted dike complexes in the oceanic crust have provided new constraints on how sheeted dike complexes form and their significance for the accretion of oceanic crust at spreading centers. Despite the general appearance of a monotonous array of side-by-side intrusions, details of sheeted dike complexes hold important keys to understanding the fundamentals of the tectonics, magma plumbing networks and hydrothermal/biological systems at mid-ocean ridges. In situ investigations of sheeted dikes and related upper crustal units in seafloor exposures provide fundamental observations that have implications for deformation during spreading, the reconstruction of ophiolite complexes, and the restoration of ophiolite structures to spreading center reference frames.

1 Introduction

About two-thirds of Earth is covered by oceanic lithosphere generated over the past ~200 Ma. Plate accretion at mid-ocean ridges (MORs) has resurfaced large swaths of the planet continuously since the inception of plate tectonics. On a planetary scale, MORs represent the most important flux of energy and matter from Earth's interior to the surface. Understanding the construction of oceanic crust and lithosphere is thus a fundamental planetary process. And yet, despite the initial recognition of seafloor spreading more than 50 years ago, many basic questions related to the geology of the oceanic crust and the processes by which it is created remain. For example, how and why does the internal structure of the oceanic crust vary with spreading rate or magma budget? Are there systematic variations in spreading processes and crustal structure along the length of discrete spreading segments? What variations occur

J. A. Karson (✉)
Department of Earth Sciences, Syracuse University, Syracuse, NY 13224, USA
e-mail: jakarson@syr.edu

R. K. Srivastava et al. (eds.), *Dyke Swarms of the World: A Modern Perspective*,
Springer Geology, https://doi.org/10.1007/978-981-13-1666-1_13

at spreading-segment discontinuities? How is the middle to lower gabbroic crust constructed? What geological features are associated with the Moho? To what extent do ophiolites reveal structures and processes that occur at mid-ocean ridge spreading centers or other spreading environments? New detailed studies on the seafloor and in ophiolites as well as new technology continually redefine and refocus investigations that shed light on these issues.

The obvious logistical limitations of investigating this vast terrane have required an indirect approach using remotely-acquired marine geophysical data and integrated with subaerial analogs. These analogs include active and ancient magmatic extensional terranes and especially, ophiolite complexes, widely considered to be analogous to crust and lithosphere generated at MORs. Thus, the large, layered ophiolite complexes and even dismembered ophiolites have played a fundamental role in the evolving perception of the geology of the oceanic crust. As direct investigations of the oceanic crust became possible using Human-Occupied Vehicles (HOVs), Remotely-Operated Vehicles (ROVs), Autonomous Underwater Vehicles (AUVs) and deep crustal drilling, new perspectives have forced a reconsideration of this view. Inherent in these efforts is the assumption that, by viewing limited areas represented by ophiolites (a few to tens of kilometers across) or outcrops on the seafloor (typically less than hundreds of meters across), very broad extrapolations can be made, globally or at least to large oceanic areas formed under similar conditions.

This paper focuses on sheeted dike complexes, a fundamental rock unit in ophiolite complexes as well as in contemporary oceanic crust. Beginning with a review of how sheeted dike complexes in ophiolites inspired early inferences about the oceanic crust and MOR processes, the history of relevant seafloor investigations leading to the discovery of sheeted dike complexes on the seafloor is considered. This is followed by an evaluation of reports of sheeted dike complexes on the seafloor. Next, salient aspects and significance of extensive sheeted dike complexes studied in deep drill holes and in situ along major tectonic escarpments are summarized. Finally, the implications of these results for ophiolites and key issues that can be addressed from these different perspectives are discussed.

2 From Ophiolites to Oceanic Crust

2.1 Sheeted Dikes as Fundamental to Seafloor Spreading

Basaltic dikes have long been known as essential components of mafic igneous terranes where they represent frozen conduits linking subsurface magma bodies to lava flows at the surface. Individual dikes and swarms of dikes in eroded volcanic systems reveal the nature of shallow subsurface magma plumbing networks. The geometry of dikes in these systems have been essential to understanding the evolution of volcanic edifices and the tectonic regimes in which they form (e.g., Walker 1992, 1993, 1999; Anderson 1951).

It was not until the late 1950s that a sheeted dike complex was first described in the Betts Cove Ophiolite of Newfoundland. Unlike swarms of dikes near intrusive centers, this sheeted dike complex has >90%, subparallel basaltic dikes that have intruded and chilled against one another and local screens of intervening country rock. Even in this early discovery, it was apparent that this distinctive rock unit must have formed in a setting in which protracted tectonic extension tapped magma from a persistent source (Barager 1954). Somewhat later, sheeted dikes in the Troodos Ophiolite were linked to the new idea (at the time) of seafloor spreading (Gass 1968). Following the correlation of ophiolites with oceanic lithosphere, several investigators were quick to correlate the layered seismic structure of oceanic crust and upper mantle (Ewing and Engel 1962) with the internal geologic structure of ophiolite complexes (Church and Stevens 1971; Dewey and Bird 1971; Moores and Vine 1971). Typically ophiolites are composed of pelagic sedimentary rocks, basaltic pillow lavas, basaltic sheeted dikes, and gabbroic rocks overlying upper mantle peridotites (Fig. 1). This recurring association was formally recognized in a report of a Geological Society of America, Penrose Conference devoted to this topic (Anonymous 1972). Detailed investigations of the major, layered ophiolite complexes provided the basis for increasingly detailed studies of the major rock units in ophiolites and the processes that created them. Sheeted dike complexes rapidly came to be perceived as one of the hallmarks of oceanic crust and inspired many detailed studies in ophiolites (Abbots 1979; Kidd and Cann 1974; Kidd et al. 1978; Moores and Vine 1971; Pallister 1981; Alexander et al. 1993; Barager et al. 1990; Nicolas and Boudier 1992; Rosencrantz 1983; Varga 1991; Nicolas 1989). The ophiolite template was widely embraced as representative of the geology of the oceanic crust and mantle, a notion that persists in research articles and textbooks today.

Since the recognition of sheeted dike complexes in ophiolites, sheeted dike units and dense swarms of parallel dikes have been documented in many different environments demonstrating that these units can form in many different tectonic settings (Fig. 2). Some of these include: island arcs (Miyashiro 1973, 1975), volcanic rifted margins (Karson and Brooks 1999; Klausen and Larsen 2002; Nielsen 1978; DuToit 1929), and ocean islands (Gudmundsson 1983; Helgason and Zentilli 1985; Walker 1964, 1986; Sillman 1987).

2.2 Matching Ophiolite and Oceanic Crustal Structures

Ophiolites are commonly folded and faulted on a large scale and in many cases highly dismembered. Although the usual rock units are present, commonly in coherent blocks many kilometers across, reconstructions of the major units and restorations of internal structures with respect to an MOR reference frame are not unique. A combination of observations and inferences from ophiolites and marine seismology provided a basis for reconstructing the observed crustal structures (Casey et al. 1983; Nicolas 1989).

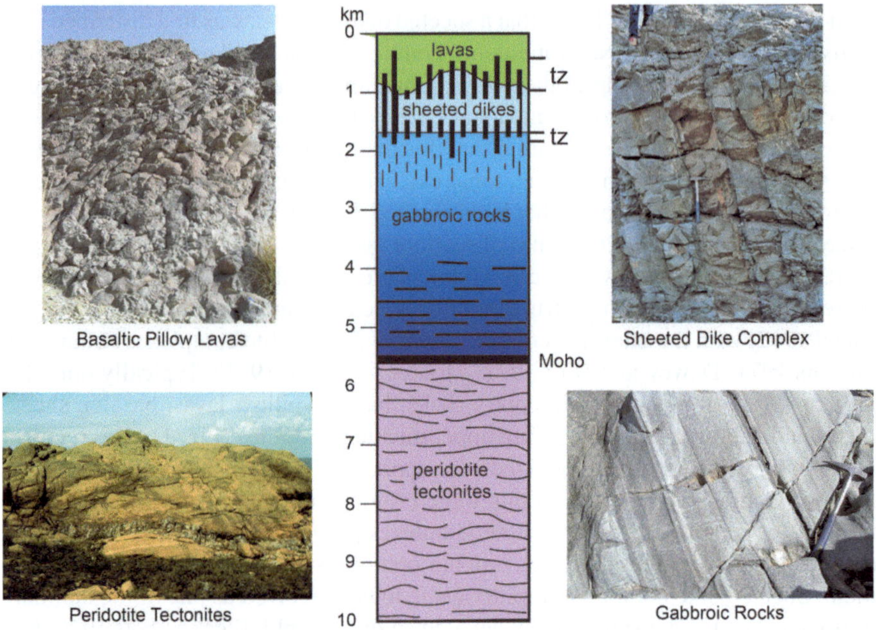

Fig. 1 Simplified schematic columnar section of rock units typically found in layered ophiolite complexes with examples of outcrop features. Unit contacts are gradational. Sheeted dike complexes are bounded by transition zones (tz) both above and below. The upper transition zone includes individual dikes and swarms of dikes intruding the lower part of the lava unit. In different ophiolite complexes, the lower transition zone is defined by a zone of brecciation, a detachment fault or intrusive gabbroic contact (from Karson et al. 2015)

Refinements in seismically-defined layers in the oceanic crust (Christeson et al. 1992; Houtz and Ewing 1976; Vera et al. 1990) were the basis for correlations with specific ophiolite rock units (Christensen and Salisbury 1982; Fox et al. 1973). Further, the horizontal seismic layers of the oceanic crust (determined over survey distances of kilometers) were correlated with sedimentary bedding, lava flows and the layered assemblage of rock units in ophiolites (Fig. 3). Based on simple mechanical models (Anderson 1951) dikes in sheeted dike complexes were assumed to be vertical structures in oceanic crust. These assumptions provided horizontal and vertical references for the reconstruction of the internal structure of ophiolites and by inference, the geometry of structures generated by seafloor spreading at MORs (Casey et al. 1983; Nicolas 1989; Pallister and Hopson 1981).

Fig. 2 Sheeted dike complexes (>90% dikes) and dike swarms in ophiolites. **a** Bay of Islands Complex, sheeted dikes, **b** Oman Ophiolite, sheeted dikes, **c** Macquarie Island Ophiolite, sheeted dikes, and **d** Troodos Ophiolite. Dense swarm of dikes intrude pillow lavas (right) in the lava/sheeted dike transition zone

2.3 *Limits of Ophiolites as Analogs for Seafloor Spreading*

The correlation of ophiolites with oceanic crust formed at MORs was disputed on the basis of geochemistry of basaltic to andesitic rocks capping the lava units of many ophiolites. Ophiolite volcanic suites were found to have both major and trace-element compositions characteristic of subduction zone environments rather than lavas known from MORs (Miyashiro 1973; Pearce and Cann 1973; Robinson et al. 1983). Further investigations of active MORs and subduction zones (forearcs, arcs, remnant arcs and back-arc basins) revealed increasing diversity of compositions of lavas in both of these settings that overlap with those of ophiolite lavas (Pearce 2003). Increasingly sophisticated studies have tended toward the view that ophiolites formed in "suprasubduction zone" environments (Moores 1982; Moores et al. 2000). The geochemistry of sheeted dike complexes in ophiolites has been less intensely studied, owing to the nearly ubiquitous effects of hydrothermal alteration, but these also commonly have compositions expected in subduction zone environments (Robinson et al. 2008).

At present, many geologists, persuaded by the internal structure of ophiolites and especially sheeted dike complexes, cling to the notion that most ophiolites formed at some sort of mid-ocean ridge spreading center or at least provide good structural

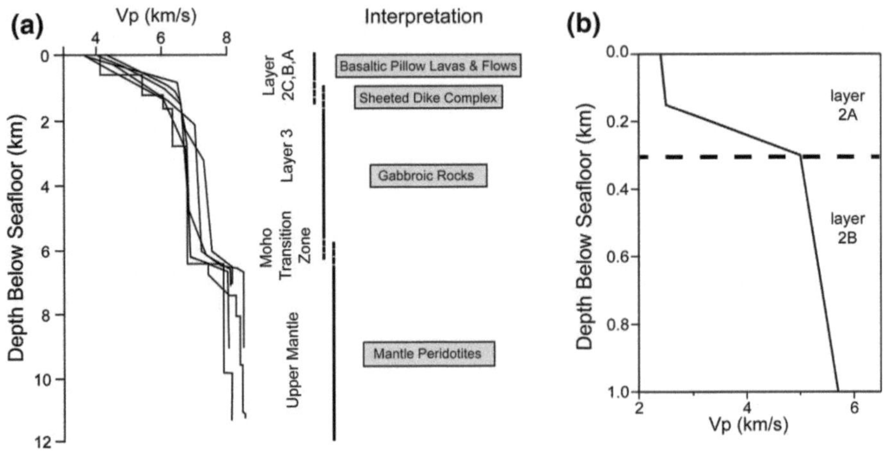

Fig. 3 The seismic structure of oceanic crust and interpretations based on ophiolite complexes and marine geology. **a** Velocity-depth profiles of oceanic (Vera et al. 1990) crust and correlations with major rock units. **b** Generalized velocity-depth profile for the upper kilometer of the oceanic crust units (Christeson et al. 2012). Layer 2A is commonly thought to correspond to lava units; deeper layers may correlate with a lava-dike transition zone or a sheeted dike complex

analogs for oceanic crust. Alternatively, geochemists tend to view ophiolites as having formed in suprasubduction zone environments. This is a potentially important distinction because crustal structures produced in island arcs, forearcs and back-arc basins may be different from those of MORs in some significant ways (e.g., Martinez and Taylor 2013). Suprasubduction zone crust will have to be investigated in situ in order to document its internal structure and evaluate how it relates to MOR crust.

2.4 *Testing the Ophiolite Model with Marine Geology and Geophysics*

Despite the intuitive appeal of the ophiolite model for oceanic crust, subsequent investigations on land and at sea paradoxically both supported and challenged this view and the assumptions upon which it was built. In the 1970s the Deep Sea Drilling Project began drilling the first reference hole into the oceanic crust, eventually penetrating >2 km into the upper oceanic crust (Anderson et al. 1982). Although continuous core recovery in this hole was poor, it confirmed the presence of pillow lavas and sheeted dikes in the upper oceanic crust. In parallel, the first direct observations of a mid-ocean ridge spreading center were being made on the Mid-Atlantic Ridge (MAR) in Project FAMOUS (ARCYANA 1975, 1978; Heirtzler and LePichon 1974) with submersible vehicles documenting the active processes on a spreading ridge. Subsequent investigations of other parts of the MAR (Stakes et al. 1984) and the East

Pacific Rise (EPR) (Ballard et al. 1981) revealed important variations in faulting and volcanism with spreading rate.

The simple view of a uniformly layered oceanic crust began to waver in the early 1980's. Although it had long been known that a diverse suite of plutonic igneous and metamorphic mafic to ultramafic rocks could be dredged near oceanic fracture zones (Shand 1949; Cann and Simkin 1971; Engel and Fisher 1975; Miyashiro et al. 1969, 1971), the first submersible investigations of rift valley and transform fault escarpments provided a new perspective. First, at the Mid-Cayman Spreading center (CAYTROUGH 1979), and later near the intersection of the MAR and Kane (Karson and Dick 1983) and Oceanographer (OTTER 1984, 1985)Transforms, crustal cross sections revealed extensive exposures of mafic to ultramafic plutonic rocks directly overlain by basaltic lavas and/or sedimentary units. The sheeted dike complexes appeared to be missing.

Submersible and camera tow studies of a section of the MAR >100 km long just south of the Kane Transform (MARK Area) showed that the spreading center is composed of distinct spreading segments with different geologic structures and implied processes (Karson et al. 1987). Segments vary from volcanically active to inactive rifts, like the FAMOUS area, to extensional core complexes with variably deformed and metamorphosed mafic to ultramafic plutonic terranes exposed in the footwall of low-angle, normal (detachment) faults (Karson 1990; Karson and Dick 1983). Superimposed on this amagmatic rift segment is a younger axial volcanic ridge (4 km wide by 400 m high by 40 km long) with an active black smoker hydrothermal system. These features demonstrate extreme variations in spreading processes on time scales of <1 Ma (Karson et al. 1987) in a single segment. Thus, both spatial (along-strike) and temporal (along the spreading direction) variations became obvious. The global perspective provided by the NSF RIDGE program of the 1990s along with more local surveys in the following decade gave birth to the current view of segmented spreading centers. A hierarchy of spreading center segments and bounding structures show systematic along-strike variations in spreading structures and active processes (Carbotte et al. 1998, 2006; Carbotte and Macdonald 1992; Karson et al. 2015; Macdonald 1998; Macdonald et al. 1988).

Near-bottom studies of major escarpments along the slow-spreading MAR with HOVs and ROVs produced more mixed results. HOV submersible investigations on the wall of the Vema Transform for the first time documented a sequence of pillow lavas, sheeted dikes, and gabbroic rocks and mantle peridotites similar to that anticipated by the ophiolite model, but with a somewhat thinner crustal section than the expected based on seismic refraction studies (Auzende et al. 1989). Meanwhile, drilling in areas considered to be "normal" oceanic crust near the MAR found many places where pillow lavas, sedimentary breccias or pelagic sedimentary rocks lie directly atop serpentinites or gabbroic rocks (Juteau et al. 1990). These results showed that oceanic crust formed at relatively slow spreading rates, and low magma budgets (volume of magma normalized with respect to spreading rate) is fundamentally different from that formed at fast-spreading ridges where the magma supply keeps pace with plate separation. The dominance of mechanical extension over magmatic construction results in an oceanic crustal structure that is significantly more complex than in the ophiolite model and that sheeted dike complexes are not necessarily part of this structure. Instead, lavas and sedimentary rocks commonly unconformably overlie variably deformed, metamorphosed and intruded assemblages of mafic and ultramafic rocks (Karson 1998). If sheeted dikes formed during spreading in these areas, they must have been removed by later spreading-related faulting. Most investigations of deeper structures have necessarily been near major escarpments at transform faults or along major rift valley walls where the magma budget during spreading is expected to have been low. It is not known if sheeted dike complexes occur in slow-spread oceanic crust generated near the centers of magmatically robust spreading segments or proximal to axial hotspots, for example, on the Reykjanes Ridge (Karson et al. 2015; Murton and Parson 1993; Searle et al. 1998).

Deep crustal drilling and investigations of major escarpments in crust formed at intermediate to superfast rates (50–150 mm/year) in the Pacific helped put ophiolite geology in a new perspective. Investigations with HOVs, ROVs, sonar and camera systems documented extensive exposures of the upper oceanic crust including pillow lavas, sheeted dikes and underlying gabbroic rocks finally verifying the ophiolite model (Francheteau et al. 1990, 1992). Deep crustal drilling at IODP Hole 1256D penetrated a continuous vertical section of lavas, sheeted dikes and underlying gabbroic rocks formed at a superfast rate (Wilson et al. 2006).

2.5 Sheeted Dike Complexes: Inspiration and Irony

Although sheeted dike units were first recognized in ophiolite complexes and provided the basis for the first detailed interpretations of oceanic crust formed at MORs, it is ironic that most ophiolites are now considered to have formed in convergent margin settings. It was only decades later that direct investigations of the oceanic crust by drilling and submersible studies showed that an upper crustal structure similar to that found in ophiolites is indeed present, and probably very widespread in the contemporary oceanic crust. Questions arising from the geochemistry of ophiolite rocks and the wide range of geologic structures found along MORs both cast doubt on aspects of the traditional ophiolite model for oceanic crust.

By investigating sheeted dike complexes in situ in the oceanic crust, where they can be unequivocally related to spreading rate, spreading direction, and horizontal reference frames, new information from the seafloor can now be re-directed back to ophiolites. Seafloor observations force the reconsideration of some of the fundamental assumptions made about sheeted dike complexes in ophiolites that may in turn reflect back on the perceptions of MOR or suprasubduction zone processes.

3 Sheeted Dike Complexes on the Seafloor

The early correlation of ophiolites with oceanic crust and mantle led to the expectation that beneath the lava units of the upper oceanic crust there might be a continuous array of sheeted dikes stretching across huge swaths of the Earth and continuously injected beneath MORs. Some workers have regarded dike intrusion as the "quantum event" of seafloor spreading (Delaney et al. 1998). As discussed above, this might be a reasonable expectation for relatively fast-spread crust with a high magma budget, but not for parts of slower-spreading ridges where faulting dominates. Nevertheless, based on many observations of dikes in oceanic crust in deep drill holes or exposed on tectonic escarpments, dike intrusion must be regarded an integral part of seafloor spreading across the entire spectrum of spreading regimes.

3.1 Dike Dimensions

The dimensions of individual dikes in the oceanic crust and most other settings, is unclear because of the relatively limited size of exposures relative to dike dimensions. However, the dimensions of dikes can be inferred by considering constraints from different perspectives. Seismic reflection mapping shows that the top of MOR axial magma chambers (AMCs) are melt-rich lenses typically at 1–2 km depth, 1–2 km wide and 50–100 m thick (Carbotte et al. 2012; Detrick et al. 1987). The melt lens is underlain by a seismic low velocity zone (LVZ) with a lower proportion of melt that extends downward to the depth of the Moho (~6 km below the seafloor) (Dunn et al. 2000). Magma chambers in slow-spreading crust are rare and reside deeper (~4 km) in the crust (Singh et al. 2006). Thus, an individual dike could be as much as a few kilometers vertically.

In seafloor and ophiolite outcrops, dikes are commonly about 1 m wide, though accurate determination of dike widths are complicated by dikes intruding the cores of earlier dikes causing separations of originally paired chilled margins. Individual dike intrusion events documented along MORs and other environments show that segmented dikes can intrude laterally for tens of kilometers (Dziak et al. 2009; Einarsson and Brandsdottir 1980; Grandin et al. 2010; Sigmundsson et al. 2015). Thus, the overall dimensions of dikes in oceanic crust should be roughly 1:1000:10,000, similar to a ribbon or knife blade (Fig. 4). Dikes intruded in slow-spreading environments may have somewhat different dimensions because of less persistent magma supply, deeper magma chambers and possibly different regional stresses.

Although dike intrusion must be an important process along all spreading centers, formation of sheeted dike complexes represents a special case. They require a persistent magma source that allows dike intrusion to keep pace with plate separation. Although sheeted dike complexes have been found in investigations of all crust formed at intermediate to superfast rates examined to date, they appear to be much less widespread in slow-spread crust (Table 1). This should not be surprising as

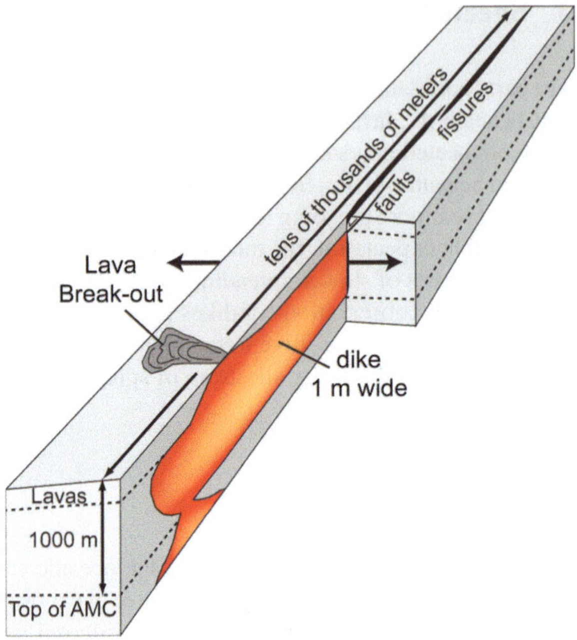

Fig. 4 Dimensions of dikes in oceanic lithosphere. Dikes in ophiolites and oceanic lithosphere are typically about 1 m wide. They intrude upward or laterally from axial magma chambers (AMC) about 1000 m below the seafloor. Migrating seismicity and periodic lava break-outs during dike intrusion events on the seafloor and on land show that they can be 10 s of km in length. Thus, dikes in oceanic lithosphere can have dimensions of about 1:1000:10,000, similar to a ribbon or knife blade (from Karson et al. 2015)

there is abundant evidence of fresh lavas and a nearly continuous magma chamber beneath MORs with fast to intermediate spreading rates. In contrast, slow- and ultraslow-spreading ridges commonly lack fresh lavas and have only isolated magma chambers. In these environments, plate separation may be almost entirely accommodated by faulting for extended periods of time. Exceptions would occur on slow-spreading ridges near hotspots such as Iceland or the Azores on the MAR, where higher magma budgets and more frequent dike intrusion would be expected.

3.2 Dikes in Slow-Spreading Crust

As noted above, many investigations of the rift and transform valley walls in relatively slow-spread crust (20–50 mm/year), document faulted crustal sections without sheeted dike complexes (Karson 1998). This has opened the door to a re-investigation of some ophiolite complexes that were previously assumed to be highly dismembered during obduction or transportation. Because of the lack of major escarpments in areas of slow spreading with a high magma budget (near some spreading segment centers or near hotspots) the upper crustal structure and style of spreading in those settings is not known at present. It is possible that sheeted dike complexes could form in these settings.

Table 1 Summary of observations of sheeted dike complexes in oceanic crust

Seafloor exposures of sheeted dike complexes				
Location	Full rate (mm/year)	Type of exposure	Notes	References
Slow-spreading crust				
Vema Transform	24	Transform valley wall	Contacts above and below are faulted; dikes vertical	Auzende et al. (1989)
MARK Area	22	Rift-parallel fault bounding oceanic core complexes	Dike margins near parallel to fault; image interpretation only	Auzende et al. (1994), Dick et al. (2008)
Kings Trough	30	Intra-Oceanic Rift	Dikes vertical; no documentation	Klitgord et al. (1990)
Intermediate-rate-spreading crust				
Blanco Transform	54	Transform valley wall	Laterally extensive exposures, intensely faulted, no lower contact observed; dikes dip moderately outward	Juteau et al. (1995), Naidoo (1998), Tivey et al. (1998), Karson et al. (2002b)
Hole 504B	60	DSDP/ODP Drill hole	No lower contact reached; dikes fractured and dip moderately outward	Cann et al. (1983), Adamson (1985), Alt et al. (1993), Schouten and Denham (2000)
Fast-spreading crust				
Hess Deep Rift	110	Upper scarp, propagating rift	Laterally extensive exposures; upper and lower contacts well preserved; dikes fractured and dip moderately outward	Francheteau et al. (1990, 1992), Karson et al. (1992, 2002a), Stewart et al. (2002)
Superfast-spreading crust				
Pito Deep Rift	142	Upper scarp, propagating rift	Laterally extensive exposures; upper and lower contacts preserved; dikes dip steeply outward	Francheteau et al. (1994), Karson et al. (2005), Chutas (2007)
Endeavor Deep Rift	200	Upper scarp, propagating rift	Local dense dike swarms; dikes weakly fractured; dip very steeply	Larson et al. (2005)
Hole 1256D	220	IODP Drill hole	Upper and lower contacts penetrated; weakly fractured dikes, near vertical	Tartarotti et al. (2006), Wilson et al. (2003, 2006)

Seafloor geology and exposures of individual dikes hint at processes along slow-spreading ridges where rifting dominates. Dikes mapped on a rift valley wall at 23°N on the MAR (SMARK Area) cut lavas and breccias and dip inward, that is, toward the axis, typically from 10 to 20° (Fig. 5). Paleomagnetic studies on oriented samples (Fig. 6) show that the dikes have been tectonically rotated into this attitude, probably from an original vertical orientation (Lawrence et al. 1998). This rotation is consistent with block rotation by slip on inward-dipping (toward the rift valley axis) normal faults that bound the rift valley.

Although some investigations have reported sheeted dike complexes in crust formed at slow-spreading ridges, well-documented cases (unambiguous images and samples) are rare. At Kings Trough, a rift in Mesozoic crust formed at an intermediate rate east of the MAR, a sheeted dike complex was reported but without documentation (Klitgord et al. 1990). Paradoxically, sheeted dikes near the MAR are best documented at the Vema ridge-transform intersection (Auzende et al. 1989) (Fig. 5), in crust formed at a spreading segment discontinuity where the magma budget is assumed to be low (Table 1). Just south of the Kane Transform Fault (24°N on the MAR) sheeted dikes are reported on a west-facing fault scarp that cuts oceanic core complexes (Auzende et al. 1994; Dick et al. 2008). None of these examples has been extensively documented but might be worth further examination.

3.3 Dikes in Crust Formed at Intermediate to Superfast Rates

The most extensive, best exposed and most intensely studied sheeted dike complexes are in crust formed at intermediate to superfast rates 50 to >150 mm/year) in the Pacific. Spectacular exposures are comparable to those of the major ophiolite complexes (Fig. 7). There are seven areas in the Pacific where deep crustal drilling or near-bottom studies of tectonic windows provide important information about sheeted dike complexes in the oceanic crust (Fig. 8). These include three deep drill holes and four tectonic windows in crust of different ages and formed at different spreading rates (Table 2). Each of these will be discussed briefly below.

3.3.1 Hole 504B

This drill site reaches a depth of >2000 m and required 7 drilling legs to achieve. It penetrates a variably altered lava unit, lava dike transition and sheeted dikes (Anderson et al. 1982). Underlying rock units were not reached. Core recovery was poor (~25%) in the highly fractured sheeted dike unit (Fig. 9); dike contacts could rarely be measured directly. If the dikes are essentially vertical, it might be expected that very few dike margins would be intersected by the vertical drill hole. However, nearly 200 dike margins or igneous contacts were recovered, thus they are not vertical. Assuming an average width of 1 m, the dikes must dip about 60° (Karson 2002). Paleomagnetic studies show that the dikes and overlying lavas have been tectonically rotated with

Fig. 5 Dikes in slow-spreading crust of the Mid-Atlantic Ridge. **a** Photomicrograph of typical diabasic (subophitic) texture in a basaltic dike with plagioclase laths partially enclosed in more equant clinopyroxene, olivine and glass. Field of view ~1 cm. **b** Fractured dike (~1 m wide, left) intruding brecciated lavas in the SMARK Area (MAR at 23°N). Dike dips steeply inward, toward the spreading axis. Individual dikes and swarms of dikes crop out in major fault zones bounding the rift valley. **c** Sheeted dikes exposed on the wall of the Vema Transform Fault (some dike margins indicated with dashed lines) (Auzende et al. 1989)

lavas dipping up to 50° toward (inward) and dikes dipping >60° away (outward) from the spreading center where they formed (Schouten and Denham 2000; Pariso and Johnson 1989). This was the first location to show that significant tectonic rotations occur during crustal accretion, a result that is echoed in all later investigations.

Fig. 6 The *Geocompass* is a relatively simple tool with a digital compass and tilt meters used to measure the orientations of structures on seafloor escarpments from HOVs or ROVs. This tool has been used to document the orientations of dikes in the oceanic crust as well as to collect oriented samples for paleomagnetic and AMS studies (from Karson et al. 2015)

Table 2 Comparison of ages and spreading rates of formation of key deep drill holes and tectonic windows in oceanic crust formed at intermediate to superfast rates

Key Study areas of oceanic crust formed at intermediate to superfast rates			
Location	Study	Full rate (mm/year)	Age (m.y.)
Blanco Transform Fault	HOV, sonar	54	0.6–1.6
DSDP/ODP 504B	Drilling	72	5.9
Hess Deep Rift	HOV, sonar	127	1–1.5
Pito Deep Rift	HOV, ROV, sonar	145	3
Endeavor Deep Rift	HOV, ROV	151	3
ODP 801C	Drilling	160	165
IODP 1256D	Drilling	220	15

3.3.2 Hole 801C

Although this hole did not penetrate a sheeted dike complex (i.e., >90% dikes), it did drill through and sample lavas and a lava/dike transition zone with mixed dikes and lavas. Progressive downward inward tilting of the lavas and outward dipping dikes show the same type of block rotation during accretion as inferred at Hole 504B (Pockalny and Larson 2003).

Fig. 7 Sheeted dikes in ophiolites and the upper oceanic crust. **a** Sheeted dike outcrop in the Oman Ophiolite. **b** Sheeted dike outcrop on a seafloor escarpment at the Pito Deep Rift. Dikes are near vertical and about 1 m wide in both areas. Continuous dike margins (arrows) bound evenly spaced columnar joints (~horizontal) within each dike (after Karson et al. 2015)

3.3.3 Hole 1256D

This hole in superfast-spreading crust penetrated all of the upper crustal units anticipated from ophiolite studies including on- and off-axis lava flows, a lava/dike transition zone, sheeted dikes and a small amount of gabbroic rock (Wilson et al. 2003, 2006). Detailed studies of the lava flows, dikes and sheeted dike complex show that these structures have also been tectonically rotated during spreading, resulting in inward-dipping lavas and outward-dipping dikes (Tartarotti et al. 2006; Wilson et al. 2003). Gabbroic rocks show contact metamorphic relationships (Wilson et al. 2006).

3.3.4 Blanco Transform Fault

The north wall of the Blanco Transform Fault (Fig. 10) exposes crust formed at the Cleft Segment of the Juan de Fuca Ridge. Northward migration of the transform zone cuts into crust formed ~10 km north of the former ridge-transform intersection (Dziak et al. 2000). Multiple HOV diving programs in the area provide detailed descriptions of the upper crustal section including a very thick lava section, lava-dike transition

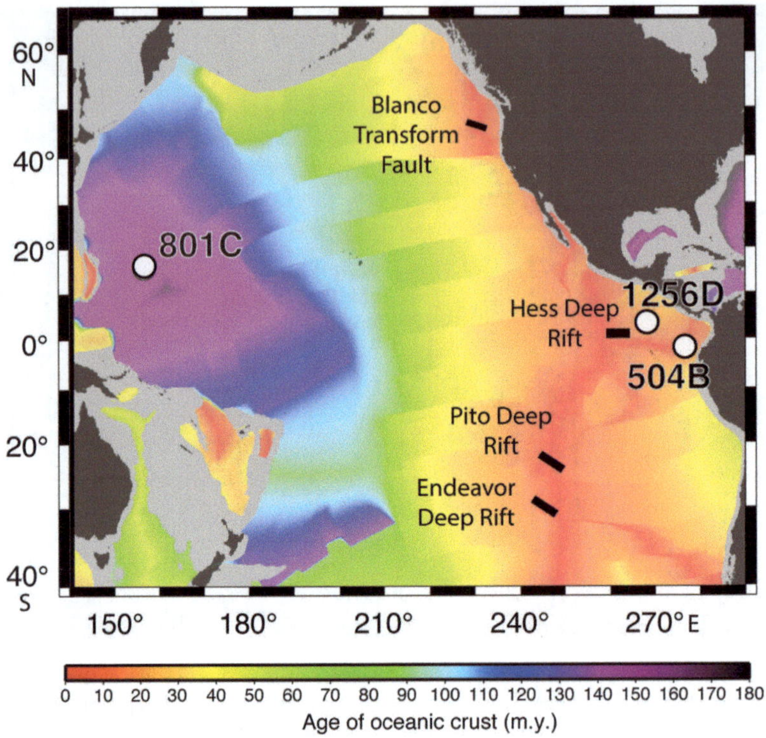

Fig. 8 Locations of key areas where the upper crustal structure of the oceanic crust has been investigated. These include three deep crustal drill holes: DSDP/ODP 504B near the Costa Rica Rift, ODP 801C in the western Pacific, and 1256D in the equatorial Pacific. Other areas are major escarpments that expose "tectonic windows" into the upper oceanic crust: Blanco Transform Fault at the south end of the Juan de Fuca Ridge, the Hess Deep Rift at the western end of propagating Cocos-Nazca plate boundary, Pito Deep Rift at propagating rift in the edge of the Easter Microplate and the Endeavor Deep Rift on the edge of the Juan Fernandez microplate. These areas are in crust of different ages and that formed at very different spreading rates (from Karson et al. 2015)

and possible sheeted dikes (Juteau et al. 1995; Naidoo 1998; Tivey 1996; Tivey et al. 1998). Although the rock units are intensely fractured, lava flows dip dominantly inward and dikes outward (Karson et al. 2002b).

3.3.5 Hess Deep Rift

The walls of the Hess Deep Rift are formed by westward, amagmatic propagation of the E-W Cocos-Nazca spreading center (Lonsdale 1988). The rift forms the northern edge of the Galapagos Microplate. The rift walls cut the abyssal hill lineaments in adjacent crust at a high angle, creating a natural cross section of the upper oceanic crust (Fig. 10). Initial reconnaissance dives discovered extensive exposures (tens of

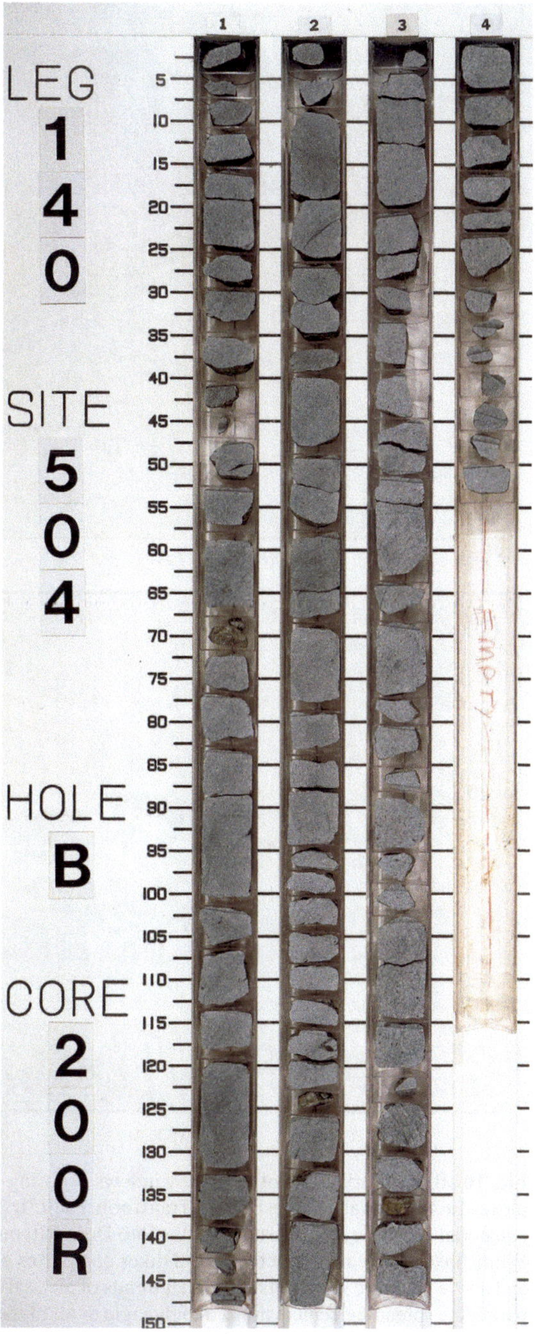

Fig. 9 A section of core from DSDP/ODP Hole 504B showing the typical, highly fragmentary recovery of material drilled from a sheeted dike complex (Alt et al. 1993). The core is about 7 cm wide

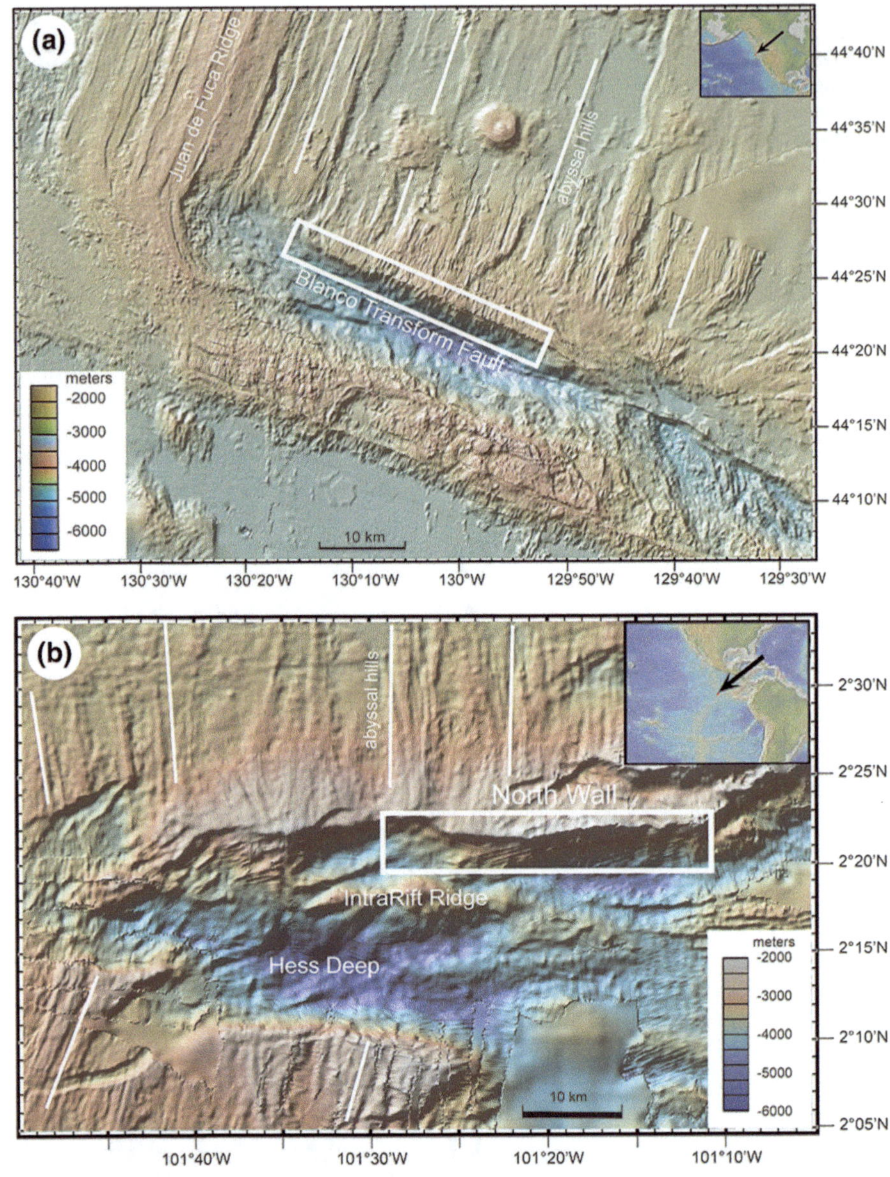

Fig. 10 Bathymetric maps of tectonic windows exposing extensive outcrops of sheeted dike complexes. **a** North wall of the Blanco Transform Fault. **b** North all of the Hess Deep Rift in the equatorial Pacific. **c** Northeast wall of the Pito Deep Rift on the NE corner of the Easter Microplate. White boxes show areas where sheeted dikes complexes are exposed structurally below lava units and above gabbroic rock units. Note that trends of abyssal hill lineaments (white lines) that formed parallel to spreading centers are at a high angle to all of these escarpments that create natural cross sections of the oceanic crust parallel to the spreading direction. All images made with GeoMap App (Ryan et al. 2009)

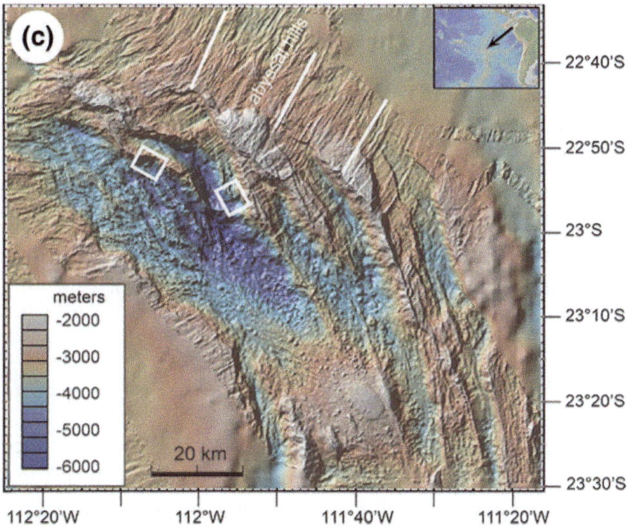

Fig. 10 (continued)

kilometers along the rift walls) of lavas, lava-dike transition zone, sheeted dikes and locally gabbroic rocks (Francheteau et al. 1990, 1992). Later more focused studies on the North Wall of the rift documented the details of the structure (Karson et al. 1992, 2002a), paleomagnetic properties (Horst et al. 2014; Hurst et al. 1994; Varga et al. 2004), igneous petrology (Hanna 2004; Stewart et al. 2002, 2005; Natland and Dick 1996) and metamorphic petrology (Gillis et al. 2001) of these rock units. These studies collectively show consistent inward tilting of lava flows and outward rotation of dikes accommodated by intense brecciation of the lava-dike transition zone and sheeted dike complex. Faulting and brecciation occur along dike margins as well as in wider areas that cut across dike margins. Progressive downward tilting of lava flows, cross-cutting dikes, and high-temperature alteration of breccias show that the tectonic rotations and brecciation of lavas and dikes occurred during subaxial subsidence during crustal accretion at the EPR (Karson et al. 2002a).

Deeper in the rift valley, below the North Wall outcrops described above, near-bottom investigations and deep crustal drilling provide additional information on the Intra-Rift Ridge, a horst block just downslope from the exposures on the north wall of the rift (Fig. 10). Two drilling legs recovered a diverse suite of gabbroic and ultramafic rocks considered to represent the lower crust and upper mantle, respectively (Gillis et al. 1993, 2013). Correlation of these units with those of the north wall relies strongly on reported sheeted dikes on the top of the ridge (MacLeod et al. 2008), which are not well documented.

3.3.6 Pito Deep Rift

Like the tectonic window at the Hess Deep Rift, large escarpments expose the upper crust at the propagating Pito Deep Rift on the NE corner of the Easter Microplate (Naar and Hey 1991). The upper crustal exposures are similar to those at Hess Deep and have been investigated extensively (Hayman and Karson 2009; Heft et al. 2008; Horst et al. 2011; Karson et al. 2005; Perk et al. 2007; Pollock et al. 2009). Like other sections described above, lava flow contacts dip inward and dikes dip steeply outward above underlying gabbro units (Karson et al. 2005).

3.3.7 Endeavor Deep Rift

Only lavas and swarms of dikes were reported from the Endeavor Deep Rift on the edge of the Juan Fernandez Microplate (Francheteau et al. 1987) (Fig. 8). The same upper crustal structural geometry of inward-dipping lavas and outward-dipping dikes occurs (Larson et al. 2005). No sheeted dike complex with >90% dikes was reached.

4 Ophiolites on the Seafloor

Although exposures of dikes and sheeted dikes in slow-spreading crust of the MAR are rare and subject to major faulting and block rotations, drilling and tectonic windows in crust formed at intermediate to superfast rates show a consistent structure that in general mimics that of ophiolites, but also suggests some important variations on long-held assumptions. Figure 11 shows schematic columnar sections for the three deep drill holes and four tectonic windows that have been investigated. The following key observations emerge (Karson et al. 2015):

(1) All sections show the general rock units anticipated from ophiolite studies to different structural depths. Contacts between rock units are all gradational and have significant (hundreds of meters) relief. All seven sites investigated show slight variations on a similar crustal structure.

(2) The thicknesses of the upper crustal rock units generally match the thicknesses of seismic layer 2 (Fig. 3) but the total lava thickness is locally comparable to but in some cases significantly greater than that of seismic layer 2A (Christeson et al. 2007).

(3) The compositions of the associated lava and sheeted dike complexes sampled in the oceanic crust are subtly different, possibly controlled by magma density (Karson et al. 2015; Pollock et al. 2009; Stewart et al. 2003).

(4) In all sections, the inward dip of lava flows (toward the spreading center where they formed) increases progressively downward. Ophiolite studies commonly assume horizontal lava flows (at least for sheet lavas). Individual dikes in the

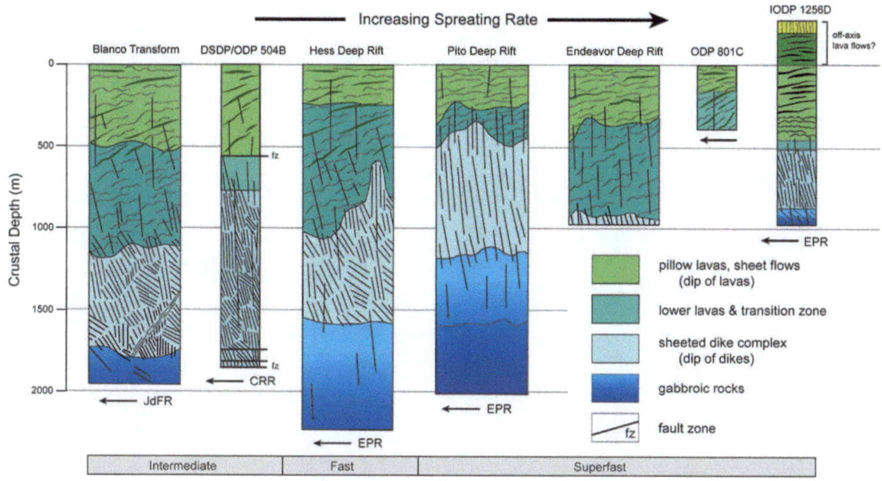

Fig. 11 Summary of upper crustal structures from 3 deep crustal drill holes (narrow columns) and four tectonic windows (wide columns) into oceanic crust formed at intermediate to superfast rates. Columns are arranged with crust formed at higher spreading rates left to right. Dip of lava flows (wavy lines) and dikes (bold lines) are shown schematically. In all cases, dikes dip dominantly (outward) away from the spreading center where they formed (to left), variably, brecciated and rotated blocks of dikes are cut by younger, undeformed dikes. Lava flow dips increase with depth (inward) toward the spreading center where they were erupted. The overall structure indicates different degrees of subaxial subsidence during crustal accretion (from Karson et al. 2015)

lava-dike transition zone dip outward. Dikes are generally orthogonal to lava flows but they are not vertical as typically assumed.

(5) Dikes in the sheeted dike complex occur in fault-bounded panels of parallel dikes (Fig. 12) but generally dip outward; paleomagnetic studies show they have been tectonically rotated (Varga et al. 2004). They cannot be assumed to provide a reliable vertical reference frame.

(6) Discrete fault zones cutting the sheeted dikes show intense brittle deformation (Fig. 13). Both inward- and outward-dipping faults occur in relatively fast-spread crust.

(7) Distributed fracturing and faulting generally increase downward through the lower lavas and transition zone and into sheeted dikes. Deformation is commonly focused in dike-parallel fault zones. Hydrothermal alteration and veining show that these fault zones control fluid flow at this crustal level (Gillis et al. 2001; Hayman and Karson 2007, 2009; Heft et al. 2008) (Figs. 14 and 15).

(8) Dikes and hydrothermal veins cut the tilted and fractured lavas and dikes (Fig. 14).

(9) Overlying upper lavas and underlying gabbroic rocks (where exposed) are much less fractured and altered than the intervening lower lavas, transition zone and sheeted dikes. Similar geological relationships can be found in some ophiolites.

Fig. 12 Sheeted dikes exposed on a vertical cliff face at the Pito Deep Rift. Individual dikes are ~1 m wide. Continuous dike margins (dipping to right) bound parallel columnar joints within each dike. As in all areas studied, dikes dip outward (away from the EPR spreading center where they formed) (from Karson et al. 2015)

Fig. 13 Faulting in sheeted dikes at the Hess Deep Rift. **a** Vertical cliff face with outcrop of intensely brecciated dike rock in the sheeted dike complex. White line shows orientation of nearby sheeted dikes. **b** Sample collected from the fault zone (~20 cm long) is fault breccia and cataclasite bounded by slickensided surfaces (right and left). **c** Photomicrograph of the faulted dike rock shows a clast of fresh diabase (left) in dark cataclasite to fault gouge (right). Field of view ~1 cm (Karson et al. 2015)

Fig. 14 Examples of typical structures found in sheeted dike complexes on the seafloor and ophiolite counterparts. **a** Outward-dipping dikes separated by intensely brecciated dike-parallel fault zones at the Hess Deep Rift. **b** Dike-parallel fault zones in the sheeted dike complex of the Oman Ophiolite. **c** Low-angle cross-cutting relations among dikes in the sheeted dike complex of the Oman Ophiolite. **d** Steeply dipping dike (center, white dashed lines) cuts moderately dipping dike (left) at the Pito Deep Rift. Steeply dipping dikes commonly cut older, dipping dikes that have been tectonically rotated. Dikes are ~1 m wide in all images. Arrows highlight selected dike margins

(10) The degree of subsidence and fracturing appears to be function of spreading rate or magma budget with less subsidence at higher rates (or magma budgets) (Karson et al. 2015). The amount of subaxial subsidence during construction must be at least as much as the depth to the deepest lava in the crustal section, that originally erupted on the seafloor.

Collectively, these relationships support a model of crustal accretion in which subaxial subsidence above an axial magma chamber and underlying low-velocity zone controls the construction of the upper crust (Fig. 16). Subsidence of lavas at the axis permits thickening of the lava units in a very narrow (~1 km) area near the axis. Subsidence is asymmetrical (greatest near the axis) resulting in tilting of lavas to inward-dipping orientations. Dikes are initially intruded vertically near the axis but rotate progressively to outward-dipping orientations as the axial crust subsides and spreads laterally. Subsidence and rotation are accommodated by pervasive brittle deformation on discrete faults and wider zones of distributed cataclastic deformation. The sheeted dike complex consists of large panels (meters to tens of meters across) of

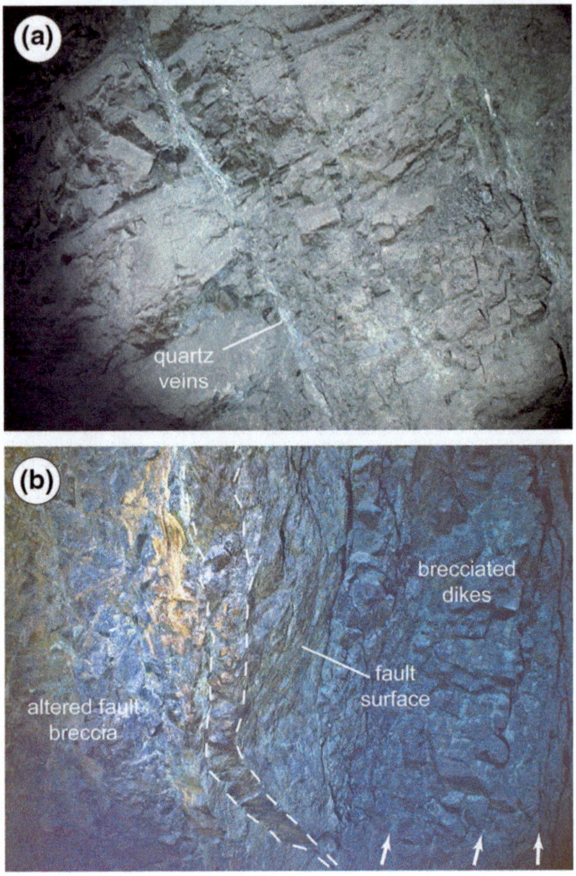

Fig. 15 Deformation and alteration in the sheeted dike complex at the Pito Deep Rift. **a** Vertical cliff face shows white quartz veins along the margins of outward-dipping (~60°) dikes. **b** Vertical cliff face with brecciated and altered dike rock with rust-colored gossan and white quartz veins in steeply dipping, fractured dikes (arrows highlight dike margins). A discrete, slickensided fault surface cuts the breccia and small black dike (white dashed lines) cuts the fault zone

parallel dikes surrounded by pervasively fractured material. Deformation is focused along dike margins and fluid flow is channeled along dike-parallel fault zones. Cross-cutting dikes and hydrothermal veins show that the rotations and brittle deformation occurred beneath the spreading axis.

This general model has its roots in older models based on relationships in ophiolite complexes from before constraints from the oceanic crust were available (Cann 1968, 1974; Dewey and Kidd 1977; Rosencrantz 1982). More recent studies have progressively refined this view (Karson 2002; Karson et al. 2002a, 2015) but it remains a work in progress. It should be noted that this model applies only to crust formed at intermediate to superfast rates. Relevant observations that could reveal upper crustal construction processes at slower spreading rates are not yet available. Observations along rifted parts of the MAR suggest that even in areas near spreading segment centers crustal accretion processes and upper crustal geology are substantially different from spreading at faster rates. At slow spreading rates, upper crustal rock units are

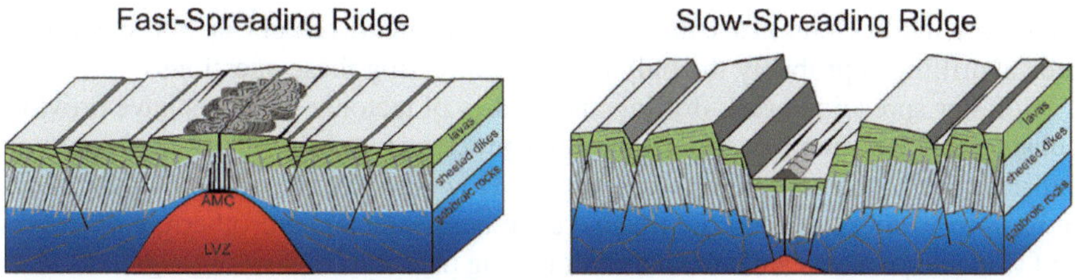

Fig. 16 Highly simplified summary of the structure of sheeted dikes and lavas in crust formed at fast and slow spreading rates. At faster rates a relatively continuous magma supply from the Axial Magma Chamber (AMC) and underlying Low-Velocity Zone (LVZ) keeps pace with plate separation. Subaxial subsidence accommodated by distributed brittle deformation (not shown) during accretion results in an upper crustal structure with inward-dipping lava flows and outward-dipping dikes. At slow spreading rates, faulting dominates over magmatic construction. Slip on normal faults results in tilting of crustal blocks with outward-dipping lava flows and inward-dipping dikes. Crustal structures formed at fast spreading rates may be similar to those formed at slower rates with high magma budgets, for example, near hot spots (from Karson et al. 2015)

likely to be subjected to extensional tectonic processes similar to those of continental rifts, for example inward-dipping normal faults and back-rotated crustal blocks (Fig. 16).

5 Discussion and Concluding Remarks: Looking Forward and Backward

Studies of ophiolites and sheeted dike complexes in particular, paved the way for understanding the internal structure of oceanic crust and processes at mid-ocean ridge spreading centers. Assumptions and imagination infused many models for spreading processes but beginning in the mid-1970s direct observations of the geology of the oceanic crust provided hard constraints that have been progressively integrated into the perceptions of seafloor spreading. Although the available data remain fragmentary, it is appropriate to consider what more ophiolites might tell us about MOR processes and, conversely, what direct investigations of oceanic crust can tell us about ophiolites and the settings in which they form.

Looking ahead to refining the understanding of seafloor spreading processes, a number of important gaps remain to be filled. One of these concerns the nature of spreading along slow-spreading ridges. Investigations over the past few decades have revealed unexpected diversity in geologic structures and implied processes, for example, oceanic core complexes (Escartín and Canales 2010). These structures probably form in areas of low magma budget, but what about the areas with high magma budgets? Do sheeted dike complexes form in these settings? How continuous are they along and across spreading directions? Given the general lack of major tectonic windows into crust that formed near segment centers or hot spots, deep crustal drilling is probably the only hope of addressing these questions.

In faster spreading crust, drilling and studies of tectonic windows have provided considerable information about spreading processes and these are in accord with expectations based on ophiolite complexes. However, significant, systematic variations in spreading such as axial magma chamber character, hydrothermal processes, and volcanic products are known to occur along individual spreading segments (Carbotte et al. 2006). Magmatism is more robust near segment centers and more limited near segment boundaries. At present, it is difficult to evaluate the available crustal structures (Fig. 11) in the context of these known segment-scale variations. It is not clear where these crustal sections formed with respect to segment centers and ends. There are more potential tectonic windows in relatively fast-spread crust and investigations of them could help assess the diversity of crustal structures and processes, but ultimately, these need to be linked to segment-scale processes. Again, the only available approach would be deep crustal drilling along an isochron-parallel transect to sample crust formed in different parts of the same spreading segment. Given the technical challenges of deep crustal drilling and the current funding climate, this type of study is not likely to happen any time soon.

Looking back on ophiolites, some new questions arise. The discovery of the complex geology of oceanic core complexes and more generally the diversity of spreading structures formed at slow-spreading ridges, has already stimulated investigations of ophiolites that lack the simple idealized layered structure of the Penrose definition (Anonymous 1972). These studies are likely to make a major contribution to the understanding of oceanic core complexes (e.g., Nicolas et al. 1999). Recognizing that most ophiolites may have formed in suprasubduction zone settings, it is likely that ophiolite geology can provide important information on crustal accretion in those environments. However, one important caution that has emerged from the perspective of sheeted dikes is that it appears that using the assumption of vertical dike orientations is unreliable. Although dikes in the oceanic crust are probably initially intruded vertically, they are rotated by subsequent crustal accretion processes. Thus, reconstructions of the ophiolite layer thicknesses and restoration of internal structures within rock units should be re-examined. Paleomagnetic data are probably the best way to test and if necessary revise these reconstructions. Seafloor and ophiolite investigations show that both inward and outward rotations are possible in different tectonic settings (Varga et al. 2004; Varga and Moores 1985).

Many authors have attempted to assign a spreading rate to ophiolite complexes on the basis of their general internal structure. For example, major faulting affecting the entire crustal section is generally thought to be more typical of slow-spreading environments. While this is probably true, it would be more accurate to interpret large-scale faulting as a result of low magma budget rather than rate alone.

The internal structure of ophiolites is also commonly used to suggest how asymmetrical structures are related to the direction back to a spreading center. At present there are no clear, consistent criteria to indicate age relations within an ophiolite [besides geochronology (Rioux et al. 2013)]. However, since all crustal sections in the Pacific show a consistent upper crustal geometry (inward-dipping lavas and outward-dipping dikes) this geometry might help identify ophiolites formed at relatively fast rates or high magma budgets.

Sheeted dike complexes in ophiolites have played an important historic role in the evolution of thinking about the oceanic crust and processes at mid-ocean ridge spreading centers. Despite geochemical data indicating that most if not all ophiolite complexes formed in suprasubduction zone environments, very similar upper crustal geological units and internal structures are now documented in the oceanic crust. This is probably less ironic than a reflection of similar geology resulting from magmatic construction and extension across a spectrum of plate boundary settings. Differences between documented seafloor outcrop relationships and expectations from ophiolites should lead to some revised interpretations based on ophiolite geology alone.

Acknowledgements Thanks to the many colleagues who have contributed so much to our investigations of ophiolites, the geology of the oceanic crust at sea, and analogous subaerial terranes. Thanks also to the National Science Foundation for the support of these diverse projects. Reviews by Craig Magee and Paul Robinson helped improve the presentation of material and interpretations in this manuscript.

References

Abbots IL (1979) Intrusive processes at ocean ridges: evidence from the sheeted dyke complex of Masirah Oman. Tectonophysics 60:217–233

Adamson AC (1985) Basement lithostratigraphy, deep sea drilling project hole 504B. In: Anderson RN, Honnorez J, Becker K (eds) Initial reports of the deep sea drilling project. 83. Washington, DC, U.S., Government Printing Office, pp 121–127

Alexander RJ, Harper GD, Bowman JR (1993) Oceanic faulting and fault-controlled subseafloor hydrothermal alteration in the sheeted dike complex of the Josephine ophiolite. J Geophys Res 98:9731–9759

Alt JC, Kinoshita H, Stokking LB et al. (eds) (1993) Proceedings of the ocean drilling program, initial reports, Leg 148. Ocean Drilling Program, College Station, TX. https://doi.org/10.2973/odp.proc.ir.148.1993

Anderson EM (1951) The dynamics of faulting and dyke formation, 2nd edn. Oliver and Boyd, London

Anderson RN, Honnorez J, Becker K, Adamson AC, Alt JC, Mottl MJ, Newmark RL (1982) DSDP Hole 504B, the first reference section over 1 km through Layer 2 of the oceanic crust. Nature 300:589–594

Anonymous (1972) Penrose field conference on ophiolites. Geotimes 17:24–25

ARCYANA (1975) Transform fault and rift valley from bathyscaph and diving saucer. Science 190:108–116

ARCYANA (1978) FAMOUS: photographic atlas of the Mid-Atlantic Ridge: rift and transform fault at 3000 m depth. Bordas, Paris, France

Auzende J-M, Bideau D, Bonatti E, Cannat M, Honnorez J, Lagabrielle Y, Malavieille J, Mamaloukas-Frangoulis V, Mével C (1989) Direct observation of a section through slow-spreading oceanic crust. Nature 337:726–729

Auzende JM, Cannat M, Gente P, Henriet J-P, Juteau T, Karson JA, Lagabrielle Y, Mével C, Tivey M (1994) Observation of sections of oceanic crust and mantle cropping out on the southern wall of the Kane Fracture Zone (N. Atlantic). Terra Nova 6:143–148

Ballard RD, Francheteau J, Juteau T, Rangan C, Normark W (1981) EPR at 21°N: volcanic, tectonic and hydrothermal processes of the central axis. Earth Planet Sci Lett 55:1–10

Barager WRA (1954) Betts Pond Area, Burlington Peninsula, Newfoundland. Newfoundland Department of Mines, Open File Report. St. John's, Newfoundland, Canada

Barager WRA, Lambert MB, Baglow N, Gibson IL (1990) The sheeted dyke zone in the Troodos ophiolite. In: Malpas J, Moores EM, Panayiotou A, Xenophontos C (eds) Ophiolites and oceanic crustal analogues: proceedings of the symposium "Troodos 1987". Cyprus Geological Survey Department, Nicosia, Cyprus, pp 37–51

Cann JR (1968) Geological processes at mid-ocean ridge crests. Geophys J Roy Astron Soc 15:331–341

Cann JR (1974) A model for oceanic crustal structure developed. Geophys J Roy Astron Soc 39:169–187

Cann JR, Simkin T (1971) A bibliography of ocean-floor rocks. Philos Trans R Soc Lond 268:734–737

Cann JR, Langseth MG, Honnorez J, Herzen RPV, White SM, et al (1983) Initial reports of the deep sea drilling project. Washington, DC, U.S., Goverment Publishing Office

Carbotte SM, Macdonald KC (1992) East Pacific Rise 8°–10° 30′N: evolution of ridge segments and discontinuities from SeaMARC II and three-dimensional magnetic studies. J Geophys Res 97:6959–6982

Carbotte SM, Mutter C, Mutter J, Ponce-Correa G (1998) Influence of magma supply and spreading rate on crustal magma bodies and emplacement of the extrusive layer: insights from the East Pacific Rise at lat 16°N. Geology 26:455–458

Carbotte SM, Detrick RS, Harding A, Canales JP, Babcock J, Kent G, Van Ark E, Nedimovic M, Diebold J (2006) Rift topography linked to magmatism at the intermediate spreading Juan de Fuca Ridge. Geology 34:209–212

Carbotte SM, Canales JP, Nedimovic MR, Carton H, Mutter JC (2012) Recent seismic studies at the East Pacific Rise 8°20′–10°10′N and endeavour segment: insights into mid-ocean ridge hydrothermal and magmatic processes. Oceanography 25(1):100–112. https://doi.org/10.5670/oceanog.2012.08

Casey JF, Karson JA, Elthon DL, Rosencrantz E, Titus M (1983) Reconstruction of the geometry of accretion during formation of the Bay of Islands Ophiolite Complex. Tectonics 2:509–528

CAYTROUGH (1979) Geological and geophysical investigation of the Mid-Cayman Rise spreading center: initial results and observations. In: Talwani M, Harrison CG, Hayes DE (eds) Deep drilling results in the Atlantic Ocean: ocean crust. Maurice Ewing series 2, pp 66–93

Christensen NI, Salisbury MH (1982) Lateral heterogeneity in the seismic structure of the oceanic crust inferred from velocity studies in the Bay of Islands ophiolites, Newfoundland. Geophys J Roy Astron Soc 68:675–688

Christeson GL, Purdy GM, Fryer GJ (1992) Structure of young upper crust at the East Pacific Rise near 9° 30′N. Geophys Res Lett 19:1045–1048

Christeson GL, McIntosh KD, Karson JA (2007) Inconsistent correlation of seismic layer 2a and lava layer thickness in oceanic crust. Nature 445. https://doi.org/10.1038/nature05517

Christeson GL, Morgan JV, Warner MR (2012) Shallow oceanic crust: full waveform tomographic images of the seismic layer 2A/2B boundary. J Geophys Res 117. https://doi.org/10.1029/2011jb008972

Church WR, Stevens RK (1971) Early Paleozoic complexes of the Newfoundland Appalachians as mantle-oceanic crust sequences. J Geophys Res 76:1460–1466

Chutas LA (2007) Structures in upper oceanic crust: perspectives from pito deep and Iceland [Masters thesis]. Durham, NC, Duke University

Delaney JR, Kelley DS, Lilley MD, Butterfield DA, Baross JA, Embley RW, Summit M (1998) The quantum event of oceanic crustal accretion: impacts of diking at mid-ocean ridges. Science 281:222–230

Detrick RS, Buhl P, Vera E, Mutter J, Orcutt J, Madsen J, Brocher T (1987) Multichannel seismic imaging of a crustal magma chamber along the East Pacific Rise. Nature 326:35–41

Dewey JF, Bird JM (1971) Origin and emplacement of the ophiolite suite: Appalachian ophiolites in Newfoundland. J Geophys Res 76:3179–3206
Dewey JF, Kidd WSF (1977) Geometry of plate accretion. Geol Soc Am Bull 88:960–968
Dick HJB, Tivey MA, Tuchoke BE (2008) Plutonic foundation of a slow-spreading ridge segment: the oceanic core complex at Kane megamullion, 23° 30′N, 42° 20′W. Geochem Geophys Geosyst 9(5). https://doi.org/10.1029/2007gc001645
Dunn RA, Toomey DR, Solomon SC (2000) Three-dimensional seismic structure and physical properties of the crust and shallow mantle beneath the East Pacific Rise at 9° 30′N. J Geophys Res 105:23537–23555
DuToit AL (1929) The volcanic belt of the Lebombo—a region of tension. Trans Roy Soc S Afr 18:189–218
Dziak RP, Bohnenstiehl DR, Matsumoto H, Fowler MJ, Haxel JH, Tolstoy M, Waldhauser F (2009) January 2006 seafloor-spreading event at 9 50′ N, East Pacific Rise: ridge dike intrusion and transform fault interactions from regional hydroacoustic data. Geochem Geophys Geosyst 10(6). https://doi.org/10.1029/2009gc002388
Dziak RP, Fox CG, Embley RW, Nabelek JL, Braunmiller J, Koski RA (2000) Recent tectonics of the Blanco Ridge, eastern Blanco Transform Fault Zone. Mar Geophys Res 21:423–450
Einarsson P, Brandsdottir B (1980) Seismological evidence for lateral magma intrusion during the July 1978 deflation of the Krafla volcano in NE-Iceland. J Geophys 47:160–165
Engel CG, Fisher RL (1975) Granitic to ultramafic rock complexes of the Indian Ocean ridge system western Indian Ocean. Geol Soc Am Bull 86:1553–1578
Escartín J, Canales JP (2010) Chapman conference on Detachments in oceanic lithosphere: deformation, magmatism, fluid flow and ecosystems (conference report). EOS Trans Am Geophys Union 92:31. https://doi.org/10.1029/2011eo040003
Ewing MA, Engel L (1962) Seismic shooting at sea. Sci Am 5:116–126
Fox PJ, Schreiber E, Peterson JJ (1973) The geology of the oceanic crust: compressional wave velocities of oceanic rocks. J Geophys Res 78:5155–5172
Francheteau J, Yelles-Chaouche A, Craig H (1987) The Juan Fernandez microplate north of the Pacific-Nazca-Antarctic plate junction at 35°S. Earth Planet Sci Lett 86:253–286
Francheteau J, Armijo R, Chiminee JL, Hékinian R, Lonsdale P, Blum N (1990) 1 Ma East Pacific Rise oceanic crust and uppermost mantle exposed by rifting in Hess Deep (equatorial Pacific Ocean). Earth Planet Sci Lett 101:281–295
Francheteau J, Armijo R, Cheminee JL, Hékinian R, Lonsdale P, Blum N (1992) Dyke complex of the East Pacific Rise exposed in the walls of Hess Deep and the structure of the upper oceanic crust. Earth Planet Sci Lett 111:109–121
Francheteau J, Armijo R, Cogné JP, Girardeau J, Constantin M, Hékinian R et al (1994) Submersible observations of the Easter microplate and its boundary. EOS T Am Geophys Un 75:582
Gass IG (1968) Is the Troodos massif of Cyprus a fragment of Mesozoic ocean floor? Nature 220:39–42
Gillis K, Mével C, Allan J et al (1993) Proceedings of the ocean drilling program, initial reports, Leg 147. Ocean Drilling Program, College Station, TX
Gillis KM, Muehlenbachs K, Stewart M, Gleeson T, Karson JA (2001) Fluid flow patterns in fast-spreading East Pacific Rise crust exposed at Hess Deep. J Geophys Res 106:26311–26329
Gillis KM, Snow JE, Klaus A, Abe N, Adriao AB, Akizawa N, Ceuleneer G, Cheadle MJ, Faak K, Falloon TJ, Friedman SA, Godard M, Guerin G, Harigane Y, Horst AJ, Hoshide T, Ildefonse B, Jean MM, John BE, Koepke J, Machi S, Maeda J, Marks NE, McCaig AM, Meyer R, Morris A, Nozaka T, Python M, Saha A, Wintsch RP (2013) Primitive layered gabbros from fast-spreading lower oceanic crust. Nature. https://doi.org/10.1038/nature12778
Grandin R, Socquet A, Jacques E, Mazzoni N, de Chabalier J-B, King GCP (2010) Sequence of rifting in Afar, Manda-Hararo Rift, Ethiopia, 2005–2009: time-space evolution and interactions between dikes from interferometric synthetic aperture radar and static stress change modeling. J Geophys Res 115(B10). https://doi.org/10.1029/2009jb000815

Gudmundsson A (1983) Form and dimensions of dykes in eastern Iceland. Tectonophysics 95:295–307

Hanna HD (2004) Geochemical variations in basaltic glasses from an incipient rift and upper level gabbros from Hess Deep, Eastern Equatorial Pacific. M.S. thesis, Duke University, Durham, NC

Hayman NW, Karson JA (2007) Faults and damage zones in fast-spread crust exposed on the north wall of the Hess Deep Rift: conduits and seals in seafloor hydrothermal systems. Geochem Geophys Geosyst 8(10):Q10002. https://doi.org/10.11029/12007GC001623

Hayman NW, Karson JA (2009) Faulting and hydrothermal alteration in superfast spread crust of the East Pacific Rise exposed at Pito Deep. Geochem Geophys Geosyst 10. https://doi.org/10.1029/2008gc002319

Heft K, Gillis KM, Pollock MA, Karson JA, Klein EM (2008) Constraints on the nature of axial hydrothermal systems from the sheeted dike complex exposed at Pito Deep. Geochem Geophys Geosyst 9(5). https://doi.org/10.1029/2007gc001926

Heirtzler JR, LePichon X (1974) FAMOUS: a plate tectonics study of the genesis of the lithosphere. Geology (June):273–274

Helgason J, Zentilli M (1985) Field characteristics of laterally emplaced dikes: anatomy of exhumed Miocene dike swarm in Reydarfjordur, eastern Iceland. Tectonophysics 115:247–274

Horst AJ, Varga RJ, Gee JS, Karson JA (2011) Paleomagnetic constraints on constructional deformation of superfast-spread oceanic crust exposed at Pito Deep Rift. J Geophys Res 116(12). https://doi.org/10.1029/2011jb008268

Horst AJ, Varga RJ, Gee JS, Karson JA (2014) Diverse magma flow directions during construction of sheeted dike complexes at fast- to superfast-spreading centers. Earth Planet Sci Lett 408:119–131. https://doi.org/10.1016/j.epsl.2014.09.022

Houtz R, Ewing J (1976) Upper crustal structure as a function of plate age. J Geophys Res 81:2490–2498

Hurst SD, Karson JA, Verosub KL (1994) Paleomagnetic study of tilted diabase dikes in fast-spread oceanic crust exposed at Hess Deep. Tectonics 13:789–802

Juteau T, Cannat M, Lagabrielle Y (1990) Serpentinized peridotites in the upper oceanic crust away from transform zones: a comparison of the results of previous DSDP and ODP legs. In: Detrick RS, Honnorez J, Bryan WB, Juteau T (eds) Proceedings of the ocean drilling program, scientific results, vol 106/109, Part B. Ocean Drilling Program, College Station, TX, pp 303–308

Juteau T, Bideau O, Dauteuil G, Manc'h G, Naidoo DD, Nehlig P, Ondreas H, Tivey MA, Whipple KX, Delaney JR (1995) A submersible study in the western Blanco Fracture Zone, N.E. Pacific: lithostratigraphy, magnetic structure and magmatic and tectonic evolution during the last 1.6 Ma. Mar Geophys Res 17:399–430

Karson JA (1990) Seafloor spreading on the Mid-Atlantic Ridge: implications for the structure of ophiolites and oceanic lithosphere produced in slow-spreading environments. In: Malpas J, Moores EM, Panayiotou A, Xenophontos C (eds) Ophiolites and oceanic crustal analogues: proceedings of the symposium "Troodos 1987". Geological Survey Department, Nicosia, Cyprus, pp 125–130

Karson JA (1998) Internal structure of oceanic lithosphere: a perspective from tectonic windows. In: Buck WR, Delaney PT, Karson JA, Lagabrielle Y (eds) Faulting and magmatism at Mid-Ocean Ridges. Geophysical monographs, vol 106. American Geophysical Union, Washington, DC, pp 177–218

Karson JA (2002) Geologic structure of uppermost oceanic crust created at fast- to intermediate-rate spreading centers. Annu Rev Earth Planet Sci 30:347–384

Karson JA, Brooks CK (1999) Structural and magmatic segmentation of the Tertiary East Greenland volcanic rifted margin. In: Ryan P, MacNiocaill C (eds) Continental tectonics, vol 164. Geological Society of London, Special Publication. Blackwell Scientific Publishers, London, pp 313–338

Karson JA, Dick HJB (1983) Tectonics of ridge-transform intersections at the Kane Fracture Zone, 24°N on the Mid-Atlantic Ridge. Mar Geophys Res 6:51–98

Karson JA, Thompson G, Humphris SE, Edmond JM, Bryan WB, Brown JR, Winters AT, Pockalny RA, Casey JF, Campbell AC, Klinkhammer G, Palmer MR, Kinzler RJ, Sulanowska MM (1987) Along-axis variations in seafloor spreading in the MARK area. Nature 328:681–685

Karson JA, Hurst SD, Lonsdale P (1992) Tectonic rotations of dikes in fast-spread oceanic crust exposed near Hess Deep. Geology 20:685–688

Karson JA, Klein EM, Hurst SD, Lee CE, Rivizzigno PA, Curewitz D, Morris AR, Hess Deep '99 Scientific Party (2002a) Structure of uppermost fast-spread oceanic crust exposed at the Hess Deep Rift: implications for subaxial processes at the East Pacific Rise. Geochem Geophys Geosyst 3. https://doi.org/10.1029/2001gc000155

Karson JA, Tivey MA, Delaney JR (2002b) Internal structure of uppermost oceanic crust along the western Blanco Transform Scarp: Implications for subaxial accretion and deformation at the Juan de Fuca Ridge. J Geophys Res 107. Paper Number 2000JB000007

Karson JA, Francheteau J, Gee JS, Gillis KM, Hayman NW, Hékinian R, Hey RN, Hurst SD, Klein EM, Naar DF, Varga RG, Pito Deep 2005 Scientific Party (2005) Nested-scale investigation of tectonic windows into super-fast spread crust exposed at the Pito Deep Rift, Easter Microplate, SE Pacific. InterRidge News 14:5–8

Karson JA, Kelley DS, Fornari DJ, Perfit MR, Shank TM (2015) Discovering the deep: a photographic atlas of the seafloor and oceanic crust. Cambridge University Press, London

Kidd RGW, Cann JR (1974) Chilling statistics indicate an ocean-floor spreading origin for the Troodos complex, Cyprus. Earth Planet Sci Lett 24:151–155

Kidd WSF, Dewey JF, Bird JM (1978) The Mings Bight ophiolite complex, Newfoundland: Appalachian oceanic crust and mantle. Can J Earth Sci 15:781–804

Klausen MB, Larsen HC (2002) East Greenland coast-parallel dike swarm and its role in continental breakup. Geol Soc Am Spec Pap 362:133–158

Klitgord K, Casey JF, Agar S, Cruise Participants (1990) Cruise report of Russian Mir dives at Kings Trough (unpublished report)

Larson RL, Popham CT, Pockalny RA (2005) Lithologic and structural observations at Endeavor Deep and their implications for the accretion process at fast to ultra-fast spreading rates. EOS, Trans Am Geophys Union 85(52):T33D–5094

Lawrence RM, Karson JA, Hurst SD (1998) Dike orientations, fault-block rotations, and the construction of slow spreading oceanic crust at 22° 40′N on the Mid-Atlantic Ridge. J Geophys Res 103:663–676

Lonsdale P (1988) Structural pattern of the Galapagos microplate and evolution of the Galapagos triple junctions. J Geophys Res 93:13551–13574

Macdonald KC (1998) Linkages between faulting, volcanism, hydrothermal activity and segmentation on fast spreading centers. In: Buck WR, Delaney PT, Karson JA, Lagabrielle Y (eds) Faulting and magmatism at Mid-Ocean ridges. Geophysical monographs, vol 106. American Geophysical Union, Washington, DC, pp 27–58

Macdonald KC, Fox PJ, Perram LJ, Eisen MF, Haymon RM, Miller SP, Carbotte SM, Cormier M-H, Shor AN (1988) A new view of the mid-ocean ridge from the behaviour of ridge-axis discontinuities. Nature 335:217–225

MacLeod CJ, Teagle DAH, Gillis KM, James Cook Scientific Party (2008) Accretion of the lower oceanic crust at fast-spreading ridges: a rock drill and near bottom seafloor survey in support of IODP drilling in Hess Deep. Unpublished Cruise Report RRS James Cook Cruise JC21

Martinez F, Taylor B (2013) Modes of crustal accretion in back-arc basins: inferences from the Lau Basin. In: Christie DM, Fisher CR, Lee S-M, Givens S (eds) Back-arc Spreading systematics: geological, biological, chemical and physical interactions, American geophysical union monograph series, Washington, DC, pp 5–30. https://doi.org/10.1029/gm166

Miyashiro A (1973) The Troodos ophiolite complex was probably formed in an volcanic arc. Earth Planet Sci Lett 19:218–224

Miyashiro A (1975) Discussion of "Origin of the Troodos and other ophiolites: a reply to Hynes". Earth Planet Sci Lett 25:217–222

Miyashiro A, Shido F, Ewing M (1969) Composition and origin of serpentinites from the Mid-Atlantic Ridge near 24° and 30°N lat. Contrib Miner Petrol 23:117–127

Miyashiro A, Shido F, Ewing MA (1971) Metamorphism in the Mid-Atlantic Ridge near 24°N and 30°N. Philos Trans R Soc Lond Ser A 268:589–603

Moores EM (1982) Origin and emplacement of ophiolites. Rev Geophys Space Phys 20:735–760

Moores EM, Vine FJ (1971) The Troodos massif, Cyprus, and other ophiolites as oceanic crust: evaluation and implications. Philos Trans R Soc Lond A268:443–466

Moores EM, Kellogg LH, Dilek Y (2000) Tehyan ophiolites, mantle convection, and tectonic "historical contingency": a resolution of the "Ophiolite Conundrum". In: Dilek Y, Moores EM, Elthon D, Nicolas A (eds) Ophiolites and oceanic crust: new insights from field studies and the ocean drilling program. Geological Society of America, Special Paper 349, Boulder, CO, pp 3–12

Murton BJ, Parson LM (1993) Segmentation, volcanism and deformation of oblique spreading centres: a quantitative study of the Reykjanes Ridge. Tectonophysics 222:237–257

Naar DF, Hey RN (1991) Tectonic evolution of the Easter Microplate. J Geophys Res 96:7961–7993

Naidoo DD (1998) Accretion of the upper oceanic crust. Ph.D. dissertation, University of Washington, Seattle, WA

Natland JH, Dick HJB (1996) Melt migration through high-level gabbro cumulates of the East Pacific Rise at Hess Deep: the origin of magma lenses and the deep crustal structure of fast-spreading ridges. In: Mével C, Gillis KM, Allen JF (eds) Proceedings of the ocean drilling program, scientific results, vol 147. Ocean Drilling Program, College Station, TX, pp 21–58

Nicolas A (1989) Structure of ophiolites and dynamics of oceanic lithosphere. Kluwer Academic Press, Dordrecht, The Netherlands

Nicolas A, Boudier F (1992) Rooting of the sheeted dike complex in the Oman Ophiolite. In: Parson LM, Murton BJ, Browning P (eds) Ophiolites and their modern oceanic analogues. Geological Society of London, Special Publication 60. Blackwell Scientific Publishers, London, pp 39–54

Nicolas A, Boudier F, Meshi A (1999) Slow spreading accretion and mantle denudation in the Mirdita ophiolite (Albania). J Geophys Res 104:15155–15167

Nielsen TFD (1978) The Tertiary dike swarm of the Kangerlussuaq area, East Greenland: an example of magmatic development during continental break-up. Contrib Miner Petrol 67:63–78

OTTER (1984) The geology of the Oceanographer transform: the ridge-transform intersection. Mar Geophys Res 6:109–141

OTTER (1985) The geology of the Oceanographer Transform: the transform domain. Marine Geophys Res 7:329–358

Pallister JS (1981) Structure of the sheeted dike complex of the Samail ophiolite near Ibra, Oman. J Geophys Res 86:2661–2672

Pallister JS, Hopson CA (1981) Samail ophiolite plutonic suite: field relations, phase variation and layering and a model of a spreading ridge magma chamber. J Geophys Res 86:2593–2644

Pariso JE, Johnson HP (1989) Magnetic properties and oxide petrography of the sheeted dike complex in Hole 504B. Proc Ocean Drill Program Sci Results Leg 111:159–166

Pearce JA (2003) Suprasubduction zone ophiolites: the search for modern analogs. In: Dilek Y, Newcomb S (eds) Ophiolite concept and the evolution of geologic thought, special paper 373. Geological Society of America, Boulder, CO, pp 269–293

Pearce JA, Cann JR (1973) Tectonic setting of basic volcanic rocks determined using trace element analyses. Earth Planet Sci Lett 19:290–300

Perk NW, Coogan LA, Karson JA, Klein EM (2007) Primitive gabbroic rocks from a tectonic window at Pito Deep: implications for the accretion of plutonic rocks beneath the East Pacific Rise. Contrib Miner Petrol 154(5):575–590. https://doi.org/10.1007/s00410-007-0210-z

Pockalny RA, Larson RL (2003) Implications for crustal accretion at fast-spreading ridges from observations in Jurassic oceanic crust in the western Pacific. Geochem Geophys Geosyst 4. https://doi.org/10.1029/2001gc000274

Pollock MA, Klein EM, Karson JA, Coleman DS (2009) Compositions of dikes and lavas from the Pito Deep Rift: implications for accretion at superfast spreading centers. J Geophys Res 114(B03207). https://doi.org/10.1029/2007jb005436

Rioux M, Bowring S, Kelemen P, Gordon S, Miller R, Dudás F (2013) Tectonic development of the Samail ophiolite: high-precision U-Pb zircon geochronology and Sm-Nd isotopic constraints on crustal growth and emplacement. J Geophys Res Solid Earth 118:2085–2101. https://doi.org/10.1002/jgrb.50139

Robinson PT, Melson WG, O'Hearn T, Schmincke H-U (1983) Volcanic glass compositions of the Troodos ophiolite, Cyprus. Geology 11:400–404

Robinson PT, Malpas J, Dilek Y, Zhou M-f (2008) The significance of sheeted dike complexes in ophiolites. GSA Today 18(11):4–10. https://doi.org/10.1130/GSATG22A

Rosencrantz E (1982) Formation of uppermost oceanic crust. Tectonics 1:471–494

Rosencrantz E (1983) The structure of sheeted dikes and associated rocks in North Arm Massif, Bay of Islands ophiolite complex and the intrusive process at oceanic spreading centers. Can J Earth Sci 20:787–801

Ryan WBF, Carbotte SM, Coplan JO, O'Hara S, Melkonian A, Arko R, Weissel RA, Ferrini V, Goodwillie A, Nitsche F, Bonczkowski J, Zemsky R (2009) Global multi-resolution topography synthesis. Geochem Geophys Geosyst 10(Q03014). https://doi.org/10.1029/2008gc002332

Schouten H, Denham CR (2000) Comparison of volcanic construction in the Troodos ophiolite and oceanic crust using paleomagnetic inclinations from Cyprus Crustal Study Project (CCSP) CY-1 and CY-1A and Ocean Drilling Program (ODP) 504B drill cores. In: Dilek Y, Moores EM, Elthon D, Nicolas A (eds) Ophiolites and oceanic crust: new insights from field studies and the ocean drilling program, special paper 349. Geological Society of America, Boulder, CO, pp 181–194

Searle RC, Keeton JA, Owens RB, White RS, Mecklenburgh R, Parsons B, Lee SM (1998) The Reykjanes Ridge: structure and tectonics of a hot-spot-influenced, slow-spreading ridge, from multibeam bathymetry, gravity and magnetic investigations. Earth Planet Sci Lett 160:463–478

Shand SJ (1949) Rocks of the mid-Atlantic ridge. J Geol 57:89–92

Sigmundsson F, Hooper A, Hreinsdóttir S, Vogfjörd KS, Ófeigsson BG, Heimisson ER, Dumont S, Parks M, Spaans K, Gudmundsson GB, Drouin V, Árnadóttir T, Jónsdóttir K, Gudmundsson MT, Högnadóttir T, Fridriksdóttir HM, Hensch M, Einarsson P, Magnússon E, Samsonov S, Brandsdóttir B, White RS, Ágústsdóttir T, Greenfield T, Green RG, Hjartardóttir ÁR, Pedersen R, Bennett RA, Geirsson H, La Femina PC, Björnsson H, Pálsson F, Sturkell E, Bean CJ, Möllhoff M, Braiden AK, Eibl EPS (2015) Segmented lateral dyke growth in a rifting event at Bárðarbunga volcanic system, Iceland. Nature 517(7533):191–195

Sillman CJ (1987) A Canary Islands dyke swarm: implications for the formation of oceanic islands by extensional fissural volcanism. In: Halls HC, Fahrig WF (eds) Mafic Dyke swarms Geological Society of Canada special paper 34. Ottawa, pp 243–255

Singh SC, Crawford WC, Carton H, Seher T, Combier V, Cannat M, Canales JP, Dusunur D, Escartín J, Miranda JM (2006) Discovery of a magma chamber and faults beneath a Mid-Atlantic Ridge hydrothermal field. Nature 442:1029–1032. https://doi.org/10.1038/nature05105

Stakes DS, Shervais JW, Hopson CA (1984) The volcanic-tectonic cycle of the FAMOUS and AMAR valleys, Mid-Atlantic Ridge (36° 47′N): evidence from basalt glass and phenocryst compositional variations for a steady state magma chamber beneath the valley midsections. J Geophys Res 89:6995–7028

Stewart MS, Klein EM, Karson JA (2002) The geochemistry of dikes and lavas from the north wall of the Hess Deep Rift: insights into the four-dimensional character of crustal construction at fast-spreading mid-ocean ridges. J Geophys Res 107(B9):2181. https://doi.org/10.1029/2000JB000051

Stewart MA, Klein EM, Karson JA, Brophy JG (2003) Geochemical relationships between dikes and lavas at the Hess Deep Rift: implications for magma eruptibility. J Geophys Res 108(B4):2184. https://doi.org/10.1029/2001JB001622

Stewart MA, Karson JA, Klein EM (2005) Four-dimensional upper crustal construction at fast-spreading mid-ocean ridges: a perspective from an upper crustal cross-section a the Hess Deep Rift. J Volcanol Geoth Res 144:287–309

Tartarotti P, Hayman NW, Anma R, Crispini L, Veloso E, Galli L, Parties IESS (2006) Structure of Hole 1256D: the role of mechanical deformation in superfast-spread crust. EOS, Trans Am Geophys Union, Fall Meeting Supplement 87(52):B31B–1091

Tivey MA (1996) Vertical magnetic structure of ocean crust determined from near-bottom magnetic field measurements. J Geophys Res 101:20275–20296

Tivey MA, Johnson HP, Fleutelot C, Hussenoeder S, Lawrence R, Waters C, Wooding B (1998) Direct measurement of magnetic reversal polarity boundaries in a cross-section of oceanic crust. Geophys Res Lett 25:3631–3634

Varga RJ (1991) Modes of extension at mid-ocean ridge spreading centers: evidence from the Solea graben, Troodos ophiolite Cyprus. J Struct Geol 13:517–538

Varga RJ, Moores EM (1985) Spreading structure of the Troodos ophiolite, Cyprus. Geology 13:846–850

Varga RJ, Karson JA, Gee JS (2004) Paleomagnetic constraints on tilt and fault models for oceanic crust from oriented samples from the Hess Deep Rift, Equatorial Pacific Ocean. J Geophys Res 109(B2102):1–22. https://doi.org/10.1029/2003jb002486

Vera EE, Buhl P, Mutter JC, Harding AJ, Orcutt JA, Detrick RS (1990) The structure of 0-0.2 My old oceanic crust at 9°N in the East Pacific Rise from expanded spread profiles. J Geophys Res 95:15529–15556

Walker GPL (1964) Geological investigations in eastern Iceland. Bull Volcanol 27(1):351

Walker GPL (1986) Koolau dike complex, Oahu: intensity and origin of a sheeted dike complex high in a Hawaiian volcanic edifice. Geology 14:310–313

Walker GPL (1992) "Coherent intrusion complexes" in large basaltic volcanoes: a new structural model. J Volcanol Geoth Res 50(1):41–54

Walker GPL (1993) Basaltic-volcano systems. In: Prichard HM, Alabaster T, Harris NBW, Neary CR (eds) Magmatic processes and plate tectonics. Geological Society of London, special publication 76. Blackwell Scientific Publishers, London, pp 3–38

Walker GPL (1999) Volcanic rift zones and their intrusion swarms. J Volcanol Geoth Res 94(1):21–34

Wilson DS, Teagle DAH, Acton GD (2003) Proceedings of the ocean drilling program, initial reports, Leg 206 [CD ROM]. Ocean Drilling Program, Texas A&M University, College Station, TX

Wilson DS, Teagle DAH, Alt JA, Banerjee NR, Umino S, Miyashita S, Acton GD, Anma R, Barr SR, Belghoul A, Carlut J, Cristpini L, Durand SR, Einaudi F, Galli L, Gao Y, Geldmacher J, Gilbert LA, Hayman NW, Emilio H-B, Hirano N, Holter S, Ingle S, Jiang S, Kalberkamp U, Kerneklian M, Koepke J, Laverne C, Lledo Vasquez HL, Maclennnan J, Morgan S, Neo N, Nichols HJ, Park S-H, Reichow MK, Sakuyama T, Sano T, Sandwell R, Scheibner B, Smith-Duque CE, Swift SA, Tartarotti P, Tikku AA, Tominaga M, Veloso E, Yamasaki T, Yamasaki S, Ziegler C (2006) Drilling to gabbro in intact oceanic crust. Science 312:1016–1020

MIX
Papier aus verantwortungsvollen Quellen
Paper from responsible sources
FSC® C105338

If you have any concerns about our products,
you can contact us on
ProductSafety@springernature.com

In case Publisher is established outside the EU,
the EU authorized representative is:
Springer Nature Customer Service Center GmbH
Europaplatz 3, 69115 Heidelberg, Germany

Printed by Libri Plureos GmbH
in Hamburg, Germany